D1492495

Plant Ecology
Origins, Processes, Consequences
Second Edition

Presenting a global and interdisciplinary approach to plant ecology, this much-awaited new edition of *Plants and Vegetation* integrates classical themes with the latest ideas, models and data. Keddy draws on extensive teaching experience to bring the field to life, guiding students through essential concepts with numerous real-world examples and full-colour illustrations throughout. The chapters begin by presenting the wider picture of the origin of plants and their impact on the Earth, before exploring the search for global patterns in plants and vegetation. Chapters on resources, stress, competition, herbivory and mutualism explore causation. After chapters on how pattern in vegetation is studied, the book concludes with a chapter on conservation that addresses the concern that one-third of all plant species are at risk of extinction. The scope of this edition is broadened further by a new chapter on population ecology, along with extensive examples including South African deserts, the Guyana Highlands of South America, Himalayan forests and arctic-alpine environments.

Paul A. Keddy has taught plant ecology for more than 30 years. He is often a conference keynote speaker, and delights in bringing science alive for his audience. Dr. Keddy's research explores environmental factors that control plant communities and their manipulation to maintain and restore biodiversity. His awards include a National Wetlands Award for Science Research, and the Lawson Medal and Gleason Prize for *Competition*, and his first edition of *Wetland Ecology* (Cambridge University Press, 2000) won the Society of Wetland Scientists' Merit Award. He has also advised organisations including World Wildlife Fund, Earthjustice and The Nature Conservancy.

Cover: The Socotra archipelago off the east coast of Africa has many endemic plants, including this dragon's blood tree (*Dracaena cinnabari*). You can read more about these trees in Section 9.4.3, and about the harm being caused by goat grazing in Box 13.2. The archipelago is a fragment of the former continent of Gondwana, which is discussed in Section 8.2.2. (Vladimir Melnik, Shutterstock).

Plant Ecology

Origins, Processes, Consequences

SECOND EDITION

PAUL A. KEDDY

CAMBRIDGE
UNIVERSITY PRESS

University Printing House, Cambridge CB2 8BS, United Kingdom

One Liberty Plaza, 20th Floor, New York, NY 10006, USA

477 Williamstown Road, Port Melbourne, VIC 3207, Australia

4843/24, 2nd Floor, Ansari Road, Daryaganj, Delhi – 110002, India

79 Anson Road, #06-04/06, Singapore 079906

Cambridge University Press is part of the University of Cambridge.

It furthers the University's mission by disseminating knowledge in the pursuit of education, learning, and research at the highest international levels of excellence.

www.cambridge.org
Information on this title: www.cambridge.org/9781107114234
DOI: 10.1017/9781316321270

Second edition © Paul A. Keddy 2017

First published 2007
Second edition 2017

Printed in the United States of America by Sheridan Books, Inc

A catalogue record for this publication is available from the British Library.

Library of Congress Cataloging-in-Publication Data
Names: Keddy, Paul A., 1953– author. | Plants and vegetation.
Title: Plant ecology : origins, processes, consequences / Paul A. Keddy.
Description: Second edition. | New York : Cambridge University Press, 2017. | First edition (2007) titled: Plants and vegetation. | Includes bibliographical references and index.
Identifiers: LCCN 2016015469 | ISBN 9781107114234 (Hardback)
Subjects: LCSH: Plant ecology.
Classification: LCC QK901 .K34 2016 | DDC 581.7–dc23 LC record available at https://lccn.loc.gov/2016015469

ISBN 978-1-107-11423-4 Hardback

The mass of vegetation on the Earth very far exceeds that of animal organisms; for what is the volume of all the large living Cetacea and Pachydermata when compared with the thickly-crowded colossal trunks of trees, of from eight to twelve feet in diameter, which fill the vast forests covering the tropical region of South America, between the Orinoco, the Amazon, and the Rio da Madeira? And although the character of different portions of the Earth depends on the combination of external phenomena, as the outlines of mountains – the physiognomy of plants and animals – the azure of the sky – the forms of the clouds – and the transparency of the atmosphere – it must still be admitted that the vegetable mantle with which the Earth is decked constitutes the main feature of the picture.

Alexander von Humboldt. 1845. *Cosmos: A Sketch of the Physical Description of the Universe*

Contents

Chapter 4   Competition   123

Boxes

Preface

Welcome to the second edition!

When planning the first edition, I wanted to write the book that should have been there to instruct me when I was a young biologist, the book I wish I had to consult when I was 18. The chapters consist of 13 topics that should be included in every plant ecology course. That would be about two weeks per topic area in a one-semester course. Each chapter begins with some inescapable basics (about one-third of the chapter), some more in-depth reading for senior students (another third), and some advanced material that might be of use later in one's career (another third). I assume of course that you will buy this book and keep it as a lifetime companion, not just rent it for a few months.[1] You can, of course, rent or borrow if you wish but good books grow along with you. I have written this book with that in mind: some basic principles are so obvious we need to hear them first when we are young and then repeatedly as we age.

At the time the first edition went to press, I was seriously ill and leaving my position in Louisiana, as well as trying to finish a guide to nature in Louisiana, and coping with unusual levels of academic perfidy. Not to mention the aftermath of Hurricanes Katrina and Rita. Louisiana was meanwhile beginning her march of folly into the ocean, amidst chants of "Drill, baby, drill!" In a perfect world, I would have had a few more quiet months for reflection and editing and revision but, of course, if you wait for the perfect time to write the perfect book, it is possible that one will die long before that state of perfection is attained. I have since had five years in the deciduous forests – living in solitude longer than Thoreau, but not so long as St. Francis. This has allowed me to reflect further on the material in the book, to consider the suggestions and opinions of published reviews, and to correspond with students.

So, you ask, what about the new edition?

My first task has been some pruning. Most plants benefit from pruning, so long as it is carried out with precision. I have taken branches out of nearly every chapter. I like the form and architecture much better now. I hope the flow of ideas is clearer, and that the remaining figures more perfectly illuminate the text. Each chapter has, more

---

1 Yes, I know good books are expensive. Particularly in hard cover. But they last a lifetime. I still have a hard cover of *Geographical Ecology* that I bought as an undergraduate. I do understand that money is in short supply for most students – I once lived in a basement apartment where I met many of the invertebrates, including scutigerans, that I was learning about at university. My suggested solution for books? Let your relatives know. Many family members are desperate for a long-lasting and worthwhile Christmas or birthday gift. Instead of an ill-fitting sweater, or a piece of kitsch that will soon be forgotten, let them choose a book from your list, and write something personal in it. Long after they have passed on, you will still have a treasured reminder of them in your personal library.

clearly I hope, those three sections: (1) basics, (2) more in-depth reading and (3) advanced material. Given this structure, beginners should remember that if any chapter seems overwhelming, it is fine to stop part way through. Even a writer has to do this. There really are parts of the book that still challenge my understanding, and I was the one who wrote it! If you merely read the first third of every chapter you will have grasped the essentials.

Having marked thousands of exams and attended an excruciating number of seminars and oral exams, I can assure you that many people who claim to have had a plant ecology course have either been badly taught or fail to remember much at all. Or both. Hence you will find certain topics continue to be emphasized simply because I found that most students I met did not know them.

Although I wrote this book for a one-semester undergraduate plant ecology course, you could also use this book in a graduate course. In this case, having laid the foundations early in the chapter (perhaps as preparatory reading assignments) you could explore the examples and discussions later in the chapter. I have continued to include important people in ecology. You will meet people including Carolus Linnaeus, Alexander von Humboldt, Joseph Banks, Alfred Wallace, Charles Darwin, Wilhelm Hofmeister, Vladimir Vernadsky, Bernhard Frank, Armen Takhtajan, Christen Raunkiaer, Fritz Haber, Ronald Fisher and Robert Whittaker. No, these are not the people who are currently promoting themselves with trumpets and headlines, nor their sycophants, but people who were devoted to botany, often overcoming great hardship to advance our understanding of plants and plant ecology. (Well, except for Haber. His wife shot herself in shame; but even so, he is part of the story.) When we forget them, we lose a part of ourselves.

I also had my fellow professors and other scholars in mind. There is so much pressure to specialize into narrow sub-disciplines that it is easy to lose track of the big picture. This book provides such a big picture as a backdrop against which your own work can be viewed more clearly. It should also help you teach topics well beyond your own specialty. The chapters end with difficult problems that remain to be solved, and so, in a general way, provide trajectories of how other people are trying to solve problems in other sub-disciplines. I hope I have dealt with your own specialty adequately but, of course, no one can claim to be an expert in every field, however widely one reads. Still, someone has to make the leap to the big picture. I often advise younger scholars struggling with their own field that one of the best ways to find new approaches and insights is to read about other fields. In this sense, deserts really have taught me a great deal about wetlands.

I have also added new material, where it was important, but very selectively. I am not of the opinion that every piece of work rushed into print in 2012 is necessarily an improvement over excellent and classic work from 1972. I have added work, much as I would have added in my lectures, to update major themes, but not to obscure the past or stroke friends' egos. Good work will stand the test of time; we ignore it at our peril. This is doubly true now that the internet and hand-held devices fragment our attention spans and reduce pages of text to a few headlines. In the past, when I had to walk across campus to the library (sometimes through the snow) and pay to photocopy articles page by page, there was an incentive to read them, not just collect them. But even then (ca. 1975) one easily slipped into the mistaken belief that if you had a copy you must have understood the content! Desks then easily accumulated piles of papers that had been copied but not read. Now that pdf files flow effortlessly over the internet, even right into my office in the forest, how easy it is to assume that the electronic acquisition of a paper, or (gasp) automatic download of a citation, is a substitute for having read and considered the contents. Increasingly, I suspect that many young scholars are citing work they have not even read. (Yes, my friends, it is far more obvious than you might suspect.) This is a dangerous route to error and ignorance. I assure you that every single article I cite in this book, I have read. Many of the books I even own. From time to time I take them down, dust them off, and re-read them. The effort is usually well rewarded.

I have also had the time to reconsider each figure. Some have been removed. I thought Lonesome George the Galapagos tortoise would have to leave, particularly as he has died, but Ole Hamman has helped me keep him in as a cheerful insert in Figure 13.8. Some figures have been revised for clarity (world floristic regions now appears in colour in Figure 2.12). A few more have been added, such as the world leaf economic spectrum (Figure 3.26). If you simply flip through the figures, you will see the improvements.

When Cambridge decided to print in full colour, I went through the entire book again. This time I was able to add in some beautiful and strange plants from around the world: the saguaro cactus, *Dendrobium* orchids, *Myrmecodia beccarii*, the ant house epiphyte, *Magnolia sinica*, the yareta, *Telopea speciosissima*, the dragon's blood tree, *Welwitschia mirablis*. There are also more colour maps, including an accurate map of world grasslands (Figure 10.7) and the latest maps of global plant diversity (Figure 12.10). The images also range across time, from early paintings of newly discovered plants to a recent highly processed satellite image from the NASA Earth Observatory. For those of you who remain intimidated by the size of the book, may I suggest you begin by just flipping through the colour images and reading those captions? I think you will find they tell quite a story in their own right.

Three technical challenges remained.

First, there was explaining what we know about the origin of plants and their impact on the Earth. I do think it is one the most important biological stories. In my experience, it is not well taught in schools, and over the last five years, I have seen important topics such as evolution and global warming systematically undermined by people being paid to shill for ignorance. I still meet students who tell me, gravely, that evolution violates the second law of thermodynamics, when, as plants brilliantly illustrate, evolution is, in fact, a direct and necessary consequence of the second law of thermodynamics. The story of plant evolution must be told, but some think it should not start the book. I have tried to juggle sections here, keeping the story near the beginning, but introducing modern plants earlier, and moving energy flow to a sub-theme.

Second, in a related way, some reviewers thought the book covered too much basic work that students should already know. I don't disagree. Early land plants and coal swamps, and the oxygen revolution and major types of plants should indeed have been taught. All I know is that I have repeatedly sat on graduate qualifying exams where students who assured us they had earned a BSc really did not know why coal was a fossil fuel, or what early land plants looked like, or what is meant by alternation of generations. We may expect that students should know such basics, but apparently they often do not.

The third issue was population ecology. Some reviewers wanted a chapter on the topic. My graduate research was in population ecology. But this experience taught me as much about its limitations as well as its strengths. In the first edition I tried to show you how populations fit into several different themes. That is, they are a tool for understanding plant communities, not an end in themselves. Still, many readers have told me that in a book on plant ecology, there should be a separate chapter on populations. I accepted this request, and have moved many population examples into a separate chapter. I also try to explain why one would study plant populations, the fundamental importance of exponential population growth to ecology and evolution, and how populations expand our understanding of processes and consequences altogether. I have also added some interesting stories about Brazil nuts and century plants.

I like this second edition better and I hope that you will agree. My aspiration in writing this book was to teach you about plants, to pass on wisdom from researchers who are now dead or retired (or both), to challenge young scientists to revisit certain problems, to share my enthusiasm for wild places, and to encourage a new generation of botanists and conservationists. I also wanted to prepare an accurate historical record of work, to counter-balance those who are deliberately distorting our field to benefit their own egos. As I worked on *Plant Ecology* I occasionally received fan mail for *Wetland Ecology*. One student recently emailed me from his hand-held device (as he studied for finals) that it had him laughing as he studied and that it was "such an enjoyable read that I hesitate to call it a text book." Based on that email, I can add one more aspiration: may you enjoy a jolly good read.

1

Plants Create the Biosphere

Plant diversity. Vegetation types. The first plant life moves onto land. Evolution and diversification of land plants. Energy flow. Membranes. The origin of eukaryotes. Photosynthesis. The oxygen revolution. The ozone layer. Plants and climate. The biosphere.

FIGURE 1.1 Flowering plants now dominate the land. Some, like this enormous saguaro (*Carnegiea gigantea*), are even able to tolerate the lack of moisture on this hot hillside in the Sonoran desert. (Photograph of saguaro cactus and author by Cathy Keddy, Phoenix, Arizona, 2007.) Saguaro cacti are introduced in Section 7.2.1, while deserts and barrel cacti are discussed in Chapter 10.

1.1   Introduction: The Importance of Plants

1.1.1 Plants Are Abundant and They Support Other Life Forms

Plants occur in almost every imaginable habitat on Earth – exposed on wind-swept mountain tops, submerged under water on lake bottoms, perched perilously on branches in the rain forest canopy or simply withstanding the sweltering desert sun. Moreover, they can be many sizes, from microscopic oceanic plankton to towering giant sequoias and eucalypts. Let me begin by introducing you to just three plants that illustrate some of the larger terrestrial species, and the topics we will explore in this book. The saguaro cactus (Figure 1.1) is seen in many films but in reality occurs only in the Madrean deserts of the New World. This one was photographed just outside of Phoenix, Arizona. The remarkable sausage tree (Figure 1.2) grows naturally only in the tropics of Africa, where it depends upon bats to pollinate its flowers. The sacred fir (Figure 1.3) is restricted to just a few mountains in Central America, where it provides a winter home for monarch butterflies. As we proceed through this book you will meet many more unusual plants and vegetation types. We will explore the factors that control their abundance and consider the challenges of managing habitats for conservation.

There is a very practical reason for learning about plants. They comprise more than 99 percent of all the Earth's living matter. That is to say, Earth is not a world of lions and whales, but a world of conifers and angiosperms. We can also say with confidence that the biosphere – including the oxygen you are breathing as you read this paragraph – is largely the consequence of the origin and diversification of plants. Without plants, conditions on

FIGURE 1.2   The world's tropical regions have an enormous number of flowering trees. One example is this sausage tree (*Kigelia africana*) on the edge of a savanna in Uganda. The flowers open at night for pollination by bats and moths. The fruit is consumed by mammals including elephants. (Photograph from Wikimedia Commons)

FIGURE 1.3 *Abies religiosa*, or sacred fir, is one of several conifers found in the mountains of Central America at altitudes of 2,000 to 4,000 metres. A single cone is typically greater than 10 cm long and typically produces 300 to 400 seeds. This is also the preferred tree in which overwintering monarch butterflies hibernate. (Worthington Smith, 1887, Peter H. Raven Library/Missouri Botanical Garden)

Earth – including temperature, types of rocks, the composition of the atmosphere and even the chemical composition of the oceans – would be vastly different. And, of course, plants provide for human sustenance, with the spread of human civilization being linked to the first discovery of agriculture. Global exploration (and wars) were driven by the search for spices, including pepper and cinnamon, which came from tropical trees. And then there are the many other products such as rubber, cotton, silk, quinine, tobacco, wine, potatoes, sugar cane, soybeans and heroin, all of which had, and continue to have, enormous impacts upon human individuals and human civilization (Laws 2010). Along the way we shall also meet asteroids, burning cliffs, dinosaurs, poisonous plants, seed-carrying ants, carnivorous plants and wild orchids. And we shall visit, briefly, locations including the

Andes Mountains, the Amazon River basin, the Sonoran desert, South African deserts, Socotra Island in the Middle East, the Galapagos Islands in the Pacific Ocean and the forested mountains of southern China. You will also meet an array of early scientists who actually did the research that made a book such as this possible.

The world will always need botanists and plant ecologists. Many of the students on my courses seem to want to use their skills to protect wild animals and improve the human condition, but I often find it necessary to explain that it is rarely possible to be effective at these tasks without some understanding of botany and ecology. If you want to contribute to ecology, or to conservation, or to many kinds of human welfare, you have to know something about plants first. Those of you planning to work in fields including forestry, zoology, fisheries management, geography, planning or environmental studies (not to mention molecular biology and medicine) may find it helpful, if not absolutely necessary, to know something about plant ecology. Indeed, one could suggest further that there is little point in going on a tropical holiday if you are unable to appreciate the remarkable plants and vegetation found there. If this book inspires you to continue with the study of plant ecology, and provides some resources to guide you in doing so, it will have succeeded. Equally, however, if it enriches another scholarly discipline that you intend to follow, or at least helps you better appreciate parts of the world that you one day visit, then it will have succeeded in another way. With the increasing specialization of many sub-disciplines in ecology in particular, and biology in general, I also think there is a need for a book that synthesizes the big picture in a way that will allow specialists to pursue their selected field more effectively. Hence this is also a book for fellow professionals and, if the early parts of each of the chapters seem somewhat basic, you will find in the later sections of each chapter enough depth and subtlety to challenge even the expert.

1.1.2 Fundamentals and Overview

While plant ecology is generally defined as "the study of relationships between plants and the environment" plants do not, as this definition implies, merely inhabit environments. Plants also create environments, and they may even control them. Where, then, should one begin a book on plant ecology? The answer is clearly genesis – the origin of plants and the processes that created the current biosphere. As we all learned in our first biology course, plants live by capturing sunlight. The first chemical process we were likely expected to memorize may well have been photosynthesis. This was a world-changing process and we will look at it in some detail.

Most newer students that I teach – including graduate students – appear to know relatively little about global processes and geological time scales. I will therefore start with the story of plants and the origin of the biosphere in a quite general way, emphasizing long-term consequences for the atmosphere, the oceans and the land. The list of further readings will allow you to pursue a deeper understanding of the impacts of plants on biogeochemical cycles, energy flow and the greenhouse effect. In Chapter 2 we will examine global patterns in plant distribution and some of the explorers who made these important discoveries, which might inspire you to visit new areas of Earth and explore them yourself. Then and only then will we encounter the material with which most text books begin: resources and plant growth. In Chapters 4 to 7 we will work our way through the processes by which plants interact with other plants, fungi and animals (including competition, herbivory and mutualism), and the ecological consequences of these interactions. In Chapter 8, we will return to time, including the impacts of meteor collisions and ice ages upon plants and vegetation. Chapters 10 through 12 have more advanced work on patterns in vegetation and how they are studied. We will conclude, in Chapter 13, with the large scale again: the growth of the human population and its consequences for the biosphere and the Earth's plants and vegetation.

But first, a brief introduction to plants and vegetation types as they exist today.

1.1.3 The Number of Species and Their Classification

There are now some 350,000 species of plants in total, spread from coastal mangrove swamps to mountain peaks. Many plants are found in deserts. Some even grow in shallow water, including shallow salt water, although none of these represents ancestral plants. Contemporary wetland plants are species that have re-invaded wetter habitats from terrestrial ancestors. In this book we shall focus a good deal of time on flowering plants, or angiosperms, because they are the most common. However, this does not mean we can safely ignore the other groups of plants. A brief introduction to the others is given, following Table 1.1, from the bottom to the top. The second most common group is the gymnosperms, with seeds, and often in cones, but plants with no flowers or fruits. They appeared much earlier in evolutionary history, and are common in the fossil record. Those of you familiar with plants will know that the gymnosperms are a somewhat artificial group, since they contain four divisions: Coniferophyta, Cycadophyta, Ginkgophyta and Gnetophyta. The latter three are quite uncommon, although of considerable botanical and evolutionary significance. They may show up from time to time in this book, but you will survive if you realize that when I refer to gymnosperms, I am mostly referring to conifers, but trying to remind you that they are not the only seed-bearing plant that lacks flowers. The last group, a small proportion of the numbers and biomass, is the spore-producing plants that represent some early evolutionary stages, pteridophytes and other early vascular plants. Some of these (such as *Lepidodenron*, see Figure 1.7) were enormous, but only relatively small species survive today, with the notable exception of the tree ferns. Lastly there is the Bryophyta. They most likely represent the earliest stage of land colonization and still remain relatively common in wet habitats, often as epiphytes. And yet there is one exception: possibly the most abundant plant in the world by weight is *Sphagnum* moss (see Figure 10.13), the plant that forms vast northern peat bogs.

Table 1.1 **The main groups of flowering plants (kingdom Plantae) and the classification used in this book.**[a] **We will return to this table again late in the book in Box 8.1.**

Group	Division
Bryophytes	Bryophyta (mosses and liverworts)
Vascular plants	
Seedless plants	Lycopodiophyta (club mosses)
	Equisetophyta (horsetails)
	Pteridophyta (true ferns)
	Psilophyta (whisk ferns)
Seed plants	
Gymnosperms	Cycadophyta (cycads)
	Ginkgophyta (ginkgo)
	Pinophyta (conifers)
	Gnetophyta
Angiosperms (flowering plants)	Magnoliophyta (flowering plants)
	Class Magnoliopsida (dicots)
	Class Liliopsida (monocots)

[a] Note that newer classifications are available, but not necessarily helpful from the perspective of plant ecology. For example, the *Tree of Life*, based on Kenrick and Crane (1997a) combines the last three divisions of seedless plants into one group, the Polypodiopsida, as a single clade. It also calls the entire group of plants in this table Embryophtes. The Angiosperm Phylogeny Website of the Missouri Botanical Garden (www.mobot.org/MOBOT/research/APweb) puts many of the above divisions into orders, and has a more complicated breakdown of the Magnoliophyta. Sometimes the name Lycophyta is used instead of Lycopodiophyta, and so on. While systematics thrives on such changes, for readers of this book and most practising ecologists, this table is quite sufficient.

Now, obviously, this is quite a broad series of generalizations. I would encourage you to revisit a book of basic botany and evolution to remind yourself of the different main groups of plants. It will make this book more interesting, and account for the occasional

digression into topics such as gametophytes. At a minimum, I shall assume that you know what an angiosperm, conifer and fern are, and I shall, when necessary, use more technical names when it is appropriate.

1.1.4 Vegetation Types and Climate

One of the most useful general principles in plant ecology is this: all the major types of plant communities on Earth are controlled by two main factors: temperature and rainfall. In general, the warmer it is, and the wetter it is, the more abundant plants will be and the more kinds of species there will be. This is why, for example, the world's rain forests are huge reservoirs of plant species. There are, as we shall see, more than ten thousand species of trees in the Amazon basin alone, not counting the orchids and bromeliads that grow on the branches of these trees. And there are more kinds of orchids than any other group of flowering plants. So Figure 1.4 is an important, one might say, foundational, figure. It was first presented by Helmut Lieth in German and later adapted in English by Whittaker in his (1975) book *Communities and Ecosystems*.

You have likely already seen a version of Figure 1.4 in a basic ecology book, but let us have a short review. If you follow the upper line in the diagram, you are moving along a gradient of increasing temperature and rainfall. You pass through four major vegetation types, tundra (arctic plants), boreal forest (conifers and cold grasslands), temperate rain forest and tropical forest. Those of you reading this book in Europe or eastern North America, for example, are in the temperate forest region of the world, a region where many tree species are deciduous.

All the regions below these four main types are produced by low rainfall. On the far right, that is in warm climates, as one moves from top to bottom it becomes drier, from tropical rain forest to tropical seasonal forest (with a pronounced dry season), to subtropical desert. Finally, in the intervening area lies a rather complex mixture of shrubland and grassland. In this region, generalizations may be more difficult. This is likely because other factors

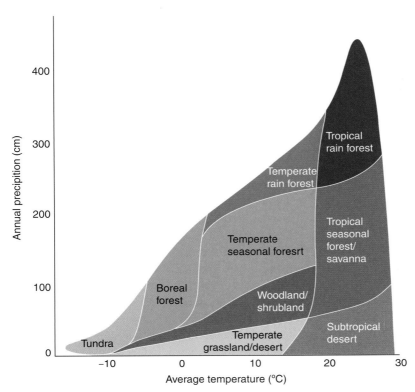

FIGURE 1.4   World vegetation types are produced by just two main factors: precipitation and temperature. We all live somewhere inside this envelope. Identify the location of your home. (From Ricklefs 2001 after Whittaker 1975)

such as fire and grazing animals have a pronounced effect on these vegetation types. Hence one has to consider factors such as disturbance (Chapter 5) and herbivores (Chapter 6) to understand what is happening. It is possible for herbivores to change grassland to shrubland by removing grasses; it is also possible for herbivores, particularly with human assistance, to turn grassland into desert (Figure 10.8). One might regard this region of the diagram as an area of lower predictability and multiple stable states. Often there can be quite abrupt thresholds to switch from one state to the other, a topic to which we will return in Chapter 13. Thus while this diagram shows the main relationships between climate and plants, other factors become important in specific situations.

1.2   The First Land Plants

The land was apparently colonized about 400 million years ago (Niklas et al. 1985; Taylor 1988; Stewart and Rothwell 1993). So far as one can infer from the known fossil record, both plants and animals colonized the land at about the same time, give or take 50 million years. It is not clear why there was a long delay before life forms were able to colonize terrestrial habitats. One hypothesis is that it took that long for there to be sufficient oxygen in the atmosphere for respiration. Another suggestion is that it took that long for ozone to accumulate and shield the Earth's surface.

Some typical early land plants are shown in Figure 1.5. These fossils of *Asteroxylon* and *Rhynia* were found in Scotland by Kidston and Lang (1921), preserved in chert that formed in the early Devonian, about 410 million years ago. These two examples appear typical of early land plants – small erect shoots

(a)

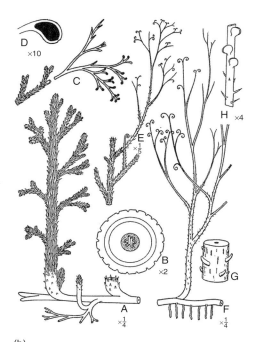

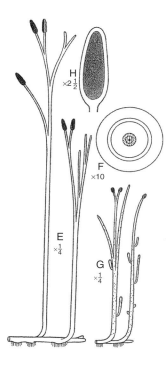

FIGURE 1.5
(a) Reconstructions of three early land plants discovered in the Rhynie chert in Scotland including an *Asteroxylon* species (left A–E), *Psilophyton princeps* (middle F–H) and a *Rhynia* species (right E–H). Sections B and F show stem cross-sections with a central vascular bundle. D and H give a longitudinal section of a typical sporangium. (From Sporne 1970) (b) Artist's impression of an early Silurian landscape (ca. 440 million years ago) showing populations of early land plants beside freshwater pools. (© Natural History Museum, London)

(b)

with horizontal rhizomes – lacking roots, leaves, seeds and woody tissues (Stewart and Rothwell 1993; Kenrick and Crane 1997a,b). Some similar genera still alive today include *Psilotum, Lycopodium* and *Selaginella*. The ancestors of such plants were probably horizontal in growth form with the gametophyte stage dominant, similar to modern liverworts. Fossilized spores of such ancient liverworts have been found in rocks approximately 460 million years old in Argentina; this could place the earliest land plants in the Ordovician era, on the western margin of the Gondwana paleocontinent (Rubenstein et al. 2010).

There is good reason to conclude that colonization of the land also required the evolution of symbiosis between plants and fungi. Much of the nutrient uptake by terrestrial plants is still accomplished by a mere 130 species of fungi in the relatively ancient genus *Zygomycotina* in the order Glomerales (Peat and Fitter 1993; Simon et al. 1993). Re-examination of fossil plants from the Devonian suggests that mycorrhizal fungi were associated with plant rhizomes as early as some 400 million years ago (Pirozynski and Dalpé 1989; Taylor et al. 1990). The fact that mycorrhizae are now found worldwide, and in groups including ferns, gymnosperms and angiosperms, is further evidence of their early origin. The fungi therefore appear to have diversified along with the terrestrial plants (Berbee and Taylor 1993; Simon et al. 1993).

The land produced intense selection upon plants. A whole new suite of traits, including a cuticle to reduce desiccation, stomata to control water loss but admit CO_2, sclerenchyma to strengthen stems for vertical growth, and water conducting tissues, arose out of the strong natural selection to cope with desiccation. Early land plants still betrayed their aquatic origins by having free-living sperm that swam from male to female organs. This is obviously workable in the ocean but not a terribly good trait for dry conditions (we return to this topic in Box 8.1). In fact, a recurring theme in many discussions of plant evolution is the way that terrestrial environments have driven modifications to plant reproductive systems to get around the constraints imposed by a terrestrial habitat (Raven et al. 2005). The gymnosperms appear to have been one of the first groups in which selection eliminated motile sperm and produced the

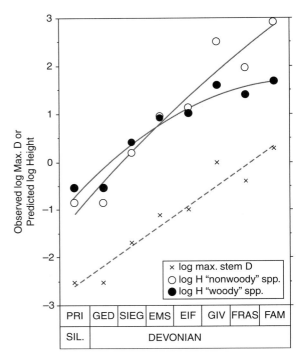

FIGURE 1.6 Plant height increased through the Devonian, as documented by stem diameters of early Paleozoic vascular plants combined with allometric equations for non-woody and woody species. (From Niklas 1994)

pollen tube (although free water is still required for the pollination droplets that capture the pollen). Further, the cycads, alone among the gymnosperms, still have sperm cells that swim down pollen tubes to fertilize the egg. The rest of the gymnosperms, and all of the angiosperms, have lost even this vestige of their aquatic origin; only nuclei move down the pollen tube.

Plant height increased steadily through geological time, presumably as a consequence of increasing competition for light (Figure 1.6). By the **Carboniferous** era, there were real forests. Trees such as *Lepidodendron* (Figure 1.7) reached a height of 30 metres with trunks a metre in diameter at the base (Bell and Hemsley 2000). Over the next 100 million years, these early vascular plants were replaced by ferns, and then by conifers (Figure 1.8). Then, just over 100 million years ago, the flowering plants (Angiosperms) arose – a topic we will explore further in Section 8.2.

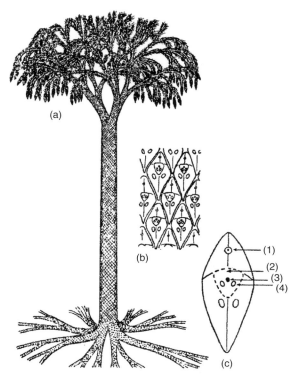

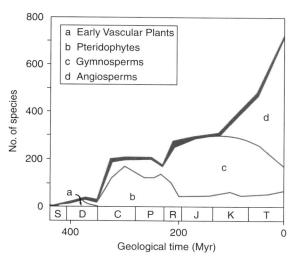

FIGURE 1.8   Trends in vascular plant diversity illustrated by the estimated number of fossil species in four major plant groups. (From Niklas et al. 1983)

FIGURE 1.7   Reconstruction of *Lepidodendron* trees: (a) whole plant; (b) leaf scars; (c) leaf base (1 ligule pit, 2 area of leaf base, 3 vascular bundle, 4 parichnos scars). (From Sporne 1970)

We should note, in passing, that the spread of land plants probably had effects on the oceans. Before land plants appeared, there would have been very rapid rates of erosion, as rainfall spilled off the naked land directly into rivers and oceans. As plants covered more land surface, erosion would have decreased. Also, biologically essential nutrients such as phosphorus would have been selectively stored, either in living plants, or in their organic debris now accumulating as soil and peat. Hence those organisms

in the ocean adapted to vast amounts of eroded material, or to higher dissolved nutrient levels, may have been replaced by other species better adapted to the new oceanic conditions.

Now that you have refreshed your familiarity with the diversity and origins of terrestrial plants, you have two choices available for the remainder of this chapter.

1. If time is limited and you want to move quickly through this book, you can wrap up the topic of plants in the biosphere by leaping ahead to Plants Affect Climate (Section 1.9).

2. If you are curious about what happened before plants arrived on land and how they created the atmosphere altogether, you can continue (Section 1.3, following) to read about the origins of photosynthesis, single-celled plants and the oxygen revolution.

1.3   Energy Flow Organizes Molecules

To understand the origin of plants, we must understand the origin of photosynthesis. For life to exist, energy flow is required. Such a requirement is met

when a planet is situated near enough to a star for sufficient energy released by solar fusion to pass the planet before dissipating into outer space. This is the case for our particular planet, situated near a star we know as the Sun. While it is not known how often life occurs, it may not be infrequent given the enormous size of the universe – our own galaxy has some 100 billion suns, and there now appears to be convincing evidence that some of these suns have their own solar systems. This provides many opportunities for other possible planets to be affected by flowing energy. Proximity to a source of solar energy is essential for life because that energy flow, by itself, organizes matter. Life, at least as it is presently understood, is matter that has been organized by energy flow. Morowitz (1968) has examined the relationships among energy flow, thermodynamics and life – asserting that in order to properly understand life one must look at the relationship between physical laws and biological systems. He demonstrates that flowing energy can create complexity out of simplicity. Once the requirement for energy is met, life then requires resources. This begs the question of what those early resources might have been. One way to answer such a question is to ask what conditions would have existed in the early Earth's atmosphere before there was life, since the early atmosphere would likely have been one source of resources for the precursors of living cells. Determining what the early atmosphere was like, however, requires considerable detective work (e.g. Oparin 1938; Strahler 1971; Levin 1994). It seems that this atmosphere would have come, in part, from volcanic out-gassings. For clues about its composition one can measure the current composition of volcanic gases. Table 1.2 shows that the early atmosphere would likely have been composed of water, carbon dioxide and sulfur. It was an atmosphere rather different from that of today. Yet, some billions of years later, these basic molecules remain as the principal constituents of cellulose, the dominant structural molecule of plants and the most abundant (by mass) molecule in the biosphere (Duchesne and Larson 1989). Morowitz (1968) presents thermodynamic calculations illustrating how energy flow stimulates chemical interactions and creates molecules with higher potential energy. From chemical interactions taking place within the volcanic

Table 1.2 The early atmosphere of Earth probably resembled the composition of gases produced by volcanoes such as these two on Hawaii. (From Strahler 1971)

Volcanic gases from basaltic lava of Mauna Loa and Kilauea	Percent composition
Water, H_2O	57.8
Total carbon, as CO_2	23.5
Sulfur, S_2	12.6
Nitrogen, N_2	5.7
Argon, Ar	0.3
Chlorine, Cl_2	0.1
Fluorine, F_2	–
Hydrogen, H_2	0.04

gas mixture given in Table 1.2, molecules such as methane and ammonia will result. These molecules are thought to have been major constituents of the early atmosphere. Morowitz demonstrates mathematically that, with energy flow and simple mixtures of gases, increasingly complex molecules are formed. For example, a gaseous mixture of carbon, hydrogen, nitrogen and oxygen at 500°C yields mostly water and CO_2 with smaller amounts of other molecules, such as methane and ethane, that have higher potential energy. The latter molecules are less likely to form because they are larger and therefore more energy is required to create them. As energy flows through the molecular system, however, the energy distribution shifts upward toward more and more complicated molecules. Morowitz postulates that energy flow through the early atmosphere yielded similar results: starting off with simple low-energy molecules such as water, CO_2 and nitrogen, more complex molecules were produced. The production of molecules was driven by the external energy source, which on Earth is the Sun. While some authors suggest that the origin of life by such means contradicts the second law of thermodynamics, what they fail to appreciate is that the second law applies to closed systems. The biosphere is an open system where, so long as energy flow occurs, organization will increase.

Another important physical condition of the early environment on Earth was the abundance of water. It is not surprising that water is still a major constituent of the bodies of living organisms. Given the probable temperatures on Earth at that time, water would be evaporating from some areas, condensing in the atmosphere and then falling as rain. As it flowed back into the sea, water would dissolve elements from the rocks – elements that would rise in concentration as water evaporated from the ocean again. These elements could interact in solution, and concentrate in locations where sea water was evaporating most rapidly. Of course, while energy flow tends to produce larger and more complex molecules, there is a natural countervailing tendency – complex molecules will also tend to fall apart into simpler molecules. But here is the crucial point – some molecules will be more stable than others. These stable ones will tend to persist and accumulate. They will steadily become more common than those other molecules that are unstable. It does not require any great scientific insight to appreciate this, nor does it require us to imagine any sort of magical complexity or life force – this process is simply a logical consequence of what we mean by the terms "stable" and "unstable." Nothing lasts forever. Some things fall apart quickly, some things fall apart slowly. So long as both kinds of things are being steadily built by energy flow, the long-lived ones will tend to become more common than the short-lived ones. It is so very simple – yet note that even at the chemical level, long before there is anything that one might be tempted to call life, there is a crude process of natural selection. Some things are surviving longer than others, and hence are becoming more common. Ammonia and methane are two such molecules that likely accumulated in the Earth's early atmosphere. Once a reservoir of larger and more stable molecules forms, these molecules can in turn interact with each other, yielding molecules with greater complexity and higher levels of potential energy. Like the simpler molecules, these more complex molecules will have varying degrees of stability.

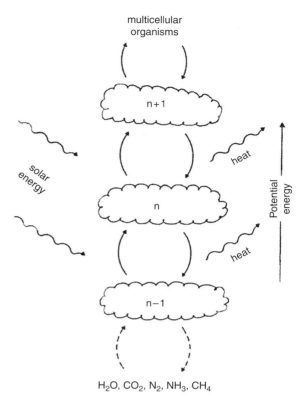

FIGURE 1.9   Solar energy creates high-energy molecules out of simpler low-energy molecules. Complex molecules and multicellular organisms are inevitable thermodynamic consequences of energy flow in the biosphere. For any arbitrary level of potential energy there is a restricted pool of substrate molecules at the next-lower level, so that even in simple molecular systems a form of resource competition can be observed. (From Keddy 2001)

Again, molecules that are unstable will fall apart and those that are stable will accumulate. Imagine this process continuing, with increasingly complex molecules forming as a consequence of external energy flow. In this simple scenario, there is ongoing natural selection for stability and persistence, even at the molecular level (Figure 1.9).

Such ideas are based upon thermodynamic calculations, simple chemistry and logic. Experimental work nicely complements them. In an early experiment, Miller and Urey (Miller 1953) set up a simple atmospheric system with a hydrological cycle (Figure 1.10). Water was evaporated and then cooled and condensed while sealed within glass tubes. Miller and Urey then let the hydrological cycle run, created electrical sparks to simulate lightning (the electrical sparks were used as an alternative to sunlight as a possible external energy source) and found that primitive amino acids formed. This classic piece of work was done in the early 1950s, and it is worth emphasizing that it was done by a graduate student. Miller was fishing around for a research project to do for graduate work and had already tried one project that did not work. Then he and his advisor heard a seminar about early conditions on Earth that stimulated them to try their experiment. This single study led to a large series of experiments wherein researchers have created all manner of artificial atmospheres and utilized different types of energy flow to explore what kinds of molecules could be produced. One could ask what factors might allow complex molecules to further increase in stability and further accumulate. Such factors would likely include: (1) protective walls, (2) the direct use of sources of energy such as sunlight, and (3) the ability to form larger aggregations to buffer against short-term periods of unsuitable conditions. Consciousness would be another step, but this is not a step that plants have taken. In *The Selfish Gene*, Dawkins (1976) argues that consciousness can be thought of as the ability to develop predictive models for future events. For example, if an organism knows that certain conditions are likely to bring winter, then it can store up food. Such ideas will not be explored further here, but Dawkins does raise other issues, one of them being the way in which molecules that copy themselves will proliferate. Returning to Figure 1.9, let us try to mentally reconstruct the circumstances on Earth some 4 billion years ago. Pools of increasingly complex molecules are accumulating as water evaporates and energy flow stimulates chemical interactions. Molecules that are stable are accumulating, those that are unstable are falling apart. Now consider the possibility

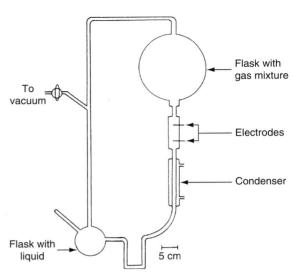

FIGURE 1.10 The original illustration of the apparatus used in the classic Miller and Urey experiment. (From Miller 1953)

of replication. Any molecule that tends to create copies of itself will accumulate more rapidly than other molecules. Dawkins suggests that the occurrence of such replicators was a critical event in the origin of life. Although he uses the word "replication," the word "reproduction" is the analogous biological term. From this perspective, then, molecular stability is survival, and molecular replication is reproduction. Thus, in a very basic and non-living molecular system, it is possible to find the sorts of ecological and evolutionary processes that occur in whole organisms. Further, one can also find larger ecological processes such as competition and predation (Keddy 1989).

Let us try to imagine a scene from this time. Here is how Margulis and Sagan (1986) describe it:

The ponds, lakes and warm shallow seas of the early Earth, exposed as they were to cycles of heat and cold, ultraviolet light and darkness, evaporation and rain, harbored their chemical ingredients through the gamut of energy states. Combinations of molecules formed, broke up, and reformed, their molecular links forged by the *constant energy input of sunlight. As the Earth's various microenvironments settled into more stable states, more complex molecule chains formed, and remained intact for longer periods. By connecting to itself five times, for example, hydrogen cyanide (HCN), a molecule created in interstellar space and a deadly poison to modern oxygen-breathing life, becomes adenine ($H_5C_5N_5$),*

the main part of one of the universal nucleotides which make up DNA, RNA and ATP.

(p. 52)

Now we will turn from this very general discussion of the origin of life and look at more specific issues such as the origin of cellular envelopes, mitochondria and chloroplasts.

1.4 Membranes Are Necessary for Life

Membranes are essential to all life as we know it, and it seems probable that they originated rather early in the history of life. In the most basic way, there is no life without a membrane to divide the world into inside (living organism) and outside (environment). The importance of membranes is emphasized by Day (1984) in his book *Genesis on Planet Earth*. One line of inquiry into the origin of membranes has examined various colloids – mixtures of finely dispersed organic matter suspended in water. Depending upon the composition and concentration, small droplets called coacervates appear.

Coacervates appear prominently in *The Origin of Life*, published in 1938 by another Russian scientist, Aleksandr Oparin. He too emphasized how energy flow drove the assembly of molecules from elements in the early biosphere and, like Morowitz, emphasized that life did not arise by chance but as a consequence of the principles of physics and chemistry. He stresses that "the formation of complex coacervates... was unavoidable because their formation requires very simple conditions, merely the mixture of two or more high-molecular organic substances" (p. 159).

A good deal of later research examines what kinds of circumstances create coacervate droplets. In some cases, coacervate droplets actually have the ability to grow by absorbing from solutions around them. Under other conditions, coacervate droplets start dividing thereby simulating a simple form of cellular replication. They also have the ability to protect their contents from ultraviolet light irradiation (Okihana and Ponnamperuma 1982). Day argues that coacervates are only pseudo cells, that they look cell-like under the microscope, but that the walls around these coacervate droplets do not behave like or have the structure of the membranes cells have today. He concludes that one must look to other examples for origins of the cellular envelope.

On natural freshwater bodies, such as lakes, one finds a surface film that tends to accumulate organic matter; this occurs because of the presence of proteins. Proteins have hydrophobic and hydrophilic components and will orient with the hydrophilic end in the water and the hydrophobic end pointing out. A thin type of membrane results. Day offers a model wherein four steps produce the modern membrane (Figure 1.11): (a) first a surface film forms with the lipid facing the atmosphere and with the protein in the water, (b) turbulence from wind causes the film to buckle, (c) eventually the film buckles so much that the edges of the film touch each other, and (d) finally the air dissolves into the water, leaving a vesicle with lipids inside and proteins outside. This lipid bilayer membrane is one of the most basic structures of life. These lipid bilayers have now been experimentally produced in laboratories and have demonstrated the ability to selectively accumulate certain substances in the surrounding water. Once membranes exist, they provide the opportunity for a powerful form of natural selection to operate, as stable entities will tend to accumulate at the expense of those that are less stable.

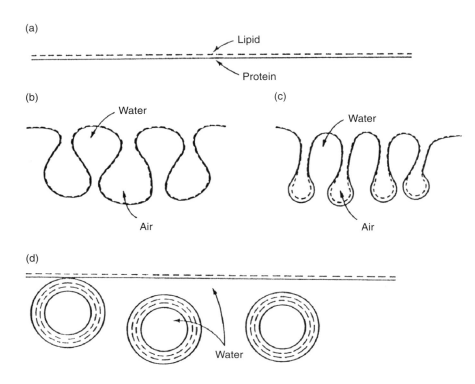

(a)

Lipid

Protein

(b)

Water

Air

(c)

Water

Air

(d)

Water

FIGURE 1.11
Membranes may have been formed by wave action acting on surface films. (From Day 1984)

1.5   Eukaryotic Cells Originated as Symbioses

Prokaryotic cells are probably rather good examples of what some of the earliest cells were like. There was considerable excitement at the discovery of possible 3.1 billion-year-old fossils of prokaryotic cells in South Africa (Schopf and Barghoorn 1967), and the fossil record for early one-celled organisms is steadily accumulating, although caution is necessary to distinguish between real fossils and microstructures that are non-biological in origin (Cloud 1976). The oldest eukaryote fossils, discovered in China, appear to date from 1.8 to 1.9 billion years BP, a date roughly consistent with that obtained from molecular clock estimates (Knoll 1992). It appears that the most primitive living eukaryotes now survive within animal hosts where guts maintain anaerobic conditions. These organisms have a well-defined nucleus and flagella but relatively simple cytoskeletons and no

mitochondria or chloroplasts (Knoll 1992). Despite the apparent simplicity of these primitive eukaryotes, one is still faced with the thorny problem of an apparent leap of complexity from prokaryotes to eukaryotes. How did one arise from the other?

Evolution does not appear to involve spectacular leaps in complexity. As so often happens, what initially appears to be a sudden step actually involves a number of unexpected intermediaries. One hypothesis for the origin of eukaryotes can be traced back to Lynn Margulis, who in 1970 proposed that eukaryotes are in fact symbiotic associations of several prokaryotes (the serial endosymbiosis hypothesis, Figure 1.12). Earlier still, in the 1920s, an American biologist, I. E. Wallin, raised this possibility and even published a book entitled *Symbioticism and the Origin of Species*, but his enthusiasm for the concept was far

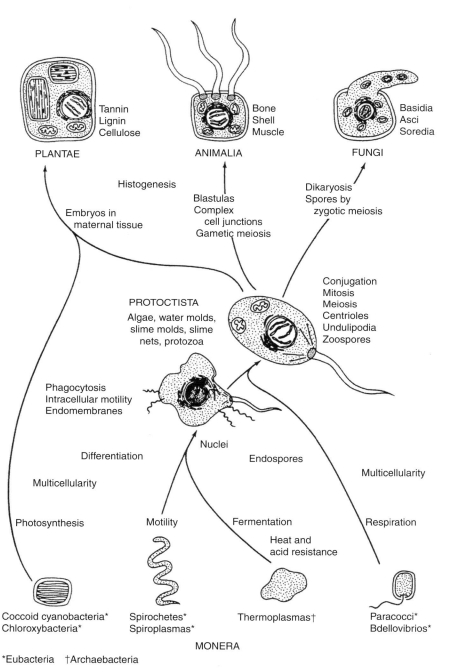

FIGURE 1.12 The serial endosymbiosis theory for the origin of eukaryotic cells. (From Margulis 1993)

ahead of the quality of his data (Wallin 1927). It is now thought that at least three structures in the eukaryotic cell had a symbiotic origin: the mitochondrion, the chloroplast and the flagellum. If an early prokaryote was invaded by a non-photosynthetic bacterium, a cyanobacterium and a spirochete, the result would be rather similar to a modern eukaryotic cell (Figure 1.12). There is a good deal of evidence accumulating to support the serial endosymbiosis hypothesis (Margulis and Sagan 1986; Smith and Douglas 1987; de Duve 1991; Margulis 1993; Roger 1999), although sceptics remain (Cloud 1976).

Finally, there is a marvellous living example of endosymbiosis: the organism is called *Myxotricha paradoxa* and it lives in the guts of termites (Smith and Douglas 1987). Why would one look for an example of an early endosymbiotic organism in animal guts? The reason is simple: guts are anerobic, and thus like the early protoatmosphere. *Myxotricha paradoxa* has three symbionts. There are no mitochondria in *Myxotricha*, instead there are endosymbiotic bacteria. *Myxotricha paradoxa* is moved around by spirochetes, and associated with each of these spirochetes is a

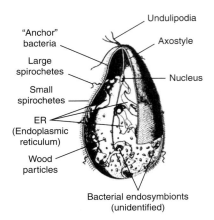

FIGURE 1.13 *Myxotricha paradoxa* is an endosymbiotic organism that lives in the low-oxygen environment found in termite guts. (From Margulis 1993)

mitochondrion-like bacterium (Figure 1.13). The serial endosymbiosis hypothesis proposes that by mixing non-photosynthetic bacteria, spirochetes or cyanobacteria one can derive most of the kinds of cells there are on the Earth today. Even cells in our bodies are symbioses of other living organisms.

1.6 The Origin of Photosynthesis

The oxygen revolution was probably the most important event in the history of life on Earth. One reads a great deal about the events that may have caused the extinction of the dinosaurs (Section 5.4.3), but the extinction of the dinosaurs pales in significance to the origin of photosynthesis. To fully comprehend the significance of the origin of photosynthesis, one must remember that Earth's earliest cells were dependent upon anaerobic metabolic processes for survival. There was no oxygen in the early atmosphere, so these cells were dependent upon breaking down the chemicals that had been created by the flow of solar energy. Neither was there an ozone layer, which, like oxygen, is a product of life, so there would have been a much more intense flow of solar

energy than there is now. The complex molecules that accumulated as a result of chemical interactions driven by solar energy were scavenged and broken down by early organisms in the absence of oxygen. This process, called fermentation, is the basic metabolic pathway upon which other, more elaborate, metabolic pathways have been superimposed.

Imagine protocells scavenging for large molecules in solution. Large molecules would be selectively absorbed and the high energy of these molecules would be used to make other needed molecules. Thus protocells would be limited in growth by the scarcity of large molecules. Over time, large molecules would probably become increasingly rare because many other protocells would also be absorbing them.

One way to avoid competition with other cells, then, would be to make the large molecules internally. Thus there would be a strong selective advantage for any cell line able to use sunlight to convert raw materials into larger molecules, particularly as conversion could occur within a membrane, thus within the control of the cellular environment. As well, the membrane would shield end products from neighbours. Thus there was natural selection for internalized processes of chemical synthesis instead of simply absorbing products formed by chemical reactions that took place outside of their membranes. How might such a transformation have occurred? Was it some rare chance event? Probably not; the transformation may have, in fact, been quite deterministic. Examine the early conditions of Earth from the point of view of physics. Is it possible to determine what type of chemical process might arise and then be selected for in order that the efficiency of early cells would be enhanced? To put the question another way, assuming that energy synthesis, of some form, must occur, is it possible to predict what kind of energy synthesis early cells would be most likely to employ? Would it be radiosynthesis, electrosynthesis, thermosynthesis, magnosynthesis or photosynthesis?

In order to determine what type of chemical process might arise and then be selected for in early cells, three basic questions must be asked. The first and most basic is, "What type of energy has the greatest flow in the atmosphere of the Earth?" The source of energy that life might exploit would likely be abundantly available. Consider the possible terrestrial energy sources. There is sunlight, electric discharges, cosmic rays, radioactivity and volcanic energy (Table 1.3). In terms of calories per square centimetre per year, sunlight would likely be the source of energy used by organisms on this planet, not radioactivity or volcanic energy. Simply based on the energy flux, one can predict that photosynthesis would be likely to arise. The second question that one must ask is, "Given that sunlight would be an obvious source of energy, is there a particular section of the electromagnetic spectrum most likely be exploited?" Table 1.4 shows a breakdown of energy versus wavelength for each part of the spectrum. Note that wavelength increases toward the infrared and energy increases toward the ultraviolet. Consider what the different

Table 1.3 Possible sources of energy for life to exploit. (From Morowitz 1968)

Source	Energy (cal cm^{-2} year^{-1})
Sunlight (all wavelengths)	260,000
<2500 Å	570
<2000 Å	85
<1500 Å	3.5
Electric discharges	4
Cosmic rays	0.0015
Radioactivity (to 1.0 km depth)	0.8
Volcanoes	0.13

Note: 1 Å = 0.1 nm.

Table 1.4 The characteristics of some regions of the electromagnetic spectrum. (From Morowitz 1968)

Region	Wavelength range (μm)	Energy range (kcal/Einstein)	Fraction of solar spectrum (%)	Molecular changes
Far ultraviolet	0.1–0.2	152.2–304.4	0.02	Ionization
Ultraviolet	0.2–0.38	75.3–152.2	7.27	Electronic transitions and ionizations
Visible	0.38–0.78	36.7–75.3	51.73	Electronic transitions
Near infrared	0.78–3	9.5–36.7	38.90	Electronic and vibrational transitions
Middle infrared	3–30	0.95–9.5	2.10	Rotational and vibrational transitions

wavelengths of solar energy do. Along the entire electromagnetic spectrum, from x-rays on the left to radio waves on the right (Figure 1.14), there is only one small portion of the entire spectrum that can cause electron orbital changes and therefore actually be directly involved in chemical interactions: that is, the visible range. At the ultraviolet end of the spectrum, the effects of electromagnetic energy tend to break molecules apart. At the infrared end the energy generates heat: that is, it causes the molecules to vibrate. Only in a narrow "visible" range does light interact with chemistry. Natural selection would therefore tend to favour cells that had internal chemistry that responded to the visible light spectrum.

The origin of chlorophyll is obviously closely related to the origin of photosynthesis. Within the photosynthetic pathway there are a number of different molecules that are sensitive to solar energy, and chlorophyll appears to be a relatively recent addition to the light-trapping system. The earlier molecules sensitive to sunlight were probably much simpler in structure and yielded a rather crude form of working photosynthesis. It seems probable that some cells synthesized porphyrins, which fortuitously had the property of absorbing sunlight in the visible wavelengths. This step probably led to bacterial photosynthesis, which requires an external source of hydrogen such as hydrogen sulfide.

At first, molecules such as hydrogen (H_2), hydrogen sulfide (H_2S) or simple organic molecules (CH_2O) might have provided the source of hydrogen for constructing more complex organic molecules, which are, after all, chains of carbon and hydrogen. As such sources of hydrogen were scavenged by more and more microorganisms, the increasing scarcity of hydrogen would have strongly selected against

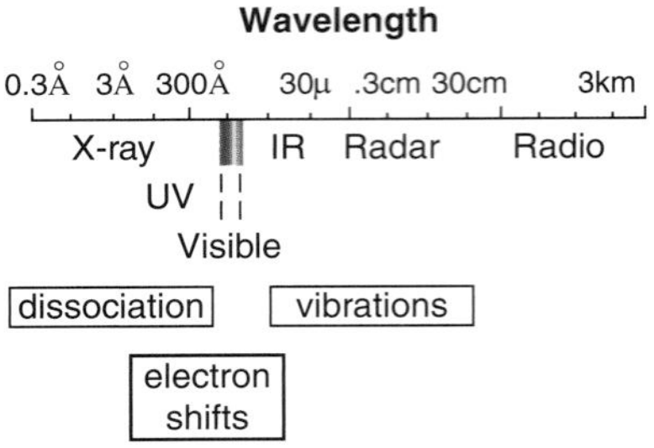

FIGURE 1.14 The electromagnetic spectrum, showing the narrow range of wavelengths causing electron shifts that is used in photosynthesis and vision.

organisms that were unable to exploit other hydrogen sources. The chemical bond holding hydrogen to oxygen in water is much stronger than the bonds holding hydrogen in the above molecules; nonetheless it seems likely that an early photosynthetic organism acquired a means to use light to split water molecules into hydrogen and oxygen. The separated hydrogen could then be added onto carbon from the atmosphere and organic matter thus synthesized. Cells with the ability to split water molecules would then have been able to extract hydrogen from a nearly inexhaustible pool. With photosynthesis achieved via hydrolysis, water and sunshine could be converted into living material. As Hutchinson (1970) describes:

The overall geochemical result [was] to produce a more oxidised part of the biosphere, namely the atmosphere and most of the free water in which oxygen is dissolved, and a more reduced part, namely the bodies of organisms and their organic decomposition products in litter, soils and aquatic sediments.

1.7   The Oxygen Revolution Was a Consequence of Photosynthesis

Photosynthesis via hydrolysis shaped the biosphere, for while it enabled cells to synthesize organic matter from abundantly available water and solar energy, it

also produced a by-product – oxygen. This gas, which was dangerously unstable and tended to react violently with other organic compounds, was released into the atmosphere. Thus the consequences of a more efficient form of photosynthesis were catastrophic for many early life forms. Three important consequences of photosynthesis are considered below, drawing largely upon accounts in Mains (1972), Margulis and Sagan (1986) and Levin (1994).

1.7.1 Ocean Chemistry Changes With Oxygen

Since the first photosynthetic organisms were aquatic, the oxygen released by photosynthesis entered sea water, and began to rust metals present in the oceans. Up until this point there were many metal ions dissolved in sea water; with the appearance of oxygen the metal ions were oxidized and precipitated. This process of oxidation formed enormous beds of sedimentary rocks that contain rusted metal ions, particularly iron oxide. Often these rocks have alternating layers of silica and iron oxide (banded iron formations) that suggests that there was some periodicity in iron precipitation (Figure 1.15). So long as

iron oxide formation used up the supplies of oxygen, however, oxygen could not accumulate in appreciable quantities.

1.7.2 Atmospheric Composition Changes With Oxygen

Once the oceans of the world had rusted, oxygen began leaking out into the atmosphere. Here the oxygen reacted with methane and ammonia, which were broken down and stripped from the atmosphere by rainfall. Once oxygen began to accumulate there was, necessarily, a growing selection for cells with cell walls that provided protection from the oxidative effects of oxygen. In the absence of a sufficiently resistant cell wall, the interior contents of a cell would oxidize. Early oxygen concentrations were initially quite low, perhaps one percent of contemporary concentrations, but for the first time the atmosphere assumed some of the properties associated with it today. For example, it oxidized metals.

(a)   (b)

FIGURE 1.15 Iron ore deposits document the emergence of the first photosynthetic organisms and the subsequent oxidation of the world's oceans. (a) Banded quartz-magnetite ore in a 1 m outcrop with (b) a boulder of banded hematite ore (23 × 53 cm) from Archean rocks in the Ikusia region in West Greenland. (Photographs by Henrik Stendal, Geological Survey of Denmark)

The processes of erosion changed with the presence of oxygen in the atmosphere. The whole physical and chemical nature of the planet slowly changed. Until this period, various kinds of fermentation had probably fuelled the cellular machinery. Now it became possible for several kinds of respiration to arise. It is interesting to note that the chemical pathways in respiration appear to have been built upon an older system that used fermentation. So respiration did not suddenly appear but was slowly modified from the existing biosynthetic pathways in fermentation. Chemical evolution, like the evolution of skeletons, is conservative and slowly modifies existing structures rather than suddenly building new ones. Consider plants and nitrogen. Modern terrestrial plants can use two sources of nitrogen: ammonia and nitrate. Ammonia was widely available early in the Earth's history, whereas nitrate is an oxidized form that would have been another by-product of the oxygen revolution.

1.7.3 The Ozone Layer Forms From Oxygen

As oxygen levels in the atmosphere continued to increase, oxygen in the upper atmosphere would have been exposed to bombardment by ultraviolet light, resulting in the production of ozone. The accumulating ozone would gradually have begun to absorb ultraviolet light. It was as if a giant window blind were slowly being pulled over the Earth. Ultraviolet light may even have produced some of the larger molecules that early microbes fed on; any such organisms would then have starved. Waves of extinction in primitive

microorganisms were likely associated with these events, just as waves of diversification were a likely consequence of the new conditions. All this raises an important question: "When did this most significant event in the Earth's history happen?"

The answer seems to be – at least 2 billion years ago (Mains 1972; Levin 1994). There are two sources of evidence. Uranium dioxide, uraninite, is deposited in sedimentary rocks only under low oxygen conditions, and the last major deposits are about this age. Extensive deposits of banded iron (Figure 1.15) also indicated early oxygen production. Different layers, sometimes only microns thick, of oxidized hematite and less oxidized magnetite may indicate seasonal cycles of photosynthesis. Oxygen-producing photosynthetic bacteria, living in warm volcanic pools in iron-rich water, may have been responsible for seasonal surges of oxygen waste. There was a surge of banded iron formation between 2.2 and 1.8 billion years BP, but some banded formations in Labrador and Greenland appear to be as ancient as 3 billion years old. These banded iron formations provide something like 90 percent of the world's sources of extractable iron (Margulis and Sagan 1986).

The presence of photosynthetic cells and an oxidizing atmosphere bring us to a situation where we can see, at least in outline, the processes we study in the field of plant ecology. Yet there are at least two more important evolutionary events from the perspective of plants. The first is the colonization of land; the second is the origin of flowers and seeds. The former is examined here, while the latter is deferred to Chapter 8.

1.8 The Cambrian Explosion of Multicellular Life

Near the beginning of the Cambrian era, about 700 million years ago, there was apparently an explosion of multicellular life. Exactly when this happened is still somewhat unclear (Day 1984), but

the timing will no doubt be better defined as more enthusiastic students work in the area of paleoecology. Why did it take so long for multicellular life to form? There are many competing hypotheses (Gould

1977). One suggestion is that it took that long for there to be sufficient oxygen in the atmosphere for respiration to evolve, and from this perspective multicellularity is a consequence of aerobic respiration. Another suggestion is that multicellular organisms lived in shallow water and that it took that much time for there to be sufficient ozone to shield the shallow water from ultraviolet bombardment. Another more interesting suggestion is that the fossil record may be somewhat misleading, and perhaps earlier multicellular life was abundant but did not have any skeletons that would fossilize. There may have been a change in oceanic chemistry, near the beginning of the Cambrian era, that permitted the deposition of calcium carbonate required to make preservable skeletons. The real explanation for the sudden diversification of life remains a puzzle, perhaps one that a student reading this book will one day solve.

The next major step was movement onto land. This brings us back to Section 1.2 and the appearance of the first land plants, such as *Rhynia*

(Figure 1.5). Now there was strong natural selection for traits that conferred resistance to terrestrial conditions. Plants began to face two contrasting types of selection. The first was for foraging and survival below ground, demanding structures such as rhizomes and then roots. The second was for light acquisition and survival bathed in the atmosphere, demanding structures such as leaves and eventually seeds. Even to this day a plant is a strangely chimerical organism: it is half selected for below-ground life, and half selected for above-ground life. Is a plant a below-ground organism with above-ground shoots to acquire light and carbon dioxide? Or is it an above-ground creature with below-ground organs to acquire water, nitrogen and phosphorus? It is neither, and it is both. Perhaps it is better to think of a plant as a organism that lives right on the soil surface, with extensions that allow it to forage in two different directions for two different sets of resources. We will look into this more carefully in Chapter 3.

1.9   Plants Affect Climate

The early atmosphere appears to have had vast amounts of CO_2. Carbon dioxide is a greenhouse gas – that is, it absorbs infrared radiation and re-radiates it as heat, raising the temperature of the atmosphere. Space probes that have visited our nearest neighbour toward the Sun, Venus, have discovered an inhospitable planet with a surface temperature of 750 K and a pressure of 90 Earth atmospheres. At least 500 K of this temperature has been calculated to be the effects of the high CO_2 concentrations of the atmosphere. This could be described as runaway greenhouse effect. Our other near neighbour, Mars, has a thin atmosphere and the concentration of CO_2 in its atmosphere is sufficient to raise the surface temperature by only 10 K, for a chilly 220 K. Earth is somewhere in between; the minor CO_2 concentration in the atmosphere is

enough to produce a greenhouse effect of 35 K and a mean surface temperature of 290 K (Table 1.5). Yet the early atmosphere on Earth almost certainly had

Table 1.5 **The Earth in context: basic comparisons with Venus and Mars. (From MacDonald 1989)**			
	Venus	**Earth**	**Mars**
Main gas	CO_2	N_2	CO_2
Atmospheric pressure (atm)	90	1.0	0.01
Mean surface temperature (K)	750	290	220
Greenhouse effect (K)	500	35	10
Δ Temperature poles to equator (percent)	2	16	40

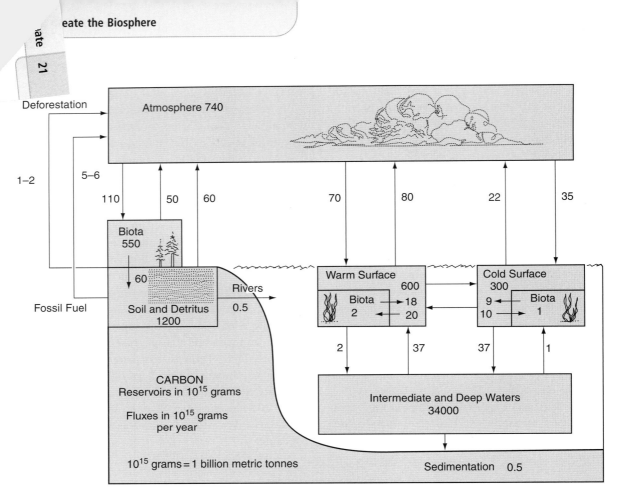

FIGURE 1.16 The carbon cycle showing storage in living plants, soils and fossil fuels. (From Moore and Bolin 1987)

more CO_2 (Table 1.2) – enough to produce much warmer conditions.

Significant amounts of carbon that have been removed from the atmosphere are now stored in the tissues of plants – some 550 billion metric tonnes (Figure 1.16, left). Further, when plants in the past died, their organic remains sometimes did not decay but instead formed thick seams of organic carbon now called coal. Ocean life forms also died and did not fully decay, which contributed to organic shales and oil-bearing rocks. Because of their organic origins, these modern fuels are called fossil fuels, and together they store vast amounts of organic carbon (Figure 1.16). While some CO_2 remains in the atmosphere of Earth, nearly all is

now trapped in living plants and fossil fuels, so temperature is moderated. Some of the greatest deposits of carbon remain in coal. There is still debate about exactly what conditions allowed such vast accumulations of plant matter. Certainly there were vast coal swamps, where inundation reduced oxygen available for decomposition, a process still found in wetlands today (Stewart and Rothwell 1993; Keddy 2010). This suggests that there were large areas of partially or seasonally flooded habitats, in which plants grew and then fell into shallow water or peat (Figure 1.17). There is a steady flow of new information on such environments from coal mines, providing new data on the plant species and environmental conditions. As but one

FIGURE 1.17 A coal swamp in the Permian era preserved by a volcano that erupted about 300 million years ago. The volcanic ash buried this ancient swamp nearly intact, until it was uncovered by an open-pit mine in the Wuda Coal Field in China. Several species of tree ferns formed the lower canopy. Towering 25 m above them were trees including a species of *Cordaites* (an extinct early conifer, shown on the left) and *Sigillaria* (an extinct Lycopod, shown on the right). (Painting by Ren Yugao, Wang et al. 2012)

example, an open-pit coal mine in China uncovered a fully preserved coal swamp landscape including ancient trees in extinct genera such as *Cordaites* and *Sigillaria*, still rooted in fossil peat that was evidently once covered by shallow water (Wang et al. 2012).

The rates at which organic carbon was buried at different periods in the Earth's history can be estimated from the chemical composition and volume of sedimentary rock of different ages. These estimates show a dramatic peak at some 300 million years BP (Figure 1.18). Moreover, deposition conditions can be estimated by ratios of pyrite sulfur to organic carbon, with lower ratios indicating that deposition took place on land. Robinson (1990) attributes the peak in coal deposition to the evolution of lignin, which is not only difficult to degrade but inhibits the decay of associated materials. Only the recently evolved basidiomycete fungi can degrade lignin, and this process is inhibited at the low oxygen levels that are typically found in flooded soils. Robinson estimates that in the Pennsylvanian, the lignin content of plants was 40 percent or higher. Her interpretation of Figure 1.16 is that, "Paleozoic forests had limited geographic extent and contained less biomass than modern

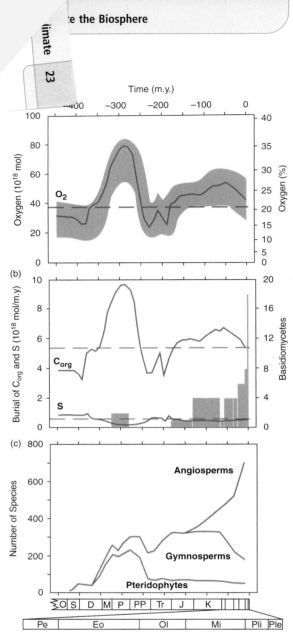

FIGURE 1.18 The chronology of four key properties through geological time: oxygen production (a), carbon burial (b), sulfur deposition (b) and plants (c) through geological history. Lower ratios of reduced sulfur (pyrite sulfur) to organic carbon indicate deposition in terrestrial conditions. Note the surge in oxygen production and carbon storage about 300 million years ago. (From Robinson 1990)

forests. By providing a rich source of lignin in an environment where lignolytic organisms were rare or absent, however, they greatly increased the C_{org} content of the infill of subsiding basins and caused the greatest bulge in terrigenous C_{org} burial in Earth history" (Robinson 1990: p. 609). This would be consistent with current understanding of the evolution of fungi, particularly the relatively recent origin of the lignin-consuming basidiomycetes (Berbee and Taylor 1993). It is also possible that there was a period of 60 to 70 million years when plants were comparatively free of herbivores, which allowed coal to accumulate rapidly (Southwood 1985).

Perhaps the story is somewhat larger still. It is not just fungi and herbivores that can affect rates of accumulation of dead plants. Many kinds of invertebrates enhance rates of decomposition by breaking down dead plant material into smaller particles themselves, and by simultaneously increasing the surface area for fungi to attack. Important modern groups of insects including beetles, flies, termites and ants did not appear until after the late Carboniferous. Hence plant debris may have accumulated to depths we rarely observe today. To try to evaluate this hypothesis, Raymond et al. (2001) compared some properties of modern peats with ancient coal peats. As one example, in a modern mangrove swamp, a leaf mat is typically less than 10 leaves thick, while in ancient *Cordaites* swamps (from four coal deposits) the leaf mats ranged from 10 to 40 leaves thick (and occasionally even more). This is consistent with lower rates of decay in Carboniferous *Cordaites* swamps.

Overall, there is a growing body of evidence that rates of terrestrial decomposition have increased since the Paleozoic. Lignin, fungi, herbivores and detritivores offer four entirely biological explanations for the accumulation of coal and, perhaps more importantly, suggest that modern conditions are not conducive to the rates of carbon storage that occurred earlier in geological history.

1.10   Sediment and Ice Cores Provide a Record of Past Environments

Let us now turn to two much shorter time periods: the past 65 million years, and the past half million years (Figure 1.19). Both graphs show merely the very tail end of the lengthy period shown at the bottom of Figure 1.18, but both are important time periods for plant ecologists to consider. It was 65 million years ago that an enormous object hit the Earth, triggering the end of the dinosaurs (the K/T extinctions) and the rise of mammals and flowering plants (Section 8.2). Thus the top graph in Figure 1.19 covers the period from the K/T extinctions to the present. And it was the past half million years that has seen plant communities exposed to at least four major ice advances

(Section 8.3). Thus the bottom graph shows the past 400,000 year time span. Since you will read more on both the K/T extinctions and ice ages in Chapter 8, let us focus here upon the theme of climate change and CO_2.

The two time scales in Figure 1.19 are studied using different sources of evidence. The top 65 million-year time scale comes from sediment cores drilled in the deep ocean floor (Zachos et al. 2001). These record the history of the Earth as measured by debris that has settled out of sea water. As Figure 1.19a shows, temperature (as measured by oxygen isotope ratios) has declined more or less steadily over this period. Note

(a) Deep sea sediment cores

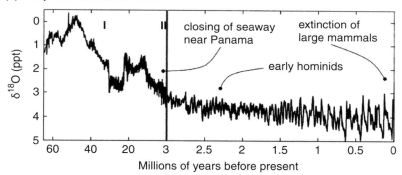

(b) Antarctic ice core (Vostok)

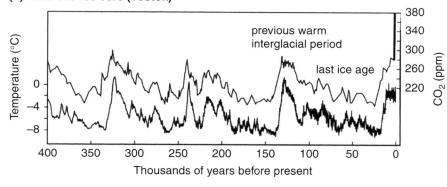

FIGURE 1.19 Changes in global climate over two time scales from two sources of data. (Adapted from Fedorov et al. 2006) (a) Global temperature for the last 65 million years inferred from oxygen isotope ratios in deep-sea sediment cores. (I) marks the onset of glaciers at the south pole, and (II) the onset of glaciers at the north pole. Note the change in time scale at the vertical bar. (b) Global temperature for the past 400,000 years inferred from oxygen isotope ratios in an ice core (bottom line), and associated CO_2 levels measured from bubbles (top line).

arth has cooled, the temperature has also
striking fluctuations. These are thought
to be caused by cyclical changes in the Earth's orbit
(precession, obliquity and eccentricity), termed
Milankovitch cycles (Imbrie et al. 1993; Muller and
MacDonald 1997), named for the Serbian mathemat-
ician and engineer Milutin Milankovitch (1879–1958)
who proposed their relationship with ice ages. This era
has seen several major changes in plants and climate.
These include the gradual rise in dominance of the
flowering plants, the expansion of C_4 grasses, the
development of ice sheets first in Antarctica (I) and
then the north (II), the diversification of grazing
animals (e.g. horses), the closing of the isthmus of
Panama, and the appearance of hominids. Overall, it
has been a period of steady cooling, ending with the
modern human era.

Ice cores drilled from glaciers provide more detailed
information on the past half million years – the
Pleistocene (Figure 1.19b). Ice cores also contain small
bubbles that hold actual samples of the previous
atmosphere. This allows scientists to track changes in
greenhouse gases such as CO_2 and methane. An ice
core taken from Vostok (Antarctica) reached a depth
of 3,623 m and provides a vivid picture of climate
change over the past 400,000 years (Petit et al. 1999).
It shows that major cool periods have occurred about
every 100,000 years. This cycle is approximately
consistent with that predicted by Milankovitch; evidence can also be found for shorter cycles of 41,
23 and 19 thousand years (Petit et al. 1999). Alternating with these ice advances were warmer interglacial periods, which are closely correlated with
peaks in CO_2 concentration (as shown by the top line
of Figure 1.19b and methane (not shown)). We are
currently in a prolonged interglacial period.

The linkages among plants, atmosphere and climate are both simple and complex. Simply, plants can
control the composition of the atmosphere, including
global CO_2 levels, and thereby affect climate. The
complexity arises because they are not acting alone.
The Earth has been through a prolonged cooling
period over the past 60 million years, variation in the
Earth's orbit having become increasingly influential

in driving cycles of glaciation (Figure 1.19a). Milankovitch cycles were not so important in the warmer
past, but appear now to be driving the onset and
conclusion of ice ages. Other factors amplify this
pattern (Fedorov et al. 2006). Rising levels of greenhouse gases near the end of glacial periods
(Figure 1.19b) appear to accelerate the rate at which
ice melts. As the ice melts, the albedo (reflectivity) of
the Earth falls, white snow and ice being replaced by
darker coloured earth and vegetation that absorb more
heat. Changes in cloud cover over the ocean may
simultaneously alter albedo. Changes in ocean circu-
lation and surface temperate may further amplify
changing climate, but this change depends in part
upon the configuration of continents. Plant communities are affected by, and yet also affect, these patterns. The vast peatlands of the northern hemisphere
arose after the last glacial period, and now continue
to store organic debris from plants, particularly
Sphagnum moss (Figure 10.13), thereby having a
cooling effect. They also emit methane, a greenhouse
gas. Peatlands now cover some 500 million ha,
nearly four percent of the Earth's ice-free land area
(Gorham 1990). Should they cease to grow, or should
they burn during droughts, CO_2 levels would increase
further.

Here we have focused upon ocean sediment cores
and ice cores to address the history of the Earth at the
global scale. You should be aware that sediment cores
taken from lakes and peat bogs record many other
more local features of the Earth's history. Such sediment cores contain many kinds of organic debris:
pollen grains record past vegetation types (Moore
et al. 1991), charcoal fragments record past fires
(Section 5.3.1) while phytoplankton remains record
aquatic events (Smol and Cumming 2000).

Let us end with the shortest time scale, the past
century, where we have various direct measurements
of CO_2 levels in the atmosphere. If you examine the far
right-hand side of the ice core record in Figure 1.19b,
you can see a sudden spike in CO_2 levels that
corresponds with humans clearing forests and
burning fossil fuels. On a geological time scale this is a
sudden event; from the short-term perspective of

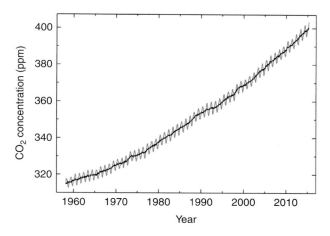

FIGURE 1.20 Monthly average concertrations of atmospheric carbon dioxide measured at Muana Lau Observatory, Hawaii (1958–2015). Current levels have passed 400 ppm. The black line is seasonally corrected data. (National Oceanographic and Atmospheric Administration, USA)

humans it has seemed more gradual. Over the last century, CO_2 levels have been rising steadily (Figure 1.20) and are now the highest recorded in the past 400,000 years.

The study of past climate may provide important information about future climates. During the warm period 40 to 60 million years ago, there were no glaciers in the northern hemisphere, and sea level was about 25 m higher than today (Fedorov et al. 2006). The last four interglacial periods have all been relatively short and the current interglacial in which we live, the Holocene, is already the longest of the four. Thus we know that it is possible for the Earth to be both much warmer or much colder than present with relatively small changes in solar input. Moreover, the current atmosphere takes humans into unknown territory, since atmospheric CO_2 levels are not only strikingly higher than in previous interglacial periods,

they are well above any level recorded in the past half million years (Figure 1.19(b)), and they continue to rise (Figure 1.20). There is therefore ample reason to suspect that rising global temperatures will have a significant effect on plant distributions in the coming century, a topic explored further in Flannery (2005) and Gore (2006).

Finally, note that the trend in CO_2 with time measured at Mauna Loa itself shows a cycle. This brings us back to plants. Each year the rapid growth of plants in the northern hemisphere draws down the pool of atmospheric CO_2 and each winter as the organic matter decays and organisms respire, that carbon is returned to the atmosphere. Figure 1.20 thus illustrates the overall theme of this chapter: that plants are not only influenced by the atmosphere, but they can change the atmosphere.

1.11 The Biosphere

The word biosphere, which refers to that relatively thin layer on the surface of the Earth within which life exists, is now rather familiar. Yet the concept, according to Hutchinson (1970), was

introduced into science rather casually by an Australian geologist, Eduard Suess, in 1875. The idea remained largely overlooked until the Russian mineralogist, Vladimir Vernadsky, published *La*

Vladimir Vernadasky Writes *La Biosphère*

La Biosphère by Vladimir Vernadsky (1863–1945, Figure B1.1.1) was published in 1929, based upon a Russian edition in 1926. In this slim book, Vernadsky publicized the term biosphere, and then laid out some basic yet visionary principles that would become important themes in ecology. These included:

1. That the rates of reproduction of organisms such as termites will lead to geometric rates of increase in population size. A queen termite, for example, produces 60 eggs per minute, or 86,400 in 24 h, in which case in a few years they could cover the entire surface of the Earth, which he estimates at 5.10065×10^8 square kilometres (pp. 39–41). (Darwin, of course, used this observation in formulating the principles of natural selection and evolution.)

2. That, as all organisms could multiply to this extent, an external force (*obstacle extérieur*) sets an upper limit to their population size. There must be, he adds, some maximum numbers for different life forms, and these abundances, N_{max}, are characteristics of different species (pp. 46–48). (Today this upper limit is termed the carrying capacity, and finds its way into the Lotka–Volterra equations as K.)

FIGURE B1.1.2 Vernadsky wrote about the biosphere when space travel was science fiction. Now it is not only possible, but routine, to note how vegetation colours the continents, and how weather systems transport water, carbon dioxide, oxygen and mineral nutrients (as dust) across the Earth. (Image by Robert Simmon, NASA Earth Observatory using Suomi NPP VIIRS data from Chris Elvidge/NOAA)

FIGURE B1.1.1 Nataliia and Vladimir Vernadsky in 1910. (From Bailes 1990)

Box 1.1 (cont.)

3. That the gases of the atmosphere are identical to those created by the gaseous exchange of living organisms. These are oxygen, nitrogen, carbon dioxide, water, hydrogen, methane and ammonia. This, he adds, is no accident (*ne peut être accidentel*). The amount of oxygen produced by plants, some 1.5×10^{21} g, corresponds to the amount of living matter that has been produced by plants (p. 57). (The origin of life on Earth is now traced back to gases such as these, leading Morowitz to observe that life is not an accident but an inevitable outcome of energy flow through the atmosphere. In spite of this, too many people still try to pretend that science says that life arose "by accident." They are out of date by about two centuries.)

Of course, Vernadsky was not always correct. For example, he asserts on p. 63 that the green microorganisms in the ocean are the principal transformers of solar energy to chemical energy on the planet. (While promises of oceans feeding the world persisted into the late twentieth century, better measures of production revealed that most of the oceans are unproductive, with pockets of high production along coasts, in estuaries and where upwellings are produced by ocean currents.) Nonetheless, one can see in this book an attempt to chart the major processes on Earth in terms of chemical and biological process calculated for the entire planet (Figure B1.1.2). The extensive modern literature on biogeochemical cycles in general, and global warming in particular, can be traced back to early work such as this.

Students can read Vernadsky with two rather different perspectives. First, it may seem remarkable that many apparently modern ideas were being discussed more than a century ago. Second, it is equally remarkable to realize just how young the science of plant ecology really is, and that we are living through a period in which an entire scientific discipline is being constructed in front of our own eyes, perhaps including our own contributions. In my experience *La Biosphère*, like *The Origin of Species*, has a strange but compelling mixture of the past and the present, which is both hard to describe and difficult to forget.

Biosphère in 1929 (Box 1.1). The word has now attained a general usage and significance that Vernadsky probably could not have imagined.

The biosphere has conditions that are rare in the universe as a whole – liquid water in substantial quantities, an external energy source (the Sun) and temperatures at which there are interfaces between solid, liquid and gaseous forms of water. Liquid water exists under a rather narrow range of conditions of temperature and pressure. It was once abundant on Mars and may still occur beneath the ice on Jupiter's moon Europa. New information is continually emerging from interplanetary space probes. At one end of the galactic temperature gradient there are temperatures of trillions of degrees inside stars and, at the

other end, there are conditions near absolute zero in the vastness of space. Neither extreme provides the conditions where biological chemistry, at least as humans understand it, can occur. The biosphere of Earth offers an intermediate set of environmental conditions.

And as for plants, they have taken these intermediate conditions and further modified them. There is an oxidized atmosphere with very low concentrations of carbon dioxide, an oxidized ocean and even immense volumes of rocks containing enormous amounts of organic matter from long dead plants. The forests of the Earth have even changed the albedo (reflectivity) of the planet, and therefore its energy balance. The biosphere, as the name indicates, is a product of life and, most particularly, plant life.

Plants have formed and transformed the physical and chemical nature of the biosphere (Table 1.6). Had plants not invaded the land back in the Silurian (Figure 1.5) and created coal swamps (Figure 1.17), you would not be reading this chapter. In the rest of this book we shall explore the great communities of plants that now rule the world. If you master a few general principles, and understand a few key causal factors, you will appreciate how they function altogether. In the next chapter we shall explore the variety of plant types and their geographical distributions. This will allow us to read about early global explorers, and many kinds of unusual plant species, and prepare us to look at general principles in the following chapters.

Table 1.6 **Some of the major effects of plants in creating life on Earth**	
Stage of plant evolution	Some principal effects
Origin of photosynthesis	1. Increased amounts of biological matter
	2. Oxidation of oceans
	a. Precipitation of metal ions
	b. Respiration
	c. Eukaryotic cells
	3. Oxidation of atmosphere
	a. Ammonia and methane removed
	b. Ozone layer formed
	c. Nitrates replace ammonia
	d. Oxidative weathering of rocks
Invasion of land	1. Reduced rates of erosion and sedimentation
	2. Removal of atmospheric CO_2 and cooling of Earth
	3. Production of coal
	4. Formation of soils
	5. Food and shelter for terrestrial animals
Origin of angiosperms	1. Flowers lead to insect diversification
	2. Fruits provide food for birds, mammals and insects

Review Questions

1. What were the consequences of photosynthesis for the environment of Earth? Explain how we can date the first occurrence of photosynthesis in sedimentary rocks. Don't forget to look again at Figure 1.15.

2. Put a mental dot on Figure 1.4 showing where you live. Where, on this diagram, did other important events in history occur? The siege of Troy? The Boer war? The Vietnam war? Where are important books located? *Heart of Darkness*? *The Gulag Archipelago*? *1984*? See if you can name a film from each region.

3. Write an enrichment box about *Rhynia* and its significance for plants. Be sure to mention where the first fossils were found, when and by whom. Describe the life cycle and habitat. Using your favourite search engine find some photos. Are there any specimens you can view in a local museum?

4. Who was the first person to write about the biosphere?

5. One frequently hears from the ill-informed that evolution contradicts the second law of thermodynamics. Explain, using the examples in this chapter, how evolution is, in fact, a necessary consequence of the second law of thermodynamics.

6. Why does the global carbon dioxide concentration in the atmosphere (Figure 1.20) show an annual cycle superimposed on the long-term trend?

7. What is a fossil fuel? How do fossil fuels regulate the Earth's temperature? (I told my classes that this would be on the final exam every year. It was. Yet every year some people still found themselves unable to answer the question! Even graduate students.)

Further Reading

Lavoisier, A.L. 1789. Elements of Chemistry. Translated by R. Kerr and reprinted. pp. xii and 1–60. In M.J. Adler (ed.) 1990. *Great Books of the Western World.* 2nd edn., Vol. 42. Chicago: Encyclopaedia Britannica Inc.

Morowitz, H.J. 1968. *Energy Flow in Biology: Biological Organization as a Problem in Thermal Physics.* New York, NY: Academic Press.

Margulis, L. 1993. *Symbiosis in Cell Evolution.* 2nd edn. New York: W. H. Freeman.

Mains, G. 1972. *The Oxygen Revolution.* Newton Abbot: David and Charles.

Niklas, K.J., B.H. Tiffney and A.H. Knoll. 1983. Patterns in vascular land plant diversification. *Nature* 303: 614–616.

Day, W. 1984. *Genesis on Planet Earth*, 2nd edn. New Haven, CT: Yale University Press.

Ferris, T. 1988. *Coming of Age in the Milky Way.* Anchor Books edition 1989. New York, NY: Doubleday.

Robinson, J.M. 1990. Lignin, land plants, and fungi: biological evolution affecting Phanerozoic oxygen balance. *Geology* 18: 607–610.

Moore, P.D., J.A. Webb and M.E. Collinson. 1991. *Pollen Analysis*, 2nd edn. London: Blackwell Scientific.

Knoll, A.H. 1992. The early evolution of eukaryotes: a geological perspective. *Science* 256: 622–627.

Schopf, J.W. and C. Klein (eds). 1992. *Evolution of the Proterozoic Biosphere: A Multidisciplinary Study.* New York, NY: Cambridge University Press.

Levin, H.L. 1994. *The Earth Through Time*, 4th edn, updated. Fort Worth, TX: Saunders College Publishing; Harcourt Brace College Publishers.

Vernadsky, V.I. 1998. *The Biosphere*. New York, NY: Copernicus (Springer-Verlag). Translated by D.B. Langmuir, revised and annotated by M.A.S. McMenamin.

Petit, J.R., J. Jouzel, D. Raynaud et al. 1999. Climate and atmospheric history of the past 420 000 years from the Vostok ice core, Antarctica. *Nature* 399: 429–436.

Bell, P.R. and A.R. Hemsley. 2000. *Green Plants: Their Origin and Diversity*, 2nd edn. Cambridge: Cambridge University Press.

Laws, B. 2010. *Fifty Plants That Changed the Course of History*. Buffalo, NY: Firefly.

2

The Search for Global Patterns

Two ways of sorting plants into groups. Function and phylogeny. (1) Functional classification: von Humboldt, Raunkiaer, Küchler. Classification of climate: the Köppen system. Biomes. (2) Phylogenetic classification: Linnaeus, Bentham, Hooker, Wallace, Darwin. Molecular systematics. Asteraceae, Orchidaceae, Poaceae. Takhtajan and world floristic regions. Ecoregions as a synthesis.

FIGURE 2.1 Botanical exploration has its risks. On his first voyage Captain Cook sailed the *Endeavour* to the coast of New Holland (now Australia), where they anchored in Botany Bay (now Sydney Harbour) on 28 April 1770, allowing Joseph Banks to begin a remarkable collection of plants entirely new to science. Only a few days later the ship was nearly wrecked on a coral reef and had to be beached on a rocky shore for repairs. (Detail from a plate from Hawkesworth's 1773 account of the voyage)

2.1   Introduction: There Are Two Ways to Classify Plants and Vegetation

We must begin with a biological paradox of sorts – the Earth is indeed mostly covered in green, but you need not one, but two, maps to adequately describe the pattern. Moreover, both maps took centuries to compose. Explorers died while preparing them, including both Cook and Magellan. Von Humboldt and Wallace also had near brushes with death, but survived to old age. Still, you would think that with just one planet, one map surely would be enough to remember. It is not, and that is probably the most interesting part of the story in this chapter. Along the way, you will find out how the notorious British penal colony of Botany Bay in Australia was named (Figure 2.1), and learn why you have to go there, and only there, to New South Wales, if you want to see wild Waratah in flower.

Here is another way to think about the underlying problem we will address in this chapter. Most of the Earth's living organic carbon is locked into the tissues of plants. The rest of the Earth's multicellular biota, including insects, birds, reptiles, amphibians and mammals, comprises a mere 0.1 percent of the carbon pool (Whittaker 1975). This living organic carbon is spread in a thin layer over the Earth's surface – thinnest in deserts and thickest in forests. To proceed further with the scientific analysis of this living organic carbon, it is necessary to cut it into smaller sensible categories. There are two tools for making this dissection – the functional and the phylogenetic. They are contradictory, yet also complementary. That is part of the paradox. Neither is entirely right, neither is entirely wrong. The functional approach aims to sort plants into groups using traits such as shape, leaf form, life history and physiology. The phylogenetic approach aims to sort plants into groups sharing a common evolutionary history. A well-trained plant ecologist should be familiar with both approaches. Too often, however, practitioners become stuck on one view, unwilling to appreciate the value of the other. We will begin here with the functional view (Section 2.2). This is a good place to start because even people with little experience of plants can recognize simple functional groups such as evergreens, or succulents or shrubs. Moreover, there is a growing body of recent work seeking predictive models of vegetation based upon plant traits (Weiher and Keddy 1999; Shipley 2009). We will then turn to the phylogenetic view (Section 2.3). This view will lead naturally to questions about plant diversity in different habitats, a topic with important implications for both theory and conservation, and to which we will return in Chapters 12 and 13.

2.2   Functional Classifications Are Based on Ecological Traits

The concern of functional classification is to classify plants and map the Earth based upon ecological traits such as morphology and physiology. The objective is not to construct groups that reflect past evolutionary relationships, but rather to construct groups that reflect evolutionary convergence. It is based on the observation that some kinds of plants look quite similar even though they have very different origins. (This, for example, allowed some western films starring Clint Eastwood, allegedly taking place on the American frontier, to actually be filmed in Spain.

Unless you are very observant, most plants of dry areas look superficially similar. But as you learn more about plants, you start to notice films where the plants in the scenes contradict the alleged setting.) Succulents, for example, all share certain traits for water conservation, but they are drawn from a wide array of plant families. Cacti are a largely New World group, while Aloes are largely African. The similarities among such succulents are the consequence of natural selection, which drives organisms to adopt similar physical forms in similar habitats, in this case deserts.

It is this kind of functional similarity we want to map in this section.

2.2.1 Functional Classifications by von Humboldt, Raunkiaer and Küchler

One of the earliest functional classifications, prepared by Alexander von Humboldt (Box 2.1), recognized 19 categories (*Hauptformen*). Most were named after a typical genus or family, such as Palmen-form, Cactus-form and Gras-form. As with many systems that are initially simple, von Humboldt's was expanded to accommodate diversity and apparent exceptions. By 1872 Grisebach's work, *Die Vegetation der Erde*, published in Leipzig, yielded some 60 physiognomic types. These were grouped into

---

Box 2.1   Alexander von Humboldt Explores the Andes

The great era of global exploration in the 1800s has passed. In today's era of satellite imagery, robot explorers on distant planets and near-instantaneous electronic communication around the world, it may be difficult to appreciate the tasks that faced early global explorers and scientists a little over a century ago. Of early scientific expeditions, Darwin's voyage of the *Beagle* down the east coast of South America and up the west coast to the Galapagos is perhaps the best known. Countless other names are less appreciated: Stanley (1874–1889, Africa), Livingstone (1849–1864, Africa), Przhevalsky (1870–1880, Mongolia and Tibet), Cook (1768–1779, Pacific Ocean and Western Coast of North America), Mackenzie (1789–1793, north western North America), Younghusband (1887–1891, Tibet). This list does not include the names of explorers of infamy who used the force of arms in their search for land or gold (e.g. Cortez, Pizzaro) and who destroyed what they found and enslaved the native inhabitants of the places that they reached; nor does it include missionaries who sought to change what they found. The names that deserve recognition are those who made contributions to humanity through their attempts to expand the frontiers of human understanding. Perhaps the greatest of these was the German naturalist Alexander von Humboldt (1769–1859, Figure B2.1.1).

FIGURE B2.1.1   Portrait from 1814 of Alexander von Humboldt, a naturalist, explorer, scientist and model for aspiring young ecologists. (From Wikimedia Commons)

(Continued)

Box 2.1   (cont.)

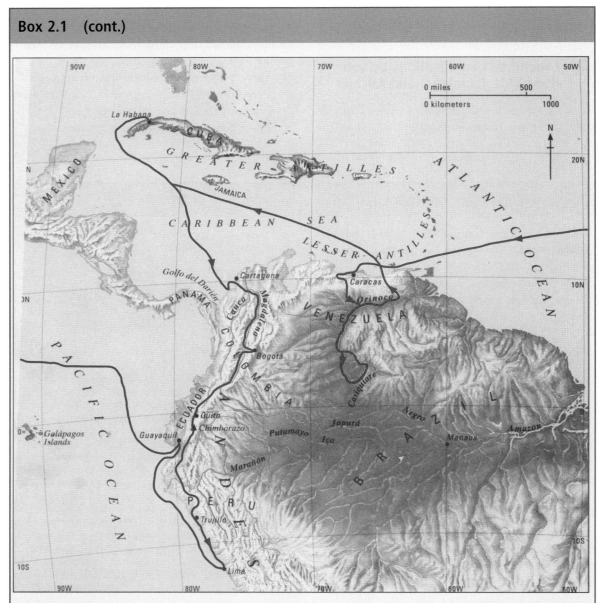

FIGURE B2.1.2   Von Humboldt's voyage from 1799 to 1803 took him from mangrove swamps and floodplain forests to deserts and mountain peaks. All the while he collected plants, made careful notes, measured temperature and barometric pressure, and later on the trip even collected data on social and economic conditions in Mexico. On the way home, he visited President Jefferson in the United States. (From Edmonds 1997)

Box 2.1   (cont.)

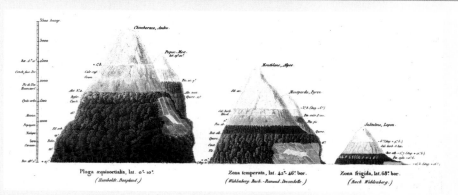

FIGURE B2.1.3   This portion of a hand-coloured plate in the first volume of *Nova Genera et Species Plantarum* (by Bonpland and von Humboldt, 1815) describes how vegetation changes with altitude at three different latitudes: equatorial (left), temperate (centre), and cold (right). The actual three mountains are Chimborazo (Ecuador), Mont Blanc (border of France/Italy) and Sulitelma (border of Sweden/Norway). The book was written in Latin and can be viewed at www.biodiversitylibrary.org/item/11233. Aimé Bonpland was a doctor and botanist who travelled with von Humboldt through South America.

Von Humboldt's family were Huguenots (French Protestants) who had fled Catholic France after Louis XIV revoked religious freedom. Von Humboldt initially thought of joining the army, began futile studies in economics and then developed a passionate interest in botany. Between 1799 and 1803 he explored the northern areas of South America, from the rain forests of present-day Venezuela to the Andes (Figure B2.1.2). He even set the world altitude limit for mountain climbing while ascending Mount Chimborazo, although altitude sickness and bad weather prevented him from reaching the peak. The years 1804 to 1827 were spent in Paris where he published the results of this expedition (Figure B2.1.3).

His fame was such that he was compared at the time with Napoleon, but in addition to being a determined explorer and an accomplished scientist he was also a decent human being. He sought out promising students and personally supported them in spite of his precarious finances; Justus von Liebig and Louis Agassiz were both launched by him. He encouraged the young Charles Darwin. Like Darwin, he denounced slavery and forced labour, but, unlike Darwin, he freely mixed with society and appeared regularly in the salons of Paris. His enthusiasm for the popularization of science led him to publish Kosmos, an overview of the structure of the universe, and he gave public lectures to audiences of a thousand at a time. He died at 90 while still working on the fifth volume of Kosmos (von Humboldt 1845–1862).

In 1831, Darwin was inspired by von Humboldt's seven-volume, 3,754-page account of his trip to South America. He is said to have read out long passages to his friend John Stevens Henslow in order to encourage a joint trip to Tenerife, but before arrangements for the trip could be made, he received a letter (29 August 1831) offering him passage on the *Beagle* in the role of a well-bred gentleman to ease the isolation of Fitzroy's command (Desmond and Moore 1991). Darwin sent von Humboldt a complimentary copy of his book on the voyages of the *Beagle*. Von Humboldt, says Darwin's biographer, was a figure of "almost mythic proportions in Darwin's mind. . . Humboldt's highly laudatory reply consequently electrified him. The grand old man behaved

(Continued)

Box 2.1   (cont.)

just as grand old men were expected to behave, and warm congratulations poured from his pen like honey, praising Darwin's 'excellent and admirable book'" (Browne 1995: p. 416).

Here is a potential model for young plant ecologists, a man brave and determined, accomplished in botany, yet compassionate and awake to the needs of his world. Compare his personality with that of Galileo, who denied Kepler a badly needed telescope out of fear of what he might discover, or of Newton, who huddled in his office at Cambridge completely cut off from the rest of the world, unwilling to share his discoveries. Perhaps even Darwin fares badly by comparison. He was reclusive and, it appears, feared the negative publicity that would arise if he published his ideas; he spent eight years studying barnacles rather than publicizing his views on topics such as slavery. Indeed, says a biographer (Browne 1995), Darwin turned his house into the *Beagle*, "self contained" and "safely isolated from the concerns of the world" (p. 530).

seven main categories (Shimwell 1971: p. 67) that would still be familiar to most field botanists today:

Woody plants (*Holzgewächse*)
Succulents (*Succulente Gewächse*)
Climbing plants (*Schlinngewächse*)
Epiphytes (*Epiphyten*)
Herbs (*Kräuter*)
Grasses (*Gräser*)
Lower plants (*Zellenpflanzen*)

To try to add increased precision, in 1903 Raunkiaer first presented his influential system, which is based largely upon the location of perennating meristems (Figure 2.2). Meristems (buds) are the basic units by which plants grow. They may be temporary or long-lived, and above or below ground. Indeed, you can argue that most plants are colonies of many meristems, a topic to which we will return in Chapter 10. So to start with how plants protect meristems seems an eminently sensible way to categorize them. To apply the Raunkiaer system, we must focus on only a subset of these meristems – those that overwinter or, to use a more general term, perennate. It is common to think of perennating meristems as the overwintering meristems, but in warmer parts of the world they may equally tolerate periods of drought, or allow regrowth after fire or grazing. Raunkiaer appreciated that the location of perennating meristems was closely

connected to climate, easily recognizable in the field, and a single characteristic that could be enumerated for statistical comparisons of geographical locations (Raunkiaer 1907, 1908). There are five main categories.

1. Phanerophytes produce their perennating meristems high above ground. You know them as trees or shrubs, and they are put into four quite artificial categories according to height.
2. Chamaephytes are also woody, but their perennating meristems are produced close to the ground, and in colder climates are often buried by winter snow.
3. Hemicryptophytes produce their meristems at the soil surface.
4. Cryptophytes produce their meristems beneath the soil surface or under water.
5. Therophytes are annuals, and pass unfavourable periods as embryos within a seed.

These terms may seem unfamiliar, but they make sense – if you know Greek. For example, *crypto* indicates concealed or hidden. (You may take it as an exercise to look up the other terms; it will help you remember them.) In practice, it is often useful to expand these five categories to include three (or four!) different kinds of phanerophytes and three kinds of cryptophytes, as well as adding stem succulents and epiphytes (Table 2.1). This conveniently produces ten categories, which is the system

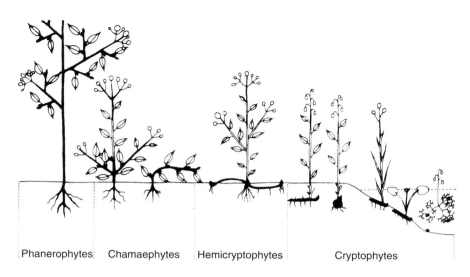

FIGURE 2.2 The Raunkiaer system classifies plants based upon the location and the protection provided to meristems. (From Goldsmith and Harrison 1976, redrawn from Raunkiaer 1907)

Phanerophytes   Chamaephytes   Hemicryptophytes   Cryptophytes

Raunkiaer used in many of his tables of life-form spectra (Table 2.2).

Raunkiaer was also astute enough to recognize the need for a null model against which world floras could be compared on the basis of meristem location (see Box 11.1). Thus he objectively selected 1,000 species from around the world to provide a random sample of life forms. It was therefore possible to use his life-form spectra not only to compare specific sites with different climates but also to compare each site to a standard, a worldwide reference spectrum, or what Raunkiaer called the "normal spectrum" (Table 2.2, bottom line). Look more carefully at Table 2.2. This world "normal spectrum" is worth considering more carefully. It answers an important general question. What are the world's plants like? The answer is that an overriding proportion of the world's plants are trees: 6 + 17 + 20 = 43 percent of them! Figure 2.3 shows one of the world's rarest species of phanerophyte. The next largest group is hemicrytophytes, comprising 27 percent of all plants.

Although many other functional classification systems exist, most involve various elaborations and expansions of the Raunkiaer system. It is therefore still a worthwhile exercise to go to a natural area and try to assign each species to one of the categories in Table 2.1. It is even better if you can compare two areas differing in altitude, soil moisture or human activity. The principal limitation of the Raunkiaer system lies in its strength: it is made for large-scale geographical comparisons. At finer scales, where larger geographic and climatic gradients are not involved, this system may lack discrimination. At finer scales, it may be other traits that become more important (e.g. Westoby 1998; Weiher et al. 1999).

Let us conclude this section with two more systems that may be considered to be extensions of the Raunkiaer system. In his monograph on the tropical rain forest, Richards (1996) introduces a classification of life forms with seven categories (Table 2.3). With a focus on traits that help produce vegetation maps, Küchler produced a rather more complicated system (Table 2.4). As you can see, this system contains many more traits, including leaf size. I would suggest skimming this table unless you are on a second or third reading! There is a further complication: "Unfortunately," says Shimwell (1971: p. 180), "like many other investigators before him, Küchler (1966) falls foul of the revision trend so that his early scheme (1949) becomes considerably altered in his 1966 paper." Table 2.4 attempts to reconcile these changes.

Table 2.1 Raunkiaer's life-form classification system slightly expanded from Figure 2.2 in order to correspond with Table 2.2, which shows the distribution of life forms in different climates. (Raunkiaer 1934) The final column shows where you can find an example of each life form in this book.

Life form	Symbol	Example in this book
Stem succulents	S	Figure 1.1, *Carnegiea gigantea*
Epiphytes	E	Figure 2.11, *Dendrobium* orchid
Phanerophytes:		
Megaphanerophytes >30 m	MM	Figure 13.1, Dipterocarp trees
Mesophanerophytes 8–30 m	MM	Figure 9.6, *Dracaena cinnabari*
Microphanerophytes 2–8 m	M	Figure 8.10, *Telopea speciosissima*
Nanophanerophytes <2 m	N	Figure 10.1, *Azorella compacta*
Chamaephytes	Ch	Figure 10.16, *Diapensia lapponica*
Hemicryptophytes	H	Figure 6.17 (bottom panel), *Sagittaria platyphylla*
Cryptophytes		
Geophytes	G	Figure 5.39, *Podophyllum peltatum*
Helophytes	HH	Figure 5.20, *Parnassia palustris*
Hydrophytes (protected by water)	HH	Figure 10.23, *Nuphar luteum*
Therophytes (annuals)	Th	Figure 5.38, *Bidens cernua*

Table 2.2 **Selected life-form spectra. (After Raunkiaer 1908) See Figure 2.2 and Table 2.1.**

Region		No. species	Percentage distribution of species among life forms									
			S	E	MM	M	N	Ch	H	G	HH	Th
Franz Josef Land, Russia	82°N, 55°E	25	–	–	–	–	–	32	60	8	–	–
Iceland	65°N, 19°W	329	–	–	–	–	2	13	54	10	10	11
Sitka, Alaska	57°N, 9°E	222	–	–	3	3	5	7	60	10	7	5
Death Valley, California	36°N, 117°W	294	3	–	–	2	21	7	18	2	5	42
Ghardaïa, Algeria	32°N, 4°E	300	0.3	–	–	–	3	16	20	3	–	58
Aden, Yemen	13°N, 45°E	176	1	–	–	7	26	27	19	3	–	17
Seychelles	58°S, 56E	258	1	3	10	23	24	6	12	3	2	16
Normal spectrum		1,000	1	3	6	17	20	9	27	3	1	13

Remember, the point of all these tables is to identify a small number of easily measured traits that allow you to sort plants whose names you don't know into functional groups. If you wish to understand more about how the Earth's climate controls plant life forms and world maps, continue. Otherwise, it is quite acceptable to leap ahead to Section 2.3 and start thinking about an alternative, that is, what phylogenetic rather than functional traits tell you about plants and plant ecology.

2.2.2 Climate Has a Major Impact on Plant Traits

Earth, like most planets, has strong temperature gradients, as illustrated by the accumulation of ice at the poles. Modifying this latitudinal temperature gradient are other factors such as winds, mountains

Table 2.3 **Seven life forms of plants in tropical rain forests. (From Richards 1996)**

A. Autotrophic plants (with chlorophyll)

  1. Mechanically independent plants

    a. Trees and "shrubs"   arranged in a number of strata (layers)

    b. Herbs

  2. Mechanically dependent plants

    a. Climbers

    b. Stranglers

    c. Epiphytes (including semi-parasitic epiphytes)

B. Heterotrophic plants (without chlorophyll)

  1. Saprophytes

  2. Parasites

FIGURE 2.3   Trees, termed phanerophytes in the Raunkiaer system, are the most common plant life form in the world. The meristems are produced high above ground exposed to the atmosphere and hence such plants are most abundant in warm climates. The name comes from the Greek word *phanero*, meaning visible, and *phyte* meaning plant. This is *Magnolia sinica*, one of the rarest species of phanerophyte in the world, restricted to a few forested slopes in Yunnan Province, China. It grows up to 40 metres tall. The flowers are 15 to 20 cm in diameter. (Photograph by Jackson Xu, Fauna & Flora International)

Table 2.4 Vegetation structure classification of Küchler. (After Shimwell 1971)

Life-form categories

Basic life forms	1949	1966	Special life forms	1949	1966
Woody plants			Epiphytes	e	X
Broadleaf evergreen	B	B	Lianas	j	C
Broadleaf deciduous	D	D	Stem succulents	k	K
Needleleaf evergreen	E	E	Tuft plant	y	T
Needleleaf deciduous	N	N	Bamboos	v	V
Aphyllous	O	O	Cushion plants	q	–
Semi-deciduous (B + D)	–	S	Palms	u	–
Mixed (D + E)	–	M	Aquatic vegetation	w	–
			Leaf characteristics		
Herbaceous plants			Hard	–	h
Graminoids	G	G	Soft	–	w
Forbs	H	H	Succulent	–	k
			Large (400 cm^2)	–	l
			Small (4 cm^2)	–	s
Lichens, mosses	L	L			

Structural categories

Height (stratification)	1966	1949		
>35 m	8	t	tall: minimum tree height 25 m	
20–35 m	7		minimum herbaceous height 2 m	
10–20 m	6	m	medium: trees 10–25 m	
5–10 m	5		herbs 0.5–2 m	
2–5 m	4	l	low: trees to 10 m	
0.5–2 m	3		herbs to 0.5 m	
0.1–0.5 m	2	s	shrubs, minimum height 1 m	
<0.1 m	1	z	dwarf shrubs, maximum height 1 m	

Coverage (1967) *Density* (1949)	1949	1966
continuous (>75%)	c	c
interrupted (50–75%)	i	i

Table 2.4 (cont.)		
Structural categories	1949	1966
parklike, patchy (25–50%)	p	p
rare (5–25%)	r	r
barely present (1–5%)	b	b
almost absent (<1%)	–	a

and large water bodies. All of these factors also influence precipitation patterns. You will recall from Figure 1.4 that the kinds of plants and plant communities that arise in any area are strongly controlled by temperature and precipitation, so that the search for patterns of plant form and vegetation type has occurred in parallel with the search for a simple method of classifying the world's climates.

And now for some history.[1] While we now accept it as obvious that temperature and rainfall control vegetation, it took several centuries of research to uncover the patterns. There were no satellite photographs – the thermometer was only invented in 1714, and von Humboldt is credited with first using it to plot isotherms of mean annual temperature and relating these to vegetation (Daubenmire 1978: pp. 280–285). In 1836, Meyen emphasized the importance of annual variation, as well as the mean temperature. Adding in precipitation was important, but it had to be connected in some way to temperature. In 1869, Lisser proposed evaluating the effectivity of precipitation by calculating the ratio of monthly precipitation to monthly temperature: "This concept of discounting precipitation in proportion to temperature was so far ahead of its time that its significance was overlooked for several decades. In fact, his contribution seems never to have been acknowledged by subsequent workers in the field" (Daubenmire 1978: p. 282). This work laid the foundations for two well-known climate classification systems, the earlier one proposed by the

European scientist, Vladimir Köppen, and a later one developed by C. W. Thornthwaite in the United States. The latter included the use of potential evapotranspiration and soil moisture storage, but Dansereau (1959) concludes that in spite of the added sophistication, the climatic types do not correspond to vegetation any better than those of Köppen, and his system (Figure 2.4) is still widely used – being found in both the *Encyclopaedia Britannica* (1991a) and the *Oxford Atlas of the World* (1997). In fact, the *Britannica* summary of climate and weather (vol. 16, pp. 436–522) makes for useful and entertaining reading.

The Köppen system assumed only five main climate types:

tropical rainy (A)
dry (B)
warm temperate rainy (C)
cold temperate rainy (D)
and polar (E).

A sixth type, highland (H) was later added to address the variation of climatic zones in high mountains. In B dryness is the key factor, and it is further subdivided into arid (BW) and semi-arid (BS) sub-types. Lowercase letters are used as modifiers for further categories of rainfall and temperature. With rainfall, for example, f denotes rainfall throughout the year, m denotes monsoon rains, s denotes a dry summer and w denotes a dry winter. Consider two examples. The Amazon River basin

---

[1] Some older and classic works, particularly those not in English, are not included in the list of references. Generally I try to indicate this in the text by putting the title directly in the text, and indicating that the work was published "in" a certain year, without using the standard form of author and date in brackets.

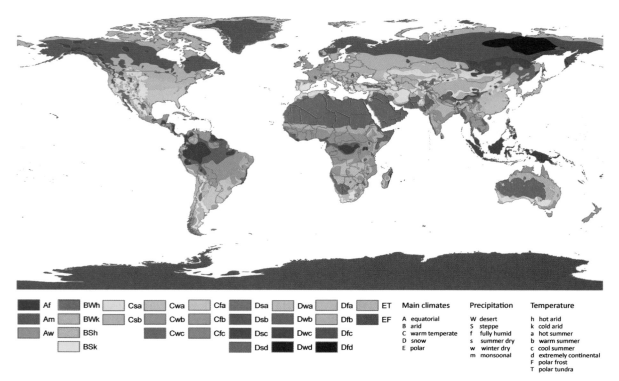

FIGURE 2.4   The Köppen climate classification system. (After Peel et al. 2007)

can thus be divided into **Af, Am** and **Aw** regions. Most of Europe, in contrast, has a **Cf** or **Cs** climate. Finally, one can add another set of lower-case letters to denote special characteristics of temperature as follows: **a** hot summer; **b** warm summer; **c** cool, short summer; **d** cool, short summer and cold winter; **h** hot dry climate and **k** cool dry climate. As an example, most of eastern north America is designated **Cfa**.

The last system for classifying climate we will consider here is the newest, the Bailey system (1998), which conveniently divides the world into only four major climate domains: Humid Tropical (**400**), Dry (**300**), Humid Temperate (**200**) and Polar (**100**). This has the advantage of being simple, and recognizing the overwhelming importance of temperature and moisture. Consider the Dry domain. "The essential feature of a dry climate is that annual losses of water through evaporation... exceed annual water gains from

precipitation. Due to the resulting water deficiency, no permanent streams originate in dry climate zones" (p. 85). Nested within each of these four large domains are smaller divisions. Dry climates can be sorted into regions of arid *desert*, and semi-arid *steppe*. Even smaller divisions exist within these, such as the Tropical Subtropical Desert division (**320**), which includes the Sahara desert in Africa, the Sonoran desert of North America, and the Great Australian desert.

2.2.3 Climate and Life Form are Summarized as Biomes

By combining knowledge about functional groups of plants, and the climates in which they occur, it is possible to divide the Earth into about a dozen ecological types, or **biomes** (Figure 2.5). The names are rather self-explanatory. Moreover, biomes are extensively described in biology text books, and introductory ecology text books, and are readily found on the internet. If you need to refresh your understanding of biomes, refer to one of these introductory sources. The main point is this: this sort of map is based upon functional types of plants. The tropical forests, for example, are dominated by phanerophytes, while the grasslands are dominated by hemicryptophytes.

2.2.4 Functional Classification Systems Have Limitations

While classifying plants according to functional type is useful for scientific purposes, there are certain problems with any system of functional classification:

1. The categories provided may be of too coarse a scale to describe many kinds of patterns, yet when the schemes are expanded they often become too complicated to be easily or usefully applied.
2. The traits used to construct the systems have to be easily assessed in the field; this, however, constrains any system to a small set of morphological attributes. Many other characteristics of plants such as physiology, anatomy and ecology cannot be included. Therefore a system may be both unrealistic and unable to adequately cover the full array of plant types that occur in nature.
3. There is no obvious way to test which system is best and therefore no way to end the proliferation of classification systems.

Let us therefore consider the alternative.

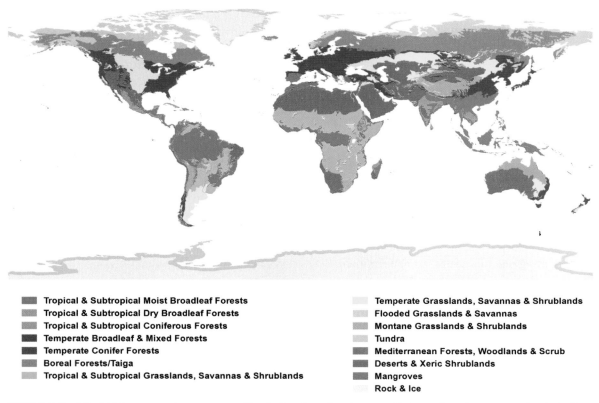

Tropical & Subtropical Moist Broadleaf Forests
Tropical & Subtropical Dry Broadleaf Forests
Tropical & Subtropical Coniferous Forests
Temperate Broadleaf & Mixed Forests
Temperate Conifer Forests
Boreal Forests/Taiga
Tropical & Subtropical Grasslands, Savannas & Shrublands

Temperate Grasslands, Savannas & Shrublands
Flooded Grasslands & Savannas
Montane Grasslands & Shrublands
Tundra
Mediterranean Forests, Woodlands & Scrub
Deserts & Xeric Shrublands
Mangroves
Rock & Ice

FIGURE 2.5 World biomes are large areas with similar functional types of plants, largely as a result of similar climates. Depending upon the source and criteria used, there are about 12 biomes. (After Olson et al. 2001, with permission of WWF-US)

2.3   Phylogenetic Classifications Are Based on Evolutionary History

2.3.1 Early Plant Classification by Linnaeus, Bentham and Hooker

The phylogenetic system is rooted in the pioneering work of Carolus Linnaeus (1707–1778) and his treatises including *Systema Naturae* (in 1735) and *Genera Plantarum* (in 1737). Since then, botanists have invested more than three centuries into exploring the world and sorting organisms into named species. Even so, there are still many species of plants unknown to science and as yet unnamed. Earlier systems of classification existed, of course, but the foundation of our modern classification is essentially Linnaean (Benson 1959; Mayr 1982). By 1764, Linnaeus had listed 1,239 genera of plants.

To put Linnaeus in context, it may be helpful to remember that at the time of his birth, Sweden was a European power; its territory extended to Finland, part of northwestern Russia, Estonia, Latvia and part of northern Germany. The Swedish monarch, Carl XII, fought ceaselessly against his neighbours from 1699 to 1718. Linnaeus was born in 1707, in the midst of these war campaigns, and named for the monarch (Black 1979). He is best known for his work in establishing the binomial system of nomenclature for species, and his books addressing both plants (*Species Plantarum*, in 1753 and *Genera Plantarum*, 5th edition, in 1754) and animals (*Systema Naturae*, 10th edition, in 1758) have been internationally accepted as the starting point for biological nomenclature. Hence, Linnaeus marks the start of the modern taxonomic era and no names published before Linnaeus are accepted, unless they were adopted by him or by subsequent authors.

While his name is now associated with a strong sense of the order in nature, during his time Linnaeus was accused of being a botanical pornographer and of introducing lewd methods into classification (Stearn 1979). These criticisms arose because of his emphasis upon using the male and female characteristics of plants as a means of classification. It probably did not help his cause that his early identification keys referred not to stamens and pistils per se but rather to characteristics (and terms) including the visibility of flowers (public as opposed to clandestine marriages), bisexual or unisexual flowers (husband and wife in the same bed or separate beds) and the number of stamens (one male, two males, up to many males). Critics were outraged; one (J.G. Sigesbeck) stated in 1737 that "such loathsome harlotry" as several males to one female would never have been permitted in the vegetable kingdom by the Creator. Linnaeus obtained revenge by naming an unpleasant, small-flowered weed after him, *Sigesbeckia*.

It is important to remember that Linnaeus' travels were confined to northern Europe, Lapland in particular, although he did encounter tropical plants later in his career when he visited Holland. As a flood of plant material from global explorers began to wash over Europe, it became clear that the 1,239 genera recognized by Linnaeus were vastly insufficient for describing the diversity of the plant kingdom. Consider the enormous number of totally new plant and animal species from Cook's voyages alone (Figure 2.1). New workers rose to the challenge. George Bentham (1800–1884) and Sir Joseph Hooker (1817–1911) were both British botanists who developed a system of classification "based upon an unexcelled study of a great section of the world's flora by detailed research upon families, genera, and species" (Benson 1959: p.464). Together they published *Genera Plantarum* (3 Vols., from 1862–1883).

Joseph Hooker also went on expeditions to New Zealand, Antarctica and India, and had a special interest in Asia. In 1849, while returning from Tibet, he was seized, along with the government agent to Sikkim, by a local anti-British ruler. Hooker was, however, allowed to collect rhododendron seeds even as they were marched south in captivity. "To stop such banditry and show – in Hooker's words – that the rajahs 'could not play fast and loose with a British subject,' southern Sikkim was promptly annexed for the Crown, with Hooker advising the expeditionary forces" (Desmond and Moore 1991: p. 370).

Hooker also became a close advisor to Charles Darwin. He was invited to examine Darwin's plants from the *Beagle* voyage "which were still lying unstudied in a room at Cambridge" (Browne 1995: p. 452). In 1844, Darwin wrote to Hooker of his ideas on the transmutation of species. The reply was positive, for which Darwin was "overwhelmingly grateful." He felt sure Hooker would be one of the best and most erudite sounding boards for his theories" (p. 453). Hooker was in attendance, with Charles Lyell, to present Wallace's paper, along with Darwin's 1844 essay, to the Linnean Society in 1858. In 1876 he was put forward for a knighthood and a very specific additional honour, the Star of India, in acknowledgement of his decades-long work on the Himalayan flora. Hooker even served as pall-bearer at Darwin's funeral at Westminster Abbey in 1882.

2.3.2 Plant Evolution: Wallace, Darwin and Bessey

Alfred Wallace (1823–1913) explored the Amazon in 1848, and published *A Narrative of Travels on the Amazon and Rio Negro* in 1853. Money was a constant problem. His companion in the Amazon, Henry Bates, had an agent who gave him fourpence a specimen (from which he took 20 percent commission) so that Bates received a mere £27 for 20 months' work (Edmonds 1997). In the end, Wallace did not even have this to fall back on: "Wallace on his return to England, was struck by catastrophe. The ship on which he travelled caught fire (August 6, 1852) and sank, with his entire magnificent collection and most of his journals, notes and sketches" (Mayr 1982: p. 418).

From 1854 to 1862 Wallace explored the islands linking Australia to Asia and during this period, while staying at Sarawak in Borneo, wrote an explicit essay on evolution, "On the Law Which Has Regulated the Introduction of New Species" in 1855. He became ill again from malaria in 1858 and, while in a fever, connected his ideas to those of Thomas Malthus (an

English economist, who you will read more about in Chapters 4 and 9): "There suddenly flashed upon me the idea of the survival of the fittest." It is well known that he sent a manuscript to Darwin, who immediately recognized the similarities to his own work. On the advice of Lyell and Hooker, Wallace's essay along with Darwin's own was read to the Linnean Society on 1 July 1858. A new era of science had begun.

Darwin seems to have received a preponderance of the credit for the discovery of evolution. It may be worth noting that Wallace was a man of modest means, while the Darwins were well-established and well-connected within English society:

> In most respects, both men were about as different as two people can be:
> Darwin, the wealthy gentleman, with many years of college education, a private scholar, able to devote all of his time to research; Wallace, a poor man's son with only a lower middle-class background (a very important factor in Victorian England), without any higher education, never particularly well-to-do, always having to work for a living, for the longest time in the exceedingly dangerous profession of a collector of birds and insects in fever-ridden tropical countries. But they agreed in some decisive points. Both of them were British, both had read Lyell and Malthus, both were naturalists, and both had made natural-history collections in tropical archipelagos.
>
> *(Mayr 1982)*

Darwin also had powerful friends in his circle of Cambridge gentlemen: friends who defended and propagated his ideas. The image remains of Darwin living in comfort, with servants on his estate, while Wallace lies sick in the Moluccas. Given that Wallace published his ideas first, it seems remarkable how much he has been eclipsed by Darwin.

The unifying framework of evolution stimulated the search for natural, that is, phylogenetic, systems of plant classification. Bessey (1845–1915) extended Bentham and Hooker's work; he postulated that the order Ranales (most closely related to the current

Magnoliid complex, Figure 2.6) represented the most primitive flowering plants. This group, he suspected, had ancestors with bisexual strobili in which the sterile bracts became a perianth, the microsporophylls became stamens and the megasporophylls became carpels (Thorne 1963). A particularly large number of primitive angiosperm traits appears to be possessed by the five genera and 90 species in the Winteraceae of the southwestern Pacific, Australia, Madagascar, South America and Mexico. The genus *Drimys* typifies a primitive type of ovary in which the margins of the carpel are not yet fused into a cylinder. The genus *Magnolia* is a better known representative (Figures 2.3 and 2.6), and hence the name Magnoliid for this group.

And now an historical tangent with a lesson about science. Adolf Engler (1844–1930) and Karl Prantl (1849–1893) were struck by the similarity between the catkins of some angiosperms and the cones of gymnosperms. *If* one postulates that catkins are primitive flowers, and derived from cones, one can offer a very different evolutionary tree from that offered by Bessey. The preponderance of evidence, particularly that from molecular genetics, does not support this interpretation. However, Engler and Prantl remain in this book for three reasons. First, you may encounter older books in which plants are still organized by their system. Second, and more importantly, they remind us that science requires different hypotheses, and that we learn about nature by collecting data to test among those hypotheses. Third, there is no shame in being wrong, so long as you present your data as honestly as possible, and accept the outcome. Engler and Prantl were wrong, but at the time no one could imagine the power of molecular systematics. Let us turn to this very topic.

2.3.3 Molecular Techniques Provide New Insights

In the late twentieth century it became possible to determine the sequences of amino acids in proteins and of nucleotides in DNA. This opened a vast new realm of information on plant evolution and a "renaissance" in studies of plant evolution (Crane in Heywood et al. 2007). New statistical techniques were developed to sort this sequence data and create evolutionary trees called cladograms (an example appears in Figure 2.7). While some of the fundamentals of the early classification schemes by Bessey and others remain intact, other views of phylogeny have been revised.

Much new data has come from direct studies of DNA. Chloroplast DNA is abundant in cells, and plant systematists have collaborated to build a large database of sequences of the chloroplast gene *rbcL*. This gene codes for a large subunit of a photosynthetic enzyme called RuBisCO (ribulose-1,5-bis-phosphate carboxylase/oxygenase). More than 2,000 species have been sequenced, and phylogenetic relationships mapped (Judd et al. 2002). The cladogram in Figure 2.7 summarizes a phylogeny of flowering plants with this input of molecular data. The study of phylogeny continues; a good reference point is the Angiosperm Phylogeny Group classification (2016). The 2003 APG classification is a partial source for the *Flowering Plant Families of the World* (Heywood et al. 2007).

So how do we now view the evolution of flowering plants? Figure 2.7 shows a set of three primitive "basal groups" on the far left – Amborellales, Nymphaeales and Austrobaileyales. The Amborellales is represented only by a small evergreen tree called *Amborella trichopoda*, endemic to New Caledonia (this island, a small fragment of former Gondwana, is considered a separate world floristic region). A data set combining five gene sequences from 51 taxa indicated that *Amborella* was a basal taxon (Parkinson et al. 1999), but a complete sequence of the chloroplast genome of this plant brings this into question (Goremykin et al. 2003). The Nymphaeales are represented by the more familiar and widespread white water lily (Figure 2.8) and include the genera *Nymphaea, Nuphar, Victoria* and *Camboba*. The Austrobaileyales includes families such as the Illiciaceae (Figure 2.9), which contains one genus, *Illicium*, with 37 species of small trees and shrubs that occur mainly in southeastern Asia

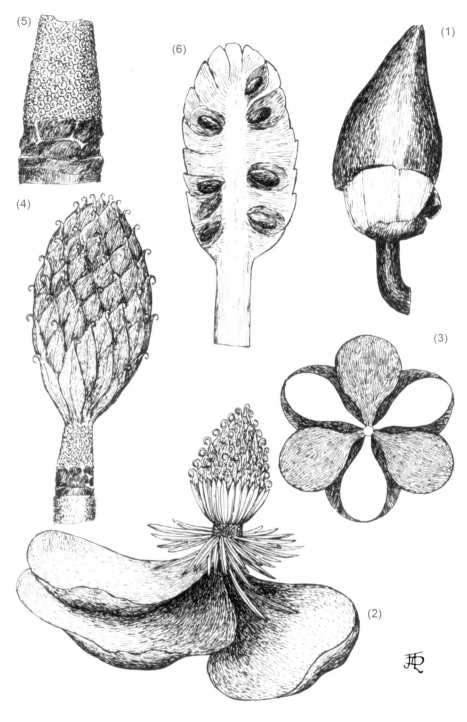

FIGURE 2.6 The cone-like flower of a primitive angiosperm (Magnoliaceae: *Magnolia grandiflora*). (1) Flower bud opening, the sepals being ruptured; (2) flower at anthesis with the upper stamens having fallen; (3) arrangement of sepals and petals in cycles of three; (4) head of fruits in side view; (5) enlargement of a portion of the receptacle (showing the scars at the points of attachment of the three sepals, the six petals in two series, and the numerous spirally arranged stamens); (6) head of fruits in longitudinal section. (Drawing by Jerome Laudermilk, from Benson 1959)

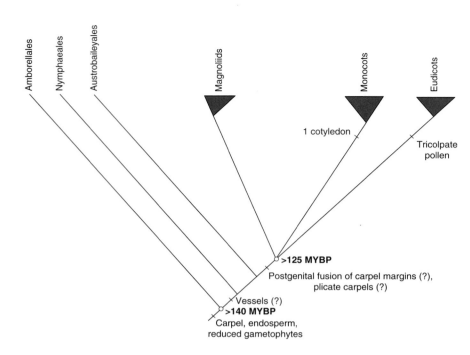

FIGURE 2.7

A cladogram showing the presumed phylogenetic relationships of the angiosperms. It was constructed using both morphological and molecular data. MYBP: million years before present. (Derived from Angiosperm Phylogeny Group IV 2016)

and the southeastern USA. Once these three basal groups are delineated, the Magnoliids can be treated as one large monophyletic group containing families such as Magnoliaceae, Annonaceae (e.g. pawpaw), Myristicaceae (e.g. nutmeg) and Lauraceae (e.g. sassafras). The remaining flowering plants fall into two evolutionary groups on the right of the cladogram in Figure 2.7, the traditional monocots, and what are now called the eudicots. For more subgroups, beautifully illustrated, you can again refer to *Flowering Plant Families of the World* (Heywood et al. 2007).

Other more traditional methods of research continue. In China, from rocks aged at 125 million years, paleobotanists have found roots, leaves, reproductive organs and pollen of an aquatic plant representing a new basal angiosperm family, the (now extinct) Archaefructaceae (Sun 2002). In Australia, an entirely new genus of conifer (*Wollemia*, Araucariaceae) was discovered clinging to the ledges in a deep sandstone gorge in Wollemi National Park in 1994 (Jones et al. 1995).

2.3.4 The Two Largest Families of Plants: Asteraceae and Orchidaceae

The families within the Magnoliid complex in Figure 2.7 "have the appearance of being 'old groups', in which evolution is proceeding at a leisurely pace and extinction has greatly affected the pattern of variation" (Dodson 1991: p. 745). You have just seen some examples in Figures 2.8 and 2.9. In contrast, the Asteraceae and Orchidaceae are two relatively homogeneous families that have recently diversified in the eudicots and monocots, respectively. Owing to their vast size (Asteraceae 32,900 species; Orchidaceae 27,800 species; The Plant List 2013) something more should be said about each of these two major angiosperm groups.

The Asteraceae (Figure 2.10) are sharply defined by having flowers grouped into a compact inflorescence surrounded by bracts (termed a capitulum). This trait is so clear that there is no species where there is doubt as to whether or not it is properly

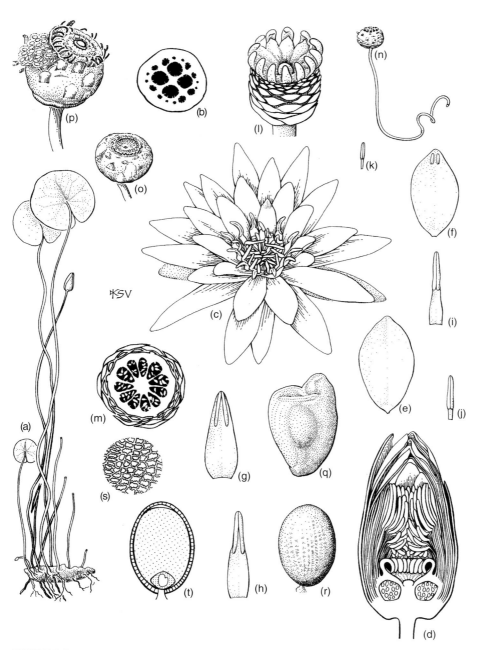

FIGURE 2.8 An example of the Nymphaeaceae (*Nymphaea odorata*). (a) Habit (× 0.3); (b) petiole, in cross-section (× 5); (c) flower (× 1); (d) flower, in longitudinal section (× 1.5); (e) petal (× 1.5); (f–h) petal-like stamens (× 1.5); (i–k) inner stamens (× 1.5); (l) gynoecium showing numerous carpels (× 2); (m) gynoecium in cross-section (× 1.5); (n) fruit, with coiled floral peduncle (× 0.5); (o) fruit (× 1); (p) dehiscing fruit (× 1.5); (q) seed, with aril (× 15); (r) seed (× 30); (s) seed coat (greatly magnified); (t) longitudinal section of seed, with embryo, endosperm, and perisperm (× 30). (Illustrations by Karen Velmure, from Judd et al. 2002)

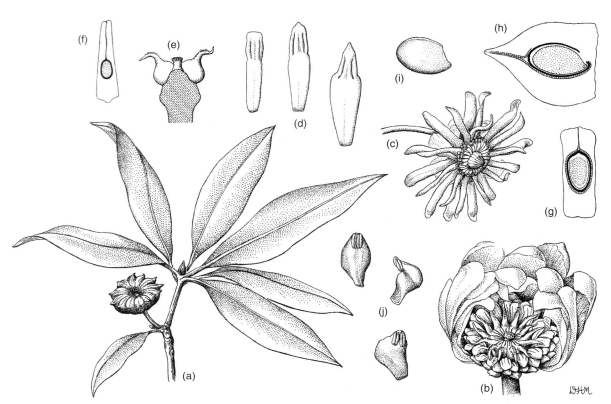

FIGURE 2.9 An example of the Illiciaceae (*Illicium floridanum*). (a) fruiting branch (× 0.5); (b) opening flower bud with receptive carpels (× 4); (c) flower, later stage at shedding of pollen (× 1.5); (d) stamens, inner, outer and an unusual subtepaloid form (× 7); (e) two carpels on receptacle (× 4); (f) carpel in cross-section, note single ovule (× 15); (g) mature fruit, with single seed, endosperm stippled (× 3); (h) mature fruit in longitudinal section (× 3); (i) seed (× 3); (j) *I. parviflorum*: stamens (× 3). (Illustrations by Dorothy Marsh, from Judd et al. 2002)

included within the Asteraceae (Cronquist 1991). Further, the species within the Asteraceae are so similar to one another that most could be placed in a single genus – partly explaining the challenges posed by this group when learning plant identification. One genus, however, would be impractical, and botanists have found it necessary to recognize subfamilies, tribes and genera to convey a sense of order. Nucleotide sequence data are now clarifying phylogenetic relationships in this enormous group (Lundberb and Bremer 2003).

The progenitor of the Asteraceae is likely to have been a small tree species from the dry highlands of Mexico (Cronquist 1991). It probably had a well-developed resin system and opposite, simple leaves. There were likely a few terminal inflorescences arranged as a cyme, each yellow flower in the cyme being subtended by a bract. The Asteraceae are not well represented in the fossil record until the early Pleistocene, about 2 million years ago, at which time climatic change may have stimulated dispersal and adaptive radiation. The Asteraceae are common in sunny places including fields associated with human agricultural activity and, indeed, the presence of pollen from the genus *Ambrosia* (ragweed) is used as a marker of European agriculture when

FIGURE 2.10 The Aster family is one of the world's largest, with some 32,900 species. All possess the distinctive capitulum of densely aggregated flowers surrounded by rows of bracts. Aster chemistry is also distinctive: many use the oligosaccharide inulin instead of starch for carbohydrate storage, and produce sesquiterpene lactones to provide defence against herbivores (Lundberg and Bremer 2003). (*Helianthus pubescens*, William Hooker, 1827, Peter H. Raven Library/Missouri Botanical Garden)

interpreting sediment cores from lakes. The family is now widespread, but still particularly common in open habitats of arid regions in North America, the Near East and South Africa. Primitive woody forms still occur in the dry highlands of Mexico.

The Orchidaceae (Figure 2.11) may have even more species than the Asteraceae, but exact numbers will remain unknown until further work is completed in tropical regions. Orchids possess three sets of distinctive traits: (1) the pollen stays in masses called pollinia; (2) the stamens and pistil are fused into one structure called the column; and (3) the tiny seeds lack endosperm and so require fungi for germination. Pollination by bees, moths, butterflies, birds and flies has been documented and the group is well known for close coevolution between flower and pollinator. Many orchids, if not all, have close associations with soil fungi that assist with nutrient acquisition even during the adult stage (see mycorrhizae Section 7.3.1, Figure 7.5). More than one hundred orchid species have no chlorophyll and are therefore completely dependent upon fungi for carbon acquisition as well as, presumably, from nearby photosynthetic plants (Dearnaley 2007). Many orchids also have a drought-tolerant type of photosynthesis known as CAM photosynthesis that you will read more about in Section 3.3.2. Similar to the Asteraceae, there is thought to have been rapid diversification of orchids in geologically recent time, making it difficult to delineate genera or higher categories. The reasons for this rapid diversification are unclear and remain a challenging question with both ecological and evolutionary components.

2.3.5 Grasses and Their Significance

Another recently evolved group is the grass family (Poaceae). There are some 11,500 species, less than half the number in the Asteraceae or Orchidaceae, but still notable because grasslands comprise the largest biome on Earth, some 41–56 million km^2. Early humans evolved in this biome, as did many of our major food crops (Gibson 2009). The ancestral Poaceae were likely species of forests and forest margins, not unlike the bamboo. Although insect pollination

FIGURE 2.11 Orchids are one of the world's largest plant families. Many orchids are epiphytes and, like cacti, have CAM photosynthetic pathways (Section 3.3.2) to minimize water loss (Silvera et al. 2009). Note also, at the bottom left, the pseudobulbs (enlarged internodes with a thick cuticle and parenchyma tissue) that can store water and nutrients, as well as allowing regeneration after fire. The genus *Dendrobium* has more than a thousand species, ranging from the high altitudes of the Himalayas to the deserts of Australia. (*Dendrobium venustum*, John Fitch, 1893, Peter H. Raven Library/Missouri Botanical Garden)

Table 2.5 The 35 floristic regions of the world mapped in Figure 2.12. (Following Takhtajan 1986)	
I. Holarctic Kingdom	D. Polynesian Subkingdom
A. Boreal Subkingdom	20. Polynesian
1. Circumboreal	21. Hawaiian
2. Eastern Asiatic	E. Neocaledonian Subkingdom
3. North American Atlantic	22. Neocaledonian
4. Rocky Mountain	**III. Neotropical Kingdom**
B. Tethyan Subkingdom	23. Caribbean
5. Macaronesian	24. Guayana Highlands
6. Mediterranean	25. Amazonian
7. Saharo-Arabian	26. Brazilian
8. Irano-Turanian	27. Andean
C. Madrean Subkingdom	**IV. Cape Kingdom**
9. Madrean	28. Cape
II. Paleotropical Kingdom	**V. Australian Kingdom**
A. African Subkingdom	29. Northeast Australian
10. Guineo-Congolian	30. Southwest Australian
11. Uzambara-Zululand	31. Central Australian (Eremaean)
12. Sudano-Zambezian	**VI. Antarctic Kingdom**
13. Karroo-Namib	32. Fernándezian
14. St. Helena and Ascension	33. Chile-Patagonian
B. Madagascan Subkingdom	34. South Subantarctic Islands
15. Madagascan	35. Neozeylandic
C. Indo-Malesian Subkingdom	
16. Indian	
17. Indochinese	
18. Malesian	
19. Fijian	

was a key factor in the origin of flowering plants altogether (Section 7.4.2), grasses have reverted to mostly wind pollination; each grass spikelet is a reduced inflorescence branch. The meristems often occur at the soil surface, making grasses one of the largest groups of hemicryptophytes, and rather resistant to both grazing and fire. The Poaceae appear to be monophyletic with a Cretaceous origin, and they probably diversified on Gondwana (Section 8.2.2). Grasses appear to have spread during drier periods of climate, which may have been accelerated by rising mountains – the Rocky Mountains, for example, created an enormous rain shadow in central North America, a landscape often too dry for trees. Likely, the spread of grasses and expansion of grasslands was accelerated by a combination of factors, including

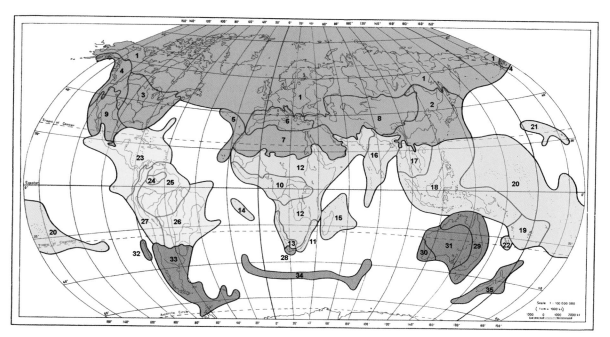

FIGURE 2.12 The 35 floristic regions of the world fall into six kingdoms. (From Takhtajan 1986) See Table 2.5 for region names. (Note that an earlier version, Takhtajan 1969, recognized 37 regions.)

drought, fire and grazing animals. The strong selective pressure for the C_4 photosynthetic pathway is demonstrated by its occurrence in about half of all grass species, and the evidence that it may have evolved up to 11 different times. The lawns in our cities and suburbs testify a continued human preference for open grasslands, while the vast fields of wheat, corn, barley, sorghum and sugar cane in the countryside illustrate the continuing economic importance of this family.

2.3.6 World Floristic Regions are Based on Phylogeny and Geography

Most plants are restricted to rather small geographic ranges. If you compile the data on their ranges, you can map which parts of the world have similar plant species, genera or families. Figure 2.12 shows the floristic regions of the world as mapped by the Russian botanist, Takhtajan (1969, 1986). Each region of the figure is mapped based on floristic similarities. To

illustrate Takhtajan's regional classification, consider two examples, one from North America and the other from Africa. The North American Atlantic Region (3) is described as follows:

> *Endemic families Hydrastidaceae and Leitneriaceae, no fewer than a hundred endemic genera (including* Sanguinaria, Sarracenia, Dirca, Neviusia, Hudsonia, Dionaea, Yeatesia, Pleea, Uvularia*) and numerous endemic species.*
>
> *(Takhtajan 1969: p. 244)*

Similarly, the Karroo-Namib Region (13) is described as having:

> *Very many endemic species;* Mesembryanthemum *and allied genera,* Tetragonia, Pelargonium, Rhigozum, Pentzia, Pteronia *and other shrubby Asteraceae are especially characteristic; in the northern parts of the Namib* Welwitschia *and* Acanthosicyos horridus *are endemic.*
>
> *(Takhtajan 1969: p. 248)*

You can see one of the characteristic genera of region 3, *Sarracenia*, if you flip ahead to Figure 3.18. You can see one of the characteristic genera of region 13, *Welwitschia*, if you flip ahead to Figure 12.14. Takhtajan's entire classification of regions is presented in Table 2.5. This also gives you a key for the map in Figure 2.12. Box 2.2 will tell you more about this remarkable man, and the plant named in his honour.

2.3.7 Some Limitations of Phylogenetic Classifications for Ecological Research

In the phylogenetic system of classification, the goal is to arrange species according to their relatedness. This perspective places a great deal of emphasis upon conserved characters, particularly those of reproductive systems, which make it possible to retrace the evolution and diversification of plants. Thus, for example, species with free-living gametophytes (e.g. Pteridophyta) are classified as entirely separate groups from those with gametophytes that are retained upon the sporophytes (e.g. Magnoliophyta) (recall Table 1.1).

So many species of plants have now been identified and cataloged that no single worker can be familiar with more than a subset. The subset you choose might be defined by taxonomic boundaries (e.g. the palm family) or geographic boundaries (e.g. the flora of the Galapagos). One of the first duties for a plant ecologist is to master the flora of the particular region to be studied; this, in itself, is no easy task. But if you learn even one new species a week, you will know hundreds within a few years.

While a phylogenetic approach is useful for subdividing the vast plant kingdom, there are at least five problems created by phylogenetic classifications:

1. Some groups of plants, in some geographic regions, particularly those from the tropics, are still poorly understood and may even be unnamed. Ecologists, however, cannot cease to study tropical areas until such time as complete classifications and phylogenetic reconstructions are available.

2. Even if the flora of an area is well studied, the flora may be so large as to present an obstacle to field identification. Further, while classification usually assumes knowledge of reproductive traits, most field studies have large numbers of pre-reproductive individuals, and even the mature individuals may lack the required structures for identification.

3. Even if the flora of an area is well known, small enough for easy identification and consists of individuals easily assigned to species, the enumeration of species names does not necessarily convey useful information to other ecologists. Papers using nomenclature from a different biogeographic region may be nearly indecipherable to those unfamiliar with that region's flora. The names, then, do not necessarily enhance communication among scientists but may even subdivide them into geographic cliques.

4. Even if the above obstacles are surmounted, the taxonomic and phylogenetic status of an individual may tell a great deal about evolutionary relationships but very little about its ecological function. In temperate climates, the Lily family may be represented by spring ephemerals of deciduous forests (e.g. *Erythronium* spp.), in boreal regions the Lily family may be found in *Sphagnum* bogs (*Smilacina* spp.) and in arid regions the representatives of the Lily family may be evergreen succulents (*Aloe* spp.). While all these species may share an evolutionary heritage, as illustrated by similarities in floral structure, and similarities in chloroplast DNA, they now occupy entirely different habitats and have different functional traits.

5. For some scientific purposes, there simply may be "too many" species in nature. For studies of biodiversity or evolutionary diversification, the detail provided by thorough taxonomic subdivisions may be necessary, but for vegetation mapping, extreme simplification is necessary. That is where simpler systems, such as Raunkiaer life forms, may be necessary.

Finally, let us say something here about the naming of vegetation types or plant communities. All of the

Box 2.2   Takhtajan Survives Stalin, Genocide and Two World Wars

FIGURE B2.2.1   The journal of the Armenian Botanical Society is called *Takhtajania*, and the image on the front cover shows *Takhtajania perrieri*, a small tree found only on the island of Madagascar. Although first discovered in 1909 by the French Botanist Henri Perrier de la Bâthiein in northern Madagascar, the exact location remained unclear, and numerous searches failed to relocate living plants. Searchers also had to deal with the densest populations of terrestrial leeches on the island. This species has now been relocated in the Marojejy National Park, which protects more than 50,000 ha of intact forest, with rainfall exceeding 3,600 mm per year. Apparently the team that first collected the plant had a malfunctioning altimeter, which sent all subsequent teams to the wrong elevations. The full story is found in Schatz (2000). This is the only species in the ancient Winteraceae family

"It was the best of times, it was the worst of times, it was the age of wisdom, it was the age of foolishness." That Dickensian description of the turmoil of the French Revolution still seems to apply to our world. Yet individuals still matter. Earlier in this chapter you met Wallace, who lived in near poverty for many years, collecting important Amazonian specimens at fourpence each (minus a 20 percent commission), and yet still leaving a remarkable legacy. You have just read about Takhtajan's work. What about the man?

Armen Leonovich Takhtajan was born in 1910 in Armenia (now in Nagorno Karabahk, a contested area officially within the nation of Azerbaijan). His main interest was to become fossil plants. Yet it would be just four years to chaos: the First World War from 1914 to 1918, the Armenian genocide in 1915, the Bolshevik revolution in 1917, the resulting Russian Civil War in 1919, not to mention the Second World War. He also survived Stalin (many didn't): Stalin's purges, show trials and labour camps killed millions of his own citizens (recall *The Gulag Archipelago*, 1973, by Aleksandr Solzhenitsyn). In 1948, Takhtajan was declared an enemy of the socialist system, and lost his position as Director of the Botanical Institute of the Armenian Academy of Sciences, and several other official roles. He left for Moscow, and then Lenigrad. He continued to study fossil plants and plant evolution.

After the collapse of the USSR and decline of Communism, Leningrad took back its earlier name of St. Petersburg. Takhtajan died there, in St. Petersburg, in 2009 at the age of 99. In addition to his many books (including *Flowering Plants: Origins and Dispersal* 1969), and the map shown in

known from Madagascar, and illustrates the rich Gondwana flora that typifies World Floristic Region 15 (Figure 2.12). (Gabrielyan, I. and J. Kovar-Eder. 2010. Obituary of Armen Leonovich Takhtajan. *International Organization of Paleobotany Newsletter* **91**: 8–9)

(Continued)

Box 2.2   (cont.)

Figure 2.12, he described many new species of fossil plants and several are named after him (Figure B2.2.1). His map continues to provide a foundation for global conservation efforts including planning by ecoregion (Figure 2.14). Moreover, his map is an important context for biological hotspots, and his own homeland now lies within one, the Caucasus biodiversity hotspot, which covers the area between the Black Sea and the Caspian Sea (Figure 13.11).

systems of plant classification that are described in this chapter assign individual plants to categories, be they phylogenetic or functional. When it comes to classifying local vegetation, as opposed to individual species, however, neither a phylogenetic nor a functional classification system is particularly appropriate. One general solution is often to select a few species that are dominant and use their names to designate the vegetation type. The phylogenetic system might yield designations such as "*Podocarpus* forest," "*Acer-Fagus* forest," "Ericaceous shrubs" or "*Andropogon* prairie." The functional system would

yield more general descriptions such as "evergreen phanerophytes," "deciduous phanerophytes," or "hemicryptophytes and geophytes." Sometimes these two types of classification are mixed. Thus Daubenmire's (1978) description of plant geography in North America includes "the *Quercus falcata* Province within the Temperate Mesophytic Forest Region" or "the *Agropyron spicatum* Province in the Steppe Region." Sometimes local names for habitats have been passed down for generations. The description of vegetation will be discussed at much more length in Chapter 11.

CONCLUSION

The culmination of two centuries of field exploration and laboratory examination can be summarized in just two maps. The first map (Figure 2.5) showed biomes. The second map (Figure 2.12) showed world floristic regions. Study these maps often. Sometimes you will see Takhtajan's map simplified to just eight regions (Figure 2.13), which is acceptable for some purposes.

Biomes and world floristic regions summarize what is known about the large-scale distribution of terrestrial organic matter. Both systems have strengths, weaknesses and value. For studies that aim to preserve the genetic diversity of the Earth's biota, the phylogenetic system would receive priority. For studies that aim to predict the consequences of climate change, the functional system would be most useful.

Of course, getting back to the paradox that opened the chapter, of needing two maps to describe one world, it may be possible to combine the phylogenetic and functional systems into one system using the concept of an ecoregion. Olson et al. (2001) define ecoregions as "relatively large units of land containing a distinct assemblage of natural communities and species, with boundaries that approximate the original extent of natural communities prior to major land-use change." Their map (Figure 2.14) recognizes a total of 867 ecoregions, nested within 14 biomes and 8 biogeographic realms (Figure 2.13). Maps of ecoregions are now used by many agencies for planning reserve systems to protect ecological communities.

Yet there may be merits to simultaneously keeping two separate systems. Carpentry projects, for example, can require both a saw and a hammer. If one asked a carpenter "Is the hammer or the saw the best tool?" one would reveal only one's ignorance of carpentry. In the same way, a plant ecologist should remain adept at using multiple systems of classification for describing global patterns in plant distribution.

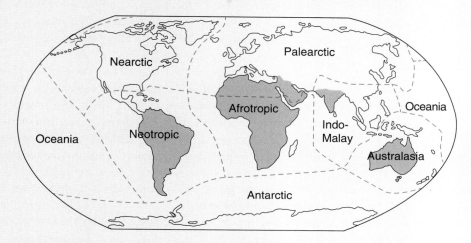

FIGURE 2.13   The eight biogeographic regions of the world. The coloured regions separated as Gondwanaland from Laurasia about 200 million years BP. The line between the Australasia and Indo-Malay regions is often called "Wallace's line." (Adapted from Pielou 1979 by Olson et al. 2001 with permission of WWF-US)

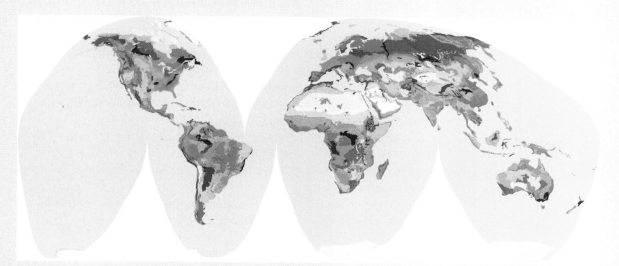

FIGURE 2.14   World ecoregions contain distinct natural communities and species. There are currently 867 of them. Ecoregions combine the phylogenetic perspective (Figure 2.12) and the functional perspective (Figure 2.5) to provide a foundation for global conservation planning and ecological research. (After Olson et al. 2001, with permission of WWF-US, online version at http://worldwildlife.org/science/wildfinder)

Review Questions

1. Contrast functional as opposed to phylogenetic systems for plant classification.
2. Visit a local natural area and tabulate the Raunkiaer life forms that are present.
3. Does the Southern Hemisphere have more kinds of plants than the Northern Hemisphere? If so, what does this mean for global conservation planning?
4. Is there a natural area near your home that has carnivorous plants or orchids? Write a short description of the habitat and offer some hypotheses that might explain the occurrence of these plants there.
5. Some iconic "American" westerns (such as *The Good, the Bad and the Ugly*) were actually filmed in southern Europe (which explains the origin of the term "spaghetti western"). Explain how this is possible, that is, why functional plant types are similar between the American southwest and the northern Mediterranean. If you watch such films carefully, can you see particular species that seem out of place? Or is the power of convergent evolution so strong that you really can't tell the difference? (Hint, look for cacti.)
6. Pick a famous explorer, and read to see what contributions, if any, they made to the study of plant distributions.
7. Briefly describe the origin of grasses from both a phylogenetic and functional perspective.

Further Reading

Raunkiaer, C. 1908. The statistics of life-forms as a basis for biological plant geography. pp. 111–147. In C. Raunkiaer. 1934. *The Life Forms of Plants and Statistical Plant Geography: Being the Collected Papers of Raunkiaer.* Oxford: Clarendon Press.

Küchler, A.W. 1949. A physiognomic classification of vegetation. *Annals of the Association of American Geographers* **39**: 201–210.

Udvardy, M.D.F. 1975. *A Classification of the Biogeographical Provinces of the World.* IUCN Occasional Paper No. 18. Morges, Switzerland: International Union for the Conservation of Nature and Natural Resources,

Takhtajan, A. 1986. *Floristic Regions of the World.* Translated by T.J. Crovello. Berkeley, CA: University of California Press.

Woodward, F.I. 1987. *Climate and Plant Distribution.* Cambridge: Cambridge University Press.

Archibold, O.W. 1995. *Ecology of World Vegetation.* London: Chapman and Hall.

Browne, J. 1995. *Charles Darwin: Voyaging.* Princeton, NJ: Princeton University Press.

Edmonds, J. (ed.). 1997. *Oxford Atlas of Exploration.* New York, NY: Oxford University Press.

Bailey, R.G. 1998. *Ecoregions: The Ecosystem Geography of the Oceans and the Continents.* New York: Springer-Verlag.

Weiher, E., A. van der Werf, K. Thompson, et al., 1999. Challenging Theophrastus: a common core list of plant traits for functional ecology. *Journal of Vegetation Science* **10**: 609–620.

Camerini, J.R. (ed.). 2002. *The Alfred Russel Wallace Reader. A Selection of Writings from the Field*. Baltimore, MD: The Johns Hopkins University Press.

Heywood, V.K., R.K. Brummitt, A. Culham, and O. Seberg (eds.) 2007. *Flowering Plant Families of the World*. Richmond Hill, Ontario: Firefly.

Resource acquisition as one logical starting point for plant ecology. The CHNOPS perspective. Costs of acquisition. The global carbon pool. Harvesting photons. Canopy architecture. Photosynthetic types (C_3, C_4, CAM). Height. The search for below-ground resources. Water. Algae in fresh water. Nitrogen. Phosphorus. Fertilization experiments. Patterns of primary production. Soils. Humans and soil resources. Resources in space and time. Gradients and patches. Conceptual classification of resources. Fluctuating resources complicate research. Chronic scarcity of resources. Nutrient conservation. The leaf economic spectrum. Plants in tropical canopies, succulents, carnivorous plants, parasitic plants. Synthesis.

FIGURE 3.1 All plants (well, nearly all) must simultaneously forage for resources above and below the ground. The leaves of this giant Amazon water lily (*Victoria amazonica*) absorb carbon dioxide and capture sunlight. Meanwhile, many metres below, the roots absorb nitrogen, phosphorus and water. Most of the plant is hidden from sight. (Photograph © Corbis)

3.1   Introduction: Plants Must Find Resources to Grow

3.1.1 The CHNOPS Perspective

Plants, like all living organisms, are constructed from relatively few elements: CHNOPS (Table 3.1). These six elements could thus be considered the fundamental resources for ecologists to study. Organisms do not use these elements in equal amounts. Table 3.2 shows that oxygen and carbon predominate, whereas sulfur and phosphorus make up less than one percent of organisms.

These elements are not equally available in time or space. In order to grow and reproduce, plants must forage for, absorb and internally transport such resources before being able to construct new tissues. They also face the challenges of conserving such resources when they are in short supply. While it is true that all life forms must go through similar steps, vascular plants face a situation very different from most other life forms. Since plants occupy the narrow interface between the atmosphere and the soil, plants must forage simultaneously in two different environments for two different sets of resources (Figure 3.1). Plants must forage in the atmosphere for carbon dioxide and light, and forage in the soil for water and mineral nutrients. Hence plants have a distinctively dual nature, possessing two kinds of architecture, morphology, anatomy and physiology. (There are partial exceptions to this specific duality, such as carnivorous plants and epiphytes that forage for nutrients with their leaves (Section 3.7.3), but these are noteworthy precisely because they deviate from the norm.) Since photosynthesis is a defining characteristic of plants, the capture and processing of light energy is normally a major part of any book on plant ecology. I assume in writing this book that you have already had introductory courses in botany and ecology. It would be tiresome of me, and a waste of your time, to repeat lengthy treatments of photosynthesis and plant nutrition already written by far more qualified authors. My approach here is to: (1) remind you of some basic principles, (2) tie these basic principles clearly to processes in plant populations and communities, and (3) introduce you to some important topics that are often overlooked in more basic books (e.g. the CHNOPS perspective, nutrient conservation in infertile habitats, gradients and their consequences, and foraging). My intention is to lay the groundwork

Table 3.1 **Major elements required by living organisms and their functions. (From Morowitz 1968)**

Element	Function
C	Structure; energy storage in lipids and carbohydrates
H	Structure; energy storage in lipids and carbohydrates
N	Structure of proteins
O	Structure; aerobic respiration for energy release
P	Structure of nucleic acids and skeletons; energy transfer within cells
S	Structure of proteins

Table 3.2 **Atomic composition of four typical CHNOPS organisms: [a]humans, [b]alfalfa, [c]copepods, [d]bacteria. (From Morowitz 1968)**

Element	Mammal[a]	Vascular plant[b]	Arthropod[c]	Moneran[d]
C	19.37	11.34	6.10	12.14
H	9.31	8.72	10.21	9.94
N	5.14	0.83	1.50	3.04
O	62.81	77.90	79.99	73.68
P	0.63	0.71	0.13	0.60
S	0.64	0.10	0.14	0.32
Total	97.90	99.60	98.07	99.72

for future chapters without repeating too much that has already been well covered elsewhere.

Before continuing with the problem of how plants obtain resources, let us briefly consider the physical properties of CHNOPS resources themselves. These elements share two chemical properties. First, all have relatively low atomic numbers (<20). Thus they are among the most common elements, since abundance varies inversely with atomic number. Further, they all have a first ionization energy between 10 and 15 eV, lower than that of the noble gases (which have low chemical reactivity) and higher than that for most other elements. Such similarities suggest that early in the history of life there was natural selection for those molecules that were relatively common and had intermediate levels of reactivity. This seems inherently reasonable. Life would be formed from the most common building blocks available. If these building blocks had too little reactivity, chemical interactions could not occur and they would rarely be incorporated into molecules. If these building blocks were too reactive, reactions would be very frequent and molecules would be easily established by small amounts of energy.

The repetitive simplicity and similarity of the basic building blocks of living organisms is illustrated by the structure of a typical amino acid in Figure 3.2. Although there are more than a hundred kinds of amino acids in nature, only 20 are common, and they differ in the chemical structures attached at the R position in the molecule. In glycine, R represents merely a hydrogen atom, H; in aniline R represents CH_3; and, in serine, CH_2–OH. Proteins, the basic structural matter of life, are composed of chains of

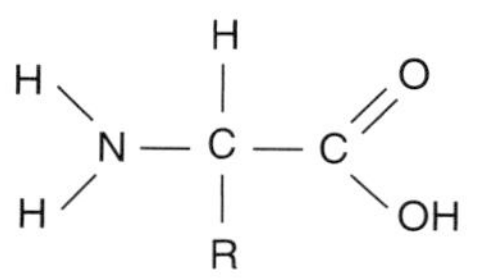

FIGURE 3.2 Every amino acid requires four elements: carbon, hydrogen, nitrogen and oxygen. They differ only in the chemical structure appended at the R position.

these different amino acids, but the similarity of their elemental composition, largely C, H, O and N, reflects the elements that were available in the Earth's atmosphere (Box 3.1).

The NH_2 in each amino acid sets the stage for a recurring issue in the struggle for survival and reproduction by plants: how to obtain those critical nitrogen atoms for protein synthesis. Every amino acid requires a nitrogen atom. On a planet with an atmosphere that is practically three-quarters nitrogen, the acquisition of nitrogen may seem to be relatively straightforward. Yet, it is not, because plants cannot use gaseous nitrogen. One of the remarkable physiological puzzles of botany is the dependence of all plants upon a relatively small subset of microorganisms that are still able to transform gaseous nitrogen into its biologically useful forms. The relative abundance of this element, and yet its great scarcity from the perspective of plants, is an overriding theme of plant ecology, and one that quite likely spills over into the animal kingdom as well (White 1993).

Phosphorus is unusual in the CHNOPS list precisely because, unlike nitrogen, it is a comparatively rare element. Phosphorus is a constituent of nucleic acids, the phospholipids of biological membranes and the phosphate esters that provide the metabolic machinery of cells (Marschner 1995). In particular it is an integral part of the citric acid cycle that extracts chemical energy from large molecules; ATP being used to store the resulting chemical energy. Each ATP molecule requires three phosphorus atoms (as well as five nitrogen atoms). Phosphorus appears to be useful because, relative to most elements, there are quite large bond energies (Dickerson 1969). This requirement for large numbers of phosphorus atoms to complete a fundamental metabolic process allows one to predict that phosphorus would be an important resource for plants.

Nitrogen and phosphorus, along with sulfur (also potassium, calcium and magnesium) are usually termed macronutrients because they are required in considerable amounts. A large number of other rare elements are required in much smaller amounts and are hence termed micronutrients; these include many

Box 3.1   The Composition and Origin of the Atmosphere

The composition and origin of the atmosphere was already a matter of systematic enquiry when Antoine Lavoisier (1743–1794, Figure B3.1.1) prepared *Elements of Chemistry* (*Traité Elémentaire de Chimie*) in 1789. He begins Chapter III:

> *From what has been premised, it follows that our atmosphere is composed of a mixture of every substance capable of retaining the gaseous or aeriform state in the common temperature, and under the usual pressure which it experiences.*
>
> (p. 16)

Not only does Lavoisier present experimental methods for determining the composition of the atmosphere, but he concludes that the atmosphere is composed of two gases, one of which supports respiration (oxygen) and the other which does not (nitrogen). (Although Lavoisier states a preference for the terms *azote* and *azotic gas*, which expresses "that property... of depriving such animals as breathe it of their lives.") He even speculates upon possible changes in atmospheric composition if the Earth's temperature were to be greatly increased by being transported into the region of the planet Mercury. Previously solid materials

> *would be changed into permanent aeriform fluids or gases, which would become part of the new atmosphere. These new species of airs or gases would mix with those already existing, and certain reciprocal decompositions and new combinations would take place, until such time as all the elective attractions or affinities subsisting amongst all these new and old gaseous substances had operated fully; after which, the elementary principles composing these gases, being saturated would remain at rest.*
>
> (p. 15)

He further speculates upon changes to be expected in the case of extreme cold, in which case "the water which at present composes our seas, rivers, and springs" would become mixed with foreign substances and become "opaque stones of various colours." New liquids would also form "whose properties we cannot, at present, form the most distant idea." (p. 16)

One of his contemporaries, Jean Paul Marat (1743–1793), also wrote several scientific works. His contribution in 1789 was not a chemistry book, but rather a radical newspaper, *L'ami du peuple*, inciting the poor to violence. These were the turbulent years of the French Revolution.

FIGURE B3.1.1   Portrait in 1788 of Antoine-Laurent Lavoisier (1743–1794), a chemist and revolutionary who explored the properties of the Earth's atmosphere. (From Wikimedia Commons)

Box 3.1   (cont.)

After narrow escapes from the police (including a well-known escape through the sewers of Paris, which left him with an agonizing skin infection from which he never recovered and which forced him to spend most of his life in a bath), Marat was elected to the Convention as a deputy for Paris. In 1791, Marat began denouncing Lavoisier for his associations with, and employment by, the former government, and in 1794 Lavoisier was tried and convicted by the Revolutionary Tribunal. On 8 May 1794 he was guillotined, his body being thrown into a nameless grave in the cemetery of La Madeleine. All the same, he outlived Marat, who was assassinated in his bath by Charlotte Corday on 13 July 1793.

metals (iron, manganese, zinc, copper and molybdenum) and two non-metals (boron and chlorine) (Larcher 2003). While a treatise on plant nutrition must address each of these elements in turn, it is reasonable to ask how many of them need to be considered for a general knowledge of plant growth and vegetation development. Is there some minimum number of resources that, by themselves, enables one to make useful predictions about the structure and function of plants and their communities? Keep this question in mind for the remainder of the chapter.

3.1.2 The Costs of Acquisition

Resources are not just of interest to ecologists. In the very first paragraph of *Das Capital*, Marx (1867) introduces the concept of commodity. A commodity, he says, "is an object outside us, a thing that by its properties satisfies human wants of some sort or another" (p. 13). In other words, it is a resource. The value of a commodity, he continues, following from Smith (1776), is determined by the amount of labour required to produce a unit of it: "If we could succeed at a small expenditure of labor in converting carbon into diamonds, their value might fall below that of

bricks" (p. 15). Commodities that appear to be very different (say, a coat, tea or gold) can be exchanged once we agree upon such a measure of value.

The energetic costs for acquisition of each resource may be one useful way of measuring their value to plants. For example, van der Werf et al. (1988) measured the respiratory energy costs of ion uptake in a species of sedge (*Carex diandra*) and reported that the proportion of total ATP demand for ion uptake ranged from 10 to 36 percent. Neighbours will tend to reduce the availability of resources, and thereby increase these costs of acquisition. Any study of a plant population or community should probably begin by explicitly considering resources. Resources will always play a role in controlling plant populations and organizing communities. Other factors such as competition (Chapter 4), disturbance (Chapter 5) and herbivory (Chapter 6), are then superimposed on the effects of resources, which is why these topics are relegated to later chapters. Since the above-ground organs of a plant are the most obvious to humans, we will first examine foraging for resources in the atmosphere and then turn to foraging in the soil. This should not, however, lead to the supposition that the below-ground nutrients are less important.

3.2   Carbon Dioxide: Foraging in an Atmospheric Reservoir

Carbon is first on the CHNOPS list. Nearly all life forms are constructed from gaseous atmospheric

carbon in the form of CO_2. Plants must forage for this atmospheric carbon in the same way they forage for water and mineral nutrients; a fact sometimes overlooked, perhaps because CO_2 is an odourless, colourless gas. The task of locating and assimilating CO_2 might at first seem to be a relatively simple task, since the distribution of CO_2 would appear to be quite homogeneous and constant. As Harper (1977) puts it: "The concentration of CO_2 in the atmosphere away from a photosynthetic surface is about 300 parts per million and shows only a slight variation from place to place" (p. 323). True, it is widespread and fairly homogeneous, but then it is also scarce everywhere. Stems and leaves are frequently viewed as organs for intercepting light when in fact they function equally in removing CO_2 from the atmosphere.

The steps for extracting CO_2 from the atmosphere evolved early in the history of life; the biochemical steps known as the Calvin–Benson cycle being ubiquitous in plants (Keeley and Rundel 2003). Foraging for CO_2 is a costly process, particularly in dry environments, because the stomata that admit CO_2 from the atmosphere into plants also allow another important resource, water, to escape. Moreover, global CO_2 levels are much lower now than during the Earth's early history (recall the immense stores of carbon removed from the atmosphere and stored as coal, Figures 1.17 and 1.18). Some of the inefficiencies inherent in the photosynthetic process may well have arisen at this time when CO_2 levels were higher – the consequence for the present being "suboptimal performance under contemporary atmospheric gas composition" (Keeley and Rundel 2003: p. S55). In general, the more CO_2 absorbed, the more water lost – a pattern to keep in mind when we discuss the C_4 photosynthetic pathway below. Moreover, all terrestrial plants, and many aquatic plants, withdraw CO_2 from one common atmospheric pool, allowing for plant–plant interactions over large distances. Further, this widespread gaseous resource can be locally depleted. Figure 1.20 shows how the levels of CO_2 in the atmosphere decline by about 6 ppm during each growing season as plants remove CO_2 by photosynthesis; the CO_2 levels then increase throughout the winter as a consequence of respiration. It is also known that these depleted CO_2 levels reduce plant performance; experiments show that plants can grow as much as 50 percent faster when CO_2 levels are increased and that the largest increases in growth occur at relatively low levels of augmentation (Hunt et al. 1991; Woodward 1992).

The importance of carbon sources for photosynthesis is illustrated by studies in which plants are grown under elevated CO_2 levels to study the possible effects of increased CO_2 in the Earth's atmosphere (e.g. Mooney et al. 1991; Porter 1993; Loehle 1995). Most plants show a surge in growth when CO_2 supplies increase (Table 3.3), although plants from infertile habitats, and plants with CAM photosynthesis, respond less than others. Of course, the observation of higher growth at higher rates at higher CO_2 levels assumes that other resources are also available; there may be situations where nutrients such as nitrogen prevent plants from growing faster at higher CO_2 levels (Hungate et al. 2003).

Table 3.3 **Responses of plants to elevated CO_2 levels. Mean weight ratio is the ratio of weight at the end of an experiment at high vs. low CO_2 levels. (From Loehle 1995)**

Group	Weight ratio	Number of species
Crops	1.58	19
Wild C_3 non-woody plants	1.35	62
Fast growing	1.54	18
Slow growing	1.23	20
Woody plants	1.41	50
Fast growing	1.73	13
Other	1.31	37
CAM	1.15	6

3.3   Light and Photosynthesis: Harvesting Photons

3.3.1   Three Measures of Photon Harvest

The foliage of plants must intercept photons. A great deal of thought has gone into the relationship between light intensity and photosynthesis. No light, no photosynthesis. Increased photon flux (light supply) will then promote increased photosynthesis until a certain threshold is reached. At this point, I_s (irradiance saturation), the photosynthetic process is said to be light saturated. The CO_2 uptake rate, maximized at this threshold, provides one useful way of describing a plant's photosynthetic character. That is to say, we can classify plants into groups based upon the point at which they become light saturated.

Another useful trait of a photosynthetic organism is its light compensation point, I_c, which is the level of irradiance at which photosynthetic CO_2 uptake and respiratory CO_2 production are in equilibrium. Plants that respire rapidly, then, require more light to achieve compensation. Once I_c has been exceeded, there is a linear relationship between irradiation and CO_2 uptake until I_s is reached.

A third measure of photon harvest is the quantum yield, Φ_A. It measures the inherent efficiency of the photosynthetic system. The greater the quantum yield, the greater the slope of the curve of CO_2 uptake plotted as a function of irradiation.

These three measurable attributes of photosynthetic systems summarize essential aspects of the capture of plant resources in the atmosphere (for further details, see Marschner 1995; Larcher 2003; Raven et al. 2005).

3.3.2   There Are Different Photosynthetic Types

The three attributes of photosynthesis described above (I_s, I_c, Φ_A) vary among plant functional types. We will look at two classifications.

One fundamental dichotomy is C_3 and C_4 photosynthesis. (Here I assume that you already know something about the process of photosynthesis, since this is taught in all introductory biology courses). Recall that C_4 photosynthesis is the more recent modification, and has an added energetic cost. The primary benefit of C_4 photosynthesis is low rates of photorespiration, thereby yielding higher rates of photosynthetic energy fixation. A second advantage is higher water use efficiency (Keeley and Rundel 2003). C_4 plants are most common in environments where light and temperature do not limit carbon uptake, that is sunny habitats with higher growing season temperatures. Tropical and subtropical grasslands are dominated therefore by C_4 grasses and sedges. The grasses (Poaceae) are the dominant C_4 family. They appear to have originated during the Cretaceous era in either South America or Australia, and have since dispersed to a global distribution. Widespread C_4 genera include *Andropogon*, *Chloris*, *Muhlenbergia*, *Paspalum* and *Sporobolus*. Other examples of C_4 plants include annual plants in deserts and certain shrubs of arid areas. This type of photosynthesis is rare in trees, which is to be expected since their growth form is ultimately a response to light competition (Keeley and Rundel 2003). One exception occurs in the Hawaiian Islands where ancient colonization by a C_4 member of the Euphorbiaceae started an evolutionary lineage of C_4 shrubs and a C_4 tree, *Euphorbia forbesii*; even so, this tree occurs in relatively open habitats.

Let us return to the important trade-off between CO_2 uptake and water loss. Higher levels of CO_2 in the atmosphere would reduce rates of water loss from plants. Hence periods with higher CO_2 would appear, from a plant's perspective, to be wetter. Conversely, as CO_2 levels in the atmosphere fall, drought becomes an increasingly important constraint upon plants. The spread of grasslands in the late Miocene (about 15 million years ago) might therefore have been triggered in part by declining levels of atmospheric CO_2, which made plants more sensitive to reduced rainfall

(Keeley and Rundel 2003). Other factors such as changing climate or increased frequency of fire are competing hypotheses for the appearance of C_4 plants. This may have implications for future changes in plant communities: higher levels of CO_2 may not only increase plant growth rates overall, but may reduce the impacts of drought and reduce the competitive advantage of C_4 growth forms. Such predictions are, however, complicated by the fact that the C_4 pathway is more successful under higher temperatures.

CAM photosynthesis is a modification of the C_4 system. The term CAM (crassulacean acid metabolism) originated with succulents in the family Crassulaceae in which the process was first found, but it has since been documented in 33 plant families, and seems to have evolved independently on a number of occasions (Keeley and Rundel 2003). Since CO_2 uptake normally leads to water loss, any photosynthetic modifications that reduce such losses may be advantageous in dry environments. In CAM photosynthesis, CO_2 is absorbed at night for use in photosynthesis during the day. Hence the stomata are open when rates of water loss are likely to be lowest. CAM photosynthesis is found in many desert plants such as the Cactaceae (Nobel 1985). It is also found in other groups where drought may be an issue, such as in epiphytes (Moffett 1994; Lowman and Rinker 2004) – in both the Bromeliaceae and the Orchidaceae, which are prominent in epiphytic communities, more than half of the species have CAM photosynthesis. Some cacti, such as the popular Christmas cactus (*Schlumbergera* spp.), are also epiphytes. The Asclepiadaceae, also common in deserts, has many CAM species.

Although in terrestrial plants C_4 is thought to be a relatively recent modification (albeit one that has evolved in multiple ways and places), it also occurs in ancient plants, such as *Isöetes*, and in aquatic plants. How does this fit into the above understanding of photosynthesis, C_4 and CAM? The answer seems to be that CAM photosynthesis is an advantage in other kinds of habitats with very low CO_2 levels (Keeley 1998; Keeley and Rundel 2003). Since gases such as CO_2 dissolve poorly in water, CO_2 levels will tend to be low, and these will be further reduced by photosynthesis. Hence, CO_2 can be critically low in aquatic

habitats, in which case it may be selectively advantageous to absorb CO_2 during the night when levels are higher due to respiration of surrounding aquatic organisms and the lack of photosynthesis, which would remove CO_2 from the water. In conclusion, foraging for atmospheric carbon has several features that distinguish it from foraging for most other resources. First, it is the only resource that occurs in a gaseous form. Second, there is a fairly well-mixed global pool of CO_2, for which plants in very different localities compete. Third, while temporary depletion zones form, they are far more transitory than depletion zones for water and soil nutrients. Finally, the absorption of atmospheric carbon has important consequences for global temperature. Were it not for plants, the Earth might have conditions like those on Venus, with CO_2 comprising some 96 percent of the atmosphere and a mean temperature of 750 K, 500 K of which is attributed to a greenhouse effect (MacDonald 1989).

3.3.3 An Exception to the Rule: Root Uptake of CO_2

The genus *Isöetes* is an obscure group of herbaceous plants thought to be evolutionary relics related to *Lycopodium* and *Selaginella*. These latter genera represent a form of plant life that once produced tree-like forms and dominated Earth. Many of the beds of coal burned today in electricity-generating plants are derived from the remains of these plants. *Isöetes* look rather like small pincushions and grow mostly in shallow water in oligotrophic lakes, although some species grow in temporary pools and a few are terrestrial. One member of this group grows at high altitudes (usually >4,000 m) in the Peruvian Andes, and the following account of the species comes from work by Keeley et al. (1994). At one time this plant was considered so distinct that it was put in its own genus (*Stylites*), but it is now considered to be more appropriately classified as *Isöetes andicola*. This bizarre plant not only has an unusual reproductive system (heterospory with free-living gametophytes), but also, over half its biomass is tied up in roots, and only the tips of the leaves emerge above ground and have chlorophyll (four percent of total biomass).

Studies with $^{14}CO_2$ demonstrated that the bulk of the carbon that *Isöetes andicola* uses for photosynthesis is obtained through the root system. There are no stomata in the leaves and the plant has a CAM photosynthetic system. Since the partial pressure of CO_2 decreases with altitude, these may be an adaptation to its highland habitat. Further studies of carbon isotopes were carried out to explore the source of carbon. Here some background is necessary. In the early 1960s, atomic bomb testing contaminated the entire atmosphere with new carbon isotopes. Studies of isotope ratios in tree rings, for example, show dramatic increases in certain carbon isotopes after the 1960s, but Keeley et al. (1994) report that the levels of CO_2 isotopes in the *Isöetes* are far below current levels and conclude that the plant is growing by fixing CO_2 from decaying peat – peat that was formed from other plants that grew long before the atmospheric testing of nuclear weapons.

The adaptive significance of this source of carbon is not immediately clear, although the lack of above-ground sources seems an inviting hypothesis. Certainly, there is an entire group of evergreen rosette-type plants called, because of their superficial resemblance to *Isöetes*, isoetids, and these plants are restricted to oligotrophic lakes where inorganic carbon levels in the water are very low. Some of these (e.g. *Lobelia dortmanna*, Figure 5.19) are also known to use their roots to take up CO_2 from sediments (Wium-Anderson 1971), and some have CAM photosynthesis (Boston 1986; Boston and Adams 1986), which indicates that isoetids have physiological as well as morphological similarities. Keeley et al. (1994) report that, in one small lake in Colombia, the CO_2 level in the sediment was $1.7\ mol \cdot m^{-3}$, whereas that in the water column was only $0.20\ mol \cdot m^{-3}$.

3.3.4 Another View of Photosynthetic Types

Larcher (2003) recognizes four distinctive functional types: C_4 plants, C_3 heliophytic herbs, trees and sciophytes (from top to bottom of Figure 3.3). Let us start

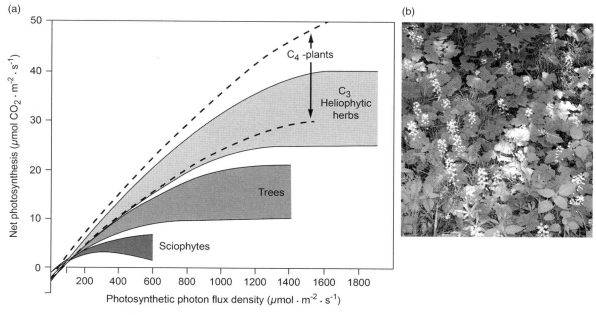

FIGURE 3.3 (a) Light–effect curves showing net photosynthesis for plants in four main functional groups, at optimal temperature and with the natural CO_2 supply. (From Larcher 1995) Note that C_4 plants have the highest photosynthetic capacity, while sciophytes have the lowest, and depend upon sun-flecks amidst shade. (b) A typical forest understorey with the sciophyte *Tiarella cordifolia* and a sun-fleck. (Photograph by Paul Keddy, Ontario)

with trees, which are almost exclusively C_3 plants. Sciophytes are plants that typically grow under trees and have adapted to cope with shade. They have a limited capacity to respond to increased light availability, but when intermittent sun-flecks strike the leaves of sciophytes, photosynthetic activities are triggered (Larcher 1995). In this way, the leaves can exploit even very brief phases of strong light. The cost of shade tolerance is that these plants rapidly reach light saturation (Figure 3.3), and their saturation rates are a fraction of those of the heliophytic herbs and C_4 plants. Heliophytic herbs have much higher rates of photosynthesis, and higher levels of light saturation – picture a marsh or grassland receiving many hours of sunlight. Finally, some grasslands have C_4 grasses, which rarely reach light saturation as such and even at moderate irradiation levels assimilate at higher rates than C_3 plants.

Another way of viewing these differences in photosynthetic response is to picture rates of CO_2 uptake during a typical day for different types of photosynthetic systems (Figure 3.4). While the heliophytes and C_4 plants steadily outperform sciophytes, during brief periods of light the latter briefly exceed the performance of the other types. Each of these light-uptake systems can often be related to the type of habitat in which these plants occur.

3.3.5 Architecture Affects Photon Harvesting

A tree can be thought of as an array of solar panels on stalks (Colinvaux 1993). Using this analogy, he challenges us to consider the way in which plants are actually constructed (multiple repetitions of small leaves on twigs) in comparison with a mechanical ideal: a large flat solar collector (Figure 3.5a). Colinvaux argues that the primary advantage of many small leaves is as follows. In the monolayer designs one huge leaf traps all incident sunlight. But in bright sunlight, photosynthesis would rapidly become saturated. In contrast, if the panel were subdivided so

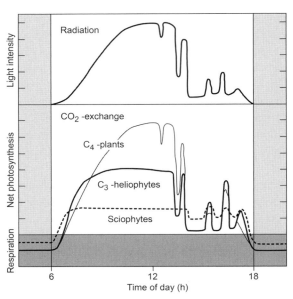

FIGURE 3.4   Schematic diagram of the daily fluctuation in CO_2 exchange as a function of the available radiation. C_4 plants can utilize even the most intense illumination for photosynthesis, and their CO_2 uptake follows closely the changes in radiation intensity. In C_3 plants photosynthesis becomes light-saturated sooner, so that strong light is not completely utilized. Sciophytes, adapted to using dim light, take up more CO_2 than heliophytes in the early morning and late evening, as well as during periods when they receive little light due to cloud cover or shade from the tree canopy, but they cannot utilize bright light as efficiently as the heliophytes. (From Larcher 1995)

that the upper layers shaded the lower layers, then the lower layers would receive diffuse light.

A three-layered device (shown in Figure 3.5b) would outperform a single-layer device and, of course, plant canopies usually consist of many more than three layers. "The payoff is increased photosynthesis in bright light" (p. 46). In such a design, the shape of the upper panels will affect the kind of shade cast upon lower panels. Irregular shapes reduce effective diameter. This may explain why so many trees have deeply notched

(a)   (b)

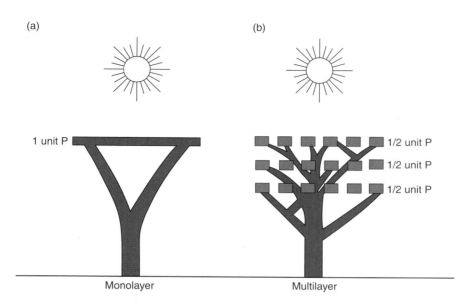

1 unit P

1/2 unit P

1/2 unit P

1/2 unit P

Monolayer   Multilayer

FIGURE 3.5 The tree on the left (a) has a single huge leaf, whereas the one on the right (b) has multiple leaves with multiple layers. What are the relative photosynthetic efficiencies of these two growth forms? Does this explain why most trees have many small leaves arrayed in multiple layers? (After Colinvaux 1993)

leaves: notches increase light transmission to lower layers of leaves. Such ideas suggest testable hypotheses; for example, canopy trees should have leaves arranged in multiple layers, whereas understorey trees should have leaves arranged in a monolayer.

Simple models help us think about nature, but it is a good practice to always consider alternative hypotheses. Several factors may be acting simultaneously. Here are some to consider:

1. The value of redundancy. In case (a) (one large layer), damage from a storm might disrupt photosynthesis in a significant portion of the leaf. In case (b) (many smaller leaves), those that are damaged are unlikely to influence those that remain intact. Think of it this way. If a brick is thrown at a window, which will be more damaged: a window with one large sheet of plate glass, or a window with a hundred independent sub-panes? Almost certainly the former, and this may be why sessile organisms, such as plants and corals, are constructed on a modular basis. Given that nature is full of forces that will damage leaves (wind, rain, hail, herbivores, falling branches) there is likely strong selection against

investing too much tissue in any large leaf. Thus canopy trees that are most exposed to storm and wind shear will have multiple small leaves, whereas more protected understorey trees can have larger leaves (a monolayer).

2. The value of increased surface area. In one large monolayer there is a minimal surface area owing to the lack of edges. As the solar panel is subdivided into progressively smaller units, there are more edges and the total surface area increases. Since CO_2 uptake is likely a function of surface area, selection for many subdivided units would enhance CO_2 uptake.

3. Nature is not perfect. Evolution by natural selection does not produce the best of all possible worlds (although it is frequently misunderstood in this way), it merely tends to select the best of available options. There are, at very least, strict biomechanical limits on the shape and size that plants can take. It is unlikely that any evolutionary pressure would be capable of producing large diameter trunks at the ends of small twigs. Similarly, leaves are produced by individual meristems. It is possible that the monolayer in Figure 3.5a could be produced by reducing the

number of meristems and increasing the amount of leaf tissue produced by each meristem, the limiting case being a single large leaf on one stalk (similar to the large single leaves in genera such as *Nymphaea*, Figure 2.8, or *Podophyllum*, Figure 5.39). But if branches and twigs are lost in this process, greatly strengthened veins within the leaf must replace them. In the case of *Podophyllum*, the leaf requires reinforced veins; in *Nymphaea* the round leaf floats on water. Even in these cases, there is not really a single monolayer leaf; while each shoot may appear to be a single solar panel, these shoots are attached to one another underground by way of rhizomes. Thus there are likely many biomechanical restraints upon the form of plants. Natural selection does not produce optimal shapes and sizes; it merely tends to perpetuate the best available options under the multiple constraints faced by many living organisms.

The issue of leaf size and shape will arise in several other contexts, later in the book, as we revisit effects of climate. Certain large-scale patterns are evident: leaves with teeth are more common in colder climates, small leaves are more common in dry climates and leaves with drip tips are more common in warm areas with heavy rainfall. Overall, however, we remain perplexed as to why leaves are the shapes and sizes they are (Horn 1971; Givnish 1984, 1987; Nicotra et al. 2011). And, in an added layer of complexity, there is also variation at the smaller scale in the density of stomata per unit area of leaf. Even among just 28 species of tropical trees in just one area of Borneo, stomatal densities varied from 27 to 380 stomata per mm^2 (Russo et al. 2010). Therefore while capacity for light gathering is certainly important, other factors may include phylogeny, biomechanical constraints, CO_2 uptake, thermoregulation, the process of leaf expansion, nutrient demands and defences against herbivores. Overall, there is no simple single explanation to account for the wide array seen in the shape of leaves.

3.3.6 The Overriding Importance of Height

Life forms on land have added problems to contend with, such as preventing desiccation and providing support to remain erect. These themes are often dealt with in introductory botany books. But there is one other consequence to living on land with a self-supported shoot: the acquisition of light now has a cost, at least when there are neighbours. This cost is incurred in constructing and maintaining the aerial shoot and leaves. In early land plants both the gametophytes and sporophytes were probably photosynthetic (as in modern-day ferns and fern allies), but as the density of plants on land increased, neighbours became more abundant, reducing the light availability to aerial shoots. As a consequence, increased height was selected for in plants (Niklas 1994), and tree-like forms arose in a number of lineages (recall Figures 1.6 and 1.7) including the Lycopodiophyta, Pterospermatophyta and Pinophyta.

Secondary growth can therefore be seen as an evolutionary response to light shortages attributable to neighbours. Since the gametophyte phase in the life cycle of each plant produces antheridia and archegonia, and since a film of free water is required for sperm to swim from the antheridium to the egg in the archegonium, there may have been selection against increased size in this life-cycle phase. This constraint upon gametophytes may explain why, in terrestrial plants, the gametophyte stage has been reduced in time, while the sporophytes have become dominant (Keddy 1981a).

The benefits from increased height are most evident when neighbours are present (Givnish 1982). Since light comes from one direction, the taller plant is always at an advantage over the shorter. This may have important implications for the nature of competition; the tall plant can shade the short one, but the short one cannot shade the tall one. As a consequence, competition for light may be strongly asymmetric, and this may in turn explain why strong competitive hierarchies occur in plant communities (Chapter 4).

3.4 Below-Ground Resources

3.4.1 Water

The second major constituent of CHNOPS life forms (Section 3.1.1) is hydrogen. Water is an abundant source of this element. Photosynthesis is the name of the process in which oxidized carbon from the atmosphere and oxidized hydrogen from water are reduced and combined to construct organic compounds of carbon and hydrogen. In most cases, water for the process is removed from soil, often from metres below the surface; hence the investment in large root systems to locate and then transport this water to shoots.

In the absence of any life at all, water would still flow within the soil from wetter areas (high water concentration) to drier areas (low water concentration), and water would still evaporate from the surface of the soil into the atmosphere. Evapotranspiration by plants, then, is not so much a new evolutionary process as a simple harnessing of the natural physical process of diffusion. That is, the soil, plants and water are three parts of one continuum (Kramer 1983). Plants exploit and enhance this natural process in at least two ways. First, the vascular system of the plant provides a more or less continuous path for the diffusion of water from the soil into the atmosphere. Second, the shoots and leaves of plants enlarge the surface area available for the evaporation of water. This combination of enhanced transport and enlarged surface area explains why evapotranspiration rates are so much higher than rates at which water would otherwise evaporate passively from a relatively flat soil surface. As water passes into roots and evaporates from shoots, it simultaneously provides a medium for the transport of other soil resources into plant tissues.

The transport of water can be described by a simple equation, the flow of water being a consequence of the driving force divided by the resistance (Kramer 1983). The driving force is the difference in water potential between the soil and the atmosphere. The resistance arises from obstacles to diffusion, such as cell walls. Through evolutionary time, the numerator in this equation (difference in water potential) has increased as leaves have become elevated further away from the soil surface. Simultaneously, the denominator of the equation (resistance) has decreased through the evolution of multicellular roots, and the change in xylem cells from thin tracheids to wide-vessel elements. The rate of movement of water through the plant is controlled primarily by the resistance to flow imposed by the cuticle and the stomata of leaves. In summary: (1) water flow is primarily controlled by the evaporation surfaces of the plant; (2) the stomata are the principal regulators of water movement; and (3) increased resistance elsewhere in the soil or roots reduces water flow indirectly by reducing leaf turgor enough to cause the stomata to close (Kramer 1983).

Water is far from homogeneously distributed in soils (Figure 3.6); its availability varies with depth, time of year and vegetation type. As this figure illustrates, tree roots must actively forage for reservoirs of water. Since a tree can transpire 200 litres of water in a single hour (King 1997), patches of water can be quickly depleted by the transpiring canopy. Two kinds of forces cause water movement into roots from the soil: osmotic movement (in slowly transpiring plants) and mass flow (in rapidly transpiring plants). In the first case, the accumulation of solutes in the xylem of the root lowers the water potential of the root below that of the soil; water then diffuses inward and root pressure develops in the xylem. As the rate of transpiration increases, the mass flow of water dilutes the root xylem sap, reducing osmotic flow. Reduced water pressure in the root from transpiration produces a steep gradient in water potential between the roots and soil.

Many plants regularly can lose from 25 to 75 percent of their saturation water content under conditions of "normal" water supply. As water content

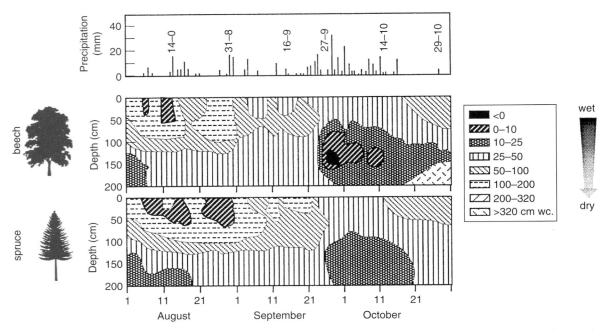

FIGURE 3.6   Changes in soil moisture over three months in beech woodland (top) and spruce woodland (bottom). Rainfall is shown above. (After Benecke and Mayer 1971)

falls, stomata close, but this has significant negative consequences for growth, since photosynthesis is dependent upon the unimpeded flow of oxygen and CO_2, and since evapotranspiration is necessary for the uptake and transport of dissolved minerals. Thus any loss of turgor will reduce rates of foraging and growth.

In many, if not most, natural environments there are periods with insufficient water for plant growth. The consequences of shortages will be explored further in Chapter 10 on stress.

3.4.2 Nitrogen

Each amino acid contains an atom of nitrogen as NH_2. Hence nitrogen is essential for plant and animal growth. Since the atmosphere is 78 percent nitrogen, plants are bathed in atmospheric nitrogen – yet nitrogen availability is one of the most important factors limiting plant growth and reproduction. How can this be? The principal reason is the structural stability of

atmospheric nitrogen, that is, the strength of the bond that links the two atoms that make the molecule N_2.

The main forms of nitrogen that plants can use are ammonium and nitrate. Ammonium and its counterpart, ammonia, are toxic at quite low concentrations; they are therefore usually absorbed and incorporated into organic compounds in the roots. In contrast, nitrate is easily transported in the xylem and can be stored in vacuoles of roots, shoots or storage organs. The cost of nitrate use lies in the necessity of reducing it back to ammonium to enable it to be further used in metabolic processes; this process consumes energy and is analogous to the process of reduction and assimilation of CO_2 in photosynthesis (Marschner 1995). These processes of nitrogen uptake are essential to our understanding of plants and plant communities, not only because nitrogen is a key resource for plants but because it is equally the key resource for animals (White 1993).

Until the industrial revolution, plants and plant communities depended upon two natural sources of nitrogen (Vitousek et al. 1997). The first source was biologically fixed nitrogen from free-living organisms (e.g. cyanobacteria such as *Anabaena*) and some symbiotic organisms (e.g. bacteria such as *Rhizobium*). These organisms produce the enzyme nitrogenase, and under appropriate (usually anerobic) conditions they can extract nitrogen from the atmosphere and incorporate it into their tissues, thereby making it available eventually to other living organisms. The second, smaller, source was atmospheric fixation. The enormous energy of lightning discharges is able to catalyze the production of ammonia from atmospheric nitrogen.

Biological fixation greatly exceeds atmospheric fixation – fixation on land alone is in the order of 100 Tg·year^{-1}, whereas atmospheric fixation is probably less than 10 Tg·year^{-1} (a teragram, Tg, is a million metric tonnes of nitrogen).

All natural plant communities, and all agricultural systems, were once dependent upon only these two sources of nitrogen. As nitrogen was depleted from European fields, it had to be augmented locally (by growing legumes and by recycling animal waste) or by importation. Ships transported sea-bird excrement (guano) from islands off the coast of Chile to supplement nitrogen supplies for European agriculture.

All this changed with the discovery of industrial methods for transforming atmospheric nitrogen into biologically usable forms (see Box 3.2). The rate of industrial fixation was 80 Tg·year^{-1} in the 1990s, and continues to grow (Vitousek et al. 1997; Matson et al. 2002). Eutrophication by surplus nitrogen is now an emerging problem for the Earth's plant communities (Ellenberg 1988a; Keddy 2010). Rainfall in industrial areas of Europe, North America and Asia now contains significant concentrations of nitrogen, and plants that normally inhabit infertile communities are disappearing as the habitats become more fertile and are invaded by species that exploit fertile soils. These topics will be explored further in Chapter 12.

3.4.3 Phosphorus

For terrestrial plants, as opposed to phytoplankton, the necessary phosphorus must be obtained from the soil (notable exceptions to this general rule include epiphytes and carnivorous plants; Section 3.7.3). There are three components of foraging by roots: (1) direct interception by roots; (2) mass flow; and (3) diffusion. For (1), as roots grow, their surfaces will naturally expand in area and encounter nutrient ions. For (2), as transpiration by shoots causes mass flow of water toward the roots, dissolved nutrients are transported. For (3), nutrients can diffuse through water in the soil to replace ions extracted by roots. Measurements of these three components of nutrient uptake (Table 3.4) show that interception is of minor importance and that the relative importance of mass flow and diffusion varies by nutrient. Phosphorus stands out because of its low values for both interception and mass flow, with nearly the entire demand being met by diffusion. The driving force of diffusion is a concentration gradient, and this gradient is produced and maintained as roots deplete the soil of phosphorus. This requires the presence of water in the soil pores. In practice, the depletion zones around plants are measured in millimetres (Figure 3.7).

The degree to which plant growth is limited by problems with phosphorus uptake is best illustrated by the exceptional means used to acquire it. Limited access to soil phosphorus has resulted in symbiotic relationships between plants and fungi, in the form of mycorrhizae (Lewis 1987; Wardle et al. 2004). The hyphae of the symbiotic fungi can absorb and translocate phosphorus to the plant from soil outside the depletion zone, increasing by two to three times the phosphorus uptake per unit of root length (Tinker et al. 1992). "Land plants need to grow towards phosphate rather than wait for it to diffuse to them" (Lewis 1987: p. 165). A delightful experiment by Li et al. (1991) measured the importance of roots and mycorrhizae in phosphorus uptake by using netting to exclude roots, and a membrane to exclude hyphae, from experimental containers (Figure 3.8). It is clear that the mycorrhizae increased the distance over which phosphorus could be extracted from the soil and therefore the total amount of phosphorus available to the plant.

In conclusion, phosphorus limitation must be an important problem for plants if it is necessary to incur

Box 3.2   Fritz Haber Changes the Global Nitrogen Cycle

Until the end of the industrial revolution, plants and plant communities depended upon two natural sources of nitrogen: biologically fixed nitrogen and nitrogen fixed by lightning discharges. All of this changed in the early 1900s. The first driving factor was gunpowder, the production of which also required nitrogen. The Germans knew that in the case of war, access to the rich beds of guano in Chile would be cut off by the British navy. The second driving factor was also related to military matters. In the case of war, German supplies of fertilizer (and hence food) would also be at risk. There had to be a way to extract nitrogen directly from the atmosphere.

This process was discovered by a German chemist named Fritz Haber (Figure B3.2.1) in 1909, and it is still named the Haber process in his honour (Oakes 2002). At high temperature and pressure, a metal catalyst is used in converting nitrogen to ammonia. So long as there are sufficient supplies of fuel, such as natural gas, the supplies of nitrogen are now nearly endless.

At very least, Dr. Haber prolonged the cataclysmic First World War by ensuring that both gunpowder and food would be amply available for the German war machine. Sadly, Haber was unable to stop there. He became further involved with chemical warfare, and introduced poison gas attacks, personally supervising them and refining the methods (Figure B3.2.2). Hundreds of thousands of soldiers from Russia, France, Britain and Canada, among others, were blinded, crippled or killed (Haber 1986; Stoltzenberg 2004). Haber married in 1901, his wife Clara being the first woman in Germany to have earned a PhD in chemistry. Horrified by his involvement in chemical warfare, she shot herself in 1915. Haber continued his poison gas research and married an apparently less squeamish woman in 1917.

FIGURE B3.2.1 Fritz Haber was awarded the 1918 Nobel Prize for chemistry for the synthesis of ammonia from its elements. (Photograph from Wikimedia Commons)

FIGURE B3.2.2   A poison gas attack on the eastern front during the First World War. As the gas drifts downwind to the left, the shadows of advancing German troops can be seen on the upper right. Fritz Haber helped develop and refine methods of gas warfare that killed or injured more than a million soldiers. (Photograph from www.ga.k12.pa.us, accessed 15 February 2005).

Box 3.2   (cont.)

The year after the end of the war, even at the time he was considered a war criminal, Haber received his Nobel Prize in chemistry (Oakes 2002). He was, however, eventually forced to leave Germany because of his Jewish ancestry, and died of a heart attack in Switzerland in 1934. The poison gas research at his institute led to the Zyklon process (Stoltzenberg 2004), which was used to produce Zyklon B, the poisonous gas used to murder Jews, political dissidents, homosexuals and other "undesirables" in their millions in concentration camps such as Auschwitz and Birkenau.

Quite apart from the dark history of chemical warfare, his other legacy, the Haber process, continues to alter the world's nitrogen cycle. The rate of industrial fixation was 80 Tg per year in the 1990s, and continues to grow (Vitousek et al. 1997). Eutrophication by surplus nitrogen (Figure B3.2.3) is now an emerging problem for the Earth's plant communities (Ellenberg 1988a; Keddy 2010).

FIGURE B3.2.3   The use of ammonia fertilizer dramatically increased during the twentieth century and became a major source of water pollution in the Mississippi River Basin. (a) Ammonia application to row crops. (Photograph from US Bureau of Labor Statistics) (b) Nitrogen fertilizer use in tons per square mile for the United States of America in 1991. (Map from US Geological Survey)

the energy costs of supporting mycorrhizae. It is logical to assume that evolutionary solutions arise only in response to a pressing problem in survival. Increasingly it appears that the symbiotic relationship between plants and fungi is not some bizarre aberration but rather a general phenomenon. Terrestrial plants with mycorrhizae might be viewed as large lichens in which the single-celled alga has been replaced by a multicellular plant. It is even possible that the association between algae and fungi was essential for the earliest invasion of land; this is supported not only by the ubiquity of the alga–fungus relationship but by the presence of apparent fossilized mycorrhizae in early land plants such as *Rhynia* (Figure 1.5), as well as the mycorrhizal gametophytes of fern allies today.

3.4.4 Experimental Tests for Nitrogen and Phosphorus Limitation

One direct way to assess the relative importance of a mineral element in controlling plant growth is to supplement nutrient supplies and see which ones cause an increase in growth (Chapin et al. 1986). In one early study of fertilization, Willis (1963) added N, P, K and NPK fertilizer to the vegetation of wet areas

among sand dunes. These areas had sparse vegetation, in which the dominant species were *Agrostis stolonifera*, *Anagallis tenella*, *Bellis perennis* and three species of *Carex*. In field trials, the addition of complete nutrients led to the production of three times as much biomass and reduced the number of plant species, after only three years. When the effects of nutrients were examined individually, the greatest response was produced by added nitrogen, with phosphorus of secondary importance. Willis then transplanted pieces of turf to the greenhouse and subjected them to different fertilization treatments. The results were similar to his field trials: complete nutrient treatment yielded 151 g of shoots (fresh weight), whereas nitrogen-deficient treatments produced 34 g, and phosphorus-deficient treatments produced 44 g. He concludes: "the sparse growth and open character of the vegetation of the Burrows are brought about mainly by the low levels of nitrogen and phosphorus in the sand." It is also

Table 3.4 The three components of nutrient uptake (interception, mass flow and diffusion) in a crop of *Zea mays*. (From Marschner 1995)

Nutrient	Demand (kg ha^{-1})	Estimates on amounts (kg ha^{-1}) supplied by		
		Interception	Mass flow	Diffusion
Potassium	195	4	35	156
Nitrogen	190	2	150	38
Phosphorus	40	1	2	37
Magnesium	45	15	100	0

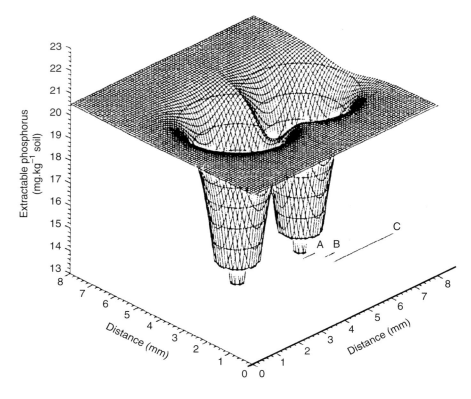

FIGURE 3.7 Profile of extractable phosphorus around two individual maize roots with overlapping depletion zones. A = root cylinder; B = root hair cylinder; C = maximal depletion zone. (Adapted from Marschner 1995)

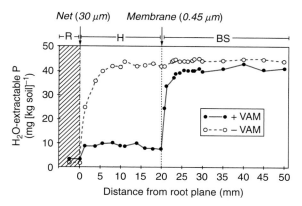

FIGURE 3.8   Depletion of water-extractable phosphorus in the root (R), hyphal (H) and bulk soil (BS) compartments of non-mycorrhizal (–VAM) and mycorrhizal (*Glomus mosseae*; +VAM) white clover plants. (From Li et al. 1991)

noteworthy that complete nutrients resulted in grasses becoming dominant, a pattern since observed in many other fertilization experiments. He suggests that increased fertility causes declines in diversity because grasses stifle species intolerant of competition from them, particularly rosette plants, annuals and bryophytes. Willis further observed that sedges and rushes appeared to be particularly successful in areas of phosphorus deficiency, suggesting that the ratio of nitrogen to phosphorus may control which particular plant group dominates a site, as Smith and Kadlec (1983) observed with algae.

A similar fertilization study in southern England used **peatland** sites including a dry *Calluna* blanket bog, a valley bog with *Sphagnum* and *Erica tetralix*, and an acid mire with *Sphagnum* and *Carex echinata* (Hayati and Proctor 1991). Fertilizer application to *C. echinata* grown in pots of peat from each site produced growth attributable to three main effects: nitrogen, phosphorus and between-site differences. Nitrogen was more limiting on the wet heath peats, whereas phosphorus was more limiting on the blanket bog. There was a minor effect of potassium, suggesting that it was present in adequate supply

everywhere except in the blanket bog. The use of peat in pots may have increased experimental control, but it is vulnerable to the criticism that it does not necessarily show that the same nutrients are important under more natural field conditions.

Let us look at one more example. Verhoeven et al. (1993) fertilized three fens with three essential nutrients (N, P, K) and a combination treatment (N + P + K). Figure 3.9 shows that the results varied among sites and plants. In fen 1, the addition of nitrogen and NPK produced significant increases in the biomass of vascular plants. In fen 2, the addition of phosphorus produced significant increases in the vascular plant and bryophyte biomass, while the addition of NPK increased only the vascular plant biomass. Fen 3 was similar to fen 1, except that the significant increase for vascular plants was accompanied by a decline for bryophytes so that total biomass was unchanged. These sorts of results might appear discouraging because each study site appears to be unique, but it is certainly clear that a vegetation type may arise either from shortages of nitrogen or phosphorus. One may not be able to tell which without doing a fertilization experiment. Or, it is possible that larger sample sizes might find patterns: in a review of 45 studies of fertilization of herbaceous wetlands, Koerselman and Meulman (1996) found an almost even split between nitrogen-limited sites and phosphorus-limited sites (Table 3.5). Co-limitation, that is, a response only to combined fertilizations, was rare. Wet heathlands all had growth limited by phosphorus, whereas fens and dune slacks could be limited by either nitrogen or phosphorus. The wet grasslands were the most complicated, with cases of nitrogen, potassium, and nitrogen plus potassium limitation.

How can we explain such patterns? Verhoeven et al. (1996) emphasize differences between the two nutrients. For phosphorus, the main source is weathering from rocks, and the main input to wetlands is from water flow. In contrast, the main source of nitrogen is fixation from atmospheric

Table 3.5 **Limiting factors in seven habitat types, as determined by biomass response in fertilization experiments. Figures indicate number of cases in which the element was shown to be limiting. (From Verhoeven et al. 1996)**

Habitat	N	P	K	N + P	N + K	P + K
Wet grassland	3	0	2	0	4	0
Wet heath	0	3	0	0	0	0
Rich fen	7	5	0	0	0	0
Poor fen	2	1	0	0	0	0
Litter fen	1	2	0	1	0	0
Bog	1	3	1	0	0	0
Dune slack	5	2	0	2	0	0
Total (45 cases)	19	16	3	3	4	0

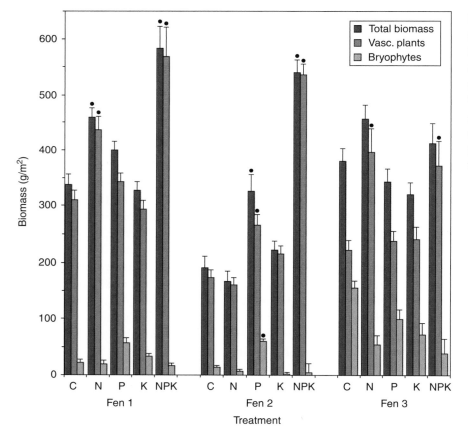

FIGURE 3.9 Response to fertilization with N, P, K and all three of these elements compared to a control (C) in three fens. Values significantly different from the control are indicated by a black dot. (After Verhoeven et al 1993)

nitrogen (and increasingly, deposition of pollutants in precipitation, Box 3.2). Thus there is likely to be a shift from nitrogen to phosphorus limitation during succession, since early in succession phosphorus is available in ground water and there is little nitrogen stored in organic matter. Consistent with this view, Table 3.5 shows that early-successional rich fens are generally nitrogen limited, while late-successional bogs and moist heaths are phosphorus limited.

3.4.5 Other Sources of Evidence for Nutrient Limitation

The concentrations of nutrients in plant tissues seem to be obvious measures of resource requirements. High concentrations of nutrients may be difficult to interpret, since they might mean that the nutrient is available in excess and is being easily taken up in luxury for future use, or that the nutrient is scarce and is being stored for future use. This is particularly problematic in infertile soils, where nutrient uptake may be entirely uncoupled from periods of growth (Grime 1977). While such nutrient data must be viewed with caution, the ratio of nitrogen to phosphorus in tissues may still be informative about the relative importance of these elements. Returning to the wetlands described in Section 3.4.4, Verhoeven et al. (1996) examined tissue nutrient concentrations for the same set of studies. The N:P ratio of 15:1 (as measured in plants from control sites) clearly separated N-limited from P-limited situations: "It can be concluded the N:P ratio of above-ground biomass at the end of the growing season (August) provides a reliable indicator of the degree to which each of these elements has been limiting plant growth in herbaceous mires." They conclude that N:P ratios greater than 16 indicate phosphorus limitation, and N:P ratios less than 14 indicate nitrogen limitation.

Many other field studies have tried to understand nutrients and plants by seeking relationships between soil nutrient levels and plant abundance or growth. Often the results suggest important roles for both nitrogen and phosphorus. Figure 3.10 shows the interrelationships of soil nutrients, biomass and primary productivity in one riparian wetland (Auclair et al. 1976a). In this case, biomass and productivity are positively correlated with nitrogen (r = 0.38, 0.39) but negatively correlated with phosphorus (r = −0.29, −0.23). Yet in a similar geographic region, but in a *Carex*-dominated ecosystem, there were no significant correlations between production and either nitrogen or phosphorus (Auclair et al. 1976b).

3.4.6 Mineral Nutrients: A Single Cell Perspective

Phytoplankton provide a useful model for studying mineral nutrients, in part because, at least in fresh water, they usually comprise single cells. Further, they represent a form of plant life that was common early in the history of the Earth. Single cells floating near the surface of an aqueous medium are unlikely to be limited by any of the three resources mentioned so far: light, water or CO_2. However, there is a striking linear relationship between the abundance of phytoplankton in lakes and the concentration of phosphorus (Figure 3.11). As well, some of the remaining variance that is unaccounted for by phosphorus concentration can be dealt with by incorporating the concentration of nitrogen.

Other nutrients can also occasionally control the distribution and abundance of plant species. In the oceans, for example, there can be relatively high concentrations of nitrogen and phosphorus yet low phytoplankton biomass. To test whether dissolved iron might be limiting growth, it was added to an unenclosed 64-km^2 patch of ocean, which was then tracked with a buoy while an inert tracer was used to monitor the shape of the patch (Martin et al. 1994; Carpenter et al. 1995).

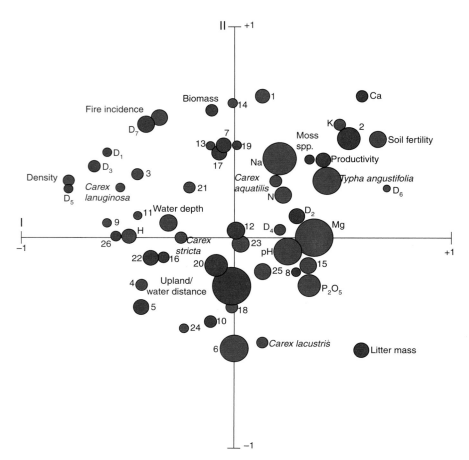

FIGURE 3.10 Interrelationships of fertility and productivity in a riparian wetland in eastern Canada (using a factor analysis model). The first and second components are shown on the horizontal and vertical axes, respectively. Factor loadings on the third component are represented by circle diameters (brown is negative, that is below the plane of the page; while green is positive, that is above the plane of the page). D_1 to D_7 are species diversity indices. Coded variables include: (1) soil organic matter, (2) soil $(Ca+Mg)/(K+Na)$, (3) date since 1 May, (4) tussock incidence, (5) distance to upland, (6) distance to water, (7) biomass, (8) average stem height, (9) *Equisetum fluviatile*, (10) *Onoclea sensibilis*, (11) *Thelypteris palustris*, (12) *Potentilla palustris*, (13) *Viola pallens*, (14) *Hypericum virginicum*, (15) *Galium palustre*, (16) *Lysimachia thyrsiflora*, (17) *Lythrum salicaria*, (18) *Cicuta bulbifera*, (19) *Utricularia vulgaris*, (20) *Impatiens capensis*, (21) *Lycopus uniflorus*, (22) *Campanula aparinoides*, (23) *Carex diandra*, (24) *Calamagrostis canadensis*, (25) *Sparganium eurycarpum*, (26) *Sagittaria latifolia*. (From Auclair et al. 1976a) You can read more about interpreting these sort of multivariate diagrams in Section 11.3.

Photosynthetic efficiency increased within the first 24 hours, and by the third day, chlorophyll concentrations had tripled. This was enough to cause a reduction in CO_2 of the surface waters, but there were no detectable reductions in nitrogen or phosphorus.

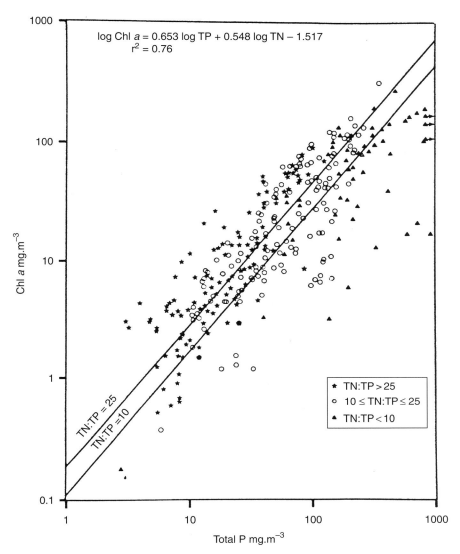

FIGURE 3.11 Total dissolved phosphorus is an important predictor of the abundance of phytoplankton as measured by the concentration of chlorophyll *a*. (From Smith 1982)

3.5 Resources Affect Entire Ecosystems

3.5.1 Primary Production Is Controlled by Resources

Now let us leap from considering individual plants to entire ecosystems. What are the patterns of energy capture at this larger scale? The capture of carbon dioxide and construction of biomass can be quantified as grams of carbon fixed per square metre per year, or as kilograms per hectare. This is known as net primary production (NPP) or, on an annual basis, ANPP. The word "net" is critical – it reminds us that plants respire some of the energy they capture in their own metabolic processes. Hence not all of the sunlight captured is converted to biomass. Roughly half is respired, and

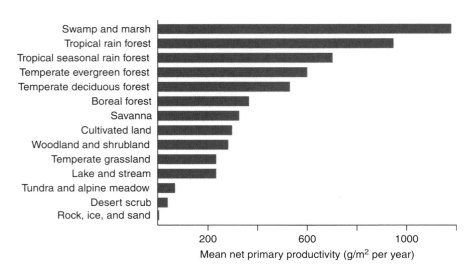

FIGURE 3.12 The availability of resources, particularly water, determines annual net primary production (ANPP) of vegetation types. Note that swamps/marshes and rain forests have the highest levels of ANPP, in the range of 1,000 g m^{-2}. (From data in Whittaker and Likens 1973)

perhaps even more in tropical forests (Amthor and Baldocchi 2001). You may recall from your introductory biology courses that different parts of the world have different rates of primary production; these are summarized in Figure 3.12. A great deal of international effort (summarized in Lieth 1975) was necessary to obtain these data. In total, ANPP was estimated to be some 160 × 10^9 t in Lieth (1975) and may be rather higher, 224.5 × 10^9 t (Vitousek et al. 1986). Note too that many writers now avoid the ambiguity in the word tons, and instead express ANPP in petagrams (Pg), 10^{15} grams, in which case, using the Vitousek numbers, the plants on Earth produce 224.5 Pg of biomass each year. Other estimates are considerably lower (Saugier et al. 2001).

Since I assume you already have access to a general book on ecology, I will not recount the basics of energy flow in ecosystems, or the structure of food webs here. That takes us beyond the boundaries of plant ecology. But I must remind you that all studies of energy flow are based on resources captured by plants, in which case the preceeding sections on plant growth can be seen to have global wildlife implications. Even the most dramatic predators in wildlife films are dependent upon humble plants. (Yet, too many wildlife biology students at the University of

Guelph could never accept this simple fact, and I recall their many machinations to study wildlife biology without taking a single course in botany!) Also recall that, in general, ANPP increases with rainfall and temperature. Thus, in general, tropical ecosystems are more productive than temperate ones, and the Köppen climate map (Figure 2.4) is useful in predicting ANPP.

Let us look at four generalizations that have important general consequences. They all start with Figure 3.12 based on a classic table in Whittaker and Likens (1973), which, in spite of much new work, is still a useful reference point.

(1) Wetlands are the most productive natural ecosystems in the world, with many swamps and marshes having an ANPP in excess of 1,000 g m^{-2} (Figure 3.12 top). This is likely because water is generally available for plant growth. Moreover, swamps and marshes receive added nitrogen and phosphorus in flood water. This is one reason why spring floods are essential in large rivers (Keddy 2010). This high plant growth in swamps is a major input to fish production, which in turn is an important source of protein in tropical regions.

(2) Tropical rain forests also have high ANPP, with a mean of 900 g m^{-2} (Figure 3.12, second bar down), which means that many areas exceed 1,000 g m^{-2}.

We may think of dense steamy rain forest as a plant paradise, but in some cases the large amount of biomass can be deceptive. The rate of production may be lower than the climate would suggest because trees may be limited not by rainfall or sunlight, but by soil nutrients. As well, higher temperatures lead to higher respiration rates (Amthor and Baldocchi 2001). The trees in fact may be growing rather slowly, and the extravagant biomass may have accumulated over many years (Forsyth and Miyata 1984; Richards 1996). When such forests are logged and cleared, the nutrients leave the site in biomass via logging trucks or run-off. Hence the resulting land may be impoverished and suitable only for rangeland, or agriculture that is dependent upon gasoline and fertilizer.

(3) Cultivated land generally has lower productivity than wetlands or rain forests, at only 290 g m^{-2} (Figure 3.12), although Leith gives somewhat higher values. Intensive modern farming can raise ANPP to, say, as high as 400 g m^{-2} (Bradford et al. 2005), but only in exceptional cases. This is usually achieved only by consuming fossil fuels to run tractors, move irrigation water, make fertilizer and apply it to fields (recall Figure B3.2.3). When ANPP figures are calculated for intensive agriculture, the energy in fossil fuel is often not subtracted from the measure of production. The fact that some crops may contain as much energy from oil as sunlight is going to be a challenge to humans as fossil fuel costs rise. Meanwhile, you should be aware that the ANPP values for cropland mostly depend upon this fossil fuel subsidy. And, over the world as whole, conversion from natural systems to agriculture almost invariably results in reduced ANPP (Vitousek et al. 1986). Again it depends on the location and the type of change in land use – in the intensively farmed Great Plains of the USA, think fields of soybeans and corn with high fuel inputs, the mean ANPP in cropland, 300 g m^{-2}, is either the same as, or slightly greater than, native vegetation (Bradford et al. 2005).

(4) Deserts have extremely low ANPP, well under 100 g m^{-2} and perhaps only a third of this. This, of course, is largely a result of low rainfall, which strictly limits rates of photosynthesis. At the same time, human activities such as population growth and overgrazing are creating anthropogenic deserts in arid areas. The process of desertification continues over large areas of the world and is likely to accelerate with increasing human populations, increasing goat populations and increasing global mean temperatures. We will return to desertification in the final chapter of the book, but meanwhile you should be aware that desertification causes a reduction in ANPP. Vitousek et al. (1986) suggest that at the global scale, desertification has already reduced global ANPP by 4.5 Pg.

(5) Root biomass is difficult to study and often overlooked. It is significant because considerable amounts of carbon are stored underground, which would mean that estimates of ANPP from aboveground parts would significantly underestimate ANPP. Yet these underground components respire and thus are a net drain on plant production. The ratio of root biomass to total biomass ranges from 0.13 in crops, through 0.21 to 0.27 in forests, to as high as 0.67 in grasslands (Saugier et al. 2001). The highest values (that is, greatest proportion of biomass stored underground) occur in biomes with severe constraints, such as drought or nutrient limitations. Including roots raises the estimates of ANPP in tropical forests by a further 20 percent, to above 1,000 g m^{-2}. Since we are already on the topic of roots, this is a good time to move to the topic of soils.

3.5.2 Soils Are Produced by Two Causal Factors

Soils provide the large-scale template for patterns of nutrient availability and for plant distribution. At the same time, plant activities contribute to soil production. There is thus an ongoing set of complicated feedbacks between plants and soils as well as among other components of ecosystems, such as fungi and grazing animals. Here we will look at some of the basic processes that produce different soil types, and some of the procedures used in classifying them. Without fertile soils to support farming, human civilization might never have progressed beyond the hunter-gatherer stage. It is no coincidence that two major centres of human culture developed on the fertile valley soils of the Middle East and Asia.

A few soil-forming processes can occur without living plants. Oxidation, for example, changes soil minerals such as iron, sulfur, uranium and manganese into oxidized forms such as hematite, pyrite, uraninite and manganese dioxide. Oxidation itself would not occur, however, without plants to produce oxygen (Section 1.6). Soils would not have formed, at least not in the conventional sense of the word, until plants invaded the land and provided a steady input of organic matter.

To start simply, the type of soil at a site can be attributed to just two causal factors, water and organic matter. Each of these can be divided between inputs and losses. This yields the following four factors that drive soil formation:

1. Rate of water infiltration
2. Rate of evapotranspiration
3. Rate of input of organic matter
4. Rate of decay of organic matter.

These factors drive processes at very small scales (centimetres of depth in the soil column) and at very large scales (thousands of kilometres across continents). At the very small scale, as water and minerals move through the soil column, and organic matter accumulates, distinctive regions form at different depths. These are known as soil horizons and are often used in classifying soils. Some examples are shown in Figure 3.13. At the large scale, the very same factors produce major soil types in response to two basic climatic variables: temperature and rainfall. Rates of chemical processes, such as oxidation, are largely dependent upon temperature and rates of leaching are largely determined by rainfall. Superimposed upon these are biological factors: rates of production generally increase with rainfall and temperature but so do rates of decomposition.

Although the above system provides a simple, mechanistic perspective it is important to note that the processes will vary with other factors. In his classic book, *Factors of Soil Formation*, Jenny (1941) listed the five causal factors of soil as parent material, organisms, climate, topography and time. That is to say, the soil arising in any particular location is a function of (1) the parent material that (2) has been acted upon by organisms and (3) climate, and (4) conditioned by topography (5) over time. Later in this chapter, we will examine the effects of topography on resource supplies for both forests and grasslands, while all of Chapter 8 will explore time.

At the smaller scale, the capacity of a soil to hold nutrients until captured by plants is determined by surface area and the surface charge of soil particles (Faulkner and Richardson 1989). Surface area, in turn, is determined by the fraction of clay-sized particles as well as by the content of organic matter. The electrical charge of the soil is also important. In most temperate zone soils, the soil particles have a net negative charge, thereby providing electrostatic bonding sites for positively charged cations. These cations, ionically bound on the surface of soil particles, can exchange with other cations that are in solution. The cation exchange capacity measures the capacity of soil to hold cations on soil particles. Metal ions (Ca^{2+}, Mg^{2+}, Na^+) dominate in mineral soils, whereas hydrogen ions (H^+) dominate in areas with high organic content, such as peatlands.

Good soil maps are vital for studying plant distributions and vegetation types in landscapes. There are, however, multiple systems for soil classification and mapping. While no single system is yet universally accepted, the World Reference Base provides a harmonized global system (Bridges et al. 1998; Deckers et al. 1998). This system recognises 30 soil reference groups. Depending upon where you live, and the scale at which you work, you may also need to learn a local or regional system until there is global consensus. (As but one example of possible confusion, heavily weathered tropical soils can be called ferruginous soils or ferralsols.)

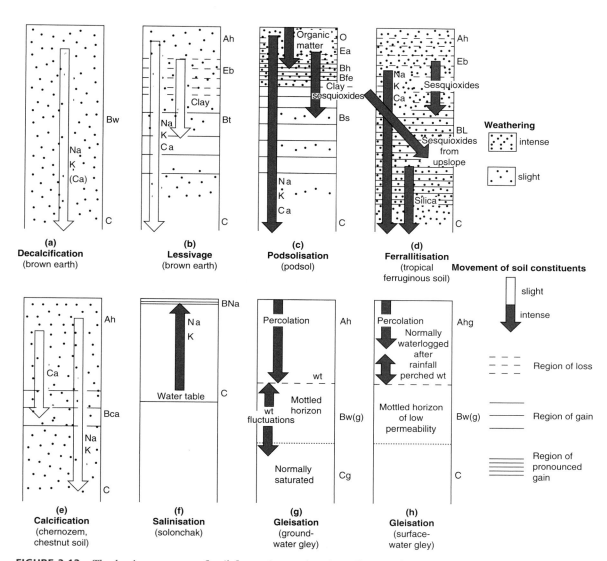

FIGURE 3.13   The basic processes of soil formation and main soil types. (From White et al. 1992)

There is little point in listing all 30 types and their characteristics here: they are described and mapped in Deckers et al. (1998) and Bridges et al. (1998). The origins of these soil types lie largely in the climate and the vegetation types mapped in Chapter 2. The many different types of soils are, however, produced principally by the four factors listed above – the effects of water and organic matter. These effects can be expanded into eight categories of process, such as

decalcification and lessivage (Figure 3.13) to produce the major soil types of the world. For a much simpler view than the 8 processes and 30 soil types, we can focus on just leaching by water and the accumulation of organic matter. Five common soil types then naturally arise: podsols, ferralsols, solonchaks, chernozems and histosols.

Certain soils are leached by excess rainfall. The typical **podsol** of northern forests has a relatively high organic content in part because of relatively low rates of decomposition. Clay particles and ions are leached by rainfall to accumulate in different soil horizons. This is very different from the typical **ferralsol** beneath tropical forests. Here high rates of decomposition produce soils with little organic matter, and heavy rainfall, in combination with the lack of organic matter, results in intense leaching. In arid climates, organic content tends to be low because primary production is low. High rates of evapotranspiration produce the opposite of leaching; ions such as Na^+ and K^+ are carried upward and deposited near the surface by evaporating water, producing a saline soil or **solonchak**. As rainfall increases, primary production increases and salinization decreases so that **chernozems** with high organic content form. Rainfall is neither so low that salinization occurs nor so high that leaching removes ions, and this balance results in soils highly productive for agriculture. **Histosols** occur where organic matter has accumulated as peat, and are commonly used in delineating wetlands.

In the United States, the soil classification system emphasizes amounts of weathering, and draws upon Latin and Greek terminology to designate ten soil orders (USDA 1975; Steila 1993). Let us consider five of these. The most recently deposited soils, called **entisols**, lack any diagnostic horizons and include young sand dunes and recent delta sediments. **Aridisols** form where conditions are warm enough for plant growth, but where a lack of moisture constrains weathering. In such soils, calcification (the accumulation of calcium carbonate) or salinization (the accumulation of mineral salts) occurs. **Mollisols** are typical of grasslands. These fertile, dark-coloured soils are thought to be formed primarily by the underground decomposition of the roots of grasses. **Spodosols** tend to form in cooler climates where conifers or ericaceous plants are found. Here, precipitation filtered through organic litter forms acids, which eventually allow clays to move downward, resulting in surface soils dominated by silicate minerals. **Ultisols** are deeply weathered acid soils formed in temperate landscapes that were not disturbed during the recent glaciations. Extremely long periods of weathering have leached soluble bases and allowed oxidation of iron and magnesium compounds.

Apart from plants and microscopic decomposers, invertebrates also play an important role in soil formation (Wardle 2002; Wardle et al. 2004). Chief among these are earthworms, of which there are some 3,000 species. Their activities include dragging plant litter down their burrows into the soil, ingesting and macerating plant litter, producing fecal material, releasing urine and glandular secretions, increasing porosity and supporting soil microorganisms (including nitrogen-fixing microorganisms) in their guts (Lee 1985). Although some earthworms in forests are confined to leaf litter or rotting logs, most excavate burrows. Up to half of the upper layers of forest soils consist of earthworm fecal material, and annual rates of production reach as high as 25 kg m^{-2}, in which case nearly one-quarter of the upper soil horizons are consumed and overturned each year. In his extensive studies on earthworms, Darwin (1881) documented how such activity results in the burial of stones at a rate of 0.25 to 0.5 mm·year^{-1}. Lee suggests that earthworms are among the most ancient of terrestrial animal groups, which seems reasonable given that a subterranean habit would allow them to tolerate two stresses that would have faced early life on land: desiccation and high levels of ultraviolet radiation. Earthworms were once uncommon in some habitats, such as the eastern deciduous forests of North America, and their introduction is now having a major effect upon soils and vegetation (Bohlen et al. 2004).

3.5.3 Life After Death: Soils, Detritivores and Decomposers

Once atoms of nitrogen and phosphorus have been used to construct plant tissues they are chemically bound and no longer available to other plants. As you will read in Section 3.7.1, plants themselves may salvage some of these atoms for use elsewhere in the same plant, particularly as leaves senesce, but this has its limits. Overall, as plants grow, increasing amounts of nitrogen and phosphorus, as well as lesser nutrients such as calcium, and potassium, become stored in plant tissues. In some areas, such as tropical forest, the amount of major nutrients in plant tissues exceeds that still available in the soil.

Once a leaf or twig falls to the ground as litter, the remaining nutrients can be reused by other plants, but only if other kinds of organisms first chew the tissue into small pieces and then break the chemical bonds that hold the nutrients in place. The process of releasing nutrients from chemical bondage within the dead plant tissues is known as **remineralization**. This process requires many intermediate stages, but it is convenient to divide it into two stages accomplished by two different sets of organisms. First, **detritivores**, particularly small arthropods, feed on dead plants and break them into smaller pieces, a process known as mechanical fragmentation. Then **decomposers**, particularly fungi and bacteria, degrade the pieces back into raw materials, releasing the nutrients. There is a complex soil food web that is associated with remineralization. Some members of this web feed directly on the plant tissues, while many others feed on bacteria and fungi that are themselves feeding upon the plant tissues. A fraction of the tissues, those most resistant to decay, accumulates as humus. The soil food web includes protozoans, nematodes, tardigrades, ants, termites, beetles, bacteria, fungi and earthworms (Brussaard et al. 1997; Wardle 2002; Adl 2003). Proteins, lipids and nucleic acids are the richest source of nutrients and most easily digested, and hence they may last only hours. In contrast, the molecules more resistant to decay including cellulose, chitin, lignin and xylan (hemicellulose) tend to accumulate (Adl 2003; Box 3.3).

You will read in Chapter 6 how many ways plants have to deter herbivores from feeding upon their leaves while they are still alive – from spines, to lignins, to individual molecules ranging from terpenes (in the conifers) to polyacetylenes (in the Asterales). This has important consequences for remineralization. In general, plant tissues that are protected from herbivores are also resistant to remineralization. And the slower the process of remineralization, the lower the re-release of nutrients for growing plants. Hence there is a complex feedback loop, with plants making resistant tissues to deter herbivores, but then slowing the process of remineralization, and having to cope with the resulting lower availability of nutrients (Wardle 2002).

In some habitats, such as northern peatlands, the detritivores and decomposers cannot process plant debris as fast as it is produced. Hence it accumulates, sometimes to a depth of many metres. You shall see more about this in Chapter 10. Meanwhile, you have already read about the Carboniferous era (Section 1.9) when organic matter accumulated as coal. During that era it is possible that decomposer food webs themselves were inadequate, since ants and termites had not yet evolved. Peatlands are often wet, which slows the activity of decomposers; but there is a further feedback loop that the resulting peat can store water and further delay decomposition.

At the other extreme from peatlands, in rain forests, the detritivores and decomposers are so efficient that very little litter accumulates on the soil surface, and equally little organic matter accumulates in the soil. The combined efforts of invertebrates and fungi rapidly degrade organic matter, and the voracious tree roots rapidly reabsorb the mineral nutrients. The understorey of a rain forest may look comparatively bare as a result (Forsyth and Miyata 1984). But even here, litter from resistant species may accumulate for a while, compared to other species that are quickly remineralized; the content of nitrogen is a good indicator of how fast a leaf decomposes (Adl 2003).

It is useful to keep these extremes in mind, peat bogs at one end of the continuum and rain forests at the other. Most other ecosystems fall somewhere in between. Rich farmland, for example, is the result of steady accumulation of nutrients and organic matter. Once the natural vegetation is removed, and crops planted, the accumulated nutrients may be depleted. This is one reason why agriculture has become increasingly dependent upon industrial sources of nitrogen and phosphorus.

Box 3.3   Big Molecules Have Big Consequences

Cellulose and lignin are two of the most abundant molecules (by mass) in the biosphere. Although their chemical nature is rather different, they are closely associated within plants, and therefore in decomposer food webs, so let us consider them together.

Structure

Cellulose (Figure B3.3.1) is the most abundant biopolymer in terrestrial environments. Each year photosynthesis creates some 10^{11} tons of dry plant material worldwide, and almost half of this material consists of cellulose (Duschene and Larson 1989; Leschine 1995). It co-occurs with two other polymers, hemicelluose and lignin, to which it is bound by non-covalent forces and by covalent cross-linkages. Cellulose is comprised of repeating glucose molecules joined end to end by β-1,4 bonds. Each glucose unit is rotated by 180 degrees relative to its neighbour. The length of this molecule can vary from 500 units in primary walls to 14,000 in secondary walls (Leschine 1995). These thin molecules are then united into microfibrils about 10 to 25 nm wide and wound together like strands in a cable to form macrofibrils. These fibrils are as strong as an equivalent thickness of steel (Duschene and Larson 1989). Cellulose molecules are even found in crystalline forms (Pérez et al. 2002). Hemicellulose contains several sugars besides glucose, including xylose, and tends to have shorter branched chains.

   Lignin is a large molecule comprised of phenols synthesized from the amino acid phenylalanine, organized in chains and networks that appear haphazard (Pérez et al. 2002; Berg and McClaugherty 2008). The most basic phenol is simply an aromatic hydrocarbon with one added –OH, for C_6H_5OH. Note, therefore, that although phenylalanine contains a precious nitrogen atom, this is removed early in the synthesis of phenols. Hence, lignin, like cellulose, does not contain nitrogen, which likely explains in part its wide occurrence in the plant kingdom (Duschene and Larson 1989). Its overall function is to give mechanical strength to plant tissues, and it is particularly associated with xylem cells and the walls of sclerenchyma cells. It is also a deterrent to many kinds of herbivores, as we shall see in Chapter 6.

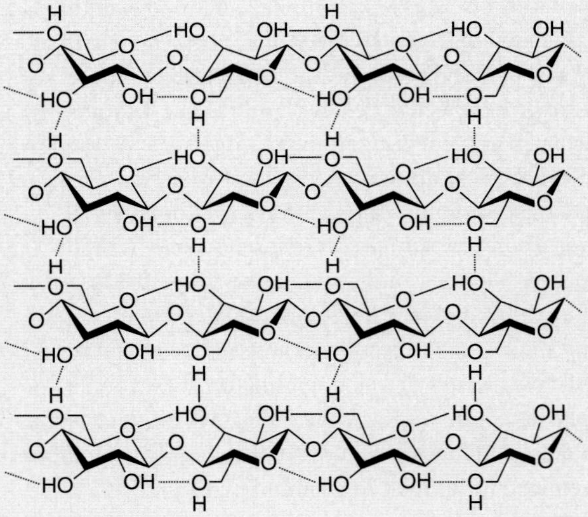

FIGURE B3.3.1   Cellulose is one of the largest and most abundant (by mass) molecules in the biosphere. It consists of long chains of glucose, which in turn are cross-linked and assembled into larger units within a plant. Cellulose, along with lignin, is the principal component of wood. Decomposition requires organisms that possess the necessary enzymes for cutting the chemical bonds to release the original glucose molecules. (Diagram from Wikimedia Commons)

Degradation

Generally, sugars and proteins in litter are consumed quickly, within months, while the degradation of cellulose and lignin in wood takes years. Although its structure is simple, even monotonous, cellulose is not soluble, so before it can be used by other organisms it must be broken down into water-soluble molecules. Cellulose can

Box 3.3   (cont.)

be degraded by both bacteria and fungi; they rely on extracellular enzymes that are secreted into their surroundings, or located on the cell surface (Berg and McClaugherty 2008).

The fungi that attack wood are often divided into white rot and brown rot, the former signifying groups that attack lignin first, the latter the group that degrade cellulose first. This is a consequence of the enzymatic systems possessed by decomposers. Overall, microorganisms have two types of extracellular enzymatic systems (Pérez et al. 2002). The first produces a suite of **cellulases**, enzymes that cleave cellulose and hemicellulose into progressively smaller pieces. The second produces a suite of **ligninases**, enzymes that similarly depolymerize lignin. These include lignin peroxidase and manganese peroxidase. Both of the latter are produced by fungi in the genus *Phanerochaete*, and these fungi are being intensively studied for their potential in degrading both lignin and some environmental contaminants (Singh and Chen 2008; Syed and Yadav 2012). Remarkably, in spite of lignin's importance in soil ecology, the global carbon cycle and its many economic implications, the actual processes involved in lignin degradation are still poorly understood at the molecular level.

In summary, cellulose and lignin are the major components of plants. They degrade slowly, and accumulate as soil organic matter or, in certain cases, peat. Litter decay rates are determined by two internal factors, the C:N ratio and the lignin:cellulose ratio (Talbot et al. 2012). The external factors include temperature and moisture availability, as well as the types of decomposers available. The most resistant organic debris accumulates as humus. Some organic soils have 50 (or even 100) kg m^{-2} of humus, which may in turn store about 700 g m^{-2} of nitrogen and 40 g m^{-2} of phosphorus (Berg and McClaugherty 2008). This takes us back to the theme of Section 3.5.3 and Figure 3.14, that key plant resources remain locked in litter until remineralization occurs.

Certain plant tissues are particularly resistant to remineralization, particularly wood. Wood is not only solidly constructed from cellulose and lignin, but it also has relatively low concentrations of nitrogen and phosphorus. As you will see in Chapter 6, most herbivores (that is, animals that feed on living plant tissue) prefer to feed selectively on plant tissues with higher levels of nitrogen and phosphorus. For this reason, wood is a relatively unpopular diet, which is one reason why large logs tend to accumulate as coarse woody debris and why a good deal of carbon remains stored in this form (Adl 2003; Barron 2003). Even wood-consuming fungi may require other sources of nutrients. Species in the genera *Arthrobotrys* and *Nematoctonus*, for example, produce sticky knobs or nooses that capture nematodes, allowing the hyphae to invade the prey and thereby supplementing nitrogen supplies (Thorn and Barron 1984; Barron 2003).

A short digression on nematodes: although this habit of consuming nematodes may seem unusual, nematodes are actually one of the most abundant groups of invertebrates in the biosphere, and are present in a wide array of natural habitats, particularly soil, where they live in thin water films and feed mainly on fungi and bacteria; 100 g of soil may yield 3,000 nematodes (Bongers and Bongers 1998). There are more than a 150 species of fungi that kill nematodes. *Arthrobotrys* fungi only produce traps when nematodes are present. The fungi can detect a family of pheromones known as ascarosides that are commonly produced by nematodes for regulating behaviour, particularly in detecting potential mates (Hsueh et al. 2013). Further, trap formation is only stimulated under nutrient-deprived conditions.

In summary, remineralization is the process that recycles nutrients out of dead plant tissues back into living plants. This process involves a wide array of

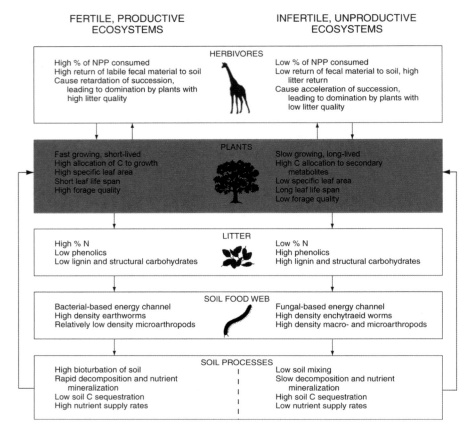

FIGURE 3.14
A comparison of soils having high (left) and low (right) mineral resources. Low levels of resources produce plants that conserve resources, being slow growing and long lived. Such plants tend to support lower populations of herbivores. These plants also drop litter that decomposes slowly, reducing soil fertility and producing microbial food webs dominated by fungi rather than bacteria. (Adapted from Wardle et al. 2004)

invertebrates including arthropods, nematodes, bacteria and fungi. The rate of remineralization can vary widely, from very slow in peat bogs to very fast in tropical rain forests. It is useful to break this continuum into two extremes, those fertile systems with rapid remineralization of tissues higher in nitrogen, and those infertile systems with slow remineralization of tissues lower in nitrogen. Figure 3.14 summarizes the many differences between these two extremes, with the former involving more bacteria and earthworms, and the latter involving fungi and arthropods.

3.5.4 Soil Resources Have Altered Human History

To a considerable extent, we can argue that human societies are dependent upon plant production, which

in turn is dependent upon the few very same resources we have considered in this chapter. Here we will consider the impact of soil quality upon the Carthaginians and Mayans in turn; in Section 13.2.2 we will return to the larger issues of deforestation of the Mediterranean landscape.

Rome and Carthage fought several wars. Rome was on the north shore of the Mediterranean while Carthage was on the south. They traded across the sea and faced each other uneasily. The Roman leader and former soldier, Cato, advocated a policy of justice and non-intervention with all foreign states – with one exception – Carthage (Durant 1944). Sent to Carthage on an official mission in 175 BC, he had been shocked by the rapid recovery of the city from the effects of preceding wars: "the fruitful orchards and vineyards, the wealth that poured in from revived commerce, the arms that mounted in the arsenals. On his return he held up before the Senate a bundle of fresh figs that he had plucked in Carthage three days before, as an ominous symbol of her prosperity and her nearness to Rome" (p. 105). In 150 BC the Roman fleet arrived offshore at Carthage. The Senate promised that if Carthage turned over 300 children of the noblest families as hostages, Carthage would remain free. Once this was done, the Romans demanded "the surrender of remaining ships, a great quantity of grain and all her engines and weapons of war." When these conditions too were met, the Romans insisted that the Carthaginians should leave their city and stay ten miles away while it was burned to the ground. The city revolted against its own leaders. Public buildings were demolished to provide metal and timber, statues were melted to make swords and even the hair of women was collected to make ropes: "In two months the beleaguered city produced 8,000 shields, 18,000 swords, 30,000 spears, 60,000 catapult missiles, and built in its inner harbor a fleet of 120 ships" (pp. 106–107). They withstood siege by land and sea for three years, but in 147 BC the walls were breached. The Carthaginians fought on street by street. When the survivors were reduced from 500,000 to only 55,000, they surrendered; they were sold as slaves and the city was turned over to the legions for pillage. The Roman Senate ordered that all of her allies were to be destroyed and, finally, that the soil should be ploughed with salt. The latter decision may be regarded as an early example of chemical warfare, since in a semi-arid climate it would be many years before more crops would grow.

In contrast to the Carthaginians, the Mayans may have been architects of their own demise for failing to conserve their soil resources. Binford et al. (1987) describe how, over a period of some 1,000 years, the Mayan population of the Americas grew rapidly. The vegetation was gradually shifted from forest to grassland, and phosphorus was leached from the soils into water bodies. Binford et al. (1987) attribute the sudden decline in the Mayan population in part to over-exploitation of the land and the exhaustion of soils.

Over human history, at least three phases of soil erosion are recognized (McNeill and Winiwarter 2004). First, around 2000 BC, soil erosion accelerated as early river-basin civilizations expanded and ascended nearby forest slopes. Second, in the sixteenth to nineteenth centuries, stronger and sharper plowshares allowed agriculture to expand into the Eurasian steppe, the North American prairies and the South American pampas. Third, in the mid twentieth century, the combined effects of agricultural machinery, industrial nitrogen fixation (Box 3.2) and exponential human population growth began driving soil erosion.

3.5.5 Two Historical Digressions: Jan Baptiste von Helmont and Titus Smith

von Helmont's Classic Experiment

Only a few hundred years ago it was still assumed that plants grow by feeding upon the soil, with roots acting in a manner similar to the stomachs of animals (Lieth 1975). This view, advanced by Aristotle (384–322 BC), persisted for more than a millenium until Jan Baptiste von Helmont (1579–1644) demolished it with an elegant experiment in the early 1600s. In Lieth's words (p. 8):

> He grew a willow twig weighing 5 lb. in a large clay pot containing 300 lb. of soil, and irrigated it with rain water. After 5 years, he harvested a willow tree of 16 lb. with a loss of only 2 oz of soil. Von Helmont concluded from this that water was condensed to form plants.

In other work, von Helmont showed the indestructibility of matter in chemical changes, invented the word "gas," and was the first to take the melting point of ice and the boiling point of water as standards for temperature (Magnusson 1990). His critical work in plant nutrition, which laid the foundation for modern studies of biogeochemical cycles, is underappreciated – a statement that can easily be confirmed by seeking his name in the indices of ecology texts. Von Helmont's willow experiment also laid the foundation for later research in agriculture, particularly Justus von Liebig's research on minerals in plant nutrition. According to Lieth, Liebig too had to fight intensely against the still widely accepted view in the mid 1800s that plants lived by extracting humus from the soil.

Titus Smith Writes About Soil Fertility in North America

Now to the New World. Titus Smith, Jr. (1768–1850) was a native of New England who travelled with his father to Nova Scotia in 1783 (Gorham 1953). Smith was commissioned by the governor of the colony to survey "the unfrequented areas of Nova Scotia, for the purpose of describing their natural resources, agricultural potentials, and suitability for settlement." The unpublished journals of these tours are deposited in the public Archives of Nova Scotia, but Smith's ecological ideas were also published in 1835 in the *Magazine of Natural History*, a report that Gorham suggests may well be the first major contribution to plant ecology in North America. This paper strongly emphasizes the contrast between vegetation of fertile and infertile habitats (an emphasis that would not surprise a visitor today interested in farming or plant ecology):

> *Upon the fertile soils the vegetation is composed of hardwood, and succulent plants with annual leaves. Their growth is rapid, and the outerbark is extremely thin... Upon a barren soil the trees are evergreen, except the hacmetac: the greater parts of the shrubs and plants are evergreens. Their leaves contain more resinous and more woody matter, than the plants of fertile soils: they also have a strong thick epidermis. The trees on this soil grow slowly, the unusual quantity of epidermis increases in an inverse proportion to the growth of the tree.*
>
> *(p. 120)*

3.6   Resources Vary in Space and Time

3.6.1 There is Small-scale Heterogeneity

We have already seen that water is far from homogeneously distributed in soils (Figure 3.6), and the same is true of other below-ground resources. Figure 3.15 depicts a small tract of prairie; even at this very restricted scale, nutrients such as nitrogen occur in patches. Further, these nutrient patches can be expected to move in time and space as a consequence of processes such as fire, and with animals adding patchiness through grazing, defecation, burrowing and death.

Even if the latter four processes were eliminated, say by removing all animals from experimental plots of prairie, other processes such as growth, death, decay, and changing rainfall and temperature would continue to alter the size, shape and duration of nutrient patches. At this scale, the essential questions may involve how organisms can best acquire and sequester nutrients when they are distributed as short-lived patches.

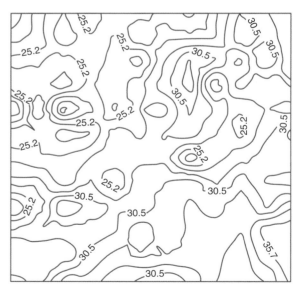

FIGURE 3.15 Spatial heterogeneity of soil nitrogen in a 12 × 12 m plot of sandy soil in a Minnesota prairie. (After Tilman 1982)

3.6.2 Resources Often Change Along Gradients

Erosion by running water is one of the dominant features shaping the surface of the Earth. As water moves from highlands to lowlands, it dissolves minerals from rocks and erodes soil. In some world drainage basins, the mass of eroded sediments exceeds 1 tonne per km^2 per year (Milliman and Meade 1983). The yield of sediment varies with annual precipitation and land use (Judson 1968). As a consequence of steady erosion and deposition, depressions in the landscape are repositories of accumulated organic matter and mineral nutrients. Deposition rates in the order of 20 to 3,000 cm per 1,000 years are suggested by palynological studies of English landscapes, with some of this being peat produced in situ (Walker 1970). Rozan et al. (1994) suggest that rates of deposition in a floodplain in eastern North America were below 10 cm per 1,000 years prior to this century but increased to nearly 1 m per 1,000 years as humans modified the landscape. This is close to the range of deposition rates reported for coastal deltas (Boesch et al. 1994).

Erosion and deposition produce gradients at a wide range of scales. In the Great Smoky Mountains, for example, rich deciduous forests develop in fertile valleys called coves, while open oak and pine forests establish on dry ridges (Whittaker 1956). Sediments from uplands are deposited in valleys to form the rich soils associated with alluvial wet prairies, marshland and swamps (e.g. Sioli 1964; Davies and Walker 1986). In peatlands, strong fertility gradients are associated with the degree to which running water, as opposed to rain water, maintains the water table, and this gradient produces striking changes in the species found. The peatlands fed by groundwater (fens) have higher pH and nutrient concentrations than those fed by rain water (raised bogs) (Gorham 1953; Glaser et al. 1990; Glaser 1992). At even more local scales, depressions in old fields contain soil with more water and nutrients than the ridges between them (Reader and Best 1989). Gradients of soil fertility typify a wide range of habitats from montane forests to wetlands to

old fields. To focus upon small resource patches, or to assume environmental homogeneity, denies one of the most natural aspects of landscapes. Therefore more attention needs to be placed upon gradients and the way in which resources and organisms become distributed along them.

The following illustrates the wisdom on resource gradients available in a traditional forestry source such as Wilde's (1958) treatise on forest soils:

> The modifying influence of the local physiographic conditions deserves particular attention in classification of forested soils in rolling or hilly topography. Such soils usually provide striking illustrations of the effects of the three major edaphic factors: water, aeration, and nutrients [Figure 3.16].
>
> The top and upper slopes of a ridge or a mound represent elements of so-called positive topography. Soils overlying this portion of the relief receive the least rain water and are subject to denudation, which decreases the supply of available nutrients. Consequently, such soils often support forest stands of low rate of growth. On the other hand, the negative topography of lower slopes receives run-off water and fertilizing products of erosion – mineral colloids, humus, and soluble salts. As the result of this enrichment of the soil, forest stands increase their height growth, or, as foresters say, "Trees level the relief with their crowns." With the descent to a depression the water accumulates in excess, and the growth of trees drops abruptly in accordance with the rapidly decreasing soil aeration.
>
> (p. 248)

In heavily developed landscapes, where pastures have replaced forests, similar resource gradients can be found. Again, erosion from upper slopes produces a soil resource gradient (Figure 3.17). Superimposed upon this, however, are grazing pressures that are most intense in the valleys. At a finer scale, local sites of erosion (eroded pasture) produce adjacent sites with deposition (very oligotrophic pasture). These local

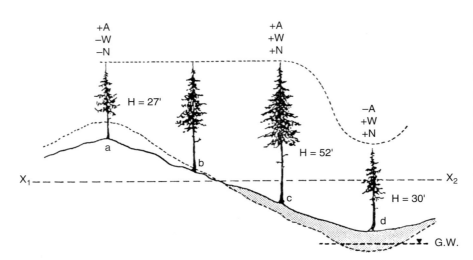

FIGURE 3.16 Effects of topography on moisture (W), aeration (A) and nutrient content (N) of soil and the resulting growth of trees. Line X_1 to X_2 delineates positive topography, subject to denudation, and negative topography, subject to deposition of eroded materials. (From Wilde 1958)

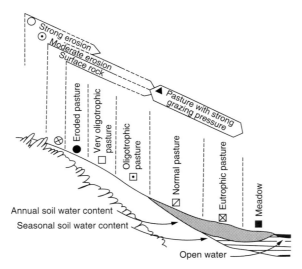

FIGURE 3.17 Typical sequence of grassland communities on a slope. The main sequence runs from "meadow" in the bottom of valleys to "eroded pasture" higher up the slope. The distribution of "strong grazing," "moderate erosion" and "strong erosion" are superimposed on the main sequence. (From Puerto et al. 1990)

gradients are also superimposed upon the longer gradient running down the elevation gradient.

Even in wetlands, which may seem to have relatively uniform topography, one can find similar

gradients. They stretch from infertile sandy shores, from which waves are eroding nutrients, to sheltered fertile bays, where silt, organic matter and nutrients accumulate. At this large scale (illustrated by, but not restricted to, wetlands) one can reasonably discuss resource gradients, because all major resources (N, P, K, Mg) are distributed along a common gradient. They are in low availability on sandy or eroding shorelines (Figure 3.18a); they are in high availability wherever sediments accumulate. In other cases, coastal plains that have been leached by rain for millenia have extensive areas of wet but infertile conditions (Figure 3.18b).

What happens if one examines resource patterns progressing from a large (landscape-sized) scale to a small (quadrat-sized) scale? Table 3.6 provides the opportunity to move down this scale, from the top matrix (marshes in eastern North America) to the bottom matrix (a single sedge meadow). In the top matrix, the soil samples are from wetlands ranging from the highly fertile (e.g. *Typha* marshes and floodplains) to highly infertile sand or gravel shorelines where insectivorous genera such as *Drosera* and *Utricularia* are common. As well, both organic matter and silt and clay content of the soil were positively correlated with nitrogen and phosphorus levels. Similar patterns occur within a single wetland

(a)

(b)

FIGURE 3.18 Infertile habitats. (a) Wave-washed shorelines are similar to hilltops, in that both are habitats from which nutrients are eroded. Note the carnivorous *Utricularia cornuta* flowering in the foreground. (Photograph by Paul Keddy, 1980, Axe Lake, Ontario) (b) Coastal plains in the southeastern United States have been leached by rainfall for millennia, and now support pine savannas. Note the carnivorous *Sarracenia alata* flowering in the foreground, along with *Drosera tracyi*. (Photograph by Robert Peet, ca. 1985, Sandhill Crane National Wildlife Refuge, Mississippi)

(Table 3.6b). At the lowermost scale (Table 3.6d), many of the correlations among resources have become negligible.

As shall become evident, assumptions about the appropriate scale of enquiry have a big impact upon one's view of competition. The assumption of nearly homogeneous environments and trade-offs among resources (e.g. Tilman 1982) may be quite appropriate for a patch such as Figure 3.15 or perhaps a single sedge meadow, whereas the assumption of resource

Table 3.6 **Resource gradients in wetlands from the large scale (top) to the small scale (bottom). Note that patterns (as indicated by the size of correlation coefficients) fade as the scale becomes smaller. (After Keddy 2001)**

(a) Marshes in northeastern North America (Gaudet 1993, her Table 1.2)

	% Organic	P	N	K	Mg	pH
Standing crop	0.77	0.76	0.66	0.58	0.67	−0.28
% Organic	–	0.77	0.57	0.50	0.51	−0.47
P	–	–	0.72	0.56	0.66	−0.13
N	–	–	–	0.53	0.63	−0.02
K	–	–	–	–	0.70	−0.28
Mg	–	–	–	–	–	−0.14

(b) One wetland complex in the Ottawa River watershed (Gaudet 1993, her Table 1.4)

	% Organic	P	N	K	Mg	pH
Standing crop	0.74	0.80	0.69	0.76	0.69	−0.45
% Organic	–	0.80	0.61	0.66	0.62	−0.61
P	–	–	0.62	0.82	0.59	−0.46
N	–	–	–	0.68	0.53	−0.18
K	–	–	–	–	0.64	−0.35
Mg	–	–	–	–	–	−0.72

(c) One vegetation zone of the St. Lawrence River (Auclair et al. 1976a, their Table 1)

	% Organic	P	N	K	Mg	pH
Standing crop	0.34	−0.29	0.38	0.49	0.17	0.21
% Organic	–	−0.27	0.37	0.75	0.59	0.18
P	–	–	−0.01	−0.48	0.33	−0.55
N	–	–	–	0.39	0.32	0.14
K	–	–	–	–	0.43	0.38
Mg	–	–	–	–	–	0.12

(d) *Carex* meadow, St. Lawrence River (Auclair et al. 1976b, their Table 1)

	% Organic	P	N	K	Mg	pH
Standing crop	0.13	−0.02	−0.02	−0.22	−0.23	−0.11
% Organic	–	−0.39	0.30	0.52	0.17	−0.14
P	–	–	−0.26	0.18	−0.21	0.03
N	–	–	–	0.24	0.26	0.04
K	–	–	–	–	0.16	−0.01
Mg	–	–	–	–	–	0.52

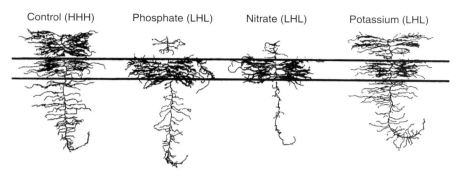

FIGURE 3.19   Responses of barley root systems to different nutrient supplies. The plants were grown in a three-chambered flowing nutrient system with the central chamber having the relevant ion at a concentration 100-fold that in the other two chambers. H, high; L, low. (From Leyser and Fitter 1998 after Drew 1975)

gradients and trade-offs between stress tolerance and resource competition (e.g. Keddy 1989) seem more appropriate to larger scales of enquiry. Further, these different scales treat resources in entirely different ways. At the small scale, resource availability is seen to be under biological control, with plants creating depletion zones around their roots. At the large scale, resource availability is viewed as a consequence of erosion and topography, with plants interacting with one another for access to different sections of this resource gradient. Perhaps greater attention to the distribution of resources in nature would allow one to better decide which scale of investigation is appropriate to a particular set of circumstances.

3.6.3 Resources Often Occur in Transitory Patches

Soil resources are not only distributed unequally in space, but patches themselves may be transitory in time. Nutrients may be available only after a transitory event such as a pulse of rainfall. Alternatively, the roots of neighbours may intercept and deplete pockets of resources. When a neighbour dies, or when surrounding biomass is damaged by an event such as freezing, the nutrients from those neighbours may also be re-released in a pulse creating a transitory patch of nutrients. Plant roots must locate such patches. While the primary root is

unaffected by nitrate, the activity of lateral root primordia is strongly influenced by the nitrate (and phosphate) concentration in the substrate (Figure 3.19). At low nutrient levels, a localized region of high nitrate will stimulate lateral root growth (Drew 1975). This root proliferation should help capture ions that have very limited mobility in the soil, such as NH_4^+ and inorganic phosphate. In contrast, there might be rather little advantage to root proliferation in response to highly mobile ions such as NO_3^- (Robinson 1996).

However reasonable such logic may seem, apparently some species such as *Arabidopsis* have a mechanism for detecting NO_3^- rather than NH_4^+ (Forde and Zhang 1998). The possible explanation lies in the relative rates of diffusion of these two ions in the soil. Under aerobic conditions, decaying organic matter will release both ions, but the relative immobility of NH_4^+ means that it is likely to be the NO_3^- ion that is the first to reach nearby roots. NO_3^- may therefore signal the presence of a nearby nitrogen source, allowing a plant detecting NO_3^- to respond before one detecting NH_4^+.

In another series of experiments (Campbell et al. 1991), two plants contrasting ecologically were exposed to resource pulses of widely different durations (from 1 minute to 100 hours out of every six days). The relatively slow-growing *Festuca ovina* showed increased relative growth rates as the length

(a)

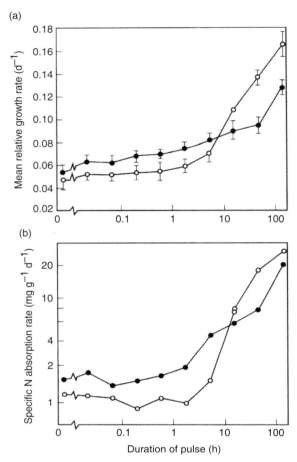

(b)

FIGURE 3.20 (a) Mean relative growth rate (+95% CI) and (b) mean specific nitrogen absorption rate of potentially rapid-growing *Arrhenatherum elatius* (○) and slow-growing *Festuca ovina* (●) plants plotted against the duration of pulses of nutrient enrichment. Pulses occurred once every six days. (After Grime 1994)

of pulse increased (Figure 3.20, solid dots). Now consider a second species, *Arrhenatherum elatius*, which has the potential to be a fast-growing species in fertile habitats. Note the top of Figure 3.20; when pulses were of short duration this species actually grew more slowly than *F. ovina*. It was apparently inferior in ability to exploit temporary pulses of nutrients, an

interpretation strengthened by measurements of nitrogen absorption rates (Figure 3.20b). Once pulses lasted longer than 10 h, the growth rate of *A. elatius* nearly doubled and exceeded that of the other species.

It is clear that the ability of plants to capture soil resources can be dramatically affected by the duration of resource pulses. Figure 3.20 does not show the consequences for competition, but it does suggest that infertile habitats with short pulses of resource might provide a refuge for the slower growing species. One can imagine that similar patterns might be found in other circumstances; say, birds exploiting nectar with different pulses of availability, or fungi exploiting detritus with different decay rates.

3.6.4 Resource Fluctuations Complicate Short-Term Ecological Studies

Resource levels (and all other environmental constraints) present two further perplexing problems. First, they fluctuate and, frustrating as it may be, may change over a wide array of time spans, from the hourly changes in soil moisture after a heavy rainfall in a forest, to the changes in rainfall patterns that occur over millennia. Such constant change is difficult enough to accommodate in ecological studies, but it is accentuated by a second problem; it may be the extreme events rather than the average conditions that control the properties observed. For example, it may not be the mean annual rainfall over the last century, but one period of drought 50 or 100 years ago that produced the mixture of species observed today (Figure 3.21). It is readily apparent that a short-term field experiment is not well suited to address causal processes in ecological systems that are largely the result of long-term fluctuations and periodic extreme events. Moreover, even if one happens to encounter an apparent extreme event during an experiment, it is unlikely that one will then simultaneously obtain the necessary comparative results from the long-term average to put the apparent extreme in an ecological context.

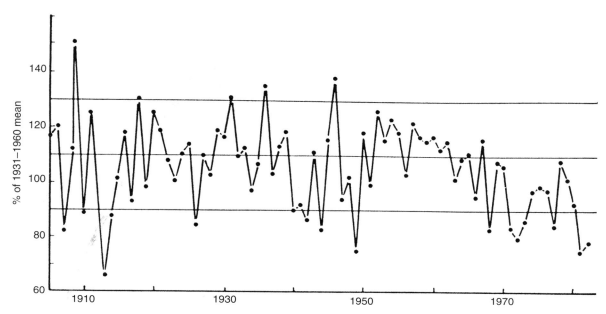

FIGURE 3.21   Annual rainfall from five stations in the Sahel expressed as a percentage of the 1931 to 1960 mean. Years that fall 20 percent above or below the mean indicate extreme events. The years of 1912 and 1949 were years of drought similar to those of 1981. These periodic droughts may have far more effect upon ecological communities than the mean value, particularly if overgrazing is an ongoing pressure. (From Sinclair and Fryxell 1985)

This conundrum apparently led MacArthur (1972) to propose that the best evidence for the existence of competition has to come from biogeographers who are accustomed to dealing with long time scales. Further, he suggested that the evolutionary responses of species, that is their ecomorphological properties (e.g. bill size and diet in birds) and geographical distributions, are the most appropriate kind of data for the study of competition. This opinion was echoed by Wiens (1977), who suggested that traits of organisms may reflect short periods of intense selection (bottlenecks) interspersed with much longer periods of weaker selection. Expanding upon this, one can imagine that any property of a species or community may be the product of short periods when resources crash below a critical level. Of course, convincing theories built upon extreme events are much more difficult to derive and to test than those built upon means.

Extreme events are, by definition, relatively rare. Further, one may be accustomed to looking for relationships among means rather than among extremes. A compensating factor may be that at least the effect of extreme events such as fire, flooding or drought may be so clear that the link between the environment and its effects is not subtle but self-evident.

The effects of fluctuating resource levels are well illustrated by the effects of drought upon two species of desert shrubs. The water potential (xylem pressure potential) of both species was measured immediately after a rainfall, and then re-measured approximately every two weeks as the soil dried (Figure 3.22). Further, some of the shrubs had neighbours whereas other shrubs did not, the neighbouring plants having been removed as a part of the experiment. Shortly after a 60-mm rainfall in mid August (the far left side of Figure 3.22), the

(a)

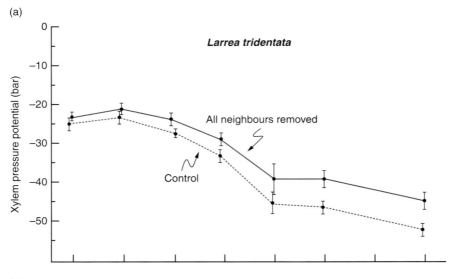

(b)

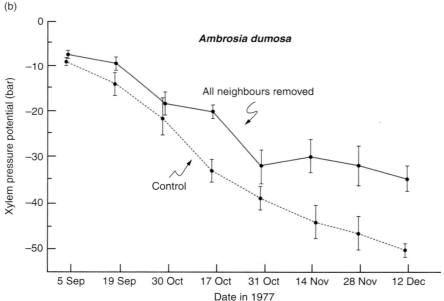

FIGURE 3.22 Declines in water potential of two desert shrubs after an August rainfall. The performance of each species is shown for control plots and for plots with neighbouring plants removed. (After Fonteyn and Mahall 1981)

water potential of each species was relatively high, and little affected by neighbours. Over the next three months, as the soil dried, water potential fell steadily. Moreover, water potential fell more rapidly in those shrubs with neighbours (dashed lines), showing that the consumption of water by neighbouring plants increased the physiological effects of the drought. This experiment has several important lessons. It shows from direct physiological measurements how plants are affected by a fluctuating resource. It also shows how neighbours can amplify the impacts of scarce resources, illustrating the effects of competition, which is the theme of Chapter 4. The experiment also shows how the results of field experiments can change with time – this case, the impacts of drought, and the effects of neighbours, only became apparent after months of measurement. Had the experiment stopped earlier, the effects of neighbours might not have been detected.

Water is an obvious resource that fluctuates over time, and rainfall data such as those in Figure 3.21 are useful because they have been routinely collected for many years. Other examples of fluctuating resources or resource patches include nectar (Feinsinger 1976), seeds in desert soils (Price and Reichman 1987), carcasses (Corfield 1973), grazing lands (Sinclair and Fryxell 1985), marshes (Keddy and Reznicek 1986), fresh-water periods in saline marshes (Zedler and Beare 1986), light gaps in vegetation (Grubb 1977) and bare areas on submerged stones (Hemphill and Cooper 1983). It would appear that relatively constant environments, or continually available resources, are exceptions and that one of the major problems faced by all organisms is coping with constantly changing environmental conditions (Levins 1968).

3.6.5 Resources Provide a Habitat Template

In order to relate the resources in a habitat to the kinds of plants found there, we might try to recognize a few fundamental types of patterns in resources (Grime 1977; Southwood 1977). Price (1984) recognized five kinds. Four of these are shown in Figure 3.23. Increasing resources are those that increase and then suddenly decline. Decreasing resources are produced suddenly at the beginning of a season and they gradually decline. Pulsing or ephemeral resources increase rapidly and then decline rapidly. Steadily renewed resources are continuously renewed over long periods. Continuous resources, not shown in Figure 3.23, are physical in nature and are largely unaffected by seasonal change. Price uses this last category to deal with space as a resource for sessile

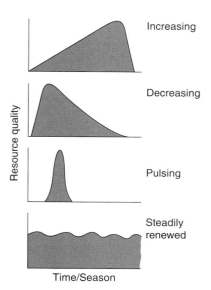

FIGURE 3.23   Four kinds of resources classified by pattern of temporal variation. (After Price 1984) This scheme does not consider spatial variation or chronically low resource availability.

organisms. Since disturbance and death continually create new spaces for colonists, space can be treated as another example of a continually renewed resource. Price's five resource types can therefore be reduced to four.

If uniform and patchy distributions of each kind of resource are recognized, then the number of kinds of resources based on Price's scheme can be doubled. Begon and Mortimer (1981) produced a more elaborate classification, which included both spatial and temporal variation. They recognized four temporal categories and three spatial categories (Figure 3.24). Long-lived plants occupy habitats with continuous or patchy resources where adult mortality is low enough that reproduction can be safely postponed. Resources that are not allocated to reproduction can then be allocated to stems, leaves and shoots. This may allow these plants to compete with neighbours for access to light and nutrients (Harper 1977; Keddy 2001).

Time

Continuous, high

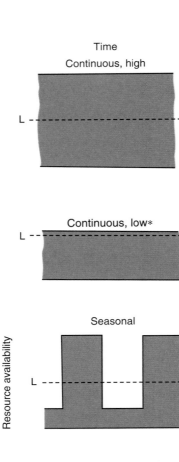

Continuous, low*

Seasonal

Resource availability

Unpredictable

Ephemeral

Space

Continuous

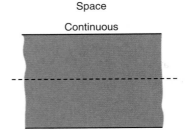

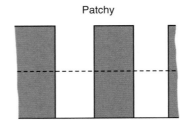

Patchy

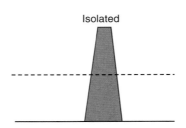

Isolated

FIGURE 3.24 A more elaborate classification of resource types, based upon variation in resource availability in time and space. There are four temporal distributions (left side) and three spatial distributions (right side) The continuous in time category (upper left) is further subdivided into separate sites with high as opposed to low resources, for a total of eight possible resource types. L indicates the level of resource necessary for reproduction. This is after Begon and Mortimer (1981) by the addition of the "continuous, low" category marked with an asterisk. For simplicity, scaling terms such as generation time and foraging range have been left off the figure; more complete accounts of scaling are given in Southwood (1977, 1988) and Begon and Mortimer (1981).

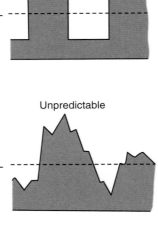

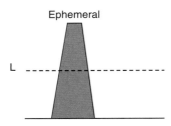

3.7   Scarce Resources Have Many Consequences

Resource acquisition is a problem that faces all plants. We have even borrowed a term from zoology, and considered it as a problem of foraging – that is, of finding and exploiting patches where resources are locally abundant. There may, however, be situations where resource levels are so low that the investment in foraging does not yield a positive return. What if there are no patches to be found or if their quality is so low that the cost of finding them exceeds the benefits obtained? Habitats with chronically low supply rates of resources may well be a distinctive category (for example Grime 1977, 1979; Greenslade 1983; Southwood 1988). The classification of resources in Figure 3.24 addresses this problem by including a category for resources that have a constant but chronically low supply rate ("continuous, low"). Grime (1977, 1979) has argued that such circumstances lead to an evolutionary strategy he terms "stress tolerance" (see Section 10.9). Under such circumstances, he argues, there is strong selection for the conservation of resources. Conservation is accomplished both by storing excess resources within the tissues of the organism and by minimizing losses (Table 3.7). This is often associated with reduction in the size and growth rate of the organism concerned. These kinds of adaptations are well known in plants from arctic and alpine habitats, arid habitats and an array of nutrient-deficient habitats such as sandy soils, serpentine soils, tree canopies and peatlands.

We will devote all of Chapter 10 to these types of extreme habitats, but to make this chapter on resources complete, we will explore evergreen foliage in more depth. We will look at global patterns in leaf architecture. And we will have an enjoyable look at five novel kinds of plants adapted to chronically low supplies of one or more resources: epiphytes, succulents, carnivorous plants, mycoheterotrophs and parasites. You can consider this last section of the chapter optional. The good news is that if you take the time, you will encounter some remarkable plants that

Table 3.7 Some attributes of stress-tolerant plants from chronically resource-limited habitats. (In part from Grime 1977; Givnish 1988)

Attribute	Presumed selective advantage
Inherently slow growth rates	Reduced demands upon resources
Evergreen foliage, long-lived organs	Reduced loss of resources
Storage of water and nutrients	Uncoupling of resource supplies from plant growth
Investment in anti-herbivore defences	Costly replacement minimized
Symbioses	Alternative paths for uptake or storage of resources
Carnivory	Alternative source of nutrients
Reduction in size	Reduced demands upon resources, low above-ground competition
Low allocation to seeds	Limited expenditure of resources, resources available for asexual reproduction

amplify and reinforce the general principles we have discussed above. All the same, if time is short, you may set this section aside and save it for future enjoyment.

3.7.1 Evergreen Plants Conserve Scarce Resources

Evergreen plants tend to have much lower tissue nutrient concentrations than deciduous plants (Aerts 1996). Evergreen leaves contain 1.4 percent nitrogen, compared to 2.2 percent in deciduous leaves. Evergreen leaves contain 1 percent phosphorus, compared to 1.6 percent in deciduous leaves. Thus from the perspective of the plant the construction costs of leaves (measured in terms of nutrients invested per unit weight of leaf tissue) is lower for evergreen than for deciduous leaves. Evergreen plants may therefore have an advantage in infertile habitats (Grime 1977; Wardle 2002). Further, evergreen plants will tend, in the longer run, to produce impoverished soils (Wardle et al. 2004).

But costs of construction are less than half the story. Ecologists, like economists (and like mining companies), often tend to emphasize extraction costs – the foraging and harvesting of resources needed for building new structures. This perspective omits the issues of conservation and recycling. Plants can recycle nutrients by removing elements from leaves before the leaves senesce and fall. This process of recycling is known as **resorption**. Completely wasteful plants would discard tissues without removing any nitrogen or phosphorus; completely conserving plants would remove all the nitrogen and phosphorus before senescence. What is the case with real plants? The proportion of nitrogen and phosphorus recovered from discarded leaves, termed **resorption efficiency**, is a measure of nutrient conservation (Killingbeck 1996). A survey of published studies on nutrient resorption in more than 200 species of plants (Aerts 1996) found that resorption efficiency differed little between nitrogen and phosphorus, or between deciduous and evergreen leaves – in all cases 50 percent resorption was a good first approximation. More precisely, nitrogen resorption efficiency was slightly lower in evergreen plants (47 percent) than in deciduous plants (54 percent), and phosphorus resorption efficiency was not significantly different between the two. The benefits of evergreen leaves would thus seem to arise from the low cost of construction rather than efficiencies of resorption.

Instead of measuring resorption efficiency (percent removal of nutrients) it is also possible to measure **resorption proficiency** (minimum concentration of nutrients in discarded leaves). In the case of proficiency, there are chemical limits on the amounts of nitrogen and phosphorus that can be extracted from discarded leaves, the lower limit being about 0.3 percent for nitrogen and 0.01 for phosphorus (Killingbeck 1996). In a study of resorption proficiency in 77 species of woody plants, Killingbeck found that evergreen and deciduous plants did not differ with respect to nitrogen whereas evergreen plants were significantly more proficient in resorbing phosphorus.

Five species of evergreen plants were able to extract nitrogen and phosphorus to the lowest recorded levels of less than 0.3 percent for nitrogen and 0.01 percent for phosphorus: *Banksia grandis* (Australia), two species of *Eucalyptus* (Australia), *Lyonia lucida* (North America) and *Pinus rigida* (North America). Plant species that have leaves with leaf nutrient levels higher than these values, then, are not fully conserving foliar nutrients.

There are some discrepancies between the two foregoing studies, probably the result of measuring resorption efficiency (percent removal) as opposed to resorption proficiency (terminal concentration). Aerts's data showed that evergreen plants had greater efficiency of nitrogen removal, whereas Killingbeck's data showed that evergreen plants had greater proficiency of phosphorus removal. At the present, it is probably best to accept that both nutrients are essential to plants, and that efficiency and proficiency are important but different measures of conservation ability in plants.

One way to measure the relationships between soil nutrients and plant responses exploits natural nutrient gradients. Relatively large-scale gradients are provided by retreating glaciers. The youngest soils occur where the ice has most recently melted. Some classic studies of ecological succession have used such circumstances (Section 8.4). A series of differently aged soils is sometimes called a chronosequence (from the ancient Greek word *chronos*, time). One such sequence in New Zealand provides a gradient of soils ranging from 60,000 to ca. 120,000 years old. The older soils are more infertile – available soil phosphorus declines from 5 mg kg^{-1} to 1 mg kg^{-1}, while available nitrogen falls from 110 mg kg^{-1} to 15 mg kg^{-1}. Richardson et al. (2005) collected leaves and recently fallen litter from eight communities arranged along this gradient.

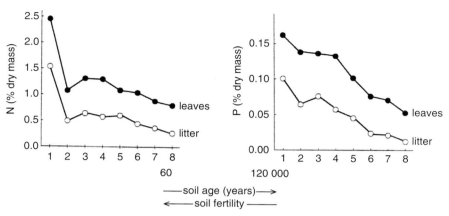

FIGURE 3.25   Concentrations of nitrogen and phosphorus in plant leaves decline along a gradient from young fertile soils (left) to old infertile soils (right). The concentrations in leaf litter (resorption proficiency) also decline. The numbers 1 to 8 indicate the rank order of site soil age from 60,000 to 120,000 years. (After Richardson et al. 2005)

Nutrient concentrations in leaves declined along this gradient (Figure 3.25). Resorption proficiency increased along the same gradient, with the plants on the oldest soils having litter that approaches the minimum values previously described, 0.3 percent for nitrogen and 0.01 percent for phosphorus.

3.7.2 Global Patterns in Leaf Architecture: The Leaf Economic Spectrum

It is hard to think about plants without thinking about leaves. As you have seen, they come in all shapes and sizes, from the enormous round and flattened leaves of water lilies (Figure 3.1), to the narrow needle-shaped leaves of the pine family (Figure 1.3) and just about everything in between. Presumably these differences in size and shape reflect different trade-offs among some of the principles described so far in this chapter. Plants need leaves mostly to provide carbon dioxide and water, but in order to construct those leaves they must also invest nitrogen and phosphorus. What type of leaf is best for particular conditions? Larger leaves may capture more sunlight but are also easily damaged. More stomata may allow more carbon uptake but also may lead to more loss of water. New leaves are an investment in an uncertain future and this investment

is more costly in infertile soils. So how do these and similar trade-offs determine the kind of leaves one sees on wild plants? The largest study of leaf architecture to date used data from 2,548 species from 219 families at 175 sites (Wright et al. 2004). Six traits were recorded overall. We will consider only three here. The most basic property was leaf mass per unit area (LMA). This is the dry-mass investment per unit of light-intercepting leaf area. Species with high LMA have a thicker leaf blade or denser tissue, or both. You can think of this as the investment in construction of the light gathering apparatus. The second trait was photosynthetic assimilation rate. This was measured under high light, ample soil moisture and ambient CO_2 levels. Higher values of this trait correspond to higher stomatal conductance and/or higher drawdown of the CO_2 levels within the leaves. The third trait was leaf nitrogen levels, nitrogen being essential for constructing proteins for the photosynthetic machinery, such as RuBisCO, and a resource often scarce in plant communities. Figure 3.26 shows the relationships among these three traits. The assimilation rate is highest in plants with low LMA and high N content. The central axis can be thought of as a leaf economic spectrum. "This spectrum runs from species with potential for quick returns on investments of nutrients

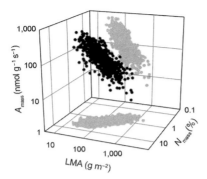

FIGURE 3.26 The relationships among LMA (leaf mass area), A_{mass} (photosynthetic assimilation rate) and N_{mass} (leaf nitrogen levels) in a set of 2,548 species of plants. (From Wright et al. 2004)

and dry-mass in leaves to species with slower potential rates of return. At the quick-return end are species with high leaf nutrient concentrations, high rates of photosynthesis and respiration, short leaf lifetimes and low dry-mass investment per leaf area. At the slow-return end are species with long leaf lifetimes, expensive high-LMA leaf construction, low nutrient concentrations, and low rates of photosynthesis and respiration" (Wright et al. 2004: p. 823).

Overall, the leaf economic spectrum shows how the availability of resources controls the size and shape of leaves. Recall that the earliest land plants, such as *Rhynia* (recall Figure 1.5), did not even have leaves. The origin and diversification of leaves is likely best understood as the result of natural selection for increased ability to gather photons and carbon dioxide, but the costs of construction mean that below-ground resources have a considerable control on leaf characteristics.

3.7.3 Bizarre Botany: Some Strange Evolution for Resource Acquisition

Epiphytes
Let us begin with an unusual example from canopy ecology (Moffett 1994; Lowman and Rinker 2004). The canopy of tropical forests supports more than 2,500 species of pteridophytes and 20,000 species of flowering plants (Benzing 1990). Such epiphytic

FIGURE 3.27 The ant house epiphyte *Myrmecodia beccarii* from Australia. Note the stem galleries inhabited by ants and the single white flowers on the shoot. This image is based on sketches by Sydney Parkinson, who was the only botanical artist on Captain James Cook's first Pacific voyage (recall Figure 2.1). This watercolour was completed in 1773, after the voyage by John Frederic Miller. (The Endeavour Botanical Illustrations, © Natural History Museum, London)

plants lack any access to mineral soil, yet must still acquire resources such as nitrogen and phosphorus.

Some epiphytic bromeliads have a bowl-like shape that traps both water and detritus, including animal excrement and dust. These plants still possess roots that securely grip the substratum, but these roots are essentially non-absorptive (Benzing 1990). The efficiency of this kind of nutrient capture is illustrated by

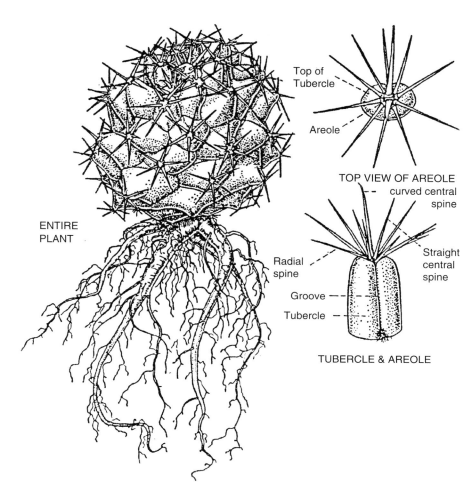

Top of
Tubercle

Areole

TOP VIEW OF AREOLE

curved central
spine

Radial
spine

Straight
central
spine

Groove

Tubercle

TUBERCLE & AREOLE

ENTIRE
PLANT

FIGURE 3.28 A barrel cactus has tissues for storing water and a reduced surface to volume ratio to minimize losses from evapotranspiration. Secondary characteristics include the array of spines to protect the resources that have been acquired, and an inherently slow growth rate. (From Benson 1950)

a study in Venezuela, where the annual input of phosphorus in rainfall was 28.4 kg ha^{-1} year^{-1}. Of this input, 21.7 kg ha^{-1} year^{-1} was absorbed by the forest canopy, leaving only 6.7 kg ha^{-1} year^{-1} to reach the forest floor (Jordan et al. 1980). Africa may actually provide phosphorus for neotropical bromeliads and forests, with dust from the Sahara Desert being transported across the Atlantic Ocean (Jickells et al. 1998). The presence of animal remains has led to claims that certain bromeliads are carnivorous, but Benzing (1990) concludes that these claims are not convincing.

Other epiphytes in tropical canopies gather nutrients by housing ant colonies, and may provide specialized organs called domatia to accommodate them. Detritus carried back to the domatia provides an important source of mineral nutrients. Some such plants will produce roots within their own domatia, in which case they are literally rooting within their own self-contained flowerpots. *Myrmecodia beccarii* (Figure 3.27) has a chambered tuber that houses ants and absorptive papillae for nutrient uptake. Similarly, *Dischidia rafflesiana* has flask-shaped leaves that collect rain water as well as detritus that is transported by the ants that live in the flasks. At least seven families of plants (Asclepiadaceae, Bromeliaceae, Melastomataceae, Nepenthaceae, Orchidaceae, Piperaceae and Polypodiaceae) produce domatia that are

(a)

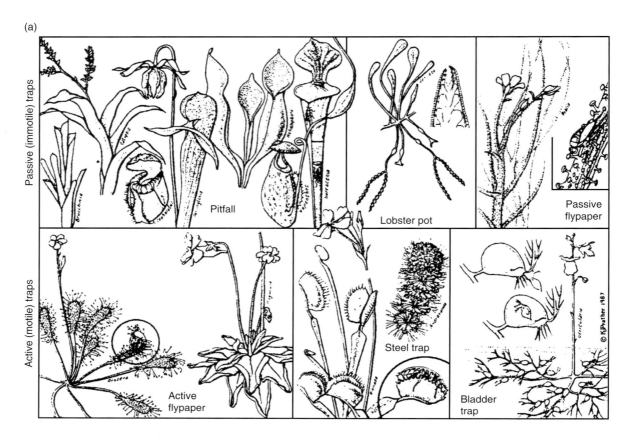

FIGURE 3.29 (a) The importance of nitrogen and phosphorus for plants is illustrated by the many kinds of carnivorous plants in the world. At least six different trapping mechanisms have been described. (After Givnish 1988) (b) The pitfall trap of *Nepenthes truncata*, endemic to the Philippines, reaches 40 cm in height, and can even capture mice. (Photograph by Anne Kitzman, Shutterstock)

occupied by ants (Janzen 1974; Huxley 1980; Benzing 1990). Less specialized cases include members of the Araceae, Bromeliaceae, Gesneriaceae, Orchidaceae and Piperaceae that root in arboreal ant nests.

Succulents

Although shortages of two essential elements, nitrogen and phosphorus, are implicated in many examples of stress tolerance, shortages of water in deserts cause some equally remarkable responses in plants. The typical barrel shape of many cacti not only allows for water storage within their tissues, but reduces surface to volume ratios to minimize water losses from evapotranspiration (Figure 3.28). Other factors may also contribute to this growth form, such as shortages of resources for constructing leaves and the inevitable losses of nitrogen and phosphorus when the leaves are shed. Reduction in leaf area and increase in leaf longevity are well illustrated by succulents, but the argument is not restricted to this growth form. The presence of evergreen leaves in chronically infertile habitats, such as tropical sand plains and temperate peat bogs, suggests that the pressures to conserve nitrogen and phosphorus are sufficient to produce evergreen plants even when water is freely

(b)

FIGURE 3.29   (*cont.*)

available (Wardle et al. 2004). Many succulents have dense arrays of spines to protect the stored water and nutrients from being consumed by herbivores.

Carnivorous Plants

In some habitats with low nitrogen and phosphorus availability, carnivorous plants exploit animal bodies as a supplementary source of nutrients (Givnish 1988). Worldwide, approximately 538 species of carnivorous plants have been described and classified in 18 genera and 8 families. Some of the better known genera are sundews (*Drosera* spp.), pitcher plants (*Sarracenia* spp.) and bladderworts (*Utricularia* spp.). In some infertile sites of South America and tropical Africa,

the genus *Genlisea* (Lentibulariaceae) actually consumes soil protozoa. These plants have no roots, and produce only a small rosette of linear leaves above ground. Below ground are bundles of hollow subterranean leaves that chemically attract, entrap and then dissolve ciliates (Barthlott et al. 1998). The intensity of the pressure to acquire nitrogen and phosphorus is illustrated by evidence that carnivory seems to have arisen independently at least six times (Figure 3.29) (Givnish 1988).

Parasites

In deeply shaded habitats, there may be no source of energy save for that captured by taller neighbours. Parasitism could provide a plant with resources without necessitating light capture. Over 4,500 species of vascular plants in 19 families are parasitic upon other plants (Nickrent 2006) from which they may take water and nutrients, and possibly photosynthates. Many parasitic plants have greatly reduced leaves and root systems since they do not need to gather resources with these organs (Figure 3.30). Among the known parasitic plants, according to Kuijt (1969), there are no seedless plants, one gymnosperm (*Parasitaxus ustus*) in New Caledonia (its host is also a member of the Podocarpaceae, Farjon 1998), and no monocots: "It is a startling and unexplained fact that the known parasitic vascular plants are limited to the dicotyledons" (p. 3). He suggests that parasitism has evolved independently at least eight times, five times in relatively large groups – Santalales; Scrophulariaceae and Orobanchaceae; Rafflesiaceae and Hydnoraceae; Balanophoraceae; and Lennoaceae; and three times in isolated genera – *Cuscuta*, *Cassytha* and *Krameria*. Evolutionary relationships of parasites are further addressed by Press and Graves (1995), Moreno et al. (1996) and Nickrent (2006).

The occurrence of parasitic plants has been documented for centuries. Some examples from the temperate zone were shown in Figure 3.30; the astonishing *Rafflesia arnoldii* from Borneo appears later in Figure 7.12. Some parasitic plants have normal leaves, like the many species of mistletoe, but in many others the leaves are greatly reduced.

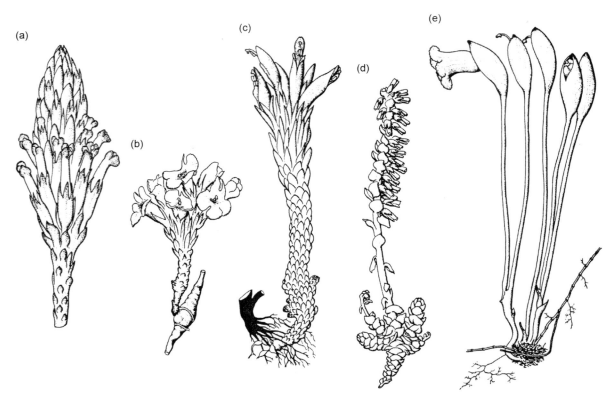

FIGURE 3.30 The broomrapes (Orobanchaceae) are most common in temperate or Mediterranean climates and generally absent from tropical regions. These five examples lack chlorophyll and show the typical parasitic growth form with greatly reduced leaves and root systems. Indeed, such plants could be described as little more than parasitic inflorescences. (a) *Harveya squamosa*; (b) *H. purpurea* on root of *Roella ciliata*; (c) *Hyobanche glabrata*; (d) *Lathraea squamaria*; and (e) *Aeginetia japonica* on *Miscanthus sinensis*. (From Kuijt 1969)

The large fleshy shoots of the Balanophoraceae look superficially like a large fungus, and it was not until 1729 that the Italian botanist Pier Micheli understood, and reported in *Nova Plantarum Genera,* that it was actually a vascular plant parasitic upon other plants. We now have the benefit of Kuijt's monograph, *The Biology of Parasitic Flowering Plants* (1969). While most parasites attach to the roots of host plants, parasites on shoots occur in three families: Cuscutaceae, Lauraceae and Loranthaceae.

One of the most widespread and familiar parasitic plants are the mistletoes that colonize the aerial branches of trees (Figure 3.31a). The evolutionary success of this group is illustrated by the diversity: 73 genera with some 900 species (Vidal-Russell and Nickrent 2008). Australia alone has 91 species of mistletoe (Watson 2011). Mistletoes are considered to be hemiparasitic shrubs. Establishing a seedling on a tree twig is no small matter, and one conspicuous feature of this group is the viscous seeds that adhere to the branches of potential hosts until roots can grow and establish a connection to the host xylem (Figure 3.31b). Most have bird-pollinated flowers and bird-dispersed

(a)   (b)

FIGURE 3.31   The Loranthaceae (mistletoe family) are parasitic shrubs with roots that penetrate the xylem of the host tree. Most are pollinated and dispersed by birds. (a) *Lysiana exocarpi* (harlequin mistletoe) showing tubular flower (ca. 5 cm length). (b) A seed (ca. 5 mm in length) attached to a branch shortly after it has passed through the gut of a bird. (Photographs by Lorraine Phelan, dry shrubland, central Australia)

fruits, and hence illustrate also mutualism between animals and plants, a topic we will explore more in Chapter 7.

In spite of the great variation in parastic plants, however, the mechanism of parasitism is similar across all groups: contact is made with the host through a **haustorium**, a specialized root that forms a connection between the vascular systems of the host and the parasite. Some species, termed hemiparasites, retain most of their photosynthetic ability and apparently take only water and mineral nutrients from hosts (e.g. some members of the Santalaceae, Loranthaceae and Scrophulariaceae). In more extreme cases, termed holoparasites, species may have no chlorophyll and be totally dependent upon their hosts for water, minerals and photosynthetic products. This is the case in some members of the Orobanchaceae (Figure 3.30), Rafflesiaceae (Figure 7.12) and Balanophoraceae.

Mycoheterotrophs: Parasites on Mycorrhizal Fungi

Another entire group of plants, typified by the ghostly white Indian pipe (*Montropa uniflora*), superficially appear to be simple parasites. They have the same reduced leaves and the absence of chlorophyll (Figure 3.32). But these plants instead are parasitic upon fungi that are saprophytic or mycorrhizal, and hence are termed mycoheterotrophs. Examples include *Sarcodes*, *Monotropa* and *Corallorhiza*.

The most extreme example of mycoheterotrophy occurs in the Australian orchid *Rhizanthella gardneri* that completes its entire life cycle underground. Hence its common name, underground orchid. The rhizome is buried from 6 to 12 cm; even the purple flowers, produced in May, remain covered by bark litter (Brown et al. 2003). This orchid is parasitic upon the fungus *Thanatephorus gardneri* that forms ectomycorrhizae on

(a)   **(b)**

FIGURE 3.32   About 400 species of plants are parasitic upon soil fungi that are in turn associated with trees. Such plants are called mycoheterotrophs. Here are two examples. (a) *Monotropa hypopithys* produces short fleshy roots in compact clumps. Basidiomycete fungi surround these roots with a hyphal net and intrude into epidermal cells. The fungi are connected to nearby conifers (Duddridge and Read 1982). DNA sequencing (nrITS) showed that the fungi belong to the genus *Tricholoma* (Bidartondo and Bruns 2002). (Note that some experts put this plant in a separate genus as *Hypopithys monotropa* Crantz.) (b) *Corallorhiza striata* is a species of orchid with no leaves and no chlorophyll. Most young orchids require fungi to germinate and establish, but some 200 species retain this juvenile dependence into adulthood. DNA sequencing (nrITS) showed that the rhizomes were penetrated by the fungus *Tomentella fuscocinerea*, which was associated with nearby trees (Barrett et al. 2010). (Photographs by Paul Keddy, plants ca. 15 cm tall)

the shrub *Melaleuca uncinata*. Both carbon and nitrogen are transferred to the orchid (Bougoure et al. 2010). It is not only leaves and roots that the plant lacks. Since there is no longer a need for genes associated with photosynthesis, more than half of the plastid genes have been lost, leading to the smallest known organelle genome in land plants (37 genes, Delannoy et al. 2011). Only 50 individuals of this orchid species are known in the wild, all in southwestern Australia shrublands.

CONCLUSION

Plants and animals share a need for the same set of resources, CHNOPS. Shortages of water, nitrogen or phosphorus frequently limit plant growth in natural habitats. There are often strong natural gradients in the availability of these resources, and they determine the distribution of plant species and the traits of these species. Figure 3.14 summarized our understanding of differences between plant communities that contrast in resource availability. With this context, you may benefit from re-reading a more thorough treatment of the physiological processes involved in photosynthesis and nutrient uptake offered in introductory biology text books.

The way in which plants forage for resources is not well understood. Part of the problem may be that most of these processes occur underground. Other factors such as competition or herbivory may override natural selection for differing modes of resource acquisition. Theoretical classifications of resource patterns (Figures 3.23 and 3.24) provide a system of organization, but few plant habitats can be accurately assigned to such categories. It may be that past work has overemphasized acquisition, while conservation of resources has been largely overlooked.

Ultimately it seems necessary to measure certain resource properties such as abundance, patchiness or predictability, and then relate these to the ecological properties of plant communities. We may learn from other disciplines. Lavoisier observed in his treatise, *Elements of Chemistry*, in 1789 (Box 3.1) that, "We must trust to nothing but facts: these are presented to us by nature and cannot deceive" (p. 2). In resource acquisition and foraging, the transition from ideas to measurable properties, to quantitative relationships, and to experiments, still largely remains to be made. Perhaps this reflects a number of historical tendencies in ecology. Since many ecologists are initially attracted by organisms themselves, the study of plant ecology naturally begins with the distribution of the organisms rather than the distribution of their resources. Resources are generally more difficult to measure than plant distributions. Much of the scientific study of plants and resources comes from relatively fertile agricultural soils. Finally, ecologists have tended to seek out homogeneous environments where patterns in resource distribution are minimized. Seeking out situations with the least possible variation, and then expecting to find general scientific relationships, is not unlike going to the bookshelves in a convenience store to uncover the great trends in literature. Future research might benefit from two perspectives. Since so much of our knowledge comes from agricultural soils, much may be learned by the study of more extreme habitats and the novel ways in which plants cope with them. At the same time, the CHNOPS perspective reminds us that all life shares the need for but a few common resources, and hence there may be remarkable similarities among very different species. Finding the right trade-off between natural historical detail and broad-scale generality is always a challenge in ecology.

Review Questions

1. The atmosphere is predominantly nitrogen, yet plant growth is often limited by lack of nitrogen. Explain this paradox.
2. What do field experiments reveal about the relative importance of nitrogen and phosphorus in limiting plant growth?
3. Since epiphytes do not have direct access to the soil, how do they obtain what we think of as soil nutrients?
4. Contrast the characteristics of plants found in infertile as opposed to fertile soils. Now contrast the characteristics of the resulting ecosystems.
5. Explain how topography can create gradients of fertility and productivity.
   Can you include examples from wetlands, forests and grasslands? What is the most common fertility gradient in your own landscape?
6. What are some advantages and disadvantages of evergreen leaves? Give some examples of habitats where evergreen leaves are particularly common.
7. Members of the Fabaceae can fix atmospheric nitrogen using symbiotic bacteria living in root nodules. Name some local examples from your own landscape. Why do all plants not have such root nodules?

Further Reading

Lieth, H. 1975. Historical survey of primary productivity research. pp. 7–16. In H. Leith and R.H. Whittaker (eds.) *Primary Productivity of the Biosphere*. New York: Springer-Verlag.

Southwood, T.R.E. 1977. Habitat, the templet for ecological strategies? *Journal of Animal Ecology* **46**: 337–365.

Smith, V.H. 1982. The nitrogen and phosphorus dependence of algal biomass in lakes: an empirical and theoretical analysis. *Limnology and Oceanography* **27**: 1101–1112.

White, T.C.R. 1993. *The Inadequate Environment: Nitrogen and the Abundance of Animals*. Berlin: Springer-Verlag.

Marschner, H. 1995. *Mineral Nutrition of Plants*, 2nd edn. London: Academic Press.

Vitousek, P.M., J.D. Aber, R.W. Howarth et al. 1997. Human alteration of the global nitrogen cycle: causes and consequences. *Ecological Applications* **7**: 737–750.

Roy, J., B. Saugier and H.A. Mooney (eds.) 2001. *Terrestrial Global Productivity*. San Diego: Academic Press

Wardle, D.A. 2002. *Communities and Ecosystems: Linking the Aboveground and Belowground Components*. Princeton: Princeton University Press

Larcher, W. 2003. *Physiological Plant Ecology. Ecophysiology and Stress Physiology of Functional Groups*, 4th edn. Berlin: Springer.

4 Competition

Neighbours consume resources. The struggle for existence. Definition. Brief history. Five kinds of competition. Intra- and interspecific competition. Competition intensity. Effect and response. Dominance. Monocultures. Asymmetry. Hierarchies. Competition gradients. Old fields, prairies, alvars, wetlands and mountainsides. Foraging ability. Mycorrhizae. Competition models. Role of models in ecology as a whole.

FIGURE 4.1 Large trees in mature forests are locked in quiet but intense competition for light, water and nutrients. The large canopy-forming trees deprive smaller trees of resources. Unless a large tree is killed by wind, lightning or insects, creating a gap in the canopy, most of the smaller ones will die silently. (Quinault Rain Forest, Olympic Mountains, Washington, by Greg Vaughn)

4.1   Introduction: Plants Struggle Against One Another for Resources

There are three principal biological interactions affecting plants: competition, herbivory and mutualism. This chapter explores the first of these. We will examine the other two later, herbivory in Chapter 6 and positive interactions in Chapter 7. We start with competition because it is so important.

Plants are mostly quiet, at least to human ears. Hence we are often oblivious to their struggles with their neighbours. To us a forest may seem relatively silent (Figure 4.1). In contrast, competition among animals can be a noisy or bloody affair. Moreover, unlike animals, individual plants cannot flee aggressive neighbours: they must simply die where they are rooted. When I worked in Louisiana there was a silly bumper sticker in our parking lot "Children are like flowers: you can never have too many of them." Whoever wrote that (and whoever stuck in on their car) illustrates my point of how oblivious people are to simple biological realities: if you see a meadow filled with wildflowers, you can be fairly certain that there are intense struggles among them, and that many will die as a consequence. ("How certain?" you ask. Well, that is why there is a full chapter on this very topic.) The point is this: habitats can indeed have "too many" flowers. Most do. In fact, when each large tree in a forest produces upwards of a million seeds each year, there is no possible outcome except that at least 999,999 of them must die. Every year. We will return to this topic under populations (Chapter 9), but my point here is that you should never underestimate the intensity of the struggle for resources among plants within plant communities. They are dying by the millions from lack of resources, but because they do so silently, most people remain unaware.

The staggering amount of death in plant communities led Darwin to describe much of life as a *struggle for existence*. For plants, the central part of this is the struggle to find, harvest, transport and retain possession of resources (Chapter 3). While resources are scarce for a variety of reasons, we will focus here on an important truth. Resources are often in short supply simply because neighbouring plants have captured them first. This contest between adjoining plants for access to the same few resources is called competition. When we visit a forest and describe it as "shaded" what we are really saying is that tall plants have already extracted the sunlight, denying smaller plants all but a few meagre photons in the occasional sun-fleck (Figure 4.1). The basis of competition is lack of resources caused by the presence of other plants vying for the same resources.

There are many kinds of competition. Introductory ecology books that deal with plants and animals together usually divide competition into resource competition and interference competition. Resource competition occurs when neighbours consume resources, thereby depleting the resources available to other plants. Interference competition occurs when there is direct interference with neighbours in order to maintain access to resources. It is safe to say that with plants, most competition is resource competition, but there may be certain cases of interference through chemical interactions or physical contact. Another common distinction is to divide competition into intraspecific and interspecific competition. Plants in a field or forest must simultaneously compete with their own species as well as with neighbouring species. This can be misleading too, because interspecific competition can occur with many different species simultaneously. There are, in fact, other kinds of competition. One source of confusion in the study of plant competition is careless use of the word competition without a modifier of some sort. So we are going to begin by showing you five kinds of competition. There are more, but if you understand these five, you will have accomplished more than enough.

But first we must have a definition. The right definition is like a sword that will clearly cleave nature into pieces that can be understood; the wrong definition is like a blunt instrument that only mashes the object of inquiry into more confusion. This is why scientists seem to spend so much time arguing about definitions. We might choose a definition that emphasizes the mechanisms of competition, its measurement by means of experiment or its long-term evolutionary consequences. It is difficult to find one definition of competition that will fit all situations. Recent text books of ecology reveal a wide variety of attempts to satisfy these conflicting objectives. Some authors have even suggested that the term no longer be used. Here I define competition as:

> the negative effects that one organism has upon another, usually by consuming or controlling access to a resource that is limited in availability.

This definition has two important parts. First, it emphasizes the main point, which is that competition is a force that has negative consequences for *both* participants. This is very important to understand; even the winner loses something in the interaction. Competition always has costs. Second, the definition suggests a mechanism, which is usually that one plant is consuming and controlling resources that the neighbouring plant also needs. Four resources – light, water, nitrogen and phosphorus – are likely to be most important.

You should appreciate that competition has a long and important history in ecology as a whole. More than a century ago, Malthus and Darwin both appreciated the intrinsic nature of organisms to multiply exponentially against limits set by resources and saw that this produces what they called "a struggle for survival." Thomas Malthus (1766–1834) was an English economist and clergyman; Charles Darwin (1809–1882) was, of course, the English naturalist who, along with Alfred Wallace, proposed the theory of evolution through natural selection. They both appreciated the power of exponential population growth, which we shall examine more closely in Section 9.2. The outcome is the important point; exponential growth means that both houseflies and elephants, given sufficient time, could multiply rapidly enough to entirely cover the land area of Earth. Given a few more generations, a ball of flies or elephants would then expand outward from the Earth's surface and eventually reach light speed; the flies, being more fecund than the elephants, would, of course, have a head start. The same is true for plants. Clearly, however, Earth is not covered by flies, elephants or, for that matter, saguaro cacti (Figure 1.1) and sausage trees (Figure 1.2). Why not? Malthus and Darwin both saw that some other factor, usually a shortage of resources, prevented most young from surviving. Although a single tree in a forest may produce millions of seeds, only one need reach adulthood for each tree to be replaced and for the forest to remain intact. The other millions of young must perish. Darwin reported that he saw, "on reading Malthus' *On Population* that natural selection was the inevitable result of the rapid increase of all organic beings."

Shortages of resources, then, limit the growth of all populations. Moreover, we saw in the last chapter how all plants depend upon rather few resources. Hence there is every reason to believe that competition among neighbours for those same resources is very important. How important? Well, that is what this chapter is about. We shall start with a number of examples that illustrate competition acting under different sets of conditions, then we shall look at some consequences for plant communities as a whole, then we shall look at some simple models that might put all the examples back together in a simple framework.

4.2 There Are Many Kinds of Competition

4.2.1 Intraspecific Competition

As the number of plants in a unit area increases, the per-capita supply of resources declines; therefore, as plant density increases, the mean plant size declines. The relationship between plant density and plant size has been extensively studied because agriculture requires some understanding of the effects of sowing density upon crop yield. Even the backyard gardener thinning a patch of radishes or peas understands intuitively that reducing the density of plants will enhance the performance of remaining individuals. Ecologists have also studied the effects of density on plants because it is relatively easy to manipulate sowing density (much easier, say, than manipulating the density of bears, birds or fungi) and thereby explore the effects of intraspecific competition. Figure 4.2 shows that eight different plant species all exhibit declining performance with increasing density. The steepness of the slope indicates how intense intraspecific competition is; the steeper the slope, the greater the effect each added individual has upon its neighbours (note the vertical scale is logarithmic). The lines in Figure 4.2 are fit by an equation of the form:

$$w = w_{m}(1 + aN)^{-b}$$

where w is the weight of individual plants, N is the density, and w_{m}, a and b are parameters (Watkinson 1985a). The form of this equation is of some interest because w_{m} can be interpreted as the weight a plant will attain if grown in isolation – where intraspecific competition is zero. The area required to supply the resources to achieve w_{m} is then a: a is the minimum area, or the neighbourhood, that a plant requires to find the resources necessary to achieve maximum growth. Finally, b can be considered a measure of the effectiveness with which resources are extracted from an area. Therefore the actual yield of an individual plant, w, will decrease as N (density) increases, as a (minimum required area) increases or as b (effectiveness of resource extraction) decreases.

The decline in reproductive output that occurs with increasing plant density is usually the result of individual plants being smaller and therefore yielding fewer seeds. In addition, the proportion of resources allocated to reproduction within an individual plant may decline with increasing density or decreasing resource supplies (Harper and Ogden 1970; Abrahamson and Gadgil 1973). To examine this experimentally, Snell and Burch (1975) grew *Chamaesyce hirta*, an annual member of the Euphorbiaceae, at four different densities combined with four levels of fertilization. Not surprisingly, mean plant size increased with fertilization and decreased with density, from a maximum size of 10 g at high fertility and low density to a minimum of <0.5 g at high density and low fertility. Over this extreme range of size variation, reproductive allocation declined from 26 to 7 percent (Figure 4.3). From one perspective, this testifies to the degree to which plants are able to maintain internal homeostasis and regulate biomass allocation in spite of an order of magnitude variation in size. From another perspective, it also appears that when resources are less freely available, a smaller proportion of them is invested in reproduction. Presumably this occurs because under stressful conditions of high density and low fertility, a greater proportion of the resources must be spent in building roots and leaves. Snell and Burch also expressed their results in calories as well as grams, although the actual limiting resources being partitioned may in fact be nitrogen or phosphorus rather than grams or calories of tissue. The actual currency by which one measures the costs of competition is still unclear.

4.2.2 Distinguishing Between Intraspecific and Interspecific Competition

The relative importance of intra- and interspecific competition can be measured using simple removal

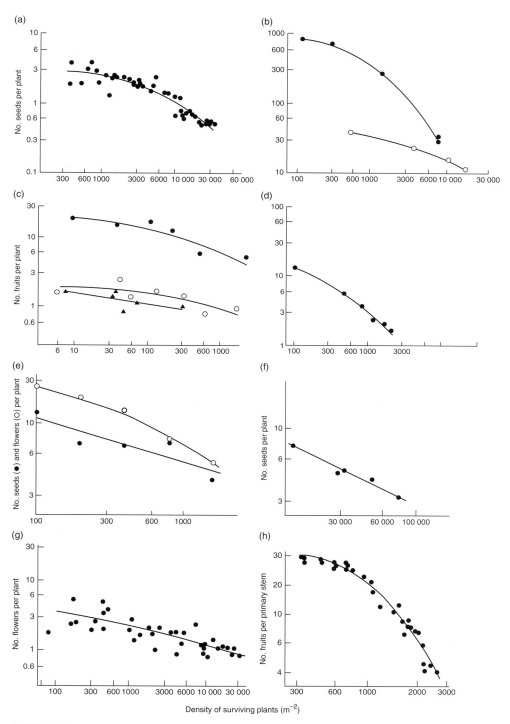

FIGURE 4.2 The relationship between reproductive output per plant and plant density at maturity. (a) *Vulpia fasciculata*. (b) *Salicornia europaea* in a high (○) and low (●) marsh. (c) *Cakile edentula* on the seaward (●), middle (○) and landward (▲) sides of a sand dune. (d) *Rhinanthus angustifolius*. (e) *Floerkea proserpinacoides*. (f) *Polygonum confertiflorum*. (g) *Diamorpha smalli*. (h) *Androsace septentrionalis*. (From Watkinson 1985a)

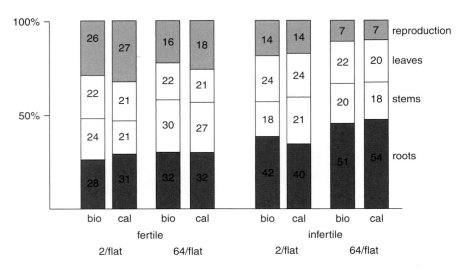

FIGURE 4.3 Allocation to reproduction is reduced by competition. Tissue allocation for the annual plant *Chamaesyce hirta* at two densities and two levels of fertility; bio = biomass, cal = calories. (After Snell and Burch 1975)

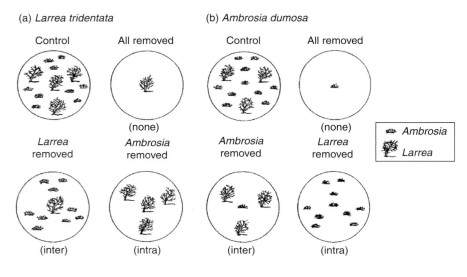

FIGURE 4.4 A removal experiment to measure intra- and interspecific competition in desert shrubs. In one treatment, both types of competition were removed (none). (From Fonteyn and Mahall 1981)

experiments. Consider the example of two desert shrubs, *Larrea tridentata* (creosote bush) and *Ambrosia dumosa* (burbage), which dominate some 70 percent of the Mojave Desert in the southwestern USA (Fonteyn and Mahall 1978). An advantage to studying desert plants is that the limiting resource is almost certainly water, and the water status of the test plants can be determined. By clipping off a branch and inserting it in a pressure bomb, one can measure how much of a water deficit exists in the branch, providing a nearly instantaneous measurement of how plants are being affected by neighbours. The lower the pressure, the greater the effect. The experiment had control plots and three different removal treatments for each of the two test species (Figure 4.4). Xylem pressure potentials were determined every two weeks through three consecutive wetting and drying cycles. Figure 4.5 shows that for

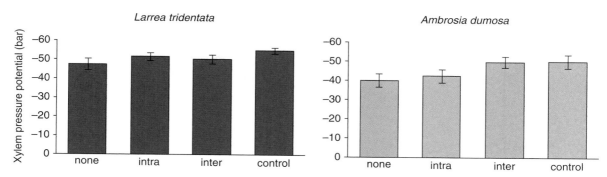

FIGURE 4.5 Effects of three kinds of experimental removals (see Figure 4.4) upon the xylem pressure potential ($\pm$ SE) of two desert plants at the end of July; these data could be re-scaled as percent reduction in performance relative to plants experiencing no competition (none). (After Fonteyn and Mahall 1981).

both species, water potential was lowered by intraspecific neighbours (that is, the same species – bar labelled intra) and by interspecific neighbours (bar labelled inter). In *A. dumosa*, interspecific competition reduced water potential more than did intraspecific competition. In *L. tridentata* the two types of competition had similar effects on water potential.

4.2.3 Competition Intensity

Competition intensity, or total competition, refers to the effects of all neighbours upon the performance of a population or an individual. This can be detected by removing all neighbours and measuring the release, if any, of the remaining population or individual relative to control plots (for example, Putwain and Harper 1970; Fowler 1981). The individuals of the test species remaining in the treatment plots serve as a "bioassay" of competition intensity (del Moral 1983; Wilson and Keddy 1986a). The difference between the performance of test individuals in treatment plots and in intact vegetation (control) measures competition intensity. Weldon and Slauson (1986) propose that competition intensity can also be measured by comparing the physiological state of organisms in plots with and without neighbours.

The mean intensity of competition faced by plant shoots in a large marsh was assessed by Shipley et al. (1991) along the Ottawa River in Canada. Ramets of three dominant plants (a sedge – *Carex crinita*, sweet flag – *Acorus calamus* and cattail – *Typha angustifolia*) were placed in cleared plots (Table 4.1, cleared) and in intact vegetation (Table 4.1, control). The clearings were maintained by weeding and by the use of plastic barriers to prevent roots or rhizomes from re-invading below ground. Further, the experiment was repeated at three elevations to test for possible changes in competition intensity with flooding. After two growing seasons the transplanted shoots were harvested, dried and weighed. The difference in weight between the transplanted shoots in cleared and uncleared plots provided the measure of competition intensity (CI). To correct for differences in test plant size, the absolute difference in weight was divided by the weight in the cleared plot, yielding relative competition intensity (Table 4.1, RCI). This shows that competition intensity was sometimes in excess of 50 percent (*A. calamus* on flooded sites low on the shore) and often greater than 30 percent (5/9 cases). Surprisingly, there was no evidence of a change in competition intensity with test species or elevation. The data in Table 4.1 also suggest that species in some habitats (e.g. *T. angustifolia* at low elevation) might be entirely unaffected, or even assisted by the presence of neighbours.

Table 4.1 **Relative competitive intensity (RCI) in a riverine wetland measured over two years by the performance (proportional decrease in plant dry weight) of three plant species at three elevations. (After Keddy 2001 after Shipley et al. 1991)**

Species	Elevation	Cleared (g)	Control (g)	CI[a]	RCI[b]
Acorus calamus	High	12.81	7.46	5.35	42
	Medium	12.18	11.13	1.05	9
	Low	7.77	2.59	5.18	67
Carex crinita	High	21.76	11.47	10.29	47
	Medium	16.44	8.85	7.59	46
	Low	23.34	19.89	3.45	15
Typha angustifolia	High	18.17	20.09	−1.92	−11
	Medium	51.42	27.66	23.76	46
	Low	22.20	26.05	−3.85	−17

Notes:
[a] CI = (cleared − control) (g).
[b] RCI = [(cleared − control)/cleared] × 100.

4.2.4 Competitive Effect and Competitive Response

When referring to the "competitive ability" of a species growing in mixture, one is actually describing something that has two separate components (Goldberg and Werner 1983; Goldberg 1990). The first is competitive effect: that is, the damage that each species can do to its neighbours. The second is competitive response: that is, the ability of each species to withstand the effects of competition from neighbours. In general use, it appears that the term "competitive ability" is frequently used as a synonym for "competitive effect"; more precision is needed.

The distinction between these components may be important in two ways (Goldberg and Werner 1983). First, competitive effects may be relatively similar among species, whereas responses may not. For example, a seedling growing in the shade is inhibited by lack of light and may be relatively insensitive to which species is actually intercepting the light. The competitive effects of all neighbours, then, may be similar. In contrast, each species of seedling might have different means to tolerate lack of light, so perhaps competitive response could be quite different. Recent experimental work, however, fails to support this suggestion; hierarchies of competitive response appear similar regardless of neighbour species (Goldberg and Landa 1991).

The categories above do not exhaust the possibilities for recognizing different kinds of competition and competitive interactions. Two others deserve consideration. First, Arthur (1982, 1987) has emphasized competition between different genotypes within populations and the evolutionary consequences of such interactions. Second, Buss (1988) has explored competition among different cell lines within individuals and its implications for the evolution of development.

4.2.5 Competitive Dominance

Competitive dominance is an outcome of interactions where one species suppresses another through resource competition and/or interference competition. It is driven by asymmetric competition, which occurs when two competing plant species have different competitive abilities. In other words, competition is asymmetric when the effect of species A upon species B is large, while the effect of species B upon species A is small (Keddy 2001). Hence one can usefully refer to one species as dominant and the other as subordinate. Asymmetric competition is probably widespread in nature. The effects of the dominant upon the subordinate increase through time by means of two positive-feedback loops (Figure 4.6). First, there is exploitation competition. The dominant lowers the resource levels for the subordinate and is simultaneously better able to forage for additional resources itself by reinvesting the newly captured resources in further growth. This foraging and growth lowers further the resources available to the subordinate. Second, there may be direct interference with neighbours to reduce their foraging and growth. The relative importance of these two positive-feedback loops is likely to vary from situation to situation, and in some cases effects may be separated into exploitation and interference only with difficulty. With plants, exploitation competition for light, water and nutrients is well documented. Interference competition is probably less important, but might involve plants poisoning their neighbours. This possibility has long been suspected (e.g. Molisch 1937; Muller 1966; Gopal and Goel 1993) but remains contentious (Muller 1969; Williamson 1990). However complicated the mechanisms, the end result of competitive dominance, however, is that one species suppresses another or excludes it from a given community. This is not a new idea. In 1929, Weaver and Clements wrote:

The plants may be so nearly of the same height that the difference is only a millimeter. Yet this may be decisive, since one leaf overlaps the other. It continues to receive light and make photosynthate, while the shaded one can not. A difference of 2 or 3 days with full or reduced photosynthetic activity is quickly shown in difference in growth. The second pair of leaves of the fully lighted plant develops earlier, the stem is thicker and can better transport

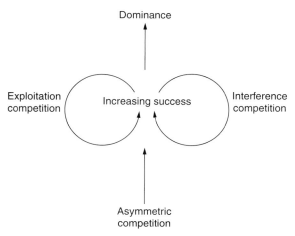

FIGURE 4.6 The positive-feedback loops that generate dominance. Success at exploitation competition increases the resources available to forage for new resources, and simultaneously reduces the resource supply for neighbours. Increased availability of resources allows some to be channelled to interference competition, damaging neighbours and leaving more resources available for exploitation by the dominant. (From Keddy 2001)

food to the rapidly growing roots. These, because of their greater food supply, penetrate a little deeper, spread farther laterally, and have a few more branches than those of their competitor. The increase in leaf surface not only reduces the amount of light for the plant beneath it, but it also renders necessary the absorption of more water and nutrient salts and correspondingly decreases the amount available. New soil areas are drawn upon for water and nutrients, and the less vigorous competitor must absorb in the area already occupied. The result is that the successful individual prospers more and more and becomes dominant.

(p. 127)

One of the difficulties encountered when discussing dominance arises from the tendency to assume that competitive interactions are symmetric and to talk loosely about "competition" between two species. As soon as there is **asymmetric competition,** the experiences of the dominants and subordinates diverge. It becomes essential to specify whether competition is being viewed from the perspective of the dominant or the subordinate. The analysis of such interactions is clarified by considering that in any competitive interaction there are both effects and responses (Section 4.2.4). The subordinate may be killed and disappear. The subordinate may tolerate the impact of the dominant, in which case it remains present, albeit at a low level.

The subordinate may escape from the competition by dispersing in space or time to another site (ruderal or fugitive species). The analysis of asymmetric interactions requires explicit consideration of the effects of the dominant and of the responses of the subordinate organisms. These issues are discussed further in Section 4.4, which looks at hierarchically structured communities.

It is important here to clarify the distinctions between competitive dominance and dominance. The word dominant is sometimes used to describe any organism that is abundant in a community. This use is misleading; abundance need not be the result of competitive dominance. Competitive dominance is abundance achieved as a consequence of exploitation of, and interference competition for, resources; that is, there is an active process of suppressing neighbours (Figure 4.7, bottom). Grime (1979) describes dominance as a process whereby one species achieves numerical dominance and suppresses others. His use of dominance is not equivalent to the term competitive dominance used here, since Grime includes a second group of effects: a species may become dominant because of inherently better abilities to withstand environmental effects such as fire, infertility or grazing. This added group of effects is shown in the upper portion of

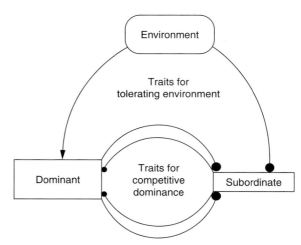

FIGURE 4.7   The possible interactions between the dominant, the subordinate and the environment. Arrows are positive effects, solid circles are negative effects. Competitive dominance refers solely to the direct links between the dominant and the subordinate (bottom). The environment (top) may independently determine which species dominates a site. In this example the environment is enhancing the effects produced by asymmetric competition, so dominance is only partly attributable to competition. (From Keddy 2001)

Figure 4.7. It seems useful to distinguish between situations where a species is dominant simply because of the inherent traits it has for tolerating the environment and situations where a species is dominant because it has traits for suppressing neighbours. The former type of dominance could occur in the absence of any competition. Experiments in which possible dominants are removed and the responses of subordinates observed are necessary to determine the kind of dominance present in a habitat or vegetation type.

4.3   Competition Has Many Consequences

4.3.1 Self-Thinning in Monocultures

One of the simplest methods for detecting intraspecific competition is to test for a negative relationship between performance and population density. Performance can be measured in many ways, depending upon the organism and the circumstances; it can be determined by examining correlations

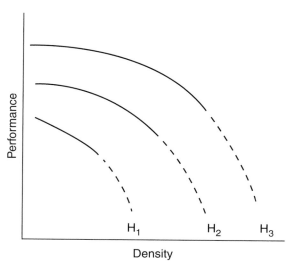

FIGURE 4.8 Performance declines with increasing density as a result of intraspecific competition. Measures of performance may include survival, fecundity or size. Each line represents a different habitat, emphasizing that performance–density relationships may differ with environmental conditions. In this case both the intercepts and shapes of the curves vary among habitats (see Keddy 1981c for other possible situations). The dashed sections of the lines represent situations not normally found in nature, which can be observed only by experimentally increasing population density. (From Keddy 2001).

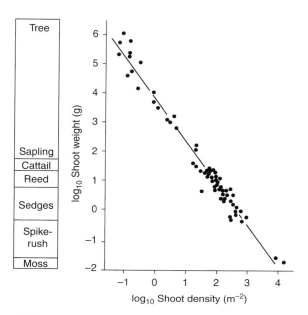

FIGURE 4.9 The relationship between performance (shoot weight) and density for 65 stands of plants representing 29 species from trees (upper left) to a moss (lower right). Intraspecific competition appears to set an upper limit to the number of shoots of a specific size that can coexist. (After Gorham 1979)

between density and measures of growth rate, survival rate and reproductive output. If the correlation is negative (higher density leading to lower performance) there is evidence for intraspecific competition. Figure 4.8 illustrates the possible relationships between performance and density and emphasizes that for any species there may be a different performance density relationship for each habitat. Also, only a portion of the line shown in Figure 4.8 may be found in nature; experimental manipulation may be necessary to observe the dashed sections.

The relationship between the size of individuals and density has been extensively studied in plant populations, partly because of the obvious

agricultural implications (recall Figure 4.2). Further, there is a well-established relationship, or "self-thinning line," between the mean mass of individuals and the density at which they are grown (Yoda et al. 1963; Harper 1977; Westoby 1984; Weller 1990), with a slope of 3:2. Most of the data regarding plant–density relationships come from the experimental study of plants in monoculture (coexisting individuals of a single species); Gorham (1979), however, has shown that there is a strong relationship between shoot weight and shoot density across 29 different plant species (Figure 4.9) ranging in size from trees (upper left) to a moss (lower right).

Such descriptive approaches have three weaknesses, which, depending upon the particular system, may be fatal flaws:

1. The most obvious is that correlation does not demonstrate cause and effect. Since different

habitats, years or quadrats provide each datum point, it is possible that the correlation is spurious. If, for example, habitats providing the most nitrogen were also most exposed to grazing or pathogens, high individual performance and low survival could be correlated in the complete absence of intraspecific competition.

2. A second, related, weakness is that such studies assume that only population density measures are needed to describe a particular habitat. Given the range of habitats that most species occupy, the performance–density relationship is likely to differ among them. Species may occur in habitats where populations have density-dependent relationships, and others where these are absent. Most species will therefore be represented by a family of performance–density response curves (Keddy 1981c; Morris and Myerscough 1991).

3. A third potential problem is that a species may have the same density in two habitats, but in one there may be a positive population growth rate and in the other a negative growth rate. The equivalence of density may therefore be transitory and misleading.

4.3.2 Dominance Patterns in Monocultures

Although it is rare in natural communities, in agricultural situations monocultures are common; cotton and wheat fields offer two examples. In a monoculture, there is no interspecific competition. Since most of the individuals are fairly similar in relative competitive ability, one might expect that the plants would all grow to be nearly the same size. This is not the case. In fact, one often finds enormous differences in size developing. It seems that small differences in initial advantages, such as slightly earlier germination or a slightly better location, multiply up to enormous differences in outcome. To study this phenomenon, one can borrow from economists who have studied income disparity in human societies.

The basic experiment is as follows. One grows a monoculture of a selected species and measures the size of each plant. The dependent variable of the experiment is size variability, which could be measured by the standard deviation or coefficient of variation in size. The usual measure of size variation is the Gini coefficient, which varies from 0 (all plants identical) to 1 (plants maximally different in performance). Weiner (1985, 1986) has examined patterns in the Gini coefficient in groups of plants grown at different densities and at different fertility levels. Let us look at them more closely.

In theory, plants grown in monoculture might all be the same size. Weiner sets out to test this theory. He begins by contrasting two models that might cause departures from size equality. The first is an exponential resource depletion model (Koyama and Kira 1956). Consider a population of seedlings that are equivalent in size or normally distributed. Allow each seedling to grow in size exponentially. The final size distribution will be a function of several variables, but the important thing is that if initial sizes are equal and relative growth rates are normally distributed, a lognormal size distribution should develop. If both the initial sizes and relative growth rates are normally distributed, a log-normal distribution will still develop (Weiner 1986). Thus even if there is no interaction at all among seedlings, a log-normal size distribution can be expected. In contrast to this exponential resource depletion model are resource pre-emption models of competition: "Such models assume that larger individuals are able to get more than their proportional share of resources," and thereby grow at the expense of smaller individuals (Harper 1977; Weiner 1986, p. 215). There are various approaches for incorporating this sort of "one sided" or "asymmetric" aspect of competition, but it is what the model predicts that is of interest. The essential point of difference between these two types of models is that resource pre-emption models all predict higher size inequality at higher densities.

Instead of comparing the results of simulation models, Weiner compared data from published studies and then did some simple but compelling experimental studies. The comparison of published studies was clear: of 16 experiments, 14 supported the hypothesis

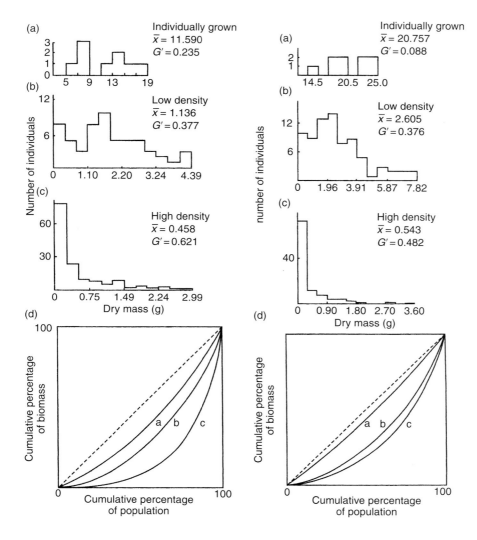

FIGURE 4.10 Dry mass distributions (with inequality measured by the unbiased Gini coefficient G′) and Lorenz curves for monospecific populations of *Trifolium incarnatum* grown in low-fertility soil (left) and high-fertility soil (right). (a) individually grown, (b) low density (200 individuals/m²), (c) high density (1,200 individuals/m²), (d) Lorenz curves for a, b and c. (From Weiner 1985)

of asymmetric competition (Weiner 1986). Then, in a series of experiments, Weiner manipulated density to explicitly test between the resource depletion model and the resource pre-emption model. The latter model predicts that size asymmetry (measured, in this case, by the Gini coefficient) will increase with density. Figure 4.10 shows the size distributions of a common herbaceous species, when grown alone, at low density and at high density, and at two different levels of soil fertility.

The results of Weiner's experiments are clear. Inequality increased from low to high density. These results contradict the resource depletion model. Finally, while both models predict increased plant size inequality with increased soil fertility, the resource depletion model predicts that inequality should be greatest in plants grown individually, and this is what occurred with *Trifolium*. Weiner concludes that size differences are continually generated by dominance and suppression but are simultaneously reduced by mortality of the smallest individuals. He suggests that the mechanism for size hierarchies is asymmetric competition for light:

> *a plant's ability to take up water or nutrients is a function of the surface area of its roots. A small plant with relatively little root surface area will not be able to absorb as much of the soil nutrients as a large plant, but it may be able to take up an amount proportionate to its root surface area, and accordingly to reduce the nutrients available to its large neighbour. However, when competition for light is intense, the effects of interference are not shared in proportion to size; a small plant will not be able to get its share of this resource, and its growth will be reduced disproportionately. Thus, if plants do not grow to the point at which the canopy becomes closed and competition for light is important, we would not expect to see dominance and suppression.*
>
> *(p. 748)*

The development of dominance hierarchies among plants as a result of asymmetric competition is clearly demonstrated by Weiner's work on monocultures. If such size irregularities and suppression can arise in simple experimental monocultures, it becomes easier to see how they can arise in mixtures of species. In mixtures one can expect species to differ both in relative growth rates and ability to suppress neighbours. These differences can be expected to generate size inequalities even greater than those found in monocultures.

4.3.3 Density Dependence in Annual Plants

The first two weaknesses of descriptive approaches in density dependence studies (Section 4.3.1) were addressed in a field study of an annual plant, *Cakile edentula*, which grows on sand dunes along the coast (Keddy 1981c, 1982). The third weakness was addressed in a re-analysis of these data (Watkinson 1985b). *Cakile edentula* can be found in a wide range of habitats and shows corresponding changes in plant size, reproductive output, survival and population density (Keddy 1981c, 1982). These habitats are arranged along a gradient. At one end one can find large plants with thousands of seeds growing amidst decaying seaweeds on open sand beaches. At the other end tiny plants with but one or two seeds can be found beneath a canopy of dune grasses. Density dependence was tested for by sowing a range of seed densities, allowing germination and growth to occur for one summer, and then testing whether performance was negatively correlated with sowing density. Two measures of performance were used: percentage of seeds sown that produced reproductive plants and mean number of fruits produced per plant. The principal results are shown in Figure 4.11. Density dependence clearly varied among the three habitats. In the middle of the gradient there was no evidence for it. At the seaward end extreme crowding significantly reduced only reproductive output, whereas at the landward end both reproductive output and survival declined with density. Survival reduction was the predominant effect at the landward end.

There are two general conclusions for studies of intraspecific competition. First, the relationship between performance and density is not a trait of a species or a population alone, but is strongly dependent upon the environment itself. Second, the dependent variable selected is extremely important: if only survival or reproductive output had been used, then entirely different conclusions would have been reached about the magnitude of intraspecific competition in different habitats. The problem with this experimental approach is that it does not yield unequivocal statements about the actual intensity of intraspecific competition in real populations unless the usual range of population density in the habitat is known. If, for example, seaward populations usually have low densities (and they do), then the intensity of intraspecific competition shown in the seaward site is a potential that is rarely realized. Thus such experimental data need to be combined with measures of population density.

One problem of interpretation still has not been removed: how is one to know that the density dependence is attributable to intraspecific

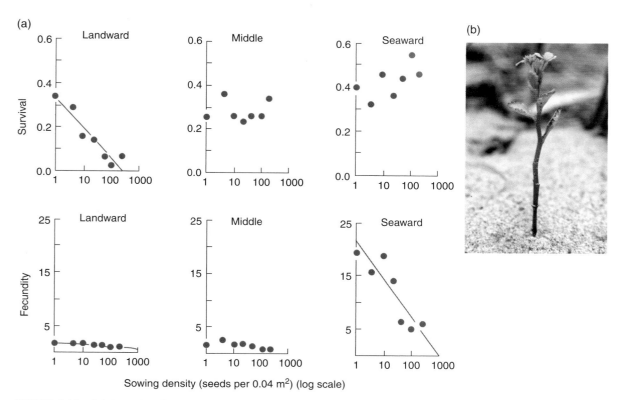

FIGURE 4.11 (a) A study of an annual plant, *Cakile edentula*, growing on sand dunes along the Atlantic Ocean, which shows how the effects of density upon survival and fecundity vary among habitats. Start at the lower left: landward plants are the smallest plants with the lowest fecundity perhaps only one or two pods; here density has little impact up fecundity but a strong impact on survival. The photo (b) shows a plant only about 6 cm tall with just two flowers. Now look at the opposite end of the gradient: seaward habitats have higher fecundity, and with strong negative effects of density. Plants here may be half a metre in diameter with some producing hundreds of seeds. There are no significant effects of density upon survival but the tendency is for positive effects. Hence the habitat and the performance measure may determine whether a researcher detects competition in the field. If, for example, the study had collected data from the middle section of the dune (which is where most of the plants are found), then no evidence for density effects would have been detected. (After Keddy 1981a; photo by Paul Keddy 1976)

competition? Many potentially confounding effects of temporal or spatial variation have been eliminated by using experimentally produced densities, so that a large number of potential alternative hypotheses have been eliminated; others, however, remain. If predation were density dependent, then the density dependence of survival could be attributed to predation, not competition. In fact, epidemics of damping-off disease (Burdon 1982) do occur in *C. edentula* populations,

although there was no evidence for it in the year in which this study was conducted. Mixed strategies can even be imagined where intraspecific competition weakens individuals that then fall prey to pathogens or are buried by drifting sand because of their small stature. One source of evidence for competition would be to augment the supply of a resource that is postulated to be limiting. By running a series of treatments with nitrogen fertilization, Keddy (1981c) showed that reproductive output of low-density landward plants (but not their survival) increased when this potential resource was supplied. This is additional evidence that intraspecific competition for nitrogen limited plant size. Since there were no effects on survival, one may postulate that an independent factor (perhaps competition for another resource, or predation) controlled survival.

The greatest weakness in this sort of experimental study is the absence of information on year-to-year variation. If storms destroy seaward populations in most years, then intraspecific competition may be much less important than the data in Figure 4.11 suggest. This criticism could only be answered by repeating the study in several years with very different weather conditions.

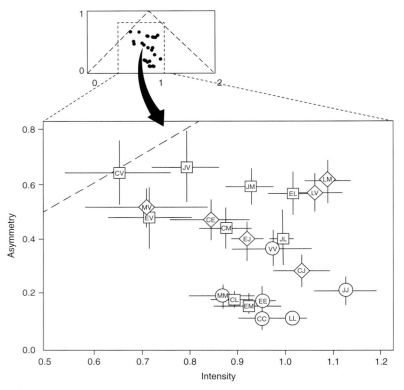

FIGURE 4.12 Phase-space diagram of asymmetry and intensity of competition for interactions between two plants of different species, where ○ = intraspecific, ◊ = intra-guild and □ = inter-guild interactions. Species names are abbreviated as follows: C = *Cyperus rivularis*, E = *Eleocharis obtusa*, J = *Juncus bufonius*, L = *Lythrum salicaria*, M = *Mimulus ringens*, V = *Verbena hastata*. For clarity, the upper small diagram shows the whole area of possible value ranges for asymmetry and intensity in a competitive interaction, with the dashed lines indicating the theoretical limits. In the lower diagram the area, which covers intensity and asymmetry values from this experiment, is enlarged. Note that, for visual expression, the intensity axis is disproportionately enlarged. (From Johansson and Keddy 1991)

4.3.4 The Relationship Between Intensity and Asymmetry of Competition

There are relatively few studies of asymmetry and intensity, and even fewer where both are measured simultaneously. Johansson and Keddy (1991) therefore grew six species of wetland plants in an additive pairwise design, so that asymmetry and intensity could be measured in each pairwise interaction. An additive design is one in which plants are grown singly to measure their potential performance and then grown with a series of neighbours to measure how each neighbour suppresses performance. Based upon a survey of traits in 43 species of wetland plants (Boutin and Keddy 1993), six species were selected for testing: three were obligate annuals and three were facultative annuals. These represent only two of a much larger number of functional groups found in wetland plants, but all of the plants selected were capable of completing their life cycle within a single growing season. Each species was grown alone to measure, Y, the yield of one individual. To assess the effects of neighbours, relative yield per plant (RYP) was calculated. When species i was grown with species j, performance in mixture (Y_{ij}) was divided by performance when grown alone (Y_i) to measure the reduction in performance from the neighbouring plant. For example, when a relatively weak competitor, *Mimulus ringens*, was grown with a relatively strong competitor, *Lythrum salicaria*, its biomass declined by 0.86 compared to its biomass when grown alone; that is, RYP = 0.14. Conversely, when the strong competitor *L. salicaria* was grown with *M.*

ringens, its performance declined by only 0.24; that is, RYP was 0.76. Therefore two RYP values are obtained for each mixture, since each species in the mixture can be compared to its growth alone.

To estimate the intensity of competition in any pot, one sums the declines in performance observed for each species. In the example above, the intensity of competition in the mixture was the sum of 0.86 (the decline of *M. ringens* owing to *L. salicaria*) and 0.24 (the decline of *L. salicaria* owing to *M. ringens*). For any pair of species, intensity was measured as $I = [(1 - RYP_{ij}) + (1 - RYP_{ji})]$. If both species are little reduced in mixture, I is near zero, whereas if both species are greatly reduced, I approaches 2.

Asymmetry can be measured by comparing the relative decline of species' performance in mixture. If both species decline the same amount in a mixture, irrespective of whether it is a large or small decline, then the interaction is symmetric. In the example above, *M. ringens* declined 0.86 in a mixture, whereas *L. salicaria* declined only 0.24; this is a relatively asymmetric interaction. One measure of asymmetry is $A = [(1 + RYP_{ij}) - (1 + RYP_{ji})]$; A can range from 0, where the species suppress each other equally, to 1, where one species eliminates the other. These two measurements, I and A, produce a phase space where intensity can range from 0 to 2 and asymmetry can range from 0 to 1. The 36 pairwise interactions measured by Johansson and Keddy (1991) occupied a small region of this space (Figure 4.12). Asymmetry values ranged from approximately 0.1 to 0.7, while intensity values ranged between 0.6 and 1.1.

4.4 Competitive Hierarchies Are Widespread

4.4.1 Methods for Establishing Hierarchies

Plant communities usually contain many kinds of species mixed together. It is not uncommon to find from 10 to 25 species in one small plot of grassland

or wetland. How do all these species interact with one another? Why are some relatively common and others relatively rare? To explore competitive interactions in the community, one could simply choose two species and measure their relative competitive ability. But which two species should you choose? If you choose the two commonest ones, it is possible, perhaps even probable, that you will also select two competitive dominants. You thereby lose information on the effects of these two dominants upon all the other plant species in the community. Arbitrarily choosing two species is clearly not a good way to study competitive inter-actions within a community.

One alternative is to include all the species in one large experiment. In such an experiment, each species is grown with every other species. There are some problems with this approach. The first is the sheer size of the experiment – if there are n species you wish to study, there are $(n^2/2) - n$ possible pairs of plants to grow in mixture. (The term n^2 is divided by two because when you grow species A with species B, you simultaneously grow species B with species A.) The second issue is a technical one. You must also grow plants of each species without interspecific competi-tion, in **monoculture**, to use as reference plants for assessing the effects of interspecific competition. Do you grow the monoculture plants singly (without any competition), or do you grow the monoculture plants paired with a second individual of the same species (with intraspecific competition)? The first case is known as an **additive** design, whereas the second case is known as a **substitutive** design. Each has its strengths. Most studies use the substitutive design, sometimes termed "de Wit replacement series" or "diallele" (de Wit 1960; Harper 1977), although it is possible that the additive design is preferable (Firbank and Watkinson 1985; Connolly 1986; Keddy 2001).

You should know how to interpret such experi-ments. Let us consider an example from a wetland plant community where seven species of plants were grown in all possible pairwise combinations using a substitutive design (Table 4.2). The table contains values of RIP_{ij} (relative increase in biomass per plant compared to biomass in monoculture) for target species (i) grown with neighbour species (j). The RIP for each species grown with itself (monoculture value; often referred to as RYP_{ii} or relative yield per plant) is assigned a score of 1.0, forming the diagonal of the matrix. The first column, Dul,

Table 4.2 Pairwise interactions among seven wetland plants. RIP is the relative increase in biomass per plant of species *i* when grown with species *j*. A row shows the effects of all neighbours on one target species and a column shows the effects of one neighbour on all target species. (From Wilson and Keddy 1986b)

Target species*	Neighbour species							Target score**
	Dul	Jun	Lys	Hyp	Rhy	Dro	Eri	
Dul	1.00	1.33	1.18	1.17	1.25	1.18	1.34	1.20^{ab}
Jun	0.63	1.00	1.34	1.46	1.46	1.52	1.49	1.28^{a}
Lys	0.88	0.87	1.00	1.63	1.63	1.78	1.57	1.31^{a}
Hyp	1.09	0.99	0.91	1.00	1.22	1.29	1.23	1.11^{abc}
Rhy	1.05	0.73	0.93	0.91	1.00	1.21	1.36	1.03^{bc}
Dro	0.98	0.91	0.93	1.02	1.02	1.00	1.11	0.98^{bc}
Eri	0.65	0.71	0.88	0.89	0.87	1.48	1.00	0.93^{c}
Neighbour score	0.89^{a}	0.93^{ab}	1.03^{abc}	1.15^{bcd}	1.21^{cd}	1.35^{d}	1.30^{d}	

Notes:

* Dul = *Dulichium arundinaceum*, Jun = *Juncus pelocarpus*, Lys = *Lysimachia terrestris*, Hyp = *Hypericum ellipticum*, Rhy = *Rhynchospora fusca*, Dro = *Drosera intermedia*, Eri = *Eriocaulon septangulare*

** Target and neighbour scores not significantly different from one another (Tukey's studentized range test, p <0.05) are denoted by common superscripts.

shows how *all target species* respond to being grown with *one* neighbour species, *Dulichium arundinaceum*, a shoreline sedge. Two species are strongly inhibited by *D. arundinaceum*, Jun (*Juncus pelocarpus*) with an RIP of 0.63 and Eri (*Eriocaulon septangulare*) with an RIP of 0.65. Two others species are less negatively affected (RIPs of 0.88, 0.98). Two species grow slightly better with *D. arundinaceum* than in monoculture, Hyp (*Hypericum ellipticum*) with an RIP of 1.09 and Rhy (*Rhynchospora fusca*) with an RIP of 1.05. The same sort of species-by-species analysis can be done with each column. In the last column, for example, all species grow better (RIP >1) with *E. septangulare* as a neighbour than when grown with their own species. In contrast to a column in the table, a row shows how *one* target species responds to *all neighbour species*.

The table thus gives the results of all possible pairwise interactions among these plants. Now let us try to sort the species from strongest to weakest competitor. One way to do this is to compare the RIPs of target species (ie. compare rows in the table). The greater the competitive ability of a target species, the fewer neighbours will suppress it (ie. the higher the number of RIPs >1). For target species suppressed by the same number of neighbour species (e.g. Lys and Hyp, each suppressed by two neighbour species), the mean RIP (target score) is used to determine competitive strength (Lys is stronger than Hyp). These two criteria were used to order the species in Table 4.2 from strongest (*D. arundinaceum*, not suppressed by any neighbours, all RIPs >1) to weakest competitor (*E. septangulare*, suppressed by five neighbour species). In a similar way, competitive ability could be determined by the ability of a species to suppress other species (the greater the number of RIPs in a neighbour column that are <1, the greater the ability of the species to suppress other species). Further, species could also be ordered from most to least competitive by target score (mean amount of suppression) or neighbour score (mean ability to suppress). In the first case, *Lysimachia terrestris* (highest target score) would be the strongest and *E. septangulare* (lowest target score)

the weakest. In the second case, *D. arundinaceum* would be the strongest (lowest neighbour score) and *D. intermedia* (highest neighbour score) the weakest.

Such tables provide important information about the multiple competitive interactions that can organize plant communities. At a minimum, they allow us to picture how each species affects the other species. These sorts of tables also provide a basis for further questions. One can test for hierarchies of sets of species, and compare one matrix to another (Box 4.1). One can ask whether the hierarchies change if you change the growing conditions (Section 4.4.2). One can ask what plant characteristics might determine position in the hierarchy (Section 4.4.3).

4.4.2 The Consistency of Hierarchies Among Habitats

If communities are composed of species with relatively similar competitive abilities and contingent competitive interactions, reversals of position in competitive hierarchies should be commonplace. But if species differ in their relative competitive abilities, if interactions are asymmetric and if competition is relatively less contingent, one may expect reversals of positions in competitive hierarchies to be relatively less common. There is a good deal of evidence to suggest that the latter situation seems more typical of real communities. Yet, in spite of the evidence of hierarchies, the view of symmetric interactions and contingent competition is firmly entrenched. In addressing the defenders of contingency and symmetry, Shipley and Keddy (1994) noted the marked tendency of the critics to prefer opinions over data.

The following examples examine three explicit tests for consistency in relative competitive ability. The first test used 23 species of wetland plants transplanted into two contrasting environments: fertile and infertile (Gaudet 1993). Relative competitive effect was measured by the ability of each species to suppress the growth of a reference species, *Lythrum salicaria*.

Figure 4.13 shows that after two growing seasons, the competitive effect of species in fertile conditions is strongly correlated with competitive performance in infertile conditions. At the same time, the relationship is imperfect ($r^2 = 0.58$); it appears that changing the environment may allow a species to slightly improve its performance relative to that of a neighbour in a similar position in the hierarchy but not significantly change its position in the hierarchy. (That is to say, an apprentice may become a journeyman, or a duke or earl, but there is no evidence that an apprentice suddenly rises to become king.)

Box 4.1   Testing for Patterns in Competitive Relationships

Experimentally derived competition matrices may contain important information about the nature of plant interactions. Keddy and Shipley (1989) explored such matrices for two properties: (1) the degree of asymmetry (non-reciprocity) of pairwise interactions, and (2) the degree to which species are arranged in competitive hierarchies (transitive networks).

Let Y_i be the yield of an average individual (ramet) of species i grown in monoculture and Y_j be the yield of an average individual of species i when grown in mixture with species j. A common measure of competitive ability of species i relative to species j is its relative yield per plant:

$$RYP_{ij} = Y_{ij}/Y_{ii}.$$

An RYP_{ij} value greater than 1 means that species i grows better in mixture with species j than it does in monoculture. If n species are grown in all pairwise combinations then an $n \times n$ matrix of RYP values is obtained. A binary matrix A can be constructed in the following way:

$$a_{ij} = \{1 \; if \; RYP_{ij} > 1; 0 \; if \; RYP_{ij} < 1\}.$$

Such a matrix records which species are competitively superior to other species. Using graph theory, one can construct paths of increasing length; the length of a path is one less than the number of species included within it. Figure B4.1.1(a) shows a transitive path with three species. The path begins with species 1, passes through species 2 and ends at species 3. The path is transitive because knowledge that species 1 competitively excludes species 2, and that species 2 competitively excludes species 3, implies that species 1 also competitively excludes species 3. Thus the species can be ranked in a hierarchy from species 1, which is capable of excluding other species, to species 3, which is capable of excluding neither of the other two species.

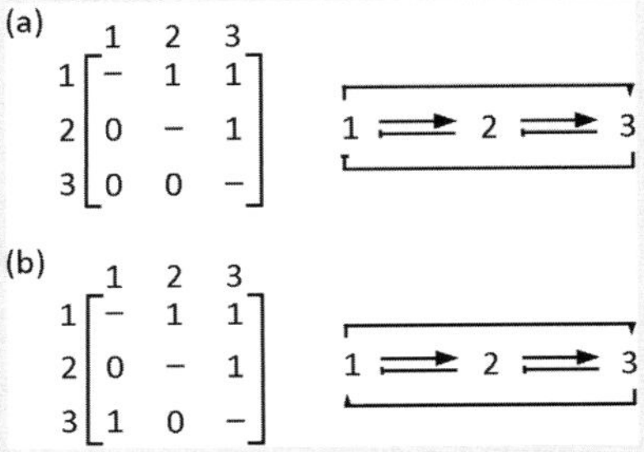

Note that there are also three transitive paths of length one, corresponding to the three cases of pairwise interactions. Figure B4.1.1(b) also shows a path of length two, but this time the path is intransitive, since although species 1 excludes species 2, and species 2 excludes species 3, species 1 does not exclude species 3. Table B4.1.1 gives three selected examples of this approach to quantifying and testing for hierarchies.

FIGURE B4.1.1   Representation of competitive hierarchies as graphs. (a) Transitive: a binary matrix and digraph with a transitive path of length two; (b) Intransitive: a binary matrix and digraph with an intransitive path of length two (from Keddy and Shipley 1989).

Box 4.1   (cont.)

Table B4.1.1 Three selected examples

L	T(L)	p	μ	m(L)	Probability ($\geq m(L)$)
Wilson and Keddy (1986a): $S = 7$, $\theta = 26/42$					
1	21	4.8×10^{-1}	4.95	16	6.6×10^{-3}
2	35	2.6×10^{-2}	0.46	16	1.3×10^{-16}
3	35	6.3×10^{-3}	0.11	7	2.0×10^{-9}
4	21	1.5×10^{-3}	0.02	1	3.0×10^{-2}
Mitchley and Grubb (1986): $S = 6$, $\theta = 19/30$					
1	15	4.6×10^{-1}	3.48	11	3.3×10^{-2}
2	20	2.6×10^{-2}	0.25	5	1.1×10^{-4}
3	15	5.8×10^{-3}	0.04	1	8.4×10^{-2}
Goldsmith (1978): $S = 13$, $\theta = 86/156$					
1	78	5.0×10^{-1}	19.29	56	$<5.0 \times 10^{-7}$
2	286	3.0×10^{-2}	4.33	90	$<5.0 \times 10^{-7}$
3	715	7.4×10^{-3}	2.68	98	$<5.0 \times 10^{-7}$
4	1287	1.9×10^{-3}	1.19	60	$<5.0 \times 10^{-7}$
5	1716	4.6×10^{-4}	0.39	19	$<5.0 \times 10^{-7}$
6	1716	1.1×10^{-4}	0.10	2	1.6×10^{-4}

The second test involved 20 species of wetland plants, grown from seed, which interacted for one growing season. The species again ranged from large aggressive species (e.g. *Typha angustifolia*) to small, rare rosette species (e.g. *Sabatia kennedyana*). Performance in infertile conditions was again strongly correlated with performance in fertile conditions (Figure 4.14).

A third test used species from different terrestrial habitats: 63 species of herbaceous plants from old fields, rock barrens and alvars in eastern Ontario (Keddy et al. 2002). In this case, the reference species was an annual plant typical of disturbed habitats. The plant species chosen were differently distributed along a gradient of soil depth and water availability, and the experimental treatments consisted of large, well-watered versus small, droughted

pots. The relative competitive performance of the plants was based on their ability to suppress the reference species. Figure 4.15 shows that competitive performance in large and small pots was strongly correlated.

4.4.3 Light and Shoot Size as Key Factors Producing Hierarchies

One of the basic characteristics of plants is their dependence upon sunlight. If two plants are neighbours and one plant is slightly taller than the other, then there are immediate consequences. The taller plant intercepts more light, as a consequence of its height, and is thereby further enabled to grow; simultaneously, it deprives the shorter plant of some photons and thus inhibits the growth of the shorter

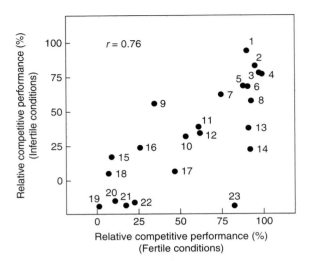

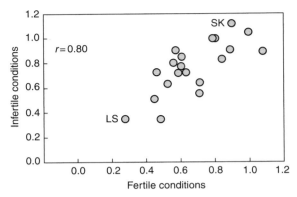

FIGURE 4.13 Competitive performance of 23 wetland plant species under low-nutrient conditions plotted against competitive performance under high-nutrient conditions. Adult plants were collected and grown together for two growing seasons. (1) *Lythrum salicaria*; (2) *Phalaris arundinacea*; (3) *Phragmites communis*; (4) *Typha x glauca*; (5) *Euthamia galetorum*; (6) *Mentha arvensis*; (7) *Acorus calamus*; (8) *Spartina pectinata*; (9) *Juncus filiformis*; (10) *Lysimachia terrestris*; (11) *Eleocharis palustris*; (12) *Eleocharis calva*; (13) *Sparganium eurycarpum*; (14) *Scirpus fluviatilis*; (15) *Triadenum fraseri*; (16) *Dulichium arundinaceum*; (17) *Carex crinita*; (18) *Viola lanceolata*; (19) *Scirpus torreyi*; (20) *Juncus pelocarpus*; (21) *Eleocharis acicularis*; (22) *Hypericum ellipticum*; (23) *Rumex verticillatus*. (From Keddy 2001 after Gaudet 1993)

FIGURE 4.14 Competitive performance of 20 species of wetland plants grown from seed in infertile conditions plotted against their competitive performance in fertile conditions. Note that the greater the competitive effect, the lower the position on the axes, since these values reflect the size of the three indicator species grown with each of the 20 test species. At the bottom left is *Lythrum salicaria* (LS), a fast-growing invasive weed, and at the very top is *Sabatia kennedyana* (SK), a rare species that in Canada is restricted to southwest Nova Scotia. (From Keddy et al. 1994)

plant. This sets up two positive-feedback loops (Figure 4.6) that can often lead to increased success for the large plant and declining success for the small one. Short plants simply do not shade tall ones to the same extent that tall ones shade short ones. Competitive hierarchies may therefore be inevitable consequences of differences in size (Weaver and Clements 1929; Weiner and Thomas 1986; Keddy and Shipley 1989). Interestingly, allometric studies of fossils reveal an exponential rate of increase in plant height during the Devonian era (Niklas 1994). This exponential rate of increase would be consistent with increasing competition for light as a result of increased terrestrial plant biomass.

Of course, plants do not compete only for light; nor do they interact only by shading. Access to light, however, can control a plant's ability to acquire other resources. Light is necessary both for constructing roots and for the physiological processes of nutrient uptake. Plants with more access to light will therefore be better able to forage for nutrients than plants with less access to light. Plants that are shaded are not only denied the photons necessary for constructing above-ground parts, but, simultaneously, their access to other raw materials is reduced and their growth is further restricted. Minor differences in height, among plants, can have a major effect on both the quantity and quality of the light available.

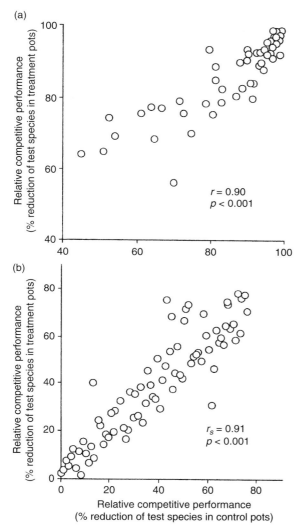

FIGURE 4.15 Competitive performance of 63 species from old fields, alvars and rock barrens of eastern Ontario in a stress treatment (small pots, drought) plotted against competitive performance in the control treatment (large pots, well watered). (a) Shows competitive performance measured as actual percent reduction in the phytometer and (b) shows only rank order (position in competitive hierarchy). (From Keddy et al. 2002)

Fitter and Hay (1983) have shown that light availability declines exponentially with distance below the top of the canopy; also, the red to far-red ratio declines below the canopy. Aware that light is an essential resource for plants, Weiner (1986) designed an elegant experiment to compare above- and below-ground competition: he found that competition for light was asymmetric, whereas competition for nutrients was symmetric. Other relevant examples are discussed in Keddy and Shipley (1989).

If asymmetric interactions for light determine a plant's position in a competitive hierarchy, then plant height should be significantly correlated with position in the hierarchy. Clements (1933) summarized the results of hundreds of transplant and removal experiments done with prairie vegetation (for example, Clements et al. 1929; Weaver and Clements 1929) and concluded that, in general, "the taller grasses enjoyed a decisive advantage over the shorter." Goldsmith (1978) studied sea-cliff plants and showed that the larger species suppressed the smaller (Figure 4.16). Wilson and Keddy (1986a) experimentally derived a competitive hierarchy for seven shoreline species. The dominants were tall species, whereas the subordinate was a small, rosette species. Keddy and Shipley (1989) showed that more than one-third of the competitive ability of these species, when grown in mixture, could be predicted based on knowledge of their heights ($r^2 = 0.37$). Similarly, in a chalk grassland study, Mitchley and Grubb (1986) derived a dominance hierarchy for six plant species. They found a significant correlation between position in the hierarchy and mean turf height of the species in monoculture; they noted that "the plants with the tallest leaves were the most effective interference." Mitchley (1988) has since shown that there is a positive correlation between the height of grassland species and their relative abundance.

Since pairwise designs increase in size by the square of the number of species examined, there are upper limits upon the number of species that can be studied if one wishes to relate traits to competitive ability. To overcome this problem, Gaudet and Keddy (1988) used a modified additive design that incorporated 44 wetland plant species that differed greatly in size, morphology and other traits. To measure

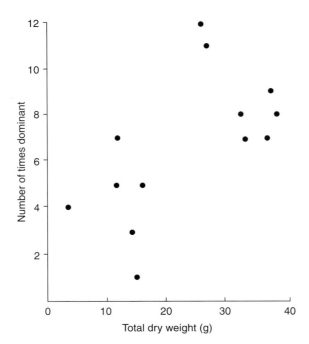

FIGURE 4.16 The relationship between position in the competitive hierarchy (measured as number of times dominant) plotted against plant size in monoculture for sea-cliff plants (data from Goldsmith 1978). Large plants suppress small ones, but the relationship is less clear for plants that are very similar in size ($r = 0.61$; $P < 0.05$). (From Keddy 2001)

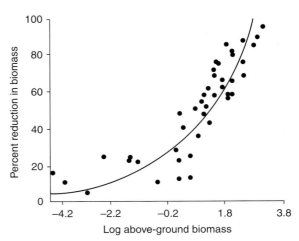

FIGURE 4.17 Screening for competitive ability in wetland plants. Percentage reduction in the biomass of *Lythrum salicaria* (when grown with different neighbours) plotted against the mean above-ground biomass of each of the neighbours (44 species, $n = 5$ replicates, $y = \exp(3.34 + 0.44x)$, $r^2 = 0.69$). The points on the left represent small evergreen species such as *Ranunculus reptans* and *Lobelia dortmanna* whereas points on the right represent large, leafy species such as *Typha latifolia* and *Lythrum salicaria*. (Data from Gaudet and Keddy 1988)

competitive ability, each plant species was grown with a reference species or phytometer (*Lythrum salicaria*), and competitive ability was measured as the ability to suppress this phytometer. They showed that simple traits such as biomass, height and canopy diameter could account for 74 percent of the measured

competitive ability. Above-ground biomass was the best predictor ($r^2 = 0.43$). A subset of the species was tested against a different phytometer and similar results were obtained. Figure 4.17 shows the relationship between the percentage reduction in the phytometer and above-ground biomass for the 44 species.

4.5 Competition Gradients Are Widespread

4.5.1 Measuring Competition Intensity Along Gradients

The intensity of competition may be one of the most fundamental properties of a community. Techniques

for measuring competition intensity have tended to view vegetation as aggregations of populations and consequently have tried to dissect these communities by studying pairwise interactions among species. In natural plant and animal communities, however, each

(a)   (b)

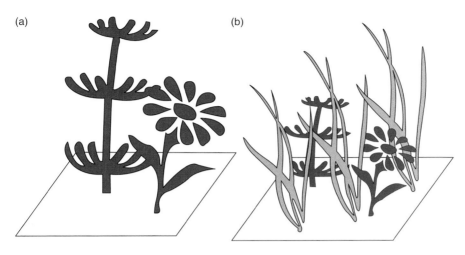

FIGURE 4.18 Indicator species, or phytometers, can be used to measure competition intensity. In this example, two test species are transplanted into a clearing (a) and intact vegetation (b). The difference in biomass of the test species after a period of time is directly proportional to the intensity of competition in the intact vegetation (courtesy of C. Gaudet).

species present experiences the cumulative effects of all neighbours more or less simultaneously (Keddy 1989; Grace 1993). In such cases, each species is said to be experiencing "diffuse competition." I prefer, and use in this section, the broader term "competition intensity" for the combined effects of neighbours; diffuse competition and predominant competition are then two ends of a continuum (Keddy 2001).

Competition intensity is measured by comparing the performance of a control organism (all neighbours present) with that of a treatment (all neighbours absent) (Figure 4.18). If the removal of all neighbours has no effect on performance, competition intensity is not detectable. The greater the difference between treatment and control, the greater the intensity of competition. This technique borrows from Clements' early work on phytometers (Weaver and Clements 1929; Clements 1935). Transplanted test individuals, he argued, were more sensitive than any environmental measurements taken at a site. Of course, it is probable that the removal of neighbours from plant communities will lead to measurable increases in light, soil moisture, and nitrogen and phosphorus levels. While measurement of these factors would be helpful for developing an

understanding of the possible mechanisms of competition, it is only by allowing one or more test plants to consume these resources that one can determine just how much performance has been reduced by surrounding neighbours. If the experimental design in Figure 4.18 is repeated in a series of habitats, it then becomes possible to test whether there is a gradient of competition intensity along the series of habitats.

4.5.2 Competition Intensity Gradients in an Old Field

Most natural gradients include a set of correlated environmental factors. In old fields, for example, depressions accumulate water and nutrients while ridges tend, by comparison, to be dry and infertile. Hawkweed (*Hieracium floribundum*) is a rosette plant that is frequently found in dry and infertile fields and pastures. Reader and Best (1989) studied the performance of this plant using demographic variables: survival, recruitment and population growth. Rather than studying a series of sites along the topographic gradient, they examined the extremes of the gradient, the tops and bottoms of depressions. Depression bottoms had deeper soil, more nitrogen and

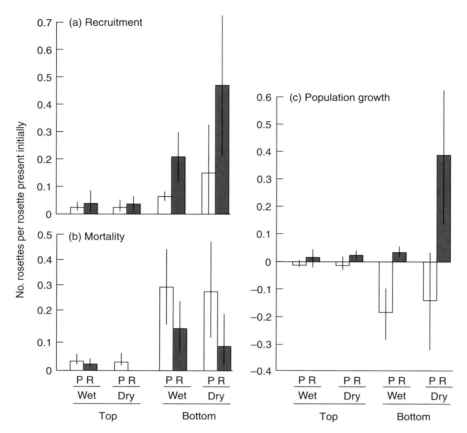

FIGURE 4.19 Mean (±1 SD) of (a) recruitment, (b) mortality and (c) population growth (i.e. recruitment minus mortality) for *Hieracium floribundum* with neighbours present (P) or removed (R) in subplots watered frequently (wet) or infrequently (dry) at the top or bottom of depressions. (From Reader and Best 1989)

phosphorus, higher above-ground biomass and lower light levels. In each habitat, the performance of *H. floribundum* was measured with and without neighbours. Only in the bottoms of depressions did competition have a significant effect on performance (Figure 4.19). Another element of this experiment was the addition of water to increase the availability of a possibly limiting resource (watered treatments are designated "wet" in Figure 4.19). Watering increased recruitment of rosettes significantly only in the bottoms of depressions. In contrast, watering had no effect on mortality. Overall, population growth was affected directly only in depression bottoms by removing neighbours and by increasing the availability of water.

Reader and Best (1989) conclude that the prostrate growth form of *H. floribundum* makes it particularly sensitive to shading. The sensitivity of rosette species to environmental gradients has been documented in habitats ranging from British grasslands (e.g. Grime 1973a, b) to North American wetlands (e.g. Keddy 1983). The study by Reader and Best (1989) illustrates several distinctive approaches to the study of competition. It used demographic variables rather than biomass to measure performance, contrasted two sites rather than a complete gradient of sites, experimentally supplemented a resource, used existing plants rather than transplanted individuals, and used a test species with a distinctive and widespread growth form.

4.5.3 Competition and Cacti

Cacti can occur in grasslands as well as deserts, particularly in arid, exposed sites with low plant cover. Burger and Louda (1995) used a factorial design experiment to test whether competition with grasses is a factor controlling the distribution of *Opuntia fragilis*, in the Sandhills prairie of western Nebraska. The experiment had plots with and without neighbour vegetation that were watered and unwatered. In each plot the performance of *O. fragilis* and damage from insects were assessed for two years. Performance of the cactus was measured by counting the number of cladodes, or pads, that comprise a plant. Cactus plants without neighbours had nearly three times the number of cladodes, which shows clearly the effects of competition, but the addition of water did not have an effect on competition. Finally, herbivore loading was higher with neighbours.

Larvae of the stem-feeding moth, *Melitara dentate*, and the weevil, *Gerstaeckeria* sp., were the major causes of cladode death, and these higher death rates may also exclude *O. fragilis* from areas of dense grass. In this case, then, the effect of herbivores appears to enhance rather than counteract those of competition with grasses in restricting the distribution of *O. fragilis*.

4.5.4 Competition Intensity Along a Soil Depth Gradient

Soil depth is a useful natural gradient on which to base a study of plant communities. Not only is soil depth easily measured, but all major soil resources (e.g. water, nutrients) are correlated with soil volume. This simplifies the complexity of multiple measurements of soil resources. Belcher et al. (1995) therefore sought out a plant community in an area where shallow soil was a predominant factor limiting growth – an alvar. An alvar is a terrestrial vegetation type that forms on thin soil over limestone (Petterson 1965; Catling et al. 1975). Alvars are similar to the cedar glades of the United States (Baskin and Baskin 1985) and are most abundant in the Baltic area of

Europe and the Great Lakes basin of North America. Along the soil depth gradient of an alvar, one may find bare limestone with shallow soil-filled depressions, grassy meadows and even mixed forest. The meadows and depressions support a rich and varied flora. In an experiment on competition in alvar vegetation, Belcher et al. (1995) examined the intensity of total, root and shoot competition along the soil depth gradient. Figure 4.20 shows that above-ground biomass was strongly correlated with soil depth ($r^2 = 0.65$, $P < 0.001$), which indicates that as soil resources increased, light availability decreased. Phytometers (the annual plant, *Trichostemma brachiatum*, in the mint family) were transplanted into areas without neighbours, with neighbours' roots only (the neighbour shoots were held aside by netting), and with neighbours' roots and shoots, in order to determine competition intensity. Figure 4.21 shows that the intensity of both total and root competition were significantly greater than zero ($P < 0.05$); intensity of shoot competition was not. Competition in this system

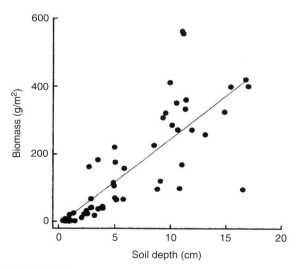

FIGURE 4.20 The relationship between above-ground biomass and soil depth within experimental plots in alvar vegetation. The relationship is described by the linear equation: biomass = 9.85 + 25.29 (soil depth) ($r^2 = 0.65$, $P << 0.001$, $n = 55$). (From Belcher et al. 1995)

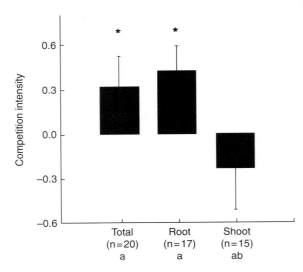

FIGURE 4.21 Mean competition intensity measured by phytometer performance for total, root and shoot competition in an alvar plant community. Bars indicate + 1 SE (sample sizes are shown in brackets). * Competition intensity was greater than 0 ($P <0.05$). Means sharing the same letter are not significantly different ($P >0.05$). (From Belcher et al. 1995)

was primarily below ground. Moreover, competition intensity did not vary significantly along the soil depth gradient.

4.5.5 Competition Intensity Gradients in Wetlands

Soil fertility and primary production are closely linked in plant communities, and resource gradients are of broad interest to community ecologists. Some vegetation types may have only narrow fertility gradients, and may therefore not be suitable for experiments involving competition intensity gradients. In wetlands, where sand dunes or sand bars grade into sediment-rich bays, a rare opportunity exists to study long, natural resource gradients. Twolan-Strutt and Keddy (1996) measured total competition intensity and its above- and below-ground components in two wetlands that represented extremes in habitat

productivity: an infertile sandy shoreline and a fertile bay. Transplants of *Lythrum salicaria* and *Carex critina* were grown with no neighbours, with roots of neighbours only, and with roots and shoots of neighbours; their growth rates during the study were used to measure competition intensity. The experiment was carried out to answer the following main questions:

1. Is there a difference in total, above- or below-ground competition intensity in two wetlands that differ in standing crop?

2. Does standing crop have an effect on total, above- or below-ground competition intensity when the data from the two wetlands are combined?

Results using both transplanted species show that total and above-ground competition intensity were greater in the high-standing crop wetland, but below-ground competition did not differ between wetlands (Figure 4.22). Moreover, when all experimental plots were examined independently, the two study sites overlapped enough to construct a resource gradient. In Figure 4.22 standing crop is used as the measure of site fertility, but independent measurements of soil nitrogen and phosphorus confirm this pattern. The top graph shows a marked gradient of competition intensity; results are near zero on sandy sites (left) and reach a maximum of about 0.6 in fertile bays (right): in fertile bays, there is a 60% reduction in growth rate owing to the presence of neighbours.

4.5.6 Competition Along an Altitudinal Gradient

Although some terrestrial experiments appear to contradict results from wetlands, a potential problem with most terrestrial experiments is that old fields are used as an experimental system. Old fields are a relatively new type of habitat created, in many cases, only a few hundred years ago and containing a mixture of native and exotic plants. Old fields might therefore not exhibit patterns found in native plant communities where species may have coevolved for thousands or even millions of years. Further, the patterns in such communities are largely ones of small-scale heterogeneity rather than of longer environmental

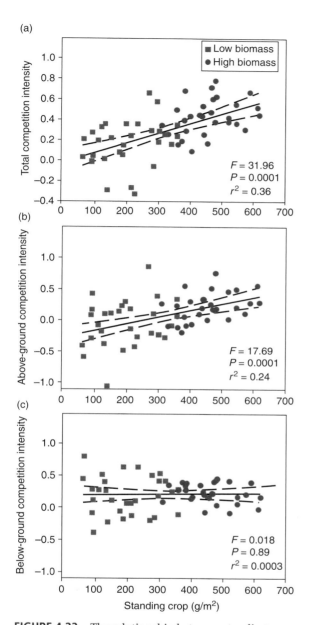

FIGURE 4.22   The relationship between standing crop and (a) total competition intensity, (b) above-ground competition intensity and (c) below-ground competition intensity for a marsh. Broken lines give 95% confidence bands. (From Twolan-Strutt and Keddy 1996)

gradients. The following example is a study of competition intensity that avoids such problems – a study of altitudinal zonation in the Snowy Mountains of Australia. These mountains have a zonation pattern typical of many altitudinal gradients: trees giving way to heathland giving way to grassland and alpine meadows. In the specific case of the Snowy Mountains, the tree line is formed by *Eucalyptus pauciflora*, which gives way to one-metre-tall heathland dominated by *Phebalium ovatifolium*, and at still higher altitudes is replaced by tall alpine herbfield, a grassland dominated by *Poa costiniana*.

Wilson (1993) measured competition intensity in the two higher elevation zones using three different species: the tree *E. pauciflora*, the grass *P. costiniana* and a rosette composite that is abundant at higher altitudes, *Celmisia longifolia*. Each species was grown in both vegetation zones under three conditions: plots entirely cleared with herbicide (NN, no neighbours), plots where neighbours were present but the above-ground parts were tied back so that only roots could interact (RN, roots of neighbours present), and natural vegetation (AN, all neighbours present). After two growing seasons, the transplanted individuals were harvested and above-ground growth rate was calculated. The effects of above-ground competition (the difference between treatments RN and AN) were typically small and did not vary between vegetation types (Figure 4.23). In contrast, below-ground competition (the difference between treatments NN and RN) was always significant and differed among the three species and between the two vegetation types.

The rosette plant, *Celmisia*, was least suppressed, and the woody plant, *Eucalyptus*, most suppressed by below-ground competition. Wilson concluded that, in these habitats, below-ground competition was more important than above-ground and that it was greater in the higher elevation grassland. Further, the ability of species to compete for below-ground resources (*Celmisia* > *Poa* > *Eucalyptus*) was the same as their distribution along the altitudinal gradient. Thus, as in Twolan-Strutt and Keddy (1996), competition intensity

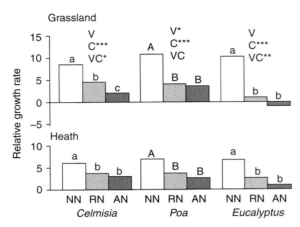

FIGURE 4.23 Growth rates of transplants of three species grown in alpine grassland and heath in three competition treatments: no neighbours present (NN), roots of neighbours present (RN) and all neighbours present (AN). The results of two-factor ANOVA are shown for each species with vegetation type (V) and competition treatment (C) as main effects (* P <0.05, ** P <0.01, *** P <0.001). Means with different lower-case letters are significantly different (P <0.05) among competition treatments within each vegetation type for each species. Means with different upper-case letters are significantly different among competition treatments across both vegetation types for each species. (From Wilson 1993)

differed between the two vegetation types. In the mountain vegetation, above-ground competition was constant and below-ground competition increased, whereas in the wetland, it was below-ground competition that was constant and the above-ground component that changed. Measurements of soil resources along the altitudinal gradient in the Snowy Mountains showed that although the grassland had intense below-ground competition, it had higher soil nitrogen levels, higher soil water levels and higher plant growth rates than the adjacent heathland.

Wilson's experiment is a model of the way in which competition experiments can be executed: it used a natural habitat, compared two major zones of an altitudinal gradient, incorporated three plant species of widely contrasting autecology, lasted two full growing seasons, partitioned competition into above- and below-ground components, and measured resources available to the organisms. There is but one limitation, and this is a feature of the analysis. Like so many competition studies, the effects of competition are partitioned into effects on a species-by-species basis. At no point does Wilson ask whether competition varies between the two zones irrespective of the test species used; note that in every case, the effects of competition seem slightly higher in grass-land but that breaking the data down into three species has two important consequences. First, it focuses one's attention upon the response of test species rather than the overall intensity of competition in a particular habitat. Second, and more importantly, it significantly reduces the number of degrees of freedom in the analysis and greatly reduces its power by splitting the data for between-habitat comparison into three groups. It is difficult to break free from a species-oriented view of nature.

4.6   Foraging Ability Might Be a Competitive Trait

The importance of size in determining competitive ability is well established (Section 4.4.3), but what other traits might control the competitive success of plants? One might suspect that a plant's ability to forage for resources is important. Early studies of resource uptake focused upon measures of the gross uptake of nutrients, as estimated by the relative growth rates of seedlings grown under standardized conditions (e.g. Grime and Hunt 1975; Grime 1979). Other studies of resource uptake have explored various factors thought to be significant in nutrient uptake such as early germination (Wilson 1988), root morphology (Boot 1989), shoot thrust (Campbell et al. 1992) and ability to exploit patches (Campbell et al. 1991).

4.7   Mycorrhizae Can Affect Competition

Might mycorrhizae influence the relative competitive ability of plants? We have already seen that mycorrhizae can increase resource uptake by roots (Section 3.4.3). Perhaps fungi can also transfer soil resources or carbohydrates from dominant plants to subordinate ones. As more experiments are done, it is becoming obvious that plants are linked underground in many complex ways. Here we will look a just three experiments that directly studied competition by comparing plants grown in cultures with and without mycorrhizae.

In the first example, seedlings of plantain (*Plantago lanceolata*) were grown either alone or with older plantain plants, and with and without mycorrhizae. The expectation was that the seedlings would suffer asymmetric competition when grown with the larger adults, and the issue to be tested was whether these effects would be smaller when the seedlings were infected with mycorrhizal fungi (Eissenstat and Newman 1990). To measure competitive performance, three characteristics of seedlings were measured: biomass, phosphorus content and nitrogen content. There were no significant benefits to seedlings when infected with fungi (Figure 4.24). In fact, Eissenstat and Newman noted that "the concentration of nitrogen and phosphorus was higher in the seedlings than the large plants... so that net transport of nitrogen and phosphorus from large to small plants seems unlikely."

In a similar sort of experiment, but focused on interspecific competition, Grime et al. (1987) grew seedlings of 20 grassland plants in mixture with larger *Festuca ovina* plants, with and without mycorrhizae. Although the mycorrhizae had little impact on the performance of *F. ovina* seedlings competing with larger plants of the same species, the balance between *F. ovina* and seedlings of other species was affected. Their results suggest that mycorrhizae played a role in interspecific competition, but a minor role that did not change the relative position in a hierarchy. Similarly,

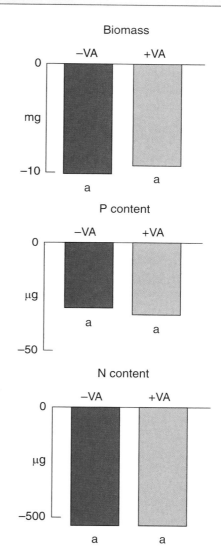

FIGURE 4.24   The reduction in biomass and nutrient content of shoots of plantain seedlings, grown with (+VA) and without (–VA) mycorrhizae, caused by competition with large plants. Means followed by the same letter are not significantly different at $P < 0.05$ (determined by competition × mycorrhizal interaction in a two-way ANOVA). (From Eissenstat and Newman 1990).

when two species of grass (*Holcus lanatus, Dactylus glomerata*) were grown in competition both with and without mycorrhizae, infection was found to increase plant weight, but any increases were offset by the higher intra- and interspecific competition that resulted from the plants being larger. As a consequence, when mycorrhizal and non-mycorrhizal treatments were compared, the benefits of mycorrhizal infection declined with increasing density (Watkinson and Freckleton 1997).

A survey of more than fifty such experiments found positive effects in half, negative in one quarter and no effect in the rest (van der Heijden and Horton 2009)! Thus generalizations remain elusive. It certainly seems that mycorrhizal networks can allow subordinates to persist in the presence of competition. Further work will, however, need to clearly distinguish between two quite different roles for mycorrhizae: they may simply allow small plants to establish and persist, or they may shift competitive relationships overall (or even both). To use an analogy from economics, mycorrhizae may act as a kind of welfare, but they do not so far appear able to alter class structure. But of course the above experiments, while difficult, are still rather simple. As you add in more species of plants and more species of fungi, the number of possible interactions increases exponentially, and one is forced to think about how multispecies networks behave (Allen and Allen 1990; van der Heijden and Horton 2009). But this takes us far beyond the topic of competition. We will return to mycorrhizae, however, in the chapter on positive interactions.

4.8   Two Competition Models

4.8.1 The Problem of Coexistence

Given the intensity of competition in plant communities, and given the tendency of a few large species to exclude many weaker ones, one could reasonably ask how weaker competitors survive at all. Why has competition not already driven most weak competitors to extinction? Perhaps it has. There are many, many species known only from the fossil record. But the world still has more than 300,000 plant species, so clearly not all have fallen prey to superior competitors. What are the possible explanations for the survival of weaker species?

There are two answers that depend primarily upon ecological time scales.

(1) Perhaps in certain cases, competition is not hierarchical but nearly symmetric, leaving no way for any single species to dominate.

(2) Perhaps weaker species can escape competition in space, by finding a habitat without the dominant competitor.

Let us begin with option (1) above. Very similar species, perhaps in some communities, such as tropical rain forest where there are thousands of species of trees, really are similar enough that no one tree species is able to dominate the others (Hubbell and Foster 1986). This option, of symmetric competition and coexistence is shown on the left side of Figure 4.25. But this explanation is not as easy as you might think, because even in the Amazon basin, where there are more than 11,000 tree species, there are still a few that are rather common and a great many that are quite rare (Hubbell et al. 2008). How are all these relatively rare species coexisting with one another, as well as with the more common species?

Many tree species depend upon light gaps for regeneration (you shall see some examples in Section 5.6 and Section 9.4.2). A light gap provides a way of escaping competition. More generally, then, it may be that some weak competitors continually evade extinction by dispersing to new habitats ahead of the competitors, that is, by finding a clearing of some sort. They complete their life cycle before dominant species invade and kill the weaker species. This takes us back to an older article yet, where Hutchinson (1961) asked about "the paradox of the plankton": how do so many

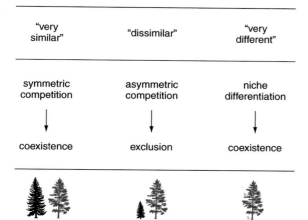

FIGURE 4.25 Three modes of coexistence in spite of competition. On the left side, competition is symmetric, so both species survive indefinitely. In the middle, there are competitive hierarchies, but the weaker species manages to disperse into gaps not yet occupied by the dominant. On the right, the subordinate species is driven to evolve away from the niche of the dominant, and occupy a new type of habitat entirely. (Adapted from Figure 6.12 in Keddy 2001, where the same principle is shown with barnacles).

species of plankton coexist in a single lake? Likely, a lake is heterogenous enough in space, and changing enough in time, that no single species can dominate.

It is relative easy to extend Hutchinson's logic to terrestrial vegetation types (Huston 1979). Hence entire books have been written about the creation of cleared patches and their implications for plant communities (Pickett and White 1985). From this perspective a clearing in vegetation provides a local opportunity to gather resources with less competition from established individuals. It may be the result of some kind of natural disturbance, which we shall consider in the next chapter. Individual fallen trees, patches of storm damage and fires are but three examples. Hence weaker competitors may survive by finding a location in time or space where they can grow. In this view of nature, there are competitive

hierarchies balanced by escape. This option is shown in the middle of Figure 4.25.

There is a third possible explanation for coexistence, but it depends upon evolution leading to niche differentiation. Thus it demands thinking at evolutionary time scales.

(3) Perhaps weaker species can escape competition in space through natural selection, by using different resources from the dominant competitor. (Or, to be more precise with language, perhaps the individuals that are most similar to the dominant die, or at least leave fewer offspring, leading to the proliferation of those individuals most different from the dominant.)

The logic here is that because competition was important in the past, species now have different habitat requirements. Borrowing mostly from zoologists, this explanation is termed "niche differentiation." From this perspective, plants may escape competition by occupying different ecological niches. This option is shown on the right side of Figure 4.25. Here one of the species has escaped from a competitive dominant by becoming different: a sciophyte and epiphyte or a parasite – each are a kind of escape from competition with the dominant. This explanation is popular because it provides a ready answer for why plant species are different from one another. The principal problem is that it assumes that competition in the past produced the patterns we see today. There is no easy way to test for competition in the past, leading Joseph Connell (1980) to paraphrase Charles Dickens and wryly observe that this explanation depends upon "the ghost of competition past."

It is likely that all three of these mechanisms occur. The problem is knowing which one applies to a particular habitat or set of species.

4.8.2 Patch Dynamics: A Model

Others have already thought about competition and coexistence. More than 50 years ago, well before Hutchinson, Skellam (1951) showed that weak competitors can indeed survive indefinitely – as long as they can disperse better than strong competitors. The argument goes like this (Pielou 1975). Imagine two competing species that reproduce once a year. Let A be the stronger competitor and B the weaker competitor. Wherever they coexist, A invariably wins. Therefore the only habitat in which B can reproduce includes those sites in which it occurs alone (Figure 4.26). Assume that the landscape has N sites, or patches of habitat, and that at equilibrium the expected proportion of sites with a single A individual at the end of the growing season is Q. This means that NQ of the sites are dominated by species A. Therefore only $N(1 - Q)$ remain for B to occupy. If we call this remaining portion of sites (those that allow B to survive) q, then q must be greater than zero for the competitive subordinate to survive in that landscape. We want to know how much better dispersal of B must be for this to occur. Therefore let F and f be the number of seeds produced by species A and B, respectively. For species B to persist, f/F must be great enough to ensure that $q > 0$. It can be shown that, for this to occur, f/F must exceed $-Q/[(1 - Q)\ln(1 - Q)]$. Provided this condition is met, species B will continue to occur in the landscape in spite of its weak competitive ability. To put it in other words, as long as there are enough disturbances, and as long as some weak competitors have good dispersal, they may coexist indefinitely.

This seems like a good time for an observation on the topic of learning ecology. One can easily be exhausted by the amount of new work appearing. On the topic of competition and disturbance, it is indeed remarkable to see how many other scholarly papers have been written on this topic, and how many new models are being trumpeted to younger readers. Many contain no reference to this most basic and convincing model. If you feel inundated by new work, this may be a sober reminder – just because something is "new" does not mean it is necessarily "better." Indeed, for students, it is often best to begin with classic studies, if only to provide a skeptical basis for evaluating new claims. One should not discount Skellam's model simply because he did it more than 50 years ago.

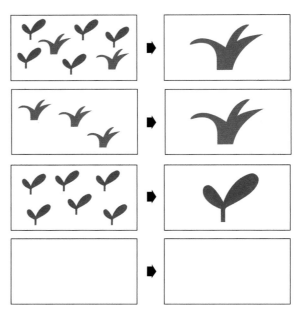

FIGURE 4.26 Competition in a patchy environment. Weak competitors (green) may survive by escaping to habitat patches that are not occupied by stronger species (brown). Four possible combinations of seedlings are shown on the left, and the outcome of adults is given on the right. (From Pielou 1975 after Skellam 1951)

The entire next chapter (Chapter 5 Disturbance) will return to disturbance, which may indeed be one of the great forces counterbalancing competition. We shall see many more examples of species dispersing into patches. Further ahead (Chapter 9 Populations) we shall encounter a large number of plant species with extremely small seeds, termed dust seeds (Figure 9.9). One potential reason for producing vast numbers of very small seeds is to provide escape to new patches ahead of competitors. Dust seeds are particularly common in orchids and bromeliads, both of which are common as epiphytes. Finding new unoccupied tree branches might be a very good way of escaping competition from neighbouring epiphytes. Overall, it seems disturbance and dispersal are rather important for escaping the negative effects of established neighbours. Indeed you could even extend this line of argument to the long-term survival of ancient plant groups such as ferns and mosses: their dispersal by means of tiny spores allows them to stay one step ahead of plants that must disperse by means of seed.

4.8.3 Gradients and Zonation: A Model

Talking about overlooked models, here is another that is directly relevant to plants. Instead of dispersing to a new patch of habitat ahead of a competitor, it is possible for species to share space along environmental gradients. You have already seen in Section 4.5 some of the many studies of environmental gradients in nature, and a full chapter on the topic lies ahead (Chapter 11). I have argued elsewhere that a gradient in ecology is like a prism in optics – it provides vital insight into underlying causes. Here let us look at a familiar model, the Lotka–Volterra model for competition, extended for a natural gradient. I am going to assume that you have already encountered this model elsewhere, but you are unlikely to have seen this extension. If you are unfamiliar with Lotka–Volterra, then either take a break to read about it, or follow along as best you can. Pielou challenges us to take the two-dimensional model and expand it to a third (Figure 4.27). The two dimensions, the axes you see on the surface of the page, are the abundances of two plant species, shown in the Lotka–Volterra framework. In this graph, each point in two-dimensional space gives the abundance of species A and B, from zero to K (carrying capacity). The lines, known as isoclines, indicate where one or the other species has zero growth rate. Given time, one of two things will happen. In the upper illustration, the pair of species will interact in such a way that either one, or the other, wins and excludes the other. That is to say, as the arrows show, the system will naturally move toward one species at K, the other at zero. That is, no coexistence. In certain situations, as the lower illustration shows, the two species will reach an equilibrium stable point of some mixture of population sizes where neither can exclude the other. This stable point, however much it may be a mathematical abstraction, has been of great interest to theoreticians as it allows for two species to coexist in spite of competition between them. However, this model, in its

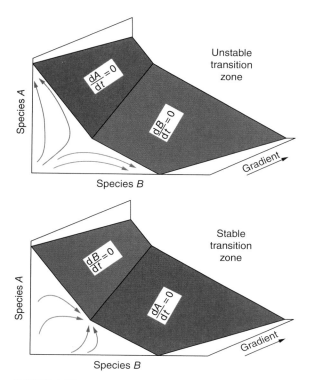

FIGURE 4.27 Competition between two species along an environmental gradient. This uses the familiar Lotka–Volterra model extended to a third dimension. Both possible outcomes yield two species occupying different locations along the gradient. (After Pielou 1975)

two-dimensional form, inherently assumes a single homogeneous environment. Such environments exist only in the minds of mathematicians, and while they may simplify the mathematics, they complicate the ecology, by asking ecologists to image a situation that simply does not occur in nature. Here is where Pielou turns this approach on its head by asking the obvious. Since nature has gradients everywhere, why not ask how this model would respond to a third dimension, one in which the competitive relationships change along a gradient. This is the black arrow going into the page. The colour surfaces now become planes where one or the other species has a zero growth rate. Thus in the upper diagram, a mixture of two species slides through time to an outcome with K,0 – one species has locally exterminated the other. But, and this is the important point, which one wins, and the trajectory they take to reach this point, depends upon where you are along a gradient. Hence, so long as you add in a gradient, it is possible for two species to coexist, since some locations will have K,0 as the outcome, while others will have 0,K. In the lower diagram, there is that stable equilibrium point, but its location drifts (along where the two plains intersect) in response to the gradient. It just depends where you are. In either case, you obtain a gradient of species composition with two zones occupied by two species, the sort of thing you see in nature all the time. There are other nuances in these outcomes, and you may wish to set some time aside for a proper exploration of the mathematics (Pielou 1975, pp. 90–99) and more on the ecology (Keddy 2001, pp. 351–354).

The Pielou gradient version of Lotka–Volterra is significant for several reasons. First, it provides an alternative way of looking at the world – as a gradient, instead of the patches we used in Figure 4.26. Second, it takes a familiar model and expands it in a novel way. And, third, it suggests outcomes that appear consistent with the natural world. All of which leads us to the topic of models altogether. Most realms of ecology have models in them. You will encounter at least one in most chapters of this book. But just what is the purpose of models? We should think carefully about this. There is a good deal of confusion out there, so do not be surprised if you feel confused when reading further. Much of the confusion arises because there are different types of models, with different purposes and there is no general agreement about how to describe them. (Indeed, some authors of models apparently do not know themselves!) It is overall much easier to make up a model than to test it. Hence our journals (and books) tend to fill up with models that only confuse newer students and exhaust more seasoned researchers. I once suggested a moratorium on publishing new models until the existing ones have been properly evaluated. After a moment of polite (or shocked) silence, the audience simply agreed to ignore my advice on this topic. So the next section will provide an overview of the role of models in ecology. You can apply this to any of the models you encounter in this book. Meanwhile Table 4.3 provides a snapshot guide to general models for competition to guide you in further exploration of this topic at your leisure.

Table 4.3 **There are many models for how competition might affect plant communities. Most have not been properly tested, and yet people keep inventing new ones. So where do you start? Probably with these six classics.**

Lotka–Volterra	Intra- and interspecific competition in two similar species. Widely found in introductory text books, but assumes two species and a homogeneous environment. (I like the presentation in MacArthur 1972.)
Patch dynamics	Patch dynamics and seed dispersal. Rarely seen in introductory books, but very sensible assumptions for sessile organisms. (See Section 4.8.2)
Gradients	Lotka–Volterra cleverly adjusted for two species along a gradient. Of great interest since zonation along gradients is widespread in plant communities. Almost ignored. (See Section 4.8.3)
Trees and light	Competitive hierarchies in forests. The shade-tolerant species should dominate, but the light-demanding species always keep one step ahead. Not unlike Skellam, in principle, but with far more data on tree biology (Botkin 1977).
Resources	Competition viewed through the depletion of resources, particularly light and nitrogen. Challenges users to think about mechanisms of nutrient uptake and depletion (Tilman 1982).
Biomechanics	Leaf height as a key factor in competition. There are costs to making taller shoots, and this has impacts on ecology and evolution of herbaceous plants (Givnish 1987).

Yes, there are many more. But these provide a foundation. I compare and contrast these models, with a few more, in Keddy (2001).

4.9   The Role of Models in Ecology

Models have many uses in ecology. You have already met some in the four chapters covered so far. More will come later. This is a good time to think about models in general. Overall, there are three main types of models, based upon their purpose: prediction, exploration and description.

1. **Predictive models** are designed to predict the future states of systems based on relationships specified within the model between predictor (independent) variables and the predicted (dependent) variables. The dependent and independent variables must be measurable. Success or failure of the model is easy to judge: the more accurate the prediction is, the better the model. Of course, it is also nice if the model is based on a few factors that are easy to measure. As an example, see Figure 3.11, where phosphorus concentrations predict phytoplankton abundance in lakes.

2. **Exploratory models** are quite different. They allow us to think through the logical consequences of certain assumptions about how the natural world works. That is, they are more of a thought experiment than an attempt at prediction. You have just seen such a model in Figure 4.27.

3. **Descriptive models** are used to summarize existing knowledge about the behaviour of a system. The picture may not be sufficiently complete to permit the model to be used to make predictions, but the summary can serve as a foundation for future work, or as a guide to possible experiments. Figure 1.4, Whittaker's summary of world vegetation types, is such a model. We shall see many more such models in Chapter 11 where the challenge is describing plant communities at different scales.

In designing models, an investigator is faced with many trade-offs. The more precise the model is made, the more it incorporates the details of a specific system, the greater the possibility that accurate prediction is possible. However, as the model is finely tuned to one situation, there is an inevitable loss of generality. The skill of the modeller determines the degree to which a model combines generality and accuracy. We will see many more models in this book, and all preserve certain parts of reality, while ignoring others. Any time you start trying to understand a model, you need to carefully assess what is being included and what is being ignored. Sometimes people who offer models are not entirely honest about this, or worse, will cavalierly ignore the important and include the trivial.

You should be aware that there are other classification systems that have their merits. Starfield and Bleloch (1986) give five reasons for constructing models: (1) to define problems, (2) to organize thoughts, (3) to understand data, (4) to communicate and test understanding, and (5) to make predictions. Pielou (1977) proposes that mathematical models can be classified into four criteria for a total of 16 possible styles of models. The criteria are (1) whether it treats time as continuous or discrete, (2) whether it is an analytical or a simulation model, (3) whether it is deterministic or stochastic, and (4) whether it is inductive (empirical) or deductive (theoretical).

Another very important classification of models comes from the realm of general systems theory (Weinberg 1975). Here we recognize small, medium and large number systems. These have very different properties, and therefore generate different approaches to modelling. Small number systems have a few components and few interactions, and these systems are amenable to precise mathematical description (say, Figure 4.27, two plants on one gradient). Large number systems lie at the other extreme, where there are so many components that the average behaviour becomes a useful description of the system (say Figure 1.4, all plants on Earth). Both types of systems are common in other scientific fields, such as physics and engineering. The problem in ecology is that ecosystems are usually neither of these. Instead, they are medium number systems. Medium number systems contain too many components to be treated analytically, and too few for gross statistical analysis. Usually some method of simplification is necessary in order to solve problems involving medium number systems. This requires carefully preserving critical interactions and components while excising or ignoring others. The inherent difficulty in doing so wisely probably explains why, at present, medium number systems require modelling approaches that are as much an art as a science. This takes us back to the three types of models offered above.

One final observation. Often, in ecology, we have some basic understanding about how a system works (say, its main components and their links to one another) but we have insufficient *useful* data with which to work. In such cases, ecologists and ecological models are presented with two daunting challenges. First, management decisions may have to be made despite the lack of data and understanding. How do we make good decisions under such circumstances? For some thoughtful opinions on this problem, you may wish to consult Holling (1978a), Rigler (1982) and Starfield and Bleloch (1986). Suffice it to say that many arguments in ecology arise because people differ in the kind of model they want, they disagree in how much complexity the model should contain, they disagree how they should measure success and they disagree how much data is needed to test the model. A model may even be an unhappy compromise among many competing interests that pleases no one! Yet, whether it is the future of forests with bark beetle populations (Section 6.7.1), or the future of individual species like the dragon's blood tree (Section 9.4.3) or the future of coastal wetlands with changing salinity (Section 13.2.3), models will always be there to guide thinking.

CONCLUSION

Given that populations can grow exponentially, whereas resources do not, most individuals are likely to be negatively affected by shortages of resources. Neighbours will usually further reduce resource levels and may also directly interfere with resource acquisition or growth. While competition for resources may be ubiquitous, the actual nature of competition (e.g. exploitation vs. interference; intraspecific vs. interspecific; monopolistic vs. diffuse; above-ground vs. below-ground) probably changes dramatically from site to site. Too many ecologists have asked "is there competition?" between a pair of selected species rather than asking what the nature of the competition is and how it varies along gradients (Keddy 1989). Soil resource gradients are widespread, in which case gradients in competition intensity, and gradients in the relative importance of above- and below-ground competition, are likely common. Models are an important tool for exploring how competition might structure plant populations and plant communities.

Review Questions

1. Explain how you use a removal experiment to test for the presence of competition.
2. Distinguish between symmetric and asymmetric competition. Give an example of each.
3. What is meant by the term competitive hierarchy? Describe the sorts of experiments that can be used to identify such hierarchies in plant communities.

4. According to the data from field experiments, how do above- and below-ground competition change along a gradient?

5. How might disturbance allow weak competitors to survive in landscapes? Use the Skellam model to explain. Can you think of any local species that might be using this mechanism (hint, look up the concept of ruderals in Section 10.8.6).

Further Reading

Clements, F.E., J.E. Weaver and H.C. Hanson. 1929. *Plant Competition*. Washington, D.C.: Carnegie Institution of Washington.

Tilman, D. 1982. *Resource Competition and Community Structure*. Princeton: Princeton University Press.

Watkinson, A.R. 1985. Plant responses to crowding. pp. 275–298. In J. White (ed.) *Studies in Plant Demography: A Festschrift for John L. Harper*. London: Academic Press.

Underwood, T. 1986. The analysis of competition by field experiments. pp. 240–268. In J. Kikkawa and D.J. Anderson (eds.) *Community Ecology. Pattern and Process*. Melbourne: Blackwell.

Gaudet, C.L. and P.A. Keddy. 1988. A comparative approach to predicting competitive ability from plant traits. *Nature* 334: 242–243.

Keddy, P.A. and B. Shipley. 1989. Competitive hierarchies in plant communities. *Oikos* 49: 234–241.

Grace J.B. and D. Tilman (eds.) 1990. *Perspectives on Plant Competition*. San Diego: Academic Press.

Wilson, S.D. 1993. Competition and resource availability in heath and grassland in the Snowy Mountains of Australia. *Journal of Ecology* 81: 445–451.

Keddy, P.A. 2001. *Competition*, 2nd edn. Dordrecht: Kluwer.

2012. Competition in plant communities. In D. Gibson (ed.) *Oxford Bibliographies Online: Ecology*. NY: Oxford University Press.

Bennett, J.A., E.G. Lamb, J.C. Hall, W.M. Cardinal-McTeague and J.F. Cahill. 2013. Increased competition does not lead to increased phylogenetic overdispersion in a native grassland. *Ecology Letters* 16: 1168–1176.

5

Disturbance

Definitions. Properties: duration, intensity, frequency, area. Fire. Erosion and deposition. Animals: beaver ponds and "gator" holes. Burial. Ice damage. Waves create chronic low-level disturbance. Catastrophic events include landslides, volcanoes and meteors. Two examples where the consequences of disturbance were measured. Gaps and gap dynamics. Buried seeds. Logging. Multiple disturbances in the Everglades. Broad-scale comparisons.

FIGURE 5.1 Fire is a natural disturbance that removes biomass from forests as well as causing shifts in species composition. (Caribbean pinelands of the Rio Platano Biosphere Reserve in Honduras, by Ronald Myers.)

5.1   Introduction: Disturbance Removes Biomass

In 1990, a bolt of lightning hit a red oak tree on the ridge near my home. It blasted much of that tree into metre-long splinters. *That* was a disturbance. Back in 1883, a volcano on the Island of Krakatau in Indonesia erupted explosively, stripping entire islands bare of forest cover. *That* was also a disturbance. In 1692 a forest fire burned through what is now the Boundary Waters Canoe Area on the border between the United States and Canada. And, yes, *that* was a disturbance too. Once you start to look for them, disturbances are really rather common. They include floods, fires, landslides, wind throws, tornadoes, hurricanes and volcanoes. So, if you live in a floodplain, or in a coniferous forest, you can be assured that sooner or later a flood or fire will come to you. Disturbance – it is only a matter of time.

Let us take a closer look at this widespread and inescapable factor that initiates, shapes and then ends plant communities. This topic follows naturally from the preceeding chapters on competition and resources. We have seen (1) that plants use a small number of resources to create new biomass, (2) that as biomass accumulates, resources – particularly light – are depleted, and (3) the result is intense competition. In principle, there should be competitive exclusion, with only those few species of plants best able to gather resources and exclude their neighbours covering any piece of landscape. Yet most habitats contain many species. Disturbance is likely the most important factor that prevents competitive exclusion. Disturbance kills plants, thereby releasing resources and making gaps within which weaker competitors may be able to survive.

Disturbance shall be defined here as a "short-lived event that causes a measurable change in the properties of an ecological community." One of the commonest properties to change is biomass. One of the commonest examples of disturbance is fire (Figure 5.1). The balance between the rate of disturbance and the rate of recovery determines the species composition of many, if not most, vegetation types. In general, disturbance is fast and recovery is slow.

This definition may seem unsatisfactory. What is short-lived? It may be useful to think in terms of the duration of plant life spans (Southwood 1977, 1988). Short-lived can be defined as an event that occurs as a pulse with duration much shorter than the life span of the dominant species in the community. According to this definition, a fire or one-year drought would be a disturbance; a slow and long-term climate change would not. Insisting upon measurable change further requires that a user of the word must identify at least one property that is measurable (e.g. biomass, diversity, species composition) and then show that it changes. No change, no disturbance (see Cairns 1980).

Another similar definition of disturbance has been offered by Grime (1977, 1979). He has suggested defining disturbance as simply a factor that removes biomass. White (1994) explains, "When the structural resistance and physiological tolerance of the vegetation is exceeded, substantial and sudden destruction of living biomass occurs; hence the recognition of... events as disturbances." This definition is satisfying because it is simple, and because it is clearly plant oriented. The reason I have offered a slightly different definition is to include factors such as soil disturbance that seem to have significant effects upon so many species of plants. Thus we might think about disturbance as having two components: the primary effect (removal of biomass) plus secondary effects (changes in the substrate). We will see many examples of secondary effects.

The word disturbance should not be used without some careful thought about just what you mean and how you intend to measure it. The word is dangerous precisely because the word is non-technical; therefore many people assume they understand it when they do not. To repeat the key points: a disturbance is short lived and it causes a measurable change in a property such as biomass. We shall look here at an array of examples to see how disturbance affects different plant species and ecological communities. As we shall see, complications can arise in applying the definition. Herbivores too can cause many kinds of disturbance, from zebras grazing on grasses to spruce budworms removing the canopy of entire forests. These too are disturbances, but they are dealt with in a separate chapter (the next one).

Disturbance has one other important consequence. It initiates the process of recovery. Many of you will know that this process of recovery is often called **succession**. Recovery (or succession, if you prefer) is usually a slow process, taking years or even centuries. It begins when plants start to produce biomass in the disturbed area. They may have survived the disturbance as buried seeds or they may arrive through dispersal. The dispersal may consist of vegetation slowly growing in from the edges, or from seeds carried by wind or animals. In this chapter we are going to focus mostly on the disturbance itself, as a factor that removes established vegetation, and less on the recovery process. We will return to recovery and succession later in the book, in Chapter 8, which deals with the importance of time in plant communities.

5.2   Disturbance Has Four Properties

5.2.1 Duration

Duration refers to how long the event lasts. A frost, fire or lava flow may last only hours; floods or herbivory may continue for days. The duration of an event may be expressed in terms of the life spans of the organisms of concern (Southwood 1977, 1988).

5.2.2 Intensity

The simplest measure of intensity would be the proportion of biomass at a site that is killed or removed. A factor that disturbs one group (e.g. trees) might not disturb another (e.g. geophytes), so the change in abundance of several groups might be measured simultaneously. Change in species composition would be an alternative measure of intensity. There is a wide range of measures of similarity between samples (Legendre and Legendre 1983). Using a standard measure of ecological similarity, one could define a range of disturbance intensities from 0 (the community is the same before and after the disturbance) to 1 (the community is completely different after the disturbance).

5.2.3 Frequency

Some events, such as hurricanes or spring floods, happen on a yearly basis. Others, such as ice storms or asteroid collisions, happen rarely. In general, the greater the intensity of the disturbance, the lower the frequency. It seems reasonable to argue that the more frequent an event is, the more likely organisms are to develop resistance to it and therefore why rare events tend to be catastrophic. In the short term, frequency can be expressed in years, but in the long term ecologists will probably find it necessary to translate this into the lifetimes of dominant organisms.

5.2.4 Area

Disturbances that affect huge areas will usually be worthy of more attention than those influencing smaller areas. Hurricanes, for example, are large-scale disturbances that are also intense. They may kill between 25 and 75 percent, and as much as 90 percent of low-lying mangal, leading Lugo and Snedaker (1974) to suggest that "hurricanes may have a very large role in determining the ratios of species within vegetation types over large areas."

5.3   Examples of Disturbance

You now have three general principles to guide your inquiry into disturbance. First, you know that disturbance removes biomass. Second, you know that there are often secondary effects upon the soil. And, third, you know that every disturbance has four properties. We are now going to enjoy a rather large number of examples that illustrate the many different kinds of disturbance in nature, and some of their primary and secondary consequences. These examples illustrate a wide range of vegetation types, intensities and frequencies. If you are pressed for time, it is quite acceptable to pick just a few to illustrate the general principles: if so, I suggest you start by reading the section on fire (Section 5.3.1), and then perhaps leap to meteor impacts (Section 5.4.3). If you have the time, however, each example has been carefully chosen to illustrate a wide range of habitats and some relevant natural history.

5.3.1 Fire Disturbs Many Kinds of Vegetation

Fire in Coniferous Forests

Forests composed of conifers are common at high latitudes and high altitudes, as well as upon infertile soils of sand plains and dunes. The vast boreal forests of the Northern Hemisphere support the largest continuous stands of conifer forest, some 15.8 million km^2, as compared with about 3.3 million km^2 in mountainous areas (Archibold 1995). Because of their high resin content and evergreen foliage, conifers burn easily. When fire occurs, the underlying soil of the forested area may also burn, since it is often comprised of peat formed from understorey mosses and tree needles. The dominant kind of fire is a high-intensity crown fire that can easily cover 10,000 ha and sometimes more than 400,000 ha (Heinselman 1981; Brown and Smith 2000). The frequency of fire changes with both vegetation and climate; cycles of 100 to 150 years are typical of lichen woodlands in

the extreme north, while those in eastern North America are longer (150 to 300 years). In drier areas of the west, surface fires of low intensity may burn every 25 years or so, whereas on floodplains *Picea alba* stands may avoid fire for 300 years.

We can recognize at least three main intensities of fire. Some fires burn only the understorey plants and accumulated litter. Other fires burn the standing vegetation, killing many of the trees and starting a new stand of forest. The most intense fires can also remove the soil organic layer. Shrubs, herbs and grasses adapted for post-fire regeneration usually re-sprout from buried meristems in the first year, while other species germinate from buried seeds. If the organic layer is also consumed, few buried propagules will remain, and most rhizomes and root systems that could sprout are also killed. Bryophytes often become abundant on moist sites. Dense regeneration of conifers usually occurs. A mosaic of different fire intensities will then produce an array of different forest types. As biomass accumulates through succession, and twigs, needles and peat accumulate under the trees, the risk of fire steadily increases. The proportion of old forests with a dense understorey of moss and peat will therefore depend on the fire frequency, which will in turn depend largely upon the climate.

A classic example of fire comes from the Boundary Waters Canoe Area on the border between Minnesota and Ontario where one can find some 215,000 ha of natural coniferous forests typical of the Precambrian shield country of central North America. Deglaciation occurred some 16,000 years BP and left behind coarse-textured soils formed on gravel, sand and bare rock outcrops. The forest types in the area include lowland conifers and spruce bogs, with pine, spruce or birch mixed with lichen outcrops in the uplands. The Boundary Waters Canoe Area was once populated by the Sioux and then by the Chippewa. By the late 1800s European settlements had begun to form. And what of the forests? Heinselman (1973) examined tree rings, fire scars and historical records to reconstruct the fire history of the area. He found that virtually all forest stands dated from one or more fires that had occurred since 1595 AD, and 83 percent of the area burned resulted from just nine fire periods: 1894, 1875, 1863–64, 1824, 1801, 1755–59, 1727, 1692 and 1681. The mean length of time between major fires was 26 years, and historical data seem to indicate that these were particularly dry periods. Figure 5.2 shows some of the earliest recorded burns. The landscape of this area is a mosaic of forest types recovering from different intensities of fire. At the time of Heinselman's study there had been a policy of fire control in place for some 60 years, but even so, the canopy trees were still the first generation of trees to repopulate burned areas.

Another way of exploring the effects of fire is to compare larger tracts of forest with those on islands where fires are infrequent (although edge effects and wind damage are likely much greater). One island in Heinselman's study had *Pinus resinosa* trees that were 378 years old. Another had *Pinus strobus* approaching 370 years old. These older stands go through a stage of senescence in which arboreal lichens and terrestrial mosses proliferate. As mosses, branches and dead wood accumulate, the area's susceptibility to fire increases steadily (Heinselman 1981). *Pinus strobus* in particular may be so long lived that it is rarely replaced before a new fire destroys the stand. Distinctive tree species' distributions may result from fire; *Thuja occidentalis* "has almost literally been driven to the lakeshores by fire" (p. 358). The species is so uncommon on uplands that one might assume that it needs abundant water to grow, but on ridges and islands where fires have been infrequent, it can occur. Many lakeshores are lined with old *Thuja* that, upon close examination, are found to be fire scarred on the side away from the lake.

Closer to the east coast of North America, where there are deeper soils, more rainfall and milder winters, a variant of the Boreal Forest known as the Acadian Forest occurs. The mixture of species here varies with local soils and climate. When Wein and Moore (1977) did an analysis of historical fires, they found much lower fire rotation periods – 1,000 and 5,000 years for the mean and median, respectively. Conifer forests had the highest fire rates and deciduous forest the lowest (Table 5.1).

It would be useful to know how fire frequencies change over longer time periods. This can be determined by measuring the abundance of charcoal in lake sediments. Figure 5.3 shows such data from the Rocky Mountains. The charcoal accumulation rate is shown on the left. Each horizontal bar in the middle shows a single fire episode. The episodes are summarized at the right as fires per thousand years. Note the low fire frequency near the bottom of the core, where the landscape was subalpine parkland. As the climate warmed, conifer forests developed, with an increase in fire frequency. The charcoal particles show that fire activity was high about 5,000 to 6,000 years ago, presumably when the climate was drier, with a downturn at 4,500 BP when the climate became wetter. The highest fire frequencies show up as a spike in charcoal input about 1,800 years ago, likely a reflection of the impacts of humans upon the landscape, both aboriginal and European.

Fire Can Kill the Cambium of Trees

In woody plants, one of the most conspicuous impacts of fire is death of the cambium, the thin layer of cells that produce both the woody tissues and the bark. When heat is applied to the surface of the bark, it is transmitted inward and the cambium temperature rises. An exposure to at least 60°C for a duration of 60 seconds or longer is usually sufficient to cause death in vascular plant tissue (Wright and Bailey 1982; Whelan 1995). Insulation by bark helps protect the cambium from reaching this temperature.

Uhl and Kauffman (1990) compared the fire resistance of different trees in rain forest just south of the equator in Para State, Brazil. A cotton rope saturated with kerosene was attached to the trunk of trees and temperature of the cambium was monitored with thermocouples inserted 10 cm above the point of attachment of the rope.

The length of rope used was one half the diameter of each tree to ensure a standard level of heat

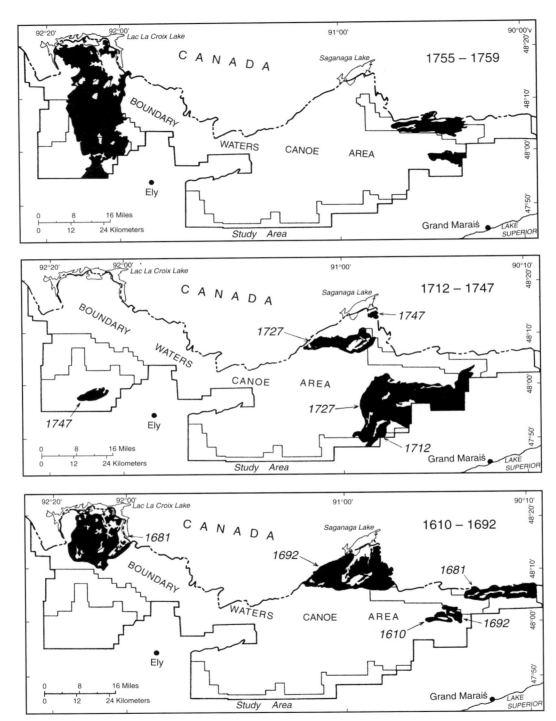

FIGURE 5.2 Shaded regions show the extent of fires in the Boundary Waters Canoe Area between 1610 and 1759. (From Heinselman 1973)

Table 5.1 **Properties of fire in the Acadian Forest of maritime Canada illustrated by the major vegetation types in eight areas of New Brunswick. (From Wein and Moore 1977)**

Vegetation type and approximate area ($\times 10^3$ ha)	Mean annual burn		Annual fire size (ha)		Mean annual no. fires
	ha	percent	mean	median	
Red spruce-hemlock-pine	2,591	5,418	0.21	681	69	7
Sugar maple-yellow birch-fir	1,655	2,569	0.16	1152	61	2
Spruce-fir coast	197	253	0.13	295	52	1
Sugar maple-hemlock-pine	1,005	731	0.07	149	49	3
Sugar maple-hemlock-pine	1,202	480	0.04	204	51	2
Fir-pine-birch	522	68	0.01	230	122	<1
Fir-pine-birch	99	2	<0.005	71	71	<1
Sugar maple-ash	305	8	<0.005	30	27	<1

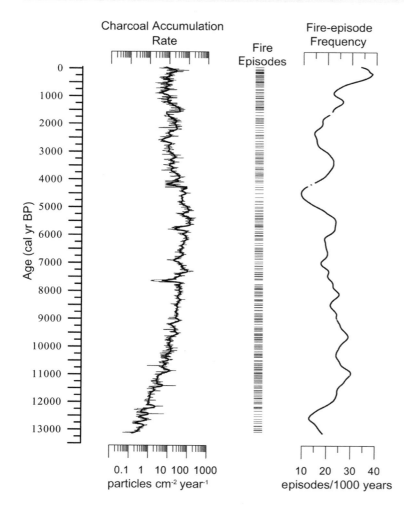

FIGURE 5.3 The abundance of charcoal in lake sediments provides a reconstruction of past fire regimes. These data come from an 11.4-metre long sediment core taken from Foy Lake, Montana. The curve on the left shows charcoal accumulation rate through time (smoothed over 150 years). When charcoal accumulation exceeds the background trend, each such peak is interpreted as a local fire (horizontal lines). The curve on the right shows the frequency of such fires per 1,000 years. (After Power et al. 2011)

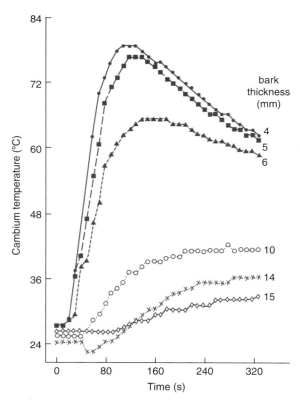

FIGURE 5.4 Thick bark protects trees from fire. In this study of the tree *Jacaranda copaia* in Brazil, standard fires were applied to bark. The thicker the bark, the lower the cambium temperature during the fire. (After Uhl and Kauffman 1990)

exposure per unit area. In *Jacaranda copaia* (Figure 5.4), the temperature reached by the cambium exceeded the critical exposure level if the bark was less than 6 mm thick; bark greater than 10 mm thick provided good defence of the cambium. Bark thickness varies among species. In the case of Amazonia, Uhl and Kauffman (1990) found that bark thickness usually ranged between 3 and 20 mm. Eleven taxa, however, had bark thickness less than 3 mm; these included species of *Ecclinusa, Apeiba echinata, Pterocarpus rohrii* and *Dialium guianense.* In contrast, five species had bark greater than 20 mm thick, including *Cecropia obtusa, Caryocar villosum* and *Laetia procera.* When an array of species was exposed

to fire, there were substantial differences in the temperature reached by the cambium, and this temperature was closely related to bark thickness.

Recurring Fire Can Create Prairies and Savannas

In areas that are frequently burned, woody plants may be unable to survive to reproductive age. In such cases, prairie or savanna vegetation may result. You saw one example earlier in Figure 3.18b, a pine savanna. Here recurring fires allow only one tree species to survive, *Pinus palustris.* High fire frequency (about once every five years) maintains a diverse understorey flora with many species of carnivorous plants and orchids (Peet and Allard 1993). Similar pine savannas occur through much of central America (recall Figure 5.1). Indeed, each vegetation type of the United States has a characteristic fire frequency and intensity (Figure 5.5); in the rainy northwest mountains, fires may occur only once every 500+ years (purple) while in the dry central prairies, fires may occur at least once a decade (yellow). The pine savannas that we just previously described (light blue) have high fire frequencies, but low intensity, as only the understorey normally burns.

Many plants of savannas have buried meristems to allow regeneration after fire (Figure 5.6). Later in this book you will encounter many other examples of fire-created vegetation types including Mediterranean shrublands and different kinds of prairies. When fire is excluded from such habitats, woody plants often invade and begin to replace the native prairie flora; vegetation managers then have to re-introduce fire to re-establish and maintain the herbaceous prairie vegetation.

Effects of Fire in Wetlands

Even wetlands can burn, usually during periods of drought. Fire is regarded as a major controller of plant diversity in both pocosin peatlands (Christensen et al. 1981) and the Everglades (Loveless 1959). Loveless concludes: "The importance of fire and its influence on the vegetation of the Everglades can hardly be over-emphasized."

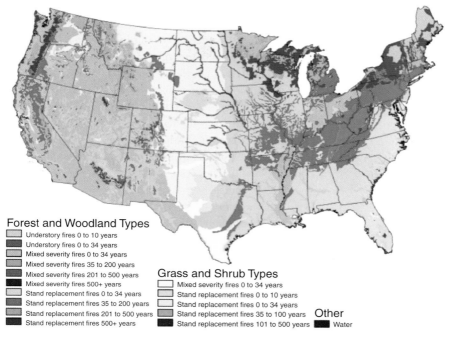

FIGURE 5.5 Every vegetation type and landscape has a characteristic fire frequency. This map shows fire frequency and vegetation types for the continental United States of America. Fires can range from understory fires every 0 to 10 years (light blue) to stand replacement fires about once every 500 years (purple). (Jim Menakis, USDA. For an online version and supporting text see Brown and Smith 2000 at www.treesearch.fs.fed.us/pubs/4554)

Forest and Woodland Types

- Understory fires 0 to 10 years
- Understory fires 0 to 34 years
- Mixed severity fires 0 to 34 years
- Mixed severity fires 35 to 200 years
- Mixed severity fires 201 to 500 years
- Mixed severity fires 500+ years
- Stand replacement fires 0 to 34 years
- Stand replacement fires 35 to 200 years
- Stand replacement fires 201 to 500 years
- Stand replacement fires 500+ years

Grass and Shrub Types

- Mixed severity fires 0 to 34 years
- Stand replacement fires 0 to 10 years
- Stand replacement fires 0 to 34 years
- Stand replacement fires 35 to 100 years
- Stand replacement fires 101 to 500 years

Other

- Water

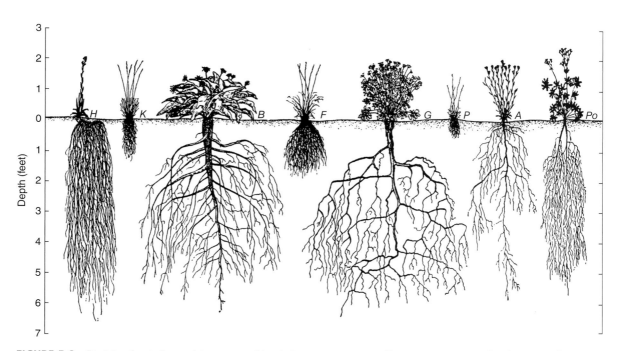

FIGURE 5.6 Prairie plants have rhizomes and buried root crowns to allow regeneration after burning. (From Weaver and Clements 1938)

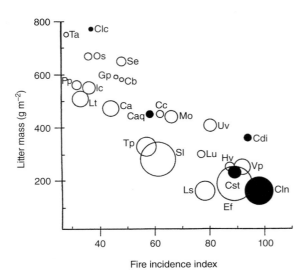

FIGURE 5.7 Litter mass and species diversity are related to a fire incidence index in a riparian wetland. Circle diameter is proportional to diversity (range 0.86 to 1.61). Principal *Carex* species shown in solid circles include: (Caq) *C. aquatilis*; (Cdi) *C. diandra*; (Clc) *C. lacustris*; (Cln) *C. lanuginosa*; (Cst) *C. stricta*. Other species include: (Ef) *Equiseum fluviatile*; (Gp) *Galium palustre*; (Hv) *Hypericum virginicum*; (Ic) *Impatiens capensis*, (Ls) *Lythrum salicaria*; (Lt) *Lysimachia thyrsiflora*; (Lu) *Lycopus uniflorus*; (Mo) moss species; (Os) *Onoclea sensibilis*; (Pp) *Potentilla palustris*; (Se) *Sparganium eurycarpum*; (Sl) *Sagittaria latifolia*; (Ta) *Typha angustifolia*; (Tp) *Thelypteris palustris*; (Uv) *Utricularia vulgaris*; (Vp) *Viola pallens*. (From AuClair et al. 1976b)

Fire becomes important during prolonged periods of drought. Low-intensity fires can simply remove existing vegetation, shift the composition of plant species from woody to herbaceous and increase plant diversity (Christensen et al. 1981; Thompson and Shay 1988). Figure 5.7 shows the effect of fire incidence on both litter accumulation and plant diversity in *Carex*-dominated wetlands along the St. Lawrence River in eastern North America. Hogenbirk and Wein (1991) have measured responses to fire in two vegetation types of the Peace-Athabasca delta (Figure 5.8). Fire reduced both the height and density of the dominant species. But

although the number of dicots increased, there was little effect upon total species richness. During the longer droughts, however, more intensive fires can burn the organic matter in the soil and create new depressions and pools (e.g. Loveless 1959; Vogl 1969).

Peatlands are particularly useful for the study of fire because, under certain circumstances, charcoal layers and macrofossils record both the fire history and the vegetation responses to the fire. *Sphagnum*-dominated peatlands are probably the most abundant peatland type in western boreal North America. Kuhry (1994) studied a series of peat cores to reconstruct fire and vegetation histories. He found that these peatlands had many macroscopic charcoal layers as a consequence of past fires. In the eight studied peat deposits, he estimated there had been one local surface fire approximately every 1,150 years. While this may be a surprisingly high rate of fire frequency, it is still an order of magnitude less frequent than estimates of fire frequency in coniferous forests in western boreal Canada (e.g. Ritchie 1987). During the hypsithermal, a period of warmer and drier climate about 7,000 years ago, fire frequencies in peatlands appear to have been twice as high as in the past 2,500 years. These fires not only burned the vegetation, but they also burned the superficial peat deposits. In spite of this, the cores suggest that the effect of peat surface fires on vegetation was short lived. This is apparently also the case in contemporary reports of peat fires. An interesting natural history story complements these findings; *Sphagnum* can apparently regenerate from stems at depths of 30 cm in the peat deposit (estimated to be 25 to 60 years old) (Clymo and Duckett 1986).

Kuhry (1994) could have stopped with his pictures of peat profiles, but in addition to these qualitative observations, he went on to test for quantitative relationships among rates of peat accumulation and fire frequency. Fire frequency was estimated as the number of macroscopic charcoal layers per 1,000 years, and peat accumulation rates were determined from radiocarbon dating. There was a negative relationship between peat accumulation rates and fire frequency (Figure 5.9). It appears, then, that the flush of nutrient-rich ash released by burning (and the

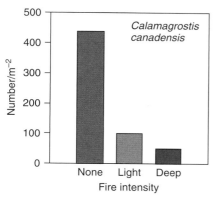

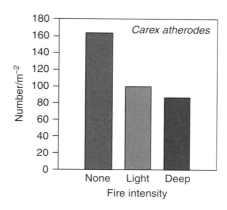

FIGURE 5.8 Effects of three fire intensities upon three properties of wetlands in the Peace–Athabasca delta. (From data in Hogenbirk and Wein 1991)

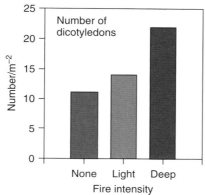

presumed higher plant growth rates) does not compensate for the loss of peat consumed by the fire. Thus fires significantly retard the growth of peatlands. This has important consequences for global warming because peatlands are an important reservoir for carbon storage. An increase in temperature would presumably lead to higher frequencies of burning, which in turn would lead to further releases of carbon stored in the peatlands (Gorham 1991; Hogg et al. 1992). This would then act as a positive-feedback loop to increase rates of global warming.

5.3.2 Erosion Creates Bare Ground

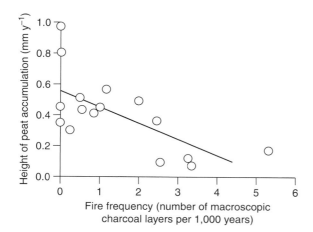

FIGURE 5.9 Peat accumulation as a function of fire frequency in western boreal Canada. (From Kuhry 1994)

Running water can create many kinds of disturbance in watersheds. Splash erosion is caused by individual rain drops, soil washes down valley slopes, while banks collapse or slump on river margins (Strahler 1971;

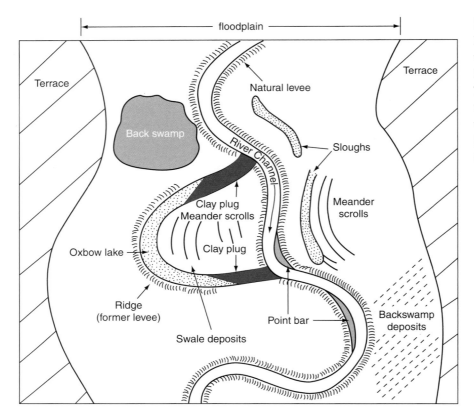

FIGURE 5.10

Disturbance creates a
wide array of wetland
habitats along rivers.
These habitats comprise
the template upon which
vegetation is formed.
(From Mitsch and
Gosselink 2000)

Rosgen 1995). In floodplains, rivers flow through valleys filled with alluvial sediments, which are continually reworked by the river, thereby destroying established vegetation and exposing new substrates for succession. Meanders are typical of many rivers, and meander systems both destroy existing vegetation and create new deposits of sediment (Figure 5.10). Once these meanders are formed, there is a gradual movement downstream ("down valley sweep") and over time all the alluvial sediments are eroded and redeposited between the limits set by river bluffs.

The result is a mosaic from open mud or sand flats to vegetation types of different ages and species composition. These mosaics can be seen along large rivers around the world including the Mississippi River, the Rhine, the Nile, the Amazon or the Yangtze.

These large-scale landscape dynamics from erosion are powerfully illustrated by the forests and floodplains of the Peruvian Amazon (Figure 5.11). Here more than one-quarter (26.6 percent) of the modern

lowland forest shows characteristics of recent erosion and deposition (Salo et al. 1986). During one 13-year period, satellite images showed that the mean lateral erosion rate of meander bends was 12 m year^{-1}. The total area of newly created land available for primary succession was 12 km^2, representing nearly 4 percent of the present floodplain area. The new substrates were first colonized by herbaceous species in genera such as *Tessaria*, *Cyperus*, *Ipomoea* and *Panicum*; smaller trees in the genera *Cecropia*, *Ficus* and *Cedrela* gradually formed a closed canopy, and eventually these became mixed with later successional species. Kalliola et al. (1991) have described the successional processes in more detail, documenting a flora of 125 plant species that colonize new sediment. Salo et al. (1986) conclude:

> *According to the repetitive nature of river dynamics, the migration of the river channel course creates a mosaic of successional forests within the present meander plain. The mosaic forest is*

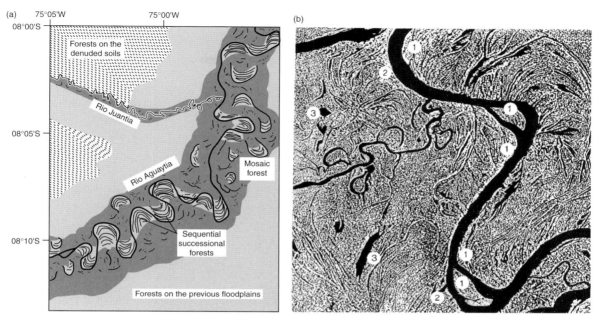

FIGURE 5.11 Lateral erosion and channel migration in Amazonian lowland forest from Landsat multi-spectral scanner images. (a) A simplified map of a meander system in a white water river (the Ucayali at Pucallpa, Peru). (b) A detailed map of forests along a meander system downstream from top figure, showing (1) areas of intense primary succession, (2) eroding forest at outer edges of meanders and (3) isolated oxbow lakes. (From Salo et al. 1986)

composed of patches of differentially aged sequential successional forest and patches of forests originating from a succession on the sites of former oxbow lakes. The annual floods further modify the mosaic pattern.

5.3.3 Animals Create Gaps in Vegetation

The entire next chapter will be devoted to effects of animals, but still something must first be said about them as a kind of disturbance. Grazing also removes biomass, and at many scales, from huge herds of herbivores grazing in the African grasslands to snails feeding in salt marshes. Animals can also create entirely new patches of habitat. It is this ability to create new patches, and their effects on plants, that we will explore here. Consider beavers and alligators, both of which are natural forms of disturbance in landscapes.

Beaver Ponds

Beavers obstruct water flow in streams thereby flooding forest and creating small ponds (Figure 5.12). Occasionally, under the right physical conditions, they can even regulate the water levels of lakes. Before the arrival of Europeans, the beaver population of North America was estimated to be 60 to 400 million individuals, with a range stretching from arctic tundra to the deserts of northern Mexico. The ponds that beavers create cause changes in forest structure, nutrient cycling, decomposition rates and the properties of water downstream (Naiman et al. 1988) as well as plant and animal diversity (Grover and Baldassarre 1995; Wright et al. 2002).

Beavers create cyclical disturbance in the landscape with two different frequencies. The short-term cycle is one of dam construction followed by dam loss. Dams may be washed out during floods, holes may be punched in dams by mammals such as otters, predators may kill

FIGURE 5.12 Beavers produce water-level fluctuations by building dams that periodically break or are abandoned. (Howard Coneybeare, Friends of Algonquin Park)

the beavers maintaining the dam. In such cases, water levels suddenly fall and many plant species regenerate from buried seeds. A long-term cycle occurs when beaver populations first colonize and then abandon sites (Figure 5.13). Building a dam changes forest to open water and wetland. Abandonment of the dam results in a short-lived mud flat, a longer period of marsh formation, and then, as the beaver meadow gradually dries, woody plants re-invade. This longer cycle of beaver ponds alternating with swamp forest probably has a frequency of centuries rather than decades.

Alligator Holes

"Gator" holes (Figure 5.14) are depressions that are either made or maintained by alligators. They are prominent enough that you can see them from airplanes should you be lucky enough to fly over the Everglades. During winter dry periods, these holes may be the only ponds remaining in a wetland (Loveless 1959; Craighead 1968). The alligator maintains ponds by pulling loose plants and dragging them out of the pool. Thicker muck is either pushed or carried to the edges of the pond.

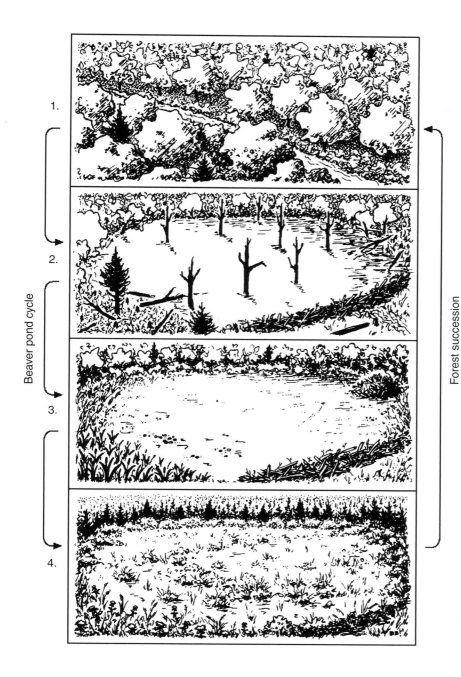

FIGURE 5.13 The beaver pond cycle going from forest with stream (1), to new pond with dead trees (2) to established pond with aquatic plants (3). When the food supply diminishes, indicated by the presence of conifers, the dam bursts and a beaver meadow forms (4). Eventually, the forest re-invades (1). Temporary events can cause a short circuit of the cycle (by Betsy Brigham, from Keddy 2000).

Gator holes were once a predominant feature of wetlands in South Florida. Craighead (1968) concludes that "in the first two decades of this century every inland pond, lake, and river held its quota of alligators." He suggests that a density approaching one alligator per acre existed in some regions. (Historical records have their limits, but the naturalist William Bartram, who travelled the St. Johns River in 1774–76, described alligators massed around his boat. He reported that when camping on beaches,

FIGURE 5.14 A "gator" hole in the Florida Everglades. The pool is visible in the centre left. Both the pool itself and the soil the alligator deposits around the edge of the pool become habitat for a wide variety of wetland plants. (By Paul Keddy, 2005)

it was necessary to keep a large fire burning all night for protection.) Gator holes, continues Craighead are "reservoirs for an amazing biological assemblage." Within them live "diatoms, algae, ferns, flowering plants, protozoans, crustaceans, amphibians, reptiles and fish." The productivity of these ponds is enhanced by uneaten food. Larger animals, such as hogs and deer, are killed by drowning but may be left for several days for ripening. The aquatic flora includes widespread genera such as *Myriophyllum, Utricularia, Potamogeton, Nymphoides* and *Najas*. The shallow water near the banks has marsh genera such as *Peltandra, Pontederia* and *Sagittaria*. Indeed, the description of this flora is remarkably reminiscent of beaver ponds. Connecting the gator holes are well-developed trail systems; some trails erode into troughs that are 15 cm deep and 60 cm wide.

5.3.4 Sediment From Flooding Can Bury Wetlands

Terrestrial communities can be buried by catastrophic events such as volcanic eruptions or landslides (Section 5.4), or burial by wind-deposited sand (e.g. Maun and Lapierre 1986; Brown 1997). Such events may be dramatic and conspicuous, but they are also infrequent. In contrast, rivers continually erode the land's surface and carry sediments that are deposited in wetlands as water movement slows. It is estimated that the world's rivers deliver in the order of 10^{10} tons of sediment per year to their mouths, which leads to the formation of large floodplains and deltas (Figure 5.15). Burial is clearly a routine experience for riparian vegetation. The importance of sedimentation as an ecological factor varies among watersheds (Figure 5.16). Southeastern Asian rivers, in general,

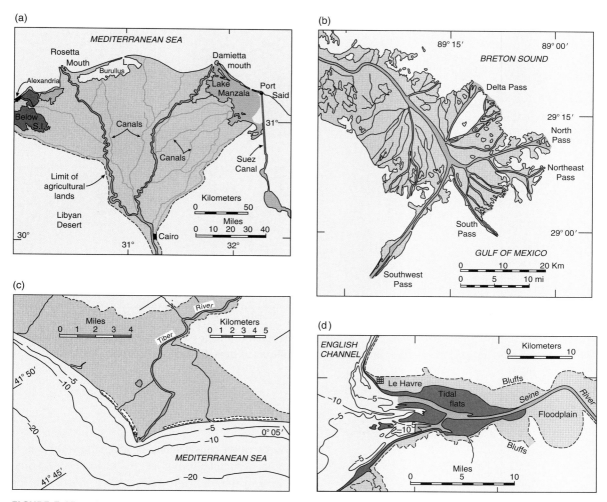

FIGURE 5.15 The world's large deltas illustrate the amounts of sediment transported and deposited by rivers. (From Strahler 1971)

are among the most prodigious transporters of sediment. Taiwan, for example, an island of 36,000 km² (roughly half the size of Ireland or the same size as Indiana), produces nearly as much sediment as the entire coterminous United States (Milliman and Meade 1983). The Yellow, Ganges-Brahmaputra and Amazon rivers have the highest annual suspended sediment loads in the world.

Burial can also occur much more slowly. Burial by locally produced organic matter such as occurs in peat bogs (autogenic burial) is much slower than burial by

externally produced sediment or organic matter such as occurs in river deltas (allogenic burial). Both kinds of burial can cause changes in plant communities, but here we are concerned with high rates of burial in deltas. These rates of burial can be even higher when storm surges or hurricanes excavate and then redeposit sediment (Liu and Fearn 2000).

Many wetland plants have well-developed rhizomes and pointed shoots (Figure 5.17). Examples include genera such as *Typha*, *Juncus*, *Scirpus* and *Carex*. Pointed shoots and underground storage

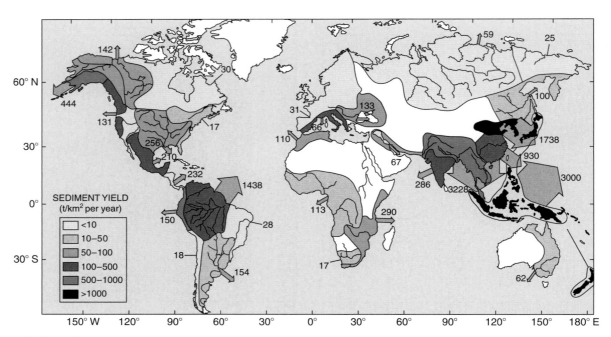

FIGURE 5.16 Annual discharge of suspended sediment from major drainage basins; arrow width corresponds to relative discharge, numbers give average annual input in millions of tons. (From Keddy 2010 after Milliman and Meade 1983)

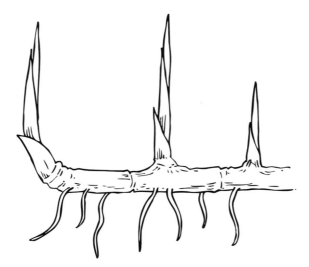

FIGURE 5.17 Rhizomes and pointed shoots allow buried plants to re-emerge. (From Keddy 2010)

structures are considered to be adaptations for penetrating accumulations of leaf litter (Grime 1979), and it is likely that the same traits also are adaptations for penetrating sediment. In contrast, evergreen rosette life forms would likely be extremely intolerant of burial, and this may be one reason why they are largely restricted to eroding shorelines (Pearsall 1920). At a larger scale, this may also explain partly why such plants are often restricted to oligotrophic lakes. Eutrophic lakes and bays with high sedimentation rates are generally occupied by larger rhizomatous plants.

How much burial does it take to change a wetland? Obviously the answer depends upon which kind of wetland (salt marsh or cypress swamp?) and how much burial (1 cm or 25 cm?). Here is one example. Allison (1995) covered salt marsh vegetation near San Francisco with 10 cm of sediment dug out of nearby tidal channels. He then followed the recovery of the plots for four years. For all species combined, vegetation cover returned to control values after only two years. Species such as *Salicornia virginica* and *Distichlis spicata* recovered quickly; other species such as *Frankenia grandifolia* and *Jaumea carnosa* recovered only when the disturbance occurred early in the growing season. There was very little new establishment from seedlings. In general, plots were

revegetated by ingrowth from adjoining plants, or else from buried rhizomes. Recovery was relatively rapid because the disturbed areas were only 1 m^2 circular plots; since most recovery was from adjoining areas, larger areas of spoil or sediment would presumably take much longer to recover. In contrast with freshwater wetlands, it would seem that seeds play a minor role in re-colonizing disturbed areas in salt marshes (Hartman 1988; Allison 1995).

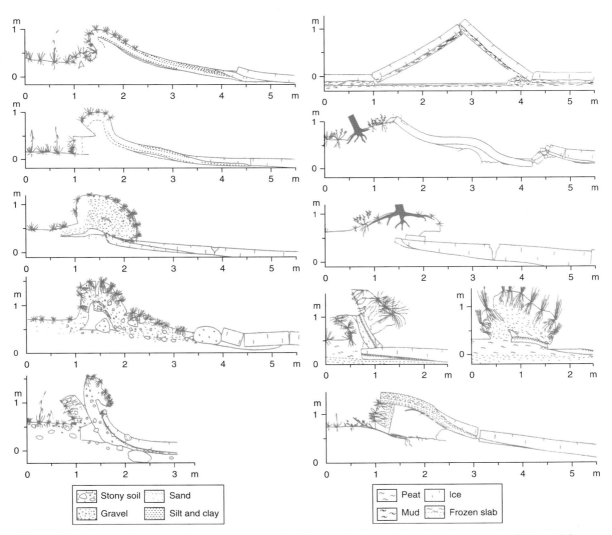

FIGURE 5.18 Ice can have a major impact upon shoreline wetlands and shoreline topography. (Adapted from Alestalio and Haikio 1979)

5.3.5 Ice Reworks Shorelines

Anyone who has watched great cakes of ice grind against a shoreline during spring flooding will be impressed by the power of ice scour to modify vegetation. In salt marshes or large lakes, one can find entire metre square pieces of marsh with 20 cm or more of substrate chopped out of the ground and moved many metres. At a smaller scale, there is the constant grinding of freshwater shorelines by the movement of ice as water levels rise and fall. Although the effects of ice cakes grinding on shorelines are visible (and audible)

during spring thaw, the processes beneath ice and snow during long winters are more difficult to study. But as Figure 5.18 shows, the results are nonetheless obvious come spring. Geis (1985) described how ice freezes onto the shoreline of lakes, forming an "ice foot." Sediments can become incorporated into this ice foot. Entire sections of shoreline are torn out of place when ice is lifted by rising water levels. According to Geis, plant biomass and diversity are reduced in the zone affected by the ice foot. Further north, ice pushing can create ridges, which produce a distinctive undulating topography along shorelines (e.g. Bliss and Gold 1994).

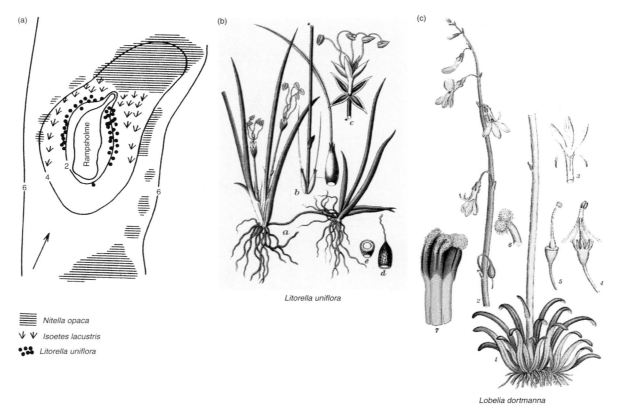

FIGURE 5.19   (a) Disturbance by waves as illustrated by an early twentieth century drawing showing the distribution of three species of rooted plants around a small island called Rampsholme in Derwentwater, a large lake in northwestern England. *Nitella opaca* (an alga in the Characeae) grows in the lee of the island, whereas *Litorella uniflora* (b) grows on more exposed sites. "*Litorella*, as in all English Lakes, is a shallow-water plant able to colonize sand of highly disturbed areas" (Hutchinson 1975, p. 424). (c) *Lobelia dortmanna* is another widespread shallow-water plant that is often found on shorelines exposed to waves. ((a) after Hutchinson 1975; (b), by Jacob Sturm; (c) by Carl Lindman)

FIGURE 5.20 A wet meadow has formed on this shoreline that is sheltered from large waves. These meadows greatly expand during low water years. In the foreground you can see the white flowers of *Parnassia palustris* and yellow flowers of *Helenium autumnale*, mixed with a variety of sedges in the genera *Schoenoplectus, Eleocharis, Rhynchospora* and *Scleria*. (Long Point, Lake Erie, by Tony Reznicek, 1979)

5.3.6 Waves

Waves illustrate events of very short duration and high frequency, almost exactly the opposite kind of disturbance from meteor impacts or volcanoes. Waves provide an opportunity for the study of the effects of chronic disturbance. It has long been observed that vegetation varies with exposure to waves, and sketches such as Figure 5.19 are typical of those in many books on aquatic botany. These effects of chronic exposure to waves are complex. Early research (Pearsall 1920) noted that there were both direct effects (e.g. biomass removal from plants, uprooting, seed dispersal) and indirect effects (e.g. erosion of nutrients, sorting of substrates, litter transport). The indirect effects create fertility gradients, so as disturbance from waves increases, fertility decreases. Species composition also changes with exposure to waves; Figure 3.18a showed the type of vegetation that develops on shorelines chronically exposed to small waves. Contrast that with the vegetation that develops in a sheltered bay, where silt and organic matter accumulate (Figure 5.20).

Much more has been written about the effects of waves on humans! Storms destroyed much of the Spanish Armada and thereby changed European history (Fernández-Armesto 1989). They also badly damaged artificial channel ports constructed for the Normandy landings in the Second World War, which nearly changed European history again (Blizard 1993). It is therefore natural that much of the work on waves can be found in manuals published by engineering agencies (e.g. US Army Coastal Engineering Research Centre 1977). Their methods have, however, been adapted for use by aquatic ecologists (e.g. Keddy 1982, 1983; Weisner 1990).

5.3.7 Storms

Storm damage is a widely studied factor in forests. It can be as small as a single tree killed by lightning or

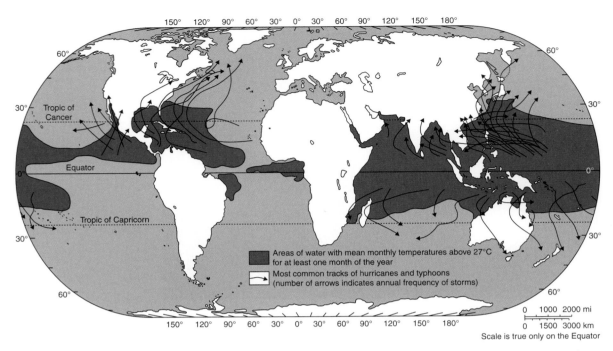

FIGURE 5.21   The distribution and frequency of hurricanes and typhoons (Encyclopaedia Britannica 1991c).

uprooted. It can be as large as an entire forest levelled by a hurricane (e.g. Denslow 1987; Dirzo et al. 1992; Merrens and Peart 1992). Hurricanes and typhoons (Figure 5.21) are frequent enough that their effects can be expected to have a recurring influence upon forests by breaking open the forest canopy and thereby allowing growth of seedlings. Anyone who has lived along the Gulf of Mexico or southeast Asia knows that hurricanes (or cyclones) are hard to predict, yet they arrive again and again. (Usually, it seems, the frequency is just long enough for people to forget and start rebuilding in flood-prone areas.) Entire neighbourhoods in coastal Louisiana were destroyed by Hurricane Katrina in 2005 (I was there) – some have been rebuilt in the identical location, while others have been abandoned. Each has its own story of folly. My own property along a small bayou had a beautiful oak forest, which was levelled, and, when I last saw it, the fallen oaks were being succeeded by thousands of seedlings of an exotic tree species, *Triadica sebifera* (Chinese tallow tree). Windthrows have a big impact

on forests. It is harder to judge their importance relative to other disturbances, such as fire. Accurate information on the frequency, intensity or proportional area of disturbance is often not available to put these events into a proper context.

Here is where long-term historical records can be helpful. A good example of a study of systematic disturbance properties used early surveyors' records of two large tracts of forest in western New York State south of lakes Ontario and Erie (Seischab and Orwig 1991). Here a total of some 25,000 km^2 was surveyed beginning in 1788. Surveyors often made notes of tree types, burns, dead trees, windthrows and old fields, and if one knows both the linear distances walked and the proportion of these lines falling into the above categories, one can reconstruct both early vegetation cover and disturbances. In a total area of >25,000 km^2, there were 140 km^2 of windthrows, 25 km^2 of agricultural fields (possibly aboriginal), 15 km^2 of dead trees and 5 km^2 of burn! The great majority of the windthrows occurred on dissected

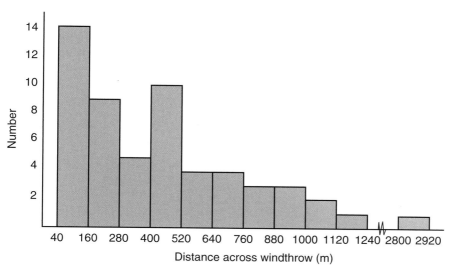

FIGURE 5.22 Size-class distribution of windthrows in presettlement forests of western New York in 1788. (From Seischab and Orwig 1991)

landscapes with steep slopes or ravines. A majority of windthrows were less than 500 m across (Figure 5.22). At the same time, while there were many disturbances, they comprised less than one percent of the study area. If one assumes that surveyors could recognize windthrows for about 15 years after the event, this yields a return time of 1720 years. Seischab and Orwig conclude that "only a small percentage of the western New York forests were subject to windthrow" (p. 119) and in general "natural disturbances were infrequent in the northern hardwood forests... steady state communities dominated the presettlement landscape" (p. 121).

5.4   Catastrophes Have Low Frequency and High Intensity

Really big disturbances can be hard to study. First, they are infrequent and unlikely to occur during the life span of an ecologist and, second, if and when they do occur, it is unlikely that an ecologist will be there to study them (or survive them). Yet, when such events do occur, they have a massive effect upon ecosystem structure, which may last for decades or centuries. They therefore cannot be ignored. The three most obvious examples are landslides, volcanic eruptions and meteor impacts. Let us look more closely.

5.4.1 Landslides

Landslides might be thought of as a relatively infrequent form of disturbance, but they are rather common in mountainous areas with heavy rainfall. Consider the Luquillo Mountains of Puerto Rico. Guariguata (1990) measured the frequency and area of landslides within 44 km² of montane wet forest here. On average, all months receive at least 200 mm of rain, but intense rains, associated with hurricanes, can deliver up to 500 mm in a day and trigger major landslides. Between 1936 and 1988, 46 landslides occurred. The most frequent size class (ca. 40%) was from 200 to 400 m², but large infrequent landslides >1,800 m² created almost 40 percent of the area disturbed (Figure 5.23). Some plant species that cannot occupy small gaps benefit almost exclusively from landslide openings. These include light-demanding ferns (*Dicranopteris pectinata, Gleichenia bifida*), herbs (*Phytolacca rivinoides, Isachne*

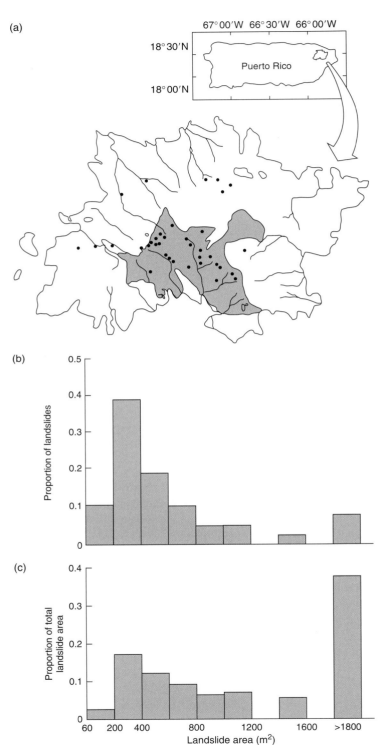

(a)

(b)

(c)

FIGURE 5.23 (a) Location of the Luquillo experimental forest in Puerto Rico and 46 landslides. The shaded area is underlain by intrusive rocks. The frequency plots (b,c) summarise the characteristics of the landslides. (From Guariguata 1990)

angustifolia) and one tree species characteristic of mature forests (*Cyrilla racemiflora*). Almost pure stands of *Dicranopteris* and *Gleichenia* ferns have also been reported in other tropical forests; both species spread by rhizomes and form thick canopies up to 1.5 m tall. Mature forests on landslides are dominated by *Calycogonium squamulosum* (Melastomataceae), *C. racemiflora* (Cyrillaceae) and *Micropholis garciniaefolia* (Sapotaceae), interspersed with patches of an emergent palm, *Prestoea montana*.

5.4.2 Volcanic Eruptions

Volcanic eruptions are less frequent than landslides, but often affect much larger areas. The largest volcanic eruption in history occurred in 1883 on the island of Krakatoa (Figure 5.24). Vegetation can be disrupted by lava, mud flows (lahars) and air-borne pyroclastic materials (tephra). Tephra deposits tend to sort by size; the fine materials are called ash, the intermediate ones lapilli and large blocks breccia or bombs. Small particulates, of course, are distributed the farthest by air currents and may spread over many square kilometres. If ejected high into the atmosphere, volcanic dust can be spread around the world. Figure 5.25 shows the distribution of volcanoes in the western United States and the results of an eruption 7,000 years BP that distributed ash over some 2500 km^2 and blanketed the hatched area with pumice more than 15 cm deep.

In the Kamchatka Peninsula of Russia there was a violent eruption of the Ksudach Volcano on 28 March 1907. Between one and two cubic kilometres of tephra was ejected, spreading pumice over more than 8,500 km^2, with deposits over a metre deep over some 50 km^2. Table 5.2 illustrates the sort of good quantitative data one needs to record for both the intensity and area of different kinds of ecological disturbance.

FIGURE 5.24 Volcanic eruptions are a large but infrequent natural disturbance. In 1883, one of the largest volcanic eruptions in recorded history occurred on the island of Kraktoa in Indonesia. It stripped entire islands of plant cover. Revegetation required long-distance seed dispersal combined with primary succession. The dust in the upper atmosphere reduced global temperatures for nearly five years, as well as producing vivid sunsets around the world. This photo shows Anak Krakatau, a new volcanic island that began forming there in 1927. (By Byelikova Oksana, Shutterstock)

Table 5.2 **The relationship between pumice depth and degree of vegetation destruction. (From Grishin et al. 1996)**

Deposit thickness (cm)	Deposit area (km²)	Nature of destruction
1–5	8,460	Destruction of some mosses, lichens, herbs and dwarf shrubs; minor damage to taller plants
5–10	1,458	Substantial destruction of some species in moss-lichen, herb and dwarf shrub layers; damage to taller plants
10–20	954	Loss of lichen-moss layer, significant destruction to herb and dwarf shrub layers; some trees die slowly by drying
20–30	228	Total destruction of lichen-moss layer, herb layer; most of shrub layer lost; significant die-back of trees
30–70	484	Destruction of all layers of vegetation and tree layer; isolated trees survive
70–100	62	Total destruction of all vegetation; reinvasion of vascular plants
Over 100	54	Total destruction of all vegetation; lichen desert persists

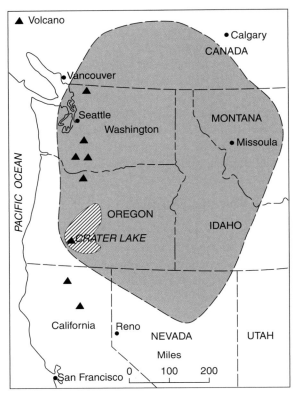

FIGURE 5.25 The area covered by pumice eruption at Crater Lake about 7,000 years ago. The shaded area shows maximum limits of ash fall and the hatched area indicates coverage by pumice of 6 inches (c. 15 cm) or more. (From Crandell and Waldron 1969)

Three different impact zones can be recognized (Grishin et al. 1996). The first received pumice deposits deeper than 100 cm. No vegetation survived. Trees still had not re-established in this zone; it remained lichen covered with sparse herbs and shrubs. The second zone received deposits of 30 to 100 cm. Occasional trees survived this disturbance and provided seed sources for re-colonization. Further, dead snags attracted birds that facilitated seed dispersal. Once plants established, their roots could reach through the ash to the buried soils beneath. As a consequence, revegetation occurred more rapidly. In the areas that received less than 30 cm of volcanic deposits, there was substantial survival, and regrowth was so rapid that there are now well-developed forests similar to those nearby that were less disturbed. Where the pumice was sufficiently thin, it may be more appropriate to describe changes as vegetation recovery rather than succession. Figure 5.26 summarizes the changes in vegetation: before, after and ca. 1995. Note just how many different kinds of vegetation resulted from this one eruption.

Sometimes rare events do happen near scientists. The eruption of Mount St. Helens in the United States

FIGURE 5.26 Vegetation maps of the Ksudach Volcano study area. (a) potential vegetation (before the 1907 eruption); (b) disturbed vegetation (a few years after eruption); (c) modern vegetation. (1) *Betula ermanii* forest; (2) *B. ermanii–Alnus kamtschatica* sub-alpine complex; (3) *A. kamtschatica* thickets; (4) high-mountain vegetation; (5) river valley forest; (6) meadows with shrubs on level sites in the forest belt; (7) moist *Calamagrostis–Carex* spp. meadow; (8) pumice desert; (9) pumice desert with scattered surviving trees; (10) pumice desert with lichen cover; (11) pumice desert sparse sub-alpine plants; (12) complex of mountain meadows with *A. kamtschatica* thickets; (13) pumice desert with lichen cover and isolated birches; (14) open young *B. ermanii* forest with dwarf shrubs-lichen cover; (15) closed young *B. ermanii* forest with isolated mature trees that survived the eruption; (16) nival belt; (17) lakes. (From Grishin et al. 1996)

provided a fine opportunity for ecologists to study the effects of volcanoes. Having been dormant for 130 years, Mount St. Helens erupted on 18 May 1980. One lateral burst levelled trees up to 20 km distant. Tephra was spewn over thousands of square kilometres, with the largest deposits on the northern slopes. Melting glaciers produced mudslides that swept downstream channels. Fine materials coated forests. Figure 5.27 shows the results. In spite of the publicity that this eruption received, and the many published studies (del Moral and Bliss 1993; del Moral et al. 1995), it was relatively small compared to other eruptions – the volume of ejecta, was for example less than five percent of the Krakatoa eruption in 1883.

5.4.3 Meteor Impacts

One only has to look at the surface of the Moon to realize that collisions among astronomical objects occur. It is easy to forget this because such events are

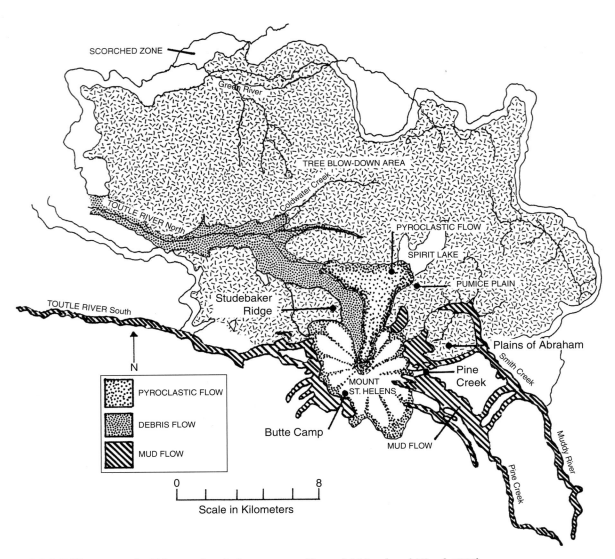

FIGURE 5.27   Mount St. Helens volcanic impact area. (From del Moral and Wood 1993)

rare (from the perspective of an ape's life span) and because erosion obliterates the effects of such collisions on Earth. Each day as the Moon rises, however, one should be reminded that catastrophes can and must have occurred on Earth too. How is one to study them? What have their impacts been?

Geologists have long recognized that a major change in the Earth's biota occurred at the Cretaceous–Tertiary (K–T) boundary. It marks the end of the Mesozoic era, known as the Age of Reptiles. The vegetation was dominated by gymnosperms, although early flowering plants (angiosperms) were also present. Then there was a sudden event. Dinosaurs disappeared; mammals become abundant. Angiosperms radiated and become the dominant plant forms. We now live in the Age of Mammals and Angiosperms. What event triggered this sudden change?

The geological record suggests that the change was abrupt. Indeed, so much so that it left a distinctive array of thin rock layers. Figure 5.28 shows a vertical

FIGURE 5.28 The K–T boundary as exposed near Clear Creek in southeastern Colorado. The white layer is ejected material with elevated levels of iridium. The dark layer, about 10 cm thick, is coal. The rocks beneath this layer contain pollen and spores from extinct Cretaceous plants, mostly broad-leaved evergreen angiosperms that comprised a coastal plain forest. (By Kirk Johnson, Smithsonian Institution)

section. The black layer is coal while the white layer underneath is stone that originated as clay. This claystone contains anomalously high concentrations of iridium (Ir), as well as other uncommon elements such as Sc (scandium), Ti (titanium), V (vanadium), Cr (chromium) and Sb (antimony). High iridium concentrations have been found in a similar layer at over 50 sites around the world. Iridium is a rare element on Earth, although common in certain asteroids and meteors. This transition is now widely believed to record the impact by an asteroid or similar object some 10 km in diameter (Tschudy et al. 1984; Wolbach et al. 1985).

Now take a closer look at evidence extracted from such rock layers (Figure 5.29). We will start below the K–T boundary and move upwards (forward in time). At the bottom there is a thick layer dominated by extinct species of broad-leaved angiosperms (Nichols and Johnson 2008). Then there is a thin layer of claystone, rich in iridium. Immediately above this there is coal, including a form known as fusinite, which is considered to be derived from charcoal. There are also abundant fossil fern spores (expressed relative to the abundance of angiosperm pollen). This shows "widespread but temporally brief dominance of ferns

followed by very rapid angiosperm recovery" (p. 1031). Since fossilized charcoal indicates fires, Wolbach et al. (1985) specifically sought out evidence on the nature of the postulated fires. They collected samples from the K–T boundary in both the northern and southern hemispheres and found 0.36 to 0.58 percent graphitic carbon, indicating a world-wide layer of soot. This corresponds to a volume of soot equal to ten percent of the present biomass of the Earth! This implies either that much of the Earth's vegetation burned or that substantial amounts of fossil fuels were also ignited. This soot would have had at least three effects (e.g. Wolbach et al. 1985; Alvarez 1998; Flannery 2001). First, it would have blocked virtually all light reaching Earth and thereby prevented photosynthesis. Second, the pyrotoxins formed during combustion would have harmed most land life. Carbon monoxide alone, if produced in the same amount as soot, would have reached a toxic level of 50 ppm (Wolbach et al. 1985). Third, the Earth would have cooled as soot blocked out sunlight.

The Chicxulub crater in the Yucatan is thought to be the impact crater from this event. You can read more about the scientific controversies in two books listed at the end of the chapter: *T. Rex and the Crater*

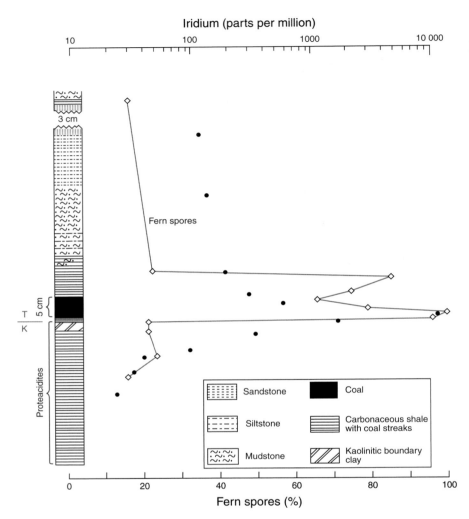

FIGURE 5.29 Data extracted from the K–T boundary interval in southeastern Colorado near Starkville. The large black dots show the variation in Ir concentration and the solid line shows the fern-spore percentages. (From Tschudy et al. 1984)

of Doom is intended for a popular audience, while *Plants and the K–T Boundary* will tell you much more about vegetation changes. Another exciting development has been the systematic exploration of sites resembling Figure 5.28 elsewhere in the world. Remarkable new and detailed descriptions of vegetation changes at the K–T boundary have emerged from locations including southern South America (Barreda et al. 2012) and Antarctica (Bowman et al. 2014).

The end of the dinosaurs and rise of the mammals has captured the human imagination. Hence the interest in the K–T extinctions. The rise of flowering plants may have been just as significant, but generally

receives less attention. Still, this was a single catastrophe. Science deals, whenever possible, in general principles rather than special cases. Table 5.3 and Figure 5.30 illustrate that there have been many more such impacts. These data have at least three constraints: (1) only larger objects make it through the atmosphere without burning up, (2) erosion buries or erases older craters, and (3) many craters likely remain undiscovered. Apparently such impacts are relatively frequent.

We do have observational data on one such impact in 1908, the Tunguska event in central Siberia. Unfortunately it occurred rather far away from

Table 5.3 **Characteristics of the ten largest (rim-to-rim diameter) terrestrial impact structures on Earth. (From Earth Impact Database 2016)**

Crater name	Location	Diameter (km)	Age (Ma)
Vredefort	South Africa	160	2023 ± 4
Chicxulub	Yucatan, Mexico	150	64.98 ± 0.05
Sudbury	Ontario, Canada	130	1850 ± 3
Acraman	South Australia	90	~590
Popigai	Russia	90	35.7 ± 0.2
Manicouagan	Quebec, Canada	85	214 ± 1
Morokweng	South Africa	70	145.0 ± 0.8
Kara	Russia	65	70.3 ± 2.2
Beaverhead	Montana, USA	60	~600
Tookoonooka	Queensland, Australia	55	128 ± 5

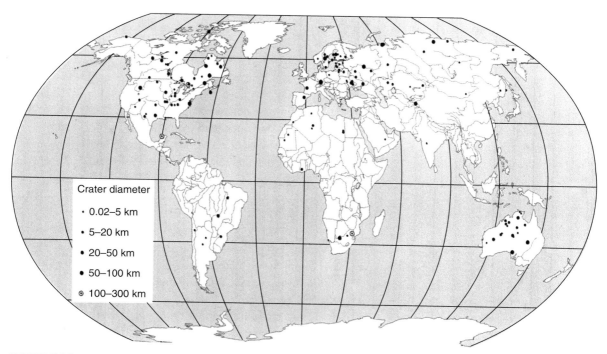

Crater diameter

· 0.02–5 km

· 5–20 km

• 20–50 km

● 50–100 km

◉ 100–300 km

FIGURE 5.30 The distribution and size of terrestrial impact structures on Earth. (From French 1998 with updates)

scientists and observing equipment. We would have had rather better information had it hit London or New York. It has been estimated that the object weighed between 105 and 106 tons and arrived at a speed of 100,000 km h^{-1}. Because of the remoteness of the site, it was not investigated scientifically until 1927. The Russian scientist L. A. Kulik found that an area of pine forest was flattened; around the epicentre everything was scorched, and very little was growing two decades later. The felled trees all pointed away from the epicentre. The absence of any object or crater suggests that the object, possibly a comet fragment, disintegrated in the atmosphere (Encyclopaedia Britannica 1991b).

Some of the interest in these events was triggered by the realization that a nuclear exchange could generate similar effects, including a nuclear winter. The immense amounts of soot entering the atmosphere would likely be enough to create a nuclear winter. (Recall how ash from a single volcano – Krakatoa, Section 5.4.2 – reduced temperature for several years.) A short historical digression, since younger readers may have no idea of the huge nuclear arsenals that faced each other in the 1970s, when two nations possessed the power to destroy each other many times over. Such was the overkill that even if the USSR had somehow attacked first and destroyed the more than 1,000 US intercontinental ballistic missiles (ICBMs), all US strategic bombers and all nuclear armed submarines in port, the United States would still have had enough nuclear warheads left

(ca. 2,400) on submarines at sea to destroy the 300 largest cities and towns in the USSR (Forsberg 1982). The same is true the other way around with slightly different figures. This capacity of mutually assured destruction (MAD) discouraged warfare, but left the fear that a war might happen by accident, or that someone might decide that launching first was still a better option than waiting. Data from other bombing campaigns, along with growing understanding of events such as the K–T, raised fears that a nuclear war would likely leave no winners and a seriously damaged biosphere. And hence added interest in the K–T event: Wolbach et al. (1985) were able to observe that the nuclear winter models then being considered might significantly underestimate the efficiency with which soot is carried into the atmosphere by fires. That is, the nuclear winter might be worse than predicted. Still a nuclear winter would not likely have extinguished all life, although as one popular writer (Schell 1982) put it, a post-nuclear biosphere would be "a republic of insects and grass." So, in actuality, the study of rare but intense impacts has had important consequences for international politics and human history.

While we are on the topic of warfare and disturbance, we should note that while humans have so far avoided a nuclear winter, there are many documented historical examples of lesser kinds of disturbance from warfare including the felling of forests to make wooden ships and the spraying of herbicides to remove hiding places for enemy guerillas (Box 5.1).

Box 5.1 Warfare as a Disturbance

It may at first seem out of place in a plant ecology book, but it would be irresponsible and inaccurate not to address the disturbances caused by human warfare. War is normally treated as just a human tragedy, but it is often also a significant cause of disturbance in natural communities. Here are just four examples.

1. **Peloponnesian wars.** The Peloponnesian wars lasted for nearly three decades during the fifth century BC, as the Athenian Empire fought the Spartan Alliance in "a terrible war that changed the Greek world and its civilization forever" (Kagan 2003 p. xxiii). In repeated naval battles, one reads of losses of hundreds of triremes, and each of these had to be replaced by the Greek taxpayers – and, more to the point, their forests. In one battle alone, at Arginusae (406 BC) the Spartans lost 77 ships and the Athenians a further 25

Box 5.1   (cont.)

(Kagan 2003, p. 458). The great forests of the Mediterranean once included the famed cedars of Lebanon. *The Epic of Gilgamesh* says "They gazed at the mountain of cedars, the dwelling-place of the gods... The hugeness of the cedar rose in front of the mountain, its shade was beautiful, full of comfort" (Sanders 1972, p. 77). Many of these were felled to make warships, and then the area was grazed by goats (Thirgood 1981). You will read more about this in Figure 13.2. You can still see such deforested landscapes in current news footage from the Middle East.

FIGURE B5.1.1   C-123 aircraft spraying defoliants over Vietnam in 1965. The four planes covered a 350 m wide swath in each pass over the forest (United States Air Force).

2. **Vietnam War.** The Vietnam War saw one of the greatest uses of bombing by aircraft in history. Explosives were dropped from bombers, creating enormous crater fields in wetlands, while napalm set forests alight. The largest scale effects, in terms of area, were likely the aerial spraying with herbicides (Figure B5.1.1), which were most intense in the southeast, particularly Dong Nai. Between 1961 and 1971 more than 6.6 million hectares of forest were sprayed, and of these, 293,000 ha were sprayed ten or more times (Stellman et al. 2003). Overall, some 2 million ha of forest were destroyed. In some areas, forests were replaced by bamboo, or herbaceous grassland. Recurring fire in these grasslands has not only prevented forests from re-establishing, but it has encroached on nearby forests (Westing 1989). On steep slopes, erosion has been severe. "It has become quite clear that, for vegetational recovery to occur in the seriously damaged inland forests, fire must be excluded and, moreover, that the worst damaged areas will require artificial planting (p. 352). Thirty years later, in 1999, one could still see sprayed swaths as straight-line clearings across the forest.

3. **Iran–Iraq War.** The Tigris–Euphrates river system supports several enormous wetlands. During the war between Iran and Iraq (1980–88), the dictatorship of Saddam Hussein drained the vast Central Marsh of Mesopotamia (al-Hammer marsh). Since the indigenous marsh Arabs were viewed as potential allies of the Iranians, Saddam Hussein diverted almost the entire flow of the Euphrates into a 560 km long drainage canal. Construction was often carried out in a brutal manner: "artillery initially bombards a district where engineering works are planned, so as to clear the local population; troops move in, to secure the district... Once a section has been completed, mines are laid to protect the embankments from attack" (Pearce 1993b). In the mid 1970s the marsh area was about 8,926 km^2 (about the original size of the Everglades), but had shrunk to only isolated patches of 1,296 km^2 by 2000. Some dikes were broken after the war to reflood selected areas, but the boundary between Iraq and Iran is still unstable, making restoration difficult (Lawler 2005).

4. **Congo Civil War.** Civil wars make government difficult or impossible. Hence the establishment of protected areas is weakened, and even when established formally, their continued use by guerillas and displaced

(Continued)

Box 5.1   (cont.)

residents means continued loss of trees cut for cooking and animals used for food. In some areas this had led to the phenomenon of empty forests (Redford 1992), areas that look green from the air but have few animals left. As you will see in Chapter 7, many of these animals are needed to either pollinate trees or disperse their fruits. In addition, illegal logging may increase either due to lack of enforcement, or to raise money for combatants. The Congo River Basin is one the world's largest wetlands (Figure 13.13), and the west coast of Africa is an area of high biological diversity (Figure 13.11). Civil war has raged there for decades. To maintain the natural plant communities of areas like Virunga National Park (established in 1925, some 800,000 ha, with habitats ranging from floodplain swamps to alpine vegetation) political stability is necessary.

The most obvious effect of warfare is the production of craters from bombing and shelling. But as you can see from the examples above, other effects such as draining wetlands or feeding guerilla armies on bush meat may in fact affect much larger areas of vegetation.

And, one more example. The red poppy, which now symbolizes the military deaths in World War I, is another plant species that survives in a buried seed bank (Section 5.6.2). In areas of France and Belgium where the plant cover was disrupted by shelling and trench warfare between 1914 and 1918, poppies emerged from buried seeds and flowered. Hence the connection between poppies and war remembrance.

5.5   Measuring the Impacts of Disturbance With Experiments: Two Examples

Thus far we have explored a wide array of observations on natural disturbances that happened as natural phenomena. To better understand how plants and vegetation respond to disturbance, one might consider creating disturbances of different types and measuring the responses. There is a growing number of such studies. Let us look at two, one from forests, the other from wetlands.

5.5.1 Forested Watersheds at Hubbard Brook

The Hubbard Brook valley in the White Mountains of northern New Hampshire is covered by second-growth northern deciduous forest growing in shallow soil over gneiss bedrock (Figure 5.31). Principal plant species of the area include *Acer saccharum* (sugar maple), *Fagus grandifolia* (beech), *Tsuga canadensis* (hemlock) and *Pinus strobus* (white pine). Variants of this type of forest stretch from Nova Scotia in the east, to Minnesota in the west, and south in the Appalachians to Virginia and North Carolina. This is also the type of forest that I can see out of my office window here in Lanark County, Ontario.

The experimental forest covers approximately 3,000 ha. In preparation for the experiments, a series of adjoining watersheds was identified, and near the base of each, a weir was constructed to monitor stream flow and water chemistry (Figure 5.32). In this way, nutrient outputs for an entire watershed could be monitored by sampling at a single point. One watershed (W6) served as a reference system; another (W2) was experimentally deforested and maintained bare for three years before vegetation was allowed to

FIGURE 5.32 The weir at Watershed 4 (Hubbard Brook Experimental Forest) showing a flow of $0.6 \ l \ s^{-1}$. Average annual flow of W4 is $9.2 \ l \ s^{-1}$. (From Borman and Likens 1981)

FIGURE 5.31 A young, second-growth northern deciduous forest at about 600 m elevation on Mount Moosilauke, New Hampshire. The forest was logged about 65 years prior to the photograph. (From Borman and Likens 1981)

regrow, while another (W4) was clear-cut in strips over a four-year period (Figure 5.33). Comparisons among these watersheds provided a great deal of data on the way in which forests cycle nutrients. The results of this work were published in a series of scientific papers during the 1970s and were summarized in Likens et al. (1977) and Bormann and Likens (1981). Here are a few of the key observations.

In the deforested watershed, annual run-off increased by some 30 percent and storm peaks were accentuated; this was the result of transpiration being all but eliminated. Loss of soil particulates increased by about a factor of ten, from 1,491 to 8,834 kg ha^{-1}.

The concentration of most dissolved nutrients in run-off increased several fold after cutting (Figure 5.34) and produced net losses of soil nutrients including nitrate (>114 kg ha^{-1}), calcium (>77 kg ha^{-1}) and potassium (>30 kg ha^{-1}). These nutrient losses reached a maximum after two years of de-vegetation, although the treatment continued for a further year, presumably the result of "progressive exhaustion of the supply of easily decomposable substrate present in the ecosystem" (Likens et al. 1977, p. 88) at the start of the experiment. This series of experiments showed the degree to which plants are able to exercise biotic control of watersheds: "These processes, integrated within limits set by climate, geology, topography, biota and level of ecosystem development, determine the size of nutrient reservoirs and produce nutrient cycles typified by minimum outputs of dissolved substances and particulate matter and by maximum resistance to erosion" (Likens et al. 1977, p. 78).

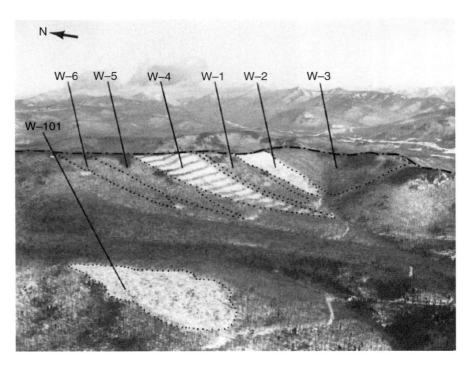

FIGURE 5.33 The Hubbard Brook Experimental Forest showing monitored watersheds 1, 3, 5, 6, and experimentally manipulated watersheds 2 (deforested), 4 (strip-cut) and 101 (commercially clear-cut). Note elevational gradient with northern hardwoods giving way to spruce-fir forest at higher elevations and on knobs. (Bormann and Likens 1981)

The Hubbard Brook study emphasized that ecosystems tend to store and recycle their nutrient capital, and one of the important detrimental effects of human disturbance is to increase the rate of leakage of nutrients. This work emphasizes the importance of declining soil fertility as a potential constraint on sustainability; that is, some landscapes may be unable to withstand repeated logging due to the combination of nutrient losses from the biomass that is removed and from erosion and leaching of the soil. Species that grow rapidly after disturbance may be particularly important for reducing leakiness since they recover and retain nutrients used in later stages of forest recovery.

5.5.2 Marshes Along the Ottawa River

Herbaceous vegetation forms along rivers where flooding or ice damage prevents woody plants from establishing. The Ottawa River in southern Canada is typical of many temperate zone rivers, with extensive wet meadows dominated by sedges (*Carex* spp., *Eleocharis* spp.) higher on the shore, and emergent marshes (with *Scirpus* spp., *Typha* spp. and *Sparganium* spp.) in shallow water. This vegetation is the result of many kinds of disturbance at many different scales (Day et al. 1988). To understand more, it is necessary to create experimental disturbances, measure the results and monitor recovery. Moore (1990, 1998) did exactly this. He artificially created bare patches in five different vegetation types along an exposure gradient. At each site, above-ground biomass was removed from 1-m² plots and the vegetation in them repeatedly compared with undisturbed controls over two growing seasons. There were two questions: (1) did the measured disturbance effects change among the particular ecological properties measured? And (2) did the effects vary among the five wetland types? The properties measured included the abundance of selected species, as well as higher-order properties including biomass and diversity. Overall, the experimental disturbances had only small consequences. A single growing season was sufficient for community-level properties such as biomass, richness and evenness to return to control levels (Figure 5.35).

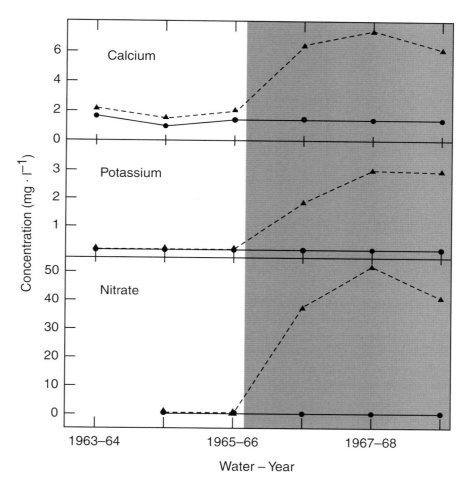

FIGURE 5.34 The annual weighted average of three dissolved ions in the water of two streams: (•) Watershed 6 was the control, while (▲) Watershed 2 was de-vegetated. The shading shows the period in which the valley was de-vegetated. (From Borman and Likens 1981)

The dominant plants that had been removed from each site tended to remain depressed for the first growing season, although by the second year effects were negligible. At the level of plant functional types (which Moore called guilds) recovery was also rapid, although there were minor changes, such as a modest increase in facultative annuals. Overall, it appeared that removing above-ground biomass had a marginal and short-lived effect on this vegetation type; this may not be a great surprise, given the dynamic nature of riparian wetlands.

The more significant parts of this story relate to the fact that Moore incorporated a naturally occurring disturbance gradient, so one could compare how experimental disturbance interacted with natural disturbance. Moreover, he used the procedure suggested above (Section 5.2) of measuring response to disturbance along a scale from 0 to 1, that is from no change in composition to complete change in composition. For this reason alone his results deserve a second look. For each property of a vegetation type (e.g. biomass, diversity) the magnitude of disturbance effects was evaluated as Z, where

$$Z = (x_0/ x_t) \times (y_t/ y_o).$$

Here x_0 is the mean value before treatment in the control sites, x_t is the mean value after treatment in the control sites, y_0 is the mean value before treatment in the disturbance sites, and y_t is the mean value after treatment in the disturbance sites (see Ravera 1989).

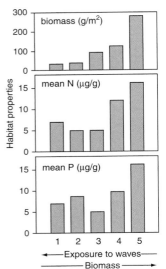

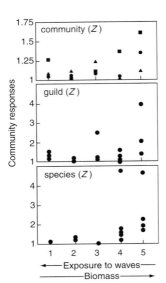

FIGURE 5.35 Marshes along rivers (left) are subject to many kinds of natural disturbance including scouring by waves and moving ice. The right-hand graphs show the impacts of an experimental disturbance – the removal of all above-ground biomass – upon five wetland communities. Z is a measure of departure from control plot values: larger Z values therefore indicate greater effects of the experimental disturbance. The effects increase from left to right; that is, habitats with higher biomass and lower rates of natural disturbance were most affected. (Quebec marshes, by Paul Keddy after Moore 1998)

The value of Z is independent of initial levels of the properties and is also independent of any ongoing temporal trends in the community. A Z value of 1.0 indicates no treatment effects, while values above or below 1 indicate increase or decrease. Figure 5.35 shows that, in each case, the experimental disturbance had the greatest effects in sheltered bays. That is, sites with higher biomass and higher fertility tended to be the most sensitive to disturbance. Perhaps this is because these are the riparian communities where disturbance is normally most infrequent.

5.6   Disturbance Creates Gap Dynamics

Sometimes disturbance causes discrete gaps to form: think of a single uprooted tree, or a pile of earth from an animal burrow or a fresh landslide. In this section we want to explore how plants respond to the creation of such gaps. You will find that this phenomenon is sometimes called gap dynamics and sometimes patch dynamics. They mean the same thing and are used interchangeably.

5.6.1 Many Kinds of Trees Regenerate in Gaps

The forested areas of the world are shaded by leaf canopies. When a gap forms in the canopy, there is increased light and tree species below begin growing upwards to fill the gap. As our example, let us spend some time in the southern regions of the Appalachian

Mountains in eastern North America. These mountains have one of the world's richest deciduous forests: there are five species of magnolia, ten species of oaks, seven species of hickory, along with a rich array of other woody plants (Stupka 1964). The other two principal areas of deciduous forests occur in western Europe and eastern Asia. Once these forest areas were contiguous, as illustrated by the species and genera that these three regions share, but drifting continents and changing climates have now isolated them from one another (Braun 1950; Pielou 1979).

The large number of tree species in the southern Appalachian forests can be accounted for by a number of factors. History is probably important. During the ice age, tree species were able to migrate southward along the mountains; in contrast, trees in Europe may have been trapped by the Alps and driven to extinction. As a result, species known from past interglacial periods no longer occur in Europe. The high number of tree species found in the Appalachians is also probably caused by variation in topography. Differences in slope, aspect, altitude and exposure provide habitat

for different species of trees (Whittaker 1956). In addition, extreme sites may provide refuge from competition (Keddy and MacLellan 1990). There is also evidence that tree species alternate with one another, saplings of one species establishing under the canopy of another (Fox 1977; Woods and Whittaker 1981). Here, however, we want to focus on one other important factor: gap regeneration. As you have read in the preceeding sections, there is constant disturbance within deciduous forests, from relatively large gaps carved out by severe storms to individual trees falling. As soon as a canopy gap forms, species beneath the gap being growing upwards to fill it. Each size of gap will also have a different light regime. Depending upon their light requirements, different tree species may succeed in filling the gap (Grubb 1977). Hence, even in flat landscapes which appear relatively homogeneous, the tree canopy and different gap sizes create a form of biological heterogeneity.

To experimentally study the role of gap regeneration in maintaining the diversity of tree species in deciduous forest, Phillips and Shure (1990) cut gaps of

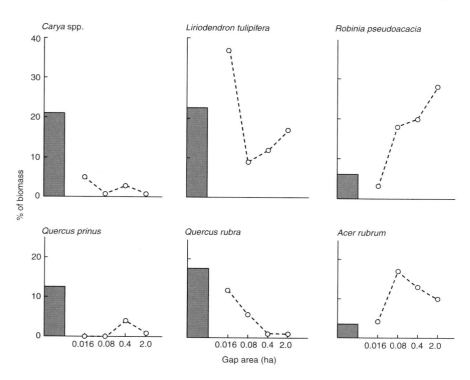

FIGURE 5.36 The regeneration of six species of tree in artificially created gaps of four sizes in deciduous forest in North Carolina. The abundance of each species is expressed as a percentage of the total biomass. Mean value before cutting (green bar), second year after cutting (○). (Data from Phillips and Shure 1990)

four different sizes (0.016, 0.08, 0.4 and 2.0 ha) in a
tract of deciduous forest in North Carolina. Then they
monitored regeneration in those gaps. Two years after
cutting there were marked differences (Figure 5.36).
On the left, *Carya* species and *Quercus prinus* showed
very low regeneration in the gaps; this may have
reflected a dependence upon simultaneous fire (not a
part of this experiment) to remove leaf litter. *Lirio-
dendron tulipifera* (a member of the magnolia family)
had highest regeneration in the smallest gaps but was
able to colonize most sizes. *Robinia pseudoacacia* (a
member of the pea family) showed a striking preference for the largest clearings. *Quercus rubra* grew best
in the smallest gaps. Many other studies show similar
patterns in a wide array of forest types from Amazonian rain forest to Australian eucalypt forests to
mangrove swamps (Lugo and Snedaker 1974; Grubb
1977; Denslow 1987). One can nearly always find a
group of species that depend upon canopy gaps
to establish. It is often a convenient place to start
learning about one's local forest ecology.

These gap-dependent species are often mixed with
species having other regeneration requirements. Some,
like the *Carya* species mentioned above, may depend
upon larger and more intense disturbance including
fires. Others may occupy extremely dry sites that other
tree species cannot tolerate. And still others may be
organized in competitive hierarchies along light
gradients (Figure 5.37). Gap dynamics may be most
important in the mesic fertile sites that otherwise would
be dominated by shade-tolerant long-lived trees.

5.6.2 Buried Seeds ("Seed Banks") Allow Regeneration After Disturbance

When a gap arises, it is often the first species to arrive
that are most successful at establishing. Many plant
species therefore have buried reserves of viable seeds
(sometimes called **seed banks**) that rapidly establish in
a gap. This is particularly common in herbaceous
species in wetlands, where seed densities in excess of a
thousand seeds per square metre are common in both
prairie marshes and freshwater coastal marshes, and

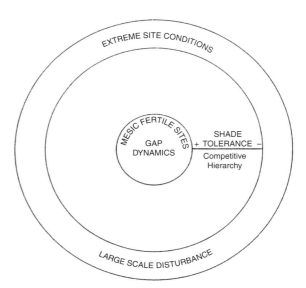

FIGURE 5.37 Disturbance may affect forests in
multiple ways simultaneously. Gap dynamics may be
particularly important for maintaining tree diversity
in mesic and fertile sites (the centre of the diagram).
Large-scale disturbances, may, like extreme site
conditions, allow species intolerant of competition,
particularly shading, to exist. In between these two
extremes, along soil fertility gradients, species may be
sorted out in competitive hierarchies. (Keddy and
MacLellan 1990)

densities in excess of ten thousand per square metre are
common in wet meadows (Table 5.4). These play a vital
role in allowing plants to re-establish after drought,
flooding or intense grazing (e.g. van der Valk and
Davis 1976, 1978; Keddy and Reznicek 1982, 1986).

How, if a seed is buried, does it become aware that
a gap has formed? Many buried seeds appear to detect
gaps from three factors: increased fluctuations in soil
temperature, increased quantity of light and changes
in the quality of light (Grime 1979). Thus most plants
adapted to exploit natural disturbances are stimulated
to germinate by a combination of high light levels and
fluctuating temperatures (Grime et al. 1981). For
many marsh and wet meadow species, regeneration in
gaps provides the only opportunity for establishment

Table 5.4 **Reserves of buried seeds are an important feature of wet meadows and marshes. (Adapted from Keddy and Reznicek 1986)**

Study	Habitat	Seedlings m^{-2}
Prairie marshes		
Smith and Kadlec (1983)	*Typha* spp.	2,682
	Scirpus acutus	6,536
	S. maritimus	2,194
	Phragmites australis	2,398
	Distichlis spicata	850
	Open water	70
van der Valk and Davis (1978)	Open water	3,549
	Scirpus validus	7,246
	Sparganium eurycarpum	2,175
	Typha glauca	5,447
	Scirpus fluviatilis	2,247
	Carex spp.	3,254
van der Valk and Davis (1976)	Open water	2,900
	Typha glauca	3,016
	Wet meadow	826
	Scirpus fluviatilis	319
Freshwater coastal marshes		
Moore and Wein (1977)	*Typha latifolia*	14,768
	Former hayfield	7,232
	Myrica gale	4,496
Leck and Graveline (1979)	Streambank	11,295
	Mixed annuals	6,405
	Ambrosia spp.	9,810
	Typha spp.	13,670
	Zizania spp.	12,955
Lakeshore marshes		
Nicholson and Keddy (1983)	Lakeshore, 75 cm water	38,259
Keddy and Reznicek (1982)	Waterline of lake	1,862
	30 cm below waterline	7,543
	60 cm below waterline	19,798
	90 cm below waterline	18,696
	120 cm below waterline	7,467
	150 cm below waterline	5,168
Wisheu and Keddy (1991)	Wilson's Lake	8,511
McCarthy (1987)	Hirst Lake	24,430
	Hirst Lake	16,626
	Goose Lake	11,455
	Goose Lake	3,743
Beaver ponds		
Le Page and Keddy (1988)	Canadian Shield	2,324

from seed. Grubb (1977) coined the term "regeneration niche" to describe the many ways in which plant seedlings and juveniles can exploit the different kinds of natural disturbance. In wetlands in eastern North America, the yellow flowers of *Bidens cernua* (Figure 5.38) are a reliable indicator that there has been a natural disturbance of some sort; after a beaver dam has collapsed, or in a drought year, they emerge en masse from accumulated seeds. During longer periods of high water, *Bidens* may flower only on the tops of beaver or muskrat houses where mud with seeds has accumulated.

FIGURE 5.38 Wetland plants such as the yellow-flowered *Bidens cernua* regenerate from buried seeds during periods of low water, in this case from a collapsed beaver dam. In the foreground are seedlings of multiple species of grasses and sedges. Table 5.4 documents the high densities of buried seeds in such habitats. (By C. Keddy, 2005)

Of course there are exceptions. In saline wetlands, seed densities are often low (<50 m^{-2}) (Hartman 1988) and a majority of the re-vegetation occurs by expansion of plants bordering the gap (Hartman 1988; Allison 1995). This exception is probably a consequence of the constraints that salinity places on the establishment of seedlings; periodic flooding with fresh water may provide the only opportunity for some saltmarsh species to establish from buried seeds (Zedler and Beare 1986).

There is one other exception. While herbaceous plants often have seed banks, it is much less common in trees (Grime 1979). Tree seeds tend to germinate and establish seedlings within a year, and it is then the seedlings that wait for a gap to open. Thus when you walk through a forest, wet or dry, you will often see thousands of young trees that have established and are waiting for a gap to form above them.

5.6.3 Rivers Create Gaps by Depositing Sediment

Gaps are also formed where rivers deposit new sediment, as you saw earlier in this chapter.

Let us just revisit the topic briefly within the context of gap dynamics. Natural processes of gap formation by erosion and deposition still occur in the valleys of some rivers such as the Amazon (Salo et al. 1986) and northern North American rivers (Nanson and Beach 1977), but overall they have been greatly reduced around the world (Dynesius and Nilsson 1994). The Rhine River, for example, has been markedly manipulated with dams, dikes and channels, whereas there were once extensive alluvial forests associated with water level fluctuations of two to three metres (Schnitzler 1995). These forests were of two main types: a "softwood" type (alder/poplar/willow) and a "hardwood" type (maple/ash/oak/elm). Newly deposited alluvial sediments are generally colonized by the softwood group, but after a period of 20 to 30 years, the hardwood species begin to establish. The challenge for conservation management of such rivers is to re-initiate the natural processes of erosion and gap formation to allow the full natural array of vegetation types to re-establish (Hughes 2003; Keddy 2010).

5.7 Logging is a Disturbance Caused by Humans

Logging is a widespread cause of disturbance in forests (Botkin 1990; Bryant et al. 1997). The most obvious consequence is the removal of trees. One way of measuring the intensity of logging is simply by the proportion of the standing biomass that is removed. By this criterion, selective removal of trees is likely to be less damaging that clear-cutting (although certain trees require large gaps, or even fires, for regeneration). Removing trees also allows light to reach the forest floor. Removing trees also extracts nutrients – the more wood removed, the more nutrients exported. Since wood itself has rather low nutrient levels, compared to branches and leaves, chipping and whole tree harvesting are more serious causes of nutrient loss. In sites with inherently low fertility, trees may be replaced by barrens. Often, it is the secondary effects that may have more impacts: the construction of roads, ruts from skidders, bark damage to trunks from skidding and later erosion.

One often reads assertions that logging is just the same as fire, and that all forests benefit from disturbance. This is a good example of the careless use of the word disturbance, which I warned against at the beginning of this chapter. "Disturbance" by fire is very different from "disturbance" by logging. For example, most fires leave standing dead trees that provide wildlife habitat. Fires may stimulate germination of plant species that require heat or smoke to break dormancy. The ashes left at the site provide nutrients for newly emerging plants. Access roads are not created by fire (although, perhaps, by fire fighters).

One area of controversy is the degree to which forests will recover after logging. From one perspective, forests are so sensitive that a single period of extraction will shift plant composition permanently. That is "primary forests" may not recover from logging. From another perspective, forests are sufficiently resilient that they will recover to their original state after a specified period of time. There is no denying

that after logging, some trees and plants return. This, after all, is the well-known phenomenon of secondary succession (Section 8.4). But are they similar to those of the original pre-logging state? Or precisely, how closely do second-growth forests come to the original old growth? The deciduous forests of the Appalachian Mountains are a good test area for asking this question. These forests are important because they are likely the largest piece of deciduous forest left on Earth, with areas protected to varying degrees within national parks and state forests (recall Section 5.6.1). Understorey herbs, particularly spring ephemerals, provide a group of target species suitable for answering this question. Duffy and Meier (1992) located nine primary ("old growth" stands) in the southern Appalachian Mountains. They then sought nine sites to pair with these, as similar as possible, except that they had been logged from 45 to 87 years earlier. The data show that every second growth site had fewer species of understorey plants. The authors conclude "even 50 to 85 years following deforestation, succession of herbaceous understorey plants in secondary mixed-mesophytic forests of the southern Appalachian Mountains resulted in only half the species richness and one-third the total cover measured in primary forests." The histogram bars in Figure 5.39 are ordered by time since logging, and there is no indication that the species richness increases over time. The interpretation of this data, of course, raises further questions. It is possible that the sites that were logged were inherently different from the primary forest sites, in spite of attempts to find equivalent forest types. It may be that the removal of trees permanently altered some physical characteristic of the site. The effects of logging roads might be one suspect; Duffy and Meier point out that fallen trees produce pit and mound topography, and clear-cutting may have reduced or eliminated this process and the diversity it creates. Another possibility is that the original woodlands established in a different period of

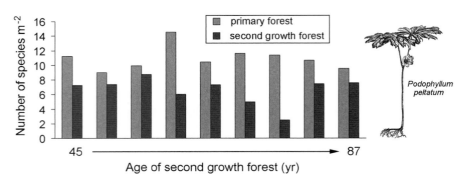

FIGURE 5.39 The number of species of understorey plants per square metre in nine stands of primary mesophytic forest in the Appalachians, and for nine nearby stands that were clear-cut from 45 to 87 years earlier (data from Duffy and Meier 1992). Note that the second growth forest becomes older from left to right. On the right is *Podophyllum peltatum*, about 25 cm tall, a common understorey species in deciduous forests.

climate, and so represent a habitat type that cannot regenerate under current conditions.

Although it was just one study of one group of plants in one forest region, this example raises profound questions that apply to many other vegetation types. How much disturbance is natural and necessary – gap dynamics (and pits and mounds) show that disturbance from single fallen trees is necessary. We also know that wind storms can fell entire tracts of forest. But the assertion that logging, selective-cutting or clear-cutting is similar is inherently doubtful, based simply upon how different the processes of logging and windfall are. And the question as to how frequently a site can be disturbed, and how intensely, without leading to irreversible changes is one that we could pose for any vegetation type with which humans interfere. At one extreme, the forests of the

Mediterranean have been abused by logging and grazing for so many centuries with so much accompanying soil erosion, it is likely that reforestation while desirable, will lead to forests that differ in significant ways from the primary forest that once grew there. There may be large areas of secondary forest, however, which are already perturbed, and with careful logging practices can continue to produce wood while maintaining the semi-natural vegetation that approximates the original.

We will return to logging in the final chapter, but first we have to look in more detail at issues such as change with time in forests (Chapter 8) and the regeneration of plant populations (Chapter 9). With these foundations established, in the final chapter we will return to a consideration of how sustainable logging might be viewed within a disturbance and recovery framework.

5.8   Multiple Factors in Plant Communities: Fire, Flooding and Drought in the Everglades

Many kinds of disturbance may operate simultaneously in one landscape. Together they can produce many different kinds of overlapping patches, each with its own species composition. Let us explore multiple kinds of simultaneous disturbances in the

vegetation of the Everglades, as interpreted by White (1994). The Everglades are not uniform: in fact, he says, they have "a great deal of structural and compositional variation, from open water sloughs with sparse macrophytes to sedge- and grass-dominated freshwater marshes, open pine stands and dense broad-leaved evergreen forest." He lists nine forces that are acting on this mosaic; they range from those that show gradual change (e.g. climate, sea level, for which change is measured on time scales greater than 10^2 years) to disturbances for which change is measured on scales of less than 10^2 years (e.g. drought, fire, storms, freezing temperatures). These latter natural disturbances tend to be short-lived events, but the communities recover from them slowly. Hence there is a basic asymmetry that we saw earlier in this chapter: disturbance is fast; recovery is slow. These lags mean that periodic disturbances can generate a mosaic representing a particular degree of recovery from the last disturbance. The rates of recovery will depend upon the amount of vegetation (if any) that persists through the disturbance, the influx of new propagules from adjoining areas and the productivity of the site.

In the Everglades, 11 of the 15 plant community types depend upon just two factors: hydrology and fire. Both of these factors are in turn connected to relative elevation. In general, as peat accumulates, the vegetation changes from sloughs to treed islands (the main sequence running diagonally from lower left to upper right in Figure 5.40). Low-intensity fires will create patches in the vegetation, but intense fires can consume peat, thereby lowering the relative elevation of the site and returning it to a much wetter vegetation type. The more intense the fire, the longer will be the recovery time. Intense fire and alligators (Figure 5.14) are apparently the only processes that can actually increase the wetness of a site. Since the mid 1980s, approximately 25 percent of the wet prairie and slough in the Everglades has been replaced by stands of sawgrass, probably as a consequence of reduced flooding and decreased fire frequency. Wet

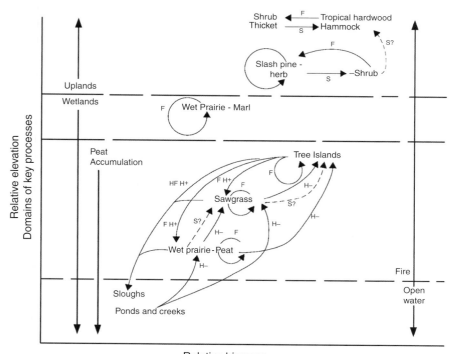

FIGURE 5.40 The vegetation of the Everglades plotted against two axes: relative elevation and relative biomass. The main diagonal from lower left to upper right consists of the change from shallow water (slough) to forest (tree islands) as peat accumulates and soil elevation increases. Disturbances such as fire push the community to the lower left, that is toward vegetation with lower elevations and lower biomass. (White 1994)

F=Fire, HF=Hot fire (peat consuming)
H+=Increased hydroperiod H−=Decreased hydroperiod
S=Succession (peat accumulation)
S?=Uncertain succession

prairies and sloughs have higher plant diversity, are major sites of periphyton production, and are important habitats for crustaceans and fish. Drainage and fire control therefore has not only changed the vegetation but also the capacity of the area to produce and support other organisms. Restoration of the Everglades will require, among other factors, more flooding and fire.

CONCLUSION

It was not my intention to catalog and describe every disturbance that can occur in plant communities. You may have skipped a few, which is fine. More could also be found. You might take it as an exercise to add one more example from your own landscape. The point of surveying so many examples was to discover how phenomena that appear very different all share a few basic properties. Each kind of disturbance has four measureable characteristics: duration, intensity, frequency and area. Each kind of disturbance has one primary factor in common – biomass is removed. Each kind of disturbance has other secondary consequences, including changes in soil organic content, nutrient levels or even elevation and moisture. Let us close simply by putting many of these different kinds of disturbance into a single figure based upon their intensity and their area (Figure 5.41).

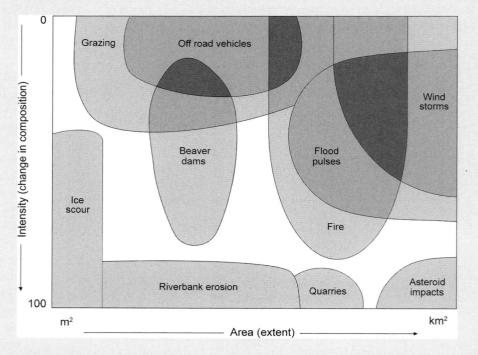

FIGURE 5.41   Some common disturbances plotted along two axes, area and intensity. Note that further down on the vertical axis means a greater decline in the amount of biomass, and greater change in the vegetation. These areas are all approximations, of course. Consider grazing. As drawn, it causes moderate changes in vegetation over relatively large areas (see Figure 6.1 to imagine the landscape). If carried out long enough, by enough animals, grazing will indeed remove all the biomass and create deserts, a topic we will return to under goat grazing (Box 13.2).

Review Questions

1. Which ecoregion do you live in, and what kind of natural disturbance was prevalent in your ecoregion 1,000 years ago?
2. What are the four properties of disturbance? Contrast an animal burrow with a landslide using these four properties.
3. What is the natural fire return frequency in your landscape? Make some predictions about what might happen if this frequency were either increased or decreased.
4. Explain how rivers create new habitats.
5. What is a seed bank? How is it related to disturbance? Can you find data for seed densities in your local vegetation types?
6. Explain how the different kinds of disturbance in the Everglades create the array of vegetation types found there (review Figure 5.40). Find a published local example illustrating similar principles. Or, create such a drawing for your own landscape.

Further Reading

Connell, J.H. 1978. Diversity in tropical rain forests and coral reefs. *Science* 199: 1302–1310.

Grime, J.P., G. Mason, A.V. Curtis, J. et al.1981. A comparative study of germination characteristics in a local flora. *Journal of Ecology* 69: 1017–1059.

Sousa, W.P. 1984. The role of disturbance in natural communities. *Annual Review of Ecology and Systematics* 15: 353–391.

Pickett, S.T.A and P.S. White. 1985. *The Ecology of Natural Disturbance and Patch Dynamics*. Orlando: Academic Press.

Salo, J., R. Kalliola, I. Hakkinen et al. 1986. River dynamics and the diversity of Amazon lowland forest. *Nature* 322: 254–258.

Leck, M.A., V.T. Parker and R.L. Simpson (eds.) 1989. *Ecology of Soil Seed Banks*. San Diego: Academic Press.

Botkin, D.B. 1990. *Discordant Harmonies: A New Ecology for the Twenty-first Century*. New York: Oxford University Press.

Kuhry, P. 1994. The role of fire in the development of *Sphagnum*-dominated peatlands in western boreal Canada. *Journal of Ecology* 82: 899–910.

Whelan, R.J. 1995. *The Ecology of Fire*. Cambridge: Cambridge University Press.

Alvarez, W. 1998. *T Rex and the Crater of Doom*. New York: Vintage Books, Random House.

Nichols, D.J. and K.R. Johnson. 2008. *Plants and the K–T Boundary*. Cambridge: Cambridge University Press.

6 Herbivory

Herbivory is a kind of disturbance. Field studies of herbivores: African mammals, beetles in rain forest canopies, tortoises on islands, sheep and slugs in moorlands. Morphological defences: squirrels and cones. Chemical defences. Nitrogen and food quality. Exclosure experiments: herbivorous insects in forests, land crabs in tropical forest, mammals in grasslands. Bottom–up or top–down? Modelling the effects of herbivory. Bark beetles in conifer forests.

FIGURE 6.1 Herbivores remove biomass from plants and from entire plant communities. Here, plains zebras and gnus in a savanna in Kenya illustrate the intensity of grazing that can occur when herbivores are abundant. (Shutterstock)

6.1   Introduction: Herbivores Have Large Impacts Upon Plants

In the previous chapter we saw how natural disturbances such as fires can remove biomass from plant communities. Now let us consider the effect of herbivores (Figure 6.1). We shall define herbivory as "the negative impacts of animals upon plants caused by the removal of plant biomass" and it is this consumption of biomass that shall be our focus in this chapter. Nonetheless, you should take note that when we talk more generally about the impacts of herbivores on vegetation, this can include other secondary effects such as trampling, and the deposition of dung and urine, as well as beneficial effects of seed dispersal. Our focus – the removal of biomass – is often quite selective, with many herbivores showing preference for meristems and seeds.

When we read the word herbivore, we are inclined to think of large mammals such as those in Figure 6.1. Indeed, some of the world's most dramatic animals, such as rhinoceros and giraffe, are herbivores – but so are a vast array of smaller animals including nematodes, arthropods and molluscs. Indeed, while images of large grazers such as those in Figure 6.1 capture our attention, it seems that herbivores, by numbers and biomass, are mostly arthropods and particularly insects (Leigh 1999). In an Amazonian forest (one of the few for which we have good data, Fittkau and Klinge 1973), the mass of ants (34 kg/ha) and termites (28 kg/ha) alone vastly exceeded that of mammals (8.4 kg/ha) and birds (3.4 kg/ha). We may watch a large mammal grazing in a forest, and be impressed to see this herbivore in action – and yet remain unaware that it is really the insects overhead in the canopy that are silently (at least to our ears) removing a greater mass of plant tissue. As we saw in the chapter on resources (Section 3.5.3), there are also a large number of animals called **detritivores** that process plant material once it has already died and, while we do not consider them, strictly, herbivores, they too are consuming plant material. It is a matter of perspective: in this chapter we are quite deliberately focusing upon a plant's point of view, where it is the removal of living tissue that has consequences for growth and reproduction.

Let us next consider a few very basic (but therefore very important) guiding principles. First, the amount of vegetation in any area of landscape will be controlled by just two factors: the rate of production of plant tissue and the rate of removal of plant tissue. The former is determined by growth and reproduction of the plants themselves. The latter is determined by disturbance. Disturbances can be either abiotic (most of Chapter 5) or biotic (most of this chapter). Herbivory is of particular interest as a disturbance because it is often so selective. Moreover, herbivores can rapidly multiply and in certain cases are able to strip the land of plant cover entirely. But not often. Plants are often better defended against herbivores than you might first suspect.

The study of plant–herbivore interactions is made difficult by the many possible ways, both direct and indirect, in which herbivores can influence plants. Eating shoots and seeds is a direct effect that is the core of our definition. The effects of trampling, urination, seed dispersal or carcasses may be more subtle and more difficult to measure. Let us consider for a moment all the possible interactions between plants and herbivores. Figure 6.2 shows the bewildering array. It also shows that microorganisms become involved as a prominent third party, through their direct effects upon plant growth and indirect effects upon nutrient recycling. This figure does not include the predators and parasites that may simultaneously influence herbivores and thereby indirectly control plant–herbivore interactions. Even so, the boxes and arrows can multiply rapidly. As an exercise, you might try to add new boxes and arrows to Figure 6.2 or even try to sketch out an alternative scheme. (Such sketches may seem elementary, but they outline how science is done: first we ask what kinds of entities there are and what linkages might occur among them.

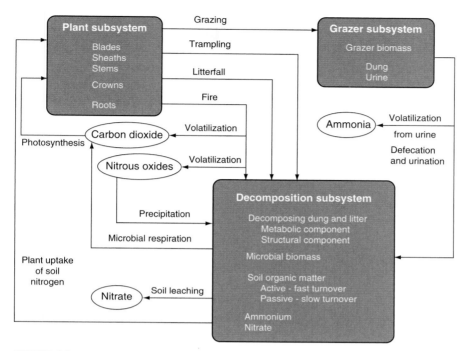

FIGURE 6.2   Another way of looking at the savanna in Figure 6.1. The living plant compartments are shown in the upper left, the grazing mammal compartments in the upper right, and the soil compartments in the lower right. Different pathways of nutrients flow from the plant subsystem to the decomposer subsystem. Their importance in any landscape depends upon the balances among grazing, fire, litterfall and trampling. Fast cycles are created where the principal pathway is through grazing; slow cycles where the principal pathway is through litterfall; and pulsed cycles where trampling or fire occurs. (From McNaughton et al. 1988)

This naturally leads to the next challenge, to measure the abundance of the entities (boxes) and the relative importance of the linkages (arrows).) Rather than elaborate on each box and arrow in Figure 6.2, I will focus this chapter on a central question regarding herbivory: *does herbivory have measurable impacts upon the properties of plant communities?* This question is a very logical starting point, for if herbivory has no effect on plant communities, or only a minimal effect, then plant ecologists can turn their attention to other factors. If, however, the effects of herbivory on plant communities are important then the challenge is to measure the effects, determine how they vary with space and time, and discern what underlying mechanisms might be at work. Note that the question being asked here is not the obvious one of whether

herbivores damage individual plants. From the perspective of the individual plant, the answer is clear: "The effects of grazing on plants growing in their natural habitats are overwhelmingly detrimental" (Marquis 1991). It is these negative effects that are the focus of this chapter.

6.1.1 Two Cautions Are Necessary

First, we should not forget the important positive impacts of animals upon plants, particularly pollination and seed dispersal, but that topic will be covered in the next chapter under the topic of mutualism. Second, while we will focus on the impacts of herbivores on plants, we cannot ignore the reverse. Plants simultaneously have impacts on herbivores.

This raises topics including animal nutrition, the evolution of teeth and guts, and various animal defences against plant poisons. While we will briefly consider these topics, for more you will need to consult books about animals.

We will begin with observations that zoologists have made upon herbivore diets. For as long as humans have been on Earth, they have likely been curious about what animals eat. (Of equal interest is what animals tend not to eat, or, from the perspective of botanists, what kinds of plants have defences to deter herbivory.) It is then logical to examine some field experiments that have tried to manipulate the effects of herbivory on plant communities. These experiments range from single fences to very large and complicated arrays of fences. Next, we will look at comparative studies that have sought to draw general conclusions about herbivory by sifting through the results of large numbers of published experiments. Finally, we will return to the topic of models and look at some challenging models that help

to clarify thinking about the dynamics of plants and herbivores in space and time.

You should be aware that this is a demanding chapter on many levels. We start with simple observations about what animals eat and how plants are defended. But, by the end of the chapter, you will be challenged by the design of enormous experiments, procedures for extracting data from collections of case studies and models that incorporate chaos theory. If you are just beginning your studies in ecology, you may find Sections 6.2 and 6.3 quite enough. As your guide, even I have to struggle with interpreting large experiments and three-dimensional models, so don't expect to grasp them on the first pass, any more than you would expect to read and remember all the characters in *War and Peace* in one weekend. Human brains have evolutionary limits too. The later sections will repay effort, particularly if you take the time to go back to some of the original papers. But, first, some interesting natural history about animals eating plants and plants defending themselves.

6.2   Observations on Wildlife Diets: Four Examples

What do wild animals eat? Wildlife biologists have investigated this question in two principal ways: they have observed feeding by wild animals and they have studied feces to reconstruct diets (Holechek et al. 1982). Consider a few examples.

6.2.1 Herbivores in African Grasslands

Large herds of grazing mammals, such as those in Africa (Figure 6.1), once lived in the arid grasslands of North and South America, Eurasia and Australia. These ecosystems, dominated by herbaceous plants and grazed by herds of large mammals, probably covered one-third to one-half of the world's surface area until as recently as the last century. The evolutionary success of large mammalian herbivores is illustrated by one mammalian family, the Bovidae (in the order Artiodactyla, containing the buffalo and

antelope), which has as many species (78) as the most diverse rodent family, the Muridae (Sinclair 1983). Over much of the world, large native herbivores have increasingly been replaced by farmland or herds of domestic cattle (McNaughton 1985; Archibold 1995).

Some of the largest remaining grazing ecosystems are found in eastern and southern Africa and much of this is in reserves that were created in the 1900s (e.g. Sinclair 1983; McNaughton 1985; Naiman et al. 1988). In such areas one can recognize three main habitats: forest, savanna and wetland. The diets of the African herbivores differ among both species and seasons (Table 6.1). The animals are selective about which plant parts they consume: most have a strong preference for green leaves rather than dead leaves while zebras apparently prefer the seeds of awnless grasses such as *Panicum coloratum*. Other species are browsers, feeding on patches of shrubs rather than grasses.

Table 6.1 **The important plant species in wet- and dry-season diets of four major herbivore species in the Serengeti grasslands. Data are percentages of total consumption. (From McNaughton 1985)**

Plant species	Animal species			
	wildebeest	zebra	gazelle	buffalo
Wet season				
Sporobolus ioclados + S. kentrophyllus	60.8	–	24.8	–
Andropogon greenwayi	11.8	41.7	–	–
Digitaria macroblephara	11.8	11.4	14.3	–
Kyllinga nervosa	–	–	12.4	–
Chloris pycnothrix	–	11.4	–	–
Themedia triandra	–	–	–	44.8
Eragrostis exasperata	–	–	–	17.0
Hyparrhenia filipendula	–	–	–	11.7
Dry season				
Themeda triandra	63.4	20.6	35.0	35.4
Sporobolus fimbriatus	6.3	–	14.8	–
S. pyramidalis	6.1	–	–	–
Hyparrhenia filipendula	–	20.7	–	26.8
Digitaria macroblephara	–	13.3	–	–
Pennisetum mezianum	–	–	14.9	–
Loudetia kagerensis	–	–	–	12.8

Herds of large grazing animals are attractive to humans and popular in game parks and zoos, probably because our minds evolved in savannas and because our ancestors found such animals to be an important source of protein. All the same, insects are potentially far more significant herbivores. So let us move to an example from the canopy of a tropical forest.

6.2.2 Herbivorous Insects in Tropical Forest Canopies

To measure rates of herbivory in the canopy of tropical rain forests, Lowman (1992) selected representative sites in the three major rain forests of New South Wales, Australia. Five tree species, largely evergreen, were selected, and leaf growth was measured monthly for five years. Most leaves were produced in a summer flush and, once expanded, leaves lasted from an average of some seven months (*Dendrocnide excelsa*) to more than five years (*Doryphora sassafras*). Mean annual leaf losses to herbivores ranged from 4.5 to 42 percent (Figure 6.3). For all five species of trees, the dominant herbivores were Coleoptera (beetles).

6.2.3 Giant Tortoises on Islands

While grazing by insects is probably the most important as measured by mass of vegetation

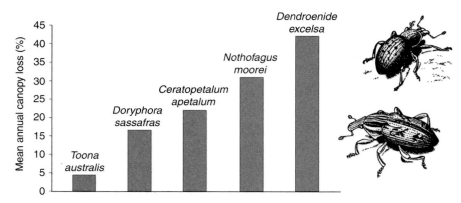

FIGURE 6.3 Herbivorous insects feed heavily on tree canopies, as illustrated by the mean annual canopy loss for five tree species in the rainforest canopy of New South Wales, Australia (data from Lowman 1992). The principal herbivores were beetles in the group Curculionidae. Right: Two representative curculionid beetles in the genus *Cleonus*. (Canstock). There are more than 40,000, perhaps closer to 60,000, species of Curculionidae; the group is one of the most important herbivores on trees and includes bark beetles (Figure 6.24), ambrosia beetles and nut weevils.

consumed, there are a few areas where large reptiles remain important.

The areas inhabited by giant land tortoises also give a glimpse of the last terrestrial grazing systems dominated by reptiles (see Figure 13.8). Such grazing systems are now restricted to the Aldabra atoll in the Indian Ocean and the Galapagos off the coast of Ecuador. Tortoises can have extremely high population densities; there are, for example, >25 per hectare on one part of the Aldabra atoll. On Aldabra, there are some 150,000 tortoises (Gibson and Hamilton 1983); in the Galapagos there are fewer, perhaps some 10,000 in total spread across a dozen populations (MacFarland et al. 1974), because of heavy predation from sailors in the 1700s and 1800s. Despite differences in tortoise population density, there are many similarities in the nature of herbivory in the two locations. At both, seasonal drought has a major impact upon the kinds of plants eaten by the tortoises and the habitats in which the tortoises feed. The favoured habitat at both places is "tortoise turf," a type of short-cropped vegetation maintained by heavy grazing. In the dry season, turf production declines and tortoises are forced to feed on other plants, particularly shrubs. In the Galapagos, cacti, as well as shrubs, are eaten during dry periods (Hamann 1993).

The usual way to discern whether a herbivore is selective is to compare the availability of food with the actual diet. If the herbivore eats food in direct proportion to its abundance in the habitat, there is no evidence for selectivity. Table 6.2 shows these calculations for Aldabra tortoises;

Table 6.2 **The overall abundance of different foods in the diets and environment of Aldabra atoll tortoises. (From Gibson and Hamilton 1983)**

Food	Percent in environment[a]	D/E[b] total
Tortoise turf	15.3	3.99
Mosaic rock	3.75	0.803
Shrub litter	38.5	0.532
Sedges	29.0	0.267
Herbs	0.06	18.3
Long grass	0.36	8.28
Sporobolus virginicus	1.76	0.789

Notes:
[a] Percentage of vegetation (excluding rock).
[b] Percentage in diet/percentage in environment.

the D/E (diet/environment ratio) is very high for herbs, long grass and tortoise turf, which shows that the tortoises feed selectively upon these components of the vegetation. Sedges, the grass *Sporobolus virginicus* and shrub litter are avoided (but note that the sedges and shrub litter together make up more than half the available food). White (1993) is of the opinion that these animals are strongly limited by nitrogen availability; consistent with his view is the observation that tortoises will also feed on carrion (including tortoise) when the opportunity presents itself (Gibson and Hamilton 1983). Moreover, when sedges are eaten, flowers and seeds are consumed, and plant reproductive structures usually have higher levels of both nitrogen and phosphorus.

6.2.4 Herbivory in Anthropogenic Landscapes

Few of us are fortunate enough to spend time in African savannas, tropical canopies or tortoise islands. Most of us are far more familiar with habitats dominated by humans (that is, anthropogenic ecosystems) where forests and their native herbivores have been replaced by pastures with exotic herbivores. Let us consider one example of such an anthropogenic habitat: the deforested uplands of the British Isles (and many parts of nearby Europe). It is only a slight exaggeration to say that there are only two major herbivores: sheep and slugs. In Great Britain alone there are more than a million hectares of such

moorland (Miller and Watson 1983). Much of this landscape and its vegetation (known for its large areas of heather, *Calluna vulgaris*) has formed in the last ten thousand years, as a consequence of deforestation and grazing, a process that began in the Neolithic but accelerated with growing human populations and the industrial revolution (Simmons 2003). The mostly treeless landscape is now maintained by grazing and fire. There are other vertebrate herbivores besides sheep: the density of vertebrate herbivores in the highlands of Scotland in the 1960s was estimated as 50 sheep, 65 red grouse, 10 red deer and 16 mountain hare per square kilometre. But invertebrates are also important. The slug populations (herds?) are less obvious to passing motorists, but species such as *Agriolimax reticulatus* and *Arion intermedius* consume approximately one gram of plant biomass per square metre per month (Lutman 1978) and can reach densities exceeding ten per square metre. Despite the abundance of herbivores in moorland, less than ten percent of the primary production of *Calluna vulgaris* is actually consumed by herbivores (Miller and Watson 1983). Hence we can think of this as a landscape mostly created by humans and their exotic herbivores. This example is important not only because it is the familiar one to many European readers, but because it illustrates what may happen to many other landscapes of the world as human populations grow and as herds of sheep and goats grow even faster, a topic to which we will return in Chapter 13.

6.3 Plants Have Defences Against Herbivores

6.3.1 Evolutionary Context

Many plants have spines, thorns and prickles – cacti are often the first to come to mind but many other plants are equally defended (Figure 6.4). Spines provide rather strong evidence that plants have evolved traits to defend against herbivores (Marquis 1991; Pennings et al. 1998; Raven et al. 2005). Pressure from

herbivorous animals can even be inferred from the fossil record. From the late Carboniferous until the late Triassic, all of the big herbivorous dinosaurs were short-necked, low browsers; by the late Jurassic, there were also prosauropod dinosaurs with long necks and anatomy that suggests that the forelegs could be raised off the ground (Figure 6.5 bottom). This coincides with the era when seed ferns became nearly

FIGURE 6.4 Spines and thorns are a deterrent to herbivores. *Gleditsia tricanthos* (honey locust) is a deciduous tree in the Fabaceae that has thorns up to 10 cm long. It grows throughout the Mississippi River valley. (Photo by Greg Hume)

extinct and conifers began to radiate (Bakker 1978; Coe et al. 1987; Wing and Tiffney 1987). By the Cretaceous, there appears to have been a shift back to big, low-browsing species such as ornithischian dinosaurs (Figure 6.5 top). As the types of grazers and browsers changed, so did the pressures of natural selection. Says Bakker (1978) "the rapid diversification of big Cretaceous low browsers changed the selective pressure on the plant communities. Intense low browsing would have increased mortality among seedlings, thinned out the forest structure, and would have favoured shrubs which could maintain a dynamic steady-state with the herbivores, rapidly regenerating the cropped foliage and colonising areas laid bare by close cropping." By and large, the

plants you see today had ancestors that survived grazing by dinosaurs.

6.3.2 Structures That Protect Seeds: The Strobilus

While all plant tissues provide potential resources for herbivores, seeds may be at particular risk because of their high protein and lipid content. Somehow seeds must be protected from herbivores. The earliest known seed plants were seed ferns (Pteridospermophyta), first found in the late Devonian, widespread in the Palaeozoic but now extinct. The arrival of the gymnosperms brought a seed-protecting structure that is still widely found today: the strobilus. You may know it as a cone. The strobilus is formed from a series of structures known as megasporophylls (thick scales that bear ovules) spirally arranged along a central axis (Figure 6.6). Strobili may have originated from shoots with widely separated branches and megasporophylls, but as the length of the axis between the branches shortened, and the branches thickened, a strobilus resulted. One can only speculate which herbivores drove the selection for the development of the strobilus (recall Figure 6.5), but the persistence of gymnosperms even today is evidence of its success.

Today, of course, it is not dinosaurs but squirrels (and birds) that are best known for feeding on cones. Let us take a closer look at the relationships between squirrels and conifers, since they illustrate herbivory and its consequences for both the plants and the plant eaters. Smith (1970) observes, "That some terrestrial animals would have been specialising as seed eaters from an early date is made more plausible by the very high nutritive value of seeds." From the perspective of a herbivore, each strobilus represents a valuable source of energy, albeit one that is well protected. In *Abies amabilis* (Pacific silver fir), for example, a single cone contains up to nearly 400 seeds, which together have a total energy value of 68,000 calories. Smith has observed that squirrels first harvest the cones of *A. amabilis*, which have the highest energy content of conifers, and then harvest from trees whose seeds have progressively lower energy content. In *Pinus*

FIGURE 6.5 Large herbivores and plants in two geological eras. The animals are fully adult individuals shown to scale. Top: (a) *Alamosaurus*, (b) *Parasaurolophus*, (c) *Pentaceratops*, (d) *ankylosaur*. (Kirtland-Fruitland-Ajo Alamo fauna, late Cretaceous). Bottom: (a) *Haplocanthosaurus*, (b) *Brachiosaurus*, (c) *Camarasaurus*, (d) *Barosaurus*, (e) *Diplodocus*, (f) *Apatosaurus*, (g) *Stegosaurus*, (h) *Camptosaurus* (the largest low-browser). (Morrison fauna, late Jurassic). The vertical scale is metres. (From Bakker 1978)

ponderosa (ponderosa pine), a cone can have up to 150 seeds and therefore contain up to 25,000 calories (Smith 1970). The high caloric value represents a high lipid content, although seeds of one pine species (*Pinus jeffreyi*) also have 20 percent protein. Smith suggests that squirrels and conifers may have co-evolved. In the Jurassic, multituberculates, a group of rodent-like mammals, appear in the fossil record at about the time that conifers with multi-appendaged dwarf shoots, as opposed to strobili, start to disappear. Many of the modern genera and species in the Pinaceae start to appear at approximately the same time as placental rodents replace multituberculates.

Certain conifers have serotinous cones in which the scales of the strobilus remain sealed by resin until seared by fire; the mature cones then open and the seeds fall onto the burned area. The world has more than 500 serotinous species of conifer. Barbour (2007) says "They occur in fire-prone, seasonally dry, and nutritionally poor habitats" (p. 296) and are distinctive in allowing for long-term seed storage in the canopy (Lamont et al. 1991). This obvious adaptation to fire means, however, that the cones and seeds retained on the tree provide a more or less steady supply of seeds for seed-eating animals. Hence canopy seed storage magnifies the selection pressures from herbivores (Table 6.3). Equally, it appears that serotinous cones have exerted selective pressure on squirrels, as is suggested by the observation that squirrels differ in the design of their jaws and the development of associated musculature (Figure 6.7); squirrels with stronger jaws occur in regions where conifers have serotinous cones. Even within a species differences are observable; red squirrels collected from forests with *Pinus contorta* (lodgepole pine) have significantly heavier jaw musculature than those in forests without this species.

(a)

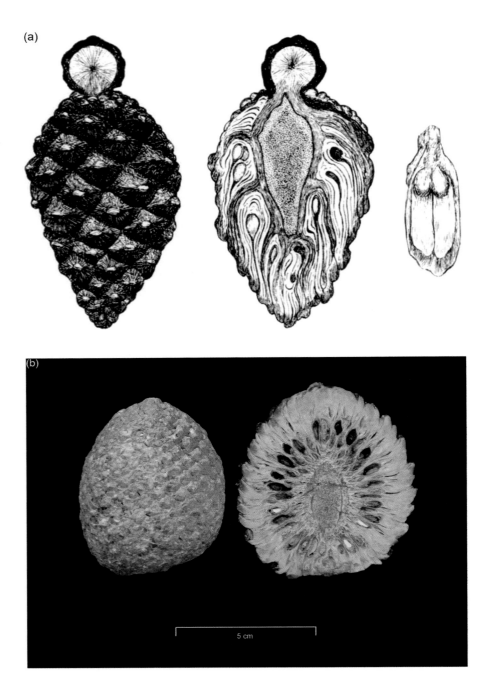

(b)

FIGURE 6.6 (a) Seed-bearing cones of a pine (Monterey pine, *Pinus radiata*, California). (Left) mature cone, external view (ca. 10 cm long), (middle) longitudinal section, (right) single cone scale with pair of seeds. (Adapted from Benson 1959) (b) A silicified cone (ca. 5 cm long) from an extinct conifer (*Araucaria mirabilis*) buried by a volcanic eruption in the mid Jurassic (ca. 160 million years ago) in what is now called Argentina (exterior (left), longitudinal-section with seeds (right)). Fossil trunks show that some trees exceeded 3 m in diameter and perhaps reached 100 m in height. Living members of the genus still occur in South America, eastern Australia and New Guinea. (Cerro Cuadrado fossil forest, © Natural History Museum, London)

Table 6.3 **The selective pressures influencing the characteristics of cones in lodgepole pine (*Pinus contorta*). (From Smith 1970)**

Characteristic of the serotinous cone	Factors selecting for the character	Factors selecting against the character
1. Hard bracts, scale bases and scale apophyses	Squirrels selectively feed on softer cones	The energy cost of building thicker cell walls in the sclerenchymatous tissue
2. Sessile cones	Squirrels occasionally fail to detach sessile cones	–
3. Asymmetrical cones	Squirrels occasionally fail to detach cones with a broad base next to the branch	The energy cost of the enlarged scales
4. Whorled cones	Squirrels occasionally fail to detach cones closely adjacent to each other	Possibly more efficient energy transport to evenly spaced cones
5. Thick vascular trace to the cone	Squirrels occasionally fail to pull off cones with thick vascular traces	Increased energy in vascular tissue
6. Few seeds per cone	Squirrels select cones with more seeds	The energy cost of sterile tissue per seed; lower reproductive output
7. Constant-sized cone crop	Maximizes seeds available for reforestation after frequent fires	Concentrated squirrel feeding pressure on cones produced in small crops

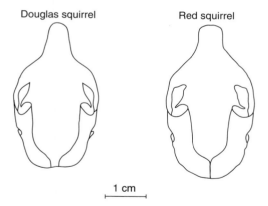

Douglas squirrel   Red squirrel

1 cm

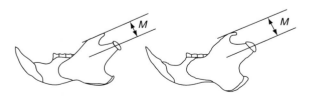

FIGURE 6.7 An example of how plants might affect animal evolution: comparison of the skull anatomy of the Douglas squirrel (left) and red squirrel (right). Red squirrels feed on the serotinous cones of *Pinus contorta*. *M* is the distance of the moment arm of force applied by the temporal muscles, showing that in red squirrels greater force can be applied by the teeth. The red squirrel skull also has a sagittal crest for attachment of well-developed temporal muscles. (From Smith 1970)

Of courses, there may be other functions to the strobilus. It certainly provides strong protection from a wide variety of seed-eating insects as well. And, as noted above, it may also protect seeds from fire. The structure may have several related functions, now, including simply reproduction, but it appears likely that herbivores have had considerable impact upon its evolution.

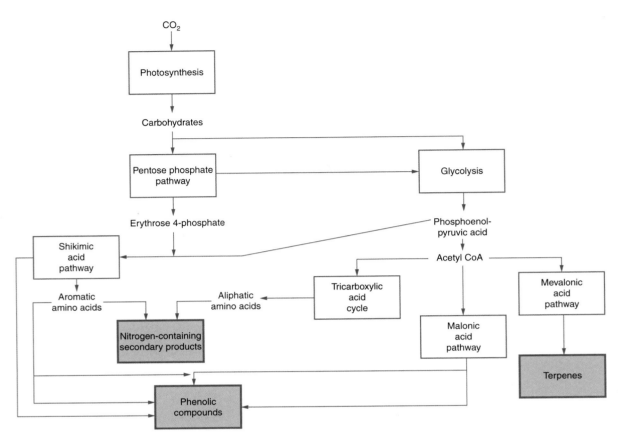

FIGURE 6.8   There are three main groups of chemical defences produced by plants. Here are the pathways that produce them. (From Taiz and Zeiger 1991)

6.3.3 Secondary Metabolites Also Defend Against Herbivores

A thorn or a strobilus is obvious to the eye, while chemical composition is not. But chemistry may be just as important in deterring herbivores. While some molecules found in plants have obvious primary roles to play in photosynthesis and growth, others do not. These latter "secondary metabolites" were once thought to be just waste products. The name has stuck, even though it is now clear that many of these compounds play active and important roles in defending plants against herbivores (Marquis 1991; Hare 2011).

There are three main groups of anti-herbivore compounds, **terpenes**, phenolics and nitrogen-containing secondary products (Figure 6.8). Note that this last group have a high cost, in that they require precious nitrogen molecules. Terpenes, in contrast, are comprised solely of carbon and hydrogen. Over evolutionary time, there appears to have been a gradual transition in the kinds of secondary metabolites involved in plant defence: terpenes can be considered relatively ancient, while steroids, sesquiterpenes and polyacetylene are known only from recently evolved and still rapidly speciating groups, such as the Lamiales and Asterales (Figure 6.9). Let us consider just three ecological examples of such chemical anti-herbivore defences. First come the conifers with their ancient terpene defence system. Next we will look again at tropical forest canopies and, finally, temperate zone wetlands.

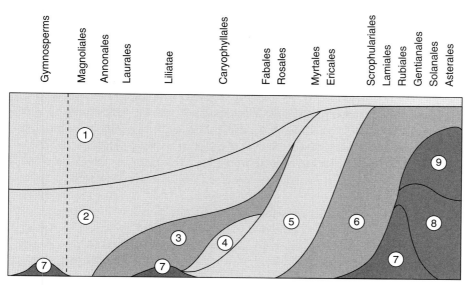

FIGURE 6.9 An overview of plant defences through evolutionary time, from primitive to derived taxa, left to right. In the course of evolution, products of the shikimate pathway (pale shading) were increasingly replaced by those of the alkaloid-synthesis pathway (medium shading) and the mevalonate-acetate pathway (dark shading). (1) lignin; (2) condensed tannins; (3) isoquinoline alkaloids; (4) betalains; (5) gallotannins; (6) indole alkaloids and iridoids; (7) steroids; (8) sesquiterpenes; (9) polyacetylene. (From Larcher 2003) Note that the recently evolved and speciose Asters (Section 2.3.4, Figure 2.10) are at the far right.

Example 1 Conifers and terpenes

In conifers, terpenes and resins are a core defence system, producing both the familiar smell of conifer foliage and providing a raw material for industrial products such as turpentine (Langenheim 2003). Terpene storage structures "can be as simple and short-lived as the resin blisters found in true firs (*Abies*), cedars (*Cedrus*), hemlock (*Tsuga*) and golden larches (*Pseudolarix*), or they can be a complex long lasting highly organized system of interconnected canals found in species of spruce (*Picea*), larch (*Larix*), pine (*Pinus*) and Douglas-fir (*Pseudotsuga*)" (Martin and Bohlman 2005, p. 41). In general, some resin is stored in resin ducts and blisters, while more is synthesized after an attack occurs. The latter is termed an "induced defence," since it occurs only after a herbivore begins eating tissue. New resin ducts may then be constructed to deliver resin to the wound. (You can see conifer resin ducts in the cross-section in Figure 8.1).

When a tree is injured, resin floods the wound, and when the monoterpenes in that resin evaporate, the rest of the resin solidifies. A vivid example is the rapid production of resins that flood tunnels being constructed by bark beetles. Overall, there are three main ways in which the terpenes and resins protect trees (Martin and Bohlman 2005). (1) The resin can provide mechanical barriers to infection and insects. (2) Monoterpenes produced by the tree can also interfere with developing insects by disrupting the endocrine system. (3) In some cases, the terpenes attract predators or parasites that then attack the herbivore.

One of the distinctive features of terpene defences is their release into the atmosphere, producing that characteristic conifer smell (mostly a kind of terpene known as pinene, $C_{10}H_{16}$). Hence terpene defences are termed volatile defences – that is, the chemicals have a low boiling point and rapidly evaporate. It now appears that a fairly wide array of plants synthesize and emit volatile compounds after being injured by

Box 6.1   Naval Stores and Resins

Naval stores refers to a once-thriving industry that supported the great sailing ships of the eighteenth and nineteenth centuries. A nation that could not build and maintain its sailing fleet could neither ensure a flow of trade nor protect its maritime sovereignty. Nor could it keep enemy armies from landing on its shores. Queen Elizabeth I, Napoleon Bonaparte and George Washington all needed sailing ships to win their wars. Wood was needed for masts, decks, keels and a thousand other parts. In addition, each sailing ship absolutely required pitch to seal the cracks between planks, turpentine for making paints and resins for treating rope to reduce decay. These three products came from pine trees. Until the advent of metal ships and diesel engines, pine forests were precious military assets with economic and strategic significance.

Why pines? Conifer trees naturally produce a sticky resin to seal off wounds caused by fire or insects. You have probably seen an injured conifer tree oozing resin that congeals into a thick amber-coloured layer. We know that trees have produced resin for millions of years, because jewellery is still being made from stones, called amber, that are fossilized resin. The remains of insects entombed in some amber testify to the power of resin in trapping insects that would otherwise injure the tree. Some entomologists have studied ancient species of insects by observing their remains inside amber.

The use of pine for naval stores is an ancient practice; the Bible records Noah using pitch to seal the seams of the ark. In the United States, the naval stores industry began in about 1600 on the east coast, based upon the pitch pine (*Pinus rigida*), which grew from Maine to New Jersey. The industry then spread south and west, following forests of other pines. In the era of the Civil War, the naval stores industry, along with agriculture and lumbering, was one of three major southern industries (Meredith 1920).

FIGURE B6.1.1   A crew collecting resin from longleaf pine trees in North America. Note the cuts in the tree bark. ("Gathering turpentine in the pine forest, Covington, La.", postcard, ca.1905, Library of Congress, P & P).

Resin was collected from living pine trees (Figure B6.1.1) and processed in much the same way that sap is collected from maple trees and turned into syrup today. The flow of resin was started by making deep v-shaped cuts into the tree bark and was collected at the bottom of the cut. Once a sufficient volume of resin had been obtained from many trees, the resin was heated over a wood fire. The fumes were driven off by the heat; what remained was pitch. Later, enclosed iron pots were introduced into the process; the fumes were condensed to form valuable turpentine. This was mostly a terpene, α pinene. Now there were two products: the more volatile spirits of turpentine, which could be used as a solvent, and the heavier resins and gums that remained in the pot.

Resin could be harvested from dead trees as well. One method required building a "kiln" – a pile of pine wood about 4 m high by 6 m in diameter, covered in earth, surrounded by a small gutter or trench. Once

Box 6.1   (cont.)

ignited, the kiln slowly burned the enclosed wood. Resin seeped from the pile and collected in the gutter. One kiln might yield 100 barrels of resin over eight or nine days of smoldering.

As the pitch pine stands in the north and east of North America were degraded, the industry moved into other pine forests, eventually reaching Texas. Remarkably, the Europeans, particularly the French, had already developed a sustainable system for producing turpentine in plantations – trees were cultivated to produce turpentine for 30 to 50 years before being harvested for timber. Yet North Americans continued using methods that killed the trees within a few years, necessitating the continual destruction of new swaths of forest.

To give some idea of the volume of resin extracted from North American pine forests, consider records from Savannah, Georgia, once a major port for the storage and export of naval stores. Between 1895 and 1902, more than a million barrels of rosin (or resin) and 300,000 casks of turpentine arrived annually (Gamble 1921). By 1914, these receipts were halved and, by 1920, fell to 223,239 barrels of rosin and 88,910 casks of turpentine.

As the pine forests reached their limits, even stumps were dynamited out of the ground to extract their valuable resin. The desperate condition of the naval stores industry was summed up by Mr. Hawley of the US Forest Service

the supply of stumps is no more inexhaustible than the supply of trees which originally grew on these stumps, and probably within 10 or 15 years after the last tree of the present stands of longleaf pine has been cut, the stump of that tree will be pulled out and distilled for the production of wood turpentine and wood rosin. What will be the condition of the naval stores industry of this country when that time comes? (Gamble 1921: p. 239)

In its short period of production, the naval stores industry claimed most of the great longleaf forests east of the Mississippi River, and one still can find depressions that testify to dynamited stumps. Many of the plant and animal species that occur in this habitat are on the rare, threatened or endangered species list, from gopher tortoises to carnivorous plants to orchids.

Turpentine remains an important solvent for a variety of chemicals, as well as for manufacturing resins (Langenheim 2003). Although live pine trees are still tapped, the dominant source for turpentine is now as a by-product of paper production from pulped conifer trees.

herbivores such as caterpillars. An undamaged plant will have a baseline level of these metabolites. Injury by a herbivore stimulates an increase in both the volume and variety of volatiles. Such emissions can attract predators that attack the herbivores. Small parasitic wasps, for example, find their caterpillar prey by smell, but mostly depend upon the smell of damaged plants rather than the smell of the herbivores themselves (Pare and Tumlinson 1999). In some cases, hyperparisitoids, that is predators upon the predators, can also be attracted by these plant emissions (Poelman et al. 2012). You can see how challenging it is to test which compounds have ecological significance given the complex mixtures of compounds that are found in nature, the general difficulty of field experiments, and the presence of so many insect species and trophic levels (Arimura et al. 2009; Hare 2011). Meanwhile, the list of volatile compounds found in plants continues to grow while rather little is understood about their significance in nature.

Of course, insects have adapted to all these plant defences. The mountain pine beetle, for example, having attacked a tree, will release pheromones to signal other mountain pine beetles to attack the same individual; if enough beetles attack at once, they can overwhelm defences. Other pheromones signal when a tree is fully occupied, discouraging more females from arriving. We shall return to bark beetles and conifers in the final pages of this chapter. Overall, insects such as bark beetles are very effective at killing conifers, yet conifers have been in the world for hundreds of millions of years, and there are still vast areas of conifer forest. This suggests that the terpene and resin system, while ancient, is still quite effective.

Example 2 Tropical forest canopies

Moving on to the defences that protect tropical tree leaves, Coley (1983) monitored 8,600 leaves on saplings of more than 40 tree species of tropical forest in Panama. The task was to assess which plant traits were associated with tissue loss from herbivores. Many different traits were measured on these leaves, including tannin content and presence of hairs. The result? Overall, young leaves were grazed far more than old leaves, probably because the young leaves were half as tough, less fibrous and more nutritious than old leaves. The only property protecting young leaves was toughness (Table 6.4, column 2). For mature leaves, most of the among-species differences in herbivory could be accounted for by differences in leaf traits (Table 6.4, right-hand column). Finally, the other noteworthy, perhaps unexpected, result was the similarity in rates of herbivory between pioneer (early-successional) trees and persistent tree species of mature forests. Grazing damage was greater during the wet than the dry season.

Example 3 Temperate zone wetlands

Although large areas of the Earth have wetlands, less is known about anti-herbivore defence compounds in wetland plants (Rosenthal and Berenbaum 1991). There is, however, a long list of candidate compounds, including glucosinolates (Louda and Mole 1991), coumarins (Berenbaum 1991), flavonoids (McClure 1970), alkaloids (Ostrofsky and Zettler 1986) and possibly iridoid glycosides (Bowers 1991), as well as fatty acids, mustard oils and steroids (Gopal and Goel 1993). Again, it is one thing to detect the presence of a secondary metabolite, quite another to show that herbivores are affected by it. This lack of information led McCanny et al. (1990) to screen 30 species of North American marsh plants to test for anti-herbivore defences against insects. Instead of measuring damage to leaves in the wild, like Coley did in the forest example above, or instead of trying to measure all the chemicals that might be present in the foliage, McCanny directly extracted an array of chemicals from a set of wetland plants and inserted this array, along with leaf fibres, into the diet of a herbivorous insect. Control insects received a normal diet, and reduction in growth relative to the controls (expressed here as a ratio) showed just how well each species was defended. In two species the values were marginally greater than one, suggesting enhanced growth rates from the wild diet, while the 28 others ranged to values less than zero, where the larvae actually lost weight while feeding. In some cases the chemical extracts and fibres had only a marginal effect – examples include iris (*Iris versicolor*, 0.94) and cattail (*Typha angustifolia*, 0.87). In others, much larger reductions in the herbivore growth rate resulted – examples include reed canary grass (*Phalaris arundinacea*, 0.58) and two species of sedge (*Scirpus cyperinus*, 0.50; *Scirpus americanus* 0.53). That is, insects feeding on extracts from the last three species grew only about half as well as those on the control diet. The two plant species capable of eliminating or even reducing larval size were both in the Asteraceae (*Corepsis rosea*, *Solidago graminifolia*). Recall that the Asteraceae are at the far right in Figure 6.9 and have an array of recently evolved defences.

Table 6.4 **Correlation coefficients between rates of herbivory and defences of young and mature leaves in tropical forests. Values are based on 22 pioneer and 24 persistent species. (From Coley 1983)**

	Young leaves			Mature leaves		
	Pioneer	Persistent	Both	Pioneer	Persistent	Both
Chemical						
Total phenols (percent dry mass)	0.014	0.030	0.040	−0.145	0.109	−0.099
Tannins – Vanil (percent dry mass)	0.278	0.182	0.190	−0.130	0.023	−0.112
Tannins – Leuco (percent dry mass)	0.220	0.207	0.209	−0.087	0.220	−0.128
Fiber – NDF (percent dry mass)	0.229	0.180	0.254	−0.119	0.056	−0.278[a]
Fiber – ADF (percent dry mass)	0.302	0.095	0.181	−0.318	−0.056	−0.424[a]
Lignin (percent dry mass)	0.213	0.148	0.133	−0.219	0.101	−0.223
Cellulose (percent dry mass)	0.340	0.036	0.195	−0.348	−0.147	−0.473[b]
Physical						
Toughness (N)	0.011	−0.473[a]	−0.045	−0.360	−0.363	−0.515[b]
Hairs – upper (no./mm^2)	0.001	−0.219	−0.146	−0.040	n/a[c]	0.291[b]
Hairs – lower (no./mm^2)	0.062	−0.306	−0.114	0.591[b]	0.089	0.635[b]
Nutritional						
Water (percent)	0.156	−0.082	0.081	−0.027	0.448[a]	0.507[b]
Nitrogen (percent dry mass)	0.028	−0.237	0.004	0.010	0.278	0.287[a]
Tannin: protein ratios						
Total phenols: protein	0.031	0.124	0.093	−0.125	0.037	−0.183
Tannin – Vanil: protein	0.289	0.111	0.167	−0.161	0.049	−0.161
Tannin – Leuco: protein	0.292	0.133	0.191	−0.182	0.145	−0.184

Notes:
[a] $P < 0.05$, one-tailed.
[b] $P < 0.01$, one-tailed.
[c] None of these species is pubescent on the upper surface.

6.3.4 Some Cautions When Interpreting Anti-Herbivore Defences

Clearly plants have defences against herbivores. You might first think that this also shows that herbivores have a major effect upon plant communities. As we shall see again and again, however, it is one thing to measure something in nature, even with carefully controlled experiments, and quite another to correctly interpret the result. Here are three cautions for you to consider.

First, it is necessary to assume that the function of the alleged anti-herbivore trait has been correctly and

FIGURE 6.10   Imagine how giant ground sloths some six metres long would change forest composition. (Hutchinson 1892, Plate XVIII by J. Smit). The mass extinction of large herbivores during the Pleistocene era has left many areas of the world deficient in large herbivores.

unambiguously interpreted. For example, some vines have thorns. Do the thorns of vines that scramble along shorelines (e.g. *Smilax* spp.) act to deter herbivores or are they a morphological device for climbing upon other plants? Or both? In the same vein, do secondary metabolites exist to protect plants from herbivores, or is their main function primarily in fact anti-microbial? After all, plants must also defend against a wide array of pathogens including fungi, bacteria and viruses. The problem of interpretation always complicates comparative evidence.

Second, there is an issue of time scales. In general, evolution is a slow process. The presence of an anti-herbivore structure, like a thorn or a strobilus, suggests that herbivores have shaped the evolution of a group, but it in no way demonstrates that it is due to a herbivore that is still active in present-day plant

communities. An example. As recently as ten to fifty thousand years ago (a very short time in geological or evolutionary context) a wave of extinction eliminated most large animals from every continent except Africa. Given the timing, and other sources of evidence, the most plausible explanation for the extinction is the arrival of tribes of human hunters and their encounter with animals that had not evolved with human hunting as in Africa (Martin and Klein 1984; Koch and Barnosky 2006). Hence, except for Africa, most areas of the world are now missing large herbivores, and great herbivores such as giant ground sloths (Figure 6.10) and woolly mammoths. Glyptodonts are entirely gone. Hence some plants may possess defences against now-extinct herbivores.

Finally, if we focus on the question of how much impact herbivores have upon plant communities,

FIGURE 6.11   Adult luna moths (*Actias luna*) do not have mouths and live only about a week. But a single female can produce several hundred eggs and the caterpillars will grow to 7 cm long, consuming the leaves of deciduous forest trees. The guts of their caterpillars contain many enzymes, including quinone reductase and glutathione transferase, that may detoxify the defensive chemicals in those leaves (Lindroth 1989). Preferred trees include the genera *Betula, Carya* and *Juglans*. (Ontario, by C. Keddy)

there is a curious bit of logic. If plants are poorly defended against herbivores, then it is likely that herbivores will have great impacts upon plants as individuals and communities as a whole. But if plants are strongly protected against herbivores, then one could argue that the impacts of herbivores will be smaller (even minimal) since plants are actively resisting the herbivory. In principle, then, one could

argue that if plants were sufficiently well defended, herbivores would have few effects at all. So far as one can tell from the evidence available, this is an extreme situation that is unlikely to occur in nature. That is, to judge from the studies of herbivore impacts that we shall see later in this book (large grazing animals in grasslands; insects in forests), most herbivores are still able to cause major changes in vegetation. At the very least, they are able to selectively remove species that are poorly defended. A classic example, which you can see from many highways, is pastures that have been overgrazed, leaving only clumps of well-defended thistles. Most herbivores are similarly selective, and so may be shifting the composition in communities. The larvae of luna moths (Figure 6.11), for example, may be changing the composition of forests by selectively feeding on certain tree species. In those cases where the species being selectively removed is a common or dominant species, then this selective removal may have dramatic effects on landscapes, as we shall see later in this chapter with spruce budworm and bark beetles. Vast areas of dying forest can raise some very challenging disagreements about vegetation management. That is why you have already read about natural disturbance cycles in landscapes. But before we continue, let us turn the inquiry on its head for a moment and ask what properties of plants might make them most desirable to herbivores.

6.3.5 Food Quality Is Predicted by Nitrogen Content

Despite having various means of defence against herbivory, plants are a source of food and nitrogen content is thought to be one of the most important factors determining the food value of plants (Lodge 1991; White 1993). The nitrogen content of many plants is often well below five percent (Table 6.5). These are very low values for supporting herbivorous animals. The struggle to extract essential resources such as nitrogen from plant tissues has likely been the major driving force in the evolution of digestive systems in terrestrial animals (Langer 1974; Janis 1976). The

Table 6.5 Composition of wetland plants. (After Boyd 1978 and Junk 1983)

Constituent (percent)	Temperate species		Tropical species	
	N	Mean	N	Mean
N	27	2.26	75	1.99
P	35	0.25	75	0.19
S	25	0.41	–	–
Ca	35	1.34	75	0.88
Mg	35	0.29	75	0.29
K	35	2.61	75	3.10
Na	35	0.51	75	0.36

presence of a rumen, the maintenance of a rich symbiotic fauna and increases in gut length are characteristics shared by most grazing animals (Smith and Douglas 1987).

Animals, then, are faced with the daunting task of extracting small amounts of nutrients from large volumes of cellulose. We have already explored in Chapter 3 how plants have used cellulose as the primary structural molecule, and some of the consequences of this for decomposers, soils and herbivores. That is to say that we could view low nitrogen concentrations in plants as one of the most all pervasive anti-herbivore traits. Because it is such a broad generalization with such a long evolutionary history, it would be easy to overlook as a defence against herbivores. But it is very effective.

One counter-measure for herbivores is to focus upon those plant tissues with the highest nutrient levels. That is, it is misleading to think just about the average nitrogen content of a plant, since certain tissues have higher nutrient levels and it is these that are often sought out by herbivores. Meristems and seeds have higher nitrogen content than most other tissues, which likely explains why herbivores selectively remove these tissues. A single grouse, for example, can harvest 1,000 dormant buds per day from deciduous trees during the winter (Huempfner and Tester 1988). Actively growing tissues also tend to have higher nutrient contents, and there are many examples of herbivores preferentially consuming new growth (Smith 1983; White 1993).

This raises the question of just how low nutrient levels can go. Wood, being mostly cellulose, certainly has extremely low nutrient levels, which is one reason why degradation of logs is such a slow process. But at the other end of the tissue continuum, nitrogen is so important for protein synthesis that meristems and seeds are likely always going to have higher nutrient levels and therefore always be at greater risk from herbivores. Which of course would explain why seeds are surrounded by tough cones, hard seed coats and even spiny fruits. While seeds can be protected in this way, actively growing meristems remain vulnerable, since actively growing tissues cannot be protected by hard tissues.

Once you begin to watch carefully for herbivore damage, you will see for yourself how herbivores select nitrogen-containing meristems. As I write this chapter in a Canadian winter forest, beavers and porcupines are stripping the bark from trees to reach the underlying cambium, which is a kind of meristem. Meanwhile grouse perch in the tops of trees and eat the dormant buds, while deer remove the tips of tree branches at lesser heights, as well as the tops of any young trees that emerge above the snow. You can make similar observations wherever you live.

6.4   Field Experiments Expand Understanding of Herbivory

Having begun with relatively simple observations on diet and morphology, it is time to examine field experiments. The principle is simple (Bender et al. 1984): exclude herbivores from some areas (these

(a)   (b)

FIGURE 6.12   An early exclosure experiment using a rabbit-proof fence across Cavenham Heath in England. (a) On the protected side (right) there are many inflorescences, while on the unprotected side there are no inflorescences, the vegetation is nibbled closely down and much bare sand is exposed. (b) Another view of the rabbit-proof fence on Cavenham Heath, this time looking in the opposite direction. Here there is a greater difference in the vegetation on the two sides. The sedge *Carex arenaria* is abundant on the protected side of the fence. (From Farrow 1917)

areas are termed **exclosures**) and compare the vegetation in these areas to the vegetation in unfenced control plots. This is not a new idea: Clements (1935) explains: "The usual type of closure is the exclosure, a fenced area of suitable size which provides protection against the coactions of one or more groups or species of animal. The enclosure is similar in design, but is used to restrict animals to a definite area to permit the direct study of their influence." One could also add other treatments that remove only selected herbivores or even add certain species of herbivores. Figure 6.12 shows an early study that put a rabbit-proof fence across an English grassland (Farrow 1917). Such experiments have been used many times since, and with increasing sophistication. You shall see some examples below, while guidelines for constructing them are given in Box 6.2.

6.4.1 Caterpillars Consume Deciduous Forest Canopies

We know that large numbers of leaf-chewing lepidopterans feed on deciduous forest trees (recall

Figure 6.11), but what are the effects on the plants? To answer this question, Marquis and Whelan (1994) selected 90 *Quercus alba* (white oak) trees in a mixed oak forest in Missouri. The trees were arranged in triplets, one served as a control, the second had native insects removed (sprayed weekly with a synthetic pyrethroid combined with hand removal) and the third was caged (a fine mesh with 3.8-cm holes provided access to insects but not to insect-eating birds). An inventory of the damage to these trees was taken each autumn for two years. Figure 6.13 shows that control plants lost roughly 15 to 25 percent of their leaf surface area depending upon the year. This, then, is the normal amount of damage that is done by herbivorous insects. Spraying and hand picking reduced the number of insect herbivores by a factor of ten, and roughly halved the amount of damage to leaves.

Now what happened in the caged plots, those without predation from wild birds? The birds in this area were typical deciduous forest species and included 17 common spring migrants and 31 spring and summer residents. The right-hand bars show that when insectivorous birds were excluded, the number of

Box 6.2   Design of Exclosure Experiments for Herbivores

Let us read from a classic exclosure experiment.

Why regeneration of beech woods in England and Scotland is so poor was determined by fencing small areas. From some areas, rabbits were excluded; from others, both rabbits and birds; and from still other, field mice as well. By placing beechnuts in each enclosure both on the surface and within the forest litter, it was found that rabbits ordinarily ate very few seeds, birds secured only those on the surface, but mice destroyed almost the entire supply, except a few overlooked in the duff. This simple experiment established the fact that, in a large measure mice are responsible for the failure of the rejuvenation of beech forests.

An exact knowledge of the amount of damage done to the range by rodents and especially of the rate and degree of recovery of the various types of vegetation after rodents have been eradicated has been obtained in a similar manner. Exclosures were made against rodents and cattle, against cattle alone, and against cattle where rodents had been killed. Clipped quadrats from these as well as from enclosures inhabited by jack rabbits and others inhabited by kangaroo rats showed the heavy toll of vegetation taken by these pests, a matter of grave importance, especially in times of drought.

(from Weaver and Clements 1929, pp. 38–39).

And now, ten considerations for designing your own exclosure experiment.

1. A well-defined question. Combine your own experience at the site with a careful study of the literature for previous problems with design. Some early examples to consult include (Farrow 1917), Tansley and Adamson (1925) and Baker (1937). Milchunas and Laurenroth (1993) have compiled a data set for 236 sites where effects of herbivory have been examined.
2. Proper randomization of treatments and controls, including stratification if necessary.
3. Appropriate design. Consider nesting each set of treatments to avoid confounding with habitat differences. Consult a standard text on experimental design such as Mead (1988). Consult a statistician.
4. Verification. Take vegetation measurements before sites are en/exclosed to test for differences before treatments are initiated. Use these data as co-variants in analysis. These data may also be used to determine the amount of replication necessary.
5. Type of fencing. The degree of permeability to grazers is determined by (a) height, (b) depth of burial and (c) mesh size. You may use several kinds of fencing to select different kinds of herbivores (yielding different treatments T_1 to T_n).
6. Confounding. Design en/exclosures to minimize shading, snow capture, protection from wind, etc. Make extra measurements to test for these effects.
7. Sampling. Avoid areas near fences to minimize problems caused by run-off, shading, toxins from metals or other fence effects. Avoid pseudoreplication (Hurlbert 1984). Maximize generality (Keddy 1989). Collect vouchers of all species and ensure that they are confirmed by an authority and stored appropriately.
8. Expansion. If resources permit, expand treatments to include resource manipulations (e.g. water, fertilizer) or manipulate other appropriate factors (e.g. fire, insects).
9. Duration. Maintain treatments as long as possible since most ecological studies are very short term. Consider what will be the criteria for ending the experiment. Is there an agency that could take over maintenance and monitoring in the long term?
10. Publish the results. Data buried in unpublished files benefit no one. Where will the raw data tables be stored for future use by other scientists?

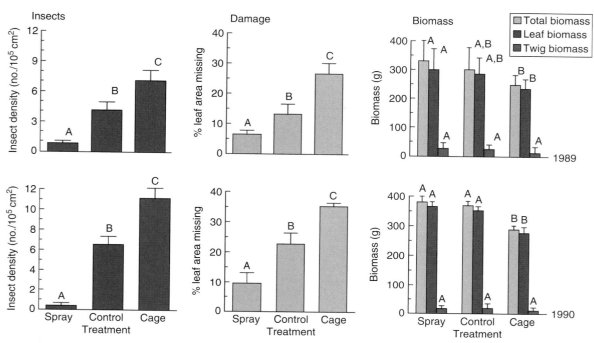

FIGURE 6.13 An experimental study of the effects of Lepidoptera on trees. The insect density (left), leaf damage (middle) and biomass production (right) is shown for 90 white oak trees divided into three treatments. In the spray treatment insects were controlled by spraying and hand picking. The controls were left without manipulation. In the cage treatment insectivorous birds were excluded. Letters denote treatments different at the $p < 0.05$ level; bars include estimates of standard error. (From Marquis and Whelan 1994)

insect herbivores almost doubled and losses of leaf area rose from 25 to 35 percent. This study therefore provides strong experimental evidence that forest-dwelling birds protect trees. If migratory bird populations continue to decline, Marquis and Whelan (1994) suggest that reductions in tree growth are likely.

This study introduces a new phenomenon in herbivory: the role predatory animals (in this case, birds) may play in protecting plants from herbivores. It is known as a trophic cascade and we shall see more examples soon.

6.4.2 Land Crabs Can Change the Composition of Tropical Forest

You have already seen that the absence of mammalian grazers on some islands has produced special situations where giant tortoises are the dominant land herbivore (Section 6.2.3). On other islands giant land crabs play a similar role. Kellman and Delfosse (1993) report that crabs can reach densities as high as one per square metre. What effects might these crabs be having on plants and forests? Green et al. (1997) tested for the effects of land crabs (*Gecarcoidea natalis*) in rain forest on Christmas Island in the Indian Ocean. Ten pairs of 5 m × 5 m plots were used, with one plot in each pair being surrounded by a semi-permeable fence. The fences allowed small invertebrates, as well as lizards, to move in and out of the enclosed areas but crabs were excluded. Seven fences were constructed under closed forest and three in light gaps. The emergence of seedlings of tropical trees was more than 20 times greater in exclusion plots (plots in exclosures), and the number of species of tree seedlings per plot was also higher (10 to 12 species in exclosures versus one in controls).

Let us look at some more natural historical details. Seedlings of trees such as *Maclura cochinchinensis*, *Planchonella nitida* and *Schefflera elliptica* were common in the exclosures, but none survived for longer than two months in control plots. Red crabs can crush propagules of many species with their claws. When propagules were placed in trays on the forest floor, 80 percent or more were removed within two weeks. Some of the propagules removed were probably consumed but at least one-quarter were dispersed, and propagules of species such as *Barringtonia racemosa*, *Pandanus christmatensis* and *Terminalia catappa* were all found inside the entrances of crab burrows. All of the exposed seeds of *Inocarpus fagifer* and *Tristiropsis acutangula* were handled by crabs, but the crabs were unable to penetrate the tough fibrous endocarps; the crabs did, however, eat the pulp of the fruits and drag the seeds to burrow entrances. It may be that crab herbivory has benefits for tropical trees (dispersal) as well as costs (loss of offspring). Crabs and squirrels, then may, unexpectedly, have a similar function in dispersing trees.

Land crabs, then, may have a substantial impact upon the forests of oceanic islands. The tree canopy on Christmas Island includes many species that are vulnerable to crabs as seedlings. This suggests that occasional periods of low crab density may be necessary for the regeneration of these trees. Overexploitation of land crabs by humans has so reduced many populations that it may now be difficult to determine what impact they once had upon forest composition. The possibility that harvesting land crabs might entirely change forest composition is yet another reminder that the consequences of human activities upon the environment can be difficult to foresee, particularly when trophic cascades occur. We will return to this important topic in Chapter 13.

6.4.3 A Large Experiment on Grasslands in Tanzania

We started with simple fences across grasslands, and then explored how experimental exclosures have been used in two cases: lepidoptera in forests and land crabs on tropical islands. I deliberately showed you examples of arthropods because they are often overlooked compared to large mammals. The final experiment we are going to examine takes us back to large mammals in the grasslands you saw in Figure 6.1. This example was chosen not only because it deals with large mammals and grazing ecosystems, but because it illustrates a wonderfully complex experimental design. It illustrates how enclosures can be used to explore quite precise questions about grazing. A word of warning: there is a lot to keep track of and a large table. You should not be surprised if you do not understand it the first time through. Focus, instead, on how the experiments were designed to answer specific questions.

The setting of this experiment is Serengeti National Park, Tanzania, a park with 13,000 km^2 of rolling grasslands, savanna and open woodlands. Rainfall in the park increases from 500 mm in the southeast to 1,000 mm in the northwest and produces a vegetation gradient of short grass (3 to 5 cm high, *Digitaria* and *Sporobolus*) in the south, through mid grass (50 to 100 cm, *Themeda, Sporobolus, Pennisetum*) up to tall grass (100 to 150 cm, *Diheteropogon, Elyonurus, Hyparrhenia*) in the north. At each site, two experimental blocks were selected, and one was enclosed by a chain link fence in order to exclude grazing mammals. Within each block, there were seven treatments: control, two disturbance treatments (deep disturbance, shallow disturbance), fire and three removal experiments (removal of dominant species, removal of subdominant species, and removal of dominant and subdominant species). Although fire is a kind of disturbance, as you know from Chapter 5, in this particular study the term disturbance was applied only to physical churning of the soil, and we will use their choice of words. At the end of each growing season, above-ground cover was measured for each species. The analysis then explored comparisons among the treatments (Table 6.6).

Consider just the responses of short-grass vegetation in Table 6.6. First, grazed plots were compared to ungrazed plots. The dominant grasses such as *Sporobolus fimbriatus* increased in the exclosure, whereas small species such as *Cyperus rotundus* declined. Next, disturbed plots were compared in grazed and ungrazed treatments.

Some species increased regardless of grazing treatment (e.g. *Melhania ovata*). In other cases, species responded positively to disturbance only in ungrazed plots (e.g. *Harpachni schimperi*), and others increased only in grazed plots (e.g. *Medicago laciniata*). Still others decreased with disturbance (*Sporobolus ioclados*), some declining only in grazed plots (*Kyllinga nervosa*) or ungrazed plots (*S. fimbriatus*). In those plots with dominant species removed, several species increased in the ungrazed plots (*H. schimperi, Digitaria macroblephara*). As the final pair of columns in Table 6.6 shows, fire had relatively few effects, although three species showed declines, two of these in ungrazed plots. Detailed results for the other communities can be found in Belsky (1992).

Since there are so many species in the study, it may be helpful to consider simplification by using factors such as plant traits. Protection from grazing increased the dominance of some species and decreased species diversity as tall species with methods of vegetative propagation shaded out shorter species and outcompeted species that reproduce only sexually. Belsky reported that tall species such as *D. scalarum, S. fimbriatus* and *Pennisetum mezianum* tended to replace short species. Even here there were exceptions; the short grass, *H. schimperi*, increased in the exclosure at the short-grass site (Table 6.6) apparently because it was shade tolerant. *Themeda triandra*, in spite of being a tall perennial, declined in exclosures in the mid-grass site, apparently because it is short-lived and has seeds that cannot germinate in shade. In disturbance treatments, in general, annual species and short species tended to increase. Fire had limited effects, its greatest being the stimulation of *T. triandra* regeneration from seed in mid-grass sites. Most other changes caused by fire were negative, presumably as a result of heat damage to shoots and buds. Finally, nearly 50 percent of the species at each site responded significantly to removals, a result that suggests that competition was controlling species' abundances. The many species that decreased when neighbours were removed suggests that commensalism also occurred or

that a third species increased after the removal and suppressed other neighbours. Surprisingly, the number of species increasing after removal of neighbours was nearly the same in grazed and ungrazed plots, and this seems to indicate that interspecific competition occurred even in heavily grazed grasslands.

To summarize further, all three types of grassland responded more or less similarly to the treatments, and the effects of grazing and disturbance were greater than those of competition and fire. The relative importance of these factors is difficult to assess, however, because the treatments represented different intensities of perturbation; comparing the intensity of perturbations is a difficult problem in all such experiments.

6.4.4 Some Lessons for Exclosure Experiments

1. You see that in less than a hundred years, ecologists have gone from the simple single-fence experiments (Figure 6.12) to complex experimental designs that require computer statistical packages and advice from statisticians (Table 6.6). It raises the question of whether experiments can become so large and complicated that we may not be able to understand the results. If so, how do we design experiments that answer important questions without overwhelming ourselves?

2. It might be tempting to avoid complexity, say, by studying only one plant species, except there would be no way to know in advance which species to use – and just think about how differently one might view this grassland had one studied just one of the rows in Table 6.6. Yet, in spite of this, there are still many single-species studies being done in ecology, and some scholars might have tried to publish the results for each species in a separate paper rather than in one table.

3. Time is an important factor in any experiment. The results in Table 6.6 suggest that five years were not long enough to find treatment effects, which emphasizes the need for more long-term studies in ecology. But five years is a long time in anyone's

Table 6.6 The effects of experimental manipulation on the cover of common grassland species for five years following the manipulations in a short-grass community in the Serengeti (+ indicates increased cover and – indicates decreased cover (p <0.05) due to the treatment). When significant interactions occurred, u indicates that significant changes occurred only in the ungrazed block and g indicates that significant changes occurred only in the grazed block. (From Belsky 1992)

Species	Initial cover of site (percent)	Species characteristics	Effects of protection from grazing in year:				Contrast 1: effects of disturbance[a] in year:				Contrast 2: effects of removal of dominant spp. in year:			Contrast 3: effects of early dry-season burn in year:	
			1	2	3	5	1	2	3	5	1	2	5	2	3
Harpachni schimperi	0.2	p,s	+	+	+	+		+u	+u	+u		+u	+u		–
Melhania ovata	0.2	p,t	+	+	+	+	+	+	+	+					
Digitaria macroblephara	3.2	p,t	+	+	+	+	–	–		–u	+u	+u	+u		
Sporobolus fimbriatus	11.3	p,t	+	+	+	+	–u								–u
Sporobolus ioclados	10.9	p,t		+			–	–	–	–u			+u		
Euphorbia inaequilatera	16.7	a,s	+	–	–	–	+	+g	+g			+g	+g		
Medicago laciniata	0.4	a,s	+							+g	–u				
Chloris pycnothrix	4.7	a,s		–	–	–	+		+g						
Cyperus rotundus	4.2	p,s		–	–	–	+	+	+u	+g			–g		
Eragrostis papposa	0.3	a,s		–	–	–	+u	+	+	+			+g		
Indigofera microcharoides	0.4	a,s						+u	+u						
Kyllinga nervosa	0.2	p,s		–	–	–	–	–g	–g	–g	+	+	+g		
Justicia exigua	0.2	p,s	–	–	–	–									
Cynodon dactylon	3.2	p,t													
Digitaria scalarum	21.7	p,t	[]	[]	[]	[]	–	–							–u
Microchloa kunthii	1.2	p,s	[]	[]	[]	[]	–	–	–	–g	–u	–			
Sporobolus kentrophyllus	7.9	p,t	[]	[]	[]	[]	–	–g	–g						
Species affected (%)			50	79	64	71	71	82	71	59	24	29	41	0	18

Notes:

a, annual; p, perennial; s, short species (ungrazed height <15 cm); t, tall species (ungrazed height >15 cm); [], effects of the grazing treatment could not be determined due to initially unequal cover values in the grazed and ungrazed blocks; [a] physical churning of soil.

career, and what of tenure committees who lack the patience for quality studies?

4. Learn from previous failures. It is important to read early work and not just decide to build a fence bought at the local hardware store. In Box 6.2 I have summarized work from a large number of studies to give you some advice on how to design exclosure experiments efficiently.

5. There was one further problem illustrated by this study. Attentive readers may have noted that there was just one large exclosure at each site. This is a mistake known as pseudoreplication (Hurlbert 1984) and one that does not allow for the proper testing of grazing effects. To correctly assess the effects of grazing, it would have been necessary to replicate the exclosures and control plots multiple times at each of the three sites. This vastly increases the amount of work necessary, not to mention the cost.

It would seem that there is some intermediate degree of enclosure complexity that would answer key questions, permit sufficient replication and continue for sufficient time. The difficulty in mastering such trade-offs among size, shape, replication, location and duration show where science remains an art. I have written more about this elsewhere, and if you wish to read more about how we might simplify complex studies in research, you can read Keddy (2010: Chapter 12).

6.5 Empirical Relationships Uncover General Patterns in Herbivory

Experiments provide a method for testing whether herbivory causes changes in vegetation, and for then exploring causal relationships in a given location or habitat. But how to move from single locations to general relationships? It is not easy. One tool is called meta-analysis. It looks at results from many experiments combined. We shall look at two examples. Both use information on the rate of production of biomass, that is, aspects of productivity.

In a remarkable piece of work similar to meta-analysis, Milchunas and Laurenroth (1993) used multiple regression analysis to synthesize a data set of 236 sites where grazed and ungrazed tracts of vegetation were compared; more than 30 journals were examined in their entirety, some 500 articles were surveyed and 97 were used in the analysis. To understand the results, some terminology is necessary. The two most important dependent variables were: (1) observed changes in species composition and (2) ANPP (annual above ground net primary production). The effects of the exclosure or, conversely, the effects of herbivory were measured using changes in species composition (Table 6.7). The measure used to assess changes in species composition potentially ranged from 0 (no differences between grazed and ungrazed sites) to 1 (completely different ungrazed and grazed sites). Annual net primary production (ANPP) was estimated as maximum standing crop. The independent variables included: (1) grazing intensity, (2) years of protection from grazing, (3) evolutionary history of grazing as assessed by a panel of experts, (4) mean annual precipitation, (5) temperature and (6) latitude. The intensity of herbivory was measured as the percentage of ANPP consumed by the grazers; it ranged from a mean of 44 percent in grasslands to 60 percent in forests. The sites covered a range of precipitation from 220 to 1911 mm year^{-1} and latitudes from 0 to 57°.

Table 6.7 shows that just three of nine potential variables explained 44 percent of the variance in composition for all vegetation types combined. Consumption (0.17) contributed about as much to the cumulative R^2 as did NPP (0.15) while evolutionary history of grazing contributed only 0.09. Species dissimilarity (differences in species composition between grazed and ungrazed areas) for all communities, grasslands plus shrublands, and grassland alone,

Table 6.7 Regression models for plant species dissimilarity of ungrazed versus grazed plant communities and the sensitivity of species dissimilarity to changing an independent variable from a low to a high value while holding other independent variables constant. All communities means grassland, shrubland, forest, mountain (alpine meadow) and desert. Consumption percent reflects the strength of herbivory. Dissimilarity values were computed as one minus Whittaker's (1952) index of association. (After Milchunas and Laurenroth 1993)

Habitat	Independent variables	Cumulative R^2	Final model		
			Slope	Sensitivity	Significance
All communities	ANPP	0.15	1.659×10^{-3}	+0.30	0.000
	Consumption (percent)	0.32	4.272×10^{-3}	+0.13	0.000
	Evolutionary history	0.41	6.569×10^{-2}	+0.20	0.000
	$(ANPP)^2$	0.44	-1.804×10^{-6}		0.033
	Regression constant		-0.214		0.003
Grasslands plus shrublands, precip. $\leq 1,000$ mm/year	Evolutionary history	0.19	6.447×10^{-2}	+0.20	0.000
	ANPP	0.32	2.139×10^{-3}	+0.37	0.000
	Consumption (percent)	0.50	4.903×10^{-2}	+0.15	0.000
	$(ANPP)^2$	0.55	-2.406×10^{-6}		0.002
	Regression constant		-0.320		0.000
Grasslands, precip. $\leq 1,000$ mm/year	ANPP	0.25	2.849×10^{-3}	+0.47	0.000
	$(ANPP)^2$	0.40	-3.260×10^{-6}		0.000
	Consumption (percent)	0.49	3.944×10^{-3}	+0.12	0.000
	Evolutionary history	0.54	6.690×10^{-2}	+0.20	0.011
	Regression constant		-0.413		0.001
Shrublands, precip. $\leq 1,000$ mm/year	Precipitation	0.62	1.071×10^{-3}	+0.53	0.000
	Evolutionary history	0.67	7.314×10^{-2}	+0.21	0.002
	Years of exclosure	0.69	2.404×10^{-3}	+0.10	0.091
	Regression constant		-0.224		0.001

increased with both ANPP and the evolutionary history of grazing (Figure 6.14): that is, the sites with the greatest history of herbivory actually showed a greater response to grazing. This at first seems counterintuitive, for one might suspect that sites with a history of herbivory would be most resistant to grazing. Perhaps it is explained as sites with a history of

herbivory having many species able to exploit the changes caused by herbivores. For all communities, species dissimilarity was most sensitive to altering ANPP, followed by evolutionary history of grazing, then consumption (Table 6.7 column 5). Remarkably, then, differences in species composition between grazed and ungrazed areas are more sensitive to

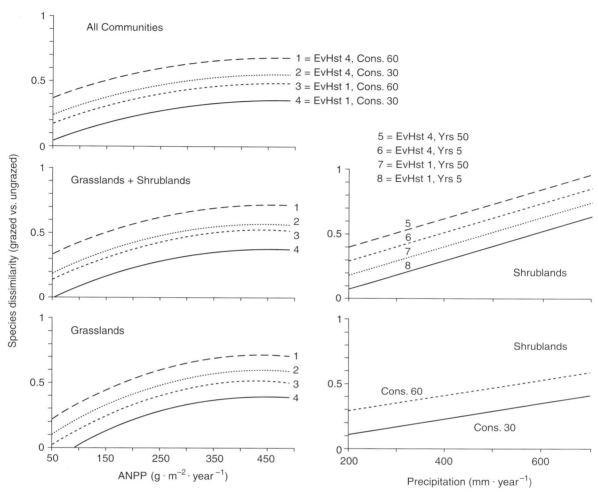

FIGURE 6.14 The search for general patterns. This is a sensitivity analysis of species dissimilarity of grazed compared to ungrazed sites in experiments from around the world. A dissimilarity value of 0 indicates no difference between grazed and ungrazed communities, and a value of 1 indicates completely distinct communities. The dependent variable was 1 minus Whittaker's (1952) index of association in regression models for all community types combined, grasslands plus shrublands, grasslands, or shrublands. EvHst, evolutionary history of grazing, ranked 1 to 4 for lower to higher values of past grazing; Cons., consumption, that is, recent grazing intensity (in percentage of annual above-ground net primary productivity [ANPP]); Yrs, years of protection from grazing. (From Milchunas and Laurenroth 1993)

physical environmental change (temperature and rainfall largely control ANPP) than to altering the intensity of grazing.

Another large group of studies was compiled by Cyr and Pace (1993) for a wide variety of habitats

with, however, greater emphasis upon wetland and aquatic habitats. The producers were phytoplankton, reef periphyton, submerged macrophytes, emergent macrophytes and terrestrial plants. Figure 6.15 shows the importance of herbivory when these producers are

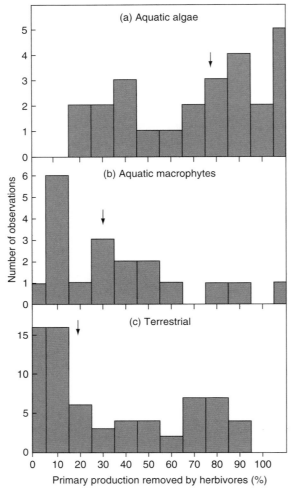

FIGURE 6.15 Impacts of herbivory as measured by the proportion of annual net primary production (ANPP) removed by herbivores for three kinds of plants. (a) aquatic algae (phytoplankton n = 17 and reef periphyton n = 8); (b) aquatic macrophytes (submerged n = 5 and emergent n = 14 vascular plants); and terrestrial plants (n = 67). Arrows indicate median values (aquatic algae 79 percent, aquatic macrophytes 30 percent, terrestrial plants 18 percent). (From Cyr and Pace 1993)

divided into three groups: aquatic algae, aquatic macrophytes and terrestrial plants. A striking result depicted in this figure is that aquatic macrophytes are much more like terrestrial plants than aquatic algae.

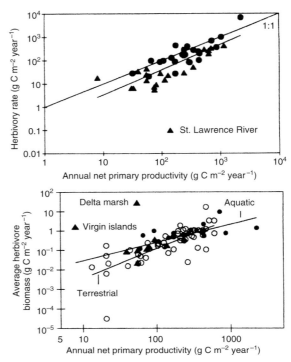

FIGURE 6.16 The rate of grazing (top) and herbivore biomass (bottom) both increase with annual net primary production. ●, Aquatic algae; ▲, aquatic macropytes; ○, terrestrial plants. (From Cyr and Pace 1993)

This echoes earlier discussions in Chapter 3 about fertility, when the question was whether wetland plants were limited by phosphorus (as with algae) or by nitrogen (as with many terrestrial plants). For aquatic macrophytes the median proportion of productivity removed by herbivores is 30 percent (compared to 79 percent for algae and 18 percent for terrestrial plants). The relationship between the rate of removal of biomass by herbivores and primary production (Figure 6.16 top) is linear with a slope not different from one and suggests that herbivores remove the same proportion of primary production across a wide range of fertility levels. The top of Figure 6.16 also shows that consumption rates are apparently an order of magnitude lower for macrophytes than algae.

In the rest of their analyses, Cyr and Pace regrettably (at least for our purposes) combine algae and macrophytes into one "aquatic" category for comparisons with terrestrial plants. Nonetheless, certain general conclusions about herbivores can be extracted. In Figure 6.16 (bottom) the biomass of herbivores is plotted against net primary productivity for both habitats. The two triangles at the upper left are submerged macrophytes (aquatic wetlands) where herbivore biomass was strikingly high. (The circle at the lower left is a terrestrial tundra site.) Excluding the two outlying triangles, herbivore biomass increases significantly with productivity, and also, excluding the outlying circle, there is no significant difference between the lines for aquatic and terrestrial habitats. Therefore, for a given level of net primary productivity, herbivores reach similar average biomass in aquatic and terrestrial ecosystems.

6.6 Some Theoretical Explorations

6.6.1 Bottom–Up or Top–Down?

The very first figure in this chapter shows grazing animals in a savanna. The most obvious interpretation of this image is that the animals are there because there is grassland to feed them. But other interpretations are possible. What if the animals are actually maintaining the grassland by preventing trees from growing there? (And just what is going on with that one lone tree? Is it the first of a forest to come? Is it the last of a forest that disappeared? Or is there some intermediate process where occasional wet years or periods without grazing allow tree seedlings to establish?) These kinds of questions lead naturally to the whole issue of the larger role of herbivores in landscapes. Are the plants controlling the animals, or the animals controlling the plants? We can recognize two possible extremes: the term **bottom–up** refers to plant communities mostly controlled by plants and resources, while the term **top–down** refers to plant communities controlled mostly by grazers (Sinclair et al. 2000; Sinclair and Krebs 2001).

Let us begin with bottom–up. At the very least, we may be certain that there is some bottom–up control, for the very simple reason that, without plants, the consumers disappear. It is therefore quite reasonable to begin with the assumption that vegetation controls wildlife, both through the creation of habitat and production of food. The presence of grasslands, for example, is generally the result of low rainfall (recall Figure 1.4). Hence the presence of grassland animals is a consequence of grassland produced by climate, a clear case of bottom–up control. Even the movement of the animals may be controlled by patterns of rainfall producing growth at different locations over seasons. (But even in this case, the particular grasses that occur may be changed by grazing, so the animals may be controlling composition at a more subtle scale.)

What about top–down? Whether the animals can influence or control the plants turns out to be much less obvious. While some areas of grassland are produced by climate alone, there are other circumstances where grazers can maintain grassland when climate alone would lead to production of forest. Grazers maintain the grassland by removing the woody plant seedlings, a topic that we shall read more about in the chapter on populations (Chapter 9). This ability to prevent tree regeneration may be one reason why the centre of the Whittaker diagram (Figure 1.4) is so problematic, since grazing may be preventing the establishment of the forest predicted from rainfall and temperature alone. Grazers may even be able to turn forest back into grassland. When grazing intensity is high enough trees may be unable to establish at all, and when the adult trees die they are replaced by grassland. This process is well-documented in Mediterranean landscapes, where forest has been converted to grassland (and even semi-desert) by goat grazing (Box 13.2). In parts of Europe and North America where closed deciduous forest is assumed to be the natural vegetation type, areas may have been kept as semi-open parkland by high grazing levels (Vera 2000).

It seems that one can find both top–down and bottom–up control in nature. At the global scale, there is no doubt that the primary cause of grassland is low rainfall. There are areas of the world that are just too dry for trees (why this should be so is an interesting question of its own, but a tangent). There are other areas where the grasslands are contingent – rainfall is high enough to allow woody plants, but grazing and fire occur often enough that woody plants cannot establish or maintain a permanent canopy. In those parts of the world where there is grassland but enough rainfall for woody plants, contingency may be all important, that is, regional changes in grazing or fire regime may be quite enough to cause landscapes to switch between forest and grassland. That is to say, bottom–up or top–down is not a question of right or wrong, but mostly an issue of location.

6.6.2 Trophic Cascades

There is one more added layer of complexity. We have discussed here only grazers and plants. There are other trophic levels on top of herbivores – large carnivores. What if the large carnivores, such as wolves, lions and alligators, are controlling the herbivores and thereby controlling the vegetation? This process is a particular kind of top–down control known as a **trophic cascade**. The presence or absence of forest as opposed to grassland may be controlled by the degree to which large predators, such as wolves or lions, are able to reduce the herbivore populations to a sufficiently low level for trees to establish. We will look at three examples, two short and one longer.

Wolves were exterminated in Yellowstone National Park in the 1920s. There was concern that large populations of elk were damaging the forests. When wolves were reintroduced in 1995/1996, there was great controversy. Some people hate wolves on principle, it seems. Many visitors, however, want a natural

landscape. And just what is natural about a landscape if the trophic cascade has been missing since 1920? There are bound to be more arguments in the future, but there is emerging evidence that after 15 years with wolves, grazing damage on trees is decreasing and new groves of young trees are establishing (Ripple and Beschta 2012). You may wish to explore the online arguments over this trophic cascade, as they will have implications for nearly every other protected area from which large carnivores are absent.

What about a smaller predator: the blue crab? There appears to be a trophic cascade involving crabs, snails and marsh plants (Silliman and Bertness 2002). When the top predator, blue crabs, were excluded from small areas of marsh, their prey, the herbivorous periwinkle, destroyed the marsh plants. In this trophic cascade, the authors suggest that commercial trapping of blue crabs could unleash overgrazing by periwinkles, thereby damaging marsh vegetation. It is possible that when you order crab for dinner, you are unintentionally contributing to allowing more snails to reach high enough population densities that they reduce the area of coastal salt marsh.

My Favourite Predator: Alligator Trophic Cascades
I spent eight years working in the coastal marshes of Louisiana and was fortunate enough to see many alligators. Most were small. Big alligators are rare because they are taken by hunters for their skins and meat. This has consequences. But, first, we must look at their diet, particularly the mammal known as nutria (or coypu).

Myocaster coypus is a large (up to 10 kg) South American rodent that was introduced into both North America and Europe (Figure 6.17, upper right). This animal is generally called coypu in Europe (Moss 1983, 1984) and nutria in the United States (Taylor and Grace 1995). Moss describes how animals introduced to fur farms in England in about 1929 escaped and multiplied to an estimated 200,000 animals by the 1960s. He observes that coypus "are extremely destructive grazers, uprooting reed and other swamp [marsh] plants to eat the rhizomes," and concludes that these animals were responsible for the loss of reed marshes

FIGURE 6.17 A trophic cascade occurs when a predator controls the abundance of herbivores that, in turn, are controlling the abundance of plants. As one example, grazing by introduced nutria (*Myocastor coypus*, upper right, ca. 0.5 m) can change the composition, biomass and area of wetlands, such as this marsh along the coast of the Gulf of Mexico. Alligators (*Alligator mississippiensis*, upper left, ca. 2 m) may protect the plants and marshes by killing nutria. (After Keddy et al. 2009.) There is a second story here: herbivores, such as *Myocastor coypus*, that feed upon both the above- and below-ground parts of plants may have higher impacts upon vegetation.

(Moss 1984). *Myocaster coypus* were introduced to Louisiana in the 1930s for fur farming, escaped during a hurricane, and reached high population levels in Louisiana deltaic marshes (Atwood 1950).

In one study with caged and uncaged plots, the biomass of dominant marsh plants such as *Panicum virgatum*, *Spartina patens* and *S. alterniflora* was higher in the caged plots (Taylor and Grace 1995).

Much larger studies have since shown that herbivory by nutria limits the production of biomass altogether – caged plots had 1.4 times as much biomass. The presence of nutria also increased the response of plants to other disturbances such as fire (McFalls et al. 2010). The changes caused by nutria likely arise from both direct and indirect effects. At least two indirect effects can be imagined. First, damaged stems and rhizomes are less able to transport oxygen below ground. Second, below-ground organs are the source of new shoots after floods and fires. Herbivores, such as *M. coypus*, that feed on both above- and below-ground parts may be particularly destructive. The presence of *M. coypus* is thought to be a major factor reducing vegetation along the Gulf Coast of North America. By reducing the amount of biomass production (and hence peat formation) in marshes, this herbivore may be causing not only changes in plant composition, but actual reduction in the area of marshes altogether. The most effective natural control agent for nutria is probably alligators. There is good evidence that alligators, by eating nutria, increase the abundance of plants in coastal wetlands (Keddy et al. 2009). Since the state of Louisiana spends millions on nutria control in order to protect wetlands (and levees), this work raises the suggestion that large alligators are worth more as large predators than they are as leather and meat.

So Is the World Bottom–Up or Top–Down?

It is tempting to ask if the world is mostly bottom–up, or mostly top–down. That is, we assume there is a dichotomy and one answer must be right and the other wrong. Too often, apparently neat dichotomies mislead us (Dayton 1979; Mayr 1982; Keddy 1989). Both processes may operate simultaneously, there may be switching among them, or other factors such as habitat productivity (Oksanen 1990), habitat heterogeneity (Hunter and Price 1992) or omnivory (Power 1992) may override or reduce their impacts. Still, it is important to know, at least in certain locations, since those who have the task of managing vegetation in large national parks must decide whether reintroducing wolves, or lions or alligators, as the case may be, will lead to restoration of natural vegetation types that declined when large predators were exterminated in the past. So far decisions have to be made on a case-by-case basis – and each new case amplifies the message of this chapter, that herbivores have significant impacts upon plants, while plants may equally have significant impacts upon herbivores.

6.6.3 Effects of Selective Herbivory on Plant Diversity

Herbivory provides an important tool for managing landscapes in general, and particularly for manipulating plant diversity (Chapters 12, 13). You may therefore be challenged one day to decide whether to increase or decrease the herbivory occurring in a landscape. There is a long list of examples to learn from – including deer in temperate deciduous forests (Kay 1993; Newman 1993; Latham et al. 2005), rhinoceros in tropical forests (Dinerstein 1992), elephants in grassland (Moolman and Cowling 1994), cattle in rangeland (Canfield 1948; Milton et al. 1994; McClaran 2003), goats in arid lands (Figure 13.2; Box 13.2) and nutria in wetlands (Section 6.6.2). Although exotic herbivores, such as goats and cattle, often have major negative consequences for the natural vegetation, in other cases, the consequences are less obvious. Predicting the effects of herbivores upon plant diversity is often complicated by the selectivity of herbivores. If herbivores feed mostly upon common plant species, rarer plant species may increase; conversely, if herbivores feed mostly upon rare species, the common species may become even more common. A simple verbal model follows, although a mathematical version is available (Yodzis 1986).

Imagine the following circumstances: a plant community with a mixture of species, some with high competitive ability and some with low competitive ability. The abundance of each species, in the absence of herbivory, tends naturally toward eventual exclusion by the species that is the competitive dominant. Over time, then, species richness will decline. Now introduce a herbivore. What will happen? The effects of this herbivore on the diversity of the plant community are impossible to predict without information about the feeding habits of the herbivore. Consider the two extremes.

At one extreme, the herbivore feeds upon the weaker competitors and avoids the dominant. In this case, the herbivore will reduce the diversity of the community. At the other extreme, imagine that the herbivore feeds solely upon the competitive dominant and avoids the weaker competitors. In this case, the herbivore will increase the biological diversity of the community.

There is third option. The herbivore may be more like a vacuum cleaner, and feed on species in direct proportion to their occurrence in nature. In this case, the effects of herbivory on diversity may be small and largely determined by the species' relative degrees of resistance to the damage of herbivory. Yodzis (1986) explores these situations mathematically. Such investigations illustrate that the effects of introducing exotic herbivores or reintroducing extirpated herbivores may be difficult to predict.

Let us conclude with an example that shows not only how grazers can influence plant diversity, but how introduced grazers may have very different effects from native grazers. The Cape Province of South Africa has a rich succulent flora that includes many hundreds of endemic species, many of which are in the Liliaceae, Asclepiadaceae, Crassulaceae, Euphorbiaceae and Mesembryanthemaceae. These landscapes have been naturally grazed by elephants for millennia; goats, on the other hand, are introduced and their populations are expanding. Moolman and Cowling (1994) used a kind of pre-existing experiment by comparing vegetation types among sites that are protected (and one assumes, with low grazing) versus sites grazed by elephants, or used for goat grazing. Figure 6.18 shows that plant diversity is deleteriously affected by grazing, but that goats have greater deleterious effects than elephants. The diversity in some families, such as the Mesembryanthemaceae, was little affected by herbivory, probably owing to the low palatability of the plants, although total vegetative cover was reduced, and many species were restricted only to the control sites. The Crassulaceae, in contrast, were strongly affected by goats but little affected by elephants. Moolman and Cowling also provide a list of species that indicates the deleterious effects of herbivory.

Mowing Machines Are Like Unselective Herbivores

Mowing by humans is, in some ways, just a form of herbivory carried out by a relatively unselective herbivore. Mowing usually increases plant diversity

(Baker 1937; Grime 1973a; Ellenberg 1988b). This appears to occur because mowing is actually somewhat selective after all: mowing tends to preferentially remove larger species with dense canopies and thereby allow smaller species, such as rosette forms, to persist. Although mowing has its disadvantages, particularly when we are trying to re-establish natural processes in plant communities, it may be one of the few ways to manage for increased diversity under eutrophic conditions.

6.6.4 A Simple Model of Herbivory

One of the simplest models adapts the logistic equation. (If you have forgotten this model, have a quick look at the beginning of Chapter 9). Recall that this model assumes that when there are few organisms and abundant resources, growth is (almost) exponential, but when population size increases and resources become scarce, the population growth slows and reaches a level known as the carrying capacity, K. This model, then, reflects the balance between exponential growth and competition for a limited resource. The logistic equation can be applied to plant populations (Noy-Meir 1975; Starfield and Bleloch 1991) as

$$\frac{\mathrm{d}P}{\mathrm{d}t} = gP\frac{(K-P)}{K}$$

where P is the amount of plant material (e.g. biomass per unit area), g is the growth rate, and K is the maximum amount of plant material that a unit area can support. Another way of thinking about this equation, if you are used to thinking about animal population models, is to think of each plant cell as an individual, in which case P is the number of plant cells and K the carrying capacity of plant cells for a particular area of landscape. Or, from another perspective, P is the number of meristems. To explore the behaviour of vegetation without herbivores, one can plot the growth rate of P ($\mathrm{d}P/\mathrm{d}t$) against biomass (P), which produces an inverted parabola (Figure 6.19a). The growth rate of the population of plant cells therefore at first increases as more and more cells are

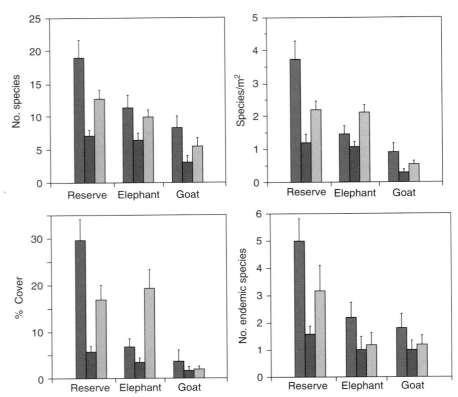

FIGURE 6.18 Grazing can change plant diversity. Here are four characteristics of vegetation types found in Addo Elephant National Park in South Africa. In each panel, the left-most cluster of bars shows ungrazed conditions, while the others show elephant grazing and goat grazing. The three sub-groups in each cluster of bars show more detail: the three vegetation types are open (brown bar), *Portulacaria* shrubs (dark green bar) and *Euclea* shrubs (light green bar). Note that grazing removes biomass, and reduces the number of kinds of plants, particularly endemic species. (From Moolman and Cowling 1994).

available for photosynthesis, and then it slowly declines as the resources available to each cell become restricted. The botanical logic behind this is simple – when plant biomass is low, each new cell will improve the photosynthetic capacity of the vegetation, but as biomass increases, more and more cells will be needed to provide structural support for photosynthetic cells, and others will be shaded, thus reducing net photosynthetic yield (we will return to this topic in Section 8.4.4). When the mean photosynthetic yield of all cells just balances their mean respiratory demands, growth will come to a halt. At this point, the level K on the

horizontal axis of Figure 6.19 will have been reached. Halfway between 0 and K the growth rate is at a maximum. This is the familiar pattern of logistic growth; the novelty lies solely in applying it to plant biomass. The level of biomass, K, will depend upon environmental factors, such as growing season, and soil fertility. Finally in the absence of herbivores, all vegetation will tend toward point K.

Now let us add in a constant grazing pressure from a herbivore. Assuming that the herbivores remove a constant amount of biomass per unit time, designated G, the equation becomes

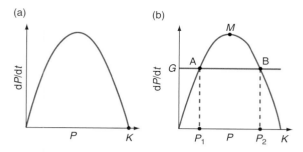

FIGURE 6.19 Vegetation growth rate dP/dt plotted in relation to plant biomass P for the logistic model: (a) no grazing and (b) constant grazing pressure G (after Starfield and Bleloch 1991).

$$\frac{dP}{dt} = gP\frac{(K - P)}{K} - G$$

Since the grazing rate is constant, that is independent of biomass, one can plot G as a horizontal line across the parabolic model of plant growth (Figure 6.19b). There is no need to solve the differential equation to learn a good deal about the behaviour of such a grazing system; this can be deduced simply from the structure of the equations and the resulting graph (May 1977; Starfield and Bleloch 1991). Returning to the growth of vegetation, it is apparent that the growth rate is positive only between points A and B, where the growth parabola lies above the grazing rate and biomass therefore accumulates. On either side of this range, the grazing rate exceeds the growth rate. At points A and B, growth just matches grazing.

The next step is to examine stability by considering the consequences of minor perturbations. Consider point B (we are still working with Figure 6.19), where the corresponding amount of plant biomass is indicated as P_2. If growing conditions improve slightly, pushing the amount of biomass to the right, the growth rate will fall below the grazing rate, and the vegetation will decline back to level P_2. If, on the other hand, drought were to reduce biomass slightly below P_2, then simultaneously the difference between the grazing rate and the growth rate increases, so that biomass accumulates, pushing the system back toward point P_2. Since the system returns to point

B when it is lightly perturbed, this is called a stable equilibrium point.

Now consider point A, with the corresponding amount of biomass P_1. This point is, in contrast to point B, unstable. If it is perturbed to the left of P_1, say by a drought, then growth rates fall further and further below the grazing rate until the plants disappear; the system slides to the bottom left and collapses. Note that even this simple model suggests that it is entirely possible for overgrazing to destroy vegetation. Conversely, if there is a surge of growth above P_1, perhaps by a few wet years, then the vegetation temporarily escapes from grazing and continues to move to the right because, as biomass increases, the difference between grazing rate and growth rate increases as well. Eventually the entire system slides over to point P_2. Over a broad range of biomass levels, this model of a simple grazing system will return to P_2, the only stable point, after perturbation. An applied way of looking at this figure is to conclude that there is a fairly wide array of plant densities that will return to P_2 (that is, there is a reasonably wide zone of stability that may be sustainable), but woe to the herder who gets greedy and allows herbivores to overgraze near P_1, near that dangerous downward spiral of collapse.

There is still more to be learned from this simple model. We change grazing intensities. If the grazing rate (G) is increased, that critical point P_1 increases (slides right), and, at the same time, the safety margin between P_1 and P_2 shrinks. Were you to allow a grazing rate higher than maximum growth rate (M) – equivalent to sliding the horizontal line up above the parabola – the herbivore would graze faster than the vegetation grew. Such a situation would be inherently unstable.

We could extend this model by changing model structure further. For example, we could allow growth rates to fluctuate in response to rainfall, or use a different model for plant growth (Starfield and Bleloch 1991). If grazing pressure is not constant, but varies with plant biomass, then a variety of outcomes is possible, depending upon the functional responses of the grazer (May 1977; Yodzis 1989). So there is much more that can be done by making the model more complicated, but you can see how even a simple model actually tells you a great deal about how plants and herbivores might interact over many years.

6.6.5 When Herbivory Becomes Catastrophe

It is now time to make that model more complicated. By making different assumptions, very different kinds of herbivore–plant dynamics can be observed. To illustrate the process of expanding the models, consider the example from May (1977) in Figure 6.20. Again, the plants are assumed to have logistic growth, represented by the parabolic curve in Figure 6.19, but instead of the herbivore being represented by a single horizontal line, it is given a more complex grazing response. This is a type III response (sensu Holling 1959) in which the herbivore feeding rate responds in a non-linear way to the resource levels. Figure 6.20 shows three different densities of herbivore; for simplicity assume that this density is something that is fixed (perhaps controlled by a farmer) rather than a population with dynamics of its own. For each herbivore density, the growth rate of the plants is given by subtracting the solid curve (grazing rate) from the parabolic curve (growth rate). If you do these subtractions, you end up with three possible situations, each shown in Figure 6.21.

Situation 1: At the low herbivore density (Figure 6.21 top), the system has a single equilibrium point that is stable at $P = A$. Here the biomass removed by the herbivore balances the growth rate of the plants and there are still many plants present.

Situation 2: At an intermediate herbivore density (Figure 6.21 middle), there are three points where the curve intersects the biomass axis (growth rate and grazing rate intersections, Figure 6.20). Points $P = B$ and $P = D$ are stable, while $P = C$ is unstable. In this case, the herbivore and plants can reach an equilibrium at either high (B) or low (D) plant biomass.

Situation 3: At a high herbivore density (Figure 6.21 bottom) there is one equilibrium point, G, corresponding to very low plant biomass.

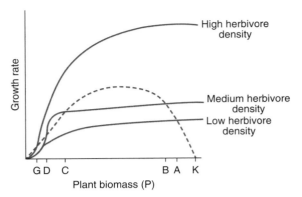

FIGURE 6.20 The qualitative behaviour of a grazing model can depend upon herbivore density. The dashed curve is the logistic growth curve of the vegetation with no herbivore present. The three solid lines are three possible grazing curves for three herbivore densities. (After May 1977; Yodzis 1989).

With just a little more effort (May 1977), all three situations can be combined into one new diagram (Figure 6.22). Here the three regions on the horizontal axis correspond to the three situations we just discussed – low, medium and high herbivore density. The threshold values that mark the transitions among these three situations are marked as either T_1 or T_2. We will let Yodzis tell us in his own words (1989) how this model applies to densities of sheep:

If, to start, there are no sheep in the field, the vegetation will eventually reach the value K of biomass density. Suppose we then stock the field with sheep, at a low or intermediate density (that is: with $H < T_2$). If we keep the flock constant for a while, the system will eventually settle into an equilibrium with the vegetation biomass somewhere on the upper solid curve, since we are starting out from a value of P (namely, K) that lies above this curve. And we can increase the density of sheep with no drastic effects, so long as we stay below $H = T_2$: each slight increase in H will just result in a slight decrease in equilibrium vegetation biomass.

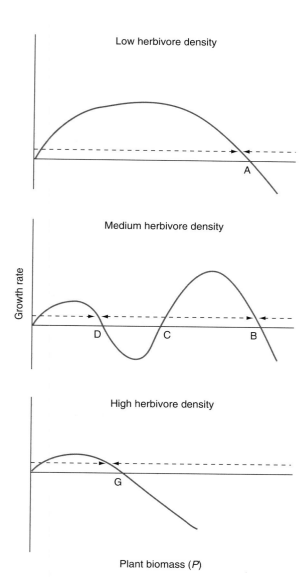

Low herbivore density

Medium herbivore density

Growth rate

D   C   B

High herbivore density

G

Plant biomass (*P*)

FIGURE 6.21   The growth rate of the vegetation plotted against plant biomass for each of the scenarios in Figure 6.20. At low herbivore density (top), there is one equilibrium point, A. At high herbivore density (bottom), there is another much lower equilibrium point, G. At medium herbivore density, however, there are three equilibrium points, D, C and B, and of these D and B are both locally stable equilibria. Depending upon circumstances, it is possible to have stable equilibria at either high plant biomass (B) or low plant biomass (D) (after Yodzis 1989).

However, if we push the sheep density above T_2, even if by only one sheep, a catastrophe will occur.

Beyond T_2, there is in the previous sketch no longer any "upper" stable equilibrium, and the vegetation biomass must plummet to the "lower" stable equilibrium curve! We no longer have a very healthy pasture: our new "high H" flock can subsist, but only rather precariously; the pasture, and our flock of sheep, may suffer severely from some relatively small misfortune such as sparsity of rain.

Worse still, we cannot easily get back to the "upper" stable equilibrium curve. If we remove a few sheep from the pasture to get back into the "intermediate H" region, the system will remain on the "lower" stable equilibrium curve, because the vegetation density is below the breakpoint value $P = C$. The only way to get back to the "upper" stable equilibrium curve is by removing enough sheep to get below the threshold $H = T_1$, into the region of "low H."

Perhaps you are thinking: isn't it still worth the risk inherent in a high H flock, because you do after all, in case there are no little disasters such as a sparsity of rain, have a big flock of sheep. The problem with this is, you've got a big flock of very scrawny sheep, for the level of vegetation biomass density associated with a high H flock is very low relative to the saturation density R_0 of the sheep. Therefore, the sheep have only enough food to barely survive: when the vegetation biomass crashes down to the "lower" stable equilibrium curve in the previous sketch, the sheep biomass will crash along with it. . . .

Notice one more thing about this system. In reality it will be subject to all kinds of unpredictable influences (weather, disease, other grazers, etc.) which will make it fluctuate around any equilibrium. Even if we are below the threshold T_2, if we are too close to T_2 these random fluctuations may bring the vegetation biomass P below the breakpoint value C, resulting in a "crash" just as if we had overstocked. If you take another look at the previous graph, you will see that the domain of attraction of

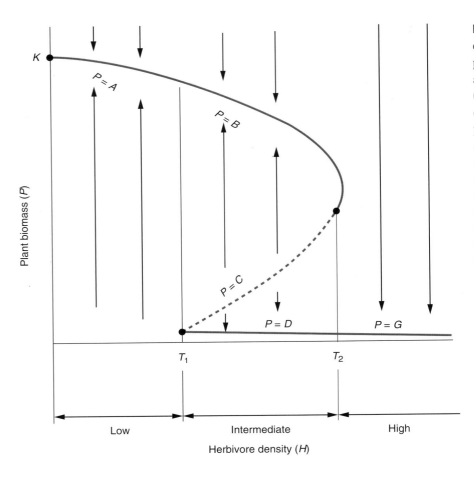

FIGURE 6.22 The equilibrium values of plant biomass (P) plotted against herbivore density (H). The arrows show the direction of change in the plant biomass at non-equilibrium, and the different values of P correspond to the equilibria shown in Figure 6.21. T_1 and T_2 are threshold values for shifts among the three models. (After May 1977; Yodzis 1989)

the upper stable equilibrium $P = B$ is smaller the closer H is to T_2, making the system more susceptible to the kind of crash just described. To be on the safe side, we should not stock too close to T_2.

This kind of possibility makes Hardin's (1968) "tragedy of the commons" all the more poignant.

(pp. 19–20)

If you are at first inclined to skip Figure 6.22 and the explanation, I understand. But it is a mistake. This is a lucid introduction to catastrophe theory, all in one figure and a few paragraphs. Sadly, Yodzis died in 2005 (you can read a summary of his contributions in Vassseur and McCann 2005), but his 1989 book on theoretical ecology is a valued legacy.

Moving From Sheep to Moths to Bark Beetles With the Same Logic

Yodzis also describes how a similar model can be formulated to describe the population dynamics of spruce budworm (*Choristoneura fumiferana*). We have already discussed both windfalls and fire as natural disturbances in coniferous forests, but there is a third factor: insect outbreaks. In eastern North America, for example, there are periodic outbreaks of a moth known as spruce budworm, whose larvae destroy the canopy of coniferous forests every 40 years or so (Holling 1978b; Ludwig et al. 1978). Unlike with fire, understorey trees of the forest remain largely intact and immediately begin to regenerate the canopy. This produces a resource management dilemma for

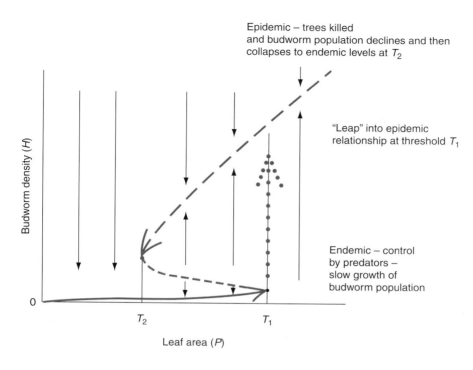

Epidemic – trees killed
and budworm population declines and then
collapses to endemic levels at T_2

"Leap" into epidemic
relationship at threshold T_1

Endemic – control
by predators –
slow growth of
budworm population

FIGURE 6.23   A simple model of spruce budworm (H, density) and balsam fir (P, average leaf area per unit area of forest). This yields a situation similar to that in Figure 6.22, where low endemic levels of budworm suddenly leap into epidemics at T_1. (After May 1977; Yodzis 1989)

humans: the pulp and paper industry would like to harvest 1/40th of the forest every 40 years to maintain steady production of pulp from its mills. The spruce budworm, however, has for thousands of years settled on a system where it harvests, instead, all of the trees once every 40 years. The natural response of foresters is to spray the forest with insecticides to keep the trees alive until they can be harvested. What then seems to happen is that the spruce budworm populations, instead of collapsing after destroying the trees (the normal cycle), remain at permanently high levels, which forces managers to continually spray forests. The results are that pesticides and their residues threaten both wildlife and humans; indeed, spraying for spruce budworm in the Acadian forests of New Brunswick was one of the classic examples of the abuse of pesticides used by Rachel Carson in *Silent Spring* (Carson 1962). Some of the battles between citizens and the forest industry have been summarized in *Budworm Battles* (May 1982). As a personal aside, I lived in Nova Scotia during one outbreak, and no doubt inhaled my share of spray, so somewhere in my

tissues there no doubt lurks one of the pesticides or its residues. A free souvenir, of sorts.

Please return to the model, this time as drawn in Figure 6.23. Here is a representation of herbivore–plant relationships that should now be familiar. In this case, the abundance of plants is represented by leaf area of the trees because this is the food factor to which budworm apparently responds. In the example of sheep herbivory, the density of herbivores is assumed to be under human control; in the case of spruce budworm it is not, and this makes the mathematics much more complicated. May (1977), however, has pointed out that the growth of budworms is much faster than the growth of trees, so for each value of P one might assume that the budworm population will be near equilibrium. Therefore one can use intuition to get at the slow dynamics of leaf area in relation to budworm density (Figure 6.23).

If one begins at a normal, low budworm density (endemic levels), the budworm is kept under control by predators and the trees are able to grow. The system then moves slowly to the right along the solid curve as leaf area increases. Eventually threshold T_1 is reached, so that the budworm population is no longer in equilibrium; it will erupt to the dashed curve (epidemic levels). The trees, however, cannot withstand the grazing rate of the upper curve, and the system slides slowly back down and to the left. When the threshold T_2 is reached, there is a final crash back down to the lower line, and the process begins anew. This cycle will repeat itself and produce the natural cycles of canopy destruction and forest regeneration that have occurred for millennia. You may wish to read Yodzis's explanation (1989, pp. 104–108) of how the model can be modified to illustrate how spraying with insecticides creates a new stable equilibrium with a fatal flaw – an equilibrium that can be maintained

only by continuous spraying. Worse, the model indicates that any cessation of spraying rapidly turns into a worst-case scenario where the forest is rapidly defoliated. (You may also wish to read Ludwig et al.'s (1978) more elaborate model of spruce budworm dynamics.) While of course any model has its limitations, it is now believed that much of eastern Canada has been brought to exactly the situation described, with enormous economic, conservation and health implications. Those who think that plant ecology is a quiet and irrelevant discipline (yes, I saw you reading the newspaper at the back of the class) need only attend public hearings on the spraying of spruce budworm to discover how relevant plant ecology really is.

6.7   Two Final Examples of Large-Scale Changes from Herbivores

6.7.1 Mountain Pine Beetles Change Conifer Forests

As you have just seen above, rather basic models of herbivory can produce complex dynamics – periods of rapid population growth can be followed by collapse. And this process can be changed, and indeed possibly amplified, by human activities. It is not always possible to identify a single cause – although forestry practices that produce monospecific forests seem to be a recurring source of problems.

Spruce budworm outbreaks were caused by a moth in the genus *Choristoneura*. Among the insects as a whole, however, it is one group of beetles (in the family Curculionidae, commonly known as bark and ambrosia beetles) that stands out for its impacts on forests (Wood 1982; Raffa et al. 2008; Bentz and Nordhaus 2009). Some feed on cones or fruits, others feed on dead twigs, but a few attack healthy living tissue. These beetles are guided in their search for injured trees by their ability to detect chemical compounds such as terpenes. Once they enter a tree, the beetles produce pheromones that can attract

thousands of other beetles within a few days, thereby exhausting a tree's resistance (Raffa et al. 2008). Beetles can create small gaps by killing single trees, or they can create large areas of dead forest that are susceptible to fires. Trees differ significantly in their susceptibility to bark beetle feeding, depending upon their site, vigour and age. The rate of resin flow is an important factor that determines rates of tree mortality, healthy trees being able to produce more defensive resin. Hence seemingly minor differences among trees may, with feeding by bark beetles, lead to changed forest composition or even the loss of certain tree species on marginal sites (Wood 1982).

The impact of beetles also varies with the type of forest. "In general, dense, older stands, where trees must compete for resources, are more susceptible to bark beetle outbreaks, whereas heterogeneous landscapes that contain many sizes, ages, and species of trees are more resistant and resilient" (Bentz and Nordhaus 2009: p. 14). In western North America, mountain pine beetles are causing widespread tree mortality (Figure 6.24). The effects include increased woody debris for decomposers, increased populations

FIGURE 6.24 The mountain pine beetle (*Dendroctonus ponderosae*) kills conifer trees in western North America. Eruptions can cause significant changes in the species composition of forests over large areas. (Map from Natural Resources Canada; beetle from US Department of Agriculture)

of understorey plants and increased tree diversity, as well as large forest fires and increased inputs of carbon dioxide to the atmosphere. Much of the information available addresses the negative effects of dying trees, without explaining the potentially beneficial, or at least neutral, long-term consequences of this kind of natural disturbance, or explaining how long-term insect population cycles (like fire cycles) may control forest composition. Of course, these population cycles themselves are affected by factors such as temperature... but that is another topic altogether.

6.7.2 White-Tailed Deer Change Deciduous Forests

With herbivores such as bark beetles, above, millions of dollars are being spent to prevent change to forests. But in other cases, with herbivores such as deer, millions of dollars are instead spent to make more herbivores that change the forest. Why the difference? Deer are also important in this book because a good many universities occur within the deciduous forest region of either western Europe or eastern North America. Apparently a lot of people therein like white-tailed deer (*Odocoileus virginianus*). In my part of the world, images of deer adorn pick-up trucks, t-shirts, baseball caps, calendars and coffee mugs. Moreoever, you can buy white-tailed deer salt and pepper sets, toilet roll holders, photo frames, clocks, hip flasks, key chains, playing cards, thermometers, bottle openers... back scratchers, candle holders, toilet seats, lamps, shower curtains, serving licence plates, weather vanes, door mats, plaques and video games. As P. T. Barnum observed "Nobody ever lost a dollar by underestimating the taste of the American public." And no small town café is complete without at least one image of an antlered buck, and perhaps several sets of antlers on the wall. This cult of deer more or less matches the boundaries of eastern deciduous forests in North America: Wisconsin to Nova Scotia south to Mississippi and Georgia. Here government agencies have entire teams of biologists devoted to the task of making white-tailed deer more abundant. Over the past decades, these teams of biologists have been raising deer populations by whatever means possible: killing predators, cutting patches in forests, winter feeding and running elaborate permit systems that protect females. They have succeeded! Forests in eastern North America likely have more deer per hectare than at any time in history.

Alas, there are unintended negative consequences for other species. Take my own acreage of deciduous forest: deer are now so abundant that many of the young trees are eaten, often leaving but one tree species, the less palatable *Ostrya virginiana*

(ironwood) to form the next canopy. Native herbaceous wildflowers, particularly spring ephemerals, are vanishing and some uncommon plants, such as the unusual gymnosperm *Taxus canadensis*, are almost extirpated. According to Alverson et al. (1988), densities as low as four deer per square kilometre may be sufficient to prevent regeneration of woody and herbaceous species. Further south, I have been told that some nature reserves can keep their wildflower populations only if they are caged to exclude deer. There is a growing awareness that large populations of deer threaten the survival of rare plant species that certain parks are intended to protect (Alverson et al. 1988; Russell et al. 2001). In Section 9.7 we will see how one rare plant (*Panax quinquefolius*) declines from deer grazing. Changes in the understorey and tree regeneration may have further unintended effects, such as declines in songbird species (de Calesta 1994; McShea and Rappole 2000). Deer also carry Lyme disease, a crippling neurological disorder caused by spirochetes in the genus *Borrelia* that are spread to humans by blood-feeding ticks. (I removed a tick just last week.) This disease has reached epidemic levels in eastern North America (Weintraub 2008). The best overview I have read about deer and forests is Latham et al. (2005), which tells the story for the state of Pennsylvania.

From the point of view of this chapter, the deer story is important. First, we learn that at least one species of herbivore can cause major changes in vegetation. But there are other lessons. The good news for conservation is that humans, with sufficient effort, can increase the abundance of selected species. Rare species might benefit if similar programs were available. The bad news is that in the context of management, deer also illustrate how human desires can lead to actions that cause a cascade of problems, a topic to which we will return in the final chapter. At the broadest level, this is a classic example of the law of unintended consequences. Since natural systems are linked together in such complex ways, as I used to warn my students, you can never do just one thing. Whatever management act you carry out will be bound to have consequences you cannot forsee. And, at the most basic level, all these negative effects arise because one group of citizens wants more hunting opportunities. So the next time you see a white-tailed deer emblazoned on a t-shirt or pick-up truck, think of the many lessons it embodies.

CONCLUSION

This chapter, like so much of biology, is rooted in the careful observation of nature. Observations on herbivores, simple questions about what is eating what, raise difficult questions about how herbivores affect plants and vegetation. This question has both short-term and long-term consequences. Such observations also lead naturally to field experiments that use exclosures. Exclosure experiments began with the use of simple fences (Figure 6.12) and advanced through the 1900s to complex multifactorial experiments (Table 6.6). Once sufficient numbers of experiments were published, it became possible to search for repeating patterns, first in a qualitative manner but then using meta-analyses such as those generated by multiple regression analysis (Table 6.7). Models then helped to further interpret the field observations and experiments.

Return to the question that was raised at the beginning of this chapter: to what extent do herbivores control properties of plant communities? The evidence would seem to suggest that herbivores can occasionally have a major effect upon plant communities but that they are often far less obvious and more nuanced. Other abiotic factors, such as drought, or biotic factors, such as competition, may be equally important. In many cases, vegetation seems to be controlled from the bottom up: that is, the composition and structure of the vegetation determine the abundance and distribution of herbivores rather than vice versa. There are also exceptions.

After all this effort, it is important to remember that herbivory is not the most important factor processing the biomass produced in plant communities. Rather, study after study over the past 50 years has demonstrated the same startling result: the vast majority of plant biomass goes straight into the decomposer food web where it is processed by small invertebrates and microorganisms (recall Figure 6.2). This generalization ranges from arid tropical grasslands (Deshmukh 1986) to temperate salt marshes (Adam 1990) to tundra (Oksanen et al. 1997) – although aquatic algae are an apparent exception (Cyr and Pace 1993). Thus the decomposer food chain is still processing more plant material than do herbivores – a reminder of the continued importance of Figure 3.14.

Thus we end on a paradox of sorts. In spite of all the efforts of herbivores, which consume almost every part of plants (with a preference for seeds and meristems), a great deal of biomass still survives to enter the decomposer food web and add to the carbon in soil. How much do herbivores affect plants and plant communities? Herbivory is an important factor, but its importance varies greatly among species and habitats. Generalizations remain elusive. Rather than asking whether herbivory is important, it seems that in the near future we will have to be satisfied with a simpler question: in which vegetation types, and with which species, are herbivores controlling species composition? And, equally, where are they having a minimal effect? These two questions, at least, should be answerable.

Review Questions

1. List some common ways that plants defend themselves from herbivores. Which seem to be prevalent in your part of the world? Can you identify several local plants that are particularly palatable or particularly unpalatable?

2. Seeds are vulnerable to predators. Explain how cones in gymnosperms protect seeds. Can you find a local museum with a display of fossil cones? Since flowering plants do not have cones, give some examples of hard structures that protect their seeds.

3. In addition to squirrels, there is another group of animals that appear to have coevolved with conifers, the crossbills. A crossbill can eat 3,000 seeds per day. Read about crossbill ecology (e.g. Benkman, C.W. 2012. White-winged Crossbill (*Loxia leucoptera*) In A. Poole (ed), *The Birds of North America Online*. Ithaca: Cornell Lab of Ornithology). How could we weigh the relative importance of rodents and birds upon conifers?

4. Explain how exclosure experiments can be used to help determine the importance of different kinds of herbivores.

5. You graduate and are put in charge of an enormous tract of forest. Local residents want you to kill the large predators in the forest. Describe the possible negative consequences of such actions upon the plants, using examples from this book or from other sources.

Further Reading

Oksanen, L., S.D. Fretwell, J. Arruda and P. Niemelä. 1981. Exploitation ecosystems in gradients of primary productivity. *The American Naturalist* 118: 240–261.

Thirgood, J.V. 1981. *Man and the Mediterranean Forest: A History of Resource Depletion.* London: Academic Press.

Coley, P.D. 1983. Herbivory and defensive characteristics of tree species in a lowland tropical forest. *Ecological Monographs* 53: 209–233.

McNaughton, S.J. 1985. Ecology of a grazing ecosystem: the Serengeti. *Ecological Monographs* 55: 259–294.

Marquis, R. 1991. Evolution of resistance in plants to herbivores. *Evolutionary Trends in Plants* 5: 23–29.

Rosenthal, G.A. and M.R. Berenbaum. (eds.) 1991. *Herbivores: Their Interactions with Secondary Plant Metabolites*. San Diego: Academic Press.

Belsky, A.J. 1992. Effects of grazing, competition, disturbance and fire on species composition and diversity in grassland communities. *Journal of Vegetation Science* 3: 187–200.

Cyr, H. and M.L. Pace. 1993. Magnitude and patterns of herbivory in aquatic and terrestrial ecosystems. *Nature* 361: 148–150.

Milchunas, D.G. and W.K. Laurenroth. 1993. Quantitative effects of grazing on vegetation and soils over a global range of environments. *Ecological Monographs* **63**: 327–366.

Fleischner, T.L. 1994. Ecological costs of livestock grazing in western North America. *Conservation Biology* **8**: 629–644.

Sinclair, A.R.E., C.J. Krebs, J.M. Fryxell, et al. 2000. Testing hypotheses of trophic level interactions: a boreal forest ecosystem. *Oikos* **89**: 313–328.

Langenheim, J.H. 2003. *Plant Resins: Chemistry, Evolution, Ecology, Ethnobotany.* Portland: Timber Press.

7

Positive Interactions

Commensalism, symbiosis, mutualism. Plant–plant cooperation (facilitation, protection, oxygen, shading). Plant–fungus cooperation (mycorrhizae, lichens). Plant–animal cooperation: pollination by birds, seed dispersal by animals (tapir, ants, rodents). Animals defending plants. Models of mutualism. Conceptual obstacles in the study of mutualism.

FIGURE 7.1 Two mutualisms for wild banana plants: the flowers are pollinated by bats and their seeds are distributed by bats. The flowers (left) open at night, have a strong odour, and abundant nectar and pollen. Note the rows of small bananas higher on the stalk. On the right, an Indian flying fox. Many tropical tree species similarly depend upon bats, sometimes for pollination, but particularly for seed dispersal (Buddenhagen 2008). (Commercial banana plants are seedless varieties propagated by clones; the genetic diversity of the species is found solely in the wild plants.) (Shutterstock)

7.1   Introduction: Plants Can Cooperate With Other Plants, Fungi and Animals

Darwin's view of the struggle for existence is often misunderstood. It is true that plants harm one another (competition, Chapter 4) and that animals harm plants (herbivory, Chapter 6), but beneficial interactions also occur. Cooperation is part of the struggle for survival. This cannot be said often enough, it seems, since many people carelessly interpret evolution through natural selection to be "every individual for himself." Some biologists still think that we must study nature solely by thinking of the individual or species in splendid isolation, when in fact nature is full of complicated entities built on cooperation. Examples? Think of your own body. Every cell is a positive interaction among both an ancient "eukaryotic" host cell and multiple prokaryotic cells (mitochondria). In the case of plants, many cells also contain a third cooperating entity; the chloroplast, which was once a free-living alga (recall Figure 1.12). Every single cell is then a mutualism. And in your body as a whole? Multiple cells with your own genotype not only compete for access to glucose and oxygen, but cooperate to build tissues and maintain consciousness. Moreover, your body contains far more bacteria than cells with your "own" genotype: there are about ten bacterial cells for every "human" cell in the human body. In the same way, a forest is not just a bunch of trees: below ground they are linked by many kinds of fungi that transfer nutrients and glucose across individuals and species. Above ground, insects, bats and birds carry pollen from one flower to another, and disperse seeds (Figure 7.1). In this chapter we want to look at positive interactions involving plants. We may be left with more questions than answers, but perhaps that will stimulate younger scholars to get on with the work of clarifying the role of positive interactions in the biosphere as a whole.

One of the peculiarities of this story is this: in spite of the many examples of positive interactions in nature, many text books all but ignore the topic and leave students with the misimpression that positive interactions are unimportant. Look at Table 7.1! I wrote about this in 1989, and have seen more unhappy examples since (Keddy 2004).

So let us set the balance straight and look at positive interactions. First, we will have some definitions, and then a little history. I will break the chapter down into sections by considering the species involved. There are four main kinds: plant–plant, plant–fungus, plant–animal and plant–microbe interactions.

I have already used two similar words, positive interactions and mutualism. There is also commensalism. For those of you who have forgotten, let us look carefully at definitions. For those of you who have not forgotten, consider this a short review that will help you design carefully thought out experiments to unravel these complex interactions.

7.1.1 Definitions

Positive interactions occur when one species benefits from the presence of another species. When both species benefit, one can use the more specific term, mutualism. Mutualism, then, is a +/+ interaction, in contrast with competition, which is the reverse, a –/– interaction (Chapter 4).

Mutualism can also be asymmetric, where one individual experiences more beneficial effects than the other. (You met asymmetric competition in Section 4.3.4.) Since it is, in fact, rather unlikely that both individuals in a mutualism will receive identical benefits from the association, most mutualisms probably are asymmetric. We won't know until benefits

Table 7.1 **The impressions given to students regarding the importance of the three major ecological interactions in the biosphere, as assessed by the number of pages on the topic referred to in the index of some text books on ecology. (From Keddy 1989)**

Text book	Mutualism	Competition	Predation
Colinvaux (1986)	1	33	70
Collier et al. (1973)	0 (1)[a]	45	30
Huchinson (1978)	0 (9)	59	6
Krebs (1978)	3	50	32
Lederer (1984)	5	21	4
McNaughton and Wolf (1979)	20	77	71
Odum (1983)	15	17	15
Pianka (1983)	3	74	41
Ricklefs (1979)	3	38	30
Ricklefs (1983)	2	11	14
Smith (1986)	2[b] (1)	19	24
Whittaker (1975)	5 (9)	18	22

Notes:
[a] The number in parentheses is symbiosis, which some authors equate with mutualism.
[b] Mutualism not in index, but present in text.

are measured in many more cases. As the asymmetry becomes more extreme, the relationship tends toward a +/0 interaction, which is a sort of limiting case where one species receives a benefit and the other appears to experience neither a cost nor a benefit from the association. This situation is called **commensalism**, although other writers also use the word facilitation. Mutualism is sometimes confused with symbiosis, but the term **symbiosis** merely denotes two species are living together in close association.

Careful study of benefits is necessary to decide whether or not positive interactions are occurring, and whether they are commensalism or mutualism. Consider three examples, beginning with two you have already seen. *Myrmecodia beccarii* (Figure 3.27) illustrates both a symbiosis and a mutualism, since the ants fertilize the plant and the plant shelters the ants. But *Myrmecodia* is simultaneously an epiphyte,

in which case it is commensal on the host tree. Mistletoe (Figure 3.31), in contrast, is a parasite on the host tree. And mistletoe is simultaneously part of a mutualism with an animal: it provides nectar and fruit for birds, and the birds disperse the mistletoe seeds. A new example: banana plants (Figure 7.1) provide food for bats, and bats both pollinate the flowers and disperse the seeds. Banana plants are simultaneously involved in a mutualism with soil fungi. One of the biggest challenges with positive interactions, then, is simply measuring which kinds of positive interactions are occurring and with which species.

7.1.2 A Brief History of Positive Interactions

One of the earliest uses of the term mutualism was in Pierre van Benden's 1875 book *Les Commensaux et les*

Parasites (Boucher et al.1982). Van Benden recognized many now classic examples of mutualism, including pollination and root-nodulation on legumes. Less than 20 years later the subject of mutualism was reviewed in *The American Naturalist* by Pound (1893). Mutualism also received prominence in *Principles of Animal Ecology*, co-written in 1949 by Warder Allee, a Quaker and pacifist, who in 1951 added a second book, *Cooperation Among Animals with Human Implications* (Allee et al. 1949; Allee 1951). More than 30 years later, Boucher (1985a) compiled an overview that included 15 papers, ranging from the natural history of mutualism (Janzen 1985) to cost–benefit models (Keeler 1985). It remains an important book.

Richard Dawkins, author of *The Selfish Gene* (1976), explains carefully how positive interactions can arise from selfish genes. This book is must reading for all young biologists. Too many of the public, and perhaps too many biologists, have read only the title and so lapse into thinking, once again, that selfishness is the antithesis of cooperation. Not at all, "selfish" gains can generate cooperation and lead to mutual benefits. So the study of selfish genes can be a foundation for the study of cooperation among plants.

Another popular book, *Gaia* (Lovelock 1979) argues that even the entire planet is a living entity. That book seems to have captured the popular imagination but, again, more have read the title than read the book. The book itself actually has some rambling attacks on conservationists since, it implies, *Gaia* is big enough to look after herself without troublemaking environmentalists annoying everyone. I have tried to write about this political side of the book but no one wants to hear – since they love the title. So I include this observation here to guide younger readers. *Gaia* can be interpreted as a marvellous happy dream about Mother Earth. In that case, it is religion, not science. *Gaia* can be interpreted as a justification for doing nothing, since nature can look after herself. In that case, it is politics not science. Or, *Gaia* might be the ultimate extension of studies of positive interactions, for example, plants cooling the Earth (Chapter 1) and plants feeding animals (this chapter). Getting back to plants, at a much smaller scale, *Positive Interactions and Interdependence in Plant Communities* (Callaway 2007) provides an entire book on positive interactions, should this chapter whet your appetite for more. (How could it not?)

7.2   Positive Interactions Occur Between Plants and Plants

Positive interactions are much more widespread than mutualisms, if only because the definition is so much wider: the term positive interaction needs only for one, rather than both, species to benefit from the interaction. Yet there are fewer examples than you might think, perhaps because plants that are growing close enough for a positive interaction may also be drawing from the same pool of resources, in which case competition may be the overwhelming effect. In spite of the negative effects of competition, there is a growing number of cases of plants acting as "nurse plants." This is where we will begin.

7.2.1 Nurse Plants

In arid environments, shade from neighbouring plants may promote the establishment of seedlings. The massive columnar saguaro cactus (*Carnegiea gigantea*), conspicuous in the Sonoran Desert of Arizona, California and Mexico (recall Figure 1.1), requires shade at the seedling stage. A study of 3,299 transplanted saguaro seedlings that were approximately 12 mm high (Turner et al. 1966) reported that all 1,200 unshaded plants died within one year, whereas 35 percent of the shaded plants survived that period. In addition, seedlings in darker soils had higher

FIGURE 7.2 Saguaro cactus grove in the Sonoran Desert (Saguaro National Monument) where cacti reach 15 m tall. (By George A. Grant, courtesy of the US National Parks Service) (See also Figure 1.1)

mortality rates than those in lighter coloured soils, probably because of the higher temperature reached by dark-coloured soil. The critical time for seedlings was the hot dry period just before the beginning of summer rain. Shrubs such as palo verde (*Cercidium microphyllum*) and creosote bush (*Larrea tridentata*) are therefore termed nurse plants because they shade saguaro seedlings until they are old enough to survive on their own in full sunlight. You can see such shrubs in Figure 7.2. Turner et al. thought that young saguaros probably become independent from their nurse plant at five to ten years of age when the saguaro would be 10 to 15 cm tall. In environments lacking shrubs, they suggest that seedlings may survive by growing on the shaded sides of rocks.

Franco and Nobel (1989) extended the work on nurse plants and saguaro in several ways. First, they showed that a second species of cactus, *Ferocactus acanthodes* (barrel cactus), is associated with the perennial bunch grass, *Hilaria rigida*.

The *F. acanthodes* seedlings could tolerate higher temperatures, however, and thus 30 percent of the seedlings were able to survive without nurse plants. (You can see an adult of this species of cactus to the left of the saguaro in Figure 1.1.) Second, Franco and Nobel provided added detail on the effects of nurse plants upon the physical environment. On a clear day in the autumn, soil temperatures exceeded 60°C compared with 47.4°C or 41.4°C under nurse plants (Figure 7.3). Daily variation in temperature was also lower under nurse plants. Total soil nitrogen levels were higher under nurse plants. Of course, nurse plants may also reduce the growth rates of seedlings through competition. Based upon effects of temperature, light and water upon net CO_2 uptake, Franco and Nobel estimated that the depletion of light and water by the nurse plant reduced the growth of *F. acanthodes* seedlings by one-third.

The beneficial effects of nurse plants have been established elsewhere. Consider four examples:

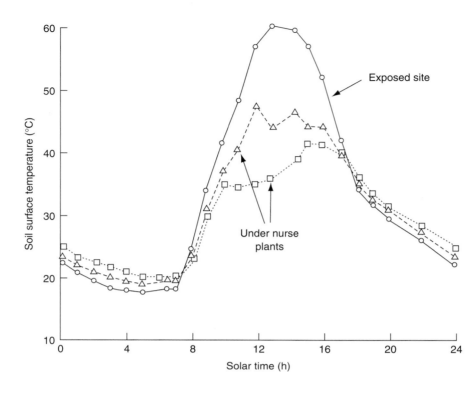

FIGURE 7.3 Effect of nurse plants on soil temperature shown by the daily course of soil surface temperatures at an exposed location (○) and at the centres of representative *Ambrosia deltoidea* (△) and representative *Cercidium microphyllum* (□) plants on a clear day. (After Franco and Nobel 1989)

desert, sand dune, savanna and evergreen scrub. (1) In the Mojave Desert, *Opuntia* cacti are associated with nurse plant species such as *H. rigida* as well as with other opuntias (Cody 1993). (2) In sand dunes around the Great Lakes, young pine trees establish beneath oak trees during sand dune succession (Kellman and Kading 1992). (3) In a savanna in Belize, Kellman (1985) transplanted seedlings of *Xylopia frutescens*, a small tree of thickets, and *Calophyllum brasiliense*, a large rainforest species, to different sites and monitored growth for three years. "In both species survivorship and growth of seedlings was superior beneath tree cover and it was unlikely that any open-grown plant would survive to reproduce" (p. 373). Thus, even when savanna habitats are protected from fire (see Figure 5.1 for such a fire), establishment of forest tree seedlings apparently requires the existence of other woody savanna plants to act as nurse trees. Those trees able to use nurse plants are likely to establish the first generation of forest after fire suppression. (4) In the evergreen scrub of central Chile ("matorral"), herbaceous plants are clumped beneath shrubs (Jaksic and Fuentes 1980).

In conclusion, neighbouring plants may produce beneficial effects, particularly in habitats where temperature or grazing is an important environmental constraint (Kellman and Kading 1992). There are likely to be strong trade-offs; any benefits from shade or protection from herbivores have to exceed the costs of increased competition for light and water. Thus it may be expected that nurse plants will be most common where environmental stress is high. The beneficial effect of neighbours may also be a common feature of primary succession. Connell and Slatyer (1977) have emphasized, following Clements (1916), that nurse plants ("facilitation") may be one of the important ecological interactions during primary succession.

7.2.2 Gradients Illustrate How Stress Affects Positive Interactions

Recall from Chapter 4 that a standard design of competition experiments is to grow plants with and without neighbours in order to measure the depression in performance caused by neighbours. The greater the depression caused by neighbours, the higher the intensity of competition. Exactly the same experiment can, however, find the unexpected: neighbours may have positive rather than negative effects. Let us return to an experiment measuring competition intensity along a gradient. Wilson and Keddy (1986a) studied an exposure gradient on a lakeshore running from open wave-washed sand beaches to sheltered bays where silt and clay accumulated. They found a significant increase in competition intensity from beaches to bays. However, the experiment showed a second unexpected result: the most extreme sites on the wave-washed beaches had negative competition intensity, suggesting that the surrounding vegetation actually protected the transplanted individuals from waves or sunlight.

When this design of experiment was repeated using 60 quadrats representing an open sunny beach and a densely vegetated bay, there was again a significant increase in the intensity of competition with increasing biomass (Figure 4.22). But look again – on the left side of the figure there are many points showing negative competition intensity, that is, a positive interaction between the transplanted individual (phytometer) and its neighbours. These results clearly show that positive interactions can be detected with field experiments, and that their importance apparently increases with increased physical constraints upon plant growth.

Other wetland gradients show similar positive effects. Bertness and Hacker (1994) used the elevation gradient in a salt marsh to test for changes in positive effects with physical stress. The typical zonation on such shorelines consists of *Iva frutescens* at high elevations and *Juncus gerardi* at low elevations, with a mixed zone in between (Figure 7.4a).

The intensity of stress is highest at the left since wave damage, hypoxia and salinity all increase with tidal flooding. In each of three zones, *Iva* and *Juncus* were planted with and without neighbours. At the high elevation (right) the growth of both *Iva* and *Juncus* was depressed by neighbours, showing the presence of competition. At the two lower elevations (left) both species benefited from the presence of neighbours. Apparently, at lower elevations, neighbours reduce evaporation and therefore reduce salinity.

7.2.3 Examples of Positive Interactions in Wetlands

You have just seen in Figure 7.4 that *Iva frutescens* (marsh elder) is a common plant at the upper border of salt marshes on the east coast of North America. Recruitment of *Iva frutescens* is influenced by both negative and positive interactions with adult plants (Bertness and Yeh 1994). Seedlings usually cannot establish because of the presence of other perennial plants. Adult *Iva* plants, however, can accumulate plant debris that kills these other perennial species, and the adults simultaneously protect *Iva* seedlings from radiation and salinity. Recruitment of *Iva* is determined by a balance between positive effects and competition.

The lack of oxygen in flooded soils is a major kind of stress in wetlands; we shall read more about this in Chapter 10. One way plants get around this stress is to transport oxygen from the atmosphere to their rhizomes and roots. In doing so, they may also assist neighbouring plants. Callaway and King (1996) showed that under cool conditions (11 to 12°C) *Typha latifolia* transported oxygen into sediments, raising oxygen levels from 1 mg l^{-1} to near 4 mg l^{-1}. Other wetland plants benefited: there was enhanced survival of *Salix exigua* and enhanced growth of *Myosotis laxa*. At higher temperatures, however, the dissolved oxygen levels fell, suggesting a shift from facilitation back to competition as temperature increased.

(a)

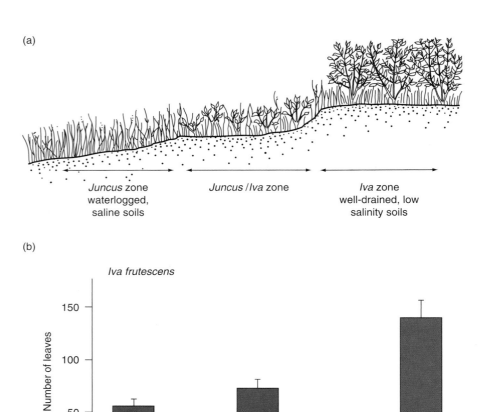

Juncus zone
waterlogged,
saline soils

Juncus /Iva zone

Iva zone
well-drained, low
salinity soils

(b)

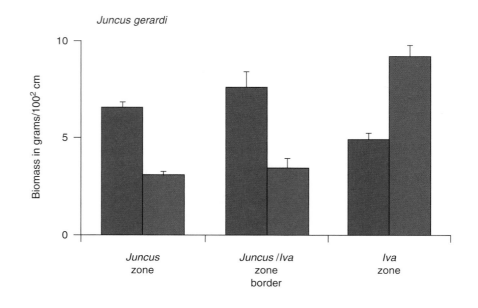

FIGURE 7.4 (a) The three vegetation zones found at higher elevations in a typical New England salt marsh. (b) Results of an experiment in which adult *Iva frutescens* and *Juncus gerardi* were transplanted with (green) and without (brown) neighbours for four months. *Iva* performance was measured as leaf counts while *Juncus* performance was measured as biomass. Each bar represents ten transplants (+/- SE). (from Bertness and Hacker 1994)

7.2.4 Commensalism May Be Common in Plant Communities

It takes experiments to test for the presence of commensalism. As such studies accumulate, it is likely that many more examples will emerge. Meanwhile, one may observe that there are other circumstances where commensalism is likely common. Since many epiphytes depend upon the tree to provide rooting space and access to sunlight, the host is providing a positive effect for the epiphyte. Thus the vast numbers of epiphytic bromeliads and orchids may well have equally vast numbers of positive interactions with trees. The effects upon the tree appear to be minimal, although there may be cases where the sheer weight of the epiphytes breaks off branches. How would you design an experiment to measure the benefits provided by the tree? Similarly, many understorey plants require shaded conditions, either because they cannot physiologically tolerate full sunlight, or because they

cannot compete with other plants better able to exploit full sunlight. You have seen (Figure 5.39) that removing the trees can have long-term negative effects upon such species.

Positive interactions between plants have some upper limit, because when two plants grow close together there is every reason to also expect strong competition for shared resources. Hence competition may often override any beneficial effects. In the case of nurse plants, if a seedling establishes and grows, the young plant will eventually begin to demand more resources from its nurse. Eventually one or the other will begin to experience negative effects. And, yet, there is one other factor to consider: when plants grow in close proximity there is an opportunity for below-ground connections through soil fungi. More opportunities for commensalism therefore arise. And many more opportunities for mutualism was well. So let us turn to plants and fungi.

7.3 There Are Many Positive Interactions Between Plants and Fungi

7.3.1 There Are Four Kinds of Mycorrhizae

Fungi associated with roots, called mycorrhizae, provide a vital means of enhancing rates of nutrient uptake, particularly in soils with low levels of nitrogen and phosphorus. In general, it seems that fungi provide phosphorus and nitrogen, and in exchange receive carbohydrates (Francis and Read 1984; Smith and Douglas 1987; Woodward and Kelly 1997). Although discovered in the late 1800s (Box 7.1), mycorrhizae were once thought to be uncommon, but it is now clear that mycorrhizae are found in most groups of plants and most habitats. There are two main groups of mycorrhizae, arbuscular mycorrhizae and ectomycorrhizae, distinguished by the degree to which roots are penetrated by the fungi (Figure 7.5 top).

Arbuscular mycorrhizae are by far the most common, occurring in about 75 percent of all vascular plants.

They are named for the distinctive storage structures that occur within the roots of the plant, arbuscules. (I regret to inform you that there is considerable confusion in terminology here, since this same association was once termed simply and sensibly, endomycorrhizae, since the roots were penetrated. As if this duplication were not enough, you will also encounter the term vesicular-arbuscular mycorrrhizae (VAM), for this same group, in reference to the vesicles that the fungus also produces inside the cells of the root. The confusion in terminology can perhaps be forgiven given the difficulty in understanding subterranean microscopic events, but, for the moment, arbuscular is the term generally used (Brundett 2009).) The arbuscular fungi are zygomycetes, belonging to the Glomales, which contains fewer than 200 mycorrhizal species in the world.

Box 7.1   Fortune Favours the Prepared Mind: Bernhard Frank Discovers Mycorrhizae

What started as a simple study of truffles in Prussia during the 1800s went on to revolutionize our understanding of trees, fungi and positive interactions in ecosystems altogether. This is a reminder to all young scientists that, in the words of Louis Pasteur, "Fortune favours the prepared mind."

You have seen that most terrestrial plants are unable to survive without the assistance provided by their associated root fungi (Figure B7.1.1). And you should even recall from Chapter 1 that mycorrhizae are associated with some of the earliest fossil land plants. But when Bernhard Frank first discovered **mycorrhizae** (and named them) he was subjected to ridicule. Fungi were known to kill plants and decompose them, but help them? Surely not!

FIGURE B7.1.1   In ecotomycorrhizae, the fungus produces a network of hyphae over thickened and shortened tree roots. This example is the fungus *Astraeus pteridis* growing on the roots of *Eucalyptus maculata* (ca. × 10). (Photo courtesy of Nick Malajczuk, Randy Molina and Jim Trappe)

Let us begin the story in Europe in the early 1800s. Albert Bernhard Frank was born in Dresden, Germany in 1839. This was a time of turmoil. Beginning in 1848, when he would have been only eleven, revolution came, in part driven by crop failures from Ireland to Russia. It was this year that Karl Marx and Friedrich Engels published the *Communist Manifesto*, ending eloquently "The proletarians have nothing to lose but their chains. They have a world to win. Workers of the world, unite." Bitter street fights broke out in Berlin, and there was outright insurrection in Paris.

Yet science, too, continued. About a decade later, in 1859, Darwin published *On the Origin of Species*. Frank earned his PhD in Leipzig in 1865, and went on to study a series of botanical problems, including heliotropism and diseases of beets and grains. He also produced a massive book, *Plant Pathology*, in 1880.

The story could end there. But his greatest discovery was yet to come. He was asked to study truffles in Prussia. Truffles, as you know, are a valuable edible underground mushroom produced by forested landscapes. Beginning in 1883, he published a series of papers that laid the foundation for the study of fungal associations with plant roots. His first work documenting the occurrence of mycorrhizae occurred in 1885. On the first page he says "certain trees... do not nourish themselves independently in the soil but regularly establish a symbiosis with fungal mycelium over their entire root system." Roots are covered by fungal hyphae, and this mantle "completely encloses the rootlet, forming a continuous cover even over the growing tip." (Of course, Frank himself wrote in German; this translation of his important paper only appeared in 2004, courtesy of Jim Trappe, himself an expert on truffles and forests, in a journal called, appropriately, *Mycorrhiza*.)

Frank's publications generated a "storm of criticism" (Hatch 1937: 29). The loud opposition to his discoveries gradually subsided, presumably as "most of those participants found, on more careful examination, that fungus roots were abundant" (Hatch 1937: 34). But it delayed scientific progress. By 1900, the year of Frank's death, another German botanist, Ernest Stahl, published "a thorough study of root condition in the entire plant kingdom" (p. 35), concluding that mycorrhizae were particularly important in infertile soils. It was only after Frank's death, then, that the true significance of his work became apparent.

Box 7.1   (cont.)

Indeed, we might speculate that the real revolution inherent in Frank's first papers has yet to play out, since most foresters still seem to know more about trees and insects than they do about mycorrhizae. Yet the key to managing and restoring the world's forests may demand that we pay far more attention to the underground network that links them into a larger biological entity. Indeed, if we start at the level of mycorrhizae, we can see forested ecosystems as some large multispecies network with an enormous number of positive interactions. Maser et al. (2008) have written an entire book on this very topic.

And, in a curiously appropriate way, Frank also expanded our understanding of natural selection. Too many followers of Darwin carelessly interpret the struggle for existence to mean a world filled with negative interactions. But mycorrhizae illustrate a profound general principle: that positive collaborative networks can also arise through natural selection.

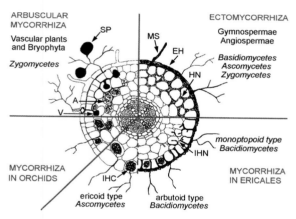

FIGURE 7.5   The four main kinds of mycorrhizae: arbuscular mycorrhizae, ectomycorrhizae, orchid mycorrhizae and ericoid mycorrhizae. MS, mycelial strands; EH, external hyphal mantle; HN, Hartig net; IHN, intercellular hyphal net; IHC, intracellular hyphal complexes; V, fungal vesicle; A, arbuscule; SP, spore. (From Larcher 2003)

Ectomycorrhizae are, as the name suggests, external, with the fungal hyphae forming a Hartig net over the root surface. These fungi are more taxonomically diverse, and most characteristic of temperate zone trees, particularly members of the Fagaceae

and Pinaceae, but also including *Eucalyptus* and *Nothofagus*. The Orchidaceae and Ericaceae are also mycorrhizal, but are distinctive enough to be put in separate categories (Figure 7.5 bottom).

A review of 10,000 plant species in 336 plant families found the relative abundances of mycorrhizae as follows: arbuscular 74 percent, orchid 9 percent, ectomycorrhizae 2 percent and ericoid 1 percent (Brundrett 2009). There were no mycorrhizae in 6 percent, particularly in families such as the Chenopodiaceae and Proteaceae. (A remaining 8 percent of families were reported to contain both arbuscular and non-mycorrhizal species, such families often coming from habitats thought to be too cold, flooded, dry or saline to support mycorrhizal fungi.)

The benefits of mycorrhizal associations to both partners are relatively clear. Hence this is a mutualism. Let us focus for the moment on the plant side of the relationship. The benefits to the plant are mainly access to nutrients that would otherwise be unavailable, or at least to increase the rates of extraction of these nutrients. There is a further benefit: by using a network comprised of fungal hyphae to extract nutrients, a plant need not allocate as much tissue to roots. Hence stubby roots are often associated with

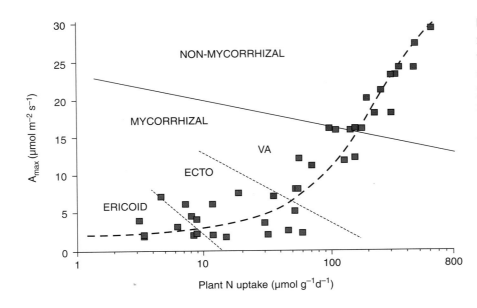

FIGURE 7.6 The relationship between the uptake rate of nitrogen and the maximum rate of photosynthesis in relation to mycorrhizal association (VA, vesicular-arbuscular mycorrhizae; ECTO, ectomycorrhizae; ERICOID, ericoid endomycorrhizae). (From Woodward and Kelly 1997) Note that the non-mycorrhizal plants (brown squares) actually have higher growth rates.

mycorrhizae while non-mycorrhizal plants have longer roots and longer root hairs. But there are also costs. The obvious one is the loss of carbohydrates to support the fungus. These costs would appear to be substantial, since in situations where nitrogen uptake and plant growth are high, plants often do not have mycorrhizae (Figure 7.6). Or, to express it in another way, the highest rates of growth are found in non-mycorrhizal plants in fertile soils. As soil resources decrease, the benefits of mycorrhizae tend to out-weigh their costs. To re-emphasize this, the presence of ectomycorrhizae, for example, does not allow plants to grow faster than non-mycorrhizal plants, but it does allow them to grow in habitats that lie outside the tolerance limits of non-mycorrhizal plants. Shrubs in the Ericaceae occupy extremely infertile soils and peat bogs where, in spite of ericoid mycorrhizae, they have extremely low rates of nutrient uptake and inherently low rates of photo-synthesis. The relative costs of mycorrhizal fungi appear to increase from arbuscular mycorrhizae through ectomycorrhizae to ericoid mycorrhizae, since each in turn is associated with more extreme conditions of infertility.

7.3.2 Ectomycorrhizae Are Vital to Forests

Ectomycorrhizae are common in forests, where they certainly affect tree growth and likely link trees into some sort of complex multispecies assemblage of plants and fungi. The species involved, and the flows among them, and the costs and benefits will take many years to unravel. All the same, if you collect wild mushrooms you soon find that certain mushrooms occur with certain trees. We are going to look at just one example, from southern Quebec, that illustrates some of the many species that may be involved in deciduous forests. Here, Nantel and Neuman (1992) sampled forest vegetation, measured the characteristics of the physical environment and also the production of basidiomycete fungi that were probably ectomycor-rhizal. Some 240 species of ectomycorrhizal Basi-diomycetes were recovered. The relationships among fungi, trees and the physical environment were then explored using multivariate methods you will learn more about in Section 11.3. There was some evi-dence that the fungi themselves were controlled by

Table 7.2 Selected fungal and tree species associations found in some forests of southern Quebec. (From Nantel and Neumann 1992)

Cluster	Basidiomycete species	Tree species
1	Amanita brunnescens	Pinus strobus
	Suillus granulatus	Populus grandidentata
2	Amanita porphyria	Picea mariana
	Russula paludosa	
	Leccinum insigne	
	Russula roseipes	
3	Amanita citrina	Acer rubrum
	Boletus piperatus	Populus grandidentata
	Russula raoultii	Populus tremuloides
4	Amanita virosa	Betula papyrifera
	Cortinarius evernius	
	Russula peckii	
	Clavulina cristata	
	Hebeloma testaceum	
5	Amanita flavoconia	Abies balsamea
	Leccinum scabrum	
	Russula fragilis	
	Russula puellaris	
6	Amanita fulva	Betula papyrifera
	Hebeloma sp.	Populus tremuloides
	Russula silvicola	
7	Hygrophorus laetus	Acer rubrum
	Hygrophorum unguinosus	Populus grandidentata
	Nolanea strictior	
	Rozites caperata	
8	Boletinus pictus	Picea glauca
	Cortinarius bolaris	Betula alleghaniensis
	Hygrophorus nitidus	
	Xerocomus subtomentosus	
	Lactarius sordidus	
9	Amanita muscaria	Acer saccharum
	Hygrophorus marginatus	Fagus grandifolia
	Hygrophorus pallidus	
	Hygrophorus ceraceus	
	Hygrophorus parvulus	
	Hygrophorus pratensis	
	Paxillus involutus	
10	Cantharellus cibarius	Populus grandidentata
	Hebeloma mesophaeum	Fagus grandifolia
	Lactarius glyciosmus	
	Russula claroflava	
	Inocybe umbrina	
	Cortinarius armillatus	

soil factors such as thickness of soil horizons, percentage of organic matter, nutrient levels of the mineral soil, soil moisture and pH. Overall, the patterns found in the fungi were best explained by the trees that were present (Table 7.2). The coniferous tree genera had fungal associates different from angiosperms and from one another. For example, the fungal genus *Hygrophorus* was found mainly in *Fagus-Acer* forest, whereas *Leccinum scabrum* was associated only with *Abies*. Other genera such as *Russula* were widely spread across vegetation types, although individual species appeared to differentiate among tree species; *Russula fragilis* was found with *Abies balsamea*, whereas *Russula roseipes* was with *Picea mariana*.

You can see that this took a lot of time and effort. Yet what is equally striking is the number of limitations on this study, which illustrate the sort of general problems that arise in the study of positive interactions.

1. Only basidiomycete fungi were included in the study.
2. Only half of the fungi could be identified.
3. Only fungi that reproduced sexually were included.
4. Abundance of fungi was estimated by dry weight of fruiting bodies (mushrooms); there may be little relationship between the physiological activity or biomass of the hyphal networks and the number of mushrooms produced.
5. The production of fungi fruiting bodies varies greatly from one year to the next, and this study examined only two growing seasons.
6. Not all sites were sampled the same year, so that among-year variation could be confounded with between-site variation.
7. There were only 11 stands, and the number of plants and environmental factors greatly outnumbered the sampling stations, which creates difficulties with data analysis and interpretation.

This list is not intended to criticize this study, but rather to illustrate the difficulties that deter the study of fungus–plant relationships. Some of these problems can be rectified in future by increased sampling effort (that is increased time and money). Identification could be enhanced with techniques such as DNA bar coding, which would also allow species in the soil that did not produce mushrooms to be identified. But this still leaves three profound questions. Which species are physically linked to which others? For each link, what is being exchanged? And what are the net benefits? You may begin to wonder how many of our cherished assumptions about forest management would change if we knew more about forest mycorrhizae.

7.3.3 Mycorrhizae May Be Less Important in Wet Habitats

In contrast with forests, mycorrhizae may be much less important in wetlands. Figure 7.7 shows the data from a review of the distribution of arbuscular mycorrhizae in 843 species (Peat and Fitter 1993). The lower bars show that mycorrhizae are relatively uncommon in wetlands. Larger global-scale surveys reinforce this pattern: "hydrophytes" are considered to be non-mycorrhizal (Brundrett 2009).

The most likely explanation for lower mycorrhizal activity in wetlands is the lack of soil oxygen. This would inhibit the growth of most fungi. In Chapter 10 we will look more closely at the importance of soil oxygen in wetlands. All the same, plants are not at the complete mercy of the habitat: non-mycorrhizal plants may also have roots that are resistant to fungal

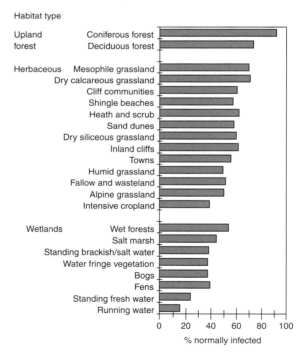

FIGURE 7.7 The relative abundance of mycorrhizal infection in different habitats. (From data in Peat and Fitter 1993)

hyphae, and so tend to remain uninfected even when the surrounding plants are mycorrhizal.

The scarcity of mycorrhizae in wetlands may have other implications. It suggests that many wetland plants may have difficulty in taking up nutrients. Hence one may speculate that soil nutrient gradients are even more important in wetlands than in terrestrial habitats. Might this account for the frequency of carnivorous plants in wet as opposed to dry soils? (Look at Figure 3.18!)

Of course, a careful reader will note two limits of literature surveys: these mostly address the presence (or absence) of mycorrhizae, not their abundance. And should one be able to measure abundance, such as percentage of root length infected, one still remains uncertain about rates of flow between the plants and fungi per unit of infected area.

Here are a few examples of mycorrhizae that do occur in wetlands. Although most aquatic plants are free of mycorrhizae, *Lobelia dormanna* has mycorrhizae on significant lengths of its roots (Nielsen et al. 2004). Of course, this is an unusual plant in other ways, a wetland plant with CAM photosynthesis and the ability to extract carbon dioxide from the soil (Section 3.3.3). An investigation of eight species of wetland plants in South Dakota (Rickerl et al. 1994) found that mycorrhizal infection declined with flooding (27 percent infection in dry areas vs. less than 1 percent in wet areas), with *Carex atherodes* and *Juncus tenuis* being entirely uninfected. *Scirpus fluviatilis* was non-mycorrhizal in wet sites, but slightly colonized (9 percent infection) in dry areas. Marsh plants in Indiana (Bauer et al. 2003) had mycorrhizal colonization levels of 3 to 90 percent, but no trend in infection with hydrologic zone. Grasses had particularly high levels of mycorrhizae.

7.3.4 Measuring Costs and Benefits

We have already seen that it is not enough to demonstrate that a plant and a fungus are connected. To demonstrate mutualism it is necessary to find benefits to both partners. This is difficult. Smith and Douglas (1987) provide a readable account of the challenges,

along with some results. They describe how radioactive-labelled sucrose has been shown to move from plant tissues to fungi in laboratory experiments, but note that field measurements are considerably more difficult to make. How much energy do plants provide soil fungi? In spruce (*Picea*) forests, carbohydrate consumption by ectomycorrhizae is some ten percent of net production. In beech (*Fagus*) forests, ectomycorrhizae account for roughly one quarter of respiration by beech root systems. And what is returned to the plant from the fungus? Again isotopes of nitrogen and phosphorus document nutrient uptake by fungi and the consequent movement of nutrients into plant tissues. When $^{32}PO_4$ (radio-labelled phosphate) is added to soil it can appear in trees within two days.

But such figures also depend upon the size of the plant and its light availability: small plants in the shade may draw carbon from the mycorrhizal network (Figure 7.8). Moreover, as we saw in Section 3.7.3, there are many plants (called mycoheterotrophs) that depend partially or fully upon mycorrhizal networks for carbohydrates as well as mineral nutrients. Consider the widespread *Monotropa hypopithys* (Figure 3.32 left). When ^{14}C-labelled glucose and $^{32}PO_4$ were supplied to neighbouring *Pinus* and *Picea* trees, these isotopes were recovered within five days from *Monotropa* plants growing one to two metres away (Björkman 1960). So here there is symbiosis, but probably parasitism rather than mutualism. The devil, as they say, is in the detail.

So what are we to conclude? At one time mycorrhizae were thought to be unusual, and experiments using them were rare. But now we understand that they are the rule rather than the exception. Hence our view of plant communities has to change. Most plant communities appear to consist of interconnected individuals and species, exchanging carbon, nitrogen, phosphorus and even water (Selosse et al. 2004; Merckx et al. 2009; van der Heijden and Horton 2009). While a majority of the plants gain carbon from photosynthesis, others are mixotrophs (that is, they photosynthesize but also draw upon neighbours for carbon) while some are apparently parasites. A few parasites are directly attached to neighbouring plants

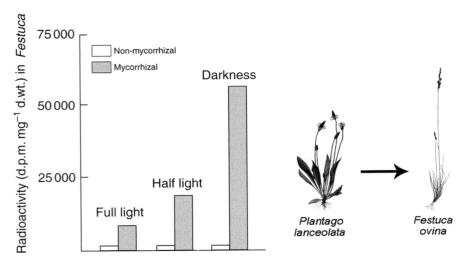

FIGURE 7.8 Movement of carbohydrate from *Plantago lanceolata* to *Festuca ovina* roots, with and without mycorrhizal connections by the fungus *Glomus caledonium*. (From Smith and Douglas 1987)

with their haustoria, but many more are attached to neighbours through mycorrhizae. A further level of complexity is being revealed by DNA techniques for identifying fungi – these reveal much more fungal diversity in mycorrhizal networks (Selosse et al. 2004; Merckx et al. 2009). In short, it may be necessary to view communities less as individuals, or populations or pairs of interacting species, and more as networks. This has profound, even troubling, consequences for the design and interpretation of field experiments. It also raises new issues for understanding coevolution. If there is one general conclusion, it is that plant communities are far more tightly linked that we once understood. It would be possible to re-read Chapters 1 to 4 and pose new questions about what such networks mean for evolution, speciation, resources and competition.

7.3.5 Lichens Are Somewhat Different, and Somewhat Similar

Lichens can be thought of as a kind of special case of plant–fungus interactions, one where the fungus has become dominant. One-fifth of all fungi, that is some 13,500 species representing 525 genera, enter into lichen associations (Hawksworth 1988). In the majority of cases the fungus forms the bulk of the thallus, and the plant (cyanobacterium or alga) is restricted to

one layer (Figure 7.9). The term **phycobiont** is used to label the photosynthetic partner since cyanobacteria and three kinds of algae (Chlorophyta, Phaeophyta and Xanthophyta) have all been reported from lichen associations (Hawksworth 1988). Normally the species of fungi are exclusively found in lichen associations and do not have other independent lives. In contrast, the algae represent free-living species, and at most only 39 genera are involved. In some cases three or four phycobiont species may occur within one thallus. In species of *Solarina*, a cyanobacterium and a green alga form separate layers, while in *Placopsis gelida* the cyanobacteria are confined to distinctive structures called cephalodia and the green alga are scattered through the main thallus.

The fungus (or mycobiont) receives carbohydrates produced by the phycobiont. In addition, in the case of cyanobacteria, ammonia fixed from atmospheric nitrogen passes to the mycobiont. The advantages, if any, that accrue to the phycobiont are less clear. It is the phycobionts, after all, that are penetrated by haustoria, and the possibility that lichens are an advanced state of parasitism should not be discounted. That is to say, returning to definitions, not every case of symbiosis should be assumed to be a mutualism. But certainly these two groups of organism have coevolved, as demonstrated by features such as the production of structures called soredia,

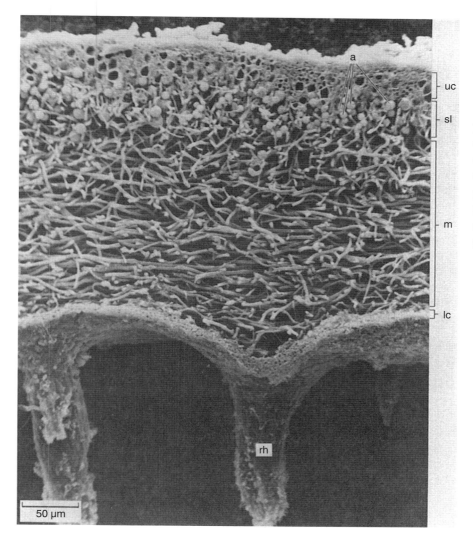

FIGURE 7.9 Scanning electron micrograph of a transverse section through *Parmelia borreri*, a foliose lichen with *Trebouxia* as the algal phycobiont. The algae cells occur in a thin layer denoted sl on the right. (uc, upper cortex; sl, phycobiont layer; a, alga; m, medulla; lc, lower cortex; rh, rhizoid. (From Smith and Douglas 1987)

which disperse both the fungus and the alga as one unit.

The photosynthetic rate per unit of chlorophyll in the alga is comparable to that of non-lichenized (that is, free-living) species (the rate per unit area is, of course, much smaller because the phycobiont comprises only five to ten percent of the volume of the thallus). When lichens are wetted at night by dew, but dry during the day, there can actually be loss of weight as respiratory CO_2 losses at night can exceed those replaced by photosynthesis during the day (Lechowicz 1981). Tracer experiments using ^{14}C show in a range of lichens that a substantial proportion of the photosynthate is released to the fungal host but estimates of quantities have been difficult to obtain; moreover, successful measurements under laboratory conditions do not necessarily provide results applicable to the field. The first lichen species in which movement was demonstrated experimentally was *Peltigera polydactyla*, which has a cyanobacterium

(*Nostoc*) as the phycobiont. Using discs of tissue exposed to a radioactive tracer, Smith (1980) found appreciable amounts of radioactive carbon in the fungal tissue within two hours. In this case glucose moves from the phycobiont to the fungus. It appears that transfer is usually in the form of a single compound, but the compound differs among the genera of phycobionts (Smith and Douglas 1987). In contrast "it has long been claimed that the fungal host provides its symbionts with mineral nutrients, but there is no experimental evidence either for the movement of nutrients or for the requirement by the symbionts for host-derived compounds. The growth of symbionts is so slow that their needs could probably be satisfied by the normal mineral content of rainwater" (Smith and Douglas 1987: p. 139). In short, this is definitely a case of symbiosis, but without better data on benefits and costs, it seems that this symbiosis is better labelled parasitism than mutualism.

7.3.6 Fungi Can Also Occur in Shoots and Leaves

While this section has focused mostly upon below-ground interactions, fungi can also live in the aerial portions of plants as endosymbionts. Such relationships are widespread, in habitats ranging from grasslands to tropical rainforest canopies (Carroll 1988). There is evidence that these above-ground fungal assocations can affect both the growth and relative competitive abilities of plants (Clay 1990; Selosse et al. 2004). Endosymbionts may also protect foliage from pathogens (Herre et al. 2007).

7.4 Positive Interactions Between Plants and Animals: Part 1 Pollination

7.4.1 Animals Pollinate Flowers

Animals Consume Pollen and Nectar

Three large groups of animals derive nourishment from flowers: insects, birds and bats (Percival 1965; Lloyd and Barrett 1996). Flowers provide two principal sources of food: pollen and nectar. Let us consider them in turn.

Pollen is a valuable food, containing protein, fats and carbohydrates. The protein content, from 25 to 33 percent (Percival 1965), is noteworthy in view of the general importance of nitrogen as an important nutrient for animals (Section 6.3.5).

Nectar is produced by glands, most of which occur within flowers, but some of which, in extra-floral nectaries, is borne on adjoining leaves or petioles. Almost any part of a flower can be adapted to produce nectar, including the sepals, petals, stamens and pistils. Nectar contains chiefly sugar, particularly sucrose, fructose and glucose. The sugar content of nectars can range from greater than 70 percent (e.g. *Aesculus hippocastanum*) to less than 20 percent

(*Prunus* spp.) (Percival 1965). Most large-flowered species with copious nectar occur in the tropics. The large flowers of the banana (Figure 7.1), for example, fill nearly to the brim with nectar. Other copious producers include *Heliconia* spp. and the striking bird of paradise flower (*Strelitzia reginae*, Figure 7.10b). In *Protea* spp., the flowers are small, but nectar from them is poured into a cup formed from adjoining bracts.

The earliest example of animal pollination likely involved beetles consuming pollen, and incidentally transferring the pollen from one strobilus to another, and from one plant to another. An interaction that began as herbivory would have had benefits for the plant such as increased seed production. As a consequence, those strobili with bracts that were more attractive to beetles proliferated. In angiosperms we call these bracts "petals." Many of the plants in the Magnolia clade (Figures 2.3 and 2.6) have a type of flower reminiscent of a strobilus adapted to beetle pollination. Over time, as other kinds of insects also began to visit flowers, there was natural selection for other kinds of floral morphologies.

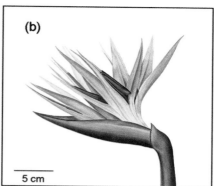

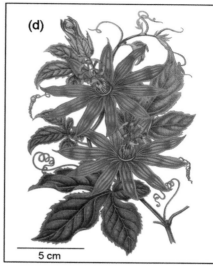

FIGURE 7.10 Four bird-pollinated flowers. (a) *Schlumbergera* sp. (South America, hummingbirds), (b) *Strelitzia reginae* (South Africa, sunbirds), (c) *Heliconia psittacorum* (Central America, hummingbirds), (d) *Passiflora vitifolia* (South America, hummingbirds). (P.-J. Redouté except for *P. vitifolia* by P. Stroobant 1867, (b)–(d) Peter H. Raven Library/ Missouri Botanical Garden)

As the insect–plant mutualism evolved, there was natural selection for flowers containing food other than pollen, particularly nectar. Not only would this increase rates of visitation by insects, but it might also reduce the amount of herbivory on the pollen itself. Plant–insect mutualisms still dominate the world today. Table 7.3 illustrates the breadth of insect types that feed upon products of flowers.

Birds Pollinate Many Tropical Flowers

Since so much is written about insects and plants, let us look at a slightly less familiar example: birds and plants. A majority of bird-pollinated flowers (Figure 7.10) occur in tropical regions, where four bird families feed on nectar: hummingbirds, sunbirds, sugarbirds and honey-eaters (Table 7.4). Unlike insects, birds take nectar, rather than pollen, as their primary food and most bird flowers therefore produce copious amounts of nectar. Large volumes of nectar need a container, so the flowers often are shaped like funnels or tubes to collect the nectar that is secreted (Figure 7.10 b and c). Vivid colours appear to attract birds by sight, pure red being most common although others are bright blue or green.

Hummingbirds are perhaps the best-known example of nectivorous birds; their ability to hover

Table 7.3 Main classes of insects that feed on the products of flowers. (From Percival 1965)

Order	Family	Genus/subgenus
Hymenoptera	Vespidae	*Vespa*
		Colpa
		Eumenes
	Bombidae	*Bombus* (bumble-bees)
		Apis (honey-bees)
	Colletidae	*Colletes*
	Andrenidae	*Andrena* (willow and golden-rod bees)
		Onagrandrena (*Oenothera* bees)
		Halictus (water lily bee)
		Halictoides (pickerel-weed bee)
	Panurgidae	*Panurgus*
	Megachilidae	*Megachile* (leaf-cutter bees)
		Osmia
	Xylocopidae	*Xylocopus* (carpenter bees)
	Anthophoridae	*Anthophora*
Diptera	Chironomidae	*Chironomus* (non-biting midges)
	Mycetophilidae	(fungus gnats)
	Ceratopogonidae	(biting midges)
	Tabanidae	*Panagonia* (horseflies)
	Nemestrinidae	*Megistorhynchus*
	Bombylidae	*Bombylius* (bee flies)
	Syrphidae	*Syrphus* (hover flies)
		Eristalis (drone flies)
Lepidoptera	Nymphalidae	*Vanessa* (tortoiseshells, peacocks)
		Pyrameis (painted lady)
		Heliconius (erato butterflies)
		Danaus (milkweed butterfly)
	Sphingidae	*Herse* (*Convolvulus* hawkmoth)
		Deilephila (striped hawkmoth)
		Macroglossa (hummingbird hawkmoth)
	Nocturidae	*Barathra* (cabbage moth)
		Triphoena (yellow underwing)
		Pluisa (silver Y)

allows them to extract nectar without landing upon the flowers on which they feed. Hummingbirds are most abundant in the tropics of South America, where approximately half of the known species (130 out of 319) occur. The bill is long and thin so that it is easily inserted into a tubular corolla (Figure 7.11 c, d). Sunbirds also have long thin bills (Figure 7.11 a, b), whereas lorys have bills more like those of parrots (Figure 7.11 f, g).

7.4.2 What Are the Mutual Benefits?

Now, all these examples are very interesting, but what of mutualism? A mutualism has to have measurable benefit to both parties. The benefit to birds (or insects, or bats) appears clear: a food source. This evidence

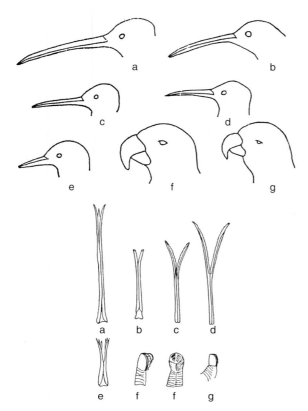

FIGURE 7.11 Bills and tongues of flower-visiting birds: (a) scarlet-tufted malachite sunbird, *Nectarinea johnstoni*, (b) west african olive sunbird, *Cyanomitra divacea guineensis* (both Nectarinindae); (c) and (d) two unidentified hummingbirds (both Trochilidae); (e) oriental white-eye, *Zosterops palpebrosa*, (Zosteropidae); (f) black capped lory, *Domicella domicella*, (g) red collared lorikeet, *Trichoglossus rubritorques* (both Loriidae). (From Percival 1965)

comes from an array of observations. The sugar content of the nectar can be measured. Four large families of birds use this source of food (Table 7.4), and their bills and tongues are shaped to efficiently extract it (Figure 7.11). Nectivorous birds prefer those flowers with the highest nectar content, as evidenced by the vigorous competition for nectar; dominant birds will set up territories and defend them against other visitors (Feinsinger 1976). The benefit to birds, then, seems clear, but what of their plant partner?

The obvious benefit is pollination. Plants that are visited by insects, birds or bats likely receive two benefits. The first is increased rates of seed set. In this case, we can think of the cost being the resources invested in flower production and the benefits being higher rates of seed production. These benefits can be measured by comparing rates of seed production in flowers that receive animal visitors as opposed to flowers that are unvisited. Unvisited flowers can be produced experimentally by various kinds of exclosures, while visited flowers can be produced experimentally by hand pollinating flowers. Thus, in the short term, there are a variety of experimental methods for measuring the benefits of pollination to flowers. The general answer is that animals benefit plants by increasing the success of seed production.

But this benefit requires further thought, because it raises the question of why plants should require pollination at all. That is, why is sexual reproduction so widespread? There is a second benefit provided by animals that is much more subtle: the benefits of outcrossing, that is, of spreading pollination not only among different flowers, but among different individuals. Since plants do not walk, they can only engage in sexual exchange with neighbours when pollen moves between individuals, and so depend upon animals to move the pollen to different individuals. The costs and benefits of sexual reproduction raise quite profound questions about mutualism and plant evolution. Hence let us have a short digression into the costs and benefits of sexual reproduction to put the animal–flower mutualism into a broader context.

7.4.3 Sexual Reproduction Has Costs

Why is there sexual reproduction anyway? The easy answer to this, and one too often heard and accepted, is that sex is necessary for evolution to occur. Certainly, sexual reproduction creates new genetic variation that includes new arrangements of genes in offspring. That is, sexuality creates novelty. While it is indeed true that one long-term

Table 7.4 **Flower-feeding birds. (From data in Percival 1965; inset: Loten sunbird, India, Arshad Ka, Wikimedia Commons)**

	Locale	Common name	Family
Major groups	Americas	hummingbirds	Trochilidae
	Africa	sunbirds	Nectariniidae
	South Africa	sugarbirds	Promeropidae
	South Pacific	honey-eaters	Meliphagidae
Minor groups	Hawaiian Islands	honey-creepers	Drepanididae
	Americas	tanagers	Thraupidae
	Australia	brush-tongued lories	Loriidae
		lorikeets	
	Africa and Australia	white-eyes	Zosteropidae
		silver-eyes	

consequence of sex is the genetic variation for evolution, this does not convincingly explain its value in the short term.

All structures have costs and benefits. The production of flowers and pollen and nectar (and seeds) have high costs. One set of these costs is obvious: the cost of resources used to produce flowers and nectar, and the cost of resources in producing fruit and seeds. In addition to these, there are opportunity costs. Each meristem that produces flowers and fruit is a meristem that could have produced more foliage to enhance photosynthesis and adult survival. Every flowering branch is a diversion of energy away from continued growth, not a small matter in a world where competition for light is often critical.

But there is a much more all-pervasive cost. The cost of meiosis (Williams 1975). Consider two individuals. Let one individual be asexual and produce dispersed meristems identical to itself. Let the other produce haploid eggs that require fertilization. The relative success of these two individuals will illustrate the relative success of asexual and sexual reproduction. Each offspring of the asexual organism will contain complete copies of the parental genotype – but each offspring of the sexual organism will contain only one-half of the parental genotype! Therefore merely by undergoing meiosis to produce eggs, the sexual individual immediately cuts its contribution to offspring by one half. Any gene that cuts reproductive fitness in half would appear to have very high costs indeed, so why do genes for sexual reproduction survive at all? From this perspective, sexual reproduction (and flowers, nectar and even seeds) becomes a problem. Why do they persist when it would be possible to double the number of genes in the next generation by simply producing asexual propagules? In the words of Williams (1975: p. 11) "nothing remotely approaching an advantage that could balance the cost of meiosis has been suggested."

It is unconvincing to try to argue that some long-term benefit arising out of the continuing potential for evolution is sufficient to balance the immediate loss of one-half of a genome each generation. Surely some short-term benefit of meiosis to individuals can be found. One proposal supposes that meiosis and sexual recombination are of immediate value in any case where offspring are likely to face a different

asexual offspring (mitosis – standardized)	sexual offspring (meiosis – diversified)
large initial size	small initial size
produced continuously	seasonal
develop close to parent	widely dispersed
develop immediately	dormant
develop directly to adult stage	develop through stages
environment predictable	environment unpredictable
low mortality rate	high mortality rate
natural selection mild	natural selection intense

Table 7.5 **Differences between asexual and sexual offspring. (After Williams 1975)**

environment from that of the adult. This explanation seems reasonable, for wherever one looks in the plant, animal or fungal kingdoms, one finds that the asexual propagules usually develop immediately and near the parent, whereas sexual propagules are usually dormant and widely dispersed (Table 7.5). The further away the offspring disperses into space or time, the lower the probability that it will encounter the same environment as its parent, and therefore the greater the value of meiosis and recombination.

Perhaps the amount of environmental heterogeneity is so great that, in plants at least, the maintenance of sexual reproduction can be justified solely as a consequence of environmental uncertainty. In this situation, the benefits of sexual variation, and the benefits of dispersal, are apparent in the short run. We could, then, tentatively conclude that the short-term benefit of sexual reproduction is genetic variation to deal with environments that are physically heterogeneous and changing with time. Hence the need for flowers to increase cross-pollination to increase the genetic variability of offspring, in the short term as well as the long term.

7.4.4 Pollination Ecology Was Founded by Sprengel and Darwin

While we now expect most people to learn about pollination in high school, it was only a few hundred years ago that biologists provided a scientific understanding of this process. Let us digress for some history.

While Linnaeus had, of course, recognized that plants were sexual organisms (much to the disgust of many of his contemporaries, Section 2.3.1), his chief objective was to use such characteristics to classify plants. Most biologists then thought that the pollen (*farina fecundus*) was carried to the stigma either spontaneously or by small vibrations, in which case insect visits to plants were by mere chance and were irrelevant to fertilization (Vogel 1996: p. 46). The colour and shape of flowers was thought to have no value other than to "manifest the solemnity of marriage." Indeed, since visiting insects did remove pollen, it was thought more likely that they were detrimental to flowers.

The process of pollination was first unravelled in Sprengel's 1793 treatise (in German), rendered in English as *The Secret of Nature in the Form and Fertilization of Flowers Discovered*. Sprengel's work remained unavailable in English until 1996 when a chapter was translated by Haase as "Discovery of the Secret of Nature in the Structure and Fertilization of Flowers" in Lloyd and Barrett (1996). It begins:

*When I carefully examined the flower of the wood cranesbill (*Geranium sylvaticum*) in the summer of 1787, I discovered that the lower part of its corolla was furnished with fine, soft hairs on the inside and on both margins. Convinced that the wise creator of nature had not created even a single tiny hair without definite purpose, I wondered what purpose these hairs might serve. And it soon came to my mind that if one assumes that the five nectar droplets which are secreted by the same number of glands are intended as food for certain insects, one would at the same time not think it unlikely that provision has been made for this nectar not to be spoiled by rain and that these hairs had been fitted to achieve this purpose.*

Alas, Sprengel was "ignored, rejected or ridiculed" when he published his book (Vogel 1996: p. 46). He left the school that employed him the same year the book appeared and spent the following years as a private scholar, supported by a small pension and housed in an attic in Berlin. His only further treatise on floral ecology was on bee-keeping. His grave is unknown, and there are no portraits of him (Vogel 1996).

Darwin revisited the topic a century later in his 1862 book on orchid pollination, titled *On the Various Contrivances by which British and Foreign Orchids are Fertilised by Insects, and on the Good Effects of Intercrossing*. The topic continues to flourish (Percival 1965; Lloyd and Barrett 1996). More than 40 books and 100 key reviews on pollination ecology appeared between 1971 and 1991 (Dafni 1992). The challenge for today's students is to acquaint themselves with the volume of work available on this topic as new work, particularly from the tropics, documents the many species of animals that pollinate flowers. Let us look at a few entertaining examples. Since you have already seen bats and birds as pollinators in this chapter, and since you no doubt heard more than enough about bees in public school, let us close with another group of flowers entirely, those that are pollinated by flies.

7.4.5 Another Example: Some Flowers Are Pollinated by Flies

One of the most remarkable groups of plants is the Rafflesiaceae, a parasitic and tropical family with 8 genera and some 50 species. The genus *Rafflesia* has 14 to 15 species in eastern Asia and Malesia (Endress 1996). The flowers are among the largest in the world (Figure 7.12) and smell like carrion. Calliphorid flies are attracted by the odour of carrion. *Rafflesia* flowers not only mimic the smell of a dead animal; the blotched, wrinkled perianth, which is borne on the ground, may visually mimic a dead animal.

Another group of carrion flowers occurs in the Asclepiadaceae where one finds "the most elaborate, complicated flowers of all the dicots" (Endress 1996: p. 302). Genera such as *Ceropegia* and *Stapelia* are

FIGURE 7.12 A *Rafflesia arnoldii* flower smells like rotten meat and is pollinated by flies. Reaching one metre in diameter, it is the largest individual flower on Earth. It is endemic to the islands of Sumatra and Borneo. As a parasitic plant, it has neither roots nor leaves, and lives inside the tissues of a host until emergence as a single flower bud. (Olga Khoroshunova, Shutterstock)

particularly adapted to fly pollination. *Stapelia*, for example, produces purplish, wrinkled, hairy flowers that smell of carrion and are borne low on the plant touching the ground. They resemble small dead rodents. (One that I grew as a houseplant had to be put on a balcony outside when it flowered because my apartment would otherwise smell of rotting flesh. Flies would then lay their eggs on the flowers, which would soon be riddled with maggots.)

In some carrion-simulating flowers then, flies are even deceived into oviposition. In such cases, it is questionable whether the flower replaces the nutrition that would be provided by a corpse. In that case, the flower is deceptive and acting as a parasite upon the pollinator population. In the case of *Rafflesia*, the male flowers provide flies with both pollen and a slimy exudate. The relative costs and benefits for the plant and the pollinator therefore depend upon just how much nutrition the flowers provide in exchange for the service of pollination.

7.5   Positive Interactions Between Plants and Animals: Part 2 Seed Dispersal

7.5.1 Animals Eat Fruits and Spread Seeds

The second great mutualism between plants and animals involves the dispersal of seeds. And this takes us to the topic of fruits, which along with flowers are one of the distinctive features of flowering plants. As noted in the preceeding section, sexually produced offspring (read seeds) generally disperse well away from parents. "The function of a ripe vertebrate-eaten fruit is to get the seeds into the right animals and keep them out of the wrong animals" (Janzen 1983: p. 232).

The mutualism is clear, in principle. The benefit to plants is dispersal of offspring. The benefit to animals is food. The food may come either from eating the fruits that surround the seed, or from eating the seeds themselves and dispersing the uneaten seeds. We will encounter some examples of both. You have already seen one example in Figure 7.1, banana seeds being dispersed by fruit bats. Figure 7.13 shows another: the tropical tree that provides chocolate has seed dispersed by monkeys and squirrels.

Other animal-dispersed fruits simply fall from the tree onto the ground where they are then consumed by mammals such as tapirs. This may also be the fate of some chocolate fruits that are not collected from the canopy. Some palms (e.g. *Maximiliana maripa*) also produce large-seeded fruits with pulp attractive to tapirs. Tapirs swallow entire fruits and defecate intact seeds as far as 2 km away from the nearest palm clumps (Fragoso 1997). Other species, however, such as rodents, peccaries and deer, eat the fruits and merely spit out the seeds, most of them falling beneath the crown of the parent plant. Seeds that are not dispersed are easily found and eaten by herbivores: bruchid beetle larvae killed 77 percent of the seeds near parent trees, but less than 1 percent of 6,140 seeds dispersed by tapirs.

Many tropical trees have fruits collected by bats. The fruits of *Andira inermis*, a medium-sized ever-green tree of the coastal lowlands of central America are eaten by two species of bats *Artibeus jamaicensis* and *A. lituratus* (Janzen et al. 1976). The bats rest in roosts, fly to an *Andira* tree to feed and then carry back single ripe fruits. These are single-seeded oblong spheres 3 to 5 cm in length, borne on long stems that project outside the tree canopy. About half of a fruit is eaten, leaving the seed and some pulp to be discarded. The fruit is rich in energy for bats. Analysis of the energetics (the energy in a single fruit and the energy costs of flight) show that a 45-gram *A. jamaicensis* could make a 1,266-metre round trip and a 60-gram *A. lituratus* could make a 951-metre round trip from one fruit (Janzen et al. 1976). To simplify it, a bat can forage on a tree 0.5 km away and recover its costs from a single half-eaten *Andira* fruit! The benefits to the tree are also substantial. Between 54 and 94 per-cent of the seeds falling under parent trees were attacked and killed by weevils, whereas under bat roosts this was reduced to only 33 percent. Seeds that were dropped in open pastures, while comprising less than 1 percent of the seed crop, had only 10 percent mortality from weevils.

These examples illustrate an important conserva-tion problem in tropical areas. Wild animals are often killed by commercial hunters and sold as "bush meat." The rate of killing is so high that entire areas of forest

FIGURE 7.13   An example of animal dispersal. Chocolate comes from the tropical tree *Theobroma cacao*, which is native to central America and the Amazon basin. The small flowers are produced directly from the bark of larger branches or trunks, where they are pollinated by a species of fly (a midge in the genus *Forcipomyia*). The pods (ca. 20 cm long) contain a fleshy pulp that attracts animals including monkeys and squirrels for dispersal. A typical tree will produce 20 to 30 fruits each year, each containing 20 to 40 seeds. (P.J.F. Turpin, 1829, Peter H. Raven Library/Missouri Botanical Garden)

are being stripped of bats and other mammals, leading to what are called "empty forests" (Redford 1992; Krause and Zambonino 2013). Fruit bats and tapirs are

not the only prey: monkeys and birds are killed as well. As a consequence, trees are being hit by two simultaneous threats: loss of pollination and loss of seed dispersal. We will return to this topic in the final chapter.

7.5.2 Rodents, Nuts and Mast Years

Other animals depend upon the energy stored in the seeds as a food source. Some trees produce edible seeds encased in dry unpalatable fruits, which we call nuts (Table 7.6). In this case, however, the cost to the plant is clear: a large proportion of the offspring, as seeds, are sacrificed so that a few are dispersed. The collective name for nuts of beech, oak and other trees is mast; this mast was once an important food for feral pigs near villages, and is even mentioned in Shakespeare's play *Timon of Athens* "The Oakes beare Mast, the Briars Scarlet Heps." Let us hear from some zoologists about squirrels and nuts.

Squirrels are strictly seed predators on those trees in the deciduous hardwood forests whose seeds are dispersed by the wind, e.g., maple (Acer), elm (Ulmus), and ash (Fraxinus). However, for the species in the families Fagaceae and Juglandaceae, squirrels are both seed predators and dispersing agents for the large, nutlike seeds. The life history of squirrels is too short in comparison to the generation time of trees and squirrels' movements are too great... for the individual squirrel to gain any advantage in leaving successful offspring by planting a new generation of food trees for the offspring. The squirrels' behaviour... is adaptive to the individual squirrel in spreading nut concentrations to prevent usage by deer. Burying the scattered nuts gets them out of sight of jays and other visual foragers. This process of scatter-hoarding nuts may also reduce the probability that other squirrels will find them, if, as appears to be the case, squirrels have a good sense of location and use it to retrieve the nuts they bury. It would have to be this advantage of reducing theft by other squirrels that would induce individual squirrels to

Table 7.6 Food value of five species of nut-bearing trees consumed by squirrels. (From Smith and Follmer 1972) Of these, the lowest value is white oak (*Quercus alba*, top) and the highest value is black walnut (*Juglans nigra*, bottom) (P.-J. Redouté in Michaux, F.A. 1865. The North American Sylva. Vol 1. Peter H. Raven Library/Missouri Botanical Garden)

Species	Lipid content (percent)	Mean dry weight (g ± SE)	Calories per nut	Proportion of nut's total energy in the kernel (percent)	Rate of ingesting energy from whole nuts (cal/s)	Rate of ingesting metabolizable energy from whole nuts (cal/s)
White oak					27[a]	20[a]
Kernel	4.6	0.40 ± 0.03	1,700	27		
Shell		0.41 ± 0.02	1,900			
Cap		0.64 ± 0.02	2,600			
Bur oak					50	41
Kernel	9.8	4.66 ± 0.19	20,200	45		
Shell		2.04 ± 0.67	7,700			
Cap		4.34 ± 0.17	17,000			
Shumard oak					42	36
Kernel	20.3	2.28 ± 0.07	11,900	66		
Shell		0.87 ± 0.02	3,900			
Cap		0.54 ± 0.03	2,200			
Black walnut					14[a]	13[a]
Kernel	23.1	2.04 ± 0.12	12,700	13		
Shell		12.20 ± 0.25	51,800			
Husk		8.62 ± 0.18	35,300			
Shagbark hickory					19[a]	18[a]
Kernel	29.3	1.01 ± 0.06	6,700	21		
Shell		2.04 ± 0.05	8,600			
Husk		3.89 ± 0.12	16,100			

Note:
[a] Overestimate because the squirrels frequently did not consume the entire kernel.

bury walnuts and the thicker shelled hickory nuts because it is highly unlikely that deer and birds can break them open... Given that dispersing and burying nuts are not altruistic acts of squirrels to benefit trees, how can the interaction between tree seed and squirrel evolve to further the trees' efficiency of seed dispersal? The tree whose seeds are too appealing will have them eaten first and leave no offspring, while the tree whose nuts are too difficult to eat will not have their seeds dispersed. If two species of trees frequently exist together in mixed stands, they will be in the same situation in which both extremes of desirability will lead to failure in seed dispersal. Therefore, it is particularly interesting that oaks and hickories, which are frequently found together in climax forests... each have a season of the year in which they are the more desirable food and are more likely to be available for squirrel use if they have been scatter-hoarded. It is possible to imagine a balance in selection on the thickness of shells of hickory nuts where too thin a shell makes the average tree too desirable and most seeds are eaten and too thick a shell where the average tree has no seed dispersal because squirrels are using neighbouring acorns. This balance coupled with the alternate-year seed production of oaks in the subgenera of red and white oaks would allow squirrels to be effective dispersing agents for three sympatric species of trees. However, the exact nature of the balance in selective pressure on the thickness of acorn and hickory nut shells is further complicated by the numerous insects which also eat the nut kernels and then bore their way out through the shell... But whatever the evolutionary influence of insects, for the various species of mast to coexist, they must reach an evolutionary balance in desirability for squirrels if all species of mast are to be effectively dispersed.

(Smith and Follmer 1972: p. 90)

Producing seeds with hard coats (think acorns or walnuts) creates an additional difficulty for the plant: while the seed must be hard enough to deter herbivores, if the seed is to germinate the embryo must be able to absorb water and split open the seed. In some cases therefore the seeds have become dependent upon being partly digested by the herbivore in order to weaken the seed coat for germination. What started out as a protective measure may end up as a dependency. We shall discuss this further in Section 7.5.4.

During mast years, seed production is synchronized, and enormous numbers of seeds are produced in one outburst. This pattern is observed in temperate conifers (Smith 1970; Janzen 1971), temperate deciduous trees (Smith and Follmer 1972; Jensen 1985; Falls et al. 2007), tropical dipterocarps (Janzen 1974; Ashton 1988; Ashton et al. 1988) and bamboos (Janzen 1976). The present-day cue that controls such mast years is still poorly understood. The evolutionary origin is more obvious: once a chance environmental event (such as an exceptionally wet or dry year) produced a degree of synchrony in seed production, it would be fine-tuned by predators eliminating the seeds of those individuals that reproduced out of phase.

7.5.3 Ants Disperse Seeds

Large fleshy fruits, or nuts, provide an obvious incentive for seed dispersal by animals. Another large group of plants depends upon ants for seed dispersal. Hence they are known as myrmecochores. Such plants produce a small ant-attracting food body called an eliasome (Figure 7.14). Ants carry the seeds back to the nest where the eliasome is removed. The seed coat may be gnawed, but the seed is generally abandoned in the nest. This appears to have at least three advantages to the plant: (1) seeds are dispersed, (2) seeds may be hidden from seed predators and (3) ant nests may provide particularly suitable germination sites, perhaps being richer in nitrogen and phosphorus than the surrounding forest soils.

Many spring flowers in the deciduous forests of eastern North America depend upon ant dispersal. In some species (e.g. *Sanguinaria canadensis*) the presence of white or cream-coloured tissue attached to the seed is obvious, but it is not always easy to judge the food value of a seed from an ant's perspective. Beattie

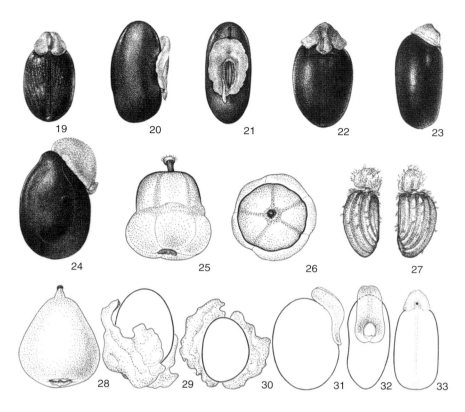

FIGURE 7.14
A selection of ant-dispersed propagules from Australia. Some are seeds and some are small fruits. (19–24, 29–33) Seeds with an ant-attracting appendage (eliaosome); (25, 26) drupe with distinct eliasome; (27) mericarps with terminal eliaosome; (28) drupe; (19) *Monotaxis linifolia*; (20, 21) *Kennedia rubicunda*; (22, 23) *Beyeria viscose*; (24) *Acacia linifolia*; (25, 26) *Leucopogon virgatus*; (27) *Xanthosia pilosa*; (28) *Monotoca scoparia*; (29, 30) *Hibbertia calcina*; (31, 32) *Daviesia mimosoides*; (33) *Bertya rosmarinifolia*. (From Berg 1975)

and Culver (1981) therefore put propagules of deciduous forest plants on the ground close to a reproducing adult plant, and watched for an hour, recording the percentage of propagules that were carried at least 10 cm by ants. Altogether they found that 21 out of 91 species were myrmecochorous including woodland genera such as *Asarum*, *Carex*, *Jeffersonia*, *Sanguinaria*, *Trillium*, *Uvularia* and *Viola*.

Arid areas also have many ant-dispersed species. There are some 1,500 myrmecochorous species reported from the dry heath and sclerophyll vegetation of Australia (Berg 1975). Ants also disperse seeds in the deserts of North America. Here the ants include both specialized genera of seed-eaters (e.g. *Pogonomyrmex* spp. and *Veromessor* spp.) as well as omnivores (e.g. *Novomessor* spp. and *Solenopsis* spp.) (Brown et al. 1986). Birds and rodents also feed on

seeds in deserts, and removal experiments show that seed densities become higher if rodents or ants are removed. As with nuts, there is likely both herbivory and dispersal occurring simultaneously.

7.5.4 Can Seed Dispersal Become an Obligate Mutualism?

There is usually some redundancy in seed dispersal. For example, if a squirrel does not collected a ripe cocoa pod, then a monkey may do so. And if neither arboreal animal collects the pod, it falls to the floor where it may be eaten by a tapir. The most extreme case of seed dispersal by animals would be an obligate mutualism. That is, a single plant species dispersed by a single animal species, with each completely dependent upon the other. You can think of this as a sort of limiting case to which evolution might take the

relationship between a seed producer and a seed disperser. In most cases studied to date there is considerable redundancy. Each tree may produce fruits and
seeds dispersed by a number of different dispersal
agents. And each disperser feeds upon fruits and seeds
from a number of different species of trees. Hence
there is a network of positive benefits, but it is not
restricted to any single pair of species.

Ecologists were therefore intrigued when a case of
a plant with a single dispersal agent was announced.
Temple (1977) reported that a Mauritian tree called the
tambalacoque (*Sideroxylon grandiflorum*) produced a
fleshy fruit with a single large seed. This fruit is a
drupe, with a large stone (a single seed encased in a
hard endocarp) covered in a 5-mm thick layer of
tenacious fleshy pulp. Temple asserted that this tree
had not reproduced for some 300 years, since the time
when dodos became extinct. When, however, the
seeds of the tree were force-fed to a substitute for the
dodo (turkeys!), some of them germinated. These three
seeds, Temple asserted, were said to be the first *S.
grandiflorum* seeds to germinate in more than 300
years, and such observations "provide empirical
support for the hypothesis that the fruits of
[*S. grandiflorum*] had become highly specialized
through coevolution with the dodo. After the dodo
became extinct, no other animal on Mauritius was
capable of ingesting the large pits. As a result [*S.
grandiflorum*] has apparently been unable to reproduce for 300 years and nearly became extinct." (I have
changed the name in the quotation and text as the
name of the tree has changed since Temple's paper
appeared.) It is an appealing story, and plausible.

Here is how the logic works. There is a coevolu-
tionary race between seed predators and their prey
(a kind of evolutionary escalation), the seed coats
become steadily thicker to protect the seed from
digestion, and eventually the seeds become dependent
upon digestion by their predator in order to escape
from the thickened seed coat. Anyone who has tried to
germinate plant seeds knows how one can enhance
rates of germination by damaging the seed coat so
water can enter the seed (a process known as scarifi-
cation). This is a good sign that one is dealing with a
species that would normally pass through an
animal gut.

But returning to the story of the *Sideroxylon* and
the dodo, there are some reasons to doubt the specificity of the dodo as a sole dispersal agent (Witmer
and Cheke 1991).

1. While *S. grandiflorum* has reproduced infrequently, there is evidence for young trees in published surveys.
2. The seed coat ruptures along a circular line, producing a cap that splits off for germination – a feature
shared with several other Mauritian tree fruits.
3. There is rather little evidence that the seeds require
abrasion to germinate, although the removal of
pulp from the fruits, which the turkeys would also
have likely done, certainly does seem to reduce
mortality of seeds.
4. Dodos were probably generalist feeders, given that
they could be kept alive on long sea journeys.
5. There are other (now extinct) species that might have
been capable of dispersing the tree, including large-
billed parrots (*Lophopsittacus mauritianus*) and two
species of endemic giant tortoises (*Geochelone* spp.).

This story shows the difficulty of making the case for
obligate relationships in seed dispersal. The occasional
establishment of wild trees without dodos does not, by
itself, indicate that dodos were unimportant, it merely
shows that dodos likely were not the sole agent for
dispersal and germination. Nor were the dodos, in
return, likely dependent solely on this seed source.

But do not forget that there can be many positive
interactions and mutualism without obligate dependency. From the perspective of the plant there is a
continuum of possibilities for dispersal: from those
with no known mechanism of seed dispersal except
gravity, through species that depend on a network of
animal dispersal agents, to species that are largely (or
solely) dependent upon a single animal. Most likely
fall somewhere in between and have a network of
dispersal agents of varying degrees of efficacy. Primates in particular are probably generalists, and while
they may have a preference for large, fleshy fruits
coloured yellow/orange/brown, their diet is determined by the mixture of fruits available at any time or

place (Chapman and Russo 2006). Primates may carry seeds hundreds of metres from the parent tree, particularly when the seeds are swallowed. There may then be secondary seed dispersal when dung beetles or rodents feed upon seeds in primate dung.

As studies of tropical seed dispersal continue, much more data will emerge. However, there will always be room for debate, since it is entirely possible that some plants may have fruits adapted to dispersal by extinct animals such as ground sloths (Figure 6.10). Hence such plants are now being dispersed, if at all, by a group of species that may have been rather unimportant in the evolution of the fruits' characteristics (Janzen and Martin 1982).

7.6   Animals Can Defend Plants From Herbivores and Competitors

The genus *Acacia* has about 1,000 species distributed throughout the drier portions of the tropics and subtropics. A small subset of this genus, the swollen-thorn acacias (Figure 7.15), are confined to the New

(a)

(b)

(c)

(d)

FIGURE 7.15   (a) *Acacia collinsii* in a pasture in southwestern Nicaragua. Note the dark bare area cleared of vegetation by the ant, *Pseudomyrmex nigrocincta*. (b) Swollen thorns inhabited by ants compared to regular thorns. (c) Raised petiolar nectary of *A. cornigera*. (d) Beltian bodies on the tips of pinnules of *A. cornigera*. (After Janzen 1966)

World. These acacias share a number of distinctive traits: (1) enlarged stipular thorns, (2) enlarged foliar nectaries, (3) modified leaflet tips called Beltian bodies and (4) nearly year-round leaf production even in areas with a distinct dry season (Janzen 1966). These acacias have an obligate mutualism with colonies of ants in the genus *Pseudomyrmex* living within their thorns.

When a queen ant colonizes a tree, she either cuts her own entrance hole to a thorn or uses one cut by a previous worker or queen. Having laid her eggs in the thorn, she forages for nectar from the petiolar nectaries (Figure 7.15c) and for solid food from the Beltian bodies (Figure 7.15d). Over time, a colony may grow as large as 30,000 individuals, at which point it may also occupy neighbouring trees within 3 to 10 m. The costs for the acacia are considerable. A two-metre acacia produces about 1 ml of nectar per day, the entire sugar source for the associated ant colony. The Beltian bodies are constricted leaflets containing large thin-walled cells apparently full of proteins and lipids; the ants harvest them, cut them up and feed them to larvae. (Beltian bodies are named for Thomas Belt, who first described their function in *The Naturalist in Nicaragua* in 1874.) The canopy of a two-metre tall acacia can bear a kilogram of swollen thorns, and a four-metre tall plant, three kilograms (Janzen 1966).

The benefits of this relationship must be relatively great to balance the costs. The principal benefit appears to be the protection of the acacia from predation and competition. The ants are aggressive and will bite and sting any herbivore that attempts to feed upon the plant; observers describe ants rushing to the ends of twigs and throwing themselves into the air to

Table 7.7 Incidence of herbivorous insects on shoots of *Acacia cornigera*, with and without the ant *Pseudomyrmex ferruginea* (Temascal, Mexico). (After Janzen 1966)

Time/occurrence	With ants	Without ants
Daylight		
Shoots with insects (percent)	2.7	38.5
Mean no. insects/ 100 shoots	3.9	88.1
Night-time		
Shoots with insects (percent)	12.9	58.8
Mean no. insects/ 100 shoots	22.6	270.1

land on visitors. Table 7.7 compares the incidence of herbivorous insects on acacias with and without ant colonies. The ants also kill the shoots of vines or other neighbouring plants that touch the acacia. There are therefore bare areas around ant-occupied acacias, whereas other species are swamped by vines and overtopped by surrounding trees. Experiments have tested for such effects: when ant colonies were killed, the acacias were rapidly defoliated by herbivores and overgrown by vines (Janzen 1967). There may be some further benefits. The shoots of swollen-thorn acacias appear to lack the protective fibrous tissues of other acacias and grow more quickly than other acacias. They also may need to allocate less energy to the production of anti-herbivore defence compounds.

7.7 Mathematical Models of Mutualism

7.7.1 A Population Dynamics Model

Just as with competition, there are many possible mathematical formulations for mutualism. We will examine only two.

The first is based upon the Lotka–Volterra models familiar from introductory ecology. In this formulation, two populations, N_1 and N_2, grow as follows:

$$dN_1/dt = r_1 N_1 [(K_1 - a_{11}N_1 - a_{12}N_2)/K_1] \text{ and}$$

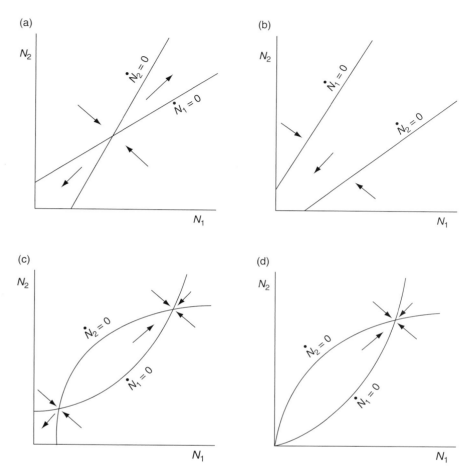

FIGURE 7.16 Models for two species having an obligate mutualism. Top, using a modified Lotka–Volterra competition equation: (a) shows a situation where the isoclines intersect at an unstable point and extinction occurs below critical densities while population explosion occurs above them; (b) shows a situation where there is no equilibrium point and both species become extinct. Bottom, using a second, more complex model: (c) shows the case where there is an unstable equilibrium point and a stable equilibrium point; species coexist provided densities are sufficiently large; (d) shows the case where there is always coexistence at a point where interspecific benefits balance intraspecific competition. (From Wolin 1985)

$$dN_2/dt = r_2 N_2 [(K_2 - a_{22} N_2 - a_{21} N_1)/K_2]$$

where the competition coefficients are indicated by **a** and the carrying capacities by **K**.

One may take these two original Lotka–Volterra equations and, instead of assuming that all neighbours cause negative effects, one can make the interspecific coefficients a_{12} and a_{21} positive to

indicate that interspecific neighbours have reciprocal beneficial effects upon each other. One can then explore how the negative effects of intraspecific competition ($-a_{11}$ and $-a_{22}$) interact with the beneficial effects of interspecific mutualism ($+a_{12}$ and $+a_{21}$), as you see in Figure 7.16. (Note: it is necessary to have as a mathematical artifact, a negative carrying capacity, because a positive density of the mutualistic partner is required for population increase (that is, $dN_1/dt = 0$ (line $N_1 = 0$) must have a positive intercept on the N_2 axis). Don't worry about this assumption now. Pay attention to the behaviour of the populations overall.)

There are two possible outcomes. First, we look for an equilbrium point, where the isoclines meet. There is one such point. The linear isoclines intersect at a single equilibrium point when $a_{12}a_{21}/a_{11}a_{22} > 1$ (Figure 7.16a). This is an unstable point: mutual extinction occurs below this point, while population explosion occurs above it. This very basic model illustrates the difficulties with many models of mutualism: since each species benefits the other, the models tend to generate a steady and unrealistic growth in population sizes (May 1973). Something else has to set a limit to growth. And, to return to the case where $a_{12}a_{21}/a_{11}a_{22} \leq 1$, both species collapse into extinction (Figure 7.16b). While this may be biologically accurate, it is rather unsatisfactory for further mathematical inquiry.

One can modify the model further. Assume that the per capita benefits of mutualism decrease with increasing population density. That is to say, as population densities increase, the benefits of mutualism are slowly outweighed by the costs of intraspecific competition. In this case, the isoclines are curvilinear (Figure 7.16 bottom). There is an asymptote at some density of partners where costs balance benefits. Figure 7.16c illustrates a scenario with one unstable and one stable equilibrium point. Figure 7.16d shows a scenario with only a single stable equilibrium point.

As with any models (recall Section 4.9) one can add further complexity (Boucher 1985b; Wolin 1985), but we will stop here. Overall, it is easier to model facultative (as opposed to obligate) mutualism since it is possible to have persistence when one species can survive in the absence of the other. As with models of competition, models of mutualism help one to think about the nature of species interactions – but may be far removed from biological reality.

7.7.2 A Cost–Benefit Model

Population dynamics models (Sections 6.6.4 and 6.6.5) are useful to help one understand how various scenarios might proceed through time but, ultimately, cost–benefit models are needed to study evolution. In fact, it is only when both costs and benefits are measured that one can be certain that there is true mutualism involved in an interaction. Let us look at one such model.

To apply cost–benefit models one needs to imagine a population of individuals in which there are three classes of individuals: successful mutualists that assist another organism and receive assistance in return; unsuccessful mutualists that provide assistance but receive no reward; and non-mutualists that make no investments in mutualism and receive no return. For mutualism to evolve, the fitness of the successful mutualists must be greater than that of either the non-mutualists or the unsuccessful mutualists. Further, the total fitness of all mutualists, successful or not, must exceed that of non-mutualists. If it does not, that trait or behaviour will be lost from the population.

The chief problems in developing such cost–benefit models include: (1) measuring costs and benefits; (2) putting all costs and benefits into a common currency; and (3) developing an appropriate quantitative framework in which to insert the measured costs and benefits. Models exist for pollination, myrmecochory, mycorrhizal fungi, ants with fungus gardens and mixed-feeding flocks (Keeler 1985).

Consider the case of animal pollination (zoophily) as opposed to self-pollination. Let w_z be the fitness of zoophilous individuals and w_s be the fitness of self-pollinators. Which will become dominant in the population? We must first account for the costs of remaining unpollinated. The fitness of zoophilous individuals is

$$w_z = pw_f + qw_u$$

where w_f is the fitness of zoophilous individuals that are pollinated (f for fertilized) and w_u is the fitness of those that are unpollinated; p and q denote their respective frequencies ($p + q = 1$).

Each state of w must have both costs and benefits. Ideally these would be measured in terms of contribution to the next generation, such as seeds produced or number of surviving offspring. Calories might be a useful choice of measurement, although grams of nitrogen or phosphorus might be more critical in some situations. To obtain the benefit one might combine the two measurements, using N, the proportion of seeds set, and v, the relative fitness of these seeds, to yield N_v. The cost I_z would be the investment in floral displays and rewards for pollinators. The variables N and p are clearly related, but Keeler (1985) separates them to emphasize the difference between plants receiving no pollen and plants with unfilled seeds in the fruits. Both N and v apply to both successfully (f) and unsuccessfully (u) pollinated zoophils, so that the equation above expands to

$$w_z = p(N_f v_f - I_z) + q(N_u v_u - I_z).$$

If the species is an obligate outcrosser, $N_u v_u = 0$, in which case

$$w_z = p(N_f v_f) - I_z.$$

For the self-pollinating individuals:

$$w_s = N_s v_s - I_s.$$

For mutualism to occur, w_z must exceed w_s, that is the benefits of the zoophilous flowers must exceed the benefits of self-pollination. That is

$$p(N_f v_f) - I_z > N_s v_s - I_s.$$

Note that since all plants produce flowers, whether they are self pollinated or animal pollinated, all will bear the cost of flower production, designated I. But since it is generally true that zoophilous species make a larger investment in flowers, then it is likely that I_z will be greater than I_s. Therefore, if there is to be mutualism, the fitness of outcrossed seeds must be substantially greater than that of inbred seeds, that is v_f must (substantially) exceed v_s.

This kind of model illustrates the careful thought necessary in considering costs and benefits of mutualism. As noted earlier in the chapter, there are definite costs to sexual reproduction, and yet it persists. This thought exercise also illustrates another benefit of models. It illustrates the challenges in testing for mutualism altogether, since one must measure costs and benefits carefully.

7.8 Mutualism Generates Complex Networks

We must now address one more source of positive interactions. We have, up until this point, considered just pairwise interactions. As soon as you enlarge your view slightly, to three or more species, large numbers of positive (or negative) interactions become possible. A good example comes from food webs, although the principle extends far beyond them. If a predator is feeding on a herbivore that is feeding on plants, and reducing the number of herbivores, then it is *indirectly* benefiting the plants. You have seen a discussion of this as top–down effects in Section 6.6.1. Or, if a parasite is feeding on a herbivore, then the parasite is also having a positive effect upon the plants. You have seen how insect parasitoids benefit conifer trees by killing bark beetles in Section 6.7.1. Overall, the

general principle is one familiar from politics: the enemy of my enemy is my friend.

Let us think back to competition. There is a view that many cases of apparent competition arise because of these higher order indirect effects. That is, you remove a plant and its neighbours grow better. We would treat this as good experimental evidence of competition. But, say the sceptics, what if the removed species was in fact attracting a herbivore, so the remaining plants grew better not because of competition, but because the herbivore was no longer visiting that patch? How many field experiments in which we remove a species and find a response are a result of such indirect effects? It is often difficult to say. But you can see that food webs of various kinds could complicate our search for competition or mutualism. Some biologists (e.g. Connell 1990; Holt and Lawton 1993) are of the opinion that apparent interactions may be particularly likely in insect food webs because parasitoids can limit their hosts to levels at which resource competition is unimportant. The same problems arise in teasing mutualisms out from complex interactions among species; in this case, it is equally possible that some cases of mutualism could be "apparent mutualism" caused by indirect effects of competition or predation.

Consider an historic example: Darwin's observation on bee pollination of red clover (*Trifolium pratense*) from *On the Origin of Species*. The plant, he says, is dependent upon humble-bees for its pollination.

Hence I have very little doubt, that if the whole genus of humble-bees became extinct or very rare in England, the... red clover would become very rare, or wholly disappear. The number of humble-bees in any district depends in a great degree on the number of field-mice, which destroy their combs and nests; and Mr. H. Newman, who has long attended to the habits of humble-bees, believes that "more than two thirds of them are thus destroyed all over England." Now the number of mice is largely dependent, as every one knows, on the number of cats; and Mr. Newman says, "Near villages and small towns I have found the nests of humble-bees more numerous than elsewhere, which I attribute to the number of cats that destroy the mice." Hence it is quite credible that the presence of a feline animal in large numbers in a district might determine, through the intervention first of mice and then of bees, the frequency of certain flowers in that district!

Darwin, 1859, On the Origin of Species

Others took the logic further.

A German scientist then continued to extrapolate that since cats were responsible for the prevalence of red clover, and since red clover was a staple food of cattle and since British sailors thrived on bully beef, one could conclude that Britain's dominant world position as a naval power was ultimately determined by the presence of cats. Thomas Huxley, tongue planted firmly in cheek, went on to note that old maids were the main protectors of cats, thus showing that the British empire owed its existence to the spinsters of England.

(Vandermeer et al. 1985: p. 326)

What these authors are saying, in a more serious vein, is that nature consists not just of pairwise interactions but of complex networks. What with mycorrhizae, pollination and fruit dispersal alone, plants are linked into a web of positive interactions. There is an entire body of mathematical theory that studies networks (Newman 2003) and graphs (Diestel 2010). It will take you into the realm of vertices, edges, paths, cycles, connectivity, spanning trees, forests, Euler tours and directed graphs. There is no room to discuss them further here, except to make you aware that they exist and have been applied to mycorrhizal networks (Caruso et al. 2012; Simard et al. 2012). Younger ecologists may find it useful to consider how one can apply such tools to real ecosystems with their complex mixture of positive and negative interactions.

CONCLUSION

Now that we have seen examples from a wide array of species and habitats, it is time to reconsider more carefully what kinds of evidence demonstrate mutualism. What can be said generally about a phenomenon that includes truffles, lichens, hummingbirds, fruit bats and tapirs?

One generalization is that mutualism is obviously very important in plant communities. A second generalization is that mutualism is routinely overlooked in text books (Table 7.1). Text books reveal, bluntly, the many attitudes, obvious and hidden, that we carry around and transmit to students. Why is a phenomenon that is so important given so little consideration? I can think of several possible reasons.

The Search for Nature Nuggets

Some people are attracted to the study of biology because they like nature. This is by no means an obstacle in itself since an intimate knowledge of nature is an excellent foundation for any aspiring ecologist. But too many fail to grow, simply collecting facts and observations as if all were of equal value. Perhaps there is a tendency to treat mutualism as simply a collection of nature stories without a theoretical context and without hypothesis-driven experiments.

The Confusion Between Mutualism and Divine Order

One might avoid studying the importance of cooperative interactions for fear of falling into a certain naive spirituality. A majority of human beings, it is safe to say, still believe in some sort of divine balance of nature. In this world view, it rains so that plants can grow. Plants produce flowers in order to make the world pretty. Birds sing in order to make humans cheerful. And so on. In this view, there is divine order and inherent purpose for everything. The study of mutualism could easily be treated as evidence of a divine order rather than science. It is not. Mutualism too arises from natural selection.

The Failure to Measure

Another obstacle is the tendency of scientists to be satisfied after showing that an apparent mutualism occurs. A similar problem occurs in the study of competition; authors seem to think that once they have found evidence of competition, their work is done. I spent the better part of a book (Keddy 1989) documenting this problem and suggesting ways to get off this dead-end path. The same analysis can be applied to mutualism. Demonstrating that a mutualism occurs is not an end in itself – it is just a beginning. The important next step is to measure the benefits accruing to each partner (just as in a competitive interaction where one measures the costs to each competitor) (Keeler 1985; Smith and Douglas 1987). The benefits might be measured in calories, grams of nitrogen, number of offspring produced or decreased rates of mortality.

One way forward is to plan more carefully to measure benefits (Figure 7.17). Ideally, these benefits should be compared to control individuals lacking their partner (in the same way, one measures competition intensity with reference to individuals without neighbours). Without measurement, and without controls, it is impossible to quantify the strength of mutualism or accurately distinguish between asymmetric mutualism and outright commensalism (Figures 7.17 and 7.18). Properties such as intensity and reciprocity may have general patterns. Another way forward is to explore the means by which such studies can be assembled into networks, and how the structure of such networks might vary among groups of species or habitats. There is much work to be done in taking the study of mutualism from the realm of natural history to rigorous and predictive science.

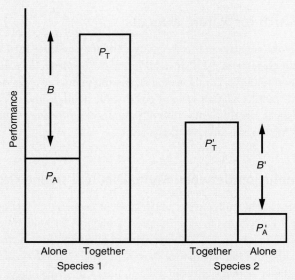

FIGURE 7.17 One way to think about measuring the relationships in a possible mutualism (P_A = performance alone, P_T = performance together).

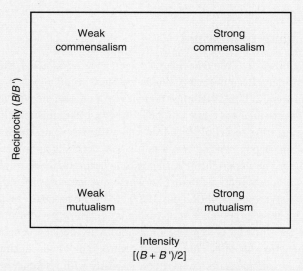

FIGURE 7.18 One way to compare mutualisms where the intensity and reciprocity of the relationships are based upon the measurements taken in Figure 7.17.

Review Questions

1. Explain what is meant by a "nurse plant" and give an example.
2. Write a short paragraph on the importance of mycorrhizae, starting with the basic resources that plants need.

3. "There is no real difference between lichens and vascular plants, except that vascular plants are larger." Consider the evidence for and against this proposition.
4. What are some characteristics of bird-pollinated flowers? Name some examples in your local flora, and identify the birds responsible for pollination.
5. What are some characteristics of ant-dispersed seeds? Name some examples in your local flora.
6. Find three examples of the following in recent scientific journals: disturbance, herbivory, competition and mutualism.
7. The angiosperms have flowers and fruits. Explain how both are the result of coevolution between plants and animals.

Further Reading

Janzen, D.H. and P.S. Martin. 1982. Neotropical anachronisms: the fruits the gomphotheres ate. *Science* **215**: 19–27.

Janzen, D.H. 1983. Dispersal of seeds by vertebrate guts. pp. 232–262. In D.J. Futuyma and M. Slatkin (eds.) *Coevolution*. Sunderland: Sinauer.

Boucher, D.H., S. James and K.H. Keeler. 1982. The ecology of mutualism. *Annual Review of Ecology and Systematics* **13**: 315–347.

Smith, D.C. and A.E. Douglas. 1987. *The Biology of Symbiosis*. London: Edward Arnold.

Williamson, G.B. 1990. Allelopathy, Koch's postulates and the neck riddle. pp. 143–162. In J.B. Grace and D. Tilman (eds.) *Perspectives on Plant Competition*. San Diego: Academic Press.

Simon, L., J. Bousquet, R.C. Lévesque and M. Lalonde. 1993. Origin and diversification of endomycorrhizal fungi and coincidence with vascular land plants. *Nature* **363**: 67–69.

Endress, P.K. 1996. *Diversity and Evolutionary Biology of Tropical Flowers*. Paperback edition 1994 (with corrections). Cambridge: Cambridge University Press.

Maser, C., A.W. Claridge and J.M. Trappe. 2008. *Trees, Truffles, and Beasts: How Forests Function*. Piscataway: Rutgers University Press.

Merckx, V., M.I. Bidartondo and N.A. Hynson. 2009. Myco-heterotrophy: when fungi host plants. *Annals of Botany* **104**: 1255–1261.

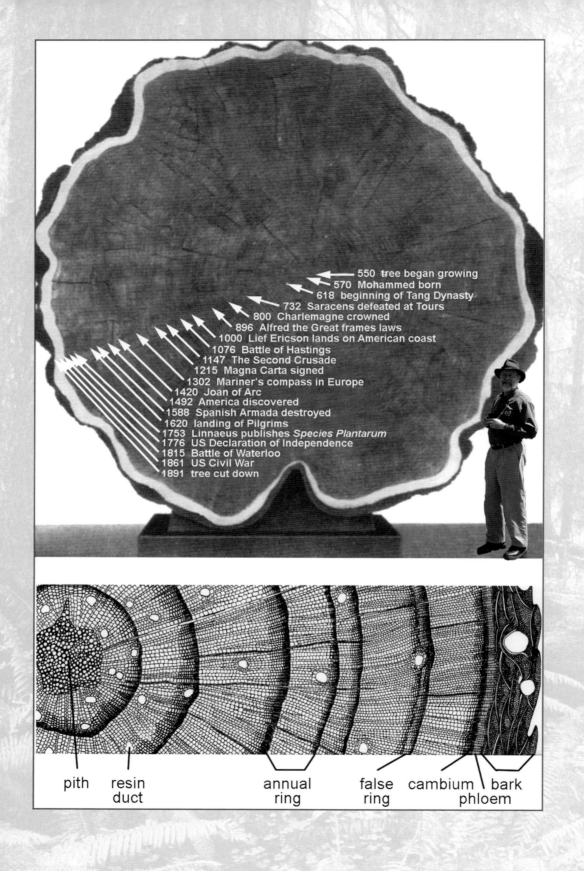

550 tree began growing
570 Mohammed born
618 beginning of Tang Dynasty
732 Saracens defeated at Tours
800 Charlemagne crowned
896 Alfred the Great frames laws
1000 Lief Ericson lands on American coast
1076 Battle of Hastings
1147 The Second Crusade
1215 Magna Carta signed
1302 Mariner's compass in Europe
1420 Joan of Arc
1492 America discovered
1588 Spanish Armada destroyed
1620 landing of Pilgrims
1753 Linnaeus publishes *Species Plantarum*
1776 US Declaration of Independence
1815 Battle of Waterloo
1861 US Civil War
1891 tree cut down

pith   resin duct   annual ring   false ring   cambium   bark phloem

8 Time

Scales of investigation. Methods. (1) The origin of angiosperms. Pangea. Gondwana. Laurasia. Nothofagus. Ginkgo. Proteaceae. (2) Glaciation. Loess. Pluvial lakes. Drought. Sea levels decrease. Migration. Hominids. Sea levels increase. (3) Succession. Conifer forests. Peat bogs. Rain forests. Sand dunes. Predictive models. Historical views on succession.

FIGURE 8.1 This giant sequoia began growing when the Romans were still ruling much of the Mediterranean, and Justinian was emperor. At the same time, in what we now call the High Sierras of North America, a ground fire allowed a seedling to establish. It was already a thousand years old when Linnaeus wrote *Species Plantarum* in 1753, and when the Bastille was stormed during the French Revolution in 1789. But it could not survive humans with metal tools, and was felled in 1891. It took eight days. A few even older trees are still alive. (You can still see cross-sections of this tree at the American Museum of Natural History in New York and the Natural History Museum in London). Below, a typical cross-section (ca. 1 cm wide) of a seven-year-old conifer. The rings are produced by cells of different sizes. Note the many resin ducts, part of the anti-herbivore defence system.

8.1 Introduction: There Are Many Time Scales in Ecology

8.1.1 Each Ecological Process Has a Time Scale

Time is easily overlooked because it is so necessary. It takes *time* for individual plants to grow. It takes *time* for competition to occur. It takes *time* to turn atmospheric carbon into buried coal. Even a humble differential equation describes how some thing, X, changes with time, as dX/dt. It is time to look at the t.

In the face of time, all phenomena are impermanent. This is a philosophical statement and a scientific observation. On the personal scale, the flowers that open on the magnolia in my back garden are present for only a few weeks each year, while the leaves last but a few months. On the scale of civilizations, sequoias and eucalyptus trees that established during the Roman empire may still be alive as adult trees. On the scale of continental drift, flowering plants from the old continent of Pangea are still drifting along and diversifying within Australia.

You have to know whether the appropriate time scale is seconds, years, centuries or eons. Consider the following. The rate at which carbon dioxide enters stomata, and the rate at which water escapes in the opposite direction, can be measured in seconds. The rate at which new leaves arise on trees, and the rate at which they are consumed by herbivores, can be measured over a year. The time it takes a giant sequoia tree to grow from a seedling must be measured in centuries (Figure 8.1). The accumulation of thick peat in wetlands to create a bog requires millennia. The spread of glaciers in an ice age, and their retreat to a new interglacial, takes closer to a million years. The rearrangement of continents through continental drift takes longer still. Somehow, practising ecologists must be familiar with all of these time scales, even if we choose to specialize in only one of them for research.

In this chapter we will somewhat arbitrarily break events into three large categories, starting with the longest periods of time first:

1. origin of flowering plants ($>10^6$ years)
2. effects of the last ice age (10^4 to 10^5 years)

3. ecological succession (10^2 to 10^3 years).

The processes that ecologists observe, and the evidence they interpret, will depend upon the scale of time investigated (Figure 8.2).

8.1.2 Some Sources of Evidence: Tree Rings, Sediment Cores and Fossils

The time problem for ecologists has a particular challenge: individuals of our species normally live for a mere "three score years and ten" (that is, 70 years), while many ecological processes take much, much longer. How might one study such long-term processes?

1. Use the age of the plants themselves. Growth rings (Figure 8.1) provide a record of events that have occurred over hundreds or thousands of years (Stokes and Smiley 1996). With care, records can be extended further back by including dead wood that still has rings. Not all plants produce easily counted rings, but even tropical trees can be aged using rings or radiocarbon dating of the wood itself (Vieira et al. 2005). Such data will, however, only take us back 10^2 to 10^3 years.

2. Use historical documents. Historical records of vegetation can be assembled and used to reconstruct changes over time. This procedure is particularly valuable in Europe where written historical accounts go back a thousand years or more. Of course, as you go back in time, documentary records are increasingly fragmentary, the names given to species may be unreliable and the observations may be qualitative rather than quantitative. You have already seen the example of using old surveyor records for documenting forest disturbance (Figure 5.22) and we shall look at ancient Greek reports on deforestation of the Mediterranean in the final chapter (Figure 13.2).

3. Use preserved plants. Peat bogs and lake sediments preserve pollen, plant and insect fragments, charcoal and diatoms from earlier times. Cores taken from

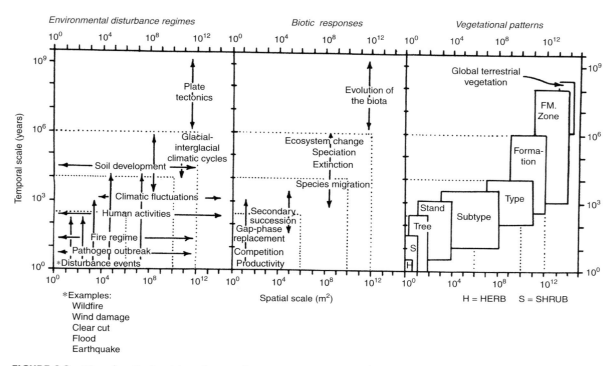

FIGURE 8.2 Time (vertical axis) and space (horizontal axis) organize a very large number of environmental processes including disturbance regimes (left), biotic responses (middle) and vegetational patterns (right). (From Delcourt and Delcourt 1988)

these habitats can provide important information on past changes in vegetation. If annual layers of sediment can be counted, these changes can be given an exact chronology; if not, radiocarbon dates can provide estimates of elapsed time since burial. The study of pollen in sediments (palynology) has been a vital tool for the study of events on scales from 10^2 to 10^5 years. You have already seen the use of sediment cores to reconstruct fire history of conifer forests (Figure 5.3) and deep ocean sediment cores for studying much longer periods of time (Figure 1.19). For longer time scales, say 10^6 to 10^8 years, fossils in dated rocks allow paleobotanists to reconstruct the evolution of plants and their changing distributions over millions of years. There are many sophisticated methods for dating rocks using rates of decay of isotopes. The use and dating of cores, pollen and fossils are discussed in more depth in Flint (1971),

Delcourt and Delcourt (1991), Stewart and Rothwell (1993) and Levin (1994).

4. Use zoned communities. Zoned communities on shorelines and mountainsides present an obvious sequence of changes in vegetation. If the spatial sequence represents a temporal sequence (and that is a big if), then measuring properties of different zones will allow description of their sequential change in time. Tree rings, or plant materials from peat, may be used to test for temporal changes. We shall see later an example of dated changes from vegetation along the edges of a retreating glacier in Alaska (Figure 8.20).

5. Of course, there is one other choice. One can simply focus on events shorter than one human life span: study population biology of annual plants or plant physiology in growth chambers. Apparently some people find this quite satisfactory. But even they might want to read this chapter. . . just in case.

8.2   Millions of Years: Flowering Plants and Continental Drift

8.2.1 Flowering Plants Appear in the Cretaceous Era

The angiosperms are now the dominant group of plants on Earth, whether you define dominance in terms of biomass, number of families or number of species. Their origin lies deep in the Cretaceous Period (65–140 million years BP), where their fossils show up in both increasing number and frequency (Figure 8.3). Where and how did the flowering plants originate? This mystery is being resolved through careful examination of various clues. To understand the significance of these clues requires some knowledge of the fossil record, geology, plant anatomy, systematics of tropical plants and DNA.

Let us review the conditions on Earth just prior to the emergence of the angiosperms (Axelrod 1970; Stewart and Rothwell 1993). At this time the single continent of the late Precambrian had split in two forming **Gondwana** in the south and **Laurasia** in the north. The dominant plants of the time were the gymnosperms – species that produce seeds but do not protect them within carpels, as the angiosperms do. True flowers and double fertilization are also absent. The diversity of gymnosperms at this time should be appreciated. Consider some examples.

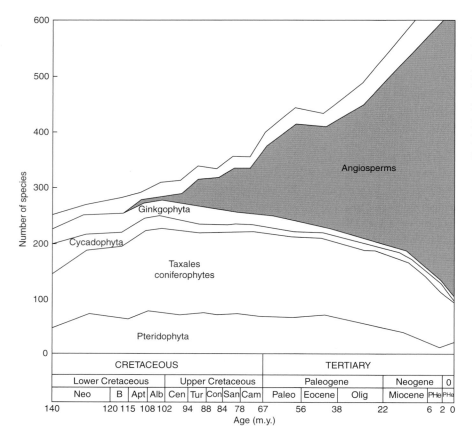

FIGURE 8.3 Changes in plant species diversity over the past 140 million years (Cretaceous and Tertiary). The data used in this graph are predominantly from the Northern Hemisphere. (From Niklas et al. 1985)

The conifers originated in the Devonian, and remained abundant during the origin of flowering plants (Figure 8.3). Currently there are 51 genera and some 500 species. Other groups became nearly extinct. The Ginkgoales were once circumpolar (Stewart and Rothwell 1993). Fossils are known from Alaska, Greenland, Scandinavia, Siberia and Mongolia as well as from the Southern Hemisphere. A single species in this group has survived to the present, *Ginkgo biloba*, known from the deciduous forests of Asia, truly a "living fossil."

Other orders of plants have disappeared. One such fossil order, the Glossopteridales, was first discovered in 1828 when certain tongue-shaped leaves were named *Glossopteris*. At first different fragments of these plants were given different names, but as the number and quality of fossils found improved, it became clear that this order consisted of woody plants with ovules (Stewart and Rothwell 1993). High-quality specimens even allowed paleobotanists to take sections of ovules and stems. These fossils are distributed across India, Australia, South Africa, South America and Antarctica, which suggests that the Glossopteridales were restricted to Gondwana. Permineralized trunks, 40 cm in diameter, have been found in late Permian rocks. The group is now extinct. Other large plant groups such as the Cordaitales and Voltziales are also extinct.

The context for the evolution of angiosperms lies among the many types of gymnosperms found in the past. Fossils of reproductive structures are critical. Paleobotanists have uncovered an entire sequence of reproductive structures ranging from leaves with attached ovules to ovules with leaves nearly fused around them (Figure 8.4), the latter giving the appearance of a carpel. The carpel, of course, is a key feature possessed by flowering plants. Even some Jurassic rocks contain angiosperm fossils (*Propalmophyllum* in France, *Sassendorfites* in Germany, *Phyllites* in England, *Palmoxylon* in Utah) and the very early Cretaceous rocks in France and California contain angiosperm fruits (Axelrod 1970). Many of the characters of modern angiosperms can be seen in

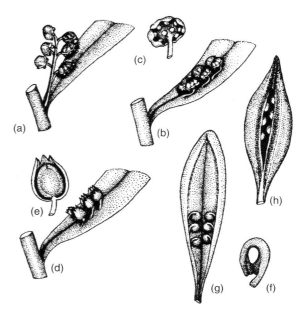

FIGURE 8.4 Fossil plants that illustrate stages in the evolution of the **carpel**. (a) Modified "gonophyll" with pinnate ovulate "sporangiophore" in the axil of a leaf. (b) Adnation of "sporangiophore" with adaxial surface of leaf leaving two rows of *Ligettonia*-like capitula. (c, d and e) Reduction in number of ovules per capitulum from several to one as in *Denkania*. (f) Fusion of cupule-like capitulum with integument of ovule to form bitegmic ovule. (f and g) Inversion of ovules to produce anatropous ovules. (h) Reflexing of leaf margins to enclose two rows of ovules in a carpel. (From Stewart and Rothwell 1993)

fossils in rocks of the middle Cretaceous (ca. 100 million years BP), including *Archaeanthus*, which apparently represents the *Magnoliales* (Figure 8.5). Stewart and Rothwell (1993: p. 452) remind us not to seek a single ancestral species:

> It is important to keep in mind that the characteristics by which we recognize angiosperms did not evolve simultaneously from a gymnosperm ancestor at the end of the Jurassic or Lower

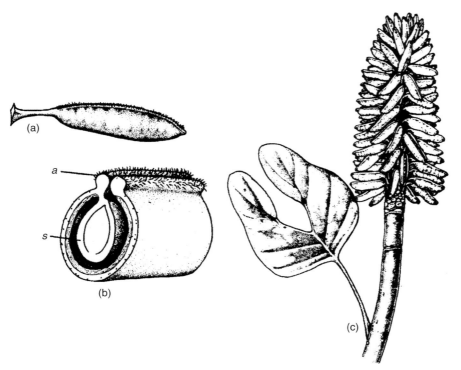

(a)

a

s

(b)

(c)

FIGURE 8.5 A mid-Cretaceous flowering plant in the Magnoliales (*Archaeanthus linnenbergeri*). (a) Reconstruction of single carpel. (b) Section through carpel: adaxial ridge (*a*), seed (*s*). (c) Reconstruction of leafy twig bearing many helically arranged conduplicate carpels. (From Dilcher and Crane 1985)

Cretaceous. Instead, many groups of gymnosperms evolved one or more characteristics that we associate with angiosperms prior to the Lower Cretaceous. Some of these groups became extinct, while others evolved new characteristics of particular selective advantage (the carpel and double fertilization, for example) in combination with already established characteristics such as reticulate venation or tectate pollen.

Some of the earliest fossil groups were the Magnoliales (containing the modern magnolias), the Hamamelidales and the Juglandales (with catkins evidently allowing for wind pollination) (Stewart and Rothwell 1993). A rich bed of early Tertiary age in southern England includes fossils of monocots, some of which can be assigned to modern genera such as *Potamogeton, Scirpus, Mariscus* and *Typha* (Collinson and Hooker 1987). New fossil angiosperms

continue to be found as researchers seek specific areas with early Cretaceous fossils (Sun et al. 2002; Friis et al. 2006).

What did the earliest flowers look like? Fossils provide some evidence, but you will appreciate that fossilized flowers are rather rare. You saw in Chapter 2 (Figure 2.7) that some living plants seem to still have traits, particularly flower types, that resemble early angiosperms. Such plants include members of the Nymphaeaceae (Figure 2.8), the Illiciaceae (Figure 2.9) and the Amborellaceae. The magnolia-like plants (Magnoliids, Figures 2.3 and 2.6) are a much larger group, again having many primitive floral traits. In contrast, groups of plants such as asters, grasses and orchids are much more recently evolved, and possess floral traits that are considered modern. In general, ancient flowers were large with many spirally arranged parts, while modern flowers are smaller and often grouped into clusters, with

Primitive flowers

parts many
parts indefinite in number
parts separate
spiral arrangement of parts
sepals, petals, stamens and carpels all present
bisexual
ovary superior (perched on receptacle)
radial symmetry

solitary on branches

Advanced flowers

parts few
parts definite in number
parts fused
whorled arrangement of parts
loss of sepals, petals, stamens or carpels
unisexual
ovary inferior (sunken into receptacle)
bilateral symmetry

grouped into spikes or clusters

FIGURE 8.6 Types of flowers associated with primitive and advanced plants. Not everyone approves of the terms primitive and advanced. Still the concept is useful, and you can to some extent substitute the words *ancestral* and *more recently evolved* if you wish, or if your instructor tells you to do so (text adapted from Heywood et al. 2007). (Left) Single flower of *Magnolia sinica*. (Jackson Xu, Fauna & Flora International) (Right) Flower spikes of *Lobelia cardinalis* and *Chelone glabra*. (Photographs by Paul Keddy)

fewer parts and an overall bilateral symmetry (Figure 8.6).

And where did the earliest flowers originate? During the Triassic and Jurassic, Gondwana separated from Laurasia as the Tethys Sea opened (ca. 250 million years BP, Figure 8.7e and f). This event appears still to exert control over the present distribution of many plant species, since the flora of the Southern Hemisphere, which is largely derived from Gondwana, is much more diverse than of those areas derived from Laurasia. We can expect angiosperms originated in tropical climates, since more than half of the current families of flowering plants occur in such a climate.

> *an analysis of the climatic requirements of ancient taxa that are still living (e.g., cycads, araucarians, tree ferns, podocarps, redwood) demonstrates that they have survived chiefly in areas of very high equability... This is also true of the more "primitive" angiosperm alliances – magnolioids, annonids, hamamelids – as well as the surviving primitive members of the major angiosperm orders.*
> *(Axelrod 1970: p. 290)*

As the flowering plants diversified, one other evolutionary change occurred. The life cycle of plants became further abbreviated as the gametophyte stage became further reduced to what is now called simply the "embryo sac" within an ovule. But the gametophyte was once a free-living life history stage, and indeed, may even have been dominant in the life cycle of early land plants. This quite remarkable story of plant evolution is worth a full chapter, but must, alas, be reduced (like the gametophytic stage itself) to Box 8.1.

Box 8.1   Mr. Hofmeister and the Vanishing Gametophyte

The Evolution of Plant Life Cycles

Sadly, plant life cycles are too often taught, if at all, through memorizing the stages and tissue types in the life cycle of a flowering plant. This is hardly an inspiration to further study in botany. Although the preface states that I assume you know some basic botany, increasingly, given the emphasis on teaching physiology and genetics, it seems that one can no longer presuppose familiarity with the alternation of generations. Yet the alteration of generations is arguably the most important theory synthesizing life history and evolution in the plant kingdom.

To understand the theory, and its implications for evolution and ecology, we need to begin, not with the flowering plants, but rather with the life cycle of primitive plants such as *Rhynia* (Figure 1.5), *Lepidodendron* (Figure 1.7), or modern seedless plants such as ferns. In these plants, we can still see clearly two entirely different life history phases, the sporophyte (which is diploid) and the gametophyte (which is haploid). The sporophyte has cells that undergo meiosis, producing large numbers of haploid cells (spores). Each spore develops into a gametophyte that produces sperm and/or egg cells. The fusion of egg and sperm then produces a diploid sporophyte. The sporophyte and gametophyte generations therefore alternate. Most vascular plant fossils (e.g. Figures 1.5 and 1.7) represent the sporophyte stage, but gametophytes, while rarer, have also been found (Figure B8.1.1).

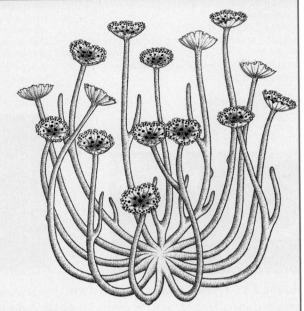

FIGURE B8.1.1   Reconstruction of a gametophyte preserved as a 400 million-year-old fossil. The eggs and sperm were produced on top of the stalked structures (termed gametangiophores) in archegonia and antheridia, which can be seen microscopically in slides made from the fossils (Taylor et al. 2005). Note the apparent similarity to modern liverworts, such as *Marchantia*, that are often studied in botany courses. (From Kenrick and Crane 1997b)

This alternation of generations would seem to be rather inefficient and improbable. Why should a plant have two life history phases at all? Why should plants that live on land have sperm that must swim through water to reach eggs? The answer illustrates a profound and frequently misunderstood point in evolution: natural selection does not create a perfect solution to environmental hazards. Natural selection can only modify an existing situation by weeding out the worst alternatives. The alternation of generations appears to have been directly inherited from the algae, where it is widespread and appears in many different guises. All early land plants, both extinct forms and those that can be thought of as living fossils, have a conspicuous alternation of generations (Kenrick and Crane 1997b; Taylor et al. 2005). In such plants, the ferns being the best-known example, both life history stages are free-living and independent of one another. In theory, one could go into a forest and seek either gametophytes or sporophytes of these species. In practice, the sporophytes are much larger and more obvious, and so they tend to receive most of the attention from

Box 8.1   (cont.)

ecologists. Gametophytes are rarely seen. Swimming sperm could severely limit reproduction, particularly in drier circumstances. Large areas of land might then have remained largely uninhabited by plants. The history of plant evolution can be viewed as a clash between the stress imposed by drought, and the inherited raw material of an algal life cycle better suited to water. Seeds and pollen arise as a consequence.

The Origin of Seeds and Pollen

Some living and many fossil plants are heterosporous – that is, the sporophyte produces two types of spores, large ones (megaspores that become female gametophytes) and small ones (microspores that become male gametophytes). *Selaginella* is the common example studied in botany courses. Should some ancestral plants have delayed dispersal of the megaspores, perhaps by something as simple as an imperfectly

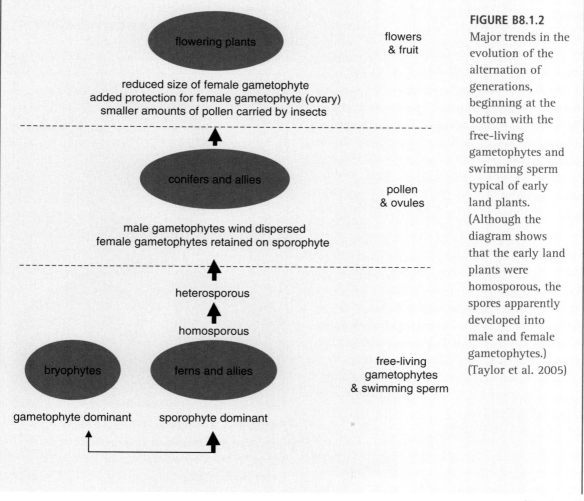

FIGURE B8.1.2
Major trends in the evolution of the alternation of generations, beginning at the bottom with the free-living gametophytes and swimming sperm typical of early land plants. (Although the diagram shows that the early land plants were homosporous, the spores apparently developed into male and female gametophytes.) (Taylor et al. 2005)

(Continued)

Box 8.1 (cont.)

rupturing sporangium, the female gametophytes could have matured while still attached to the sporophyte. This minor aberration could have had multiple short-term advantages including: (1) protection of the gametophyte from predators, (2) continued provision of water and nutrients by the sporophyte, and (3) a general reduction in the time required for a complete life cycle. A female gametophyte retained upon the sporophyte is the precursor to the ovule and seed.

Microspores that would otherwise have dispersed to the ground, developed into male gametophytes, and released sperm, could equally have been dispersed by wind to adjacent shoots possessing female gametophytes. Male gametophyte dispersal through the air would have been the precursor to pollen. Most of the vascular plants, including the conifers and the flowering plants, now have seeds and pollen.

The Vanishing Gametophyte

The gametophyte stage is still prominent in the life cycle of most primitive plants. It remains the dominant life history phase in mosses. Early fossils of gametophytes are even larger than those of present plants, with antheridia (sperm-producing organs) and archegonia (egg-producing organs) borne on stalked gametangiophores (Figure B8.1.1). The evolutionary history of plants illustrates continued pressure to reduce the size and complexity of this reproductive stage.

The stages in the reduction of gametophytes are represented by varying living plant groups (Figure B8.1.2). In the seedless vascular plants, represented by ferns and their allies, gametophytes are still free living. In general we can reconstruct the following sequence of events: female gametophytes were at first retained within the sporangia of the sporophyte, then reduced to ovules protected within cones and, finally, in the case of flowering plants, reduced to a mere handful of cells in what is cryptically named the "embryo sac." In a less dramatic trend, male gametophytes became reduced in size until, in the case of flowering plants, they become a mere vegetative nucleus within a pollen grain. One can view this trend in two ways, both of which have value. If we focus upon the extreme modification and reduction in the gametophytes achieved by natural selection, we see a strong and consistent evolutionary trend through the entire plant kingdom. If we focus instead on the continued occurrence of alternation of generations, we see a conserved and unifying algal life cycle in spite of the enormous evolutionary diversification that has occurred in vascular plants.

FIGURE B8.1.3 Wilhelm Hofmeister (1824–77) uncovered the alternation of generations in plants and published a monograph on the topic in 1851. (From Goebel 1905)

Box 8.1   (cont.)

A Unifying Theory

Remarkably, this essential unifying theory of plant evolution was largely unravelled by a single botanist, Wilhelm Hofmeister (1824–77). Hofmeister (Figure B8.1.3) was largely self-taught, having left school at the age of 15 to work as an apprentice in a book and music shop in Leipzig. Yet he became "one of the most notable scientists in the history of plant biology" (Kaplan and Cooke 1996). Note that while he did this vital research on plants, Mr. Hofmeister was not a tenured professor – he supported himself, and carried out most of his research in his own time, largely from 4 am to 6 am before going to the store (from Goebel 1905). He was only 27 when he published his ground-breaking monograph that documented the alternation of generations in plants. This appeared in 1851, a full eight years before Darwin published *The Origin of Species*. Not until 1863 (at the age of 39) was Hofmeister employed as a professor.

8.2.2 Continents Derived From Gondwana Have Remarkable Plant Diversity

Many ancient plant groups are found today in regions of the world that derived from Gondwana: South America, Africa, Australia, New Zealand and India (Figure 8.7). Let us consider four important plant groups that follow this pattern.

Forests on Drifting Continents: The Strange Case of *Nothofagus*

The southern beech (*Nothofagus*) today is found exclusively in the Southern Hemisphere, in areas such as Tasmania, the temperate montane forests of New Guinea and New Caledonia, and the Tierra del Fuego area, although the fossil distribution of *Nothofagus* extends to Antarctica (Humphries 1981; Figure 8.8). *Nothofagus* tends to form nearly pure stands on the east coast of the Andes, for example, although in other cases it grows mixed with gymnosperms such as *Araucaria*, *Podocarpus* and *Dacrydium* (Archibold 1995). The gymnosperms in the Podocarpaceae and Araucariaceae also have a similar distribution (Farjon 1998). Podocarps are noteworthy for another reason, although they are gymnosperms, they lack a strobilus and instead bear ovules singly on the tips of stalks.

Some Deserts Have an Ancient Gondwana Flora

Although deserts are defined by climate (recall Figure 2.4), the origins of desert plants themselves are more complicated. The desert floras of the Southern Hemisphere contain two groups of species. One group has been derived from adjacent mesic floras that progressively invaded and evolved as new desert areas formed. That is to say, the ancestors of such species can be found in adjoining habitats. But a second group of species appears to include much older kinds of plants that originated in the ancient desert flora from Gondwana. To see two examples, flip ahead to Figures 9.6 (*Dracaena cinnabari*) and 12.14 (*Welwitschia mirablis*). Here is how one paleobotanist has described this feature of southern hemisphere deserts (Axelrod 1970: 311).

Relatively free and isolated from competition by northern shrubs, and surviving under a more equable climate, some bizarre ancient taxa have persisted in larger numbers in austral lands, not only in the semidesert to desert areas, but also in the dry tropical forest, savanna, and thorn scrub vegetation adjacent to them. But in North America and the Eurasian region, they were subjected to more extreme climates and were unable to compete with the new xerophytes that originated there and have largely supplanted them.

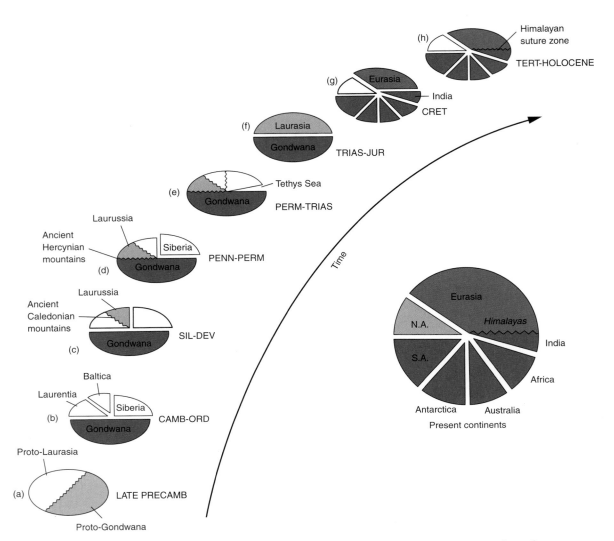

FIGURE 8.7 The fragmentation of continents from the Cambrian to recent time, with particular reference to Gondwana. (After Levin 1994)

Tropical Floras Were Divided and Isolated by Continental Drift

Many tropical plant groups are spread across South America, Africa and Asia – 54 forest plant families are pantropical (Figure 8.9). At one time it was postulated that these species somehow spread across the ocean by means of birds or ancient land bridges; in light of continental drift, however, we now understand that these groups were once widespread on Gondwana. The distribution of the subgroups, such as the eight families shared by South America and India/Australia (Figure 8.9), can be explained by allowing for extinction of these groups in Africa. As Gondwana split into smaller fragments, which

FIGURE 8.8   (a) The current distribution of the genus *Nothofagus* (southern beech) spreads on both sides of Antarctica. Fossil plants and pollen occur in Antarctica. (b) The photos show a fossil leaf of *N. beardmorensis* from Oliver Bluffs Antarctica (left) compared to a modern leaf of *N. pumilio* from South America (right). (Map after Pielou 1979; photographs by Allen Ashworth)

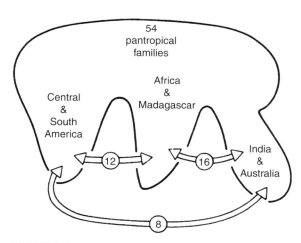

FIGURE 8.9   Schematic representation of the ranges of families of tropical forest plants. (From Pielou 1979)

drifted across climatic zones, it is reasonable to expect that extinctions occurred and account for the variation in the groups that now occur in these regions. Here is how Axelrod and Raven (1972: p. 229) describe changes in Australia as it drifted away from the other continents:

> ...as the Australian plate moved into lower middle latitudes following Eocene time, it entered the permanent high-pressure belt of low precipitation at the south margin of the tropics. As a result, the temperate rain forest, composed of southern beech [Nothofagus], araucarias, podocarps, proteads, and other evergreen dicots that had covered much of the continent, was progressively restricted to moist equable southeast Australia – Tasmania. During this movement, many taxa underwent severe restrictions in range, and others must have become extinct. Some that were in the region survive as relicts on offshore lands such as New Caledonia, Fiji, and New Zealand.

Some groups diversified into new species: the remarkable Proteads and Eucalypts. As a continent

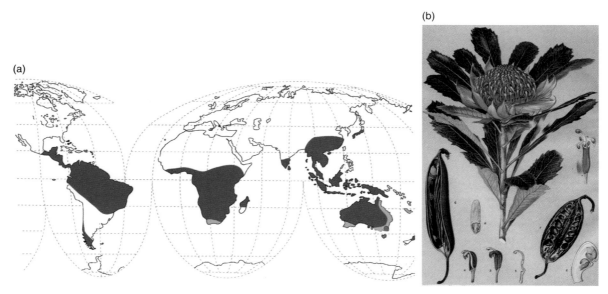

(a)

(b)

FIGURE 8.10   (a) The distribution of the plant family Proteaceae. The light-green areas are regions of high generic diversity. (b) This painting is a representative protea (*Telopea speciosissima*, Waratah) that is restricted to a small region in southeastern Australia (●). The vivid red flowers are up to 10 cm across and the shrub may grow to 4 m. It is bird pollinated but dispersed by wind. (Map adapted from Johnson and Briggs 1975; painting by Edward Minchen 1985, courtesy of Australian National Botanic Gardens)

drifts and climate changes, opportunities may arise for speciation. The distribution of the family Proteaceae still shows its origins on Gondwana (Figure 8.10), but it has expanded to 68 genera and 1,252 species (The Plant List 2013). Another Gondwana family, the Restionaceae, has expanded to some 47 genera and 482 species in the Southern Hemisphere. Again, let us hear from Axelrod and Raven (1972: p. 230):

Rafting lands to new climatic belts has provided new opportunities for evolution. As Australia moved into a zone of warmer drier climate, plant genera such as Acacia, Eucalyptus, Casuarina, Melaleuca, Hakea, *and* Eremophila... *proliferated into scores of new species adapted to progressively drier, more continental climates.*

8.3   Thousands of Years: The Pleistocene Glaciations

The last great ice age ended a mere ten thousand years ago, and vast areas of the Earth's surface still show its effects. That last ice advance, of course, was just the most recent in a long series during a period of climatic instability that began several million years ago, a period during which extensive ice sheets and other glaciers formed repeatedly (e.g. Delcourt and Delcourt 1991; Levin 1994). No student of plant ecology can afford to be ignorant of the effects that the last ice advance had upon the landscape. The effects are often very local and their results may vary from heavily scoured landscapes lacking soil, to great terminal moraines, to wind-deposited soils, to large lakes, depending upon location. This section will give an overview to put local details in a global context.

First some terminology. The last ten thousand years, an interglacial period with human activity, are called the Holocene. The preceding era, called the Pleistocene, is recognized by the repeated advance of large sheets of ice. The Holocene and the Pleistocene together comprise the Quaternary Period. At one time, the advances and retreats by ice were assumed to consist of four major advances interspersed with three major warm periods. The names of the advances differed depending upon where the field-work had been done. In North America they were named, from youngest to oldest: Wisconsinan, Illinoian, Kansan and Nebraskan; in Great Britain: Devensian, Wolstonian, Anglian and then a further series of at least four other advances. In Europe they were Weichselian, Saalian, Elsterian and then a different set of at least three advances. It has proven difficult to put these together in one large picture, particularly as the numbers of advances and retreats have expanded with our knowledge of glacial geology. In fact, marine sediment records suggest that a more complicated series of advances and retreats occurred (recall Figure 1.19).

8.3.1 Erosion and Deposition Were Caused by Glacial Ice

Substantial portions of the Earth were covered by ice as little as 25,000 years ago. As the ice retreated, it left behind the present-day topography. Soil was scoured away and pushed south of the ice margin. (American farmers now grow corn using the soil that was once on my Canadian land.) After the ice melted, the landscape was left covered with a variety of glacial deposits including moraines, eskers, lakes and clay plains (Figure 8.11). At a slightly large scale entire drainage basins changed – consider the quite remarkable changes in the shape and drainage of the Great Lakes during this period (Figure 8.12). The distribution of many vegetation types around the Great Lakes can still be best explained by past events.

Consider just three examples. The distribution of the nationally rare plant, *Rhexia virginica*, is still associated with areas once covered by water (Keddy 1981b). The arctic species *Saxifraga aizoon* still grows in a canyon that formed when once the Great Lakes drained through Algonquin Park. The extensive pine forests of the upper Ottawa Valley occur on an old delta where a glacial river drained into the Champlain Sea. Each post-glacial landscape has its own story, and that is why you should own a physiographic map of your own regional landscape.

8.3.2 Loess Was Deposited by Wind

Wind moved large volumes of dust and soil that was exposed along the ice margin and in floodplains. Deposits of this wind-blown soil, called loess, cover large areas of the world: the central plains of North America, the north European plain, central Asia, northeast China and the pampas of Argentina. The deep deposits of loess in the United States appear to have been carried east from the floodplain of the Mississippi River. The deep deposits of loess in China appear to have been swept in from the Gobi Desert; in places they are 100 metres thick.

8.3.3 Pluvial Lakes Expanded

Changes in water balance south of ice sheets caused the accumulation of water in previously arid areas (Flint 1971). Large lakes formed in areas such as the southwestern United States, central Australia and Africa (Figure 8.13). Many of these lakes are now entirely dry or greatly reduced in size.

8.3.4 Drought Affected Tropical Forests

Other areas became drier. Since tropical rain forests occur in warm areas with abundant precipitation, it seems reasonable to ask how these forests might have changed during glacial periods. There are two main possibilities. Perhaps changes in these rain forests were minimal, since this area is near the equator and

(a)

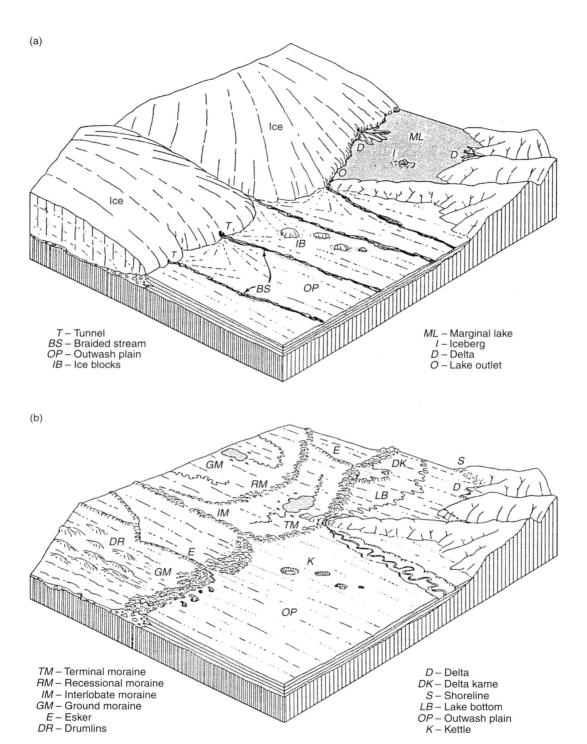

T – Tunnel
BS – Braided stream
OP – Outwash plain
IB – Ice blocks

ML – Marginal lake
I – Iceberg
D – Delta
O – Lake outlet

(b)

TM – Terminal moraine
RM – Recessional moraine
IM – Interlobate moraine
GM – Ground moraine
E – Esker
DR – Drumlins

D – Delta
DK – Delta kame
S – Shoreline
LB – Lake bottom
OP – Outwash plain
K – Kettle

FIGURE 8.11 Landforms produced near the margin of an ice sheet. (a) Ice margin in almost stagnant condition. (b) Ice entirely gone, revealing subglacial forms. (From Strahler 1971)

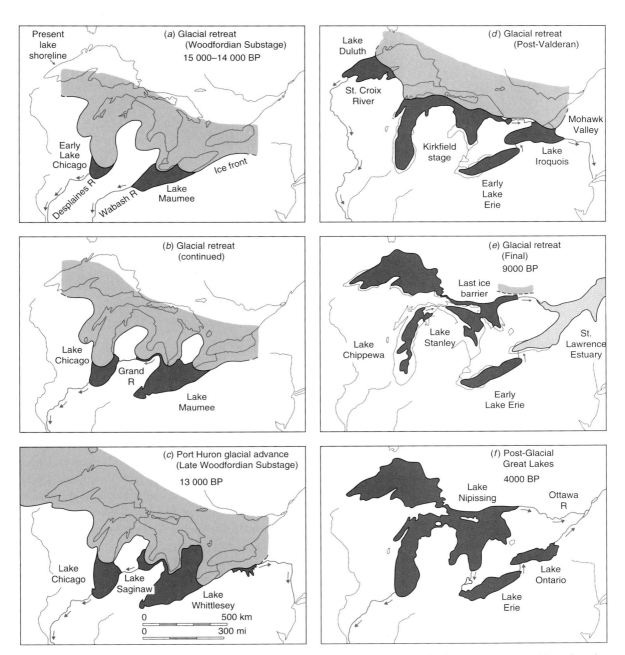

FIGURE 8.12 Sequence of stages in the development of the Great Lakes over the last 20,000 years. Note that the sequence (a–f) runs top to bottom, left side, then right side. (From Strahler 1971)

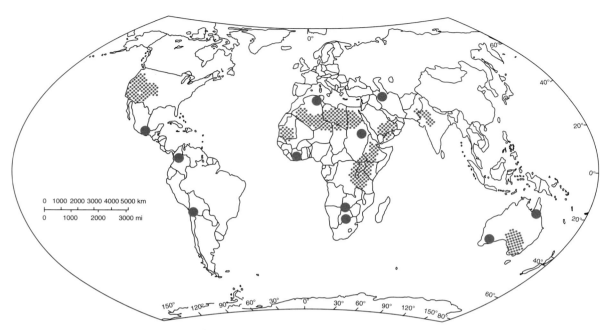

FIGURE 8.13 Over the last 30,000 years pluvial lakes have formed and disappeared from the shaded regions of the Earth. Dots show isolated lakes. (After Street and Grove 1979)

far from advancing ice sheets. Alternatively, perhaps drier periods allowed savanna to invade and replace forest. Let us consider the rain forests of the Amazon. There has been considerable debate about how they were affected by northern ice sheets (Pielou 1979; Colinvaux et al. 1996).

Here is a challenge: how does one test whether savanna invaded the Amazon basin during drier periods tens of thousands of years ago? There are two main sources of biological data available: sediment cores from the Amazon delta and pollen cores from lakes.

First consider the Amazon delta. If the rain forests periodically receded, this should be mirrored in changes in the characteristics of sediments in the delta. Samples of Amazon River detritus taken from coastal sediment indicate that the Amazon basin remained continually forested, at least up to 70,000 years ago (Kastner and Goñi 2003).

Now consider pollen cores from lakes. Periods of savanna should be marked by layers having pollen grains from grasses and other savanna plants. Finding good pollen records is much more difficult in the Amazon than at higher latitudes (Colinvaux et al. 2001). The Amazon basin has few old lakes with long sediment records, and further, many rainforest trees produce little pollen that could drift into lakes, since the pollen is dispersed by animals. These problems have been slowly overcome, and a number of cores taken for study (including a 40,000 year pollen record from Lake Pata on a 500 to 600 m tall **tepui**, a rock outcrop that emerges above the rain forest (Colinvaux et al. 1996)). This core, and others like it, show no evidence of periods of savanna expanding into the rain forest (Colinvaux et al. 2000, 2001).

Savannas are widespread elsewhere in South and Central America and in the world (look ahead to Figure 10.7), and their distribution is very sensitive to drought and fire. In drier climates, or drier locations, savanna can replace forests. Figure 8.14 shows that on a global scale, areas of forest were much reduced during the last ice age maximum.

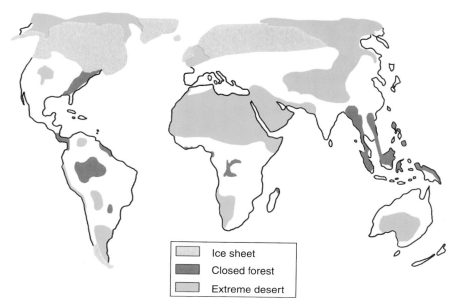

FIGURE 8.14 The distribution of extreme desert and closed forest during the last glacial maximum 18,000 years ago. This map does not show changes in sea level – particularly the extensive land bridge between Australia and New Guinea. For a complete global vegetation map from this period, see Ray and Adams 2001. (Adapted from Adams 1997)

Ice sheet
Closed forest
Extreme desert

In contemporary times, recurring fire is a natural part of the balance between forest and savanna (recall Figure 5.1 and Section 5.3.1), and the type of savanna that forms will depend upon fire frequency (Myers et al. 2006). In tropical landscapes with high fire frequencies, most species of trees survive only in narrow strips of forest along water courses (termed riparian forest or gallery forest). During wetter periods with lower fire frequency, the forest may expand into upland areas, while during drier periods with higher fire frequency, the forest may shrink to the margins of rivers. In order to investigate the potential importance of riparian forests as refugia for forest plants, Meave and Kelmman (1994) explored an area of Belize that was mostly savanna, and enumerated the tropical tree species they encountered in riparian forests. A total of 292 species were found in a cumulative sample of only 1.6 ha. Some tracts of riparian forest had as many tree species as continuous forests elsewhere in Central America. Moreover, small changes in flood frequency and soil fertility can produce different types of riparian forest (Veneklaas et al. 2005), which may further increase the number of species of trees that can survive in landscapes that are largely savanna.

We will return to the climatic control of forest and grassland boundaries in Section 10.2.2.

8.3.5 Sea Levels Fell as Ice Sheets Expanded

The enormous amounts of water locked into glacial ice lowered sea levels by as much as 100 metres. This would have opened up new areas of habitat along previously submerged coastal plains, as well as created land bridges between some continents. The continental margins of eastern North America, for example, appear to have had a rich wetland flora appropriately called the Atlantic Coastal Plain flora; peat balls recovered from offshore illustrate how well developed the peatlands in these areas were. When the sea level rose as the ice melted, most of this habitat was re-submerged, which left widely spaced fragments of this distinctive wetland habitat along the eastern coast of North America. The distribution of *Sabatia kennedyana* in Cape Cod and southwestern Nova Scotia suggests that at one time these areas were connected by land, but when the sea level rose the distribution became fragmented (Figure 8.15).

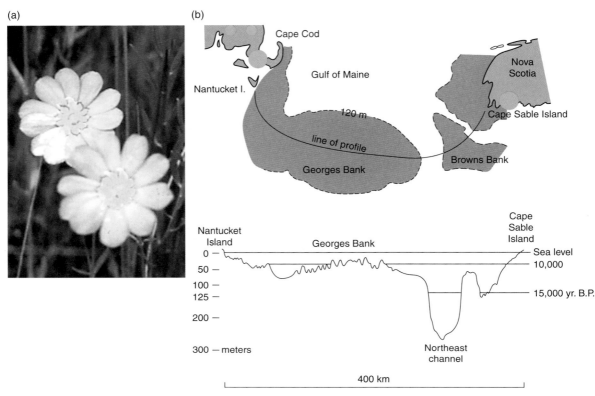

FIGURE 8.15 Falling sea levels during the last ice age opened a migration route between Nova Scotia and New England by exposing Georges Bank and Browns Bank (now famous fishing areas) as dry land (b). This probably explains disjunct plant distributions such as that of (a) the threatened *Sabatia kennedyana* (Plymouth gentian) •. (Photo by Paul Keddy; map adapted from Ogden and Harvey 1975)

Another isolated population of this species occurs further south in the Carolinas.

8.3.6 Plant Distributions Changed

Changes in sea levels, pluvial lake levels and ice cover would have forced major changes in the distributions of plant and animal species. Picture the world's vegetation types at the maximum of the last ice advance: Figure 8.16 shows that the vegetation of eastern North America at the peak of the ice age (bottom) was vastly different from that today (top).

Similarly in Europe, at the maximum of the ice age, plant distributions were shifted far south, with deciduous forests restricted to small areas in the south east along what is now the Mediterranean, one on the west coast of Italy and the other north of the Adriatic Sea (Figure 8.17). In addition, some temperate zone tree species in genera such as *Carya, Liquidambar, Robinia* and *Tsuga* became extinct around this time, perhaps because the Alps prevented southward migration as the ice advanced (Daubenmire 1978).

As the climate warmed, tree distributions began to shift north. Pollen cores confirm that there was rapid re-colonization northward (Figure 8.18). Such reconstructions have now been given an added impetus by fears that we are altering Earth's climate (Section 1.10). Establishing a connection between

(a) Present

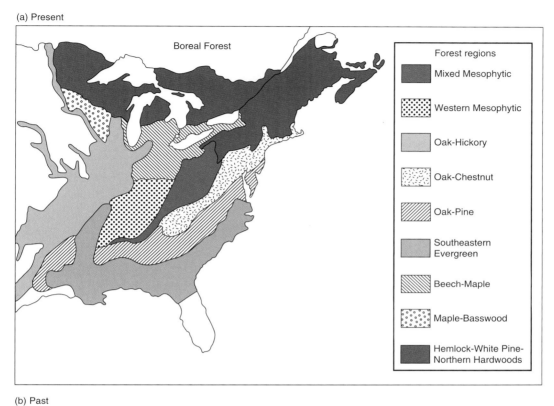

(b) Past

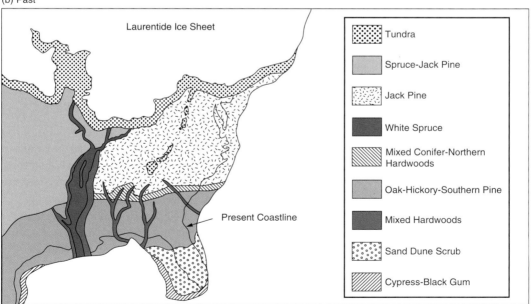

FIGURE 8.16 (a) The present deciduous forests of eastern North America. (b) A reconstruction of the past vegetation during the last glacial maximum about 18,000 years BP. (From Archibold 1995)

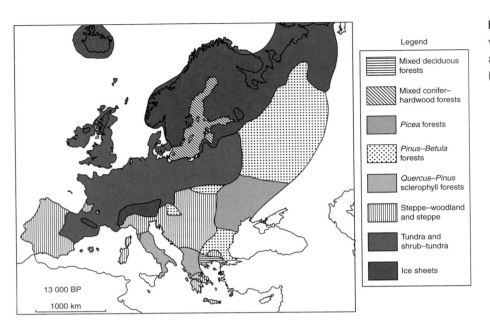

FIGURE 8.17 The vegetation of Europe about 13,000 BP. (From Huntley 1990)

Legend

	Mixed deciduous forests
	Mixed conifer–hardwood forests
	Picea forests
	Pinus–Betula forests
	Quercus–Pinus sclerophyll forests
	Steppe–woodland and steppe
	Tundra and shrub–tundra
	Ice sheets

13 000 BP
1000 km

climate and vegetation in the past provides a vital tool for forecasting changes in vegetation.

8.3.7 Humans Appeared and Spread to New Continents

Modern humans (*Homo sapiens*) appeared in Africa about 200,000 years ago and spread to Asia 80,000 to 60,000 years ago, near the end of the last Pleistocene interglacial. At this time Neanderthals (*H. neanderthalensis*) were already in Europe and parts of Asia. It is thought that *H. sapiens* arrived in North America at least 10,000 years ago by crossing the Bering land bridge, although archeologists are still struggling to date the exact time of arrival (Pringle 1996; Jenkins et al 2012). As *H. sapiens* spread across North America, there was a massive die off of large mammals, and paleo-ecologists still argue about whether climate change or overhunting was responsible for this wave of extinction (Martin and Klein 1984; Koch and Barnosky 2006). The fact that similar waves of extinction are associated with the arrival of humans in other areas including Australia and Easter Island suggests that the end of these large mammals was caused by newly arrived humans (recall Figure 6.10). Apparently our ancestors were busily changing the new lands they encountered. In the short period since the last major glaciation, our species has spread around the world and multiplied until its numbers (in 2016) exceed 7.5 billion. These human populations are now a major factor changing plant distributions, vegetation types, and even global cycles for elements such as carbon and nitrogen. We will return to this topic in the final chapter.

8.3.8 Sea Levels Rose as Ice Sheets Melted

As the ice sheets melted, sea levels began to rise again, creating the more familiar outlines of lakes and continents. But there was one other brief period of intense flooding, since the weight of the ice had actually depressed the land. Hence sea water flooded inland much further than it occurs today. For example, the land near my own home in Ontario was flooded beneath the Champlain Sea that extended inland up the St. Lawrence River (Figure 8.12e), and, indeed, one old shoreline of this sea winds through my property. Further inland, the enormous Lake Agassiz formed in central North America (Teller 2003), and periodic

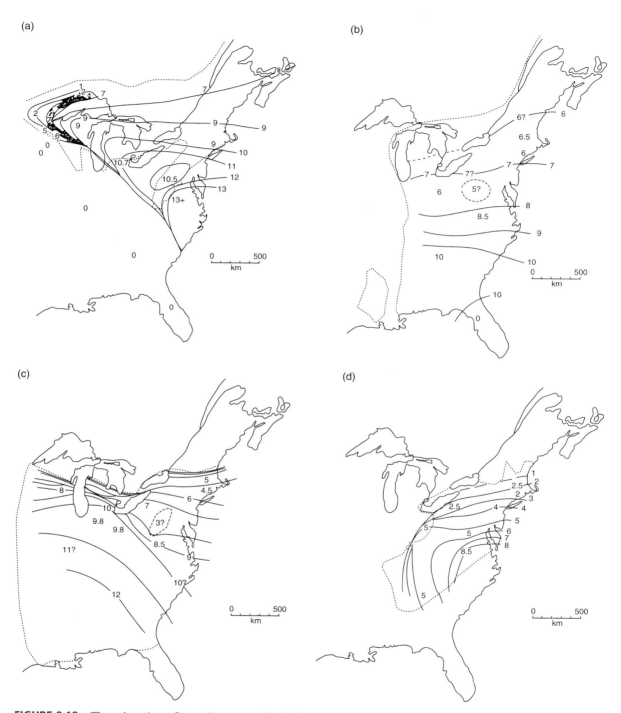

FIGURE 8.18 The migration of trees in eastern North America as the continental glaciers receded: (a) white pine, (b) chestnut, (c) hickory, (d) beech. Note that each species apparently re-invaded from slightly different refugia, some of which may have been on the now-flooded continental shelf. Solid lines mark the advancing frontiers at 1,000-year BP intervals and dotted lines surround the modern ranges of the species. (From Davis 1976)

floods from Lake Agassiz began building the modern delta of the Mississippi River. Most of the Hudson Bay lowlands were flooded, and covered with a layer of clay, laying the substrate for what is now one of the world's largest peatlands (Abraham and Keddy 2005). Similar changes occurred in northern Europe. However, these inland extensions of the ocean lasted only a few thousand years, for as the land began to rise again, these incursions from the sea retreated. As northern coastal land rebounds from the weight of the ice, lines of raised beaches continue to appear (Stevenson et al. 1986; Bégin et al. 1989).

Water levels in the Black Sea also changed after the ice age, particularly during periods when the lake was unconnected to the Mediterranean and lowered by evaporative losses. Some of you may be familiar with the popular book *Noah's Flood* (Ryan and Pitman 1999), which suggests that Black Sea levels were low from evaporation and that large areas were being cultivated by Neolithic farmers. Some 7,600 years ago, they suggest, the Mediterranean reached a high enough level to pour eastward into the Black Sea. Flow rates may have greatly exceeded those of Niagara Falls, leading to inundation of more than 150,000 km^2. To the peoples of that time, it might indeed have appeared as if the entire world had flooded, and Ryan and Pitman suggest that this event

gave rise to many stories about great floods, including those of both *The Epic of Gilgamesh* and the *Book of Genesis*. It may also have triggered migration of Neolithic farmers toward central Europe. Not surprisingly, there is disagreement about almost every aspect, including timing, depth and effects! Yanko-Hombach et al. (2007) report that there were two floods, an "intense and substantial" inundation in the Late Pleistocene (ca. 15,000 BP) and a lesser inundation in the early Holocene (about 10,000 BP). Giosan et al. (2009) used radiocarbon data from molluscs to conclude that the flooding would have been 30 m rather than 80 m. New geological and archaeological work will no doubt refine the story. Perhaps the Black Sea is not the source of the flood stories. But one should not forget that the enormous effects of changes in climate, ranging from entirely new fluvial lakes to great floods during melting glaciers, may have been retained in human memory in various stories.

And nearer the equator? In South America the entire Amazon valley appears to have been drowned about 15,000 BP. During this period, a large freshwater lake 1,500 km long and up to 100 km wide may have extended inland from the mouth of the Amazon (Irion et al. 1995; Müller et al. 1995). Sediments deposited in this lake would have produced deltas in the middle Amazon.

8.4 Hundreds of Years: Succession

8.4.1 Succession Is Directional Change in Vegetation

Succession refers to the sequence of changes in vegetation that occurs after a site is disturbed – a sequence of events that normally leads to the re-establishment of the vegetation that existed before disturbance. If the disturbance was minor, and the soil and buried propagules remain, the recovery of the vegetation is usually rapid. This rapid re-colonization is termed secondary succession. Examples of secondary succession include the forests colonizing gaps

in forest created by storms or logging. If the disturbance is more severe, and the soil and propagules are also destroyed, a much longer series of events must occur. The soil itself must be re-established and propagules may have to invade from distant sources. This process is called primary succession. Examples of primary succession include the sequence of changes in which sand dunes become forest, shallow water becomes peat bog, or sites deeply buried by volcanic debris become forest.

Frequently it is possible to distinguish a number of distinctive stages through which recovering communities pass; these are termed seres. When the process of change ends, the species in the final sere are capable of maintaining themselves and restricting further invasion. This situation is known as a climax. If you revisit the chapter on disturbance, you can see many cases where a disturbance leads to succession, a sequence of seres and a climax. The concept of succession is vital to understanding change in plant communities.

Succession is one of the key concepts in ecology, yet one that has at times been overused. As a consequence, it has occasionally become fashionable to move to the opposite extreme and deny that succession occurs. Before considering the debates that have occurred regarding succession, let us examine some real examples.

8.4.2 Four Examples of Succession

Succession After the Retreat of Glaciers

The Glacier Bay region in Alaska provides a 200-year-long perspective on succession. Retreating ice has left behind a series of plant communities of known ages (Figure 8.19). The dates are based upon evidence such as the number of rings in trees growing on moraines, written records and photographic surveys. The type of vegetation changes with site age. The youngest sites have mosses in the genus *Rhacomitrium* along with low-growing vascular plants such as *Equisetum variegatum*, *Dryas drummondii* and *Salix arctica*. The oldest sites are forested by *Picea sitchensis* along with *Tsuga heterophylla* and *T. mertensiana*. This shift from mosses to dense forest is typical of pioneer succession in other deglaciating sites.

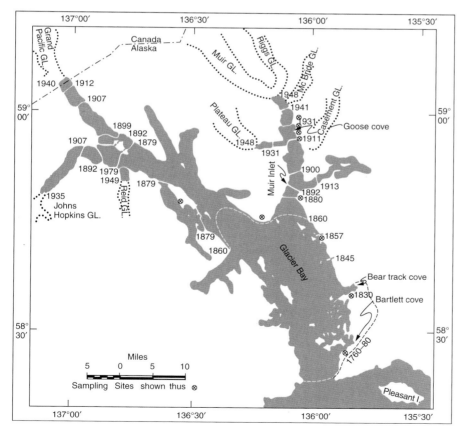

FIGURE 8.19 Ice recession at the Glacier Bay fiord. (From Crocker and Major 1955)

Crocker and Major joined an expedition to the site in 1952 to study how soil changes along this successional gradient. After deglaciation, the amount of organic carbon in the soil increased steadily (Figure 8.20). Total soil nitrogen also increased over the first 50 years (Figure 8.21), likely due to nitrogen fixation by the shrub *Alnus crispa*. Soil nitrogen later declined as spruce forest established. Acidity also declined: over the first 100 years, the pH of the upper 5 cm of soil decreased from >8.0 to 5.0. The rate of decline was most rapid during the first 50 years and changed only slightly after 100 years.

In 1984, Bormann and Sidle (1990) revisited this area and sampled the biomass and nitrogen content of the vegetation at four seral stages from *Alnus*-dominated sites (ca. 60 years old) to *Picea* forest (ca. 210 years old). Biomass reached a peak of 300 tonnes ha^{-1} after a century and a half (Figure 8.22). Total nitrogen showed similar patterns but reached an asymptote half a century sooner.

For more on work at this classic site you can read Crocker and Major (1955) and Kershaw and Looney (1985).

Succession in Peat Bogs

In wet locations, the accumulation of partially decayed plant debris, particularly *Sphagnum* mosses, can eventually produce deep accumulations of peat (Figure 8.23). Over time, as the peat deepens, rooted plants lose contact with underlying mineral soil. The landscape becomes dominated by ericaceous shrubs or even coniferous trees rooted in peat (Figure 8.24). Eventually, peat accumulates to such a depth that the vegetation is little affected by the underlying topography and instead becomes largely controlled by climate (Foster and Glaser 1986). The climax stage can be an ombrotrophic bog – one in which the peat is so deep that rainfall provides the only source of nutrients.

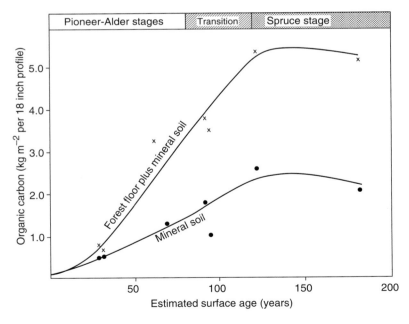

FIGURE 8.20 The accumulation of organic carbon in the mineral soil and on the forest floor under *Alnus crispa* and *Picea* in the chronosequence at Glacier Bay, Alaska. (From Crocker and Major 1955)

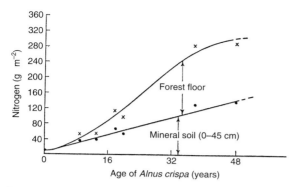

FIGURE 8.21 The accumulation of nitrogen in the mineral soil and on the forest floor under *Alnus crispa*. (From Crocker and Major 1955)

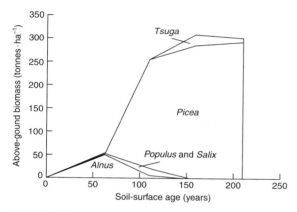

FIGURE 8.22 Distribution of above-ground biomass by species in the chronosequence at Glacier Bay, Alaska. (From Bormann and Sidle 1990)

This general process of peat bog succession has been understood and described for more than a century (Gorham 1953, 1957; Gore 1983; Keddy 2010). Many cores have been taken from bogs. Such cores provide information on plant composition from both pollen and partially decayed plant remains, along with larger pieces of wood that can be radiocarbon-dated. Since peat accumulation is relatively slow, bog succession takes much longer than forest succession. But compared to other geological processes it is

relatively rapid: many northern areas now covered with peat were uncovered from glacial ice as recently as 10,000 years ago.

Forest Succession After Fire

We looked at effects of fire on conifer forests in North America in Chapter 5. Let us consider another example from the opposite side of the world. It is significant for at least four reasons: it comes from an isolated fragment of Gondwana, the island of Tasmania, which is also an area with globally important forests. It is an example of succession in a rain forest. It includes interesting trees such as southern beech, *Nothofagus*, and the world's tallest angiosperm, *Eucalyptus regnans* (Box 9.1). And it illustrates the effects of fire, which we previously considered in the rather simpler forests of central North America (Section 5.3.1).

The dominant trees of these rain forests are *Nothofagus cunninghamii* and *Atherosperma moschatum*. After an intense fire, the normal sequence of succession progresses through at least three seral stages: from (1) sedges and shrubs, through (2) a wet sclerophyll forest dominated by species of *Eucalyptus* and *Acacia* to (3) a mixed forest with *Eucalyptus* towering over a rainforest understorey (Figure 8.25). If no fire occurs, the *Eucalyptus* cannot regenerate, and so they are ultimately lost from the community, leaving (4) *Atherosperma* and *Nothofagus*. (This is rather similar to the situation in Great Lakes area (Section 5.3.1), where pine trees from earlier successional stages can survive, towering over deciduous trees, but without further disturbance, they eventually disappear from the climax vegetation.)

The sequence of events after a fire depends upon the life history traits of the trees and local conditions. Consider just four species of woody plant.

Acacia dealbata reaches reproductive maturity early and has a short life span of some 70 years. It then survives only as seeds, which remain viable in the soil for about 400 years. It thus appears early in succession and depends upon recurring disturbance.

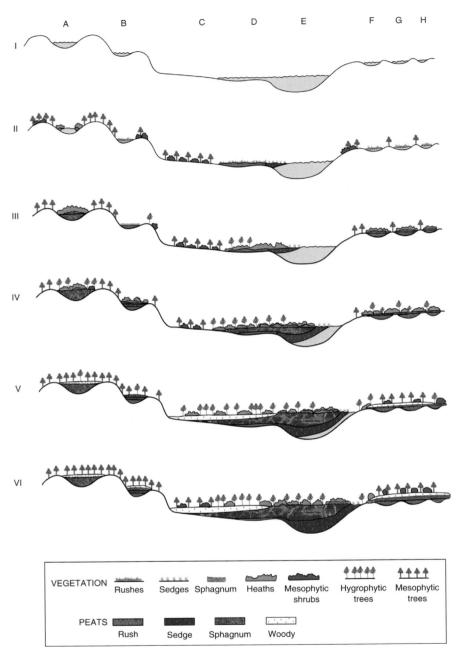

FIGURE 8.23 The development over time of peatlands in landscapes on the Precambrian shield, showing both the events in closed drainage (A, B, F, G, H) and situations where there is more seepage or water movement (C, D, E). (From Dansereau and Segadas-Vianna 1952)

FIGURE 8.24
A peatland in Algonguin
Park, Ontario. Peat bog
succession occurs in wet
locations, where rates of
decomposition are slow
and peat accumulates.
Sphagnum mosses are
particularly common in
such locations. As
succession continues, the
peat slowly fills in wet
depressions and may
become deep enough to
support conifer forest.
(Photograph by Paul
Keddy)

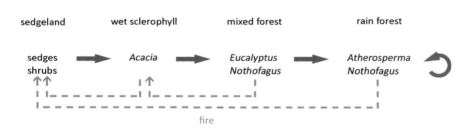

FIGURE 8.25 A model
for succession in the
forests of Tasmania,
greatly simplified from
Noble and Slatyer (1980).
The green arrows show
progressive seral stages,
while the red arrows
show the effects of fire.

Eucalyptus regnans establishes rapidly after fire, but may persist as adults for some 400 years. It produces abundant seeds, but there is no seed storage in the soil. Thus the seed source disappears when the adults disappear. If the adults are entirely removed, long-term dispersal is necessary for the adults to re-establish.

Atherosperma moschatum is tolerant of shading. This small evergreen tree is a typical climax species that can regenerate in its own shade.

Nothofagus cunninghamii is also tolerant of shading, and may establish somewhat sooner forming a *Eucalyptus/Nothofagus* seral stage.

Figure 8.25 shows a very simple version of what happens after a fire occurs in a mixed species stand. At the far left, all species are assumed to be present as juveniles. A transition occurs after 10 years when *Acacia* and *Eucalyptus* reach reproductive maturity and can produce seeds. After 30 years the rainforest species also reach maturity. Reproduction of *Acacia* ceases; after about 70 years living plants will have senesced and died. The forest will now be a mixed forest with *Eucalyptus* emergent above a rainforest understorey. After 400 years the *Eucalyptus* is lost. If the rain forest is undisturbed for a further 100 years the result is pure *Atherosperma moschatum* rain forest or a mixed *Atherosperma/Nothofagus* forest. The dashed arrows show the effects of fire on each stage in the replacement sequence.

The general story is this: the climax vegetation is evergreen rain forest and the existence of rain forest largely depends upon locations that are protected from fire (Bowman 2000). Across much of Australia, *Eucalyptus* forests and interspersed sedgelands or grasslands are a consequence of recurring fire. There are many other nuances of natural history, which you can further explore in Nobel and Slatyer (1980) and Bowman (2000). Of course, there are exceptions. Ridges may retain sedgeland owing to the extremely infertile soils. Floodplains may have rain forest owing to lower fire frequencies in wet habitats. The different forest types of this part of Tasmania (Figure 8.25) are therefore controlled in part by fire frequency, but other issues include soil fertility, fire intensity and the life history characteristics of each plant species. This example illustrates why theories of succession can be controversial: while fire undoubtedly is a key factor in creating recurring disturbance, and while certain climax species can reproduce in their own shade, there will always be distinctive environments and distinctive species that add complexity.

Sand Dune Succession

Another classic sequence of vegetation types in primary succession occurs on sand dunes. The sequence begins with open sand near the water, has a variety of grass-dominated seral stages and eventually ends as forest. Many common examples come from coastal areas of the Northern Hemisphere, where it typically takes several thousand years for forest to establish. You will no doubt have come across this typical sequence in an introductory text book already. Here, I want to show you a much longer sequence of events, and from a warmer climate. We will look at a large sand dune system on the subtropical coast of Queensland in extreme eastern Australia (Walker et al. 1981). Six of the dunes that form this system depict a successional sequence that extends through the late Quaternary Period. The younger dunes, as you might expect, show a typical progressive sequence ending in forest with trees including *Banksia serrata*, *Eucalyptus intermedia* and *Angophora woodsiana* (Figure 8.26a). Also as expected, biomass increases with succession from open sand to closed forest.

But there is more. Eventually the forest begins to degrade again. There are smaller species of trees, and these are replaced by shrubs and sedges (Figure 8.26b).

(a)   (b)

FIGURE 8.26   The late stages of forest succession on dunes. (a) Mature forest established on young sand dunes ca. 10,000 to 20,000 years old. (b) Degraded forest on ancient sand dunes ca. 200,000 to 400,000 years old. (Dune system at Cooloola, Queensland, photograph by Bob Peet)

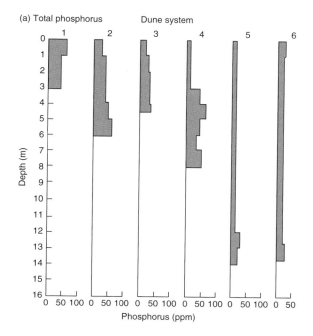

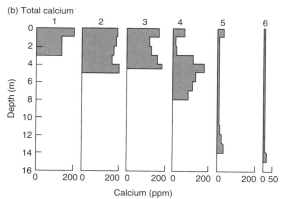

FIGURE 8.27 Two soil nutrients in six dune systems (1 youngest to 6 oldest) at various depths. (a) Total soil phosphorus (ppm); (b) total soil calcium (ppm). (From Walker et al.1981)

This change is apparently caused by rainfall leaching nutrients such as phosphorus and calcium out of the upper layers of soil (Figure 8.27). It therefore appears that along this dune sequence there is at first a succession to forest cover, but then a regression to an infertile open shrubland. Thus the younger dunes show a familiar progressive sequence, while the older dunes show a regressive sequence (Walker et al. 1981). The term regressive is slightly misleading, however, because the vegetation on the oldest dunes is not the same as the vegetation on the younger dunes. Biomass and soil nutrients may show regression, but the species composition continues unidirectional change to sedges (*Caustis recurvata, Coleocarya gracilis*) and shrubs (*Acacia ulicifolia, Aotus lanigera*) that occur only on the oldest dunes.

The degree to which this model applies to other tropical systems and other infertile soils remains to be seen. Similar patterns likely occur elsewhere where sandy soils and high rainfall coincide. Near the Gulf Coast of Veracruz, Mexico, there is an extensive system of Pleistocene-age dunes covered by low semideciduous forest; rainfall exceeds 1,300 mm, most of which falls between June and September. Measurements, by Kellman and Roulet (1990), show high infiltration rates and leaching of soil nutrients: "The successional pattern that emerges from these data is of an early phase of succession in which biomass increases, and increased quantities of nutrients are cycled, but during which absolute loss of nutrients increases even more rapidly. This is followed by a much longer phase in which soil mineralogical changes are induced by weathering, and result in a less leaky system" (pp. 673–674). These later successional forests cycle large quantities of nutrients and have a much lower output of nutrients. At this stage of succession, fine roots are concentrated in the upper 20 cm of the soil, presumably to enhance the recapture of nutrients from litter.

There are multiple reasons for reflecting on the example from the Cooloola dunes.

1. It illustrates the importance of time scales, that a vegetation sequence occurring over tens of thousands of years (progressive) may be replaced by a different sequence over hundreds of thousands of years (regressive).

2. It is a reminder of how many examples in ecology come from recently glaciated landscapes in the Northern Hemisphere where the soil is still developing after the last ice age.

3. It is a reminder of the struggle plants have to gather soil resources, and to replace nutrients lost in fallen leaves (Chapter 3).
4. It is a reminder of how logging might lead to losses of nutrients (Figure 5.34), and cause long-term changes in forest composition and biomass.

8.4.3 Predictive Models for Plant Succession

Matrix Models

If we know the seral stages in a succession, and if we know the rate at which vegetation moves from one stage to the next, we should be able to put this information into a matrix for a simple predictive model (Usher 1992). Imagine a landscape with n different vegetation types. Further imagine that information is available from two time periods, perhaps from aerial photographs or even old maps. It is then possible to assign a probability for each vegetation type turning into another vegetation type. Each such transition can be represented as an element in this $n \times n$ matrix. Using the very oversimplified model of Tasmanian rain forest the matrix would look like Figure 8.28. Note that even this very simply model has three important parts.

1. The upper right, in green, shows progressive succession, with the probability that a vegetation type

From state	To state			
	sedgeland	sclerophyll	mixed forest	rain forest
sedgeland	p_{11}	p_{12}	p_{13}	p_{14}
sclerophyll	p_{21}	p_{22}	p_{23}	p_{24}
mixed forest	p_{31}	p_{32}	p_{33}	p_{34}
rain forest	p_{41}	p_{42}	p_{43}	p_{44}

FIGURE 8.28 Succession can be represented by a matrix, where each element (p_{ii}) gives the probability of transition from one seral stage to another in a specified period of time. This is a mathematical representation of Figure 8.25. The green elements are for succession, the diagonal is vegetation that does not change and the red elements indicate regression to an earlier seral stage due to fire.

will advance to the next seral stage in the specified time.
2. The lower left, in red, shows disturbance, the probability that a fire will return the vegetation to an earlier seral stage in the specified period of time.
3. The diagonal, in black, shows the probability that a vegetation type will remain unchanged in the specified period of time.

Up to this point the matrix describes the changes observed in one time period. If one is prepared to make the assumption that future changes in an area will be similar to those that occurred in the past (and obviously this assumption requires careful evaluation), then one can assume that the probabilities of the vegetation changes represent a stationary Markov chain, in which case two predictive capacities arise (Waggoner and Stephens 1970). First, multiplying the matrix by itself will predict the vegetation components in the next time period. Second, it is possible to evaluate the steady state, when further transitions will result in a constant proportion of vegetation types in each of the n classes. This is a great deal of information on forest change from such a simple data set.

Consider another example, this time from a peatland in Estonia. Aaviksoo et al. (1993) classified Estonian wetlands into 17 different vegetation types, such as sedge fen, swamp forest, reeds and bogs. Peat cores were used to estimate transition probabilities (that is the values of p_{ii}) among the 17 classes (yielding a 17×17 matrix). Each 25 cm of peat corresponded to approximately 100 to 300 years. In this case, many different vegetation transitions were documented, and no single successional sequence could be found. When projected forward, however, the model showed that the proportion of willow, reed and sedge wetlands declined, and raised bog with *Sphagnum fuscum* steadily increased as peat accumulated.

Usher (1992) explores matrix models further with some worked examples.

Individual-Based Models of Forest Succession

Instead of working with vegetation types, you can build models using mechanistic assumptions about the interactions of plants. This approach has often been used in forests to produce predictive models for landscapes with names such as FORET and JABOWA (Botkin 1977, 1990; Shugart et al. 1981; Urban and Shugart 1992; Austin et al. 1997). Such models may be particularly successful where forests have few species and where a few ecological characteristics, such as shade tolerance, are useful for predicting species interactions. Such models can also be modified to include the effects of different kinds of disturbance, such as fire or logging. Both FORET and JABOWA simulate the behaviour of temperate forests in eastern North America by simulating the growth of individual trees in 10 × 10 m plots. Trees grow by gathering resources within the plot (light is assumed to be a key resource), and the model explores competitive interactions among the trees (Botkin 1977). Tree growth on large areas of landscape is determined by dividing the area into plots 10 m on a side.

Owing to the size of these models, there is room here to explore only the basic conceptual structure. For example, a tree growing in the open collects radiant energy in proportion to its leaf area, and its growth will also be proportional to leaf area. The equation for growth can then be modified to incorporate the effects of shading by neighbours, as well as allow for different degrees of tolerance to shade. For regeneration to occur, patches must arise, their frequency being determined by mortality rates of adults. New saplings are added each year, based upon the amount of light available and relative tolerances to shade (and with temperature and soil moisture considered as well). If, for example, light levels are high, shade-intolerant species such as cherry are added. If light is very low, only shade-tolerant species such as beech are added.

Urban and Shugart (1992) will tell you more about such models, including 30 named models for forests ranging from FORET (Appalachian hardwoods) to KIAMBRAM (subtropical rain forest).

8.4.4 More on Mechanisms of Succession

It is likely that a few simple mechanisms underlie most successional trends. I wish to draw two of them to your attention.

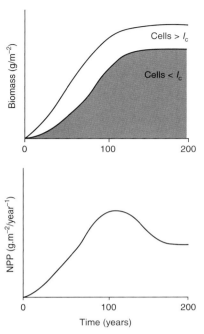

FIGURE 8.29 Biomass and net primary productivity change through succession as a direct consequence of the relative numbers of cells above and below the compensation point, I_c.

We can view succession as a very simple issue of light as a resource. Consider the progressive increase in plant biomass that occurs after a disturbance. This can be evaluated in terms of light compensation points for individual leaves or even cells (Figure 8.29). Early in succession, almost all cells are exposed to periods of intense light ($I_c \gg 0$). Biomass will accumulate and, over time, competition among plants results in larger species replacing smaller ones. This has two consequences. First, some leaves (generally those of small plants or the lower leaves of larger plants) are shaded. Second, for increasing numbers of cells in increasingly more leaves, rates of respiration then exceed rates of photosynthesis ($I_c < 0$). Further, larger plants must allocate progressively more tissues to support functions. Woody stems and branches do not photosynthesize, but they do respire. For non-photosynthetic support tissues, respiration exceeds photosynthesis, ($I_c \ll 0$). Thus as time passes the number of cells with $I_c < 0$ increases absolutely and as a proportion of the

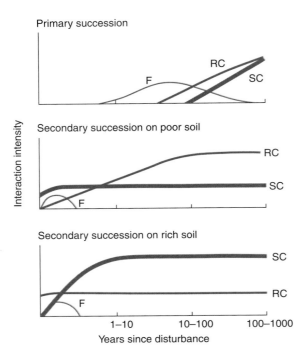

Interaction intensity

Primary succession

RC
F
SC

Secondary succession on poor soil

RC
F
SC

Secondary succession on rich soil

SC
F
RC

1–10   10–100   100–1000

Years since disturbance

FIGURE 8.30 Trends in facilitation (F), shoot competition (SC) and root competition (RC) during primary and secondary succession. (From Wilson 1999)

cells present in the forest. When respiration balances photosynthesis, accumulation of biomass ceases.

Succession is also driven by interactions between plants, mostly by **competition**, but also by positive interactions, or **facilitation**. There are three general patterns (Figure 8.30): (1) facilitation is relatively uncommon except in the most stressed sites in early primary succession; (2) competition is an important force that occurs soon after disturbance when early successional plants begin to influence one another; and (3) the intensity of competition increases through time. Therefore trends known from natural spatial gradients (Chapter 3) appear similar to those found along temporal gradients. Perhaps this should not seem too surprising since many of the same factors change, including soil organic matter and biomass. Figure 8.30 also shows three scenarios: primary succession, secondary succession on poor soil and secondary succession on rich soil. In all three cases, facilitation occurs early in succession when physical

stress is most prevalent. In all three cases, competition intensity increases through time and root and shoot competition increase simultaneously. On poor soils, however, root competition eventually becomes predominant, whereas on rich soils shoot competition may remain predominant.

8.4.5 There Are Disagreements About Succession

Succession is a perfectly useful idea that appears to describe natural processes rather well. But the idea can also be abused. And therein lies a rather long debate. I could ignore the debate, and send you on your way. In that case, you may very well make mistakes that are already a hundred years old. Instead, I could tell you more about the debate, at the risk of confusing you in an already demanding chapter. To venture further, we must review the last hundred years of writings on succession, including ideas that did not stand the test of time, and eddies of conflict that generated more heat than light and more confusion than clarity. Sometimes it is best to just let these past misunderstandings fade from memory! My suggestion is that if you are on a short course and need to learn a little bit about succession, you can safely skip this section. But if you are a professional, or wish to become one, and plan to work in zoned communities or successional landscapes, you must understand this body of work.

To start at the very beginning, there are many situations where one may observe the **zonation** of plant communities: along altitudinal gradients on mountainsides, around lakeshores, along the edges of old fields and across sequences of sand dunes. Sometimes, but not always, these sequences represent patterns of ecological succession. We will focus on shorelines, although one could follow the same arguments using any zoned vegetation. The sequence of plant communities along shorelines (Figure 8.31) appears to show a successional trend – the same sequence that would occur as a wetland gradually filled with detritus and turned into land. "Zonation, therefore, is taken to be the spatial equivalent of succession in time, even in the absence of direct evidence of change" (Hutchinson 1975: p. 497). Zonation

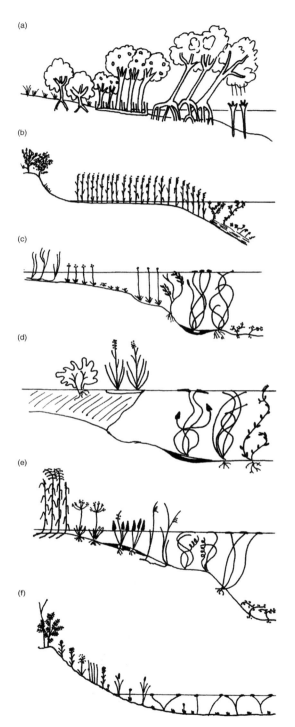

can be viewed as succession in both peatlands (e.g. Dansereau and Segadas-Vianna 1952) and small marshes along lakes (e.g. Pearsall 1920; Spence 1982). In each of these circumstances, the organic matter produced by the wetland, combined in some cases with sediment trapped by the vegetation, gradually increases the elevation of the substrate and turns shallow water into marsh and marsh into land. This successional sequence was termed a hydrosere by Tansley (1939). Better evidence for such sequential change comes from cores taken in wetlands. Using a sequence of 20 sediment cores, divided into 12 different vegetation types, Walker (1970) found that there was a temporal sequence from open water to mixed marsh (Figure 8.32).

The importance of succession as a general phenomenon in plant communities was repeatedly emphasized and systematized by Clements (e.g. Clements et al. 1929) and the terms such as sere and climax are essentially his. In his treatise, Clements (1916) recognized a number of mechanisms and processes involved in succession. Succession depends upon the kind of disturbance that initiated it (nudation), the propagules that remained in the soil (residuals), the arrival of migrants at the site (migration), success of the migrants at establishment and growth (ecesis) and how the migrant plants altered their biotic environment (reaction). As the species grow, interactions become important (competition). These successional processes continue until an equilibrium is reached between the species on the site and their environment.

But staying with the topic of shoreline zonation, while it might be true that zones in small pools, or peatlands, represent an ordered sequence through time, it seems rather unrealistic to say that zoned communities on large water bodies illustrate succession! Neither Lake Superior nor the Baltic Sea are likely to fill with organic matter in the near future. The over-use of the

FIGURE 8.31 The zonation of vegetation on shorelines has been described in many studies. (a) Mangrove swamp of the Caribbean (after Bacon 1978); (b) eastern shore of Lake Kisajno, northeastern Poland (after Bernatowicz and Zachwieja 1966); (c) sandy shoreline (after Dansereau 1959); (d) bog (after Dansereau 1959); (e) St. Lawrence River (after Dansereau 1959); (f) Wilson's Lake, Nova Scotia (after Wisheu and Keddy 1989b). But when does zonation show succession as opposed to simple flood tolerance?

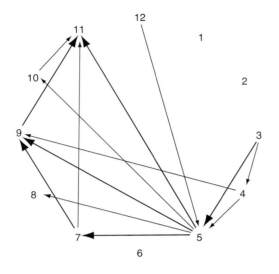

SUCCEEDING VEGETATION

	1	2	3	4	5	6	7	8	9	10	11	12	T
1	·	·	·	·	·	·	·	·	·	·	·	·	0
2	·	·	·	1	1	·	·	·	·	·	·	·	2
3	·	·	·	2	5	·	·	·	·	·	·	·	7
4	·	·	1	·	3	·	1	·	2	1	·	·	8
5	·	·	·	·	·	1	4	3	5	2	4	·	19
6	·	·	·	·	·	·	1	·	·	·	·	·	1
7	·	·	·	·	·	·	·	·	4	1	3	·	8
8	·	·	·	·	·	·	1	·	1	1	·	·	3
9	·	·	·	·	1	·	·	1	·	1	5	·	8
10	·	·	·	·	·	·	·	·	·	·	4	·	4
11	·	·	·	·	·	·	·	·	2	1	·	·	3
12	·	·	·	·	2	·	1	·	·	·	·	·	3
T	0	0	1	3	12	1	8	4	14	7	16	0	66

(Left axis label: ANTECEDENT VEGETATION)

FIGURE 8.32 Transitions over time among 12 stages of wetland vegetation in 20 pollen cores from a range of wetlands in the British Isles. The vegetation seres include open water (1) to reeds (5) to bog (11) to mixed marsh (12). Top, transition diagram; bottom, tabulated frequencies. (From Walker 1970)

concept of succession has led to a certain amount of opposition to the entire idea.

Gleason was a well-known opponent, an early reductionist who felt that the emphasis should be upon the individual responses of plants to their environments. In 1917, Gleason laid out 28 numbered statements to summarize his individualistic views. Portions of three of them are offered below:

> The development and maintenance of vegetation is therefore merely the resultant of the development and maintenance of the component individuals, and is favored, modified, retarded, or inhibited by all causes which influence the component plants. [1]
>
> The actual mature immigrant population of an area is therefore controlled by two sets of factors: the nature of the surrounding population. . . and the environment, selecting the adapted species. [9]
>
> The common cause of succession is an effective change in the environment. [25]

A huge amount has since been said and written about "Clements versus Gleason" and what this means for our understanding of succession in general and plant communities a whole, a topic to which we will return in Section 11.5.1. Population biologists, who view nature as a collection of populations (as we will do here in the next chapter) have tended to embrace the Gleasonian view (Harper 1977). But trying to create this division in ecology is somewhat misleading. Take those three points above. It is hard to imagine that Clements would seriously disagree with any of them. The debate, if there is one, is really about subtle differences in emphasis. Students of plant ecology are at some risk of having to read great volumes of words about succession and of being misled by many "modern" opponents of succession, who would have one believe that they have something new and clever to say on the matter (when in fact they have not read the original work, a point made independently by both Jackson (1981) and Booth and Larson (1999)). McIntosh (1985) has outlined Clements' views as well as those of his supposed opponents, and he observes that: "Clements's ideas, like those of prophets generally, have been widely used and misused by his adherents and detractors; and careful re-reading of his major accounts of succession. . . is often required to distinguish his actual ideas from what they are said to be" (McIntosh 1981: p. 10). One might therefore be best advised to read Clements (1916), Gleason (1917) and then (with a great deal of scepticism) a modern synthesis (like this book, or Glenn-Lewin et al. 1992).

Here is my synthesis. Succession may be ubiquitous. It is closely related to zonation in situations such as peatlands, where organic matter accumulates, but even early writers understood that the generalization did not apply to all sites. Moreover, modern ecologists tend to place greater emphasis on the many natural disturbances that delay, or even restart, proposed successional sequences. As the effects of fires, floods, storms and droughts were better documented and understood, many successional sequences could perhaps be more accurately understood as dynamic balances between succession and disturbance (e.g. Pickett and White 1985). Simultaneously, population biologists began placing increasing emphasis on the mechanistic interactions among species, leading Horn (1976, 1981) to suggest that succession was best understood as a "statistical result of a plant-by-plant replacement process."

Along with a gradual shift in emphasis toward the study of the dynamics of individuals in patches within landscapes, the ubiquity of buried seed reserves became evident and their importance a subject of investigation. It became apparent that buried seeds were more than mere detrital accumulations; seed reserves held the potential for community regeneration, and in many cases disturbance triggered the re-emergence of species from their reserves of buried seed (e.g. Harper 1977; Leck et al. 1989). Charles Darwin himself had commented on the remarkable number of seedlings that emerged from a spoonful of mud and, increasingly, soils have been found to be vast repositories of buried seeds (e.g. Salisbury1970; van der Valk and Davis 1976, 1978; Leck et al. 1989). Thus many zonation patterns are not successional sequences so much as short-term responses of plant communities to local changes in the environment.

In just 50 words, you ask? Succession is a vital concept for understanding changes in vegetation over short periods of time. It should be in every ecologist's tool box. But it becomes a problem if you have only one tool. Recall the old warning: to a person with a hammer, every problem looks like a nail.

CONCLUSION

Each and every piece of vegetation you visit is the product of many overlapping time scales.

First, the evolutionary time scale determines which plants are available to form the pool of species for an ecological region. Even a simple piece of remnant forest on a campus in the United States may have ancient gymnosperms mixed with more recently evolved angiosperms. Within the angiosperms, a single mountain-side may support ancient angiosperm groups such as magnolias with nearby clearings having recently evolved grasses and asters. And if you are exploring a forest that was once part of Gondwana, you will encounter a very different group of species than if you were on part of Laurasia. These past events, then, spill forward into the future, determining the raw material or species pool from which plant communities are formed.

Second, the last ice age has had a major impact on the world's landscape and vegetation. Was your home under the last big ice sheet? If you live in a glaciated area, you can still see the effects in lakes, eskers, drumlins, clay plains, peat bogs and other physiographic features that control modern vegetation. If you live in an unglaciated area, the effects will have been different: you may see salty plains that were once pluvial lakes, old beaches that are now dry ridges, deep deposits of loess with steep-sided valleys or deltas that were formed by rivers once laden with glacial silt. You should know about which of these occur in each landscape you visit, since they provide the template for plant vegetation that is now present.

Third, natural disturbance followed by vegetation recovery produces recognizable patterns and sequential changes in vegetation. In many cases you are encountering vegetation types that are not stable, but merely in transition from one phase to another. It is quite possible to walk into a piece of young forest and make reasonable predictions about how the forest will appear in one or two centuries. In many landscapes it is important to maintain biodiversity, and this means maintaining the full array of seral stages in the landscape. This in turn challenges you to maintain the disturbances that create those seral stages. You may require simple models for vegetation dynamics such as those introduced in Section 8.4.3.

These three time scales should be constantly in the back of your mind, whatever teaching or research or management (or travel) you do. In the next chapter we will continue with time, but at mostly shorter time scales still: the length of time it takes an individual plant to grow to maturity and reproduce.

Review Questions

1. Where is the nearest location to you that has fossil plants? What age are the fossils? Describe the general type of plant community represented by these fossils.
2. What is a gametophyte? Describe the life cycle of *Rhynia* and then explain how this very same life cycle is still present in flowering plants.
3. How did glaciation over the past 100,000 years affect your region of the world? Make the list as long as possible. If you live near an ocean, where was the shoreline during the last glacial maximum?
4. Explain how the study of pollen grains in lake sediment allows plant ecologists to reconstruct the history of plant communities, including the frequency of fire. Find an example from your ecoregion.
5. What is meant by the term succession? Give an example from this book. Give an example from your ecoregion. What is meant by the term "climax vegetation"?
6. Will forest succession after logging be the same as succession after fire? Explain why, or why not. (There is more on this topic in Chapter 5.)
7. Can plants lower sea levels? Explain the linkages that make this possible. (Consult Lackner, K. S. 2003. A guide to CO_2 sequestration *Science* **300**: 1677–1678 for more information.)

Further Reading

Clements, F. E. 1936. Nature and structure of climax. *Journal of Ecology* 24: 254–282.

Dansereau, P. and F. Segadas-Vianna. 1952. Ecological study of the peat bogs of eastern North America. *Canadian Journal of Botany* 30: 490–520.

Walker, D. 1970. Direction and rate in some British post-glacial hydroseres. pp. 117–139. In D. Walker and R. G. West (eds.) *Studies in the Vegetational History of the British Isles*. Cambridge: Cambridge University Press.

Horn, H. 1976. Succession. pp. 187–204. In R. M. May (ed.) *Theoretical Ecology: Principles and Applications.* Philadelphia: W. B. Saunders.

Connell, J. H. and R. O. Slatyer. 1977. Mechanisms of succession in natural communities and their role in community stability and organization. *The American Naturalist* **111**: 1119–1144.

West, D. C., H. H. Shugart, and D. B. Botkin (eds.) 1981. *Forest Succession: Concepts and Application.* New York: Springer-Verlag.

Delcourt, H. R. and P. A. Delcourt. 1991. *Quaternary Ecology: A Paleoecological Perspective.* London: Chapman and Hall.

Pielou, E. C. 1991. *After the Ice Age: The Return of Life to Glaciated North America.* Chicago: University of Chicago Press.

del Moral, R. and L. C. Bliss. 1993. Mechanisms of primary succession: insights resulting from the eruption of Mount St. Helens. *Advances in Ecological Research* **24**: 1–66.

Petit, J. R., J. Jouzel, D. Raynaud et al. 1999. Climate and atmospheric history of the past 420,000 years from the Vostok ice core, Antarctica. *Nature* **399**: 429–436.

Wilson, S. D. 1999. Plant interactions during secondary succession. pp. 629–650. In L. R. Walker (ed.) *Ecosystems of Disturbed Ground.* Amsterdam: Elsevier.

Laws, B. 2010. *Fifty Plants that Changed the Course of History.* Buffalo, NY: Firefly.

9 Populations

Plant populations. Seed production. The fate of seeds. Competition among seedlings. Seed size. Clonal integration. *Panax quinquefolius* and deer grazing. *Cakile edentula* and wind transport. Plant life spans. Age-structured populations.

FIGURE 9.1 Most plants produce large numbers of seeds, as illustrated by this fruit-laden date palm (*Phoenix dactylifera*). Likely native to Mesopotamia, this palm is now widely cultivated in the Middle East. In the wild state, it would be wind pollinated and animal dispersed. (Shutterstock)

9.1 Introduction: Working With Single Species

It is sometimes useful to focus on a single species of plant in isolation from the many other species around it. That is, one can treat this species as a population of individual plants. From this perspective, the abundance of a particular plant in nature is a function of its population dynamics. This requires studying factors such as the rate of production of new offspring (Figure 9.1), the rate of mortality of individuals with age, as well as immigration or emigration through the movement of seeds.

The principal advantage to this approach is that it focuses our thinking on individual species and their change in abundance through time. The principal disadvantage is the inherent difficulty in deciding what comprises an individual plant. In some cases, say annual plants in a freshly tilled field, or elm trees in a forest, this is relatively easy. But in many other cases it is not, since most plants have a wide array of styles of asexual reproduction and many apparent individuals are in fact linked underground. Thus a single shoot of an ostrich fern may be part of a vast interconnected network of rhizomes, while a hundred-year-old poplar tree may be little more than a short-lived shoot of a large multistemmed tree that has occupied a valley for millennia.

The remarkable capacity for vegetative reproduction is one of the noteworthy ways in which plants differ from most kinds of animals. In animals (with certain exceptions, such as corals) individuals are obvious. In plants (with certain exceptions such as annual plants and single-stemmed trees) the individual is as much an exception as the rule. Nonetheless, the advantages of focusing upon individuals and populations sometimes outweigh any costs. The population approach may be most appropriate when (1) discrete individuals are obvious, or (2) when a species is rare and must be studied for purposes of conservation, or (3) when a single species is of enough commercial value to require demographic information. Even those interested primarily in causal factors

may find it convenient to look at a habitat from the perspective of a single species. The processes in a tract of rain forest, for example, may be very different depending upon whether one looks at populations of rainforest trees, or populations of epiphytes on their branches or populations of herbs growing in their shade.

Looking at populations also is one way of reminding oneself that vegetation is not static in time. "Plant ecologists," say Harper and White (1974: p. 419), "have concentrated on the physiology, structure and taxonomy of vegetation, while largely ignoring the population phenomena that constitute the underlying flux." Harper (1967) advocated that plant ecologists should adopt the perspective of zoologists, and focus upon the quantitative study of single populations of species of plants. This approach, sometimes called the Darwinian approach (Harper 1967) or plant demography (Harper and White 1974), explores how life history characteristics such as birth and death rates interact to influence the abundance of plants in specific habitats. Each population of plants exists, after all, only so long as its rate of reproduction exceeds, in the long run, its rate of mortality. Population ecologists study how characteristics such as rates of flowering, seed production, seed dispersal, germination and survival vary among habitats (Harper 1977; Silvertown and Charlesworth 2001).

In principle one could begin at any life history stage, but it is traditional to begin with studies of seed production and dispersal. Plants can produce vast numbers of seeds (Figure 9.1, Table 9.1), of which only a small fraction will survive to become new adults. The fate of seeds and seedlings has therefore attracted considerable interest. Questions might include: how far do seeds move from adults, what eats the seeds and does the probability of establishment differ among habitats? This is how we shall begin. First, we shall look at some factors that

Table 9.1 Annual seed production per tree for 15 selected species of trees. (Adapted from Table 1 in Greene and Johnson 1994; *Betula papyrifera* shoot by C.E. Faxon in Sargent, C.S. 1902 *The Silva of North America*, Peter H. Raven Library/Missouri Botanical Garden)

Tree species	Seeds yr^{-1}
Betula papyrifera	27,239
Liriodendron tulipfera	13,509
Betula alleghaniensis	12,158
Picea glauca	7,202
Abies concolor	5,196
Picea rubens	5,051
Picea engelmanni	3,671
Pinus ponderosa	2,667
Fagus grandifolia	1,892
Acer saccharum	1,751
Pinus lambertiana	1,159
Quercus coccinea	525
Quercus prinus	107
Quercus velutiana	93
Carya glabra	40

Betula papyrifera

affect seed production and seedling establishment. Next, we shall look at the range of variation in charactersitics such as seed size. Then we shall look at clonal growth. Then we shall look at two examples: the population ecology of a rare deciduous forest plant and the population ecology of a common sand dune species. Finally, we shall look at plant life spans and reproduction.

9.2 Population Models and Exponential Growth

The most basic question in population biology is "How many are there?" Many years ago when I worked at the front desk of the Algonquin Park Museum, the most common question was "How many bears are in the park?" Of course, the point is that we did not know. It is no small challenge to count all the bears in a park. In the same way, it would be no small matter to count all the pine trees or white trilliums. Only in the case of plants with very small and restricted populations are we ever likely to know precisely how many there are – assuming it is a species to which the concept of the individual applies.

The second basic question in population ecology is "How is the population changing with time?" This assumes that we have carried out a census not only once, but enough times to plot a trend with time. Again, from the front desk of the Algonquin Park Museum, "Why are there fewer deer than I saw here last time?"

One of the most basic population equations involving time is simply

$$N_1 = N_0 + \text{births} - \text{deaths} + \text{immigration} - \text{emigration} \qquad (9.1)$$

where N_0 is the number of individuals at the first census, and N_1 is the number at the second. It is usually convenient to assume that a census is carried out yearly. In many cases, one can assume that immigration and emigration are minimal or, barring that, that at least they are approximately equal. If one is counting the plants in a small patch of forest surrounded by similar forest, such equality is likely. If one is counting the number of plants on an isolated island, it is likely that emigration will exceed immigration, dependent upon the number of offspring that disperse into the surrounding water. In some habitats, like populations of plants growing on cliffs or branches, the rate of emigration will be exceptionally high and most offspring will be lost entirely. This may explain why so many plants in such habitats also have methods of vegetative propagation.

Another way to look at population growth is to start with the number of individuals one has, N_0, and use estimates of rates of reproduction to attempt to forecast future populations. Often it is feasible to count, or at least estimate, the number of seeds a mature plant will produce by counting the capsules or cones on the plant, and subsampling to find out how many seeds each contains. A single giant sequoia tree can produce 4.5 million seeds in a single year, and even the common white birch can produce more than 25 thousand (Table 9.1). In such cases, the population growth model is simple. The number in the future will be the number we have now, multiplied by the number of offspring they produce, r. The inherent growth rate is then

$$N_1 = rN_0 \qquad (9.2)$$

If nothing changes, the next year, N_2, will be rN_1 or r^2N_0. By this logic, the number of individuals will increase each year as r^tN_0 where t is the amount of time elapsed.

Of course, we also have to allow for some rate of death in the population. We can modify r to reflect not just births (seeds), but the difference between observed births and deaths, that is, $r = b - d$, where r is the production of offspring and d the number of individuals that die in a given period. The death rate may vary from high (in the case of annual plants) to very low (in the case of canopy trees). It will also vary with the habitat in which we measure b and d. Generally speaking, even when we add in the effect of deaths of mature individuals, most populations have an inherent capacity to increase exponentially (that is they have a large inherent r). This will generate the powerful equation for exponential growth, where the number of individuals at time t is simply

$$N_t = N_0 e^{rt}. \qquad (9.3)$$

Death rates themselves will vary with the age of the individual (seeds, seedlings or adults) and with habitat. Such considerations are explored in Harper (1977) and Silvertown and Charlesworth (2001). In general, rates of mortality are very high, particularly for seeds and seedlings.

This is a powerful and important equation. It is so simple that it is often overlooked. But this is the engine for the vast populations of plants that cover much of the Earth and also for the overpopulation of Earth by humans. It also provides the raw material for evolution, since it guarantees that vast numbers of individuals must die in the struggle for existence. Indeed both Malthus and Darwin were greatly impressed by this equation and its implications. In Kingsland's (1995) words, it is "the hidden power of nature." As Darwin calculated, an annual plant producing only two seeds could produce a million plants in 20 years. And a single pair of elephants would have at least 15 million descendants after five centuries. It is generally true that if you take almost any organism, including elephants, and allow exponential growth, the Earth will soon be covered with them. The fact that that Earth is not covered by elephants or, for that matter, saguaro cacti or sausage trees, raises the more perplexing questions of population ecology. For example, why are there not more of certain species, given the vast number of seeds some can produce? Generally it comes down to factors that increase death rates – consumption of seeds by herbivores, death of seedlings from drought, or death of seedlings from shading by adults are three common ones. And we have already seen examples of such mortality in the chapters on competition and herbivores. If the size of a population is found to be remaining more or less constant, it indicates that somehow an enormously high death rate must be balancing the inherent potential for exponential growth.

If the population is declining, as in the case of the growing number of threatened and endangered plant species, some causal factor must be determined. Often it is simple destruction of habitat – you can't have wild orchid populations in shopping mall parking lots.

Other times it is increased mortality, such as increased rates of logging of adults, or grazing by herbivores. In other cases, lack of pollinators may be interfering with reproduction. One of the challenges of population ecology is to determine the causes of such population declines and suggest remedial measures. We shall see an example of this with wild ginseng in a later section of this chapter.

Instead, the population may be increasing, as with many introduced species. The rapidity with which a single invasive species can spread and dominate a landscape is often difficult for non-biologists to appreciate. People tend to think linearly, whereas population growth occurs geometrically. Every time a foolish person introduces a plant to a region in which it does not normally occur, there is a small but finite risk that the plant's offspring will explode by means of exponential growth to become an expensive nuisance.

9.3 How Many Seeds Will a Plant Produce?

Bigger plants will generally produce more seeds than smaller plants. To explore such patterns, Shipley and Dion (1992) collected data on 285 herbaceous plants representing 57 temperate zone species, and made measurements of characteristics including the number of seeds produced, the mean weight of seeds and the mean above-ground weight of shoots. In this collection of species, a shoot (ramet) was usually a functional individual with its own roots, leaves and flowers. The number of seeds produced by a shoot was positively correlated with the size of the shoot (Figure 9.2) suggesting, reasonably, that larger plants have more resources to expend on seed production. The number of seeds also was negatively related to the weight of a seed, as would be expected if resources were limiting seed production (Figure 9.3). Evolutionary options appear to be constrained along a continuum of either producing a few large seeds or many small seeds.

More specifically, the allometric relationship for seed number per ramet (N_s), vegetative ramet weight (W_v in grams) and individual seed weight per ramet (W_a in grams) was:

$$N_s = 1.4 \, W_v^{0.93} W_a^{-0.78}$$

That is, an increase in weight of above-ground tissues in a ramet leads to an almost proportional increase in seed production, since the allometric slope of W_v (0.93) is not significantly different from 1. In contrast, the increase in average seed weight leads to a less than proportionate decrease in seed number, since the allometric slope of W_a (–0.78) is significantly less than 1. These two factors alone accounted for 81 percent of the variance in seed number among ramets. In other words, all of the other aspects of plant reproductive systems from phylogenetic history to pollination mechanisms can together explain only a further 19 percent of the variation in seed production among plant species.

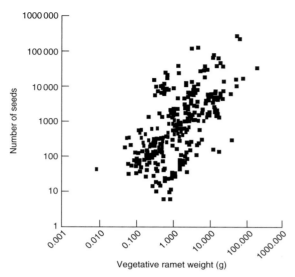

FIGURE 9.2 The number of seeds a plant produces is positively correlated with plant size ($r = 0.62$). In this example plant size is expressed as ramet (shoot) weight. The data represent 285 ramets from 57 species from habitats including wetlands, old fields and woodlands in the vicinity of Montreal, Canada. (From Shipley and Dion 1992)

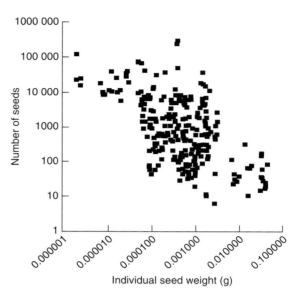

FIGURE 9.3 The number of seeds produced by a plant is negatively correlated with individual seed weight ($r = 0.60$). In this example, seed number is expressed per ramet (shoot). The data come from the same source as Figure 9.2. Note the logarithmic scale. (From Shipley and Dion 1992)

9.4 The Fate of Seeds

9.4.1 A Typical Type III Survival Curve

The quite remarkable reproductive capacity of plants is balanced by a precipitous rate of death of seeds and seedlings. If you plot the number (or proportion) alive as a function of time, the general pattern is a steep decline known as a type III survival curve described by a negative exponential curve (Figure 9.4). Such a curve is intended to show how many are alive from one initial cohort, that is, one year of reproductive output. You can, however, also think about it as some sort of overall average that you might expect to find if you were able to follow many cohorts in many years. We have no full cohort data for long-lived plants.

True, one can follow an entire generation of plants if they are short lived, say annual plants, and I have done so myself, but annuals are an exception in the plant world. Long-lived perennials and trees predominate in many landscapes. How are we to think about their populations both within and across generations?

Figure 9.4 shows a typical survival curve for a long-lived species. Let us use this as a framework for most of the rest of this chapter. The figure has no data precisely because no one has yet produced a survival curve of such dimensions. Even so, we know the shape it takes: negative exponential. It is convenient to think of it as comprising six phases, although you could also argue the need for at least one more, the

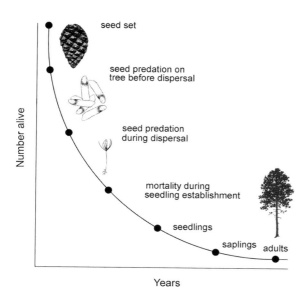

FIGURE 9.4 Most, if not all, plants have a type III survival curve in which large numbers of young die early. Only a few become adults. The main sources of mortality include seed predators, pathogenic fungi, foliage-eating herbivores and competition from established individuals, as discussed in the text. The curve is actually much steeper than shown, and the left side could even be a logarithmic scale.

survival of fertilized embryos until the seed is mature. For simplicity, let us include this in phase 1.

1. Seed mortality while still attached to the parent
2. Seed mortality during dispersal
3. Seedling mortality during establishment
4. Seedling mortality during early years of growth
5. Sapling mortality
6. Adult mortality

Seed mortality while still attached to the parent

Figure 9.1 and Table 9.1 illustrate the extreme upper left of the curve. Most plants are copious seed producers; a giant sequoia may produce more than 4 million seeds in a year. But, even here, mortality may already have occurred. Seed production may have been reduced due to early mortality from herbivores eating young seeds or fruits. As just one example, in a multi-year study of sugar maples, the percent of viable

embryos was usually much less than one half (Hett 1971). Many kinds of beetles, for example weevils, oviposit on nuts; the larvae then bore inside, feeding on the seed while protected by the tissues of the parent plant (Crowson 1981). Many arboreal mammals feed on seeds, such as squirrels gathering acorns (Section 7.5.2). It is possible, with work, to estimate the percentage of seeds containing weevils, or measure the rate at which tagged acorns still attached to a tree disappear.

Seed mortality during dispersal

Plants have relatively little control over where their young end up. Those seeds released from a giant sequoia cone drift in the wind to an unknown destination. Large numbers are dispersed to the wrong habitat, to locations where germination may not be possible. In the case of animal dispersal, a proportion may be digested by the dispersal agent, while others may be deposited in the wrong habitat – think of a banana seed carried deep into a cave by a fruit bat. In Table 9.2 below you shall see a study where thousands of acorns were place in different habitats – and in some habitats, none remained after just weeks. In Figure 9.5 you shall see that small mammals were the major cause of seed mortality for ash trees.

Seedling mortality during establishment

The production of the first root and leaves is a very vulnerable phase for most plants, since the protective tissues have been breached. Many seedlings are killed by slugs or fungi. In one study of tropical tree seedlings, mortality usually exceeded 90 percent within the first year, with pathogens the largest source of mortality (Augspurger 1984). In general, it seems likely seedlings from larger seeds have more energy reserves, and are better able to withstand adversity (Figure 9.8). At the other extreme, orchids with dust seeds (Figure 9.9), no development will occur unless the seedling encounters an appropriate type of mycorrhizal fungus.

Seedling mortality during early years of growth

Probably the main risks to small seedlings are desiccation during periods of drought, and consumption by herbivores. Growth is certainly slowed by the presence

Table 9.2 **The survival of acorns in and around oak woods in the English Pennines during the winter and spring for two years (1958–60). Small rodents were the main predators. (From Cavers 1983 after Jarvis 1964)**

Situation	Total no. acorns distributed	Sound acorns remaining after two weeks	
		No.	Percent
On surface of litter of:			
Deschampsia flexuosa (grass)	3,350	0	0
Pteridium aquilinum (bracken)	1,225	0	0
Holcus mollis (grass)	325	0	0
Calluna vulgaris (heather)	50	0	0
Oak leaves	475	4	0.8
Beneath surface litter of:			
D. flexuosa and leaves	165	42	25.5
P. aquilinum and leaves	120	11	9.2
Oak leaves	665	127	19.1

of neighbouring plants as well, which has the disadvantage of keeping the plant small and vulnerable, as you saw in the earlier chapter on competition. Tallying the causes of death is a challenge, and it is not always possible to say why each seedling died, merely that it is no longer there.

Sapling mortality

It is possible to tag saplings and follow their fate for a period of time. We have, in general, more data of this sort, since it can involve just a few years. Even so, such studies have a general problem: there is no way of being certain that the short period of study is representative. Perhaps it is decadal droughts that kill most saplings. We can say with some confidence that saplings, generally, are playing a waiting game, just staying alive until an adult near them dies, opening a gap that they can then exploit. At this phase in their life history, survival rates may be more linear than exponential (Hett 1971). In general, the small saplings in some forests may be much older than their size suggests.

Adult mortality

It is rather easier to tag adult trees in a forest and count how many are alive each year. Except for one thing. By this time, the few remaining individuals are well established and survival rates are rather high. For a student doing a four-year degree, or a professor with a thirty-year career, there can be rather little mortality to report. Hence other sources of data, such as tree rings, are sought to try to estimate mortality rates. Alas, many tropical trees do not produce rings, so one has to make do with size classes. And even in trees with well-marked rings, while one can age every tree in a stand of forest, one can only infer that these are survival curves, as many other factors can affect age distribution.

Overall, most researchers have to do the best they can looking for a phase of this long curve that is feasible to measure for a short period of time. Since there are so many studies, and so many of them are fragmentary, I suggest the best approach is to first appreciate the overall curve and its life history phases. Then, for any particular study, it is possible to see where it fits into the big picture of the life span of a species and the survival of populations and cohorts. Let us now turn to some selected examples.

9.4.2 Quantitative Studies of the Fates of Seeds

Once a plant has produced seeds, the environment begins to select which ones will survive. Given the vast number of seeds that a plant can produce, studies of their fate are difficult. One good source of data comes from the production of seeds by trees; these are also known as mast. Botanists have, in fact, rather

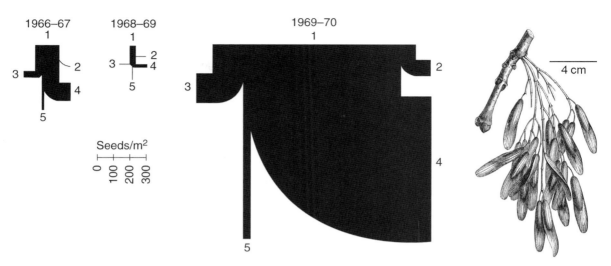

FIGURE 9.5 The fate of ash fruits (*Fraxinus excelsior*) up to germination during three different years: (1) total fruit production, (2) fruits with non-viable seeds, (3) destroyed by insects, mainly moths, (4) eaten by small mammals, and (5) germinating. (After Cavers 1983)

carefully studied the fate of mast, partly because of the economic and cultural value of woodlands in the European landscape. Seed production was studied in the Prussian forests in the late 1800s (Grubb 1977) and the British ecologist A. S. Watt is well known for his careful quantitative studies on the seeds of both oak and beech (Watt 1919, 1923). Watt's important new procedure was to increase control by placing out counted numbers of seeds in different habitats, and then censusing them on repeat visits. This enabled him to determine how long acorns survived in each situation, the value of leaf litter or burial in preventing destruction, and the different kinds of damage in different habitats (Cavers 1983). This work was extended in two studies, which are briefly examined here. In the first case, Jarvis (1964) continued to study the regeneration of oak trees, but on the Pennine hills of England. He had observed that oak seemed unable to regenerate when the ground was covered by the grass *Deschampsia flexuosa*. Jarvis confirmed Watt's finding that acorns could survive only under a cover of leaves, and no leaves would accumulate in the grazed turf of *D. flexuosa*. His sample sizes were also much larger than those of Watt. The effects of surface litter are apparent in Table 9.2. Gardner (1977) carried

out similar studies upon the reproduction of ash trees using nets to catch seeds and estimate seed production. Figure 9.5 shows the results for three years. Although the number of seeds produced varied by several orders of magnitude, a large proportion was consumed by small mammals each year. But also, in each of these years, some seeds escaped predation and survived to germinate. In the fourth year of his study, all the seeds were either damaged by moth larvae (75 percent) or eaten by small mammals (25 percent). One year, however, is a short period of time given trees that can live several centuries. For a tree population to remain stable, it is only necessary that one seed per tree of millions survives to adult size.

And now an example from tropical forest, of seedling survival in nine species of trees in Panama (Augspurger 1984). All nine species were canopy trees with wind-dispersed seeds. The study lasted one year. For the most common species (*Platypodium elegans*) 500 seedlings were tagged, while for the least common (*Cavanillesia platanifolia*) 25 were tagged. To set the scene, consider the most common species, *Platypodium elegans*. It is in the legume family, with large compound leaves, which are deciduous, being dropped during the dry season, which is also when the seeds are dispersed. The flowers are orange and produce wind-dispersed pods with one or two seeds. To return to the fate of seedlings across nine different species, most of the seedling mortality occurred within two months of germination. In seven species, more than 90 percent of the seedlings were dead within a year, with the exception of *Aspidosperma cruenta* with only 67 percent dead and *Triplaris cumingiana* with only 85 percent dead. Two more general patterns emerged. First, seedling survival was higher in plants that established at a greater distance from parents (eight of the nine species). Second, species grew better when transplanted into light gaps.

9.4.3 Dragon's Blood Trees in Deserts and Seedlings in Forests

See Figure 9.6 – it looks beautiful – but something is wrong with this scene. In arid environments many species of plants, such as this dragon's blood tree (*Dracaena cinnabari*), have difficulty in establishing offspring. We have seen a similar example with saguaro cacti (Section 7.2.1), where establishment is infrequent and dependent upon occasional wet years and nurse trees.

The dragon's blood tree is endemic to the Socotra archipelago in the Arabian Sea, just off the Horn of Africa, in Yemen. Socotra Island is noteworthy for

FIGURE 9.6   A beautiful plant (ca. 4 m), but something is still wrong with this scene. Can you see it? Think about the population perspective. Where are the seedlings? Once the adults die, this hillside will be bare. In arid environments, such as the Socotra archipelago between Somalia and Yemen, the dragon's blood tree (*Dracaena cinnabari*) has difficulty in establishing offspring. A few may establish in the right site or wet year, but once goats are allowed to graze, it is often enough to tip the balance from forest to desert. (Vladimir Melnik, Shutterstock)

having an eight percent forest cover, uncommon on the Arabian mainland (Adolt and Pavlis 2004). The tree itself is considered as "vulnerable" (IUCN 2013), which means the species is in decline although not at immediate risk of extinction.

The trees are difficult to age since they are in the monocotyledons and do not produce the rings typically found in conifers (Figure 8.1). Observations on size and branching patterns do confirm that a majority of trees are more than 200 years old (Adlot and Pavlis 2004).

Where are the young trees? There are constraints on both seed production and on seedling establishment. Regarding seed production, many plants do not flower, and only a small proportion of these produce berries. Inhabitants may strip inflorescences and berries from the trees to feed goats and livestock, particularly during periods of drought. Regarding establishment, a few seedlings may establish in the right site or wet year, but goat feeding is often enough to kill those seedlings. Most regeneration is occurring only in places inaccessible to goats, such as cliffs. It has been suggested that regeneration may be restricted to rare events, either wet years or periods of low goat density. Some hypothesize that the climate is becoming drier (IUCN 2013). If indeed the climate is becoming more arid, this could be part of the reason for a decline, but of course it also increases the significance of grazing. The hypothesis that reproduction occurs only during years of low goat populations re-emphasizes the negative role of grazing in general and goats in particular. Anyone familiar with the deforestation of the Mediterranean region – from Spain to Lebanon – under the steady pressure from goat grazing, would probably see the dragon's blood tree not as a special case of a rare tree in decline in Yemen, but as just another symptom of a widespread and recurring problem with large herds of goats in semi-arid climates (Thirgood 1981; Figure 13.2).

Whatever the precise causes are for regeneration failure of dragon's blood trees on Socotra, the situation does illustrate how sensitive tree populations are to herbivores, in the seed production stage, during dispersal and in the early establishment phase. Even that enormous giant sequoia shown in Figure 8.1 was once a small seedling, and one can argue that in the fate of such seedlings lies the future of forests.

9.4.4 More on Saguaro Seedlings

We started the book with a saguaro (Figure 1.1), and have already discussed the difficulty of establishing new seedlings in deserts. This is why nurse plants may be so important (recall Section 7.2.1).

But other factors besides nurse plants may be the real culprits in limiting populations of saguaro. Seed predators and rainfall have to be considered too.

Saguaro regeneration is also reduced by seed predators. Steenbergh and Lowe (1969) observed that a large proportion of the seed crop is consumed in June by birds while the fruit is still on the plant. Doves, thrashers, cactus wrens, woodpeckers and flickers were all conspicuous consumers. By July, ripe fruits had fallen to the ground where birds, mammals and insects consumed them. In the early summer, coyote droppings consisted almost entirely of saguaro seeds. Ants also collected seeds. (To what extent these agents disperse seeds rather than destroy them was not discussed.) After germination, seedlings are killed by weevils, cutworms and moth larvae. Beyond these effects of predators and herbivores, drought (80 percent), frost (12 percent) and erosion (8 percent) were the main causes of seedling death.

The importance of drought and predation will vary among sites and years. One must not forget that the saguaro is very long-lived. Most studies of nurse plants last for only one or a few years. An age-distribution analysis of saguaro plants in the Sonoran Desert (Turner 1990) showed that during the period 1790 to 1960 there were three surges of establishment. Not surprisingly, perhaps, these were periods of abnormal wetness. The most recent surge of establishment, detected in 1974, can be traced back to two unusually intense rainfalls, associated with Tropical Storm Norma (1970) and Hurricane Joanne (1972), rains which in some cases produced temporary lakes and berms of organic debris on receding shorelines. Similarly, the declines observed in saguaro populations during the 1950s were likely related to a quarter century of drought, during which *Larrea* and *Cercidium* cover as well as saguaro establishment declined.

It would therefore be possible to argue that the size of saguaro populations is largely a result of long-term and large-scale climatic factors – death during long droughts and brief episodes of establishment during moist periods. In this scenario, nurse plants may be a secondary phenomenon in which ecologists observe seedlings establishing next to neighbours during their brief periods of research, but where a vast majority of the recruitment actually occurs as rarely as once or twice a century.

9.5   What Determines the Size of Seeds?

There is enormous variation in seed size among plants, some of the smallest being the dust-like seeds of orchids (ca. 0.0001 mg) to the enormous 20-kg seeds of the double coconut (Figure 9.7). Earlier in this chapter we saw that there is an inherent trade-off in allocating resources to reproduction: the production of many small seeds as opposed to the production of a few large seeds. There is a growing consensus that plants with seedlings that establish under stressful conditions will allocate more energy to individual seeds to better support the early establishment phase of the seedling (Westoby et al. 1997). Figure 9.8 shows evidence consistent with this view. Nine woody plant species with an array of seed sizes were established in shade and their death rates then monitored (Grime and Jeffrey 1965). The trees with the larger seeds (e.g. *Quercus rubra*, *Castanea mollissima*) had much lower death rates than those with smaller seeds (*Betula lenta*, *B. populifolia*).

But there are many plants with very small seeds. After making an exhaustive study of seed production, Salisbury (1942) observed that several groups of plants occurring in stressed habitats have extremely small seeds (Figure 9.9) in which "the food provision

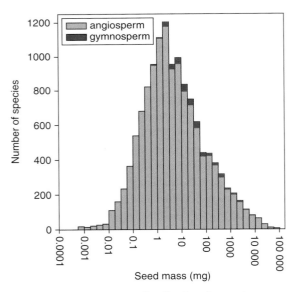

FIGURE 9.7 Frequency distribution for seed mass across 12,987 seed plant species (From Moles et al. 2005). At the left are the 0.0001 mg dust seeds of orchids. On the far right is the 20 kg fruit of the double coconut (*Lodoicea maldivica*), a palm that is endemic to only two islands in the Seychelles.

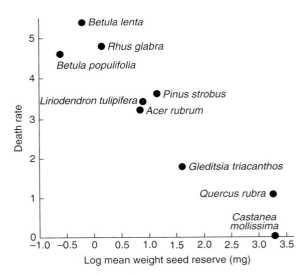

FIGURE 9.8 The relationship between death rate in shade and seed size for nine tree species. (From Grime and Jeffrey 1965)

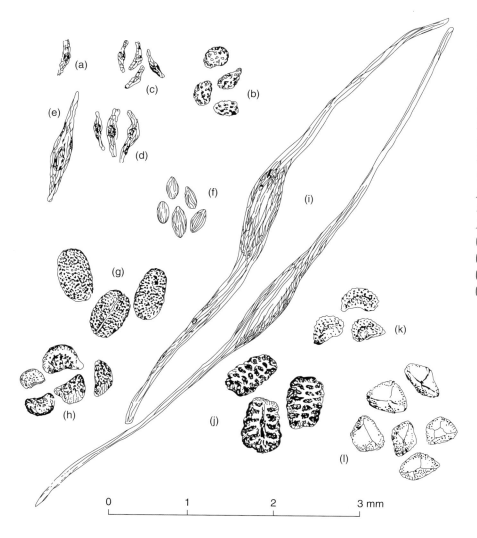

FIGURE 9.9 A selection of dust-sized seeds found in the British flora. (a) *Gymnadenia conopsea* (an orchid); (b) *Orobanche elatior* (a parasite); (c) *Pyrola secunda*; (d) *Pyrola media*; (e) *Drosera anglica*; (f) *Crassula tillaea*; (g) *Digitalis purpurea*; (h) *Polycarpon tetraphyllum*; (i) *Narthecium ossifragum*; (j) *Scrophularia nodosa*; (k) *Spergularia rubra*; (l) *Samolus valerandi*. (From Salisbury 1942)

for the germinating seedling is meagre in the extreme" (p. 5). This is particularly characteristic of the families Orchidaceae, Pyrolaceae, and Ericaceae. "It may be as an epiphyte, in the dim light of a tropical rain forest; as a terrestrial orchid, in the close turf of a chalk down; or as a pyrola in the dull illumination of a calluna heath or a pine wood" (p. 5), but the seedling is able to capture sufficient resources through mycorrhizal fungi. Recall that many plants are mycorrhizal (Section 7.3). Parasitic plants also have small seeds, and again have an external source of resources while establishing.

9.6 Clones and Genets

Plants need not reproduce sexually, at least for the short term. One of the remarkable features of plants is their capacity for clonal growth – growth without sexual reproduction. The bodies of animals reach a rather well defined size and then stop growing. Further resources are largely used for sexual activity and raising young. Most plants, in contrast, can grow nearly indefinitely. Plants have multiple growing points called meristems, and if some meristems are damaged, the others will continue growth. Plants can therefore grow to indeterminately large sizes – either in height, as in trees, or laterally, as in many herbaceous plants. Often what we call a plant – a "frond" of bracken fern, a flowering "stem" of cattail, or even a poplar "tree" – may be no more than a single shoot of a much larger and more long-lived plant. The world's largest plant (by mass) is a single clone of aspen (*Populus tremuloides*) with 47,000 stems connected below ground, and covering more than 42 ha (Grant 1993). Even in the common bracken fern (*Pteridium aquilinum*), clones may exceed 1,000 m in diameter (Parks and Werth 1993).

This requires us to introduce a new set of terms for plants: the distinction between a ramet and a genet. To paraphrase from the classic *Population Biology of Plants*, the **ramet** is the unit of clonal growth, the module that may often follow an independent existence if severed from the parent plant (Harper 1977: p. 24). Many ramets may eventually arise from what was once a single seedling. A **genet**, in contrast, is the inclusive term we can apply to all the ramets that arose from that one seedling, the word reflecting an identical, or nearly identical, genetic identity. A plant community therefore has two levels of structure: genets and ramets. Generally there are far more ramets than genets. In the case of the poplar, mentioned just above, there is one genet with 47,000 ramets. This is probably an extreme case, but there is often no way to tell. Imagine the below-ground connections in a huge cattail marsh, or a mangrove swamp or grassland (Figure 9.10). There may be a huge number of ramets belonging not only to a single species, but in fact to a single genet. In such a case the concept of the individual becomes of doubtful significance. For most practical purposes when we apply the term individuals to plants, we are referring to the ramet, which is a sort of functional

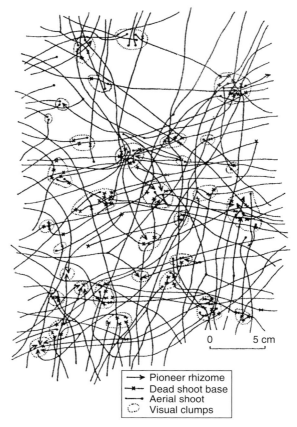

FIGURE 9.10 The arrangement of aerial shoots and rhizomes in an established stand of *Calamagrostis neglecta* in Iceland. (From Kershaw 1962).

individual, while keeping an open mind as to how many genets may be involved.

Harper (1977) does not like the common term *vegetative reproduction* at all (pp. 26–27). Fair enough. The production of clones, that is of new ramets, is more like growth than reproduction. Here, I will mostly use the term *clonal growth* to describe a plant producing nearly identical offspring of various kinds, and the term *sexual reproduction* for offspring produced via meiosis and, mostly, seeds. It is the production of genetically identical offspring, that is, clonal growth, that I want you to focus on here. There are a vast number of shapes and sizes of ramets, and hence a vast number of terms. To illustrate: there are bulblets produced on fern leaves, gemma cups produced by liverworts, runners produced by strawberries, rhizomes produced by heliconias and turions produced by water lilies. Further, adventitious buds may produce new shoots on stumps, or otherwise damaged trees. Finally, a plant may simply fall apart, with each fragment growing independently – duckweeds are an obvious example, but many other plants may fragment underground. A few species even appear to flower and set seed, but in fact bypass the sexual process and produce apomictic offspring. All of these provide a means to avoid the many costs and risks associated with sexual reproduction. Even the century plant, famous for one burst of reproduction by flowers and seeds, often puts out a few ramets just to make sure. The most obvious advantage of producing ramets rather than seeds is that the larger offspring are more likely to survive. But there is more, much more, than this simple advantage. Although much thought has been put into the advantages and disadvantages of clonal growth (e.g. Jackson et al. 1985; Harper et al. 1986), I still like the elegant, if older, models offered by Williams (1975). The names used for his models include animals, and I have not tried to rename them in strictly botanical terms, although you may find this a useful, if challenging, exercise. Here are his three models that address clonal growth, sexual reproduction and environmental conditions.

9.6.1 The Strawberry–Coral Model

Strawberries and corals both produce colonies by clonal reproduction, but can also produce large numbers of small, sexual offspring. Clonal reproduction may be advantageous within relatively desirable patches of habitat, whereas sexual reproduction may be reserved largely for dispersal among patches. When a young strawberry seedling first becomes established, it develops stolons that asexually produce new individuals. This process could continue without limit – as long as the favourable patch is very large. Yet the more the plant expands, the greater the number of

shoots that will arise ever further from the location where the plant first established as a seedling. Hence, as the plant expands, the odds increase that a shoot will encounter less favourable conditions. At some point, further allocation of resources to new shoots at the edge of the clone will not result in more offspring. At this point, sexual reproduction allows dispersal to new sites.

Occasional sexual reproduction is also advantageous when patch characteristics change as described for the previous model. No patch will survive forever. The change may be physical, such as a drought or fire that erases the patch and replaces it with a different set of conditions. The change may also be biological, since new genotypes and species are continually dispersing seeds into the patch. The bigger the clone, the more reproductive propagules it can produce, and so there may be a selective advantage to those clones that delay sexual reproduction until the local patch is filled.

Since so many plants have this combination of sexual and asexual propagation, we may think of this as being the fundamental model for plants.

9.6.2 The Elm–Oyster Model

In some cases, clonal reproduction is unlikely to be advantageous and there may be strong selection for sexual reproduction. The model name refers to those species that can be thought of as having no clonal reproduction – single individuals hold one small piece of space and flood adjoining areas with sexually produced offspring. Elms, like many other species of trees, produce vast numbers of seeds, and it should be readily apparent that one elm tree can produce vastly more seedlings than can possibly establish as adult trees. The space occupied by each adult elm tree is not unlike the single patch occupied by the clone of a strawberry. Large numbers of seedlings may colonize a patch, but owing to intense competition among them, a site will tend to be occupied by the one individual that is marginally better at exploiting resources, suppressing neighbours, and minimizing losses to pathogens and herbivores. Over evolutionary time, those trees that survived were not those most successful at making more copies of the same genotype, but rather those producing offspring with many different genotypes with differing competitive abilities.

Many trees conform rather well to this model. Other trees have varying degrees of asexual propagation, and may, in the case of species such as *Populus*, form huge clones.

9.6.3 The Aphid–Rotifer Model

The model name refers to life cycles that alternate between periods of asexual reproduction in the summer and periods of sexual reproduction toward the autumn. Clonal reproduction may be advantageous during periods with relatively stable environmental conditions, whereas sexual reproduction may be reserved for times when the environment begins to change. Future environmental conditions that will arise from the change are unknown. Sexual reproduction, which results in greater variation among offspring, increases the likelihood that some will be better adapted to surviving under the new conditions. Continued clonal reproduction would produce only

individuals adapted to the environment before the change occurred.

As Williams' name suggests, this is a common life cycle in animals where the concept of the individual is rather fixed. In plants, we may think of the many species of duckweed (including *Lemna*, *Spirodela*, *Wolffia*), while allowing that some also overwinter with vegetative turions. This model may be most applicable to plants when we consider each meristem as an individual. Annual plants, then, have long periods when each meristem produces more identical individuals (new shoots), but each autumn the meristems switch to producing flowers and then seeds.

In addition to the potential importance of clonal growth as a reproductive strategy, discussed above, it may have one other advantage in areas where water or nutrients are in very low supply. The interwoven webs of rhizomes (Figure 9.10) may permit shoots to share resources. Shoots that are located in stressful patches may be able to use resources transferred from adjoining shoots in richer patches. The branching pattern of a plant may reflect its style of foraging in an array of such patches (Bell 1984; Huber et al. 1999).

9.7   A Population Study on the Effects of Herbivores

The woodland plant American ginseng (*Panax quinquefolius*, Figure 9.11) is best described by the elm–oyster model, making it useful in population studies. The species is becoming increasingly rare from the combination of commercial collecting and overgrazing by deer. Ecologists have constructed demographic models that include rates of seed production and new plant establishment, as well as rates of adult survival for this species (Charron and Gagnon 1991; Nantel et al. 1996). Overall, populations grow and decline slowly, with the largest plants being most important for population growth rate because of their larger seed production. Population viability estimates

can be made by projecting population size into the future with a stochastic model, asking how large a population has to be to ensure a 95 percent probability of survival for 100 years (McGraw and Furedi 2005). Applying this model to their data on existing population sizes in central Appalachia, current rates of deer browsing are forecast to eliminate 29 out of 36 known populations within the century with a 99 percent probability. Even the largest population, 406 plants, has a 43 percent probability of becoming extinct before 100 years. Figure 9.12 shows population viability as a function of population size for five different levels of deer browsing.

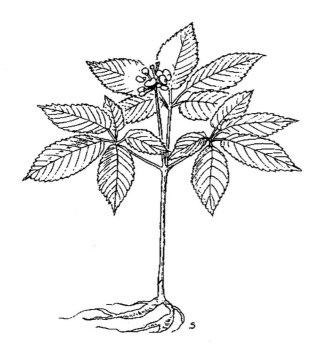

FIGURE 9.11 American ginseng (*Panax quinquefolius*) grows in deciduous forests. It is threatened by overharvesting for medicinal purposes, and deer browsing. (Illustration by Marion Seiler from Radford et al. 1968)

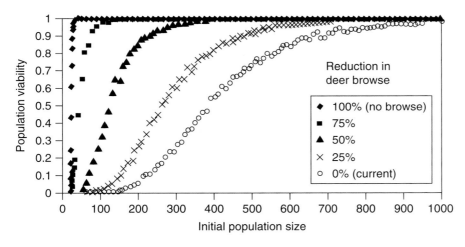

FIGURE 9.12 The population viability of American ginseng increases with initial population size and increases (that is, shifts up and to the left) if deer browsing is reduced. (From McGraw and Furedi 2005)

9.8   A Population Study on the Effects of Seed Transport Along a Gradient

Short-lived annual plants are convenient for studies of demography since they live but a single year. Consider the case of the plants in the genus *Cakile* that grow on sand beaches and shingle bars along sea coasts (Rodman 1974). On a typical sand dune system *C. edentula* plants are distributed from the open sand

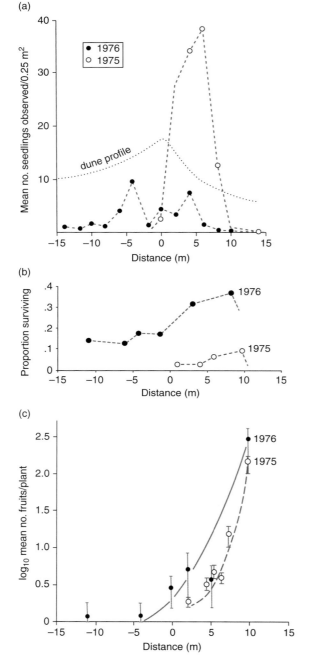

FIGURE 9.13 The seedling density of the annual plant *Cakile edentula* observed for two years. The plant grows along a sand dune gradient stretching from vegetated dunes (a) to open sand beach (c).

beach to areas covered by dense grass many metres from the sea. A majority of the seeds is produced by a tiny fraction of the total population found on the open beach; here large plants surrounded by decaying seaweed can produce hundreds of seeds, whereas the plants inland produced as few as one or two seeds (recall Figure 4.11). Both survival and reproductive output decline with distance from the sea (Figure 9.13). In habitats near the sea there are high losses to dispersal because every year onshore winds combined with storm waves move a majority of seeds inland into areas with high mortality and low survival rates. As with many cases of dispersal, not all seeds arrive in good habitat for growth and survival.

There is a fail-safe mechanism of sorts. The fruits of *Cakile* are two-parted. The distal part, which has a thick buoyant seed coat, breaks off the parent plant, while a smaller proximal section of the fruit remains attached. Each part normally contains a single seed. Thus one might conclude that there has been strong selection to invest half the offspring in the same habitat in which the parents grew. In the case of open sand beaches, the system may not work as intended because large waves tear up the adult plants and disperse the fragments along with the distal seeds. Even here, however, the proximal fruit segments with attached fragments of parent plants may tend to resist landward movement in the wind.

On shingle bars, the reproductive system of the plant has different consequences (Keddy 1980). First, because of the gaps between stones, distal seeds roll in the wind only a short distance before falling into a crack. Second, waves tend not to tear up shingle, so that the dead stalks of plants remain in situ, with the proximal seed still attached. As a consequence of these two features, each spring one finds circular seed shadows around the still-rooted remains of dead

Both the proportion surviving (b) and number of fruits per plant (c) increased toward the ocean. Note the log scale for reproductive output. (From Keddy 1980)

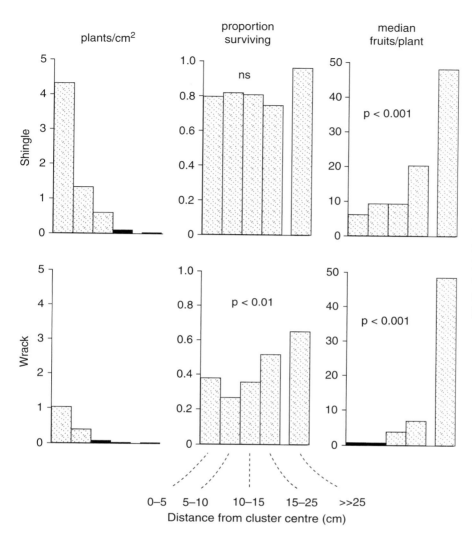

FIGURE 9.14 Seedlings of the annual plant *Cakile edentula* grow in dense clusters on gravel bars near the ocean. Each cluster has, at its centre, a dead parent plant from the previous year. The proportion surviving increases with distance from the centre of the cluster if there is wrack on the gravel. The number of fruits per plant increases with distance irrespective of the presence of wrack. (From Keddy 1982)

plants from the previous year. The third feature of this habitat is large patches of wrack (largely dead leaves of *Zostera marina*); some plants therefore grow in open gravel and others in patches fertilized by wrack.

Where one is able to measure the dispersal from parent plants (Figure 9.14), few seedlings are found further than 15 cm away. There is a suggestion in both habitats that survival increases with dispersal distance, but this trend is statistically significant only in wrack. Near the parent plant, where densities are high, a fungal damping-off disease attacks seedlings at the soil line, and the damp environment provided by wrack not only more than doubles mortality rates (from 0.2 to 0.5) but also produces significantly higher mortality nearer the parent. Thus there is a measurable survival benefit to dispersal away from the parent plant, at least in habitats with high soil moisture. But survival is less than half the story. In both habitats there are dramatic increases in reproductive output with distance from parent plant. Even a few centimetres of added dispersal can lead to a doubling of reproductive output.

9.9   Plant Life Spans

Life span can range from a few months to a few millennia, depending upon the species and habitat. Think *Cakile edentula* in the first case, and *Sequoiadendron giganteum* in the second. The short-lived plants, often called ruderals, or annuals, allocate all of their reproductive energy to seeds all at once. They typically occupy habitats in which resources are freely available, but the probability of surviving to reproduce in future years is small, usually as a result of recurring threat, such as fire, cold or drought. Such habitats would be the "seasonal" or "ephemeral" habitats in Figure 3.24. Many ruderal plants occur as weeds in agricultural fields, where the plough annually destroys biomass (Harper 1977). A more unusual case is the annual plants that occur in deserts, where the plants exploit short wet periods between long droughts (Venable and Pake 1999). At the other extreme from deserts, some annual plants also occur in wetlands during temporary periods of low water (van der Valk and Davis 1978). In general, any short-lived habitat selects for plants with short life spans and relatively rapid growth rates (Grime 1977, 1979). Ruderals often produce large numbers of relatively small seeds. These seeds may remain dormant in the soil, producing large concentrations of buried seeds known as seed banks (Leck et al. 1989). (Table 5.4 showed you some examples of densities of buried seeds in flooded habitats.)

Long-lived plants, usually called **perennials**, delay reproduction until they achieve relatively larger sizes, a process that may take many years. Trees are the most common example. Giant sequoia trees (*Sequoiadendron giganteum*), for example, can live over 1,000 years, and once one reaches reproductive age, it can produce 4.5×10^6 seeds per year (Harper and White 1974). Long-lived plants occupy habitats where adult mortality is low enough that reproduction can be safely postponed. Resources that are not allocated to sexual reproduction can then be allocated to stems, leaves and shoots. This may allow these plants to compete with neighbours for access to light and nutrients. Most perennial plants reproduce for many years upon reaching adulthood, and are termed iteroparous. These include most trees, such as those in Table 9.1.

There is a small subset of perennials that flower and reproduce only once before dying, termed

FIGURE 9.15 *Agave americana* (century plants) do not really take a hundred years to flower, but certainly decades. Then they die. Thus they provide a classic example of a semelparous plant. The flowering shoot may be up to 8 m tall. (Arizona, Anton Foltin, Shutterstock)

semelparous plants. The latter group includes foxgloves (*Digitalis purpurea*, Sletvold 2002), bamboos (Janzen 1976), some yuccas (Young and Augspruger 1991) and a mountain lobelia (*Lobelia telekeii*, Young 1990). Under what circumstances should a species terminate life with a single massive reproductive episode? There are multiple competing hypotheses to account for this "big-bang" mode of reproduction. Three include:

1. Large inflorescences may be necessary to attract pollinating insects.
2. Large numbers of seeds produced in one burst may satiate and overwhelm seed predators.
3. Unstable environments may select against longer lived plants.

It is noteworthy that these hypotheses involve quite different mechanisms – pollinators, herbivores and physical conditions. It may even be that there is a different reason for different species. Although it is the long-lived species, such as the so-called century plant, that often attract human attention (Figure 9.15), the actual age of reproduction for semelparous plants often varies with the conditions of the habitat. A few plants normally take just two years to reproduce (e.g. *Campanula medium, Daucus carota, Digitalis purpurea, Echium vulgaris, Verbascum thapsus*) and are termed biennials. However, this term may be a misnomer. True, in gardens they may flower in the second year, but in the wild it may take many years to accumulate enough resources to flower. For example, in teasel, the size of a rosette will predict both overwinter survival and the probability of flowering, and only the largest will flower with certainty (Werner 1975). Hence it is probably best to think of semelparous plants as a heterogeneous assemblage. Annuals are a fairly natural subgroup, and then there are species that take two or more years to flower, depending upon the environment and the genotype.

Even semelparous plants can have a back-up plan. Agaves, for example, can reproduce not only by seed but by (1) basal shoots, (2) rhizomatous suckers and (3) bulbils – small aerial rosettes that fall from the plant to establish new clones of the parent (Arizaga and Ezcurra 1995).

Box 9.1 Really Big Trees and Really Old Trees

At this point you may be wondering just how big and how old plants can get. But this raises thorny questions such as: what do you mean by big, and what do you mean by an individual?

Really Big Trees

The world's biggest trees are conifers.

The tallest tree in the world is the redwood (*Sequoia sempervirens*), found in coastal northwestern California, more than 100 m tall (a little over 115 m, to be exact).[1]

The giant sequoia (*Sequoiadendron giganteum*), found in the Sierra Nevada of California, is also big. Although these trees may not be as tall as redwoods (the tallest is 95 m), their larger girth makes them the world's biggest single tree by weight.[1] You can see a cross-section in Figure 8.1.

Two other conifers also reach heights of 100 m.

Douglas fir (*Pseudotsuga menziesii*) ranges from British Columbian south in the mountains to Mexico. Logging has removed most of the bigger trees throughout the range. The highest concentration of trees over 100 m tall is in northwestern California among the redwoods.[1]

Sitka spruce (*Picea sitchensis*) extends from Alaska to California; some of the big trees are found in British Columbia and Washington, but most are in northwestern California among the redwoods.[1]

(Continued)

Box 9.1   (cont.)

The tallest flowering plant in the world is a species of eucalyptus tree (*Eucalyptus regnans*), which occurs in southern Australia, including the island of Tasmania (Figure B9.1.1). There are many over 60 m tall, and a least one 100 m tall.[1] For more on their ecology, see Figure 8.25 and associated text.

Really Old Trees

The world's oldest living individual conifer is a bristlecone pine (*Pinus longaeva*) in California, where one tree is older than 5,060 years.[2]

But now that we enter the world of clones, you will know why the whole issue of ramets and genets (Section 9.6) complicates the situation immensely. Once you look for old genets, one finds many that are thousands of years old, including a 12,000-year-old creosote bush (*Larrea tridentata*) and a 13,000-year-old huckleberry (*Gaylussacia brachycera*).[4] Many large genets likely remain to be documented and aged.

The world's oldest living conifer clone may be a Norway spruce (*Picea abies*) in the southern mountains of Sweden. It is 9,550 years old. This tree has expanded clonally, so the older shoots are dead and the current upright trunk dates back only to the 1940s.[2,3]

FIGURE B9.1.1   *Eucalyptus regnans*, the tallest flowering plant in the world, occurs in southern Australia including the island of Tasmania. As with the big conifers, these huge trees depend upon occasional fires to germinate new seedlings (Figure 8.25). These particular trees established after a stand-replacing fire in 1926, and now have an understorey of other trees, shrubs and tree ferns. (Institute for Redwood Ecology)

Biggest and Oldest?

The world's oldest living tree is an enormous clone of aspen (*Populus tremuloides*) with some 47,000 stems covering more than 40 ha. The age is estimated to be 80,000 years.[4,5] If you count the weight of the entire aspen clone (the genet), then this aspen plant could be the largest living creature in the world. In which case it has to be moved ahead of the redwoods and giant sequoias. It even has its own name, Pando.[4]

1 Institute for Redwood Ecology, Humboldt State University, www.humboldt.edu/redwoods.
2 List of oldest trees, www.en.wikipedia.org/wiki/List_of_oldest_trees.
3 BBC, Swedes find 'world's oldest tree',17 April 2008.
4 Sussman, R. 2014. *The Oldest Living Things in the World*. Chicago: University of Chicago Press.
5 Grant, M. C. 1993. The trembling giant. *Discover* 4(10): 82–89.
    Note to big tree enthusiasts. I have converted to metric, 300 feet (ca. 100 m) being an important reference point. This list is provisional, since more large species of trees may yet be found, say, in South America in the Andes, while those already on the list will grow (and, sooner or later, die). Case in point: vandals killed the world's largest pond cypress (*Taxodium ascendens*) on 16 June 2012.[4]

9.10   Population Ecology of the Brazil Nut Tree: A Size-structured Model

We will now look further at size-structured populations, since most long-lived plants represent a wide array of sizes (ages). Our working example will be a tropical tree, which we can say with some certainty is one of the most difficult groups for such analysis. The difficulties include enormous size and age, as well as problems with aging trees. However, we have some reasonable data from Brazil nut trees (*Bertholletia excelsa*). We will first look at the importance of the tree, then some basic ecology, mostly drawing upon Mori (1992), and then we will look at the challenges of producing a population model for such a species using Zuidema and Boot (2002).

9.10.1 Economic Importance

The Brazil nut tree is of some general interest because it illustrates how tropical forests may be more valuable for their products than as pastures. Indeed, you may have some Brazil nuts in your house right now, perhaps in a tin of mixed nuts or in a rain forest chocolate bar. This species also illustrates themes covered earlier in the book, including insect pollination and mammal dispersal. Although we will focus on Brazil nuts, you should be aware that another wild species of great value is the rubber tree (*Hevea brasiliensis*).

A single mature tree can produce 100 kg of nuts in a year and annual production averages some 40,000 tons. In 1986, the value of nuts exported from Manaus alone was more than 5 million dollars. Although there are attempts to grow the species in plantations, a majority of the crop is still gathered from the wild, and there is no guarantee that this species can be forced into plantations. Hence it may provide a long-term source of financial return from natural rain forests.

9.10.2 Ecology

The family to which this tree belongs has about 200 species in the Neotropics, distributed from Mexico to southern Brazil. Although it grows in the Amazon basin, the habitat is upland not flooded. The adults emerge from the canopy above many of the other tree species, and may reach heights of 50 m and diameters of 2 m. They do not reproduce until more than 100 years old, and may live for 600 to 1,000 years (Vieira et al. 2005). The flowers are pollinated mostly by large-bodied bees. Each flower lasts only a few hours. In contrast, it may take 15 months for the fruit to mature, and take a further 12 to 18 months to germinate. The fruits (pods) are more than 10 cm in diameter, and can weigh up to 2 kg; and each one contains some 10 to 20 seeds with stony coatings (Figure 9.16 right). Each fruit, when ripe, falls to the ground, not unlike a cannon ball. Fruits normally fall during the rainy season in January and February. The seeds are then removed and dispersed by rodents, particularly agoutis. Agoutis are a long-legged species of ground-dwelling rodent that feed on fallen seeds and fruits in tropical forest; some of the larger species may reach weights of 3.5 kg and lengths of 50 cm. Agoutis, like squirrels (Section 7.5.2), are known to be important in seed dispersal, except that agoutis usually forage on the ground unlike squirrels, which often forage in the canopy itself. In the case of Brazil nuts, agoutis may eat some of the nuts immediately, but others are stashed under a few centimetres of soil, and eventually a few of these germinate to establish new trees.

9.10.3 A Size-structured Model Using the Lefkovitch Matrix

Now let us turn to the question of how you might quantitatively describe a population of these trees, and make some forecasts about their future. To start, it is useful to know the number of individuals of each size. Since some kinds of plants, including many tropical trees, do not produce age rings, it is often convenient to use size classes, from seedling to adult, in regular increments. This is also sensible for another reason: size is the best predictor of reproductive output. For a population that is replacing itself, the typical shape of the age or size class distribution is the

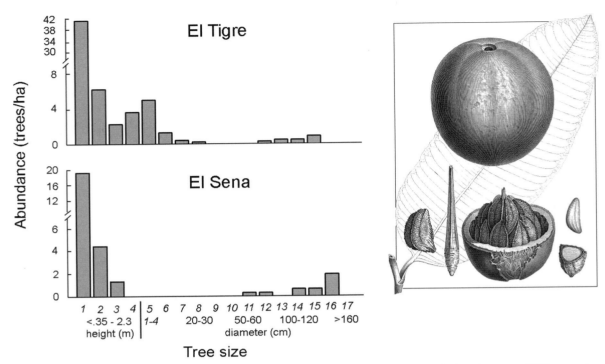

FIGURE 9.16 Left: Population structure of Brazil nut trees in two sites of the Bolivian Amazon, El Tigre and El Sena. Size class 1 has seedlings, size classes 12 and above are mature trees. Note the relative abundance of young trees compared to the old ones. (After Zuidema and Boot 2002) Right: Leaves, fruit and seeds of this species from von Humboldt's expedition to the Andes. (P.J.F. Turpin, 1808, Peter H. Raven Library/Missouri Botanical Garden). The fruit is about 10 cm in diameter.

type III survival curve you saw in Figure 9.4. For comparison, Figure 9.16 (left) shows two sites where all individual Brazil nut trees were tallied, showing many young trees on the left and a few large ones on the right.

It would now be very useful to know how this size class structure might change with time. At what rate do small trees become big trees? At what rate are new offspring produced? Is the population self-sustaining? This information is summarized in a mathematical form as a Lefkovitch matrix (Lefkovitch 1965) shown in Figure 9.17.

To predict future population structure with a Lefkovitch matrix, you need to know four more things. Three have to do with survival. First, you need to know the proportion of individuals of a given size that will grow enough to move from one size class to the next.

Second, you need to know the proportion of a particular size class that will die. Third, you need to know what proportion will simply remain in the size class (that is, do not grow much, but remain alive). Obviously, if you have measured any two of these, the third can be calculated, since they must add to 1. For each size class:

$$\text{proportion growing} + \text{proportion dying} + \text{proportion remaining} = 1.$$

Each year (or some other agreed upon interval of time) the population will move to the right (as ages increase and sizes increase), but the height of each bar will decrease owing to deaths.

So we can predict the future for each size class, as individual trees either die or slowly move from left to right across size classes. Now what about the far-left bar, the smallest size class, recruitment? We must be

Size structure at time t+1	Size structure at time t																
	1	2	3	4	5	6	7	8	9	10	11	12	13	14	15	16	17
1	0.455	0	0	0	0	0	0	0	0	0	0	**12.3**	**14.6**	**16.9**	**19.3**	**22.3**	**26.6**
2	0.091	0.587	0	0	0	0	0	0	0	0	0	0	0	0	0	0	0
3		0	0.147	0.780	0	0	0	0	0	0	0	0	0	0	0	0	0
4	0		0	0.134	0.821	0	0	0	0	0	0	0	0	0	0	0	0
5	0	0		0	0.167	0.941	0	0	0	0	0	0	0	0	0	0	0
6	0	0	0		0	0.044	0.938	0	0	0	0	0	0	0	0	0	0
7	0	0	0	0		0	0.047	0.961	0	0	0	0	0	0	0	0	0
8	0	0	0	0	0		0	0.034	0.946	0	0	0	0	0	0	0	0
9	0	0	0	0	0	0		0	0.049	0.940	0	0	0	0	0	0	0
10	0	0	0	0	0	0	0		0	0.055	0.937	0	0	0	0	0	0
11	0	0	0	0	0	0	0	0		0	0.058	0.936	0	0	0	0	0
12	0	0	0	0	0	0	0	0	0		0	0.059	0.966	0	0	0	0
13	0	0	0	0	0	0	0	0	0	0		0	0.029	0.968	0	0	0
14	0	0	0	0	0	0	0	0	0	0	0		0	0.027	0.971	0	0
15	0	0	0	0	0	0	0	0	0	0	0	0		0	0.024	0.965	0
16	0	0	0	0	0	0	0	0	0	0	0	0	0		0	0.020	0.967
17	0	0	0	0	0	0	0	0	0	0	0	0	0	0		0	0.018	0.985

FIGURE 9.17

A Lefkovitch matrix for the Brazil nut trees of El Tigre (upper histogram in Figure 9.16). Note on the extreme lower right diagonal that the probability of a tree surviving to the next year is very high, greater than 0.95. These trees certainly live to 300 years, and sometimes to 1,000. (After Zuidema and Boot 2002)

able to estimate the number of young that will be added to the far left (smallest) size class in each time interval. Typically, one first measures reproductive output for each adult size class. If you then multiply the number in each size class old enough to reproduce by the number of offspring that they produce, you know how many new young to add at the far left.

With many fewer words, this is tidily represented in the Lefkovitch matrix. It may look complicated (Figure 9.17), but it has only three parts. The main diagonal (numbers in colour) is the proportion of individuals that simply remain in each size class for another time interval. That is, they go from size class x to size class x. In this example, in the smallest size category just about half remain each year (upper left, 0.455), while in the largest category nearly all stay that size for another year (lower right, 0.985). This makes general sense. Small trees must grow, while large trees stay large and produce seeds. Just below the main diagonal is a second diagonal that describes the movement to the next higher size class. These

plants go from size class x to size class x +1 during the same time interval. Finally, at the upper right (numbers in bold) is the reproductive output for each size class. All this recruitment from all the upper size classes is tallied, and then put into the upper-left category of each time interval. To keep it simple, we are not going to worry about whether it is seeds, seedlings or small trees that enter here – that is why I have used the general word recruitment. As you know, enormous numbers of seeds are killed before becoming seedlings, and many seedlings are killed before becoming saplings; somehow, these losses need to be accounted for and included. Note that in Figure 9.16, the four smallest size classes consist of height classes, while the others are diameter classes.

So, if you put the number of individuals in each size class into a vector to describe your population at time t, and multiply it by the Lefkovitch matrix, you obtain the population structure at time t + 1. You can continue this as long as you like as far into the future as you wish. You now have a tool for predicting, or at least for forecasting. The word forecasting is more realistic, since there is no guarantee that the growth rates or reproductive rates will not themselves change, say if the climate changes or if pollinators disappear. The point is that if you have a description of the current population structure, and a Lefkovitch matrix, you have a means for exploring potential future populations. You can even change growth rates or reproductive outputs in the matrix, and see what might happen should these sorts of changes occur.

A more general model still is the Leslie matrix (Leslie 1945), in which one needs only a single diagonal, since all individuals move from one category to the next in lockstep. In general, this approach is more useful for short-lived organisms, particularly animals, and is the one you are more likely to have previously encountered. For size-structured populations, like most woody plants, the Lefkovitch matrix is likely to be more useful.

One of the more remarkable things about this kind of matrix is that if you keep multiplying it forward, year after year, the population size class structure reaches a steady state or stable size structure (Caswell 2001). This stable age distribution is of considerable interest since it describes some sort of natural equilibrium age class structure to which the population will return if perturbed. In practice, of course, so many other factors change with time over the life span of a tree that stable age class structure is unlikely to be attained, but still the stable age class structure is a useful tool for thinking about populations all the same. It is also a foundation for the topic of population viability analysis, which asks about the future survival probabilities of a population. For many threatened species, this is a useful management tool (recall Section 9.7), and you may wish to pursue it further in Morris and Doak (2002). As for age-structured matrix models in general, we may close with an observation by Caswell (2001: p. 6): "For all their power, these models are among the most intuitive and easy to grasp in all of population biology."

CONCLUSION

Populations are ultimately controlled by many of the same factors that control vegetation overall. Thus if one is studying the demography of saguaro cacti, one could think of it as a study of how stress restricts establishment, and put it in the stress chapter, or one could think of it as a case study of how mutualism affects establishment, and put it in the positive interactions chapter (as I did in this book, Section 7.2.1). Or one could think of it as a population study and put it in the population chapter. None of these would be wrong, just different ways of looking at the same issues. Overall, after many years of teaching, I am inclined to think increasingly that most students will benefit most from thinking of controlling factors in order of their likely importance, and how such factors restrict the distribution of species found in local floras (species pools). Hence I have retained some examples of population studies in other chapters. But the ultimate argument in favour of thinking from the point of view of populations, at least from time to time, is that it reminds us how prolific most plants are – there typically are vast numbers of offspring with low probability of any one becoming a well-established individual. If there is any generalization possible, it is that a vast majority of the world's plants are iteroparous perennials with high rates of juvenile mortality.

Review Questions

1. Exponential population growth is one of the most powerful biological forces on the planet. Explain why it makes invasive species so very dangerous.

2. A pond has a small population of floating aquatic plants. The number of plants doubles every week. After 40 weeks, the pond is covered in these plants. When was the pond half covered?

3. Why do some plants, such as orchids, have very small seeds? Why do some plants, such as oaks and walnuts, have rather large seeds? What are the trade-offs between the two?

4. Contrast the type III survival curve of plants with the type I survival curve of most mammals.

5. One giant sequoia tree can produce millions of seeds. If exponential growth is as important as this chapter says, why is the world not covered in giant sequoias?

6. What is the difference between a ramet and a genet? Why does it matter? Explain using the example of Pando (Box 9.1).

7. Explain how molecular techniques might be used to determine whether an area dominated by clonal plants (say a cattail marsh, a papyrus marsh, or a poplar forest) contains more than one genet. Find a recent example in a scientific journal.

8. What is a Leslie matrix and what is its purpose?

Further Reading

Harper, J. L. 1977. *Population Biology of Plants*. London: Academic Press.

Begon, M. and M. Mortimer. 1981. *Population Ecology: A Unified Study of Animals and Plants*. Oxford: Blackwell.

Keddy, P. A. 1981. Experimental demography of the sand dune annual, *Cakile edentula*, growing along an environmental gradient in Nova Scotia. *Journal of Ecology* **69**: 615–630.

Jackson, J. B. C., L. W. Buss and R. E. Cook. 1985. *Population Biology and Evolution of Clonal Organisms*. New Haven: Yale University Press.

Young, T.P. 1990. Evolution of semelparity in Mount Kenya lobelias. *Evolutionary Ecology* **4**: 157–191.

Kingsland, S. E. 1995. *Modeling Nature: Episodes in the History of Population Ecology*. Chicago:University of Chicago Press.

Nantel, P., D. Gagnon and A. Nault. 1996. Population viability analysis of American ginseng and wild leek harvested in stochastic environments. *Conservation Biology* **10**: 608–621.

Venable, D. L. and C. E. Pake. 1999. Population ecology of Sonoran Desert annual plants. pp. 115–142. In R. H. Robichaux (ed.) *The Ecology of Sonoran Desert Plants and Plant Communities*. Tucson: University of Arizona Press.

Caswell, H. 2001. *Matrix Population Models: Construction, Analysis, and Interpretation*, 2nd ed. Sunderland: Sinauer.

Silvertown, J. and D. Charlesworth. 2001. *Introduction to Plant Population Biology*, 4th edn. Oxford: Blackwell.

Morris, W.F. and D.F. Doak. 2002. *Quantitative Conservation Biology: Theory and Practice of Population Viability Analysis*. Sunderland: Sinauer.

10

Stress

Definitions. Lack of water: deserts, grasslands, Mediterranean shrublands (maquis, kwongan, fynbos), rock barrens (rock domes, tepui, alvars, serpentine). Resource unavailability (peat bogs). Cold as a regulator: Arctic and alpine plants, early spring leaves. Salinity as a regulator: salt marshes and mangal, Extremes: lichens in rocks, plants in wetlands. Pollution: acid precipitation in the Smoking Hills, radiation in forests. More on definitions: stress and strain. Stress has metabolic costs. Measurement of stress in experiments. Evolution in stressed habitats. The CSR perspective. Scale.

FIGURE 10.1 Some species of plants can tolerate extremely adverse conditions. *Azorella compacta* (Yareta), the world's largest species of cushion plant, grows in the Andes Mountains at altitudes greater than 3 km, where the main stresses include low temperature, high winds and seasonal drought. It grows only about 1.5 cm in height and diameter each year (Kleier and Rundel 2004), and large plants may be several thousand years old. The inset shows a single twig (ca. 2 cm), the first scientific drawing of this species, from von Humboldt's expedition, published in 1821. In this chapter we will explore some different kinds of environmental stress, and consequences for the morphology and physiology of plants. (Photo by Pierre-Jean Durieu, Shutterstock; inset P.J.F. Turpin, 1821, Peter H. Raven Library/Missouri Botanical Garden)

10.1 Introduction: Stress Constrains Growth

It is time to revisit the topic of resources. A second chapter on resources, you ask? Yes, but this time we will focus on habitats that lack essential resources. Although there can be many reasons why resources are in short supply, there are important similarities among species and habitats that arise. Shortages of resources will tend to reduce rates of growth, thereby reducing attributes such as shoot size, root length, and allocation to flowers and seeds (Grime 1979; Levitt 1980; Larcher 2003). We will use the word stress to describe "any environmental factor that reduces the rate of production of biomass".

It is important to understand that stress is fundamentally different from disturbance. Stress prevents plant growth in the first place; while, in contrast, disturbance removes biomass that has already been produced. You will recall from Chapter 3 that plants require a small set of resources for growth, principally water, light, nitrogen and phosphorus. Habitats that are deficient in one of these will reduce plant growth. So will a collection of other factors including cold, salinity and unusual soils such as serpentine.

Not everyone likes this word, stress. Some botanists think that we already have a word for resources, so why coin another for the opposite, lack of resources? The answer is mainly that sometimes it is useful for thinking, writing or explaining, to have two words that are opposite in meaning. We talk about good and bad, for example, or black and white, not unbad or unblack, respectively. So it is with stress. A desert is stressful because of an absence of resources, mainly water, which reduces the rate of production of plant biomass. Another reason we need a word like stress is that in certain locations, factors that are not resources can inhibit plant growth. Cold, for example, can act

directly upon plants by freezing cells and injuring the plant, or by inhibiting the supply of resources (Figure 10.1). Hence cold, like salinity, is not a resource but a source of stress we shall term a regulator. Finally, in this chapter, I want to look more closely at a few habitats in which plant growth is difficult, or "stressful". In Chapter 3 we saw how some plant traits are affected by low productivity, the result being parasitic and carnivorous plants. Here we want to continue that line of inquiry but at the larger scale of entire habitats.

There is a vast and growing scientific literature describing the physiological mechanisms by which plants respond to stress, and it would be possible to devote a lifetime to the study of how selected species respond to selected environmental constraints. Levitt (1980), for example, has written a two-volume review on environmental constraints, excluding nutrient deficiencies, the latter being covered in Marschner (1995). Further, Larcher (2003) has provided an overview of physiological ecology. This chapter draws upon these three sources where necessary. The emphasis in this chapter is, however, upon general principles that apply to a broad array of types of plants and plant communities. The intention is to describe the evolutionary consequences of survival under conditions that continually constrain growth, and the types of vegetation that arise under such conditions. Given the vast number of plant species in the biosphere and the large number of constraints that can affect them, some such generalizations must be made. We will therefore leave many of the fine-scale mechanisms by which such effects arise at the molecular, cellular and tissue levels for books addressing environmental physiology.

10.2 Habitats That Lack Resources: Drought as a Widespread Example

10.2.1 Deserts

Water is an essential constituent of life. A good place to begin the exploration of stress is in habitats that lack water (Figure 10.1). Shortages of water have been a constant stress on plants ever since their ancestors first occupied terrestrial environments more than a quarter of a billion years ago. The absence of water is the key factor that produces the world's deserts (Figure 10.2). Deserts occupy 26 to 35 percent of the land surface of the Earth; the largest example is the Sahara, which covers some

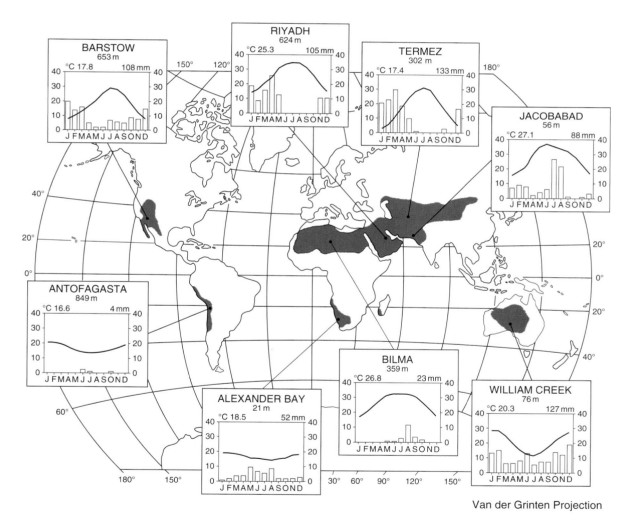

Van der Grinten Projection

FIGURE 10.2 Distribution of arid regions and representative climatic conditions. Mean monthly temperatures are indicated by the line and mean precipitation for each month is shown by the bars. Station elevation, mean annual temperature and mean annual precipitation appear at the top of each climograph. (From Archibold 1995)

FIGURE 10.3 Some examples of different growth forms allowing plants to survive drought. (a) Deciduous "bottle" trees with water-storing trunks (*Adansonia/Chorisa* type); (b) succulents storing water in the stem (Cacti/*Euphorbia* type); (c) succulents with water-storing leaves (*Agave*/Crassulaceae type); (d) evergreen trees and shrubs with deep tap root systems (sclerophyll type); (e) deciduous, often thorny shrubs (*Capparis* type); (f) chlorophyllous-stemmed shrubs (*Retama* type); (g) tussock grasses with renewal buds protected by leaf sheaths, and with wide-ranging root systems (*Aristida* type); (h) cushion plants (*Anabasis* type); (i) geophytes with storage roots (*Citrullus* type); (j) bulb and tuber geophytes; (k) pluviotherophytes (annual plants); (l) desiccation-tolerant plants (poikilohydric type). (From Larcher 1995)

9 million km^2 of North Africa (Archibold 1995). In deserts rainfall is often less than 100 mm per year. The coastal Atacama Desert in Chile averages only 0.7 mm per year, and several years can pass with no measurable precipitation. In other cases, areas in South Africa and Australia may receive more than 500 mm per year yet be classed as deserts because the rain falls over such short periods that organisms still face long periods of drought (Archibold 1995).

Plants that tolerate these environments must have a suite of traits that reduce strain from water loss. To maintain internal water concentrations that allow chemical processes to occur, plants must take up water when it is available and reduce rates of loss when it is not. Even small losses in water potential, a few bars (8 to 10 percent of saturation), is sufficient to decrease rates of cell division. Since most water arrives in short bursts of precipitation (Noy-Meir 1973), desert plants must either absorb water from brief surface pulses or else construct deep roots (Schwinning and Sala 2004).

Desert plants provide clear examples of traits that reduce water loss (Figure 10.3). More than a hundred years ago Kearney and Shantz (1912) divided desert plants into three functional groups: drought-escaping, drought-evading and drought-enduring species. Drought-escaping plants are largely annuals that remain as buried seeds until stimulated to germinate by rainfall. Drought-evading plants are perennials that remain dormant during dry periods but rapidly produce shoots during short-lived wet periods. The best known desert plants are, however, the drought-enduring succulents (Figure 10.4; recall also Figures 1.1 and 3.28). Some of the typical traits of these species are shown in Table 10.1.

One set of traits reduces the loss of water. Since water is lost mostly through the epidermis, and in particular through stomata, desert plants have a thick

FIGURE 10.4 *Pachypodium namaquanum* (Apocynaceae) in Richtersveld National Park, South Africa, in a vegetation type known as succulent karoo. The plant typically occurs in colonies of 30 to 1,000 on steep granite slopes up to 4 m. According to Rapanarivo et al. (1999), there are 23 species in this genus, 5 in southern Africa and 18 in Madagascar (Colour photo by Winfried Bruenken, Wikimedia Commons; black and white photo from Rundell et al. 1995; illustration by Robert Jacob Gordon, ca. 1778, courtesy of Rijksmuseum, Amsterdam)

layer of wax covering the epidermis and greatly reduced numbers of stomata. In non-desert species the number of stomata per square millimetre of leaf surface ranges from 800 in some trees to 100 to 200 in most annuals, but in desert succulents there are only about 20 (Marschner 1995). Further, surface to volume ratios are dramatically reduced. Consider three examples. *Pachypodium namaquanum* has cylindrical stems, with the leaves reduced to small tufts at the summit of the plant (Figure 10.4). Many members of the Cactaceae are reduced to spheres, which brings surface to volume ratios to an absolute minimum

(recall Figure 3.28). The stone plant, *Lithops salicola*, can be considered to have taken this "minimum" to a further extreme, reducing its exposed surface area further by being entirely buried except for two flattened lobes at the soil surface (Figure 10.5).

If evapotranspiration were entirely prevented, however, photosynthesis would also cease. Many desert plants therefore have a distinctive type of photosynthesis called crassulacean acid metabolism (CAM) photosynthesis. Stomata open only during the night, when rates of water loss from evapotranspiration are at their lowest (Figure 10.6). Carbon dioxide is absorbed and then stored in vacuoles as malate. During the following day, the stomata close to conserve water, but photosynthesis can proceed using the stored carbon. As a consequence, desert succulents such as *Agave deserti* transpire only 25 g of water for each gram of CO_2 taken up by leaves, and the barrel cactus (*Ferocactus acanthodes*) loses 70 g (Nobel

1976). Equivalent values for C_3 and C_4 plants are 450 to 600 g water/g CO_2 and 250 to 350 g water/g CO_2, respectively (Szarek and Ting 1975).

Preventing water loss is only half the story. There is also the issue of gaining water for growth. Water is taken up rapidly after rainstorms. During dry periods roots have few, if any, lateral branches, but upon wetting rain roots rapidly form. By the time the soil is dried again, these rain roots have been shed, and the main roots have become sealed for the drought (Gibson and Nobel 1986). Once absorbed, the water can be stored for long periods in the spongy tissues of cacti. In the barrel cactus (*F. acanthodes*), stored water can extend photosynthesis for up to 50 days (Nobel 1977). A specimen of *Opuntia* cactus (*O. bigelovii*) was still alive three years after being completely severed from its roots and suspended on a post in the desert (Gibson and Nobel 1986).

The presence of spines is also characteristic of desert plants, but why should this be the case? Some spines may indeed minimize the direct effects of drought – spines can provide shade to the plant and reduce heat loading and evapotranspiration. In most cases, however, spines are probably best understood as a secondary adaptation to drought. Recall Chapter 6 on herbivores, where spines are a

Table 10.1 **Traits associated with plants tolerant of drought. (From Gibson and Nobel 1986 and Archibold 1995)**

Relative importance	Trait
Primary	Reduced surface to volume ratio
	Wax coating
	Reduced density of stomata
	Spines/hairs (for shading)
	CAM photosynthesis
Secondary	Spines (defence against herbivores)
	Secondary metabolites (defence against herbivores)
	Camouflage (defence against herbivores)

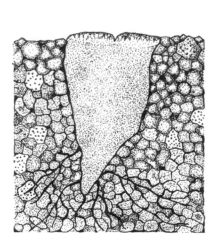 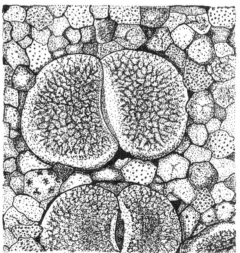

FIGURE 10.5 Stone plants such as *Lithops salicola* (Aizoaceae) reduce the surface area for water loss by means of burial. This habit may also reduce exposure to herbivores, since the two leaves may mimic the colour and even the texture of surrounding stones. Left, growth form; right, surface view of expanded leaves. (From Cloudsley-Thompson 1996)

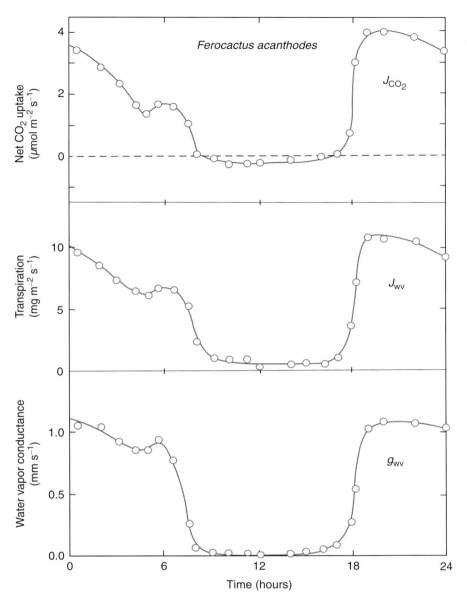

FIGURE 10.6 Gas exchange measured in the field for *Ferocactus acanthodes* over a 24-hour period. (From Gibson and Nobel 1986)

common means to deter grazing animals. It would appear that the cost of producing tissue in deserts is so high that there is also strong natural selection to defend those tissues. And, following another theme from Chapter 6, secondary metabolites are also present. The mescaline produced by the peyote cactus (*Lophophora williamsii*) can be regarded as an alkaloid that deters herbivores. It is even possible that some dessert plants use camouflage to avoid detection, such as the aptly named stone plants (Figure 10.5).

10.2.2 Grasslands

Drought also produces grasslands, the largest biome on Earth, some 41 to 56 million km^2, about one-third

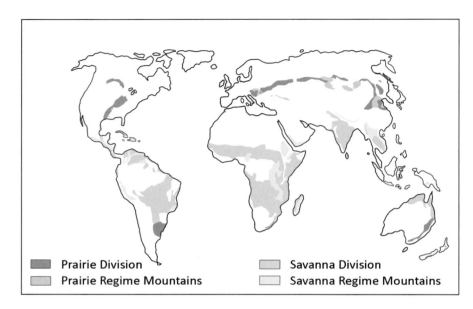

FIGURE 10.7 The global distribution of grasslands. There are two main types: prairie (green, in more temperate areas) and savanna (yellow, in more tropical areas). Each can also occur at higher elevations on mountains. This map was produced using the world ecoregions map in Bailey (1998).

Prairie Division
Prairie Regime Mountains
Savanna Division
Savanna Regime Mountains

of the Earth's land surface area. They are named for the dominant group of plants, grasses, which evolved from woody ancestors during the Cretaceous era and spread into a wide array of arid environments (Gibson 2009). Grasses are distinctive for their reduced flowers (associated with wind pollination), their herbaceous growth form and their capacity to rapidly produce new shoots from the basal **meristems**. Much of the biomass in grassland is stored below ground in the form of rhizomes and roots (recall Figure 5.6). The basal meristems and below-ground storage allow grasses to quickly recover from both fire and grazing. Grasslands occupy an intermediate set of moisture regimes, where annual precipitation ranges from 15 to 120 cm, neither so dry as to produce deserts, nor so wet as to allow forests. Otherwise, grasslands can occur in a wide range of temperatures, from the arctic to the tropics. In more temperate regions, wetter conditions produce tall-grass prairie while drier regions produce short-grass steppe (Sala et al. 2013). In hotter climates, they are often called **savannas**, and may have scattered trees; the tree species differ with region, including acacia trees in Africa, pines around the Gulf of Mexico, and oaks in Europe and North America.

Grasslands are widespread (Figure 10.7). There are many more grassland maps available – and often they are quite contradictory. Some maps show nearly all arid areas as grasslands (including regions that are more accurately named shrublands or deserts), while other maps show only certain types of grasslands (thereby showing smaller areas of grass-dominated habitats than there really are). A certain amount of disagreement is reasonable: how many trees make an area a savanna as opposed to grassland, and how many shrubs make a shrubland rather than a prairie? And how should one show small areas of open grassland that are interspersed with desert or forest? Experts still disagree, but the map in Figure 10.7 is my choice, and is based on a world ecoregion classification (Bailey 1998). If you plan work in grasslands, you should also seek out a map compiled for a smaller area, such as your own continent or island.

Climate is not the only factor producing grasslands. In areas where the rainfall is actually sufficient for forests, recurring fire and grazing can prevent tree establishment. Other local factors that may influence the occurrence of grassland include soil depth, soil texture, aspect and flooding (Anderson et al. 1999). Humans often increase levels of both grazing and fire, so some grasslands are likely anthropogenic (Vera 2000). Grasslands, including pastures, can also arise after logging, and may persist and even spread from recurring fire and grazing.

Overall, one can think of grasslands as arising from a nested series of causes. The primary cause for most grasslands is an arid climate. Secondary effects arise from disturbance (fire and grazing), both of which can modify species composition and extend grassland area. Tertiary effects arise from humans, who can further modify composition and extend area. Determining the relative importance of these three factors can be difficult, particularly in areas with a long history of human activity such as in Europe and Africa. Indeed, in some arid climates, it is also possible for grazing to trigger a loss of grassland – that is, to trigger conversion to shrubland (Archer 1989; Sayre 2003). Hence if you recall Figure 1.4, it may be useful to think of the grassland area as being a rather large and indeterminate region (Figure 10.8). In some areas, climate alone is responsible, but in many other areas, combinations of fire, grazing and human activity are also active. Hence when reading about grasslands and the factors that cause them, one must think carefully about the causes and effects as they operate at local scales.

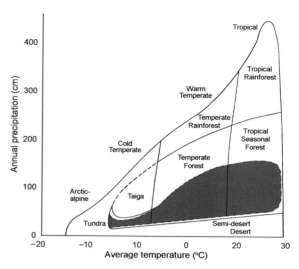

FIGURE 10.8 Grasslands can be produced by climate alone, or by climate and secondary factors such as fire, grazing and humans. Hence it may be helpful to think of the classic Whittaker diagram (recall Figure 1.4) as having a large central indeterminate area between forests (top) and deserts (bottom) in which grasslands can arise under the right circumstances. The closer one gets to the margin of the green area, the more sensitive the grasslands may be to changes in grazing or fire.

10.2.3 Mediterranean Shrublands

In some arid parts of the world, drought is more subtle. Regions with a Mediterranean climate have warm dry summers that alternate with cool moist winters (Archibold 1995; Cowling et al. 1996). These conditions occur in five widely separated geographic regions totalling about 1.8 km^2 (Figure 10.9). In order of decreasing size these are the Mediterranean basin in southern Europe, coastal California (chaparral), southwestern Australia (kwongan), central Chile (matorral) and the cape of South Africa (fynbos).

The dominant life form in all these regions is sclerophyllous shrubs (Mooney and Dunn 1970; Archibold 1995). These shrubs have many traits associated with drought tolerance: (1) small

heavily cutinized leaves; (2) small stomata (although relatively large numbers of them, up to 1,000/mm^2, roughly ten times that of succulents); (3) vertically oriented leaves to reduce heat loading and promote stem flow when rain occurs; and (4) vigorous resprouting after fire. Each of the five Mediterranean regions has evolved in isolation from one another, providing a good example of convergent evolution in life form (Table 10.2) combined with local speciation.

In Africa and Australia, the family Proteaceae (recall Figure 8.10) is very well represented in both the fynbos and kwongan. Other genera have rapidly diversified into this habitat including *Eucalyptus* (>300 species), *Acacia* (>400 species) and *Erica* (>500 species). The fynbos of South Africa has a pool of more than 8,500 species, with

FIGURE 10.9 The global distribution of Mediterranean vegetation types. (After Cowling et al. 1996)

Table 10.2 **Life-form spectra (recall Figure 2.2, Table 2.1) for various Mediterranean shrub communities. Ph, phanerophyte; Ch, chamaephyte; H, hemicryptophyte; C, cryptophyte; Th, therophyte. (After Archibold 1995)**

Location	No. species	Life forms (%)				
		Ph	Ch	H	C	Th
California						
Chaparral	44	41	16	18	11	14
Coastal scrub	65	17	19	20	3	41
Chile						
Matorral	108	24	14	20	6	36
Coastal matorral	109	12	15	17	9	46
Southern Africa						
Fynbos	448	34	31	16	15	4
Coastal renosterveld	127	12	14	19	45	10
Australia						
Mallee-broombush	288	57	19	9	8	7
E. diversifolia heath	274	56	19	10	7	8
E. behriana-herb alliance	50	48	24	12	2	14
Israel						
Quercus–Pistachia association	206	47	14	20	8	11
Pinus–Juniperus association	73	8	37	20	15	20
Arbutus–Helianthemum association	138	20	23	28	17	12

FIGURE 10.10 (a) This Mediterranean sclerophyllous shrubland, termed chaparral, occurs in California. (Montara Mountain, San Mateo County, Tom Parker) It is dominated by shrubs in three genera, *Adenostoma* (Rosaceae), *Ceanothus* (Rhamnaceae) and *Arctostaphylos* (Ericaceae). *Arctostaphylos*, known collectively as manzanita, is the most diverse woody genus in the California Floristic Province with over 100 taxa; the inset shows *A. nummularia* in flower. (January, Mendocino County, photographs by Tom Parker)

a species density greater than 95 per 1,000 km², a value exceeded only by tropical rain forests with >100 species per 1,000 km² (Bond 1997). These are arranged into many different types of plant communities at local scales (Table 10.3). And, in keeping with the theme of stress, this diversification has occurred in a particularly adverse environment for plant growth: drought combined with infertile soils and frequent fire.

The Mediterranean region of California has vegetation known as chaparral (Figure 10.10). The state vegetation manual (Sawyer et al. 2009) recognizes more than 50 different types of chaparral;

these arise from differences in rainfall, altitude, exposure and soil. Common genera include *Arctostaphylos* (ca. 100 species) and *Ceanothus* (ca. 50 species) (Parker 2007, 2011). Fire plays a role in determining the vegetation composition. Some of the shrubs resprout after fire, while other species form persistent seed banks and are well known for having seeds that require smoke or heat-shock to germinate (Keeley 2007: Table1). However, natural fires caused by lightning likely occurred only once a century; more frequent burning regimes are leading to loss of chaparral and replacement by invasive species (Keeley 2007).

Table 10.3 **Plant community and soil characteristics of fynbos vegetation on the Agulhas Plain, South Africa. (From Cowling 1990)**

Site/Community	Dominant species	Soil (parent material)	Soil fertility index[b]
Geelrug transect			
G1 Proteoid fynbos[1a]	*Protea compacta, Elegia filiacea*	Deep, colluvial podzol (TMG sandstone)	−0.25
G2 Proteoid fynbos[2]	*Leucadendron platyspermum, Restio similis*	Gravelly, leached lithosol (TMG sandstone)	−0.19
G3 Proteoid fynbos[3]	*Leucadendron elimense, Blaeria klotzschii*	Lithosol (Sil-ferricrete)	−0.17
G4 Asteraceous fynbos[4]	*Disparago anomala, Thoracosperma interruptum*	Duplex-loam on clay with ferricrete stoneline (Bokkeveld shale)	0.59
G5 Renoster shrubland[5]	*Elytropappus rhinocerotis, Ischyrolepis caespitosa*	Alluvial sand over transported clay	0.36
Hagelkraal transect			
H1 Proteoid fynbos[1]	*Protea compacta, Willdenowia glomerata*	Deep, colluvial podzol (TMG sandstone)	−0.22
H2 Proteoid fynbos[6]	*Protea susannae, Euchaetes burchellii*	Deep, colluvial podzol (Bredasdorp limestone)	−0.14
H3 Proteoid fynbos[7]	*Leucadendron meridianum, Phylica sp. nov.*	Calcareous lithosol (Bredasdorp limestone)	0
H4 Asteraceous fynbos[8]	*Passerina paleacea, Euclea racemosa*	Deep, calcareous dune sand	0.16
H5 Forest[9]	*Sideroxylon inerme, Stipa dregeana*	Deep, colluvial, calcareous sand (Bredasdorp limestone)	0.69
Soetanysberg transect			
S1 Proteoid fynbos[1]	*Protea compacta, Staavia radiata*	Deep, colluvial podzol (TMG sandstone)	−0.23
S2 Ericaceous fynbos[10]	*Syndesmanthus globiceps, Erica coccinea*	Leached lithosol (TMG sandstone)	−0.12
S3 Restiod fynbos[11]	*Leucadendron linifolium, Chondropetalum deustum*	Shallow, calcareous sand (TMG sandstone)	−0.05
S4 Asteraceous fynbos[8]	*Agathosma collina, Ischyrolepis eleocharis*	Deep, calcareous dune sand	−0.03
S5 Thicket[12]	*Olea exasperata, Euclea racemosa*	Deep, calcareous dune sand	0.31

Table 10.3 (cont.)			
Site/Community	Dominant species	Soil (parent material)	Soil fertility index[b]
Miscellaneous plots			
HN1 Proteoid fynbos[7]	*Leucadendron meridianum, Thamnochortus paniculatus*	Calcareous lithosol (Bredasdorp limestone)	0.02
S6 Ericaceous fynbos[10]	*Blaeria ericoides, Thamnochortus lucens*	Leached lithosol (TMG sandstone)	−0.09

Notes:
[a] Sites with the same superscript numeral belong to the same community.
[b] Derived from covariance biplot analysis of soil data.
TMG: Table Mountain Group

In the European Mediterranean basin, the effects of drought have been increased by centuries of forest clearance and overgrazing, and have produced dense shrub communities known as "maquis" or "garrique." Hence, as you shall see in Chapter 13, the European Mediterranean is a classic example of human alteration of natural vegetation patterns.

10.2.4 Rock Barrens

Wherever there is shallow soil, it is likely that plants will be under stress. Shallow soil stores less water, thereby increasing the probability, intensity and duration of drought. As well, essential nutrients will be in short supply since the volume of soil available to any plant is a function of soil depth. Consequently, areas with rock near the surface tend to have distinctive vegetation, including plant species tolerant of infertility and drought (McVaugh 1943; Anderson et al. 1999). Since dry vegetation burns easily, recurring fire is also often associated with such conditions. Because the vegetation is slow to regenerate after disturbance, barrens of all kinds are particularly vulnerable to human disturbance.

Rock barrens are common at high latitudes or altitudes where glaciers have removed most of the surface soil. Alvars are a special kind of glaciated

rock barren formed over limestone (Figure 10.11). Alvars are globally restricted habitats, being common only in the Baltic areas of northern Europe and in the area north of the Great Lakes (Petterson 1965; Catling and Brownell 1995). Soil depth and vegetation biomass are the primary gradients that produce different vegetation types (Belcher et al. 1992), and in those sites studied to date, below-ground competition exceeds above-ground competition (Belcher et al. 1995). In Europe, grazing has been an important historic factor in alvars, whereas in North America there is evidence that fires were important (Catling and Brownell 1998). Periodic drought may also greatly inhibit invasion by forest, mortality of rates of 60 to 100 percent having been observed in woody plants during one dry summer (Stephenson and Herendeen 1986).

Some rock barrens can be found outside glaciated regions. In the southern Appalachians, outcrops of metamorphic rocks can form smooth domes rising as much as 200 m above the surrounding terrain. Rainfall drains off such domes, and may also carry away organic matter that might otherwise form soil. McVaugh (1943) described a series of such outcrops, and their distinctive flora and vegetation. Amidst expanses of bare rock, small depressions accumulate water and organic matter, and thereby support islands of vegetation. Each depression develops a

FIGURE 10.11 Limestone pavement in an alvar on the south coast of Manitoulin Island. Some of the world's largest alvars occur on this island and on the adjoining Bruce Peninsula. Grasses such as *Sporobolus heterolepis* and *Deschampsia caespitosa* grow in the cracks between the pavement blocks. The distinctive yellow "flowers" (inset) are capitula of *Hymenoxys herbacea* (lakeside daisy, Asteraceae), a species restricted to alvars around the Great Lakes. Each capitulum is about 3 cm in diameter. (Photographs by Ryan Gardner, Misery Bay, Ontario Parks)

series of distinctive vegetation zones. At the very margin are crustose lichens or mosses (*Grimmia* spp.); these give way in turn to fruticose lichens (*Cladonia* spp.) or fern allies (*Selaginella* spp.), then large mosses (*Polytrichum* spp.) and, finally, near the very centre, vascular plants including grasses (*Andropogon* spp., *Panicum* spp.), rushes (*Juncus* spp.) and sedges (*Rhynchospora* spp.). These domes have presumably been isolated from the surrounding forest vegetation for a considerable time since, in the southern United States,

at least 12 plant species are endemic to such outcrops (Shure 1999).

In equatorial regions, rock outcrops can have a remarkably rich flora. One of the best examples is the tepui (or "inselbergs") of the Guyana highlands in northern South America. Here rock outcrops rise steeply from amidst tropical forest (Figure 10.12). The flora on these outcrops tends to be dominated by four families: Rubiaceae, Melastomataceae, Orchidaceae and Cyperaceae (Prance and Johnson 1991). Many plants have a bizarre growth form

(a)

(b)

FIGURE 10.12   (a) Tepui are flat-topped sandstone mountains edged with sheer cliffs. Some are surrounded by forest. (b) The tops of tepui have infertile wet barrens and pools with many endemic species including carnivorous plants in the genera *Utriculara*, *Drosera* and *Heliamphora*. (Venezuela, Vadim Petrakov, Shutterstock)

characterized by "thick, sclerophyllous, highly reduced, glossy, waxy or revolute leaves, often crowded into tufts of rosettes or covered by a sericeous, gray, white or brown tomentum. Frequently the stem becomes conspicuously shortened or elongated, simple and virgate producing a weird appearance in the landscape" (Steyermark 1982: p. 205).

Certain rock outcrops impose further constraints upon plant growth through the lack of major nutrients, absence of micronutrients such as calcium, and the presence of higher than normal concentrations of elements such as nickel or chromium. These rocks are collectively known as serpentine. They are widespread, with significant outcrops in locations including California, Newfoundland, central Europe, Scandinavia, Russia, Tibet, China, Japan, Brazil, Zimbabwe, New Zealand and Australia (Whittaker 1954a, b; Coleman and Jove 1992). Since the composition of serpentine rocks is quite variable, some suggest the term is of doubtful value, but recurring themes of inadequate soil nutrients and

high concentrations of metals seem worthy of recognition. Further, there is now evidence that serpentine rocks have a distinctive geological origin (Coleman and Jove 1992; Roberts 1992). They appear to form when slabs of seabed are lifted onto continental margins instead of being subducted and destroyed along the margins of oceanic plates. Thus the regions of serpentine are associated with the margins of ancient oceans.

Serpentine rocks may have particularly distinctive vegetation when other kinds of stress are superimposed. In northern areas, such as Newfoundland, Roberts (1992) added, stresses include (1) drought, (2) wind, (3) erosion from lack of vegetative cover, and (4) cryoturbation (disturbance by recurring freeze–thaw cycles). Here, serpentine rocks support arctic–alpine plant communities in exposed locations, peatlands and sedge meadows in less exposed sites, and occasionally trees and shrubs in sheltered areas. In tropical areas, leaching may increase the deficiency of nutrients. In equatorial areas such as Brazil (Brooks et al. 1992), such rocks yield sparsely vegetated domes covered in *cerrado* (grasses, shrubs and low trees) giving way to *campo rupestre* (grasses and shrubs).

Serpentine rocks have been studied for many years, in part because they are visually distinct in many landscapes, and tend to have endemic plant species. Simple pot experiments using serpentine soil (Kruckeberg 1954) showed that fertilization with NPK increased the growth of plants in serpentine but only if the plants already possessed the ability to grow on serpentine. Non-serpentine plants did not respond to NPK, which indicates that some other factor was limiting their growth in these soils. Only the serpentine plants could grow on serpentine soil alone, but if calcium was added as gypsum, then both serpentine and non-serpentine plants were able to grow. Such experiments suggest that it is the low calcium level rather than absence of the macronutrients NPK that prevents many plants from growing on serpentine. Of course, increasing the soil pH will also reduce the solubility of possibly toxic metal ions such as nickel and aluminum. This also implies the importance of competition. Serpentine ecotypes cannot spread to sites with normal soils because other plants are better able to exploit these soils. This appears to be another case where stress-tolerant species are competitively excluded to inferior habitats.

10.3   Habitats Where Resources Are Present, Yet Unavailable: Peatlands

In some habitats, the resources are physically present, but stored in a form that plants cannot use. From the perspective of the plant, strain is again induced, but the cause is subtly different. A particularly good example of this is provided by peat bogs (recall Figure 8.24). Here the plants are rooted in a matrix of organic matter, but because of low decay rates the essential elements such as nitrogen and phosphorus remain chemically bound within those partially decayed plant remains. In extreme cases the vegetation depends upon the dilute nutrient solution provided by rain water.

Peatlands are flooded more or less permanently with the water table near the soil surface. Under these conditions decomposition rates are reduced, and organic matter accumulates as peat. Once organic matter has accumulated to a depth of about 10 cm, the plant roots are increasingly isolated from access to the mineral soils beneath the peat. Over time, then, plants become more dependent upon dilute nutrients deposited in rain water (Gorham 1957; van Breemen 1995) and have distributions strongly related to nutrient levels in the groundwater (e.g. Gore 1983; Glaser et al. 1990; Vitt and Chee 1990). Fertilization experiments have shown an array of types of limitation involving nitrogen, phosphorus and potassium (Section 3.4.4).

Adaptation to these infertile conditions requires a variety of unusual plant traits. The most visible is the tendency to sclerophyllous regular type foliage. It is puzzling to see wetland plants showing leaf forms typically found in arid conditions. It appears to be a consequence of low nutrients. Deciduous leaves require conditions of relative fertility, since a plant must continually replace the nitrogen and phosphorus lost in deciduous foliage (Grime 1977, 1979; Chapin 1980; Vitousek 1982). Thus evergreen shrubs in the Ericaceae and evergreen trees in the Pinaceae dominate peatlands. Recall that both of these groups are well known for the use of mycorrhizae in nutrient uptake.

Peatlands are also distinctive in the abundance of bryophytes, and one genus, *Sphagnum* (Figure 10.13), is dominant in bogs. There may be more carbon incorporated in *Sphagnum*, dead and alive, than in any other genus of plant (Clymo and Hayward 1982). The success of *Sphagnum* in dominating large areas has been attributed to at least three characteristics (Clymo and Hayward 1982; van Breemen 1995; Verhoeven and Liefveld 1997). First,

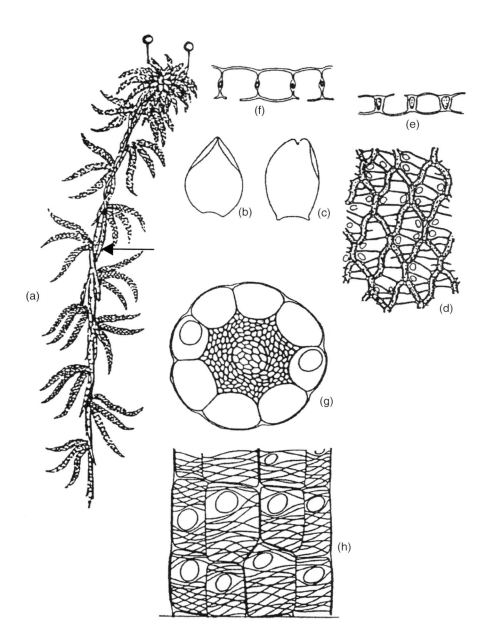

FIGURE 10.13
Morphology and anatomy of *Sphagnum*. (a) to (e) *S. papillosum*: (a) sporophyte-bearing shoot (with pendent branches; see arrow); (b) branch leaf; (c) stem leaf; (d) network of chlorophyllose (stippled) leaf cells, surrounded by porose hyaline cells; (e) cross-section of leaf; (f) same for *S. magellanicum*; (g) and (h): cross-section and external view of stem of *S. papillosum*, showing the large, porose hyaline cells, with fibril thickenings of cell walls at the outside. (From van Breemen 1995)

the morphology and anatomy ensure that *Sphagnum* carpets act like sponges to create permanently wet conditions; about 98 percent of a living *Sphagnum* carpet is pore space: 10 to 20 percent in the hyaline cells of the moss tissue and the rest on the outside of the plant (van Breemen 1995). Second, the cell walls have a high cation exchange capacity that both acidifies the local environment and retains nutrients. Third, organochemical compounds such as phenolics and uronic acids may contribute to suppressing the growth of vascular plants. The establishment of *Sphagnum* may be one of the most important factors in triggering succession in peatlands (Section 8.4.2). As peat accumulates and absorbs water, the diminutive *Sphagnum* moss can actually flood and kill forests, a process known as paludification (van Breemen 1995). The three largest peatlands in the world are the West Siberian Lowland, the Hudson Bay Lowland and the Mackenzie River Basin, comprising more than 3 million square kilometres (Fraser and Keddy 2005).

Box 10.1   The Discovery of Carnivorous Plants

In 1776, as the war for independence raged in eastern North America, biologists were still exploring the continent (Egerton 2007). William Bartram (1739–1823) is probably best known for his book *Travels Through North and South Carolina, Georgia, East and West Florida, the Cherokee Country, the Extensive Territories of the Muscogulges, or Creek Confederacy, and the Country of the Choctaws* (Bartram 1791). This book records his observations while making a journey on horseback 6,000 miles from east to west through the American South from the Carolinas to Louisiana between 1772 and 1777.

Bartram later wrote (1791: p. xx) that *Sarracenia* (pitcher plants), *Drosera rotundifolia* (sundew) and *Dionaea muscipula* (Venus flytrap) are all insect catchers. Hence he appears to deserve credit for being one of the first to describe plants that capture animals. But most extraordinary, he says, is the *Dionaea muscipula*, the Venus flytrap (Figure B10.1.1):

> there behold one of the leaves just closed upon a struggling fly, another has got a worm, its hold is sure, its prey can never escape – carnivorous vegetable!

Darwin himself picked up this thread nearly a century later. Before we examine his treatise on sundews, consider the times in which he worked. Despite his wealth and privilege – as described by Desmond and Moore (1991), two of Darwin's many biographers – the productive period of Darwin's writing life (ca. 1840–80) was a time of both political and personal turmoil. It is true that the British Empire was near the peak of its power and prestige, and that Darwin himself was wealthy, but Europe was in chaos. There was, for example, famine in Ireland, and falling corn prices hit farm incomes, including those of Darwin's own tenants; Darwin was forced to reduce rents. At the same time, his "old sickness was as virulent as ever... Friends thought him a hypochondriac, because he routinely trotted it out as an excuse... But he was no malingerer. The sickness was real and distressing, although no one knew what caused it" (p. 335).

In 1848, an insurrection swept Italy. Barricades went up in Paris at the same time, protesters were shot, the troops mutinied and the French king abdicated; 30,000 communists were said to be planning a revolution. Panic spread among the wealthy, even in England, where 150,000 Chartists were expected to converge in London to support their demands for land taxes, property taxes, and wealth taxes – all of which would have hit Darwin hard. Eighty-five thousand special constables were sworn in "to quell the insurrection by force" (p. 353).

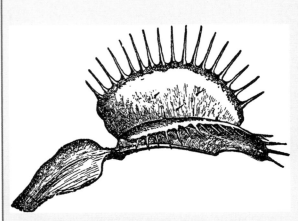

FIGURE B10.1.1   Leaf of *Dionaea muscipula* (from the 1893 *Encyclopaedia Britannica*). For the full plant, see Figure 12.13.

Box 10.1   (cont.)

Meanwhile, Darwin's health continued to fail. "Nine months of nagging fears and obsessive work had taken their toll, leaving Charles chronically depressed. Waves of dizziness and despondency swept over him. Through the winter he suffered dreadful vomiting fits every week. His hands started trembling and he was 'not able to do anything one day out of three.' There were disquieting new symptoms: involuntary twitching, fainting feelings, and black spots before his eyes" (p. 361). He was convinced that death was approaching. Yet he continued to study sundews.

> *Darwin sprinted on with* Insectivorous Plants, *plagued by the interminable manuscript. The prose was muddy, and by February he was bogged down and gasping... He was even "ready to commit suicide," a startled Hooker heard, and the death of an old, sad Lyell on the 22nd [of February] left him feeling "as if we were all soon to go".*
>
> *(Desmond and Moore 1991: p. 613)*

Insectivorous Plants appeared that year and sold out quickly; "in July a 1,000 copy reprint vanished within a fortnight. The name Darwin was a draw now however odd the subject. Who could imagine a 450-page catalogue of plant experiments selling faster than the *Origin of Species*?" (p. 616).

The book begins simply (p. 1): "During the summer of 1860, I was surprised by finding how large a number of insects were caught by the leaves of the common sun-dew (*Drosera rotundifolia*) on a heath in Sussex." In his memoirs, edited by his son Francis (Darwin 1950), he describes this incident more fully.

> *In the summer of 1860 I was idling and resting near Hartfield where two species of* Drosera *abound; and I noticed that numerous insects had been entrapped by the leaves. I carried home some plants, and on giving them insects saw the movements of the tentacles, and this made me think it probable that the insects were caught for a special purpose. Fortunately a crucial test occurred to me, that of placing a large number of leaves in various nitrogenous and non-nitrogenous fluids of equal density; and as soon as I found that the former alone excited energetic movements, it was obvious that here was a fine new field for investigation.*
>
> *During subsequent years, whenever I had leisure, I pursued my experiments, and my book* Insectivorous Plants *was published in July 1875 – that is, sixteen years after my first observations.*

Returning to *Insectivorous Plants*, Darwin summarizes the benefits arising from carnivory (p.14):

> *The absorption of animal matter from captured insects explains how* Drosera *can flourish in extremely poor peaty soil, – in some cases where nothing but* Sphagnum *moss grows, and mosses depend altogether on the atmosphere for their nourishment.*

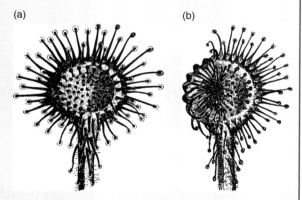

(a)   (b)

FIGURE B10.1.2   (a) The upper surface of the leaves of *Drosera rotundifolia* (round-leaved sundew), Charles Darwin determined, is covered by an average of 192 tentacles. (b) When he placed a piece of meat on the disc, the tentacles on that side inflected over it as they would with an insect. These drawings were made by his son, George. (From Darwin 1888)

(Continued)

Box 10.1   (cont.)

The leaves of *Drosera* are round and covered in gland-bearing tentacles that secrete a sticky fluid (Figure B10.1.2). The capture mechanism is relatively simple (p. 13).

When an insect alights on the central disc, it is instantly entangled by the viscid secretion, and the surrounding tentacles after a time begin to bend, and ultimately clasp it on all sides. Insects are generally killed, according to Dr. Nitschke, in about a quarter of an hour, owing to their trachea being closed by the secretion.

In the second edition of *Insectivorous Plants* completed by his son, Francis Darwin, in 1888, we find the description of additional experiments carried out by Francis (pp. 15–16):

My experiments were begun in June 1877, when the plants were collected and planted in six ordinary soup-plates. Each plate was divided by a low partition into two sets, and the least flourishing half of each culture was selected to be 'fed'. While the rest of the plants were destined to be 'starved'. The plants were prevented from catching insects for themselves by means of a covering of fine gauze, so that the only animal food they obtained was supplied in very minute pieces of roast meet given to the 'fed' plants but withheld from the 'starved' ones. After only 10 days the difference between the fed and starved plants was clearly visible: the fed plants were of brighter green and the tentacles of a more lively red. At the end of August the plants were compared by number, weight, and measurement, with the following striking results:

	Starved	Fed
Weight (without flower-stems)	100	121.5
Number of flower-stems	100	164.9
Weight of stems	100	231.9
Number of capsules	100	194.4
Total calculated weight of seed	100	379.7
Total calculated number of seeds	100	241.5

These results show clearly enough that insectivorous plants derive great advantage from animal food. It is of interest to note that the most striking difference between the two sets of plants is seen in what relates to reproduction – i.e. in the flower stems, the capsules, and the seeds.

10.4   Habitats Constrained by a Regulator: Cold

10.4.1 Arctic and Alpine Plants

Temperature gradients are found on all known planets. Earth and Mars both have ice caps at the poles, although Earth has only water ice, whereas Mars appears to have both water ice and carbon dioxide ice. High mountain tops have similar extremes of cold, and so the vegetation of arctic and alpine regions is usually treated jointly as arctic–alpine vegetation (Figure 10.14). These environments share

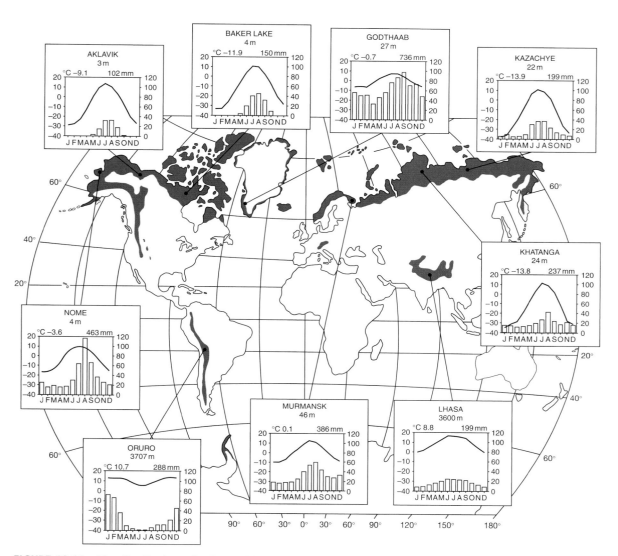

FIGURE 10.14 The distribution of polar and high mountain tundra ecosystems and representative climatic conditions. Mean monthly temperatures are indicated by the line and mean precipitation for each month is shown by the bars. Station elevation, mean annual temperature, and mean precipitation appear at the top of each climograph. (From Archibold 1995)

a number of environmental characteristics that put stress on plants (Billings and Mooney 1968; Savile 1972): (1) low winter temperature, (2) low summer temperature, (3) short growing season, (4) strong winds, (5) long photoperiod, (6) low light intensity, (7) low soil nitrogen, and (8) low precipitation. The flora of the arctic and alpine regions is drawn

from many different plant families, but all species must be able to cope with these environmental conditions. Recall the opening illustration in this chapter (Figure 10.1) for an extreme example.

Low temperatures appear to be the overriding factor. Life as we know it requires liquid water and, if the water is frozen, metabolic processes cannot occur. Further, when ice crystals form in tissues, cell walls and membranes are ruptured (for a graphic demonstration, take a piece of soft plant tissue like a fruit, freeze it, thaw it, and note how the tissues are softened and how much water leaks out). The exact temperature at which the liquid content of cells will freeze depends upon the concentration of solutes. In some cases cell fluids may be super-cooled, meaning that they can be cooled to a temperature below freezing without ice crystals forming. Solute concentrations can be increased by the accumulation of soluble carbohydrates, polyols, amino acids, polyamines and water-soluble proteins (Larcher 2003). These and other aspects of cold tolerance are summarized in Figure 10.15.

The predominant effect of low temperature is the reduction in acquisition of resources through lower rates of photosynthesis and nutrient uptake (Woodward and Kelly 1997) – this is why cold is considered a

regulator. While the exact mechanisms will vary, it should again be evident that chemical reactions, in general, are a function of temperature and, furthermore, that most enzymes have certain temperature ranges within which they function most efficiently. While the cold conditions can be modified by increasing the temperature of tissues, or reducing the temperature of optimum enzyme function, there is no way to avoid the basic laws of chemistry. Plants can avoid cold stress by growing only during warm periods, or they can tolerate cold, say, by increasing the concentrations of compounds within cells that act as anti-freeze, but both of these strategies have costs.

Stress avoidance, that is, restricting growth to warm periods, usually incurs the high costs of a greatly reduced growing season. Creeping and rosette growth forms are less exposed to damage from cold; their meristems are protected in at least three ways: they remain close to ground that is heated by the sun,

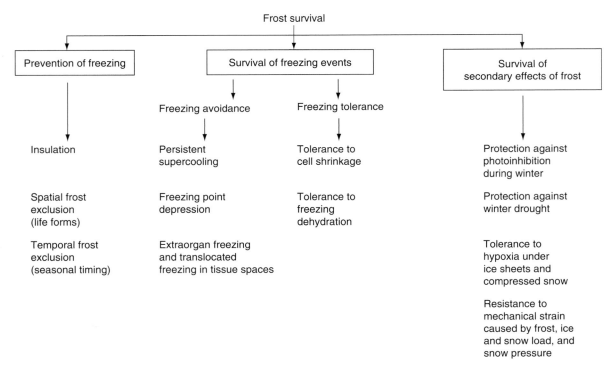

FIGURE 10.15   How plants survive frost and winter stress. (From Larcher 1995)

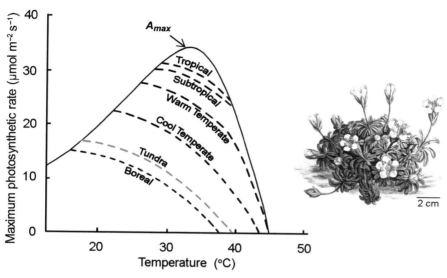

FIGURE 10.16 Maximum photosynthetic rate by biome against temperature. A_{max} is the general maximum photosynthetic rate that cannot be exceeded in any climate (after Woodward and Kelly 1997). The higher line for tundra as opposed to boreal plants may seem unexpected, but this is probably related to the dwarf stature of the tundra plants and the higher near-ground temperatures experienced by leaves during periods of high irradiance, compared with cooler leaves in the tall forest canopy of the boreal forest. Right: A typical cushion plant, *Diapensia lapponica*. (S.T. Edwards, 1808, Peter H. Raven Library/Missouri Botanical Garden)

they are less exposed to wind and the dense (and often hairy) foliage may trap warm air (Archibold 1995). Some arctic plants also have deep anthocyanin pigmentation, which appears to increase rates of absorption and therefore the temperatures of tissues. Savile (1972) suggests that deeply pigmented plants can even extend the growing season by absorbing enough light to commence growth while still buried under snow in the spring. A few arctic and alpine plants also have flowers that track the sun and are shaped like parabolic reflectors (e.g. *Dryas integrifolia*, *Papaver radicatum*). The higher temperatures that are produced by this combination are thought to both attract pollinating insects and enhance maturation of the seeds by maintaining higher temperatures in the tissues of the ovary (Kevan 1975).

Stress tolerance incurs two costs. First, there is the metabolic expense of constructing anti-freezing compounds. Second, there are many possible ways for such compounds to interfere with other aspects of cellular metabolism (Loehle 1998a). As a consequence of such constraints, the maximum photosynthetic rates of arctic and alpine plants are at best only a half of those of plants in warmer areas (Figure 10.16). Note, then, that not only is the growing season often short, but rates of biomass production are inherently low.

Abrasion may be a less appreciated consequence of cold. Savile (1972) observed that: "The most serious form of winter injury to arctic plants is unquestionably that due to abrasion by wind-driven snow particles" (p. 15). Rather than the large soft flakes of snow typical of the temperate zone, much of the arctic snow consists of small, hard and sharp crystals, which, when driven by winter gales, are strongly abrasive. Valleys and the lees of hills provide shelter where dwarf phanaerophytes are found, but each winter the new shoots may be trimmed back by winter

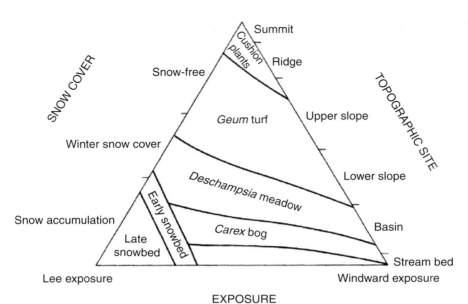

FIGURE 10.17 Alpine vegetation along environmental gradients in the Rocky Mountains. (From Archibold 1995)

gales. Woody plants may be abraded an entire metre above the snow.

These effects of wind abrasion may further illustrate why Raunkiaer placed emphasis upon the location of meristems in his physiognomic classification of plants (recall Figure 2.2). Traits that are interpreted as adaptations to cold may in fact be adaptations specifically to reduce abrasion. Many genera of arctic plants (e.g. *Empetrum, Salix, Vaccinium, Arctostaphylos*) have a prostrate, creeping growth form in spite of being woody plants. Others (e.g. *Draba, Diapensia, Cassiope, Saxifraga*) form densely packed shoots, termed cushion plants, as does *Azorella compacta* (Figure 10.1).

For all these reasons, topography has an enormous effect on the composition of plant communities. Any variation in topography will modify exposure to wind speed and snow depth (Johnson and Billings 1962; del Moral 1983). Hence topography produces a wide array of plant communities in arctic and alpine areas (Figure 10.17). Areas that are buried by snow are protected from abrasion, but there is a cost – they also have a very short growing season since the ground is snow covered for a longer period of time. These snowbeds (Figure 10.17 lower left) tend to have

very few plant species. In the Olympic Peninsula of western North America (del Moral 1983), examples include *Viola glabella, Luzula campestris, Ranunculus eschscholzii* and *Carex spectabilis*. Note too that in the same Figure 10.17 cushion plants are found on snow-free ridges.

Extreme environments, such as cliffs, may have arctic plants well beyond what is considered their normal geographic range. These disjunct populations presumably reflect distributions that were once more extensive during the ice age. The cliff creates a cold and wet environment that simulates some arctic conditions. It may also simply provide a refuge from competition with temperate zone species. In eastern North America, for example, disjunct populations of arctic and alpine plants are found along the north shore of Lake Superior (Figure 10.18)

10.4.2 Deciduous Forests

Cold conditions can occur well outside the arctic, of course. Large areas of deciduous forest experience winter cold and snow each winter. Hence their life cycles are driven by the alternation of summer growth and winter stasis. Many drop their leaves in

FIGURE 10.18 (a) The distribution of an artic–alpine plant, *Castilleja septentrionalis*, in eastern North America shows two disjunct populations far to the south: the northern shores of Lake Superior and the White Mountains of New Hampshire. (b) Photo of *Castilleja septentrionalis*. (Map from Given and Soper 1981; photo by Christopher Mattrick, United States Forest Service)

the autumn. The climate, however, also affects the shape of leaves. Botanists have long noticed that leaf shape varies with latitude: tropical plants tend to have comparatively large leaves with entire margins, but as mean annual temperature drops, the proportion of species with toothed or lobed leaves increases. This change in leaf shape with climate is so diagnostic that it allows paleobotanists to infer the climate associated with fossil floras (Wing 1997). But why should plants in colder areas have toothed leaves?

To explore a possible function of lobed leaves, Baker-Brosh and Peet (1997) used radioactive-labelled carbon to measure sites of photosynthetic activity in the juvenile leaves of 18 woody plants from temperate deciduous forests. Radiography showed that marginal teeth were far more photosynthetically active than the rest of the leaves (Figure 10.19): "Our data indicate that leaf teeth and lobe tips of some species mature early, producing swollen and photosynthetically active tissue while the leaf is quite young and otherwise not photosynthetically active... The deciduous nature of temperate forests creates a situation where precocious photosynthetic structures may be beneficial to trees" (pp. 1253–1254). Lobes and teeth, then, may be leaf organs that mature early to take advantage of the precious long days of sunlight in the spring.

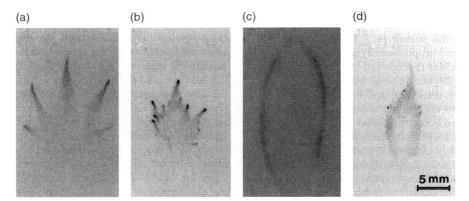

FIGURE 10.19 Autoradiography of immature tree leaves. The dark regions in the marginal teeth and lobes indicate ^{14}C assimilation from heightened photosynthetic activity. (a) *Liquidambar styraciflua*; (b) *Acer rubrum*; (c) *Carya alba*; (d) *Ulmus alata*. (From Baker-Brosh and Peet 1997)

10.5 Habitats Constrained by a Regulator: Salinity

10.5.1 Salinity, Plant Zonation and Physiological Drought

Salinity, like cold, can restrict plant growth, producing distinctive vegetation types such as salt marshes (recall Figure 7.4 top) and mangal (recall Figure 8.31 top). To avoid the negative effects of salinity, plants may exude salt using specialized glands or hairs. They may also shed those parts of the plant in which salinity has reached high levels. Other species tolerate the effects of salinity by maintaining growth in spite of disrupted ion levels in their tissues (Larcher 2003). The effects of small changes in soil salinity on plant growth is vividly illustrated by the visible changes in composition of plant communities in salt marshes and saline lakes. Plants occur in discrete zones associated with specific elevations and salinity (e.g. Poljakoff-Mayber and Gale 1975; Tomlinson 1986; Adam1990).

To understand the negative effects of salinity, it is important to remember how water is extracted from soil: evapotranspiration in the leaves creates osmotic gradients within the tissues of the plant. The water deficit in the leaves is transmitted down the plant through the xylem, causing water to diffuse into the roots (Salisbury and Ross 1988). The greater the salinity, the stronger this osmotic gradient must be to extract water from the soil. These water deficits in photosynthetic tissues can be measured with a pressure bomb, and Figure 10.20 shows that plants growing in salt water have much greater negative tension in their xylem. Hence we can think of salinity as creating a form of drought.

10.5.2 Stress, Zonation and Competition

Here is a good point to remind oneself that the effects of physical factors may act in an indirect way. It might be convenient to assume that physiological adaptations to salinity explain plant zonation in a simple way: plants grow where their physiological tolerances best match the environment. End of story. This assumption, however, requires us to set aside other factors that we already know are important in vegetation, such as competition, herbivory and positive interactions. The real story

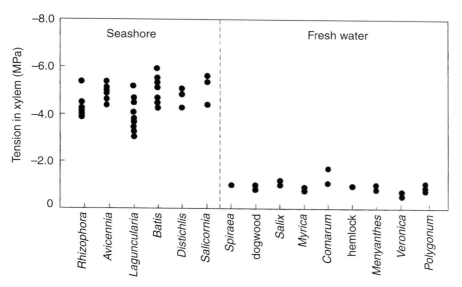

FIGURE 10.20 Xylem tension is higher in wetland plants from habitats that are saline (left) as opposed to fresh water (right). (After Scholander et al. 1965)

appears more complicated: there is growing evidence that species zones are in fact controlled by two interacting causes: stress at one end and competition at the other. Let us look at an example from the salt marshes along the coast of Alaska. Here there are four zones that can be delineated as elevation increases (Jefferies 1977; Vince and Snow 1984): outer mud flat (*Puccinellia nutkaensis),* inner mud flat (*Triglochin maritima*), outer sedge marsh (*Carex ramenskii*) and inner sedge marsh (*C. lyngbyei*). At the lower extreme, the outer mud flat with *P. nut-kaensis* is flooded some 15 times per summer for periods of two to five days each, creating soil water salinity of 15 to 35 percent. At the upper extreme the sedge marsh with *C. lyngbyei* is flooded only twice per summer, when a new or full moon coincides with the perigee (although this single flood may last more than five days), and soil salinity is only 6 to 11 percent, slightly below the 12 percent of flooding sea water. Even further upslope one finds freshwater riverbank levees with *Poa emenins.*

What happens if you move species from one habitat to another? Reciprocal transplant experiments showed that all species could grow in all habitats – so long as neighbouring plants were removed (Snow and Vince 1984). Further, the

P. nutkaensis from the outer mud flats actually grew nearly four times larger when transplanted upslope to the inner mud flat than when transplanted within its own zone. The two species from the highest elevations (*C. lyngbyei* and *P. eminens*) did, however, show reduced growth when transplanted to the outer mud flats. When the same five species were grown at different salinities in pots, all grew best in waterlogged but low-salinity conditions. Thus, in spite of conspicuous zonation, the limited distributions of these species cannot be accounted for simply by their physio-logical tolerances to stress from salinity and flooding. Competition is also a factor. Snow and Vince (1984) suggest that "species occurring in zones along a physical gradient are often limited by physiological tolerance toward one end of the gradient, and by competitive ability towards the other." This important general principle has now been demonstrated in an array of other habitats (Keddy 1990; Pennings and Callaway 1992).

10.5.3 Salinity and Pulses of Regeneration

In other saline wetlands, such as those in Mediter-ranean climates, periods of salinity occur cyclically

(Zedler and Beare 1986). During dry years, hyper-saline conditions develop. Marsh species such as *Spartina foliosa* and *Typha domingensis* cannot tolerate these levels of salinity, and are slowly replaced by salt-tolerant species such as *Salicornia virginica*. This process is reversed during abnormally wet years when higher stream flows and longer rainfalls flush accumulated salt from the soil. If the wet period is short (three to six weeks), only salt-tolerant species such as *Salicornia foliosa* can re-establish, but if the wet period is longer, brackish and freshwater marsh species can re-establish as well. The duration and intensity of freshwater periods determine which species can regenerate. The duration and intensity of the following hypersaline periods act as a filter to determine which species persist. There is thus a constant cycling through different vegetation types driven by changes in moisture supply (Figure 10.21). This example illustrates another important principle regarding stress: the patterns that one observes at any particular point in time can often be interpreted only by understanding how stress changes with time.

10.6   Two Extreme Cases of Stress Tolerance

10.6.1 Endolithic Plants

Some plants may even grow within rocks. Extreme temperatures and wide temperature fluctuations are thought to be the principal factors rendering rock surfaces uninhabitable by plants (Bell 1993). Endolithic floras have been reported from sites in Europe, North America, South Africa, Australia and Antarctica. We will consider two examples.

Consider the Antarctic. Some areas of this continent are ice-free deserts: temperatures range from $-15°C$ to $0°C$ in the summer to as low as $-60°C$ in winter. There is no visible plant or animal life, and even lichens are rare. A flora, including lichens and cyanobacteria, is found growing either in cracks (chasmoendoliths) or structural cavities in porous rocks (cryptoendoliths). Only translucent rocks can provide a habitat for such species. Figure 10.22 shows a cross-section through such rocks and a diagramatic interpretation of growth and exfoliation.

Rock outcrops and cliffs are common in many landscapes. Rock-inhabiting plants have been found to occur on a limestone escarpment in southern Canada. Endolithic life forms were sought by measuring photosynthesis and respiration of rock fragments, assaying ground rock samples for chlorophyll, and using randomly located microscope fields to count visible organisms (Matthes-Sears et al. 1997). The surface of the rocks showed cover of 26 percent by cyanobacteria, 20 percent by lichens and 3 percent by green algae. In the endolithic samples, 50 percent of the random microscope fields contained fungi, 4 percent contained green algae and 1 percent contained cyanobacteria.

10.6.2 Flooded Plants

Too much water can cause stress. The effects of flooded soils upon plant growth are obvious in wet locations (recall Figure 8.31). Roots of actively growing plants require oxygen, and in well-drained soils, aerobic respiration of plant roots and soil microorganisms typically consumes 5 to 24 g oxygen per square metre (Jackson and Drew 1984). When the soil is flooded, pore spaces become filled, and gas exchange between the soil and the atmosphere is virtually eliminated. Microorganisms and roots consume any remaining oxygen. Toxic compounds such as ammonia, ethylene, hydrogen sulfide, acetone and acetic acid are then formed from the anaerobic decomposition of organic matter (Ponnamperuma 1984).

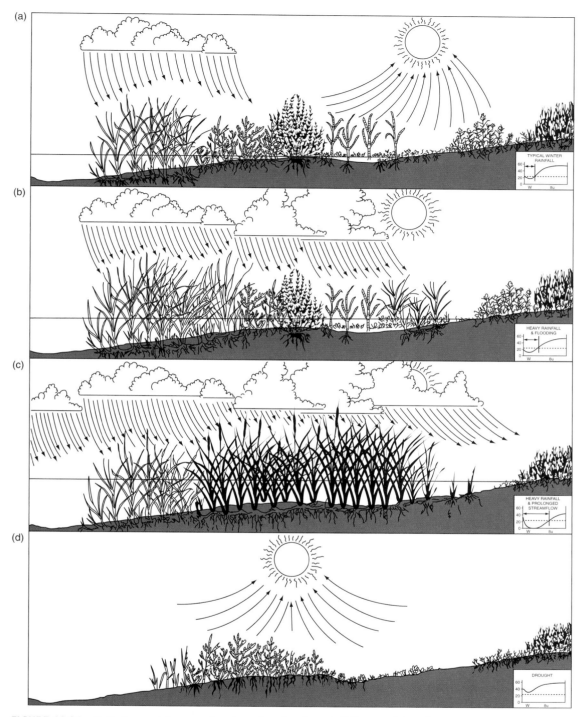

FIGURE 10.21 Salinity, and its occasional moderation by rainfall, may drive changes in coastal marshes in arid climates. (a) Typical saline situation. (b) Floods reduce salinity and allow expansion of *Spartina foliosa*. (c) Prolonged flooding eliminates salt marsh vegetation and allows brackish marsh species to establish. (d) Periods without rainfall or flooding create hypersaline conditions that kill all but a few highly salt-tolerant species such as *Salicornia virginica*. (From Zedler and Beare 1986)

(a)

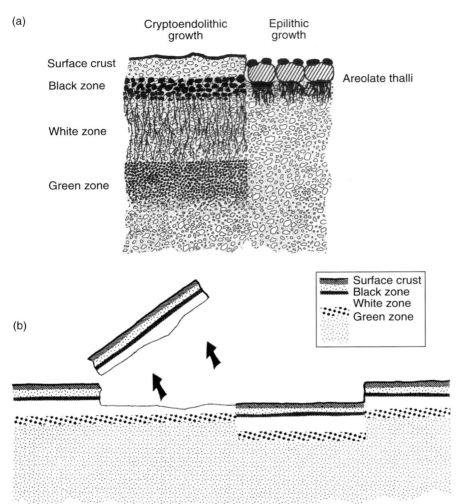

Cryptoendolithic growth

Epilithic growth

Surface crust

Black zone

White zone

Green zone

Areolate thalli

(b)

Surface crust
Black zone
White zone
Green zone

FIGURE 10.22 Lichens in Antarctic sandstone with (a) cryptoendolithic and epilithic growth contrasted and (b) exfoliation and regeneration. (From Friedmann 1982)

The principal mechanism of stress avoidance is air spaces, known generally as aerenchyma, that link roots to leaves. The presence of aerenchyma is one of the most obvious characteristics of wetland plants. In some cases, it appears that the roots of aerenchyma-bearing plants can oxidize their surroundings (Hook 1984) and even provide oxygen for the respiration of roots by neighbouring plants (Callaway and King 1996).

In the absence of other evidence, it is assumed that oxygen diffuses down aerenchyma. But in water lilies, and perhaps in other species as well, there is an internal pressure gradient that causes mass flow. Gases are pressurized in young leaves, thereby forcing air down through the petiole into the rhizome, back up old petioles, and finally escaping through older leaves (Figure 10.23). As much as 22 litres of air a day can enter a single floating leaf and flow to the rhizome

(Dacey 1980; Dacey in Salisbury and Ross 1988: pp. 68–70).

Most trees do not show such obvious features for withstanding flooding, but there is one conspicuous exception – some woody plants produce above-ground extensions of their roots called pneumatophores. These may allow roots direct access to atmospheric gases, but the function has yet to be convincingly demonstrated. It may simply be that pneumatophores vent toxic gases from the flooded soil. Since pneumatophores occur in both major groups of woody plants, gymnosperms (e.g. *Taxodium*) and angiosperms (e.g. *Avicennia*), there would seem to be some consistent evolutionary advantage to their possession in wet soils.

Besides avoiding stress through the development of aerenchyma, there is also good evidence for tolerance to lack of oxygen in some aquatic plants. It has long been known that some contain ethanol and that the rhizomes of plants such as *Nymphaea tuberosa*, *Sagittaria latifolia* and *Typha latifolia* can live anerobically for extended periods (Hutchinson 1975). This remarkable tolerance of rhizomes to anerobic conditions was demonstrated by Laing (1940, 1941), who grew rhizomes from genera including *Acorus*, *Nuphar*, *Peltandra* and *Scirpus* in water through which nitrogen was bubbled. The rhizomes were able to respire anerobically, producing ethanol, with three percent or less of oxygen. Both *Pontederia* and *Typha* showed long persistence even in pure nitrogen!

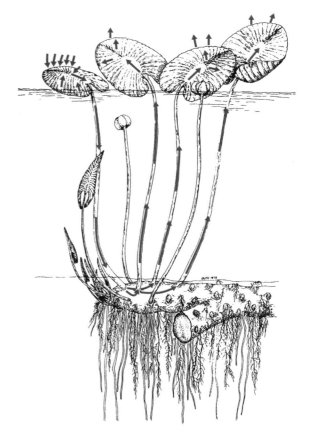

FIGURE 10.23 The movement of air through aerenchyma in water lilies (*Nuphar luteum*). (From Dacey 1981)

10.7 Pollution Is a Source of Stress for Plants

Pollution is a source of stress for plants. It has been a problem at least since Roman lead mining in southern Spain, starting around 600 BC, left traces that are still recorded in cores taken from Greenland glaciers (Rosman et al. 1997). Pollution is certainly a story of the modern industrial era, as illustrated by my own body, containing as it does lead (from leaded gasoline), strontium (from atmospheric nuclear tests), mercury (from coal-burning power plants), DDT (from forest insect spraying), PCB (from electric transformers), BPA (from plastics) and many more that they will likely find should you wish an autopsy. Younger readers will have a different mix, of course, including radionuclides from Chernobyl, neonicotinoids and microplastics. Many of these kinds of pollution also affect plants. Forest die back from acid rain is one of the more visible examples (Little 1995). It is beyond the scope of this book to discuss pollution and plants, except to remind you that it falls into the category of stress, and to share two compelling

FIGURE 10.24 Smoke rises from smouldering bituminous shale, creating a plume of acid rain on the shores of the Beaufort Sea east of the Mackenzie River delta. The cliffs were already on fire when first mapped in 1826. (Angar Walk, 2010, Wikimedia Commons)

FIGURE 10.25 Six months of exposure to gamma radiation created the circular bare patch near the centre of this aerial photograph. The forest is on the property of the Brookhaven National Laboratory on Long Island. (From Woodwell 1963)

examples: an ancient source of acid rain and a classic study of radiation impacts upon forests.

10.7.1 Acid Rain: Lessons From the Smoking Hills

East of the delta of the Mackenzie River in the Northwest Territories of Canada, bituminous shale in 100-metre-high sea cliffs has ignited naturally (Figure 10.24). Explorers document burning as early as 1826, but large piles of ash and oxidized shale at the base of the cliffs suggest that burning has occurred for a much longer period. The prevailing wind carries sulfurous plumes inland where the tundra is fumigated with sulfuric acid mists. Along a gradient of increasing exposure to smoke, concentrations of soil contaminants rise and the number of plant species declines (Freedman et al. 1990). In the most intensely affected sites, all terrestrial vegetation has been killed. Near the cliffs are pollution-tolerant species including *Artemisia tilesii*, the grasses, *Arctagrostis latifolia* and *Phippsia algida*, and lichens such as *Cladonia bellidiflora*. These species are rarely found elsewhere, except at sites where disturbance by erosion is frequent. The least polluted sites have typical tundra vegetation including lichens, bryophytes, *Carex* spp., *Eriophorum* spp., *Dryas integrifolia*, *Luzula arctica*

and *Salix glauca*. These effects rival the worst anthropogenic cases of ecological damage in the vicinity of metal smelters.

10.7.2 Radiation: Lessons From the Brookhaven National Laboratory

Long Island lies on the eastern coast of North America near New York, and is covered with oak–pine forest with *Quercus alba*, *Q. coccinea* and *Pinus rigida* (Figure 10.25). During the Cold War that followed the Second World War, in the 1950s and 1960s, there was great concern about the effects of the fallout that might occur after a nuclear war. These worries were amplified by the frequent testing of atomic bombs in the atmosphere, and by the growing risks posed by commercial nuclear reactors. To study the effects of radiation, a source of caesium 137 was exposed on a post in a stand of Long Island forest (Woodwell 1962, 1963). Rates of exposure varied from several thousand roentgens (R) per day within a few metres of the post, to about 2 R per day at a distance of 130 metres. After six months of irradiation, effects were obvious as far as 40 metres from the source (Figure 10.26). Six zones could be delineated, ranging from total kill of all higher plants to intact forest:

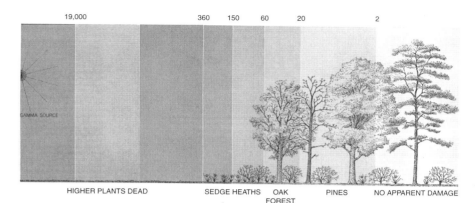

FIGURE 10.26 A radial section showing the effects of chronic irradiation upon species composition and physiognomy of the oak–pine forest shown in Figure 10.25. Numbers indicate radiation level (roentgens/day). (From Woodwell 1963)

1. Devastated zone (>360 R day^{-1}) where all woody and higher herbaceous plants were killed, leaving mosses and lichens.
2. Sedge zone (150–360 R day^{-1}) where all woody plants were killed and the sedge *Carex pennsylvanica* increased in cover.
3. Heath zone (60–150 R day^{-1}) where the trees were killed and ericaceous shrubs were the dominant plants.
4. Oak forest (20–60 R day^{-1}) where the pines were killed leaving an oak canopy.
5. Pine zone (2–20 R day^{-1}) where pine shoot expansion was greatly reduced.
6. Oak–pine forest zone (<2 R day^{-1}) where there was no apparent damage.

Individual tree trunks provided sufficient shielding to produce green "shadows" of living *Carex* where most other vascular plants were killed. Lichens were most resistant to radiation, 11 species surviving on trees at exposures exceeding 2,200 R day^{-1} after 32 months. Pine trees were more sensitive than oaks. It appears that larger plants were more sensitive than small ones. Further, the relative sensitivities of tree species were inversely correlated with the mean volume of chromosomes at interphase. The diversity of plant species also decreased with increasing radiation.

This experiment provided early evidence of the severe ecological effects to be expected from nuclear war. In popular terms, it led to predictions that the post-nuclear world would be a "republic of insects and grass" (Schell 1982). Combined with other impacts, such as those of nuclear winter (Turco et al. 1983), scientists concluded that any large-scale nuclear exchange would probably damage the victor nearly as much as the country that was defeated. Such experiments may have contributed to the ending of atmospheric testing of weapons, to nuclear non-proliferation agreements and even to treaties that have reduced the risk of nuclear war.

10.8 Some Theory

10.8.1 Concepts of Stress and Strain

We have used a rather simple (one might even say simplistic) definition of stress given in Section 10.1, any environmental factor that reduces the rate of production of biomass. Other books (e.g. Levitt 1980; Larcher 2003) use the term differently.

Levitt (1977, 1980) uses two terms: stress and strain. Stress is the external pressures put on an organism by its environment, and strain is the degree to which the physiology of the organism is damaged by that stress. An organism can therefore adapt to a stress by either (1) increasing the stress necessary to produce a specific strain, or (2) decreasing the strain produced by a specific stress (Levitt 1980). In the first case, while the organism cannot alter the external stress, it can prevent the penetration of the stress into its tissues; this is called **stress avoidance**. Some kind of barrier is generally used to prevent the stress from causing strain. Alternatively the organism may allow the stress to enter its tissues but all the same avoid the internal damage. An organism with **stress tolerance** is able to prevent, decrease or repair the injurious strain induced by the stress. In the case of flooding, for example, you have already seen that the strategy of stress avoidance supplies oxygen to the tissues (using aerenchyma) while the strategy of stress tolerance adapts a metabolism to withstand low oxygen concentrations.

10.8.2 Competition Is a Source of Stress

Competition from other plants can be considered a kind of stress, one that adds to that being imposed by abiotic conditions. Stress is measured in costs to the individual plant. These costs might be immediate (the need for increased root construction to locate water) or they might be longer term (the energy required to produce bark to reduce evaporation). Some environments are so far removed from an organism's requirements that they are lethal. One can measure the degree to which any environment departs from the optimum conditions by measuring what it costs the organism to survive under these suboptimal conditions. Or, in the same way, one can measure the amount of stress in an environment by comparing the growth of a plant in that environment with the growth of a plant given optimal growth conditions. The lower the rate of growth relative to the optimum, the greater the stress.

Imagine growing plants along a gradient of resource levels (Figure 10.27), from optimum (left) to lethal (right). The metabolic costs to the plant will increase as stress increases (Figure 10.27, lower curve). Now add in competition. If neighbours reduce

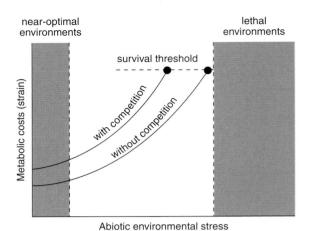

FIGURE 10.27 Suboptimal environments produce strain within organisms. At some extreme combination of conditions, the plant can no longer maintain homeostasis and it dies (•). Competition increases the abiotic stress by reducing resource supplies. (From Keddy 2001)

resource levels further, they will increase the strain imposed by the resource gradient. Thus the growth rate (the inverse of the lines in the figure) will fall more rapidly with competition, and the survival threshold will be lowered. Competition has therefore been defined as "The induction of strain in one organism as a direct result of the use of resource items by another organism" (Welden and Slauson 1986).

10.8.3 Stress Creates Metabolic Costs

In Chapter 3 we saw that foraging for resources incurs costs. These costs might be measured as the energy expended to transport ions across root membranes, as the energy invested in constructing roots to forage for nutrients, or as the nitrogen invested in constructing new leaves to forage for light. The general effect of stress is to reduce the return per unit of investment. This generalization applies to all plants (and all living organisms) whether typical of stressed or productive habitats. The lower the availability of resources, the lower the return for any investment and the higher the cost of survival. Recall from Chapter 3 that the respiratory energy costs of ion uptake in a species of sedge

(*Carex diandra*) was as high as 36 percent of available ATP (van der Werf et al. 1988).

To survive, any individual must maintain a positive energy balance; that is, the rate of photosynthesis must exceed the rate of respiration. Shortages of resources tend to simultaneously reduce rates of photosynthesis and increase rates of respiration. Consider a plant growing in a productive site that is temporarily droughted. Closed stomata will reduce the rate of water loss, but will simultaneously inhibit rates of photosynthesis. Since living cells continue to respire, the ratio of respiration to photosynthesis climbs. If respiration exceeds photosynthesis, the plant is living on stored resources, and if this continues for sufficient time the plant will exhaust these stored reserves and die. Further, in order to obtain water, particularly if neighbours are present depleting the same scarce water, energy must be expended to build new roots to forage for water. Hence just when respiration rates are already high, and photosynthesis rates are already low, added resources may have to be expended for survival. The first definition of stress (Section 10.1) may therefore be rephrased: **stress** is any factor that increases the metabolic costs of survival, and the intensity of the stress is measurable by the amount of energy consumed to deal with it. This accords nicely with the view that energy is the common currency with which to measure the ecological consequences of various phenomena.

One general experimental procedure for examining stress would be to grow individuals along a gradient of resource availability and measure their performance under different conditions. A physiologist might measure metabolic effort as, say, grams CO_2 emitted per unit of time, at each position along the gradient. A less elegant, but usually satisfactory, alternative is to use biomass as an indicator of performance. Reductions in biomass represent the net costs imposed by each set of conditions.

10.8.4 Evolution and Risk Aversion

When thinking about effects of stress, it is important to separate short-term physiological effects from long-term evolutionary effects. In the short term,

drought may cause stomata to temporarily close. In the long term, drought may select for plants having fewer stomata altogether. There is good evidence that plants in stressed environments are selected to have inherently slower growth rates (Grime 1977; Loehle 1998a). The previous example of reduction in density of stomata illustrates how natural selection can produce traits that adapt a plant to stress but will simultaneously limit growth rates under better conditions. More generally, it seems that there are at least three potential causes of low growth rates in plants tolerant of stress. Let us focus on one particular well-known source of stress, cold, while realizing that similar principles apply to other stresses. Three reasons for inherently slow growth rates are: higher production costs, higher operating costs and risk aversion.

1. Higher production costs. Plants from cold regions have thicker leaves, cell walls and cuticles. The increased strength of leaves likely reduces deformation when the leaf is frozen. The cost of this tolerance to freezing is incurred in the added costs of energy and nutrients needed to produce such strongly lignified leaves.

2. Higher operating costs. Many northern leaves contain a variety of chemical compounds including lipids and sugars that act as anti-freeze, allowing leaves to reach lower temperatures without physical damage from frozen and ruptured cells. Anti-freeze compounds, however, not only have production costs, but high concentrations may interfere with other aspects of cellular metabolism. Such interference is likely to reduce growth rates.

3. Risk aversion. Periods of extreme stress are likely to select for traits that minimize risk rather than traits that maximize growth. That is, even if production costs and operating costs were zero, there might still be selection for cautious growth. A northern plant that produces shoots rapidly in early spring may be killed by a late frost. The advantage of growing earlier than most neighbours would seem to be substantial, but this advantage may be balanced by the high costs of damage during unfavourable years. Hence, the logic goes, plants from stressful habitats minimize

risk by cautiously investing in production of new roots and shoots.

Risk aversion is an important factor in evolution. To draw upon a human analogy, risk aversion is the equivalent to investing one's retirement money in low-yielding bonds, rather than putting it all in the stock market. True, the stock market can produce much greater returns – but it simultaneously increases the risk of catastrophic losses. For plants, catastrophic loss usually means death and a failure to contribute to future generations. Hence, the logic goes, those plants present today are the progeny of the cautious investors, since the risk-takers have all died during bad periods. Or, in other words

> The net result of these factors is a conservative growth strategy that tracks the long-term average or the worst conditions and fails to respond to enrichment or favorable growing conditions.
>
> *(Loehle 1998a: p. 736)*

10.8.5 Plants in Stressed Habitats Have Low Growth Rates

Stressed habitats present particular challenges for plants, for plant ecology and for conservation management. They will therefore return for consideration in future Chapters 12 (Diversity) and 13 (Conservation and Management). Let us summarize three points to lay a foundation for those chapters.

1. Since plants of stressed habitats have inherently low growth rate, stressed habitats will, in general, recover more slowly after disturbance. Hence grazing, logging or the use of off-road vehicles may have to be strictly controlled or entirely prohibited. Collection of biomass for fuel from plants like Yareta (Figure 10.1) is almost certainly non-sustainable.

2. Some sites with chronic stress seem to accumulate large numbers of plant species. The fynbos of South Africa (Sections 10.2 and 12.5) and the tepui of the Guyana highlands (Section 10.2) are apparently inhospitable habitats, yet areas with extremely high plant diversity. These areas may require special conservation measures to maintain their high biological value.

3. When a stressed site is fertilized, plants with inherently low growth rates will likely be displaced by common species with higher growth rates. In many cases, this results in rare plant species being replaced by widespread and common species (Section 12.10). More generally, eutrophication (recall Box 3.2) is likely to have negative effects upon many species adapted to infertile habitats. Examples of the negative effects of nutrients on vegetation include chalk grassland (Grime 1973), pine barrens (Ehrenfeld 1983) and wetlands (Keddy 2010).

There may also be a few other generalizations we can make about plants of stressed habitats:

4. Wet areas with low nutrient availability often have carnivorous plants (recall Section 3.7.3, Box 10.1).
5. The kinds of mycorrhizae change with the intensity of stress (recall Section 7.3).
6. Stress seems to increase the likelihood of positive interactions among plants (recall Section 7.2).
7. Stressed habitats seem to have higher below-ground competition than above-ground competition (Section 4.5).
8. Plants in stressed conditions may also have higher investments in anti-herbivore protection (recall Sections 3.7 and 6.3).

10.8.6 The CSR Synthesis

At this point some synthesis may be desirable, if not absolutely necessary. The synthesis that follows was initially developed to examine the evolutionary strategies that might occur in different types of living organisms. This framework can also be used to describe the processes that predominate in creating different plant communities. In order to summarize a large volume of work (e.g. Grime 1973a,b, 1974, 1977, 1979) in a few sentences, let us return to the topic of resources, as presented earlier in Chapter 3. Plants require resources to grow. The availability of resources will determine which of three types of plants occur: competitors, stress tolerators or ruderals (hence the C–S–R acronym).

Competitors. In the absence of external environmental constraints, plants will interact strongly with one another for access to these resources. Roots and shoots will grow rapidly so that a plant may acquire resources; simultaneously these resources will be denied to neighbouring plants. Plants that possess traits that confer success in this active contest for resources are considered to be competitors.

Stress tolerators. If resources are scarce, the returns from active foraging will be minimal and quite possibly well below the costs of building elaborate roots and shoots. Instead, there will be slow growth and, in addition, growth will be uncoupled from periods of resource availability. This will produce a conservation strategy rather than an exploitation strategy and stress tolerators will occur.

Ruderals. If there is frequent disturbance, intense interactions will be prevented. Instead, there will simply be a race to capture as many resources as possible before the next disturbance kills the members of the community. Under these conditions, short-lived and fast-growing species will occur, a group called ruderals. Those familiar with the r–K continuum will recognize ruderals as being the typical r-selected species and competitors as more like the K-selected species.

Characteristic	Competitive	Stress-tolerant	Ruderal
Morphology of shoot	High dense canopy of leaves; extensive lateral spread above and below ground	Extremely wide range of growth forms	Small stature, limited lateral spread
Leaf form	Robust, often mesomorphic	Often small or leathery, or needle-like	Various, often mesomorphic
Litter	Copious, often persistent	Sparse, sometimes persistent	Sparse, not usually persistent
Maximum potential relative growth rate	Rapid	Slow	Rapid
Life forms	Perennial herbs, shrubs and trees	Lichens, perennial herbs, shrubs and trees (often very long-lived)	Annual herbs
Longevity of leaves	Relatively short	Long	Short
Phenology of leaf production	Well-defined peaks of leaf production coinciding with period(s) of maximum potential productivity	Evergreens with various patterns of leaf production	Short period of leaf production in period of high potential productivity
Phenology of flowering	Flowers produced after (or, more rarely, before) periods of maximum potential productivity	No general relationship between time of flowering and season	Flowers produced at the end of temporarily favourable period
Proportion of annual production devoted to seeds	Small	Small	Large

Table 10.4 Some characteristics of competitive, stress-tolerant and ruderal plants. (From Grime 1977)

The characteristics of these three functional types of plants are summarized in Table 10.4.

The three groups of species, suggested Grime, can be positioned along two axes: disturbance and stress (Figure 10.28). Habitats with both high disturbance and high stress cannot support plants, or vegetation. How can one measure disturbance and stress to position species, or habitats, along these axes? As a first approximation, the position of each species in this phase space can be determined by two measurements: the position along the C–R axis can be determined by height and capability for lateral spread, and the position along the C–S axis determined by maximum relative growth rate. Neither of these may be perfect, but the point is to have some measurable attribute. This is a good example of the pragmatic, or empirical, approach to plant ecology (Rigler 1982; Keddy 1987). Southwood (1977, 1988) presented a very similar analysis of life history types, with the exception that he used the word adversity rather than stress.

The two-dimensional model in Figure 10.28 can be represented as a triangle (Figure 10.29). Some thoughtful criticisms have been raised (Taylor et al. 1990), particularly the suggestion that triangular ordinations obscure relationships that might be orthogonal (Loehle 1988b). Whether or not you use the triangular representation, the point is that this approach sorts both plants and vegetation into three main types based upon plant traits and their functions in communities. The Raunkiaer system (Section 2.2.1), by contrast, yields 9 to 13 types. One could ask which is more accurate. It depends on the amount of variability in a particular area, and the purpose of the classification. Three types might be enough for a grassland; a dozen might be more realistic for the world flora. Other classifications may use elements of both. There are now functional

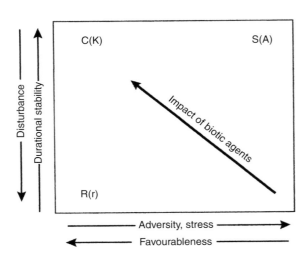

FIGURE 10.28 Two major forces that influence vegetation structure and the evolution of life-history traits are disturbance and stress (or adversity). The letters outside the parentheses follow Grime (1974, 1977): R ruderals; C competitors; S stress tolerators. The letters inside the parentheses follow Southwood (1977, 1988): r r-selection; K K-selection; A adversity-selection. (From Keddy 1989)

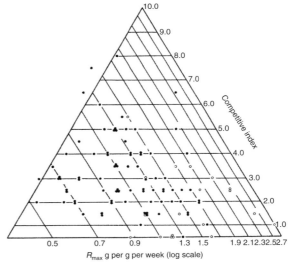

FIGURE 10.29 Triangular view of the CSR model of plant evolutionary strategies. Plant species are arrayed along two axes. The first is a competitive index (based upon height and capacity for lateral spread) and the second is maximum relative growth rate (based upon screening of seedlings for R_{max}). (From Grime 1974)

Box 10.2 Stress Acts as a Filter Upon Species Pools

Although plants produce and disperse vast numbers of seeds, most fail to produce new adult plants. Some are eaten before germinating. Others decay before germinating. Even if they germinate, many kinds of environmental stresses kill young plants. Drought and cold can kill outright. Low nutrient levels and shading can slow growth, making the seedlings vulnerable to further threats. Very few plants live to maturity. By killing plants, the stresses introduced in this chapter often control which plants occur in which locations.

There is a more general way of describing this. We can refer to all the species that occur in a region as a **pool** of species that could potentially occupy any particular piece of habitat. These species rain propagules into a patch of habitat, and then stress acts like a filter, removing all the species that are unsuited to that location. From this perspective, the biological factors such as competition and herbivory are also filters. Hence one begins with a large number of species and individuals, the number of each being determined by the surrounding adults and their ability to disperse propagules into a patch. Environmental filters then subtract and subtract and subtract from the pool until only a few individuals remain. Those are the ones that you see when you visit that location. You see the few survivors, not the vast number of failed immigrants.

The study of pools and filters is one general approach to understanding and predicting plant communities. In principle, if you know the stresses that occur in a certain location, and you know the ecological traits of the species in the pool, you should be able to predict which species will eventually survive. Wet habitats, then, have wetland plants because flooding kills species lacking the ability to tolerate standing water. Arctic habitats have cushion plants because cold kills species that do not have this growth form. And so on. To be more formal still, one begins with a vector of the abundances of S_n species. After a filter kills some of them, S_{n-x} remain (Figure B10.2.1). Add another filter, and another subset dies. Thus, rather quickly, a large regional pool becomes a small subset of species in a specific location.

Let us return to some examples you have encountered so far. The presence of aerenchyma would predict occurrence in wet sites. The presence of thick bark would predict occurrence in regularly burned sites. The presence of a cushion growth form would predict occurrence in cold and windy sites.

You will see in the next chapter how Raunkiaer first attempted to create null sets of species from a pool. With computers and modern statistics, this process has been greatly refined. Large data sets containing the traits of an entire pool remain less common. In one recent study, Laughlin et al. (2011) used 10 life history traits from 15 species of mostly coniferous trees in the southern Rocky Mountains. Traits included bark thickness, height, wood density, specific leaf area (SLA) and leaf nutrient levels. To determine which traits are most important, one begins with the assumption that the most successful species (that is, those with the most biomass) will possess the most appropriate traits for a specified environment. How can we predict which

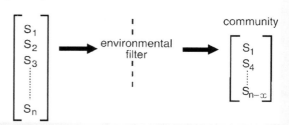

FIGURE B10.2.1   A filter takes a pool of S_n species and reduces it by x species to S_{n-x}. Although a region can have a very large pool of species (what is often called the local flora) a few filters are often sufficient to produce the small number of species found in a typical ecological community.

(Continued)

Box 10.2   (cont.)

species those will be? The general procedure is to try and solve a system of equations in the following general form

$$\sum_{i=1}^{S} t_i p_i = \overline{T}$$

This equation states that the linear combination of species traits (t_i) and unknown species relative abundances (p_i) is equal to the constraint $\overline{T}$, where $\overline{T}$ is the trait value of an average unit of biomass in a community. This approach used a model called MaxEnt, the maximum entropy model for plant communities (Shipley 2009). An exploration of the above forest data set showed that from one third to three quarters of the trait values were related to measured environmental factors. The relationships between traits and environments allow predictions of vegetation changes expected with warmer climate: *Pinus ponderosa* forest can be expected to change to *Juniperus monosperma*, depending upon rates of migration.

There are other models for linking plant traits to environments, such as Traitspace (Laughlin and Laughlin 2013). The challenge for applying such tools is often still a practical one – accurately measuring large numbers of traits on large numbers of species. Which traits should one use, and what is the best way to measure them (Weiher et al. 1999)? There is also the challenge of obtaining accurate measures of the environmental conditions that produce those traits. As you have seen in this chapter, there are strong relationships between environmental constraints (stresses) and the traits that plants have. But much of the information we have addresses only one or a few selected species. The challenge now is to advance from measuring single traits on single species, to measuring sets of traits in entire pools of species (Weiher and Keddy 1999). This should allow us to make an important leap in plant ecology, from merely describing vegetation to predicting how plant communities are assembled in different habitats with different sets of constraints.

classifications available for many of the world's vegetation types and groups of species such as herbaceous understorey plants in forests (Givnish 1987), wetland plants (van der Valk 1981; Boutin and Keddy 1993) and benthic marine algae (Steneck and Dethier 1994). These fall somewhere in between the CSR and Raunkiaer system in terms of scale and number of functional types. Once one begins to think in terms of functional types, it seems reasonable to ask whether there are general rules about plant communities that apply to functional groups as opposed to species. Given the vast number of plant species, it may be that combining them into a smaller number of functionally similar groups will be useful. This takes us back nicely to Chapter 2, where we contrasted the classification of plants by phylogenetic and functional methods.

10.9   Stress Acts at Many Scales

You have seen earlier in this book that the major biomes such as desert, rain forest and boreal forest can be understood and explained by different amounts of rainfall and different temperatures (recall Figure 1.4). It seems likely that these same two factors act at many different scales, all the way down to the distribution

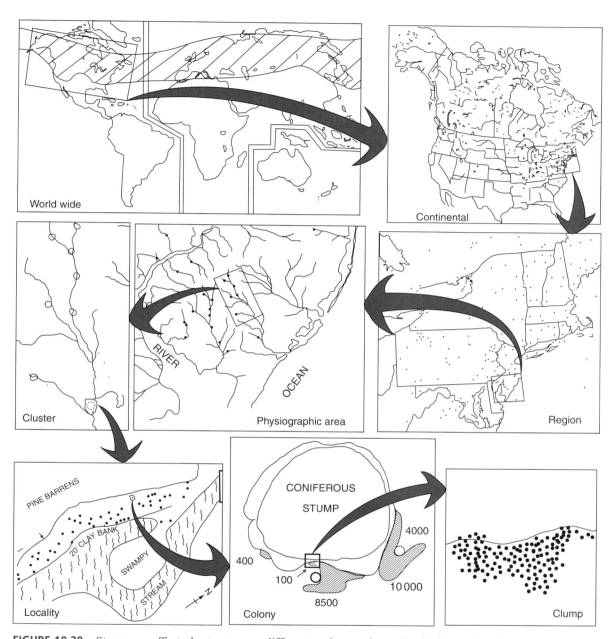

FIGURE 10.30 Stress can affect plants at many different scales, as shown by the hierarchy of distribution of the moss *Tetraphis pellucida*. (From Forman 1964)

of individual mosses on boulders. That is, there is a hierarchy of scales that determine the distribution of any species of plant. To emphasize this obvious point, Figure 10.30 shows the hierarchical distribution of a moss, *Tetraphis pellucida*, from its global distribution at the upper left to its distribution on a single stump at the lower right. Forman (1964) measured the growth of several life history stages of *T. pellucida*, in 504 different experimental environments, and then tried to use these measurements to predict the plant distribution at each scale. The results were often unsatisfactory.

You will read many, many studies explaining how the physiology of a particular species is adapted to its habitat. Not surprisingly, few conclude that the physiology of the selected organism is *not* related to its environment! As a consequence, a good deal of the work written on stress can be criticized as little more than physiological studies combined with entertaining stories. Forman's study was prescient in that in 1964 he had the courage to both make predictions from his physiological measurements and to admit failure when the predictions failed. There is still not a single species of plant for which the causal factors of distribution, at each of these scales, are yet understood.

CONCLUSION

There is no entirely successful categorization of kinds of stress. The three-category classification used in this chapter (absence of resources, unavailability of resources, presence of a regulator) is artificial and not very satisfying – a conclusion you may already have reached from the selection of examples. Consider the limitations. The distinction between absence and unavailability is somewhat artificial since, from the plant's perspective, the situation is the same: a shortage of a resource is preventing growth. From an agricultural perspective, the distinction may retain its utility. In the first case, "absence of resources," one can increase productivity just by supplying the necessary resource – irrigation of deserts being an obvious example. In the case of "unavailability of resources," one needs instead to modify the environment in order to release the resource from its chemical bonds – afforestation is often accomplished by draining peatlands thereby allowing decomposition to occur and releasing nutrients for tree growth. The category of "regulators" is also problematic. Salinity is primarily a stress on plants, to be sure, yet sea water does contain many minerals essential for plant growth (Na, K, Mg, Cl); it is only the high concentrations of the minerals that create stress. In the case of flooding, a major part of the stress is caused by depletion of oxygen, which is a resource. Yet even if one were to somehow provide the roots of submersed plants with oxygen, the physical damage from moving water and the filtration of sunlight by water would remain as stresses (Larson 2001).

These three categories of stress are helpful if they illustrate the array of ways in which stress can occur; the same classification, however, can become an obstacle if one applies it mindlessly or if one attempts to fit every example encountered into a single neat category. Perhaps some bright young reader will produce a better system.

However you classify stresses, you will see that they produce some very distinctive vegetation types. They also produce distinctive plant traits, such as inherently low growth rates. The CSR model suggests that stress tolerators are a rather distinctive group of plants that can be recognized in many different locations.

Overall, stressed vegetation types provide a natural experiment where ecologists can study the evolutionary consequences of chronic shortages of resources. These consequences apparently include evergreen foliage, low growth rates, high below-ground competition, carnivory and commensalism. Further, humans are increasing the stress on many vegetation types from heavy metal pollution, acid rain, nuclear accidents and salinization. It would appear that there will always be work for ecologists willing to study stress in plant communities.

Review Questions

1. Where is the nearest rock barren to your home or university? Where is the nearest area of serpentine? Can you find a list of rare or endemic species for either one?
2. Describe some traits typical of plants that occupy arid environments. Now describe some traits of plants that occupy flooded environments. Note that these can be thought of as the two ends of a long gradient.
3. Explain how fire and grazing expand the areas of the Earth that are covered by grasslands. Without fire and grazing, where would grasslands most likely occur?
4. Why do plants capture small animals? What is the nearest example? See if you can arrange a field trip to see the species in its natural environment. Can you find a list of the animal species it captures?
5. How do peatlands form, and why are nitrogen and phosphorus usually scarce?
6. What do stressful environments tell us about the likely consequences of pollution?
7. Describe the three main plant strategies in the CSR model.

Further Reading

Woodwell, G.M. 1962. Effects of ionizing radation on terrestrial ecosystems. *Science* **138**: 572–577.

Grime, J.P. and R. Hunt. 1975. Relative growth-rate: its range and adaptive significance in a local flora. *Journal of Ecology* **63**: 393–422.

Grime, J.P. 1979. *Plant Strategies and Vegetation Processes*. Chichester, UK: Wiley.

Levitt, J. 1980. *Responses of Plants to Environmental Stresses*, Vols. I and II, 2nd edn. New York, NY: Academic Press.

Dacey, J.W.H. 1981. Pressurized ventilation in the yellow water lily. *Ecology* **62**: 1137–1147.

Friedmann, E.I. 1982. Endolithic microorganisms in the Antarctic cold desert. *Science* **215**: 1045–1053.

Rapport D.J., C. Thorpe and T.C. Hutchinson. 1985. Ecosystem behaviour under stress. *The American Naturalist* **125**: 617–640.

Anderson, R.C., J.S. Fralish and J.M. Baskin (eds.) 1999. *Savannas, Barrens and Rock Outcrop Communities of North America*. Cambridge: Cambridge University Press.

Keeley, J.E. and P.W. Rundel. 2003. Evolution of CAM and C_4 carbon concentrating mechanisms. *International Journal of Plant Science* **164** (Supplement): S55–S77.

Larcher, W. 2003. *Physiological Plant Ecology: Ecophysiology and Stress Physiology of Functional Groups*. 4th edn. Berlin: Springer.

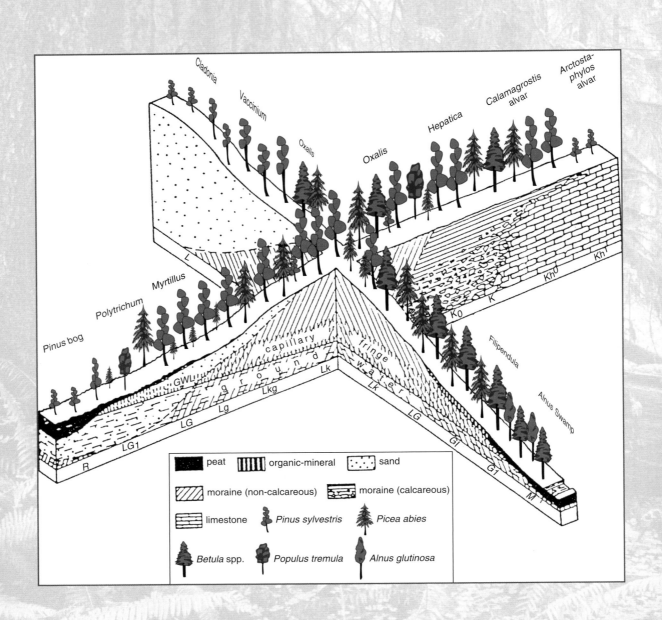

11 Gradients and Plant Communities

Vegetation description. Block diagrams. Profile diagrams. Gradient diagrams. Summary diagrams. Indirect gradient analysis. Use and abuse of multivariate models. Measurement of similarity. Ordination techniques. Sea cliff vegetation. Riverine wetlands. Functional summaries. Vegetation classification. Phytosociology. Site classification and land management. Direct gradient analysis. The importance of null models and tests. Salt marsh zonation. Freshwater shoreline zonation. Emergent wetland zonation. On the existence of communities: the null model perspective.

FIGURE 11.1 Plants are distributed along gradients at many scales, from entire continents, to mountains, to shorelines. How do we meaningfully describe these patterns? This example shows two intersecting gradients found in Estonian forests. (Frey 1973)

11.1   Introduction: Gradients Create Pattern in Plant Communities

Accurate description is essential for both scientific research and conservation. Even large experiments need description – the selection of each dependent variable requires a decision about which properties are most appropriate to describe the possible responses of manipulated communities. In the case of conservation applications, accurate description provides methods to divide landscapes into ecologically similar units, which can then guide the selection of new protected areas, or suggest those areas most appropriate for urban development. Further, once each vegetation unit in a landscape is delineated, it is then possible to select the most appropriate management activities such as grazing or passive recreation. A perusal of any recent journal of plant ecology is therefore likely to yield at least one paper on techniques of vegetation description. While it is of course important to use the right tool in the right way, it also seems that some writers forget that statistical methods are just that – a tool to answer a question. The question must come first. At the same time, one cannot deny that we continue to need good tools for describing vegetation, although not all tools need large statistical packages on computers, at least not if you can draw with pen and ink (Figure 11.1). In a sense, all of the tools we have for analyzing vegetation try to produce a visual simplification of nature so that we can see the essential patterns. The challenge of description has at least four sub-questions.

1. Which measurable properties are most useful for describing vegetation?
2. How do these properties change in space and time?
3. Which patterns in these properties generate important questions about mechanisms?
4. Which patterns in these properties usefully synthesize essential aspects of vegetation?

In this chapter we will survey some of the methods used to describe vegetation, explore their value in scientific research and landscape management, and then revisit their value in basic scientific research.

11.2   Describing Pattern Along Obvious Natural Gradients

The description of species and vegetation distributions in relation to natural gradients is an obvious starting point for examining pattern in local plant communities. An observer can often readily see the changes in vegetation with altitude, amount of flooding, soil depth or salinity. The task then is to choose a method for describing the vegetation and plot the changes in vegetation along this gradient. This approach is often called **direct gradient analysis** (Whittaker 1973a). In contrast, **indirect gradient analysis** uses statistical methods to extract gradients from vegetation data (Section 11.3). There is a long tradition of direct gradient analysis, from Tansley and Chipp (1926) and Weaver and Clements (1929) to Whittaker (1973b) and Orloci (1978). Let us consider a few examples.

First, consider forest gradients in northern Europe, more precisely Estonia, where soil depth, soil moisture and pH are three created gradients along which plant species are distributed. A two-way **block diagram** (Figure 11.1) provides an excellent, if rather qualitative, introduction to the landscapes here. Different segments of the gradient may represent different forest conditions that may require different forestry practices.

In some landscapes, a single gradient may suffice. In his classic book on tropical rain forests, Richards (1996) advocates making **profile diagrams**: "A narrow rectangular strip of forest is marked out with cords, the right angles being obtained with the help of a prismatic compass. In Rain Forest the length of the

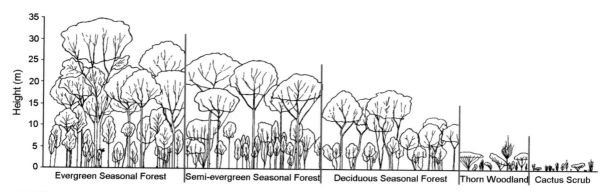

FIGURE 11.2 The distribution of vegetation types along a gradient of increasing dryness in Trinidad. (From Beard 1944)

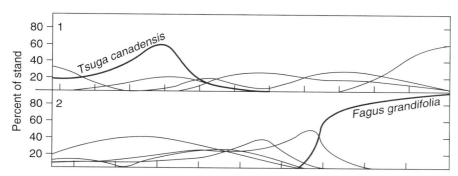

FIGURE 11.3 Distribution of selected tree species in the Great Smoky Mountains, USA. Relative density (percent of stems 2 cm and over) is shown for each species. (1) Dominance of *Tsuga canadensis* along a moisture gradient at 1070 to 1380 m elevation. (2) Dominance of *Fagus grandifolia* along an elevation gradient in mesic sites. (From Whittaker 1956)

strip should not usually be less than 200 ft. (61 m); 25 ft. (7.6 m) has proved a satisfactory width. All small undergrowth and trees less than an arbitrarily chosen lower limit of height are cleared away. The positions of the remaining trees are then mapped and their diameters noted..." Profile data from several sites can be arranged along a gradient such as rainfall, as in Figure 11.2. These sorts of rainfall effects can be found at quite local scales in mountainous areas or on islands. This tool, while laborious, has the advantage of producing a visual representation of the vegetation community, including obvious changes in physiognomy. It does not, however, show changes in taxonomy or diversity in any quantitative way. Similarly,

the gradient along which the plants are arranged is again somewhat qualitative.

More quantitative information about the vegetation can be added by replacing the sketches of plants with measured levels of abundance. Depending upon the study, these measurements might include the cover, density, biomass or canopy cover of a species. There are many examples of such gradient diagrams. For example, Whittaker (1956) described tree distributions along gradients of altitude and exposure in the Great Smoky Mountains by plotting the percentage of the stand comprised by each species (Figure 11.3). While some detail about the physiognomy is lost in such diagrams, information on

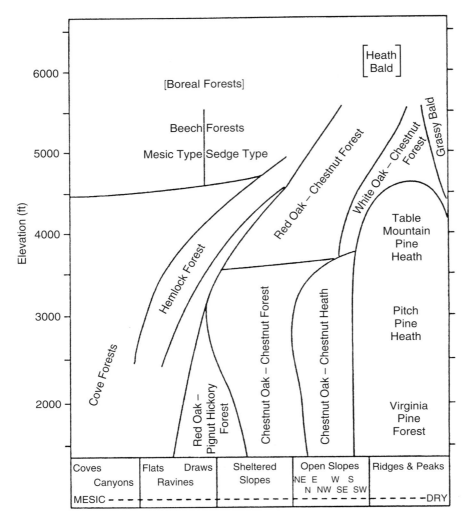

FIGURE 11.4 Non-boreal vegetation types in the Great Smoky Mountains, USA, displayed with respect to elevation and topographic-moisture gradients. (From Whittaker 1956)

individual species and diversity is captured. Comparable examples can be found in Whittaker (1960) and Peet (1978). Different vegetation types can also be plotted along gradients (Figure 11.4). Frequently the delineation of vegetation types is made rather subjectively, based upon the familiarity of an ecologist with the vegetation of the region, who, one trusts, can distinguish between, say, beech forest and hemlock forest. But, of course, these forest types do intergrade, and you shall later see some quantitative methods for making such diagrams.

Diagrams of vegetation types along environmental gradients need not be ends in themselves – they can

also be used as vegetation templates upon which other data are plotted. Figure 11.5 shows an example for the Colorado Front Range of western North America. Look more closely at the axes and the fine dashed lines. It begins with data on gradients. If one were to start at low elevation in a ravine (origin of the graph), and begin climbing, one would start in riparian (floodplain) forest and eventually reach mixed mesophytic (deciduous) forest at an altitude of about 2,200 m. If one continued climbing to 2,500 m one would then enter coniferous forest with *Picea engelmannii* and *Abies lasiocarpa*. Above 3,400 m one would encounter alpine vegetation. If one tired of climbing after

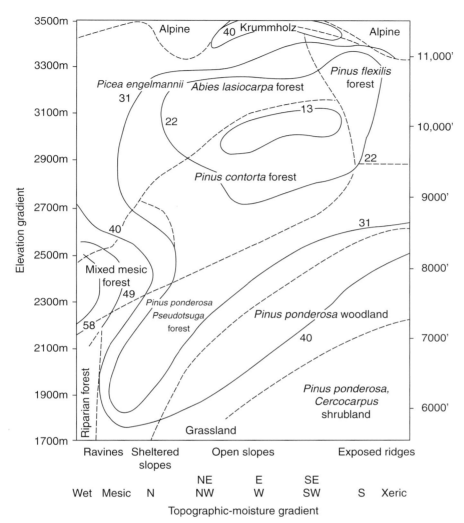

FIGURE 11.5 Species richness in relation to elevation and topographic-moisture gradients in the Colorado Front Range. Solid lines are species richness isopleths starting at 13 species per 0.1 ha and increasing by increments of 9 species. Dashed lines denote boundaries of forest types. (From Peet 1978)

reaching the mixed mesophytic forest, and began walking along the 2,400-metre elevation contour (that is, to the right on the graph), one would reach open slopes and pass into a coniferous forest comprising *Pinus ponderosa* and *Pseudotsuga menziesii*. If one continued to walk along this contour out onto an exposed dry ridge top, one would find oneself in *Pinus ponderosa* woodland. To expand the information in this diagram, Peet also tabulated the species richness of the vegetation and plotted this information as isopleths (the solid lines) on the same figure. The mixed mesophytic forests have the highest richness

(approaching 60 species), whereas the open slopes at 3,000 m have the lowest richness (less than 13 species). Such a template would be a valuable tool for describing vegetation in a park guide and for showing the distributions of other organisms, in the same way as Peet plotted richness.

Finally, the data on vegetation can be simplified still further with the objective of producing a summary diagram of the environmental factors that control the vegetation types of an entire region. In this case, the vegetation gradients may not be actual gradients studied step by step in the field; rather, the

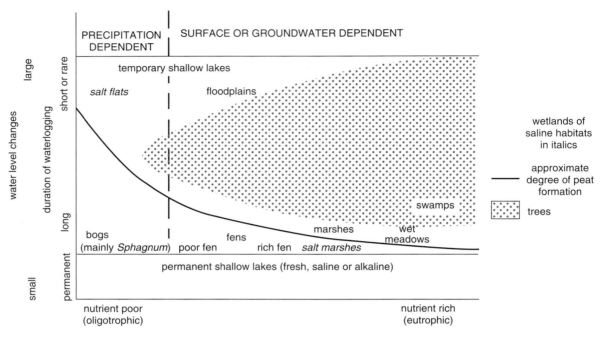

FIGURE 11.6 Wetland type related to water-level changes and nutrient supply. (After Gopal et al. 1990)

display is a means of abstracting and depicting the kinds of vegetation that would be found if one were able to sort all of the existing vegetation types along a few axes. These sorts of summary displays can be very helpful in condensing large amounts of information about landscapes into one diagram; a good analogy might be the periodic table of the elements. For example, Gopal et al. (1990) bravely tried to represent all the world's types of wetland vegetation in one diagram (Figure 11.6). The two main axes were duration of waterlogging and fertility. Superimposed upon duration of waterlogging was the amplitude of water-level changes, and superimposed upon fertility was whether the water came from precipitation or groundwater. The major vegetation types (bogs, fens, marshes, swamps, salt flats and floodplains) were then plotted. Further, the region dominated by trees was shaded in. Unlike a profile diagram, a diagram such as this conveys little information about any particular fen or bog; the strength lies rather in the context into which a particular bog is placed.

The examples above illustrate some of the common ways in which natural gradients have been used to describe vegetation. They are most useful in locations where vegetation gradients are conspicuous (Figure 11.7). These visual methods differ from one another in two ways: (1) the dependent variables used to represent the vegetation, and (2) the independent variables used to describe the environment. The appropriate variables very much depend upon the question being posed by the researcher and the needs of his or her audience. A formal test of the community unit hypothesis might require considerable statistical sophistication, whereas a report for the citizens managing a local nature reserve may require self-explanatory pictorial clarity. One might combine several approaches; a profile diagram through a swamp might be combined with a summary diagram of the entire landscape for example. Although many important questions in ecology now require quantitative methods and field experiments, for many other purposes such as conservation and inventory work,

FIGURE 11.7   Some landscapes have strong natural gradients that produce visible changes in vegetation. Higher elevations may be too dry, cold, windy, rocky or acidic for most plant species to tolerate. Superimposed upon changes in altitude are other factors including soil moisture, aspect and rock type. Such circumstances are illustrated by the granite peaks in the Huangshan Mountains of eastern China. The *Pinus hwangshanensis* (Huangshan pines) have inspired more than a thousand years of Chinese brush paintings, and are now included within a UNESCO World Heritage site. (Shutterstock)

clear descriptions are still invaluable. Before continuing with more sophisticated versions of gradient analysis, it would be helpful to make sure that you completely understand each of the figures above. You should be able to place yourself mentally in the vegetation at any point in any of the above figures and then imagine moving along an axis, visualizing the vegetation as you do so.

Through the early history of ecology, the emphasis was very much upon describing such vegetation patterns. However, along the way a subtle change began to occur. From such representations of species along gradients, scientists began drawing inferences about the underlying mechanisms causing the distributions. This is a dangerous (if necessary) step, because causation cannot be deduced from simple observations. (Were this to be so, we could prove to our satisfaction that having breakfast causes the Sun to rise, because one seems always to occur with the other.) Yet these sorts of distribution diagrams were soon being applied to one of the most contentious debates in plant ecology: whether or not vegetation communities exist. We shall return to this topic in the final section of this chapter.

11.3   Multivariate Methods for Pattern Detection

Sometimes the patterns in plant distributions and vegetation types are less obvious than those we have seen above. When patterns are not obvious, or when you want a quantitative rather than qualitative view of the pattern, there are two multivariate statistical tools that you can apply: ordination and classification. The objective of **ordination** is to create a simplified representation of the pattern of continuous variation in vegetation. The objective of **classification** is to sort different vegetation types into a limited number of groups. In both cases one begins with descriptive data from a large number of sample units. Typically, one has a list of s plant species occurring in a long list of q quadrats (that is, an $s \times q$ matrix of observations). This type of data set is very common in plant ecology. The matrix can be very large, with hundreds of species or quadrats, in which case it may not even be possible to guess the patterns that exist without a statistical tool. We will begin with ordination, reserving classification for Section 11.4. In practice, however, both begin with the same type of data matrix, they are often similar in approach and often both can be found in a single statistical package.

Ordination does not extract something new from the data; rather it is a means of reducing the dimensionality of the data so that previously hidden patterns become evident. In general this means taking the data from an n-dimensional cloud of points (where each dimension represents the abundance of one particular species and each point in the cloud represents a quadrat) to two dimensions (or rarely three), which can be neatly laid out on the page of a journal. Although the complex statistics involved provide ordination with a superficial appearance of rigour, a simple analogy may be the process of taking a complex three-dimensional object, like a frog or a porcupine, and reducing it to a two-dimensional array by driving a large truck over it. In doing so, certain aspects of the shape are emphasized, and certain others are distorted. The objective of the multivariate

technique is to squash the data in a way that emphasizes the patterns and minimizes the distortion, but fundamentally there is no way to reduce the dimensionality of the data (or a frog) without introducing some degree of artificiality.

11.3.1 The Data Matrix

To appreciate the results of an ordination, one must first clearly visualize the data set itself. In the simplest possible case, the data consist of a rectangular matrix of s species abundances in q quadrats (Figure 11.8); such a data set might be used in any number of analytical procedures. A more elaborate matrix can be created when extra rows are added for environmental measurements made in each quadrat. Part of the difficulty of combining environmental measurements with

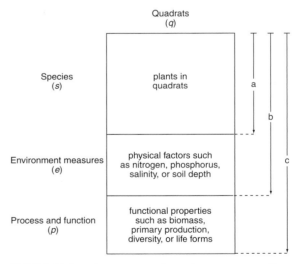

FIGURE 11.8   The raw data for ordination normally consist of a matrix of s species measured in q quadrats (region a). Increasingly, environmental measures are analyzed simultaneously with the species data (region b). Data on process and function can also be incorporated (region c).

species data is that their differences in statistical structure risks further distortions during data analysis. There is also a basic conceptual issue: does the ecologist want the physical factors to influence the patterns that are extracted, or only to add these factors in after the vegetation patterns have already been extracted? In the first case the physical factors are included to create the patterns; in the second, they are applied only to interpret possible causes of the patterns. This conceptual dichotomy illustrates again why it is important to determine your objectives clearly before embarking on data collection, much less choosing the technique for ordination of the resulting data. Finally, Figure 11.8 shows that it is also possible to add rows in the matrix for other properties of vegetation such as primary production, biomass or diversity.

11.3.2 Measuring Similarity

Having selected the kind of raw data to be used in the data matrix (Section 11.3.1), the next step in ordination is to calculate the similarity of every possible pair of quadrats. There is a large number of possible measures of similarity, and these have been conveniently tabulated by Legendre and Legendre (1983), who advise that the choice of a particular measure depends largely upon the objectives of the researcher. We will touch lightly on a few examples here, but it is imperative to consult a more authoritative source (e.g. Orloci 1978; Gauch 1982; Legendre and Legendre 1983; Digby and Kempton 1987; Krebs 1989) before making your decision. Here I want you to focus upon why one does ordination (or classification), and how it fits into the larger strategic picture of plant ecology. These are valuable tools, but they should be used to accomplish a purpose in a research program, not just because the technique happens to exist.

There is one fundamental dichotomy in measures of similarity. The appropriateness of a measure depends upon whether the data matrix has only presence or absence data, or whether the data matrix has measures of abundance for each species.

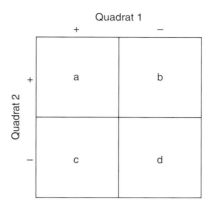

FIGURE 11.9 Similarity in species composition between two quadrats can be expressed in a 2 × 2 contingency table where a is the number of species present in both quadrats, d is the number of species absent from both quadrats, b is the number of species present only in quadrat 2, and c is the number of species present only in quadrat 1.

Presence/absence data

With presence or absence data, the usual measure of similarity for a pair of quadrats is based upon the number of species that are found in both quadrats and the number that are absent from both quadrats. To picture this, imagine that a pair of quadrats has been cast into a 2 × 2 contingency table (Figure 11.9). If the number of joint occurrences of species, a, and joint absences of species, d, is divided by the total number of species being examined, the result is the simple matching coefficient:

$$\frac{a + d}{a + b + c + d}$$

In many cases, the joint occurrences of species will be more informative than the joint absences. Similarity may then be expressed simply as the number of joint occurrences of species, a, divided by the number of species in the two quadrats combined. This yields the Jaccard coefficient, which, although it dates back to 1901, is still a common and valuable measure of similarity for ecological purposes:

$$\frac{a}{a+b+c}$$

Abundance Data

If the species abundances are known for each quadrat, say as counts of individuals or measures of biomass or cover, it is then possible to use another class of measures of similarity. One of the most obvious will be familiar from secondary school analytical geometry courses: the Euclidean distance between the two sample units in n-dimensional space, where each dimension represents one of n species. If the difference in abundance along each of these n axes is measured, the degree to which the two quadrats, i and j, differ in species composition is

$$\sum \sqrt{\left(x_{ik} - x_{jk}\right)^2}/n$$

where $\sqrt{\left(x_{ik} - x_{jk}\right)^2}$ is the Euclidean distance between quadrats i and j for species k, and n is the number of species.

Another common measure of similarity is the correlation coefficient, where the similarity of quadrats is assessed by the degree to which species abundances in the two quadrats are correlated. If both samples have the same species with the same abundances, the similarity is 1. The correlation coefficient, which would be familiar to most of you is, however, susceptible to problems caused by joint zeros (species absent from both quadrats), or when a few common species swamp the others. As far back as 1949, Cole noted that simple presence or absence data are often more informative than measures of correlation based upon species abundances.

The point here is not to summarize all the measures of similarity, since they have been fully enumerated elsewhere (e.g. Legendre and Legendre 1983; Digby and Kempton 1987). Rather, the point is to understand the tactical objective: to measure the similarity of each pair of quadrats sampled. Whatever the final decision on the measure of similarity, the result will be a matrix of similarities for each pair of quadrats. This similarity matrix is the intermediate step between the raw data matrix and the ordination. Since there are q quadrats in the data matrix, the similarity matrix will

be $q \times q$. Since the measure of similarity is usually symmetrical (the similarity of quadrat 1 compared to quadrat 2 is the same as quadrat 2 compared to 1), only the values on one side of the diagonal need to be reported. The diagonal, which consists of similarities for each quadrat compared to itself, can be omitted.

In summary, we begin with a raw data matrix. A measure of similarity is then used to transform these data into a similarity matrix. The similarity matrix is the input for the next step: ordination or classification. The objective of the ordination or classification is to take the information in the similarity matrix, and present it in a pictorial manner that is easier to comprehend. Since we find it difficult to think in more than three dimensions, the usual objective is to find a means to represent the patterns of similarity in two or three dimensions.

11.3.3 Ordination Techniques

The final decision required of the investigator is to select a technique for ordination. This will in part be determined by the kind of raw data available for preparing the similarity matrix. The techniques available for ordination include Bray–Curtis ordination, principal component analysis, factor analysis, reciprocal averaging, canonical correspondence analysis and detrended correspondence analysis, listed roughly in order of their historical use by plant ecologists. The trade-offs among these approaches can be explored in papers such as Gauch and Whittaker (1972), Gauch and Wentworth (1976) and Gauch et al. (1977). One of the biggest problems with many techniques is their tendency to take linear relationships in the data and bend them into curves in the ordination; this happens even when entirely artificial data are used (Gauch and Whittaker 1972). The second axis of the ordination therefore frequently reflected not a real environmental gradient but simply the curvature created by the technique itself (rather like the way in which flattening an animal will generally create a protuberance to one side or another). Newer techniques such as reciprocal averaging (RA) and detrended correspondence analysis (DCA) tend to minimize this curvature but then so too does the earliest Bray–Curtis ordination (Gauch et al. 1977). Since new techniques are continually being developed, it is important to consult the most recent literature before choosing a technique. At the same time, the most recent techniques are often the least well understood; for all their weaknesses, the Bray–Curtis approach and principal components analysis are at least relatively simple and their deficiencies well known. By using such basic techniques, one also avoids being seduced by a technique with such complex algorithms that one can no longer understand them entirely.

11.3.4 Ordinations Based Upon Species Data

Figure 11.10 shows the results of an ordination of sea cliff vegetation in England using the principal components analysis technique (Goldsmith 1973a). While this technique has now been replaced by more

elaborate ones, mentioned above, the patterns are still evident. The array shows a distribution of data points, each representing a quadrat, plotted according to the first two principal components (Figure 11.10a). Note that in this example, the data matrix consisted only of species abundances in each quadrat; the pattern reflects the changes in composition of the vegetation compressed into two dimensions. To interpret the possible environmental causes of these patterns, one might return to one's knowledge of the natural history of these species. For example, *Armeria maritima* is found in exposed and saline sites, and so its abundance in one portion of the ordination space might suggest that these are the quadrats from exposed and saline cliff faces (Figure 11.10c). By plotting the abundance of species for each point, one can see that different points represent dominance by different plant species. Alternatively, if one wanted to explore environmental factors more directly, one could superimpose measurements of environmental factors, such as soil

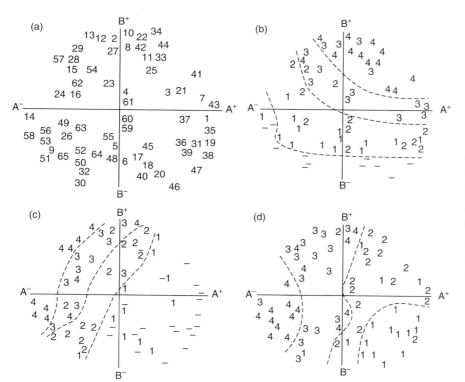

FIGURE 11.10 An ordination of sea cliff vegetation using principal components analysis (PCA). (a) distribution of 65 quadrats along the first two PC axes; (b) abundance of *Festuca rubra*, a dominant grass species; (c) abundance of *Armeria maritima*; (d) salinity of each quadrat measured as conductivity. (After Goldsmith 1973a)

salinity, on each point. Gradients in soil salinity could then be visualized (Figure 11.10d), and one might tentatively conclude that soil salinity is an important factor controlling plant distributions on sea cliffs.

Another reason for beginning with Goldsmith's work, apart from the simplicity and clarity of the patterns he found, is the remarkable fact that he did not stop his enquiry after doing the ordination. Based upon these patterns he hypothesized that competition was a major factor controlling species distributions and that salinity controlled the outcome of competition. Then he actually did the experiments (Goldsmith 1978) to test these hypotheses! This is such a rare feature of ordination studies that it deserves recognition. Another example can be found in the work of Larson on lichens and their distribution on cliff faces; the initial ordination work (Larson 1980) led to hypotheses about physiological responses to temperature, moisture and drought, and these hypotheses were tested with an extensive series of growth chamber experiments (e.g. Larson 1982, 1989).

11.3.5 Ordinations Can Combine Species and Environmental Data

In Goldsmith's (1973a) work (Figure 11.10), the ordination was based only upon the plant species, and the physical factors were then explored in a second stage. Alternatively, one might combine physical factor data and species data in one large matrix, or even expand the data further by adding in factors such as plant diversity or productivity. Recall Figure 3.10 that showed the results of a factor analysis of an expanded data matrix for a freshwater marsh in eastern Canada. Since little was known about relationships among species, biomass, diversity and fertility, Auclair et al. (1976a) sought patterns by combining all of these different measures in one data set. The particular technique, factor analysis, is no longer widely used, but in practical terms, it still illustrates rather well the nature of the results of ordination. In Figure 3.10 at the upper right, it is evident that soil fertility, productivity and the abundance of *Typha angustifolia* are all closely associated

in this wetland. The upper left shows that fire incidence, species density and the abundance of *Carex lanuginosa* are associated. This figure also incorporates a third dimension by using different types of dots, with the green circles being positive (raised above the surface of the page) and the brown ones being negative (falling below the surface of the page).

One of the strengths of ordination is its ability to incorporate and display many environmental factors and vegetation characteristics that are correlated to varying degrees at different scales. One frequently attends seminars where someone will point out that the factor under study is actually correlated with others, as if this observation were a clever and original criticism of the research. In fact, the world of plant ecology is filled with strongly correlated factors. This is not a problem so much as a simple statement of reality. On a hillside, say, where moisture, soil depth, temperature and vegetation may all be correlated, it may not even make much sense to ask about the independent effects of each of these factors, for even if one factor could be isolated from all others, the fact remains that the others co-vary in nature.

To emphasize this point, and further illustrate the strength of ordination, consider the practical example of herbaceous vegetation on rock outcrops within forest (Figure 11.11a). At the large scale, there is variation caused by geology and climate. At more local scales, aspect, slope and elevation will vary among sites. Superimposed upon this variation are biotic processes such as dispersal, competition, mycorrhizae and grazing, and correlated physical factors such as soil moisture, depth of snow cover, growing season, soil temperature, wind speed and so on. To study any one of these factors in isolation would be misleading because none of them occurs in isolation. Therefore the best procedure seems to be to describe the multivariate variation in vegetation and relate it to multivariate variation in the environment. Therefore Wiser et al. (1996) collected species presence and cover data for 154,100 m^2 samples from 42 Appalachian peaks from 11 locations in the mountains of North Carolina. A total of 294 vascular plant species was recorded. Environmental factors

(a)

(b)

FIGURE 11.11
Multivariate analysis of
high-altitude rock
outcrops in the
Appalachian Mountains.
(a) The *Selaginella
tortipila/Carex
umbellata* vegetation
type on Hawksbill
Mountain (Photograph
by Bob Peet).
(b) Hierarchical
classification of
154,100 m^2 plots into
nine vegetation types
using TWINSPAN. (c)
DCA ordination of all
plots coded by the nine
TWINSPAN
communities. (d)
Environmental
correlations
superimposed over plot
locations in DCA space.
(From Wiser et al. 1996)

Lower elevations

Higher elevations

Felsic rock
with seepage,
or mafic rock

Felsic
rock

Mid – high
elevations

Higher
elevations

Southern

Northern
& central

Abundant
seepage

Little
seepage

Somewhat
lower
elevations,
or less
exposed sites

Highest
elevations,
exposed sites

Low sol.rad.,
amphibolite
or metabasalt

High sol.rad.,
other rock
types

Anakeesta
slate

Other
rock, or
anakeesta
slate >1980 m

1. Coreopsis/
Schizachyrium

2. Selaginella/
Carex umbellata

3. Aronia/
Kalmia

4. Chelone/
Oxypolis

5. Paronychia/
Polypodium

6. Deschampsia/
Angelica

7. Picea/
Leiophyllum

8. Calamagrostis/
Rhod. carol.

9. Aster/
Menziesia

(c)

FIGURE 11.11   *(cont.)*

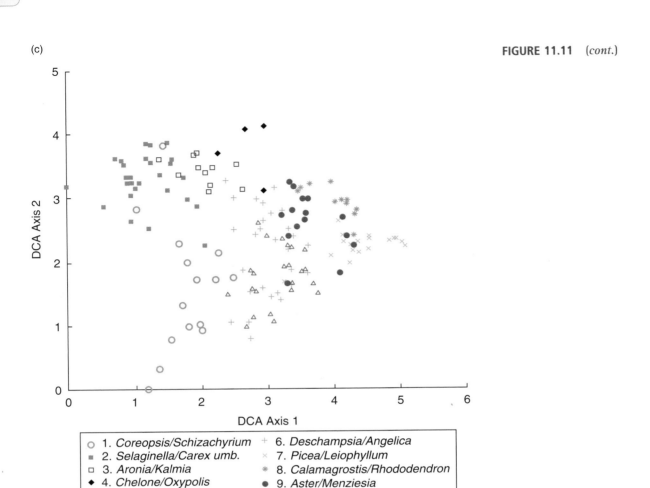

included topographic position, vertical relief, surface fracturing, exposure, rock type (mafic, felsic, intermediate) and outcrop area. In smaller plots, lichen cover, seepage, pH, organic content and soil nutrients were recorded.

A hierarchical classification using the now standard technique of TWINSPAN (Section 11.4.2) was first used to sort the vegetation samples into nine different community types (Figure 11.11b). Only vegetation data were used in the analysis, but note that the dominant vegetation types are associated with different elevations at first, and then with rock type and exposure. The *Aronia arbutifolia –Kalmia latifolia* community, for example, occurs on shallow slopes over felsic bedrock (granite domes) at low to mid elevations. In contrast, the *Selaginella tortipila–Carex umbellata* vegetation occurs below 1,600 m on felsic bedrock in the southern part of the study area. When the vegetation samples are ordinated using detrended correspondence analysis (DCA), the nine community types can then be superimposed on two DCA axes (Figure 11.11c). The *Selaginella–Carex* community falls lowest on axis 1, for example, and the *Coreopsis–Schizachyrium* community falls lowest on axis 2. Finally, superimposed upon the same ordination axes are the environmental factors. Each factor is labelled and oriented to indicate its relative contribution to the two axes; the longer the line, the greater the effect. Thus plant communities to the left, low on axis 1, have higher solar radiation and higher pH, whereas those to the right, high on axis 1, have higher elevation and fracturing. The combination of all these

(d)

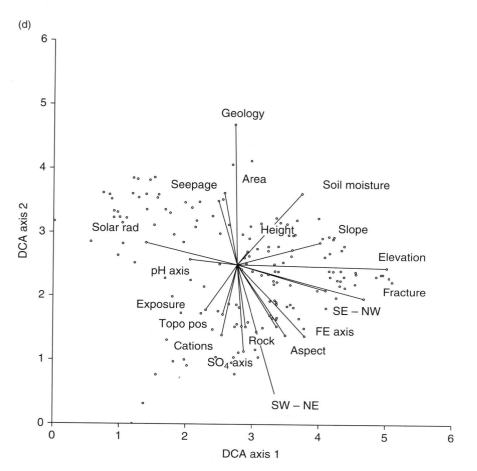

FIGURE 11.11   (*cont.*)

analyses, as Figure 11.11 shows, provides a detailed snapshot of the vegetation and environmental factors on high-elevation rock outcrops and a solid footing for conservation planning, or future experimental work aimed at unravelling cause and effect.

11.3.6 Functional Simplification in Ordination

When the results of ordination are presented, species names are often used. The problem with this approach is that there are so many plant species in the world that most of us know only a limited flora. The names on the axes are therefore largely meaningless to many readers. Even if one recognizes the names, they are of limited use in drawing ecological conclusions from the data. Why should this be surprising? After all,

names are something put on plants by humans, they are not natural properties of plants or ecosystems. When one tries to interpret ordination axes using species names, one is usually drawing upon some additional knowledge about each species: its size, its tolerance of salinity or some other trait. The problem is that the reader who is unfamiliar with the species does not know the traits unless they are explained on a case-by-case basis. This suggests that it might be very useful to plot plant traits upon the ordination; these might include competitive ability or relative growth rate (as measured in independent experiments). This is rarely done. Even when factors such as fertility and litter mass are included, as in Figure 3.10, there is still a good deal of effort involved in interpreting the results, particularly for the non-specialist. One could, however, simplify the results by

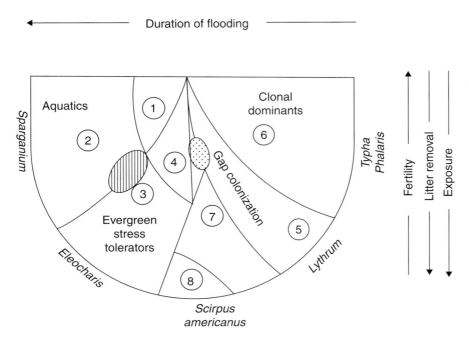

FIGURE 11.12

Summary of dominant plant species and processes along shorelines of the Ottawa River, Canada, derived from an ordination of plant species overlaid by environmental factors. The hatched and stippled regions show the locations of *Scirpus acutus* and *Spartina pectinata*, respectively. (From Day et al. 1988)

subjectively extracting the key factors and plotting them in simplified form. One would then combine the main species associations with the key factors thought to be producing them. (Of course, such a figure would be placed in the discussion section of the paper, not the results.) Another study of riparian wetlands used this approach to produce a functional summary of herbaceous vegetation types, where both dominant species (e.g. *Sparganium eurycarpum*) and processes (e.g. gap colonization) were combined (Figure 11.12).

11.4   Vegetation Classification

Classification divides landscapes and vegetation into discrete groups. Vegetation types are often delineated on the basis of species composition, but classification schemes vary depending upon their purpose. In selecting a classification scheme one needs to consider objectives carefully. As a research tool, classification may have limited value, tending to obscure the inter-actions that occur among species along gradients. One is reminded of Pielou's (1975) warning about ecologists who are more interested in finding the elusive "homogeneous ecosystem" than in studying the gradients that occur in real systems. Further, classification may suffer from the same problems as ordination – being applied mindlessly too often as a substitute for a

properly thought out research question. When it comes to the practical task of managing real landscapes and plant communities, or planning reserve systems, some method of vegetation classification is essential. For example, if one is managing large tracts of mixed deciduous forest, those regions dominated by oak may need to be burned to produce oak regeneration; if one does not have a vegetation map indicating areas of oak forest, one cannot proceed with appropriate manage-ment. Similarly, in designing a reserve system to pro-tect the different ecosystems of the world, one cannot know how many reserves are needed without first identifying ecosystem types. A list of major vegetation types provides an objective template for judging when a reserve system is complete. As well, procedures such as gap analysis, to detect ecosystem types not yet included in the reserve system (Section 13.3.1), are entirely dependent upon the adequacy of the classification scheme (Noss and Cooperrider 1994). At present, most vegetation classification schemes are developed by individual land management agencies, according to agency objectives. Gradually, consistent global systems such as maps of ecoregions (Olson et al. 2001; Figure 2.14) are emerging.

No single classification scheme will fill every need; reserve selection may emphasize the natural diversity of vegetation and land forms, whereas management of large mammals may require knowledge of only a few cover types, and management of an endangered species may require mapping of some very specific feature significant only to the particular species. In designing and applying vegetation classification schemes, we should strive to make the best possible use of existing data and to incorporate the latest understanding of vegetation dynamics.

11.4.1 Phytosociology

Phytosociological systems of vegetation classification are particularly well developed in Europe. Here there has been an emphasis upon classifying stands of vegetation rather than studying changes along gradients, and a number of classification schemes have been developed. The most widely used system is that of the Zurich–Montpellier school of phytosociology (Westhoff and van der Maarel 1973; Mueller-Dombois and Ellenberg 1974; Beeftink 1977). This system is based entirely upon the plant species cover in standard sample areas. The vegetation therein is then classified hierarchically, going downward through classes, orders, alliances, to associations. Each level in the hierarchy is identified using parts of the names of the defining species along with a special suffix to indicate the level (Table 11.1). Thus a tract of salt marsh dominated by *Spartina maritima* would be in the Class Spartinetea, Order Spartinetalia, Alliance Spartinion and

Table 11.1 **Levels and units of the formal hierarchy of the Zurich–Montpellier school of phytosociology from highest (division) to lowest (subvariant), with suffixes and examples of the construction of names based on denominating taxa. (From Westhoff and van der Maarel 1973)**

Syntaxon	Suffix	Examples	Denominating taxa
Division	-ea	Querco–Fagea	genus *Fagus*
Class	-etea	Phragmitetea	*Phragmites australis*
		Querco–Fagetea silvaticae	*Quercus robur, Fagus sylvatica*
Order	-etalia	Littorelletalia	*Littorella uniflora*
		Festuco–Sedetalia	*Festuca* and *Sedum* L. div. spp.
Alliance	-ion	Agropyro–Rumicion crispi	*Agropyron repens* (syn.; *Elytrigia repens*) and *Rumex crispus*
		Alnion glutinosae	*Alnus glutinosa*
Suballiance	-ion (-esion)	Ulmion carpinifoliae (Ulmesion)	*Ulmus carpinifolia*
Association	-etum	Ericetum tetralicis	*Erica tetralix*
		Elymo–Ammophiletum	*Elymus arenarius, Ammophila arenaria*
Subassociation	-etosum	Arrhenatheretum elatioris brizetosum	*Briza media*
Variant		ibid., *Salvia* variant	*Salvia pratensis*
Subvariant		ibid., *Bromus* subvariant	*Bromus erectus*

Table 11.2 Zurich–Montpellier classification of salt-marsh communities in the southwestern Netherlands based on

Classes Orders	Thero–Salicornietea Thero–Salicornietalia	Spartinetea Spartinetalia		Asteretea tripolii Glauco–Puccinellietalia	
Alliances	Thero–Salicornion	Spartinion		Puccinellion maritimae	
Associations	Salicornietum strictae	Spartinetum maritimae	Spartinetum townsendii	Puccinellietum maritimae	Halimionetum portulacoidis
Column	1	2	3	4	5
Number of relevés	14	24	30	124	40
Character-taxa of the associations					
Salicornia europaea coll.[1]	100(1--3)	33(+-2)	14(+-2)	78(+ -2)	40(+-1)
Spartina maritima	14(+)	100(2--4)	–	6(+)	–
Fucus vesiculosus f. volubilis	–	79(+-5)	7(2--3)	4(+ -2)	–
Spartina townsendii agg.	86(+-1)	79(+-2)	100(3--5)	50(+-2)	35(+-2)
Puccinellia maritima[2]	25(+-1)	33(+-1)	30(+-1)	100(3--5)	92(+-2)
Halimione portulacoides	8(+)	–	33(r--1)	83(+-2)	100(3--5)
Artemisia maritima	–	–	–	5(+-1)	–
Armeria maritima	–	–	–	6(+-1)	–
Carex extensa	–	–	–	–	–
Puccinellia distans	–	–	–	–	–
Puccinellia fasciculata	–	–	–	–	–
Puccinellia retroflexa	–	–	–	–	–
Scirpus maritimus var. compactus[3]	–	–	27(+-2)	–	–
Faithful taxa of Puccinellion maritimae					
Bostrychia scorpioides	–	17(+-4)	14(+-2)	21(+-4)	37(+-4)
Character-taxa of Armerion maritimae					
Juncus gerardi	–	–	–	2(+)	–
Festuca rubra f. litoralis	–	–	–	14(+-2)	62(+-1)
Glaux maritima	–	–	10(r--1)	50(+-2)	15(+-1)
Parapholis strigosa	–	–	–	–	–
Agrostis stolonifera var. compacta subvar. salina	–	–	–	–	–
Character-taxon of Puccinellio-Spergularion salinae					
Spergularia salina	–				
Character-taxa of Glauco-Puccinellietalia					
Spergularia media	–	–	7(r-+)	73(+-2)	42(+-2)
Limonium vulgare ssp.vulgare	8(+)	8(+)	30(r-+)	73(+-2)	60(+-2)
Character-taxa of Asteretea tripolii					
Aster tripolium	50(+-2)	33(+-2)	77(r--2)	98(+-2)	97(+-2)
Triglochin maritima	8(+)	–	27(r--2)	86(+-4)	62(+-2)
Plantago maritima	–	–	20(r--1)	65(+-4)	65(+-2)
Other taxa					
Suaeda maritima	50(+-1)	12(+-2)	37(+-1)	64(+-2)	60(+-2)
Atriplex hastata	–	–	77(+-2)	25(+-2)	5(+)
Elytrigia pungens	–	–	7(r-+)	1(+)	22(+)
Lolium perenne	–	–	–	–	–
Plantago coronopus	–	–	–	–	–
Phragmites communis	–	–	–	–	–

Addenda

Column1: Zostera noltii 29(+-2); Column 7: Centaurium pulchellum 17(+-2), Carex distans 5(+), Sagina maritima 2(+), Solanum dulcamara 2(+); Column 8: Centaureum 14(r), Trifolium repens 14(r), Centaurium littorale 14(+)[0]; Column 9: Polygonum aviculare 41(+-2), Elytrigia repens 22(+-1), Potentilla anserina 8(+-1), Plantago major 16(+-1)[0], Poa annua 11(+-2), Cochlearia officinalis 11(+), Festuca arundinacea 8(+), Cirsium arvense 5(+-1), Poa trivialis 3(3), Hordeum secalinum 3(1), Taraxacum Centaurium pulchellum 10(+), Plantago major 20(r)[0], Matricaria inodora 10(r)[0], Bromus mollis 10(+), Sagina maritima 40(+-2), Juncus bufonius 40(r--2), littoralis 5(+).

[1] In the alliances Thero–Salicornion and Spartinion represented by S. stricta Dum.

[2] Preferential character-taxon of the association; also selective character taxon of the alliance Puccinellion maritimae and exclusive character-taxon of the order Glauco–Puccinellietalia.

[3] Also character-taxon of the alliance Halo–Scirpion.

N.B. The superscript[0] is a convention meaning that the taxon is represented by stunted individuals.

Braun–Blanquet species abundance measures. (From Westhoff and van der Maarel 1973)

Armerion maritimae			Puccinellio–Spergularion salinae			Halo–Scirpion
Artemisietum-maritimae 6 61	Juncetum gerardii 7 64	Junco–Caricetum extensae 8 7	Puccinellietum distantis 9 37	Puccinellietum fasciculatae 10 10	Puccinellietum retroflexae 11 12	Halo–Scirpetum maritimi 12 19
$30(+-1)^0$	$42(+-1)^0$	$14(+)^0$	$35(+-2)^0$	$70(r--2)^0$	$100(1--2)^0$	–
–	–	–	–	–	–	–
–	–	–	–	–	–	–
13(+)	20(+−2)		$8(+)^0$	–	17(r)	37(+−2)
43(+−2)	31(+−2)	14(+)	62(+−1)	90(r−−3)	25(r−−1)	26(1−−3)
98(+−3)	62(+−1)	–	5(+)	–	–	–
93(+−3)	59(+−1)	–	8(+−1)	–	–	–
2(+)	90(+−3)	14(r)	–	–	–	–
–	–	100(1−−4)	–	–	–	–
–	–	–	100(1−−5)	$20(r--2)^0$	17(r−+)	–
–	–	–	–	100(2−−4)	17(r)	–
–	–	–	–	–	100(1−−4)	–
–	–	–	$32(+-2)^0$	$10(+)^0$	$8(r)^0$	100(3−−5)
5(+ −2)	–	–	–	–		
–	86(+−5)	100(2−−4)	14(+−2)	30(+)	–	5(+)
100(3−−5)	97(+−5)	86(+−3)	35(+−3)	10(r)	–	
33(+−1)	98(+−3)	100(2−−3)	35(+−3)	30(r−−2)	–	11(+)
5(+−1)	64(+−3)	43(+−2)	19(+−2)	10(+)	8(+)	–
–	11(+−1)	100(+−3)	65(+−2)	30(+−2)	–	42(+−4)
–	–	–	97(+ −3)	80(+ −2)	100(r− −2)	–
72(+−2)	47(+−1)	14(+)	19(+−1)	20(1−−2)	8(+)	–
66(+−2)	91(+−3)	57(r−−2)	$3(+)^0$	–	–	–
100(+−2)	72(+−2)	71(r−−1)	65(+−3)	100(r−−3)	100(+−4)	74(+−2)
29(+−1)	77(+−2)	29(+)	5(+−1)	50(r−−2)	8(r)	16(+−1)
72(+−3)	98(+−3)	100(+−2)	11(+)	–	8(+)	–
31(+−1)	$20(+-1)^0$	–	$38(r--1)^0$	$30(r--1)^0$	$8(1)^0$	–
11(+)	$5(+)^0$	–	$62(+-2)^0$	$40(r)^0$	–	63(1−−3)
31(+−2)	28(+−2)	–	30(+−2)	–	–	32(+−2)
–	2(+)	–	30(+−3)	10(r)	–	–
–	2(+)	29(+)	11(+−1)	10(1)	$25(+)^0$	–
–	2(+)	$43(r-+)^0$	30(+−3)	$10(r)^0$	$67(r--2)^0$	11(+−2)

pulchellum 57(r−−2), *Carex distans* 29(r−−2), *Juncus maritimus* 29(r−−1), *Lotus tenuis* 29(2), *Hippophae rhamnoides* 29(r−+), *Trifolium fragiferum* 14(r), *Sonchus arvensis* 24(+−2), *Leontodon autumnalis* 5(+), *Trifolium repens* $11(+)^0$, *Coronopus squamatus* 8(+−2), *Matricaria inodora* $11(+)^0$, *Bromus mollis* 8(+), *Ranunculus sceleratus* sp. 5(+), *Sonchus arvensis* 3(+), *Poa pratensis* 3(+), *Solanum nigrum* 3(+), *Senecio vulgaris* 3(+), *Anagallis arvensis* 3(+), *Leontodon nudicaulis* 3(+); Column10: *Hordeum marinum* 10(r), *Samolus valerandi* 10(r); Column 11: *Bromus mollis* 8(r); Column12: *Ranunculus sceleratus* 5(+), *Cochlearia officinalis* 5(+), *Atriplex*

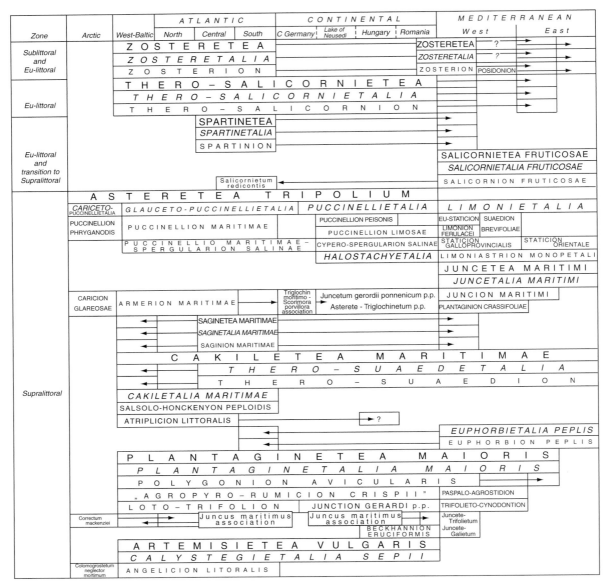

FIGURE 11.13 Geographical distribution of the European higher salt-marsh syntaxa. (From Beeftink 1977)

Association Spartinetum maritimae. Table 11.2 illustrates the application of this technique to salt marsh communities in the Netherlands. Once such tables are prepared for different sites, it becomes possible to recognize the geographical distribution of specific vegetation types and the relationship between these vegetation types and the environmental gradients (Figure 11.13). In this classification

system, a plant community is termed a "phytocoenose," a quadrat becomes a "relevé" and communities become "syntaxa." The use of this approach may be most suitable in European landscapes that are largely mosaics of vegetation types resulting from past agriculture, grazing and forestry regimes. It seems less useful in those areas of the world where natural gradients remain. All students of plant ecology should be familiar with this system since it is widely used in the European scientific literature.

11.4.2 Classification for Land Management

There are many statistical methods for classifying vegetation data. Sneath and Sokal (1973) provide a particularly lucid introduction to the topic, although there are many other sources you could also consult (e.g. Legendre and Legendre 1983; Pielou 1984; Digby and Kempton 1987; Gnanadesikan 1997). All have the objective of organizing a set of quadrats into clusters having similar species composition. Just as with ordination, one begins with the species by quadrats $(s \times q)$ data matrix shown in Figure 11.8, one selects a measure of similarity and then one applies a statistical technique (algorithm) to obtain the clusters. You may have noticed that I slipped a common classification method into the ordination discussion in Section 11.3.5 (Figure 11.11b), in order to present the Appalachian Mountains example in its entirety. In that case, the classification tool was TWINSPAN (two-way indicator species analysis), which not only sorts quadrats into clusters, but organizes them hierarchically and identifies indicator species for each cluster. In the Appalachian example, the resulting clusters were then superimposed upon the ordination in Figure 11.11c. Note in this example that the habitat interpretations (e.g. "lower elevations") in the TWINSPAN diagram (Figure 11.11b) were added later to assist with interpreting the data, and are not a part of the TWINSPAN output.

To illustrate the application of classification to land management, we will focus upon the boreal forests near Hudson Bay in Canada, as summarized in a document called *Field Guide to Forest Ecosystem Classification for the Clay Belt, Site Region 3E* – developed by a consortium of government agencies concerned with forest management in northern Ontario (Jones et al. 1983b). This report begins: "If management knowledge and experience are to be organised, communicated and used effectively, a practical, clear system for classifying [forest] stands

(ecosystems) is needed to ensure that each manager knows what the others are talking or writing about" (p. 1). The first step in understanding the value of this document in particular and classification in general is to explain the title. What is the clay belt, and what is a Site Region? In 1961, Ontario was divided into 13 distinctive Site Regions, based upon landform and climate (Hills 1961). Each Site Region was divided, in turn, into smaller regions termed Site Districts. For decades, this was the system used to classify Ontario ecological regions. In order to make this system consistent with more modern terminology, as presented in Chapter 2 (recall Figure 2.14), these Ontario Site Regions have since been reworked into Ecoregions, while Site Districts have become Ecodistricts. Ecoregion 1E, for example, refers to tundra-like lowlands along the coast of Hudson Bay; Ecoregion 7E refers to rich deciduous forest along the north shore of Lake Erie. The region we are focusing on here, Ecoregion 3E (Figure 11.14), was formed at the end of the last ice age when Lake Barlow–Ojibway formed at the margin of the retreating ice (Baldwin1958). The lake lasted about 2,000 years and apparently drained some 11,000 years BP, leaving 70,000 square miles between James Bay and the Great Lakes with a thick layer (ca. 70 m deep) of lacustrine clay. Thus the origin of the popular name, the clay belt. Since it is so flat, the drainage is poor, and extensive peatlands now cover much of the area. In

FIGURE 11.14 Ontario has been divided into 13 ecoregions, each of which has a characteristic landform and climate. Ecoregion 3E, shown here, typically has flat clay plains covered by peatlands with *Picea mariana* (black spruce) forests.

these peatlands, *Picea mariana* is the dominant tree. More well-drained areas including rock outcrops and sandy ridges have a fire-dependent tree, *Pinus banksiana*.

To construct the guide, a large number of vegetation samples were collected, ordinated using detrended correspondence analysis, and then classified into 23 vegetation types using TWINSPAN. These were simplified to 14 "operational groups" for management purposes: "No classification is of practical use if the management staff cannot allocate any stand to its class quickly in the field using only a few easily

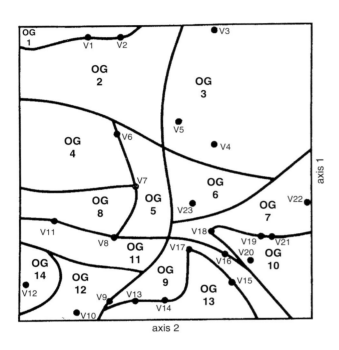

OG 4 FEATHERMOSS-COARSE SOIL
Sb (&/or Pj)-Feathermoss on Fresh-Moist Sandy or Coarse Loamy Soil

OG 5 FEATHERMOSS-FINE SOIL
Sb (&/or Pj)-Feathermoss on Fresh-Moist Fine Loamy-Clayey Soil

OG 6 LYCOPODIUM
Lycopodium-Herb Poor on Fresh-Moist Loamy Soil

OG 7 MIXEDWOOD-HERB RICH
Hardwood/Mixedwood-Herb Rich on Fresh-Moist Fine Loamy-Clayey Soil

OG 8 FEATHERMOSS-SPHAGNUM
Sb-Feathermoss-Sphagnum on Moist Fine Loamy-Clayey Soil with 20–39 cm. Organic Matter

OG 9 CONIFER-HERB/MOSS RICH
Conifer Mixed-Herb Rich on Moist Fine Loamy-Clayey Soil

OG 10 HARDWOOD-ALNUS
Hardwood-Alnus-Herb Rich on Moist Fine Loamy-Clayey Soils with Thick Black Organic-Mineral Forest Humus Form

OG 11 LEDUM
sb-Ledum on Wet Moderately Decomposed Organic Soil with Thick Surface Fibric Horizon

OG 12 ALNUS-HERB POOR
sb-Alnus-Herb Poor on Wet Moderately Decomposed Organic Soil with Thick Surface Fibric Horizon

OG 13 ALNUS-HERB RICH
sb (&/or Ce/L)-Alnus-Herb Rich on Wet Well Decomposed Organic Soil with Thick Surface Fibric Horizon

OG 14 CHAMAEDAPHNE
sb - Chamaedaphne on Wet Poorly Decomposed Organic Soil with Thick Surface Fibric Horizon

Operational group names

OG 1 VERY SHALLOW SOIL OVER BEDROCK
Sb &/or Pj-Feathermoss-Lichen on Very Shallow Soil over Bedrock

OG 2 VACCINIUM
Pj (& Sb)-Vaccinium-Feathermoss-Lichen on Dry or Fresh Sandy Soil

OG 3 DIERVILLA
Hardwood/Mixedwood-Vaccinium-Diervilla-Herb Poor on Fresh Sandy or Loamy Soil

FIGURE 11.15 Ordination and classification divides a region of northern Ontario (Site Region 3E in Figure 11.14) into 14 vegetation types, or operational groups, associated with different soil types and water availability. (From *Field Guide to Forest Ecosystem Classification for the Clay Belt, Site Region 3E*, Ontario Ministry of Natural Resources, 1983)

recognised diagnostic features. Further, there should be only a sensible number of classes to cover the whole range of forest conditions if the number of management prescriptions is to be reasonable" (Jones et al. 1983a, pp. 1–2). The classification was combined with the ordination to produce an array of the 23 vegetation types in the two-dimensional phase space defined by the ordination axes. Operational groups were overlaid on the results (Figure 11.15). Each operational group has a name, ranging from the driest, OG1, "Very Shallow Soil Over Bedrock" to a peatland, OG14, "Chamaedaphne," referring to the dominant ericaceous shrub in this habitat. The guide provides a short key using indicator species to enable

a manager to rapidly assign a site to one of these 14 operational groups.

Figure 11.16 summarizes the vegetation and physical factors in this classification. In the upper left, OG1 is dominated by Pj (jack pine, *Pinus divaricata*) whereas OG14 (lower left) has Sb (black spruce, *Picea mariana*); the codes are those in standard use by Ontario government foresters. The understorey vegetation changes from feathermoss (*Dicranum* spp.) in OG1 to *Sphagnum* in OG14. Soil texture is bedrock in OG1 and organic matter (peat) in OG14. Finally, the soil moisture regime ranges from dry–fresh in OG1 to very wet in OG14. In short, the array in Figure 11.15 summarizes the vegetation types found in the clay belt of

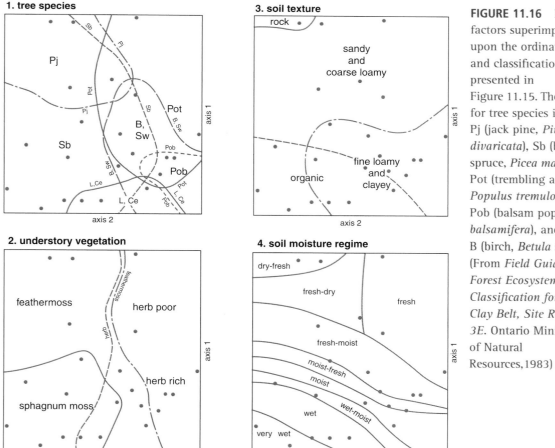

FIGURE 11.16 Four factors superimposed upon the ordination and classification presented in Figure 11.15. The codes for tree species include Pj (jack pine, *Pinus divaricata*), Sb (black spruce, *Picea mariana*), Pot (trembling aspen, *Populus tremuloides*), Pob (balsam poplar, *P. balsamifera*), and B (birch, *Betula* spp.). (From *Field Guide to Forest Ecosystem Classification for the Clay Belt, Site Region 3E*. Ontario Ministry of Natural Resources, 1983)

very shallow soil over bedrock

Sb &/or Pj-Feathermoss-Lichen on Very Shallow Soil over Bedrock

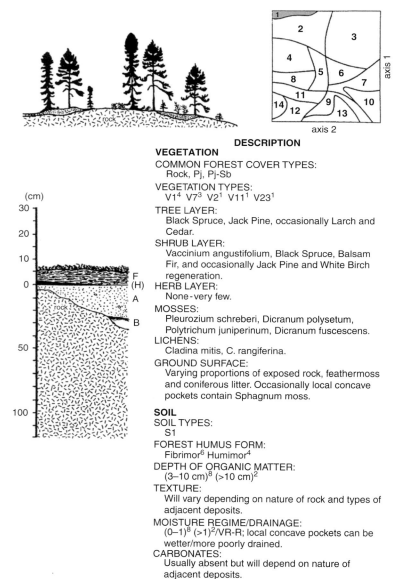

DESCRIPTION

VEGETATION

COMMON FOREST COVER TYPES:
  Rock, Pj, Pj-Sb

VEGETATION TYPES:
  V1[4] V7[3] V2[1] V11[1] V23[1]

TREE LAYER:
  Black Spruce, Jack Pine, occasionally Larch and Cedar.

SHRUB LAYER:
  Vaccinium angustifolium, Black Spruce, Balsam Fir, and occasionally Jack Pine and White Birch regeneration.

HERB LAYER:
  None-very few.

MOSSES:
  Pleurozium schreberi, Dicranum polysetum, Polytrichum juniperinum, Dicranum fuscescens.

LICHENS:
  Cladina mitis, C. rangiferina.

GROUND SURFACE:
  Varying proportions of exposed rock, feathermoss and coniferous litter. Occasionally local concave pockets contain Sphagnum moss.

SOIL

SOIL TYPES:
  S1

FOREST HUMUS FORM:
  Fibrimor[6] Humimor[4]

DEPTH OF ORGANIC MATTER:
  (3–10 cm)[8] (>10 cm)[2]

TEXTURE:
  Will vary depending on nature of rock and types of adjacent deposits.

MOISTURE REGIME/DRAINAGE:
  (0–1)[8] (>1)[2]/VR-R; local concave pockets can be wetter/more poorly drained.

CARBONATES:
  Usually absent but will depend on nature of adjacent deposits.

FIGURE 11.17   A typical one-page biography of an operational group (OG), in this case the example of black spruce and jack pine in shallow soil on rock ridges (OG1). Each OG requires a different set of management guidelines. (From *Field Guide to Forest Ecosystem Classification for the Clay Belt, Site Region 3E.* Ontario Ministry of Natural Resources, 1983)

northern Ontario, from jack pine on rock ridges to ericaceous peat bogs with black spruce in the wet sites. In between one finds other conditions, such as Sw (white spruce, *Picea glauca*) on loam with a rich herbaceous understorey (e.g. OG7).

The guide then provides a short biography of each operational group (Figure 11.17) with a profile of the community and a summary of the vegetation and soils. Note that it is keyed back to the ordination in Figure 11.15 with a shaded region in the small ordination diagram at the upper left. Simply flipping through the 14 pages with the biographies provides a rapid reconnaissance of the communities to be expected in the vast area of site region 3E. Managers can therefore prescribe different land-use strategies according to the operational group. Jack pine on rock ridges might, for example, be clear-cut to enhance regeneration, whereas wet peatlands might be left untouched. Intermediate sites, such as OG6, might be clear-cut and then replanted. Similarly, if one has the objective of setting up a park system to represent the vegetation diversity of this region, one would want

to ensure that the park system contained large enough examples of these 14 operational groups so that they would persist through time. If a gap analysis of the existing system were conducted one might similarly discover that one OG was not represented, and this would guide one in seeking a representative example of this OG to add to the park system.

Figures 11.15 to 11.17 not only illustrate the procedure of classification used in one section of the boreal forest of Ontario, but they also provide a series of steps that could be used in any other vegetation region. The field guide from which these figures are reproduced contains other elements, including keys to ecosystem types based upon plant species, descriptions of the soils and a section on plant recognition with line drawings so that field workers can quickly assign sites to OGs. I describe this example in some detail not only to illustrate the role of classification, but to provide some inspiration to young ecologists who may face a similar task in another part of the world.

11.5 Gradients and Communities

11.5.1 Clements and Gleason

We have now seen how natural gradients provide a useful starting point for describing pattern in plant communities (Section 11.2). We have also examined how gradients can be detected by applying ordination techniques to species × sample matrices (Section 11.3). Gradients also figure prominently in one of the major debates of plant ecology (Whittaker 1962; McIntosh 1967; Colinvaux 1978; Shipley and Keddy 1987), usually framed as a debate between two American ecologists, Frederick Clements (1874–1945) and Henry Gleason (1882–1975). Since both Clements and Gleason wrote extensively, any summary of their work is bound to be an oversimplification, and text books are forced by space limitations to present caricatures of their views. Essays that attempt more detail often fall into using quotations from different periods

of their careers, which usually generates more confusion and leads away from the original topic of debate (the structure, if any, of plant communities) toward separate topics – how the views of two scholars changed with time, and even how they understood or misunderstood the meaning of words such as "random" (e.g. Nicolson and McIntosh 2002). Before we know it, an important topic – the nature and meaning of pattern in ecological communities (a topic that could inspire new work) – becomes the frustrating attempt to try to interpret 50-year-old quotations from long-dead scholars whose views likely changed during their careers.

Yet one cannot ignore this topic. It involves two important historical figures in ecology. It raises the important issue of how structured ecological communities actually are. Moreover, those of you who are already familiar with the topic from introductory texts may expect a more nuanced treatment of the debate. Further, the topic reminds us that, contrary to the impressions left by some books (e.g. Strong et al. 1984; Gotelli and Graves 1996), the issue of structure in communities, and the use of null models to examine it, is a traditional approach, not a radical new topic in ecology (Box 11.1). There is also an important lesson for future work. The debate illustrates the folly into which we fall when we argue about topics without explicitly describing what measurements are needed to distinguish between alternatives. It therefore issues a challenge to us: either frame an argument in a measurable way, or ignore it and move on to something more fruitful. So let us revisit the topic with the advantage of nearly a century of hindsight.

In general, Clements is presented as an ecologist who argued for strongly integrated communities with consistent, recurring species composition (community units) that were somewhat like super organisms:

> The developmental study of vegetation necessarily rests upon the assumption that the unit or climax formation is an organic entity... As an organism the formation arises, grows, matures, and dies. Its response to the habitat is shown in processes or functions and in structures which are the record as well as the result of these functions [p. 3]... The climax formation is the adult organism, the fully-developed community, of which all initial and medial stages are but stages of development [p. 125]... Finally, all of these viewpoints are summed up in that which regards succession as the growth or development and the reproduction of a complex organism [pp. 3–4].
>
> *Clements (1916)*

Gleason is presented as his antagonist, who in 1917 published a counter-treatise: "The structure and development of the plant association." In one familiar sentence, Gleason says:

> The development and maintenance of vegetation is therefore merely the resultant of the development

and maintenance of the component individuals, and is favoured, modified, retarded or inhibited by all causes which influence the component plants.
> (p. 464)

Who could disagree? I'm not convinced that if Clements were available for interview, that he would argue with this statement. If there is a difference between the two, it seems to be a matter of degree (see Box 11.2).

11.5.2 The Temporary Victory of the Gleasonian View

Many ecology texts give the impression that the matter has been resolved in favour of Gleason's individualistic approach (see Colinvaux 1978 for a particularly readable summary). This "individualistic" approach is based on each individual species having its own distribution, independent of other species with which it may occasionally occur.

Two types of evidence are usually presented to support this view. The first is the use of ordination (Section 11.3) to describe species distributions along gradients; ordination often seems to show continuous variation in species associations (individualistic/continuum approach). The second is the description of species distributions along natural gradients such as those in Section 11.2. When plotted and viewed by eye, few distributions (e.g. Figure 11.3) show strong evidence of discrete communities.

What has been missed in relating this story are two simple counter-observations:

1. Ordination is a technique designed to array species along gradients of vegetation composition. The fact that one uses a gradient-seeking technique and then (surprise!) finds gradients is not particularly conclusive evidence for continuous variation in species composition. Not surprisingly, if instead one uses techniques of classification, one can recognize apparently discrete communities. Ordination and classification each find what they are designed to look for – vegetation gradients and vegetation types, respectively. Indeed,

Box 11.1 Getting the History Right: Null Models in Ecology

How might we determine whether plants are organized into communities, or merely randomly distributed with respect to one another? A review (Harvey et al. 1983) and an entire book (Gotelli and Graves 1996) have dealt with this topic. Unfortunately for historical accuracy, the book asserts that the term "null models" was invented in 1981 by American zoologists meeting in Florida. At best, this fundamentally misrepresents the short-term historical record (Keddy 1998), since other ecologists had already developed null models for plant communities. For example, decades earlier, P. Greig-Smith, and later, E. C. Pielou, presented many null models for the organization of communities, and discussed methods to test them (e.g. Greig-Smith 1952, 1957; Pielou 1975, 1977). However, even these writings on null models were built upon a foundation created by statisticians for whom the concept of a null model (e.g. the binomial, the Poisson and the normal distribution) stretches back at least to the pioneering work (*Statistical Methods for Research Workers*) of Sir R. A. Fisher (1925) (Figure B11.1.1).

One of the earliest null model seems, in fact, to be attributable to Raunkiaer, whose global study of plant growth forms we encountered in Chapter 2. He understood the need for a random model against which he could judge observed patterns in growth form. Long before the age of Monte-Carlo computer programmes, he describes how he selected 1,000 species at random from the world's

FIGURE B11.1.1 Ronald Fisher (1890–1962) laid the foundations for many modern techniques of statistical analysis (1946, Walter Stoneman, IM/GA/WS/1174 © Godfrey Argent Studio).

flora, and constructed a life-form spectrum for this group, which he then used as a reference spectrum against which to compare the life-form spectra of specific climates (recall Table 2.2). Since my reference copy of Raunkiaer's original work (an English translation from the Danish, edited by Tansley) is a badly tattered photocopy obtained on interlibrary loan (and since his original papers were published in Danish, German or French), I assume that many readers are unlikely to have access to the originals. Moreover, many readers or writers may accept without evidence the impressions provided by Gotelli and Graves. I will therefore quote from Raunkiaer's work at more than usual length.

Raunkiaer (1908) introduced the classification of plants into life-forms, and tabulated these life-forms for an array of different climatic types. He begins his discussion with the observation that Nanophanerophytes are

(Continued)

Box 11.1   (cont.)

abundant in the Seychelles. Recall that Phanerophytes are trees and shrubs with buds (meristems) borne above the surface of the ground (Figure 2.2). Also recall that Nanophanerophytes are the smallest of the Phanerophytes being less than 2 m tall (Table 2.1), thereby including a wide range of shrubs typical of Mediterranean climates, fynbos, semi-deserts and peatlands.

But what do these numbers mean? Are we to conclude from the fact that the Nanophanerophytes are the best represented life-form that it is the Nanophanerophytes that are particularly characteristic of the humid and hot tropical regions? By no means! The large number of Nanophanerophytes [sic] might perhaps mean that this life-form is very common in the world taken as a whole... what we lack is a standard, a 'normal spectrum' with which to compare the spectra of the various regions, and by means of which the value of the individual numbers can be determined. It is most reasonable to suppose that a normal spectrum of this kind might be found in the spectrum of the whole world, that is to say the percentage of each life-form in the flowering plants of the world.

(Raunkiaer 1934: p. 115)

He goes on to describe the difficulty of obtaining a random sample of the world's flora, particularly given that the Compositae (now Asteraceae) alone then had some 13,000 species. In 1908, he had a spectrum for 400 species "even though 400 be too small a number... I shall use a spectrum founded on this number as a preliminary normal spectrum" (p. 16). By 1916, he had completed the monumental task (Raunkiaer 1918). He first describes the difficulty in obtaining a list of the species in the global pool from which he could make a random selection.

I came upon the difficulty that there was no comprehensive descriptive list of Phanerogams of the whole earth which represented a conglomeration approximately as uniform and homogeneous as a mass of seeds which one can thoroughly mix by shaking. If there were available a catalogue of all the Phanerogams in which the different species were arranged alphabetically according to their scientific names without reference to the genus it might be assumed that we should have such a mixture; but a catalogue of this sort does not exist.

(Raunkiaer 1934: p. 429)

The need for a list without reference to genus is critical, "since the species of the same genus very often belong, all or in great part, to the same life-form..." (p. 429). Raunkiaer has already glimpsed, then, the problem of defining what random sample truly means when some groups of species are closely related to one another. He uses the analogy of trying to mix seeds thoroughly when some of them are stuck together.

He describes the problem of selecting 1,000 species from the catalog of vascular plants called the *Index Kewensis*, which, at that time, recognized some 140,000 plant species (although the accepted figure today exceeds 350,000; The Plant List 2013). To get around the problem that these species are organized by genus:

I therefore took groups of species with certain intervals between them and between the species of each group. I chose ten groups, each of one hundred species, the ten groups being so distributed in the Index Kewensis *that the first began on p. 150, the second on p. 400, the third on p. 650, &c., that is at intervals of 250 pages, and I chose in each of the hundred columns beginning at the above designated starting-points the last cited species... It is very possible that another method of selection would have been better... In the year*

Box 11.1   (cont.)

1908 I determined the life-forms of the first 400 species, and in the same year published the normal spectrum calculated from them... In the autumn of 1916 I determined the life-forms of the remaining 600.

(Raunkiaer 1934: p. 430)

He compares his normal spectrum based on a sample size of 400 with that of 1,000, reproduced here in Table B11.1.1. Next, he sorts the 1,000 species used to define the normal spectrum into four groups (Gymnosperms, Monocotyledons, Choripetalae and Gamopetalae) and calculates the percentage found in each group. Using another compendium, Engler and Prantl's *Natürlichen Pflanzenfamilien* (as well as other sources), he then determines the total number of species known at the time (139,953) and the percentage of these species found in each of the four groups. Finally, he compares the calculated percentages based on his sample of 1,000 to the actual percentages determined using the entire flora and finds that "The correspondence between the actual and calculated numbers in the two series of percentages must be considered very good" (p. 433). The Choripetalae showed the greatest difference: "the difference between the calculated and actual percentage (49.8–46.4) amounts to 3.4, and is thus greater than twice but smaller than three times the mean error [± 1.6], quite a satisfactory approximation" (p. 433).

Table B11.1.1

	Plant life-form									
No. species in sample	S	E	MM	M	N	Ch	H	G	HH	Th
400	1	3	6	17	20	9	27	3	1	13
1,000	2	3	8	18	15	9	26	4	2	13

although Whittaker's data are often used to support the individualistic approach, Whittaker himself (1956) divides forest vegetation into discrete types.

2. Examining species distributions along gradients by eye is inconclusive. Without criteria of falsification, and without null models, debates can continue indefinitely. Certainly Whittaker's data (Figure 11.3) do not show clearly demarcated plant communities along elevation gradients. But are these species distributions actually random? Perhaps they are even overdispersed from interspecific competition (Weiher and Keddy 1995b).

Or perhaps they are clustered but the data are noisy and a statistical test is needed to find pattern within the noise. There are other possibilities too: real communities might be obscured by "blended" or "blurred" boundaries (sensu Pielou 1975).

Much effort could have been saved if ecologists had consulted James' (1907) essay on pragmatism:

The pragmatic method is primarily a method of settling metaphysical disputes that otherwise might be interminable... whenever a dispute is serious, we ought to be able to show some practical

Box 11.2   Gleason, Clements and a Community Structure Continuum: A Possible Synthesis

It is likely that Clements and Gleason now suffer from misrepresentation, their names now providing convenient labels for two possible situations that need names. At one extreme (Figure B11.2.1 left) lie situations where species have almost no interactions with other species, or at least no more than expected by chance; such random aggregations of species probably should not be called a community at all. At the opposite extreme (Figure B11.2.1, right) lie groups of species with tightly integrated relationships, a situation that would deserve the term community. Neither end of the continuum is likely to be found in nature. The terms Gleasonian and Clementsian really just refer to different ends of this organizational continuum. There may be historical reasons for holding on to the terms Gleasonian and Clementsian as opposing regions of the continuum. You should know that other scholars (Nicolson and McIntosh 2002) believe that my view continually misrepresents Gleason as being too far to the left in Figure B11.2.1.

   The position of a community along the structure continuum might be assessed using two different kinds of data – descriptive or experimental. In the first case, the matrices can represent measures of interspecific association using standard association measures such as χ^2. Alternatively, a thorough set of experiments, such as we saw in Chapters 4 and 7, would measure the amount of competition or mutualism between each pair of species. Data on pattern are far more available, and easier to collect, than data on species interactions, but in the long run, it is likely to be the latter that will be needed to arrange communities along this continuum.

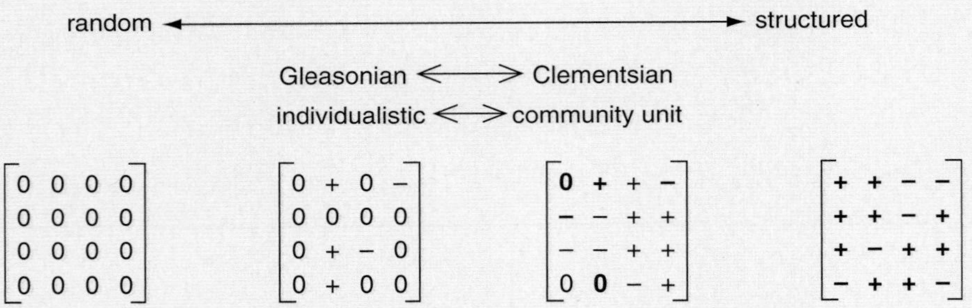

FIGURE B11.2.1   The strengths and types of association between species in a community can be shown in a matrix. Some may lack any interactions (far left) and might not even be called a community. Others may have intense positive and negative interactions and be considered a tightly structured community (far right). In this context, Gleasonian and Clementsian views merely represent different regions of this continuum, though neither occupies the extreme.

difference that must follow from one side or the other's being right.

(p. 10)

James' essay is still delightful reading. He was a philosopher who wrote to be understood rather than to impress others by his scholastic virtue, and his style still has lessons for young scholars.

11.5.3 Null Models and Patterns Along Gradients

Oddly, many of the assessments of communities and patterns seem to have overlooked the enormous body of research that used statistical techniques to appraise the degree to which species are positively and negatively associated with one another, or with particular habitats. (Early examples included Cole 1949 and Greig-Smith 1957). The null model, that there are no positive or negative associations (that species are randomly distributed with respect to one another), has now been repeatedly falsified.

Patterns of association are, of course, only pieces of evidence for communities, but many authors (e.g. Diamond 1975; Strong et al. 1984; Gotelli and Graves 1996) have minimized this body of studies, and overlooked just how thorough the studies of pattern have been (e.g. Cole 1949; Greig-Smith 1957; Agnew 1961; Kershaw 1973; Dale 1999). A quite separate body of studies also grew up using data from species along gradients (Pielou 1977; Dale 1999). In her short academic career, Chris Pielou produced a stream of papers and books addressing the search for pattern in nature. In her books one finds many methods for testing null hypothesis of community structure. Although she does not appear to have been familiar with James' (1907) ideas about pragmatism, her approach is what he advocated more than half a century earlier. Do communities exist? Well, what can be measured along gradients that would support or refute the idea of discrete communities? How does one assign probabilities to the outcomes? Since Pielou's work is so often either overlooked or misunderstood, let us trace the history of the gradient approach to testing for the existence of discrete plant communities.

11.6   Empirical Studies of Patterns Along Gradients

The observation of plants along gradients usually reveals distinctive "zones" where different species apparently dominate sections of the gradient. This is called zonation. There are three major uses for quantitative studies of zonation.

1. Although there are many pictures of zonation (recall Section 11.2) and a vast literature describing zonation, there is no way to summarize or compare such descriptive studies except with measurable properties.

2. There are many theories of resource and habitat use (e.g. Miller 1967; MacArthur 1972; Pianka 1981; Weiher and Keddy 1995b) that may be evaluated using zoned plant communities.

3. The question about whether discrete communities occur has raged on for decades, and there is no way to slay this dragon and end the debate conclusively except to actually measure the degree to which plant distributions exhibit individualistic or community patterns.

It is this third use that we will explore here. Since statistical analysis of zoned communities is likely less familiar, and has fewer monographs to consult than ordination, I will give rather more detail in this section.

From the perspective of distributions along gradients, the "community unit" concept proposes that when species distributions are plotted along some gradient or gradient-complex whose rate of change is constant, there exist groups of species, i.e. "communities," which occur in sequence along the chosen gradient (Whittaker 1975). Within each grouping most species have similar distributions and the end of one group coincides with the beginning of another. The "individualistic" concept, in contrast, proposes that "centres and boundaries of species distributions are scattered along the environmental gradient" (Whittaker 1975) and no distinct groups of species are predicted to exist. These alternatives are illustrated in Figure 11.18a.

Following Pielou (1975, 1977), explicit hypotheses regarding these two concepts can be formulated using upper and lower boundaries of species along gradients (Figure 11.18b). The **community unit hypothesis** states that:

(a)

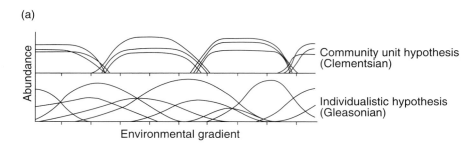

FIGURE 11.18   (a) The individualistic and community unit hypotheses; (b) recast into a testable form. (After Shipley and Keddy 1987)

(b)

	Boundaries	
	Clustered	Random
Coincide	Community unit hypothesis	?
Independent	?	Individualistic hypothesis

Upper and lower boundaries

1. There should be significantly more boundaries (both upper and lower) in some intervals of the gradient than in others, i.e. boundaries are clustered.
2. The number of upper and lower boundaries per interval should increase and decrease together along the gradient.

The **individualistic hypothesis** states that:

1. The average number of boundaries (both upper and lower) in each interval of the gradient should be equal except for random variation about the mean.
2. The number of upper boundaries per interval of the gradient should be independent of the number of lower boundaries.

At one extreme (Figure 11.19 left), distributional limits of species may be overdispersed, like the shingles on a roof; at the other extreme (Figure 11.19 right) they are clustered (Pielou 1975; Underwood 1978; Weiher and Keddy 1995b). The middle case is a random distribution.

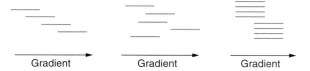

FIGURE 11.19   Species may be distributed along gradients in a manner that is overdispersed (left) like shingles on a roof, random (middle) or underdispersed (right) like pages of a book. Underdispersed boundaries are usually called clustered boundaries. Statistical tests can distinguish among these possibilities. (From Keddy 2001)

Now to some examples. The first comes from salt marshes. Pielou and Routledge (1976) collected data on species distributional limits in five sets of salt marshes at different latitudes in eastern North America. They found in many transects that species boundaries were significantly clustered – there were zones of species with similar distributional limits. Salt marsh zonation therefore looks similar to the right

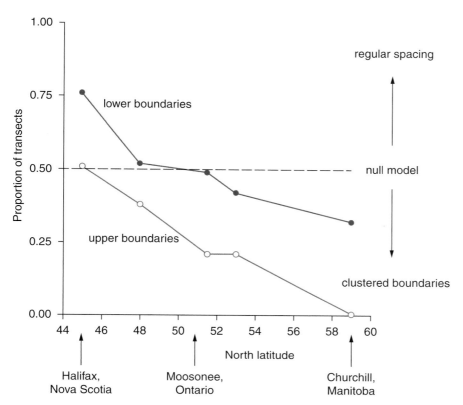

FIGURE 11.20   The clustering of species boundaries in relation to latitude for salt marshes. The higher the latitude, the lower the proportion of transects with clustered boundaries. Note: the lower the proportion, the greater the degree of boundary clustering. (After Pielou and Routledge 1976)

side of Figure 11.19. Moreover, the upper limits clustered more than the lower limits, irrespective of latitude (Figure 11.20). The study clearly showed that with proper sampling methods and appropriate null models, it was possible to find measurable patterns in zoned vegetation.

The causes of such patterns cannot be deduced solely from statistical analyses of pattern. Nonetheless, Pielou and Routledge did find evidence that biological interactions were responsible for some of these patterns. Their logic was as follows. If zonation was solely the result of physiological responses to salinity and inundation, then species' upper and lower distributional limits would be independent. If, however, one species set the limits of another through competition, then there would be a tendency for the upper limits of one species to coincide with the lower limits of the other. Using a set of 40 transects near Halifax, Nova Scotia, they found that distributional limits of species tended to coincide (p <0.001). Therefore they concluded that competition could

produce some of the observed patterns in salt marshes. Regrettably, the test was too crude to compare the intensity of competition among latitudes.

The next example comes from a lakeshore – data on zonation from a small, sandy lake near the Great Lakes. Axe Lake has an array of zonation patterns, from those associated with open sand beaches, to those of fertile bays, to floating bogs (Keddy 1981b, 1983). The flora of this lake and its array of vegetation types appear in many ways typical of the northern temperate zone. The following patterns were found:

1. Both the upper and lower boundaries of species were clustered. Just as Pielou and Routledge (1976) showed, there were certain elevations where more species reached their limits than would be expected by chance alone. This is shown in Figure 11.21 where the measures of boundary of clustering for all 25 transects fall below zero.

2. The degree to which species distributions were clustered (the intensity of the zonation on a

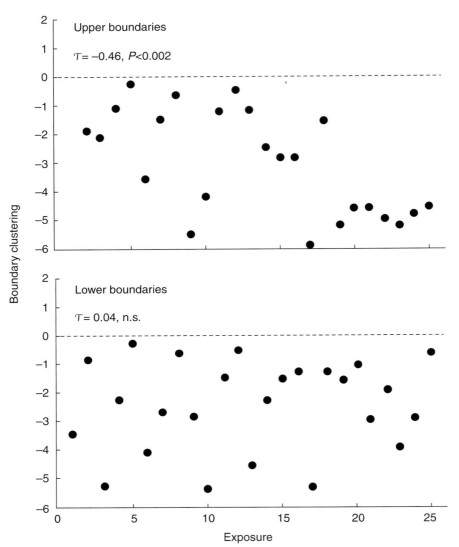

FIGURE 11.21 The clustering of species boundaries in relation to exposure to waves for a small lake. The more negative the measure of clustering, the more clustered are the boundaries. The dashed line presents the null model. (From Keddy 1983)

shoreline) increased with exposure to waves. This occurred because exposure to waves increased the clustering of upper boundaries (Figure 11.21, top); lower boundaries were unaffected (Figure 11.21, bottom).

3. The locations of boundary clusters were pushed up the shoreline as exposure to waves increased. This can be seen in the field distributions of species, where aquatics such as *Lobelia dortmanna* moved up the shoreline (Figure 11.22). It showed up in the

joint distribution of species as a landward shift in the distributional limits of all species (Figure 11.23).

Our third zonation investigation was carried out in a freshwater riparian marsh. It explored all four possibilities for patterns along a gradient that are shown in Figure 11.18b and were pointed out by Whittaker (1975). Shipley and Keddy (1987) collected data on species boundaries from 13 transects in the marsh. As with the example from Axe Lake,

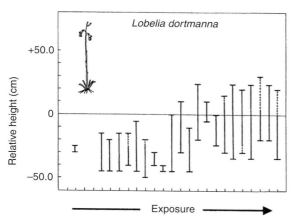

FIGURE 11.22 The relative height occupied by a shoreline plant as a function of exposure to waves; zero marks the August water line. (After Keddy 1983)

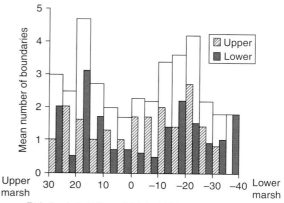

FIGURE 11.24 Zonation in a riparian marsh. The mean number of species boundaries in each 5-cm height interval is plotted against relative height. Within each bar the mean number of upper boundaries (hatched) and lower boundaries (brown) are shown. (From Shipley and Keddy 1987)

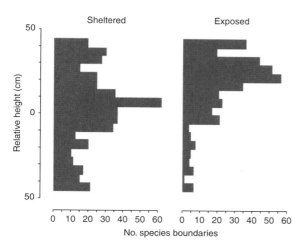

FIGURE 11.23 The relative height of species distributional limits (upper and lower boundaries combined) for shoreline plants in ten transects sheltered from waves (left) and ten transects exposed to waves (right). (From Keddy 1983)

the distribution of species boundaries was tabulated for 5-cm increments of elevation. Along this gradient the dominant species changed from *Carex crinita* to *Acorus calamus* to *Typha angustifolia* with increasing water depth. These data were analyzed

using analysis of deviance, which is analogous to analysis of variance, but does not assume normality in the error structure of the model. They found that both upper and lower boundaries were clustered (Figure 11.24). This was clearly contrary to the individualistic concept. But they also found that the pattern of clustering was different between upper and lower boundaries – a result inconsistent with the community unit concept. They therefore concluded that rather than a simple dichotomy between two models, there was a need to erect multiple models for species relationships that occur in nature. In other words, more than 50 years of debate about patterns had dragged on (Box 11.2), in part because the patterns were not expressed in clear testable form. This example illustrates the power of zonation pattern analysis as a research tool in ecology.

The study described above had two significant weaknesses. First, it tested a broad general model with data from a single wetland. Second, it used only data on the distributional limits of species. In our final example of zonation studies, Hoagland and Collins

(1997) tried to rectify these deficiencies. First, they collected data from 42 wetland sites. Second, they measured three attributes of zonation patterns: (1) boundaries of species distributions; (2) modes of species response curves; and (3) nested structure. The use of three properties not only provides a more powerful way to test among competing models, but it also allows the creation of new kinds of distribution models. Hoagland and Collins examined their results in light of the following four contrasting models of zoned vegetation:

1. The highly deterministic community unit model of Clements (1936) could be interpreted to imply that plant communities are comprised of distinguishable associations of species with little overlap in species distributions among associations. This model can be portrayed as a series of species response curves in which the starting and ending points of species distributions are clustered (Figure 11.25a).

2. Other interpretations of this community unit model are possible. Clements (1936) described the occurrence of "predominants," species that were dominant and spanned one or more associations. Figure 11.25b shows a model in which boundaries and modes of response curves are clustered yet some species response curves are nested within the curves of other, more dominant, species.

3. The individualistic distribution of species (Gleason 1926) and the continuum concept of vegetation (Whittaker 1967) are represented in Figure 11.25c as a series of broadly overlapping species response curves with randomly distributed starting and stopping boundaries, and modes, along an environmental gradient.

4. Dominant species may be regularly spaced and encompass several curves of subordinate species; the hierarchical continuum model predicts that modes and boundaries of species response curves are random, but because distributions are hierarchical, it predicts that species distributions are nested (Figure 11.25d).

Three test statistics were used to discriminate among these models in the 42 wetland sites. The three test statistics were as follows:

1. Morisita's index (Hurlbert 1990) was used to determine whether or not species boundaries were clustered:

$$I = Q \sum_{i=1}^{Q} \left(\frac{n_i}{N}\right) \left(\frac{n_i - 1}{N - 1}\right)$$

where Q is the number of quadrats, n_i is the number of starting and stopping boundaries in the ith quadrat, and N is the total number of boundaries.

2. The degree of aggregation (P) of species modes was determined using the sample variance of distance between modes (Poole and Rathcke 1979):

$$P = \frac{1}{k+1} \cdot \sum_{i=0}^{k} \left\{ y_{i+1} - y_i - [1/(k+1)] \right\}^2$$

where k is the number of species, $y_{i+1} - y_i$ is the distance between modes, and $1/(k + 1)$ is the mean of $y_{i+1} - y_i$. If $P = 1$, modes are randomly distributed; if $P < 1$, modes are regularly distributed; and if $P > 1$, modes are aggregated.

3. Nestedness was determined by using the index of Wright and Reeves (1992):

$$N_c = \sum_{i=1}^{K-1} \sum_{m=i+1}^{K} \sum_{j=1}^{S} X_{ij} X_{mj}$$

where S is the total number of species, K is the number of quadrats, and species richness of quadrat i > quadrat m. $X_{ij} = 1$ if species j is present in quadrat i and 0 if it is absent (same for X_{mj}, quadrat m, species j). This index counts the number of times that the presence of a species in a quadrat correctly predicts its presence in quadrats that are more species-rich. The value of N_c was then used to calculate a relative nestedness index:

$$C = \frac{N_c - E\{N_c\}}{\max\{N_c\} - E\{N_c\}}$$

where $E\{N_c\}$ is the expected value and $\max\{N_c\}$ is the value of N_c for a perfectly nested matrix. C ranges from 0 (complete independence) to 1

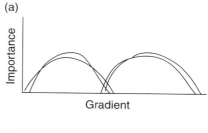

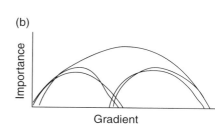

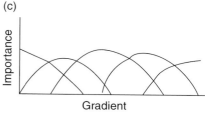

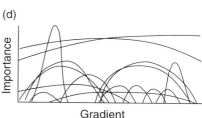

FIGURE 11.25 Four possible patterns of species growing along gradients. The top two (a, b) represent the community model; whereas the lower two (c, d) represent the continuum model. The right hand pair (b, d) possess the additional feature of being nested. (From Hoagland and Collins 1997)

Table 11.3 **Summary of six models of distribution along gradients (based on the distribution of boundaries of species response curves, modes of species response curves and degree of nestedness of species distributions) and the prevalence of these models in a set of 42 transects from wetlands in Minnesota and Oklahoma. (After Hoagland and Collins 1997)**

	Boundaries clustered	Modes clustered	Distributions nested	Examples found
Community-unit	yes	yes	no	0
Nested community-unit	yes	yes	yes	3
Alternative model	yes	no	yes	7
Alternative model	no	yes	yes	16
Continuum	no	no	no	0
Nested continuum	no	no	yes	16

(perfect nestedness). Cochran's Q was used to test for significance of nested species distributions.

All 42 transects were nested (Table 11.3). This is an important generalization. Hoagland and Collins interpret this as evidence for "hierarchical" community structure. Given the many uses of the word hierarchy, it may be more useful to simply use the descriptive result: nested patterns are the rule in zoned vegetation. Clustering of boundaries occurred in only 10 out of 42 transects; thus the continuum model is more prevalent in the wetlands they studied than

indicated by Pielou and Routledge (1976), Keddy (1983) or Shipley and Keddy (1987). Unfortunately the use of Morisita's index, rather than previously used indices, raises the possibility that the prevalence of the continuum model in Hoagland and Collins' (1997) data may be an artifact of the test used. Such problems emphasize the need for methodological consistency.

More than half of the transects did not fit any of the four models described above (Table 11.3). Seven had clustered boundaries but unclustered modes, whereas 16 had clustered modes but unclustered boundaries. This work shows the merit of applying a battery of tests to analyze zonation patterns. The differences among the transects and among published studies suggest that ecologists require several different models to describe zonation patterns in nature; Dale (1999) provides the most recent compendium of approaches.

CONCLUSION

Here, in summary, are the main questions that ordination and, to a lesser extent, classification, can answer:

1. Are there any patterns or trends in the data?
2. Which sets of species tend to co-occur?
3. Which environmental factors tend to be associated with these patterns?

The principal objective of this chapter was not to authorize you to carry out different kinds of gradient analyses, but rather to equip you to understand research papers or consulting reports that include them. The secondary objective was to advise those of you who use such tools to do so wisely. You should think about what these techniques can and cannot do, and use them sparingly. Too often, they are used unnecessarily, perhaps just to create a false sense of sophistication. Perhaps a simple sketch like a profile diagram (Figure 11.2) will suffice, or perhaps there is another tool, such as statistical analysis of patterns along gradients (Section 11.6), that would answer your question more clearly. Your research, and your sampling, should always be driven by one or a few clearly stated questions.

You should know what you plan to do with the results of your work before collecting the data. If you want to explore patterns of continuous variation, you will likely want to use ordination. If you want to assign the sample units to discrete groups, you will likely want to use classification. You may even want to use both – but do not do so simply because of the easy availability of computer software. In other cases, neither may help answer your question. There are also multiple types of data that can be collected (presence/absence? biomass? visual cover?), multiple ways of comparing quadrats (matching coefficient? Jaccard? Euclidean distance?) and multiple techniques to consider (principal component analysis? factor analysis? multidimensional scaling? structural equation modelling?). All of these decisions should be made before you collect data. There is a growing number of books that deal solely with the advantages and disadvantages of different approaches (e.g. Whittaker 1967, 1973b; Orloci 1978; Gauch 1982; Legendre and Legendre 1983; Digby and Kempton 1987; Tabachnick and Fidell 2001; Grace 2006) – the manual that comes with a particular software package should never be your sole reading in this field.

It may be useful to remember that the mere application of computing power does not guarantee good science. Nor is it a substitute for being able to identify plants and their habitats. Although it pre-dates computers and multivariate statistics, you might keep in mind an observation by Tansley, one of the founders of plant ecology: "Besides stimulating many good biological minds, ecology had a great attraction for weaker students, because it was so easy to describe particular bits of vegetation in a superficial way, tending to bring the subject into disrepute" (Tansley 1987: p. 6).

While multivariate descriptive techniques are sometimes overused, there remain many cases of complex plant communities where patterns that are now described subjectively might be better summarized and communicated with multivariate tools. Every plant ecologist should therefore have an understanding of the value of these techniques and what multivariate analysis can and cannot tell us. When circumstances demand ordination, then one should carefully specify the nature of the sampling, choose the most appropriate technique, seek advice from someone already familiar with the technique, fully describe the patterns and be explicit about the hypotheses that are generated.

Review Questions

1. Explain, using the material in Chapter 3, how moisture gradients form along topographic gradients.
2. Visit a local protected area and make a sketch that shows the distribution of plants along one important natural gradient.
3. Explain the difference between direct and indirect gradient analysis. Find an example of each in a recent scientific journal.
4. Visit a local mountain and make a sketch of the mountain in the style of Humboldt's drawing of Mt. Chimborazo (Figure B2.1.3).
5. How do you measure if plant distributional limits are randomly arrayed along a natural gradient? What does it mean if they are clustered?

Further Reading

Gleason, H. A. 1926. The individualistic concept of the plant association. *Bulletin of the Torrey Botanical Club* **53**: 7–26.

Mueller-Dombois, D. and H. Ellenberg. 1974. *Aims and Methods of Vegetation Ecology*. New York: John Wiley and Sons.

Goldsmith, F.B. and C.M. Harrison. 1976. Description and analysis of vegetation. pp. 85–155. In S.B. Chapman (ed.) *Methods in Plant Ecology*. Oxford: Blackwell Scientific.

Gauch, H.G. Jr. 1982. *Multivariate Analysis in Community Ecology*. Cambridge Studies in Ecology. Cambridge: Cambridge University Press.

Legendre, L. and P. Legendre. 1983. *Numerical Ecology*. Amsterdam: Elsevier.

McIntosh, R.P. 1985. *The Background of Ecology: Concept and Theory*. Cambridge Studies in Ecology. Cambridge: Cambridge University Press.

Digby, P.G.N. and R.A. Kempton. 1987. *Multivariate Analysis of Ecological Communities*. London: Chapman and Hall.

Shipley, B. and P.A. Keddy. 1987. The individualistic and community-unit concepts as falsifiable hypotheses. *Vegetatio* **69**: 47–55.

Wilson, E.O. 1993. *The Diversity of Life*. New York: W.W. Norton.

Wiser, S.K., R.K. Peet and P.S. White. 1996. High-elevation rock outcrop vegetation of the southern Appalachian Mountains. *Journal of Vegetation Science* **7**: 703–722.

12 | Diversity

Three main patterns: area, habitat variation, latitude. Some examples of diversity: Mediterranean climates, carnivorous plants, deciduous forests, endemic species. Four models describing diversity at smaller scales: intermediate biomass, competitive hierarchies, intermediate disturbance/gap dynamics, centrifugal organization. Relative abundance patterns. Evenness and diversity. Laboratory experiments. Field experiments. Conservation.

FIGURE 12.1 There are so many kinds of trees! The Amazon basin alone has more than 11,000 of them (Hubbell et al. 2008). And the world has even more kinds of plants – more than 350,000 species, with some groups still poorly studied (The Plant List 2013). Why do some areas of land have so many more plants than others? How can one predict how many plants a given area will contain? You will be able to answer both these questions (at least in part) by the end of this chapter.

12.1 Introduction: Why Are There So Many Kinds of Plants?

How many kinds of plants are there altogether? And why do some areas of the world have more plant species than others (Figure 12.1)? In too many text books and reviews, students new to these questions are referred to articles such as Hutchinson (1959) and May (1988) – both of which conspicuously ignore plants! Hutchinson (1959) in his essay entitled "Homage to Santa Rosalia" says "Why are there so many kinds of plants? As a zoologist I do not want to ask that question directly, I want to stick with animals but also get the answer." Nearly thirty years later, May (1988) proposed, in a leading scientific journal, to address how many species are on Earth – giving only one single statement (without a refer- ence!) on plant diversity! Apparently, to some ecologists, plants are a mere afterthought. Yet plants constitute over one-third of a million species (The Plant List 2013) and over 99 percent of the Earth's biomass (Whittaker 1975). Of course, the fungi also receive superficial treatment in many papers, while the number of species of fungi has been estimated at 1.65 million (Hawksworth 1990). Needless to say, students are not being well served with such articles on diversity. So let us do it right and get the topic back on track.

In this chapter we will explore the factors that allow so many different species of plants to occur and to coexist. The approach emphasizes two questions:

1. What environmental factors are correlated with plant diversity at specified scales?
2. What methods have provided evidence about the nature and causes of these patterns?

Once again definitions matter. The chapter is written in such a way that you likely do not need to know these definitions in advance: you can confidently start with Section 12.2 and learn the terminology as you go. But for those of you who like definitions first, note that the title of this chapter, diversity, is a commonly used word. Hence it is often used carelessly. We ran into this same problem with the word disturbance in

Chapter 5. It is easy to use the word mindlessly and create confusion. The topic of this chapter, overall, is indeed, diversity. But by the end of the chapter you will see that this word, when properly used, means rather more than just the number of species at one location.

We will start by generally exploring the number of plant species found at one location. This is often termed species richness. In using this word, we mean to focus on the number of species (and not their abundance). The term also assumes we have a speci- fied area, whether it is a metre squared quadrat of grassland, a hectare of forest or an entire island, along with a list of all the species found there. So it is necessary to pay attention to the sampling scale when you refer to species richness. The first part of the chapter will look at patterns in species richness at large scales (Sections 12.2 to 12.5) and at smaller scales (Section 12.6).

Later in the chapter we will focus attention on the fact that species are not equally common: some are abundant, while others are quite rare. When we con- sider the number of species in a sample as well as their relative abundance, it is better to use the term diver- sity. One quite remarkable pattern in ecology is this: any sample is likely to be dominated by a few common species, and almost always contains many rare ones. We often use the word rare to mean the species for which we can find only a few individuals. It is entirely possible for a species to be rare in a particular sample – say in one particular piece of forest or on one particular island – without being rare elsewhere.

It is important to understand that the word rare also has a scale. This issue becomes important because species that are rare at rather large scales are also often at risk of decline or outright extinction. Hence many regions have official lists of rare plants. If such species are declining in population size, they may have more precise designations, such as threatened or endangered species. The latter are precise terms that are both legal and biological.

This chapter will focus mostly upon two basic scientific issues. First, it is puzzling that some areas have so many more species than other areas, and we will look at the patterns. Second, it is puzzling why some species are so much more abundant than others. These are the central issues that we explore in this chapter. Such questions, however, do lead naturally to the question of why species are disappearing, what criteria are used to designate

them as rare or threatened or endangered, and what might be done to protect such species. I will have less to say about that topic. My view is that once you have the scientific principles in place, you will be able to apply them to the legal frameworks that exist where you happen to work. And, of course, Chapter 13 will focus more upon plant conservation.

If you feel confused at this point about the different meanings of richness, diversity and rarity, don't worry. That is why this chapter exists.

12.2   Large Areas Have More Plant Species

One of the most fundamental observations in ecology is the increasing number of species encountered with increased time spent searching for them, or with increased area of habitat explored. (Time and area are often related, in that the more time you spend searching, the more ground you are likely to cover.) The relationship between species and area can be quantified as:

$$S = cA^z,$$

where S is the number of species, A is the area and c and z are constants (Rosenzweig 1995). This exponential relationship is conveniently made linear by taking logarithms of both sides

$$\log S = \log c + z \log A;$$

in which case the constant ($\log c$) represents the intercept of the line and the slope is given by z. This relationship was first quantified more than a hundred years earlier using plant species in England (Figure 12.2). The English countryside is a particularly good region in which to study such patterns because so many amateur naturalists there have studied plant distributions, providing a rich and relatively accurate database. Now this linear relationship has been documented for many kinds of plant and animal species and for many natural areas of landscape. These studies have revealed that

typical values of z for large, contiguous areas of habitat range from 0.12 to 0.18. For isolated patches, such as islands, z tends to be much higher, from 0.25 to 0.35.

The mechanisms that produce such patterns likely change with area, which may also explain why the values of z change. At the small scale, say within a square metre, the number of species is likely in part determined by the size of the plants (which will determine the number of individuals that can fit into a quadrat), the amount of competition among species and the degree to which they partition the habitat.

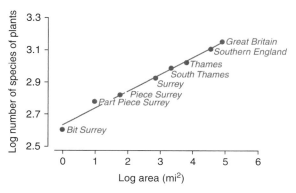

FIGURE 12.2   A species–area curve for the plants of the Surrey region of Great Britain constructed in 1859. (Rosenzweig 1995)

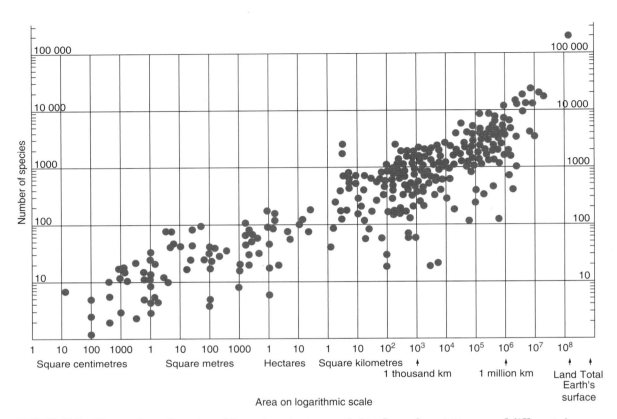

FIGURE 12.3   The number of species of flowering plants recorded in floras from 244 areas of different sizes throughout the world. (After Williams 1964)

At larger scales, differences among quadrats become more important, with areas having many endemics, or areas having more variable environments likely to have higher z values. To explore such issues, Williams (1964) compiled data from no less than 244 sites differing in size and location in one figure, finding, not unexpectedly, that the number of species increased with area, and the relationship was particularly strong at scales above 1 km^2 (Figure 12.3). More recent work (which, typically, has overlooked Williams' pioneering contribution) suggests that z is low at small scales ($z = 0.1–0.2$ at scales less than 100 m^2), increases at intermediate scales ($z = 0.4–0.5$ at scales of 1 ha to 10 km^2), and then falls back to 0.1–0.2 at even larger scales – at least in the east of the English county of Berkshire. The reasons for such changes remain unclear, although some hypotheses

are offered in Williams (1964), Rosenzweig (1995) and Crawley and Harral (2001).

One of the great values of the species–area model is its immediate applicability to the conservation of biological diversity. The model says quite explicitly that the greater the area of habitat protected in a reserve system, the greater the number of species that will be included. Moreover, it tells us what this relationship is: for every doubling of area there will be an increase of roughly 20 percent in the number of species. There is a commonsense logic to this: some species are naturally rare, have very small geographic ranges, or require very specific kinds of unusual habitat, and so will be encountered only by searching relatively large areas of landscape.

Viewed from the other direction, the species–area model has a depressing reverse logic. As the area of natural landscapes decreases (as forests are logged, prairies are ploughed, wetlands are drained and rangeland is overgrazed), species will disappear with mathematical certainty. Exactly how fast will they disappear? Every time the area of natural habitat is reduced by one-half, the number of species will fall by one-fifth. We will return to rates of extinction in natural habitats in the final chapter, but it is sobering to appreciate just how striking the reduction has already been in many natural ecosystems – the longleaf pine savannas of the southeastern United States have lost more than 95 percent of their original geographical area (Platt 1999). Too many natural vegetation types have

already been pushed to the extreme lower left of Figure 12.3.

But why do species become extinct in smaller areas? The model does not say. In each case there is a special cause. The last individual might be felled by a logging company, or eaten by a goat, or killed by an exotic insect, or flattened under a new highway, or desiccated when its habitat is drained for development. The number of explanations can be nearly as large as the number of species. But in the broad general sense, the reason for rising rates of extinction is the steady loss in area of natural vegetation in the world. In Chapter 13 we will explore this problem further and examine some possible courses of action.

12.3   Areas With More Kinds of Habitat Have More Plant Species

In his 1931 treatise, du Rietz observes "The more different habitats there are in a country, the greater will be the number of species; the more uniform the habitats, the smaller will be the number of species. A plain has not as great a chance of attaining wealth in species as a mountain country" (p. 15). That is, the more kinds of habitat there are, the more kinds of plants there tend to be. This is largely because each species of plant has physiological and morphological adaptations that are of benefit in specific sets of environmental conditions, so plants specialize (to varying degrees) on particular habitats – marsh plants are found in marshes, cliff plants are found on cliffs, carnivorous plants are found in infertile soils and so on. There is an additional reason why habitat and plant diversity tend to be correlated. Some plants have relatively weak abilities to compete with other plants, and so are restricted to those habitats that the stronger competitors cannot occupy. These weaker competitors have been called "fugitive species" (Horn and MacArthur 1972), "interstitial" species (Grubb 1987) or "peripheral" species (Keddy 1990b).

Hence plants are distributed among different habitats according to their physiological requirements

and their relative competitive abilities. Although the species–area pattern provides a good general model, in most cases there is a rather complicated set of factors controlling the occurrence of each species. Unusual microhabitats will often enhance local diversity. A good example comes from my early days as a park naturalist in Algonquin Provincial Park in Ontario. Here, amidst a largely forested landscape with acidic bedrock and soils, there was one isolated location for a small arctic plant called *Saxifraga aizoon* – the north-facing walls of one canyon where calcium-rich water seeped out of the rock. This canyon was once a river produced by melting glaciers at the end of the last ice age. The presence of this plant in Algonquin Park therefore involved multiple factors including dispersal by an ancient river and survival since the last ice age on cold north-facing rock walls that trees could not occupy. No cliffs, no alkaline water seeping from the cliff, no *S. aizoon*. It is this way with many other unusual species, which is what inspires many field botanists in their search for rare plants.

For similar reasons, in any fixed area of landscape, the greater the range in altitude, the more different kinds of vegetation there will be. This is an old idea, and so I illustrate it with an old figure reproduced from Oosting's 1956 text book, *The Study of Plant Communities*, which shows how vegetation zones change with altitude in the arid mountains of southwestern North America (Figure 12.4). Note the alpine tundra at the highest altitudes. Alpine tundra is found even on mountains near the equator if they are sufficiently high; in East Africa, for example, alpine tundra with giant senecios and lobelias is found above 3,700 m, with permanent snow above 4,600 m (Richards 1996). Earlier (recall Figure 11.4) we saw another example of changes in vegetation with altitude in the Great Smoky Mountains, where the highest elevations have heath areas called "balds." While the most obvious factor influencing species distributions is the lower temperature at higher elevations, this is not the only factor. Ridges drain quickly and tend to be dry, whereas valleys accumulate water and tend to be wet. Soil erodes from the tops of slopes and accumulates in valleys, producing fertility gradients. Slopes that face the

south are warm, whereas slopes that face north are cool (hence the diagonal zonation lines in Figure 12.4). Often, too, mountains have exposures of different kinds of rock types that form different soil types (Figure 11.11, also from the Great Smoky Mountains). And earlier still (in both this book, and in history), recall von Humboldt's illustration of the plants and vegetation of Mount Chimborazo in the Andes (Figure B2.1.3). All these examples illustrate how the number of kinds of habitats for plants (and hence the number of kinds of plants) increases with topographical variation. This is one reason why mountainous areas, such as the Himalayas, the Andes and the Caucasus, tend to be biological hotspots of plant diversity at the global scale (Myers et al. 2000; a topic of focus in Chapter 13).

It might be reasonable to raise an objection now – surely some relatively flat areas, such as the Amazon floodplain, also have a lot of plant species. Does this not contradict the above rule? It might. But we can rescue the idea of habitat variability by arguing that flooding creates many different moisture regimes, and that each moisture regime also has a characteristic set of plant species (Figure 12.5). Perhaps a few centimetres of water in a floodplain is the equivalent of many metres of elevation on a mountainside. Or perhaps the constant disturbance from erosion and deposition creates a mosaic of new habitats, each of which has its characteristic set of species (recall Figures 5.10 and 5.11). Or perhaps the Amazon is not as diverse as it first seems – studies of botanical diversity suggest that the adjoining Andes and the coastal Atlantic Forests have even more plant species (see Figure 13.11). Or perhaps over such vast areas, even though there are few kinds of different habitats, there has been insufficient time for any one plant species to force another into extinction. While there are many hypotheses, the answer for lowland tropical rain forest is as yet unresolved (Hubbell and Foster 1986; Richards 1996). But even if we do not understand the exact reasons why there are so many kinds of plants in lowland tropical forests, the species–area relationship still works there just as it does everywhere else.

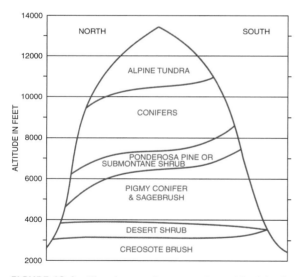

FIGURE 12.4 The changes in vegetation with altitude on mountainsides show how topographical variation can increase the number of vegetation types, and hence the number of plant species. (Southwestern North America, from Oosting 1956, after Woodbury 1947)

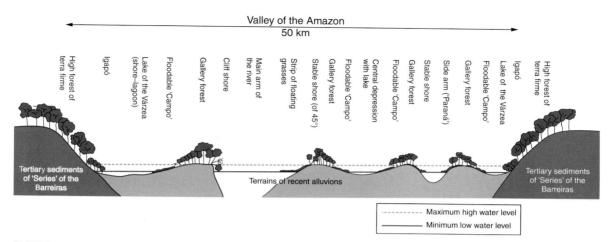

FIGURE 12.5 This cross-section through the Amazon valley shows how the physical conditions provide a template that generates different vegetation types with different plant species. (From Keddy 2010 after Sioli 1964)

But then the Amazon is not alone – there are other flat areas with large numbers of species. Two of the best known are the heathland plant communities of South Africa (fynbos) and southern Australia (kwongan), where large numbers of similar-looking species co-occur. Rosenzweig (1995) uses the term "species flocks" to designate many apparently redundant species within one functional type.

12.4 Equatorial Areas Have More Plant Species

Most plant species occur in tropical regions (Table 12.1; Figure 12.6). Three main groups of vascular plants have been tabulated (The Plant List 2013). The Pteridophytes, most of which are native to the moist tropics, include some 10,600 species. The gymnosperms, although locally important in terms of biomass in selected types of conifer forests, include probably 1,000 species. The angiosperms have the rest – some 300,000 species.

When Linnaeus first began classifying plants he started with the northern temperate flora and by 1764 had already listed 1,239 genera. He had hoped that his classification system would apply to the tropics as well (Mayr 1982). But as great expeditions such as those of von Humboldt (Box 2.1) returned with cargoes of pressed specimens, it became clear that the tropics contained an array of species far greater than anyone had anticipated (Edmonds1997).

There is still no general agreement on the factors responsible for high tropical diversity (Huston 1994; Rosenzweig 1995; Rohde 1997; Gaston 2000; Willig et al. 2003). It seems likely that the two overriding factors for plants are warmth (in particular, absence of freezing conditions) and moisture. One way ecologists have sought to understand causes of diversity is to explore relationships between plant diversity and measured climatic factors. Australia is a particularly good example, since it contains many different vegetation types, and a rich flora derived from Gondwana. Figure 12.7 illustrates the changes in the number of overstorey species with both latitude (e.g. warm temperate, subtropical, tropical) and moisture availability. Irrespective of latitude, drier environments have fewer overstorey species and all three latitudes converge where the evaporative coefficient falls to 0.03. Moreover, within the subtropical and tropical climate types,

Table 12.1 **Distribution of higher plant species (pteridophytes, gymnosperms, angiosperms) by continent. (After Groombridge 1992)**

Area	No. plant species
Latin America (Mexico through South America)	85,000
Tropical and subtropical Africa	40,000–45,000
North Africa	10,000
Tropical Africa	21,000
Southern Africa	21,000
Tropical and subtropical Asia	50,000
India	15,000
Malaysia	30,000
China	30,000
Australia	15,000
North America	17,000
Europe	12,500

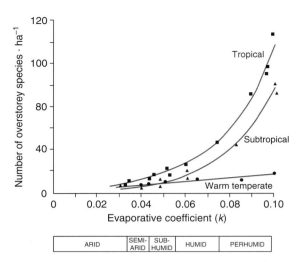

ARID	SEMI-ARID	SUB-HUMID	HUMID	PERHUMID

FIGURE 12.7 The number of overstorey species (trees and tall shrubs) per hectare increases with moisture availability in Australia. (From Specht and Specht 1993)

the evaporative coefficient is able to account for more than 90 percent of the variation in overstorey species. This coefficient is derived from the equation:

$$MI = E_a/E_o,$$

where MI is the monthly moisture index, E_a is the actual evapotranspiration (cm·month^{-1}) and E_o is the pan evaporation (class A pan; cm·month^{-1}) (Specht and Specht 1993). The greater the MI, the more actual evapotranspiration approaches that measured experimentally:

$$MI = k(P - R - D + S_e),$$

where P is precipitation, R is run-off, D is drainage (all measured in cm·month^{-1}), and S_e is extractable soil water (cm at the beginning of the month). Measuring the values of k allows climates to be divided into arid ($k < 0.035$), semi-arid, subhumid, humid and perhumid ($k > 0.075$) (Specht and Specht 1989).

In the tropics and subtropics, the ratio of understorey to overstorey species remains fairly constant with k values above 0.03 (Figure 12.8a, b). In the temperate zone, the proportion of understorey species drops as k increases, although the diversity of

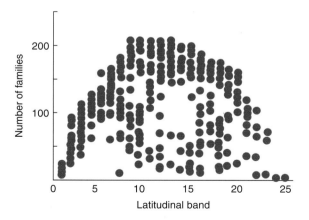

FIGURE 12.6 The number of families of seed plants plotted for 24 bands of latitude from north (left) to south (right). The equator lies between bands 12 and 13. (From Gaston et al. 1995).

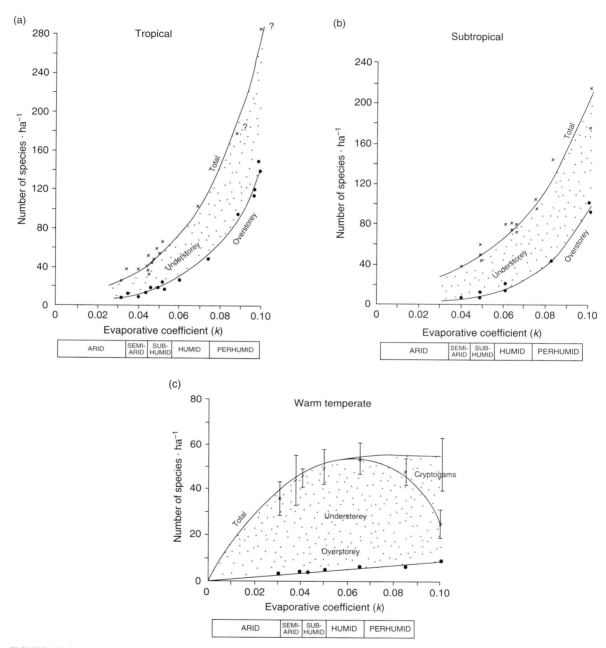

FIGURE 12.8 The number of overstorey and understorey species plotted against evaporative coefficient for (a) tropical, (b) subtropical and (c) warm temperate plant communities in Australia. (From Specht and Specht 1993)

cryptogams (non-seed-producing plants including ferns and bryophytes) somewhat compensates (Figure 12.8c). The diversity of the understorey is controlled by the structure of the overstorey since it regulates the amount of solar radiation that penetrates the canopy. Thus the annual production of the over-storey canopy may be an excellent predictor of diversity. Consider the difference in the nature of forest canopies: at one extreme, in tropical monsoon climates, the growing season extends for ten months and total shoot growth can exceed 11,000 kg ha^{-1}, whereas at the other extreme, semi-arid temperate climates, the growing season may be a few months long and total shoot growth an order of magnitude lower at some 1,100 kg ha^{-1}. A complicating factor is the effectiveness of the canopy at intercepting direct sunlight; this actually increases at higher latitudes because the angle of elevation of the Sun's rays declines with increasing latitude, so a given area of canopy can actually intercept direct sunlight over a larger area of understorey. This may help explain the precipitous drop in seed plant diversity and their

replacement by cryptogams (mostly ferns and bryophytes) in Figure 12.8c.

In another analysis of Australian forests, Austin et al. (1996) used multiple regression models to seek predictors of tree diversity. The order of variables entered in the model was: temperature, plot size, top-ography, rainfall, radiation, seasonality of rainfall and a soil nutrient index based upon the phosphorus content of the bedrock. Figure 12.9 shows how the total number of tree species can be represented in a phase space whose outer limits are set by sampled rainfall and temperature combinations. Austin et al. (1996) concluded that temperature (vertical axis) seemed to be more important than rainfall (horizontal axis) in controlling how many species of trees occur.

In summary, we can begin exploring diversity guided by three provisional rules:

1. The larger the area, the greater the number of species.
2. The larger the variation in environmental condi-tions, the greater the number of species.
3. The closer to the equator, the greater the number of species.

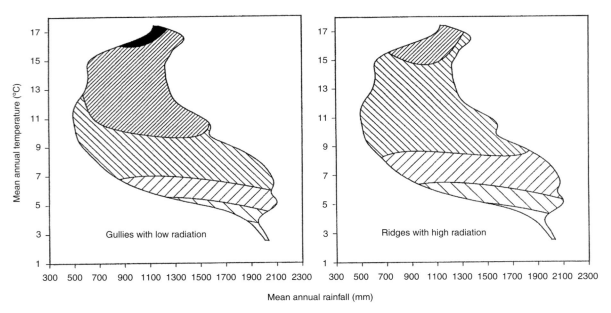

FIGURE 12.9   The number of tree species per 0.1 ha in southwestern New South Wales, Australia in two topographic positions (■) above 8.00; (▨) 4.0–8.0; (◩) 2.0–4.0; (▱) 1.0–2.0; (▱) 0.5–1.0; (▱) below 0.5. The region outside the envelope represents unsampled rainfall and temperature combinations. (From Austin et al. 1996)

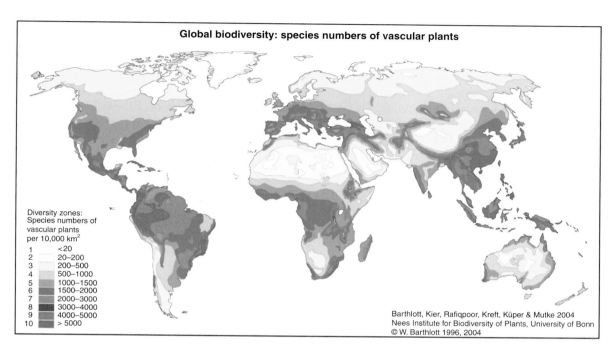

FIGURE 12.10 The number of species of vascular plants at the global scale. (From Mutke and Barthlott 2005)

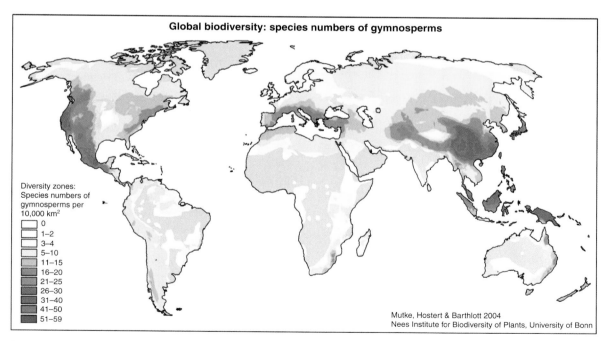

FIGURE 12.11 The highest number of species of gymnosperms is centred in the Himalayas. (From Mutke and Barthlott 2005)

Let us close this topic by returning to the largest scale of all, the distribution of all the plant species on Earth. Just where do you find the most plants? As you now know, from the three numbered principles above, it is likely to be large areas near the equator with mountains. Where might we find a map? Simply obtaining the data for it is a great logistical challenge. It means compiling data on plants and plant distributions from around the world into one usable format. One team of scholars (Mutke and Barthlott 2005) used more than 3,270 data sets showing richness to produce the map in Figure 12.10. There are two main centres of diversity, South America and southeast Asia. More specifically, there are four large centres of plant diversity:

the Andes in northern South America, the coastal areas of Eastern Brazil, the Himalayas including southern China, and the archipelago that includes Sumatra and Malaysia. A fifth region extends from southern Mexico to Costa Rica. If you look more closely, you will also see the Cape of Good Hope area and eastern Madagascar.

What if we pick out just the gymnosperms? The task here is somewhat easier since there are only about 1,000 species of gymnosperms in the world. Figure 12.11 shows that the Himalayas stand out distinctly, although the authors note that Mexico and California combined do have almost as many species as Southeast Asia.

12.5   More Examples of Plant Species Diversity

12.5.1 Mediterranean Climate Regions

Five areas of the world have a Mediterranean type of climate with warm, dry summers and cool, wet winters (recall Figure 10.9). In order of decreasing size these are the Mediterranean basin, coastal California, southwestern Australia (kwongan), central Chile and the Cape of South Africa (fynbos). The species pool for each region has been estimated by

Cowling et al. (1996) in Table 12.2. Fynbos has one of the largest species pools in the world. Many of the species are sclerophyllous shrubs that appear superficially similar, but Cowling et al. (1996) attribute the high plant diversity to relatively low growth rates and the reshuffling of competitive hierarchies after fires. There are some distinctive features of species pools in these areas. The Southern Hemisphere family Proteaceae is very well represented in both the fynbos and kwongan. These two areas also appear to have converged in a number of other ways: they share high diversity, a high incidence of species with obligate dependence upon fire for reseeding, serotinous seed storage in the canopy and seed dispersal by ants (myrmecochory, Section 7.5.3). Other genera have rapidly diversified into this habitat including *Eucalyptus* (>300 species), *Acacia* (>400 species) and *Erica* (>500 species). Of course fire is an important factor in other environments where species richness is comparatively lower (e.g. boreal forest). The difference in diversity between these areas influenced by fire is latitude, with higher diversity found at lower latitudes. Across all five regions there is a clear relationship between species richness and area (Figure 12.12). The southwestern

Table 12.2 **Plant species diversity of Mediterranean climate regions. (From Cowling et al. 1996)**

Region	Area $(10^6 km^2)$	Native flora (no. species)
Coastal California	0.32	4,300
Central Chile	0.14	2,400
Mediterranean Basin	2.30	25,000
Cape	0.09	8,550
SW Australia	0.31	8,000

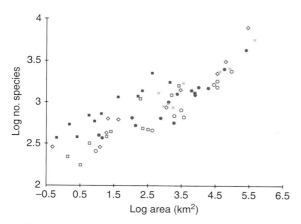

FIGURE 12.12 Species–area relationships from Mediterranean-climate regions. ◊ SW Australia, ■ Cape (SW) (South Africa), □ Cape (SE) (South Africa), × Mediterranean Basin, ○ central Chile, ● coastal California. (From Cowling et al.1996)

region of the Cape has, on average, 1.7 times the diversity of southwestern Australia, about 2.2 times the diversity of the southeastern Cape, coastal California, and the Mediterranean basin, and 3.3 times the diversity of central Chile.

12.5.2 Carnivorous Plants

Extreme environmental conditions provide opportunities for the diversification of species over evolutionary time. In very infertile soils, for example, plants have evolved a wide array of devices for trapping insects in order to compensate for severe shortages of nitrogen and phosphorus. In fact, carnivory evolved in seven different plant families including about 600 species (Table 12.3; Pietropaolo and Pietropaolo 1986; Givnish 1988), and Givnish estimates that carnivory evolved independently at least six times.

Unless extreme conditions persist through time, the plants that have adaptations to these conditions will be lost from the global pool of species. Carnivorous plants are now threatened by a number of factors that simultaneously reduce the area of infertile and wet habitat. Vast areas of wetland are lost through drainage, infilling or the construction of dams (Dugan 1993; Dynesius and Nilsson 1994; Keddy 2010). In the remnant areas of wetland, nutrients are accumulating through run-off from agricultural activities, discharge from sewers and deposition in precipitation (recall Box 3.2). As the area of suitable habitat declines, it is inevitable that the number of species in the habitat will also decline. Superimposed upon this is the market for these plants among collectors. Perhaps the most threatened carnivorous plant is *Dionaea muscipula*, the Venus flytrap (Figure 12.13), which is highly valued by collectors and greenhouse owners because of its active mechanism of trapping insects. The single member of this genus is endemic to the infertile coastal plains of North and South Carolina where it occupies peatlands on sand plains ("pocosins"; Richardson 1981) and is often associated with other more common carnivorous plants such as *Drosera*, *Pinguicula* and *Utricularia* species (Estill and Cruzan 2001). The combination of habitat loss in the pocosins and poaching by unscrupulous collectors has threatened the survival of this species in the wild.

12.5.3 Deciduous Forests

The southern Appalachians support one of the world's richest deciduous forests, particularly in the region known as "mixed mesophytic forest" (Figure 8.16a). There are five species of magnolia, ten species of oak and seven species of hickory, along with a rich array of other woody plants (Braun 1950; Stupka 1964). The other two principal areas of deciduous forest occur in western Europe and eastern Asia (Archibold 1995). Once these forest areas were contiguous, as illustrated by the fact that the species and genera still have distributions encompassing these three regions, but continental drift and climate change have now isolated them from one another (Braun 1950; Pielou 1979).

The large number of tree species in deciduous forests in North America can be accounted for by a

Table 12.3 Summary of diversity in the global pool of carnivorous plants. (From Pietropaolo and Pietropaolo 1986)

Family	Genus	No. species	Geographic distribution	Type of trap
Byblidaceae	*Bybis*	2	Australia	Passive flypaper
Cephalotaceae	*Cephalotus*	1	SW Australia	Passive pitfall
Dioncophyllaceae	*Triphyophyllum*	1	West Africa	Passive flypaper
Droceraceae	*Aldrovanda*	1	Europe, Asia, Africa, Australia	Active
	Dionaea	1	North and South Carolina, USA	Active
	Drosera	120	omnipresent	Passive flypaper
	Drosophyllum	1	Morocco, Portugal, Spain	Passive flypaper
Nepenthaceae	*Nepenthes*	71	area surrounding and including the East Indies	Passive pitfall
Sarraceniaceae	*Darlingtonia*	1	California and Oregon, USA and western Canada	Passive pitfall
	Heliamphora	6	northern South America	Passive pitfall
	Sarracenia	9	North America	Passive pitfall
Lentibulariaceae	*Genlisea*	14	tropical Africa, tropical South America, Madagascar	Passive lobster
	Pinguicula	50	NorthernHemisphere, South America	Passive lobster
	Polypompholyx	2	Australia	Active mousetrap, suction type
	Utricularia	ca. 300	omnipresent	Active mousetrap, suction type

number of factors depending upon the breadth of space and time being considered. History is certainly important (Braun 1950; Latham and Ricklefs 1993a, b). During the ice age, tree species in North America were able to migrate southward along the mountains; in contrast, trees in Europe may have been trapped by the Alps and driven to extinction. Temperate-zone trees such as *Carya*, *Liquidambar*, *Robinia* and *Tsuga* became extinct in Europe around this time (Daubenmire 1978); as a result, these species, known from past interglacial eras, no longer occur in Europe. The high number of tree species presently found in North American deciduous forests is also probably due to the variation in topography – differences in slope, aspect, altitude and exposure provide habitat variation.

12.5.4 Diversity, Biogeography and the Concept of Endemism

A species is **endemic** when it is restricted to a small geographical range. One example is the Venus

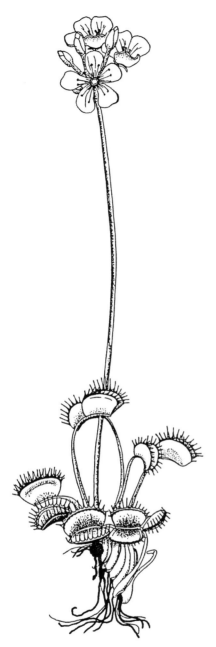

FIGURE 12.13 Venus flytrap, *Dionaea muscipula,* with inflorescence. (From Pietropaolo and Pietropaolo 1986)

flytrap, mentioned above, which occurs only in a small area of southeastern North America (Estill and Cruzan 2001). Another example is the gymnosperm *Welwitschia mirabilis*, which occurs only in the Namibian desert (Figure 12.14). Although endemism is a relative term, it is useful since it draws attention to species with relatively small geographic ranges. Areas may have large numbers of endemic species because of either high rates of evolution of new taxa (neoendemics) or high rates of survival of old taxa (paleoendemics). Sites with many endemics often have high diversity and are therefore important target areas for conservation. To study endemism, Cowling and Samways (1995) collected data on endemism in 52 regions from tropical rain forest to polar deserts. As possible causal variables they considered area, latitude, altitudinal range, mean annual rainfall and mean annual temperature. Latitude was the strongest single predictor of the number of endemics, accounting for more than half of the variation. Areas nearer the equator tended to have more endemics (Figure 12.15a). Mean annual rainfall was also important, but contributed little to the final multivariate model because log mean annual rainfall is correlated with latitude ($r^2 = 0.58$). The second most important variable in the analysis was therefore area, with larger areas tending to have more endemic species (Figure 12.15b). This is not surprising given the general relationship between species richness and area (Section 12.2). In addition, larger areas might also provide a wider array of habitats for speciation of neoendemics or refuges for survival of paleoendemics. The best fit model ($r^2 = 0.63$) was:

$$\log \text{endemic species} = 1.54 - 0.042 \, \text{latitude} + 0.466 \, \text{log area}.$$

A regression model such as this one is useful for further investigation. Not only does it show that diversity generally increases in the manner indicated, but also that residuals can be examined to find anomalies. Strong positive deviation from the relationship occurred in seven sites including: four Mediterranean climate regions (southwestern Australia, coastal California, southwestern Cape, central

FIGURE 12.14 This classic endemic plant, *Welwitschia mirablis*, grows in a narrow strip of desert on the west coast of Africa. It is surely one of the strangest plants on Earth. A gymnosperm (note the cones), the plant has only two leaves (with continuous basal growth) that can reach 9 m long and 2 m wide, often shredded by the strong desert winds. A thick cuticle reduces water loss. It is dioecious (separate male and female plants), and both sexes produce nectar to attract insect pollinators. The photo shows a young female plant more than a metre across. Like saguaro cacti (Section 7.2), seedlings establish only during infrequent rainy periods. This plant is the sole surviving representative of an entire family of vascular plants and is thought to be a relic from Gondwana and the Jurassic era. It was named for Fredrich Welwitsch who first found it while exploring the area in 1859 (Bustard 1990). (Namibian desert, Patrik Mezirka, Shutterstock)

Chile) and two Asian regions (Caucasus, Tien Shan-Alai). The middle-Asian regions are still poorly known, although the Tien Shan-Alai region alone has 3,370 endemic plant species (Major 1988). Clearly such areas deserve more attention in conservation planning. Strong negative deviation occurred in seven areas, five of which were polar desert and tundra, which had been glaciated during the Pleistocene Epoch. The other two were temperate forests in the Carolinas of southeastern North America and tropical forests in the Ivory Coast of West Africa. Perhaps the explanation for the latter two areas lies in their history: there is no source of tropical forest species adjacent to the Carolinas, and during the Pleistocene rain forest was eliminated from much of West Africa.

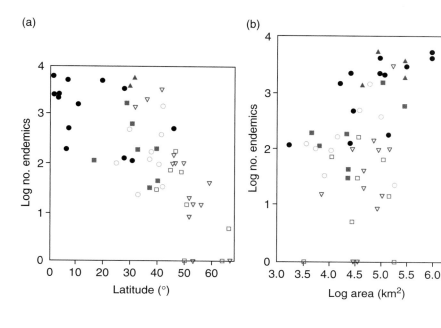

(a)

(b)

FIGURE 12.15 The number of endemic species plotted against (a) latitude and (b) area for 52 regions on continental land masses. (●tropical and subtropical forest and savanna, ○ temperate forest and woodland, ▲ Mediterranean climate shrubland and woodland, ■ warm desert and steppe, □ cold desert and steppe, (▽) boreal forest and tundra. (From Cowling and Samways 1995)

12.6   Models to Describe Species Diversity at Smaller Scales

We have seen above that area, habitat variation and latitude can explain much of the large-scale variation in plant diversity. When an ecologist goes out to a particular location or habitat, however, it is immediately evident that no single habitat contains all the species known from the region. Each habitat contains only a small subset of the regional flora. This leads to a different set of questions. What determines the number and kind of species in one location or habitat? In this case, we can consider the flora of the region to be fixed, and it is the differing composition of habitats that requires explanation. In this field of inquiry, the number of species in a region is usually called the species pool, as it provides the pool of raw materials from which ecological communities are assembled (Keddy 1992; Eriksson 1993; Pärtel et al. 1996; Grace 1999, 2001; Weiher and Keddy 1999).

Ecologists have therefore sought models to describe plant diversity at these smaller scales and there is an enormous number of such models (Keddy 2001, 2005b). Some models are entirely theoretical – their

task is to explore the consequences of certain assumptions – and while they may be elaborate, they may have little or nothing to do with reality. Pragmatic models, by contrast, are based upon actual patterns in data. We may not yet understand why these patterns occur, but that is the task of future experimenters. Here I introduce four pragmatic models that apparently describe diversity at local scales.

12.6.1 Intermediate Biomass

Grime (1973a, 1979) observed that British grasslands with intermediate levels of biomass appeared to have the largest numbers of plant species. He postulated a general unimodal relationship, the "humped-back" model, between plant species richness and above-ground biomass (Figure 12.16). In addition to suggesting the pattern, he suggested two possible mechanisms: stress and disturbance. Species richness is low at low biomass because of high levels of

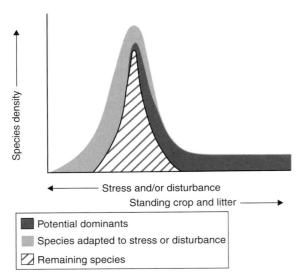

FIGURE 12.16 A model illustrating the impact of stress and/or disturbance on species richness (density) and species composition. (After Grime 1973a,b, 1979)

stress or disturbance, and only a few plants can tolerate these extremes. Species richness is low at high biomass because of dominance by a few strong competitors that create closed canopies.

Ecologists have since tested for this pattern elsewhere. A useful model applies to more than one location. Figure 12.17 shows its application to Mediterranean grasslands. In European grasslands (Figure 12.18) phosphorus alone apparently has an enormous influence on diversity, although the maximum richness is strongly skewed to the left.

A useful model also applies to more than one habitat. The intermediate biomass model has been explored in fens (Wheeler and Giller 1982), lakeshores (Wisheu and Keddy 1989a), interior wetlands (Moore et al. 1989) and coastal wetlands (Gough et al. 1994). For a large sample of interior wetlands, Figure 12.19a shows that plant diversity is highest in low biomass sites, although the pattern is strongly skewed to the left, and may be more of an envelope with an upper limit rather than a simple line. If rare species alone are considered (Figure 12.19b), the vast majority occur in those sites with very low standing crop (less than

100 g/0.25 m²). For this reason, Moore et al. (1989) argue that infertile sites are of particular significance to the conservation of plant diversity.

This is likely to become a larger issue with time, as nutrient-laden precipitation, run-off and groundwater fertilize these habitats that were once infertile (e.g. Ehrenfeld 1983; Ellenberg 1985a; Newman et al. 1996; Matson et al. 2002; Keddy 2010; Box 3.2). The presumed mechanism is the increase in biomass with fertility, and the resulting rise in competition intensity, which leads to the replacement of smaller and more slow-growing plant species by tall canopy-forming dominants. Eutrophication, then, pushes the plant community from the left to the right of Figures 12.16 to 12.19. Although there are exceptions to this pattern (e.g. at small scales, Moore and Keddy 1989a; in saline marshes, Gough et al. 1994), overall, the "humped-back" model provides a useful tool for describing vegetation and, perhaps more importantly, for predicting response to management.

12.6.2 Intermediate Disturbance

Competition is intense in many plant communities as we saw in Chapter 4. Hence a few species typically dominate any particular habitat. So how do so many other plant species manage to survive? One answer seems to be that recurring disturbance creates gaps in this matrix of strong competitors, and many (if not most) other species survive in such gaps. This idea occurred to an entire cohort of ecologists nearly simultaneously in the late 1970s and early 1980s – with a set of papers including Grubb (1977), Connell (1978), Grime (1979), Huston (1979), White (1979), Pickett (1980) and Sousa (1984). Of course, as you saw in Figure 4.26, Skellam explored how disturbance allows coexistence nearly a generation earlier! In Chapter 5 we explored many aspects and examples of disturbance.

Intermediate disturbance models (Figure 12.20) postulate that diversity is controlled by two contradictory forces: rate of disturbance and rate of recovery from disturbance. Both of these control the rate at which strong competitors exclude weak ones. Think

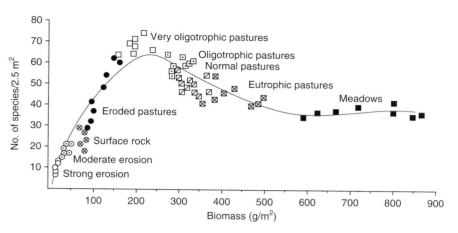

FIGURE 12.17 The relationship between the number of species and biomass for grasslands in the Mediterranean region of Spain. (From Puerto et al. 1990)

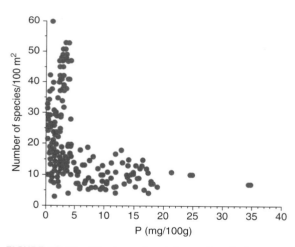

FIGURE 12.18 Plant species richness in relation to soil phosphorus for a set of European pastures. (From Janssens et al. 1998)

about disturbance from grazing animals. If grazing is very light, a few canopy-forming plants may shade out all neighbours – and diversity will be low. If grazing is very intense, nearly all species will be eaten, leaving only a few very unpalatable species (or even bare soil). Both extremes have low diversity. But now imagine intermediate amounts of grazing that are sufficient to reduce, but not eliminate, the dominant competitors, and create gaps in which other species can survive. This intermediate level of disturbance will

allow more species to coexist, and there will be some specific rate of disturbance that will allow the greatest number of species to coexist. Hence the name – intermediate disturbance. The second axis on the figure is necessary because the rate at which plants grow new biomass will determine how much disturbance will be needed to maximize diversity.

When a disturbance produces relatively small gaps still surrounded by intact vegetation, the term gap dynamics is often used to describe the disturbance. Many kinds of trees will regenerate only in gaps that are created when one or more mature trees are killed (e.g. Shugart et al. 1981; Denslow 1987; Botkin 1993). Which species colonize a gap often depends as well upon the size of the gap (recall Figure 5.36). Even small gaps can enhance recruitment in grasslands (Brewer 1999; Jutila and Grace 2002). Diversity of landscapes can be increased by gaps created by landslides (Guariguata 1990), storms (Seischab and Orwig 1991), volcanic eruptions (del Moral and Wood 1993) and floodplain deposition (Salo et al. 1986).

This model has important implications for plant conservation. If the rate of disturbance is altered (say, by changing fire regimes in a landscape), or if the rate of recovery is altered (say, by increasing eutrophication), diversity will change. There is one complication in applying the model to a natural habitat. If we ask the question of whether a disturbance, such as fire or grazing, will increase or decrease diversity, it is

(a)

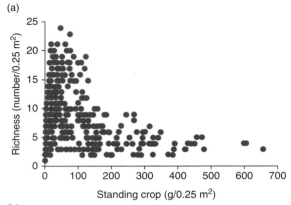

(b)

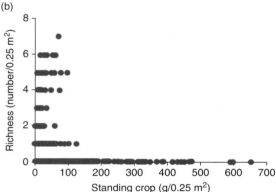

FIGURE 12.19 The relationship between plant species richness and soil fertility (measured as above-ground biomass) for shoreline wetlands based on 401, 0.25-m² quadrats from wetlands in Ontario, Quebec and Nova Scotia, Canada. (a) All species. (b) Species considered nationally rare in Canada. (From Moore et al. 1989)

obvious from inspection that the answer depends upon where the system is located in two-dimensional space. Grazing can increase diversity if the system is at the upper left of Figure 12.20, or it can decrease diversity if the system is at the lower right. Thus one often cannot give a simple answer to the question of whether a particular factor will increase or decrease diversity – it depends upon the history and current state of the system. This is called contingency.

The conservation implications of the intermediate disturbance model are illustrated by the next example, which introduces a study at one of the largest scales of which I know – fire and plant diversity in boreal

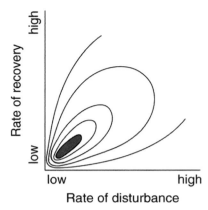

FIGURE 12.20 The number of plant species in a habitat (and the biomass) is a consequence of a dynamic equilibrium between the rate of disturbance and the rate of recovery from disturbance. The inner ellipsoid has the highest number of plant species. (After Huston 1979, in which these axes are designated "frequency of [biomass] reduction" and "rate of [competitive] displacement")

forests. (The other large-scale application of this model may be the Amazonian floodplains discussed in Section 5.3.2, Figure 5.11.) Boreal forests are heavily dependent upon fire for regeneration and landscapes often consist of mosaics of forest dating from fires of different ages (Heinselman 1973, 1981; Rowe and Scotter 1973; Wein and Moore 1977; Archibold 1995). Suffling et al. (1988) set out to test whether an inter-mediate frequency of forest fires would lead to the highest vegetation diversity at the landscape scale in accordance with predictions of intermediate disturbance models. Their study area was a large tract of forest in northwestern Ontario. The forests here are dominated by fire-dependent species such as *Pinus banksiana* and *Picea mariana* in the rock barrens of the west and grade into *Abies balsamea* and *Picea mariana* in the east. Forest fire records showed an increase in fire frequency from east to west, with the exception of a fire-prone area on shore deposits, eskers and rock barrens along the shoreline of former Lake Agassiz (Figure 12.21). The diversity of forest types (as measured by the Shannon index of diversity) was strongly related to fire history (Figure 12.22).

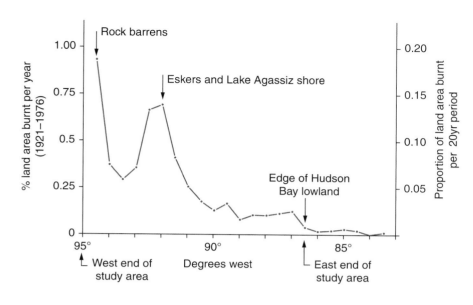

FIGURE 12.21 Regional variation in forest fire occurrence in the boreal forest of northern Ontario. The climatically induced trend is complicated by relatively dry geomorphological features that increase local forest fire occurrence from 91° to 93° longitude. (From Suffling et al. 1988)

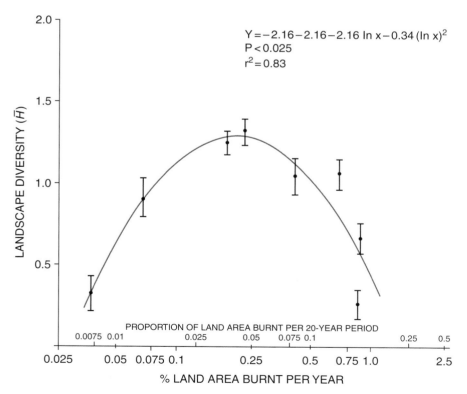

FIGURE 12.22 The relationship between landscape diversity (H) and disturbance by forest fires in eight 250-km^2 samples. (From Suffling et al. 1988)

Maximum diversity occurred at intermediate fire frequencies of 0.25 percent of the land burnt each year.

Forest fire control might lead to increased diversity in some fire-prone habitats (at the far right of the figure), but overall it will reduce landscape diversity in areas with intermediate to low disturbance (Suffling et al. 1988). They might have added that, at a larger scale, there is a further diversity issue. If all vegetation types are maintained with the same fire frequency, then similar diversity patterns will be found along the entire transect they studied. Maintaining low, intermediate and high fire frequencies in both the eastern and the western portions of the region will maximize diversity at a still higher scale. Suffling et al. (1988) emphasize correctly that fire control will force the entire landscape to the far left side of Figure 12.22, reducing the diversity of vegetation types (and presumably the number of component species). Maximum diversity is attained by having some sites at the maximum along with sites representing both extremes. Managing for diversity within a single park, then, is not the same as managing for landscape and vegetation diversity over a park system as a whole.

12.6.3 Centrifugal Organization

We know from Section 12.3 that heterogeneous environments support more kinds of plants. One model that can summarize these patterns is called the centrifugal model (Keddy 1990b; Wisheu and Keddy 1992). Consider a simple proposition – that any landscape contains multiple biomass gradients such as you saw above in Figures 12.16 to 12.19, with each biomass gradient controlled by a different limiting factor. At one end of each gradient, which we shall call the core habitat, nutrients and water are freely available, which allows a closed canopy to develop and here light becomes a limiting resource. Think, for example, of floodplains where the core habitat is dominated by *Phragmites* and *Typha*. At the low biomass end of each gradient, however, quite different factors may limit plant growth – say, drought, flooding, high or low pH, low nitrogen, low phosphorus or grazing. In order to tolerate such limiting factors, plants require different kinds of adaptations, and thus the low biomass end of each gradient supports a different set of plant species. Consider the different kinds of species found in low biomass habitats such as chalk grassland, sea cliffs, alvars, sandy shorelines, rock outcrops, etc. The resulting pattern can be represented by a set of gradients radiating outward from the core habitat (where light is the main constraint) to peripheral habitats (where a wide array of factors can be limiting) (Figure 12.23). Many unusual genera occur in peripheral habitats, including *Armeria*, *Asplenium*, *Castilleja*, *Coreopsis*, *Drosera*, *Eriocaulon*, *Isoetes*, *Lobelia*, *Lycopodium*, *Ophioglossum*, *Parnassia*, *Pinguicula*, *Sabatia*, *Saxifraga* and *Utricularia* (e.g. Fernald 1921, 1922, 1935; Peattie 1922; Soper and Maycock 1963; Goldsmith 1973a,b; Moore et al. 1989; Wisheu and Keddy 1989b).

Several predictions can be made from this model. First, while the intermediate biomass model (Section 12.6.1) focuses upon a corridor of diversity at intermediate levels of biomass, the centrifugal model emphasizes that over multiple gradients, it is the lower biomass habitats that likely support the largest number of plant species. Thus the larger the number of extreme environments, the higher the total diversity is likely to be. In terms of reserve design and selection, the centrifugal model draws attention to the need to ensure that extreme or unusual kinds of environmental conditions are included. It may be the atypical habitats that contribute most to total plant diversity – rock outcrops (Wiser et al. 1996; Shure 1999), alvars (Belcher et al. 1992; Catling and Brownell 1995), wet meadows (Peattie 1922; Hill and Keddy 1992; Sorrie 1994) or cliffs (Soper and Maycock 1963; Goldsmith 1973a).

The centrifugal model also predicts probable responses to eutrophication. As noted earlier (Section 12.6.1), eutrophication is a growing threat to plant diversity, as it allows larger canopy-forming plants

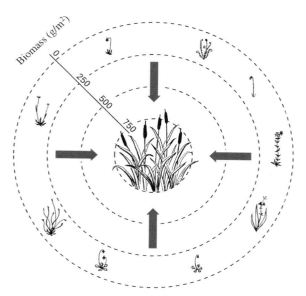

FIGURE 12.23 In the centrifugal model, biomass gradients radiate outward from one core high-biomass habitat. The biomass scale is shown at the upper left. There are many different low-biomass peripheral habitats, each with distinctive plant species, often rosette growth forms. Examples here include (from the top right, clockwise) *Drosera, Utricularia, Rhexia, Sagittaria, Drosera, Parnassia, Eleocharis, Sabatia* and *Pinguicula*. Other similar genera from eastern North America and western Europe might include *Armeria, Castilleja, Eriocaulon, Hieracium, Isoëtes, Lobelia* and *Saxifraga*. Eutrophication (green arrows) increases biomass, causing the loss of peripheral species and convergence in composition upon a few common canopy-forming species typical of core habitats.

to exclude smaller, stress-tolerant plants (Keddy 2010). Eutrophication will force plant communities that are initially different to converge into one light-limited habitat, thereby leading to even larger losses of plant diversity (Figure 12.23, thick arrows) – a growing problem worldwide (e.g. Ehrenfeld 1983; Ellenberg 1985; Newman et al. 1996; Keddy 2010). A specific kind of plant is usually eliminated by

eutrophication – creeping or rosette species that depend upon sites with low biomass and high light availability. Examples of such sensitive species can be found in many vegetation types: *Poterium sanguisorba* and *Carex flacca* in chalk grassland (Austin 1968), *Lobelia dortmanna, Eriocaulon septangulare* and *Sabatia kennedyana* in wetlands (Wisheu and Keddy 1989b), *Armeria maritima* on sea cliffs (Goldsmith 1973b) and *Hieracium floribundum* in old fields (Reader and Best 1989).

One of the most intensively studied cases of eutrophication damaging native species comes from the Everglades (Loveless 1959; Davis and Ogden 1994; Newman et al. 1996). These were once a vast rain-fed wetland, with extremely low nutrient levels, producing a unique vegetation type of species adapted to wet, infertile conditions. As the sugar industry began to exploit the northern Everglades, vast amounts of nutrient-rich water poured south, fertilizing these wetlands. The consequence was the invasion of a tall, rapidly growing wetland plant – cattail (*Typha*). Cattails changed not only the plant species, but also the suitability of the wetlands for indigenous wildlife. Plumes of invading cattail now mark where nutrients flow into the Everglades. Although court cases brought by environmental groups have led to some efforts to reduce nutrient input, it remains to be seen whether this process can be halted or reversed (Sklar et al. 2005). Currently, the Florida sugar industry provides a text book case of eutrophication.

J. S. Beard (1944) described centrifugal patterns in tropical forests. Rain forest, the core community, occurs where conditions are optimal for growth.

> We may begin by envisaging a mesic or optimum habitat where availability of moisture – and thus every condition for plant growth – is as ideally favorable as it can be in the tropics... the land must be well-drained, the soil deep and permeable, moisture must be available in sufficient quantity throughout the year and the situation sheltered from violent winds. There must be neither inundation nor seasonal drought, evaporation

must be moderate and frost unknown. Such conditions naturally favor the tallest, most luxuriant and most complex type of vegetation in the American tropics, the vegetal optimum, rain forest. Rain forest is by no means so common in the tropics as is popularly supposed, for these ideal growth conditions are rare.

(p. 134)

Five vegetation types are produced by different kinds of constraints: the seasonal formation (seasonal lack of rainfall), the dry evergreen formation (constant lack of available moisture), the montane formation (cold and exposed), the swamp formations (flooding) and the marsh or seasonal swamp

formations (alternating inundation and desiccation). Each of these five, he notes, is the product of an environmental factor that inhibits the occurrence of rain forest.

These series are additional to the single optimum formation, rain forest. Each series consists of stages between the optimum and the extremely adverse and thus the head of each series approached closely to the optimum... the series may be regarded as radiating outwards from the optimum like the spokes of a wheel or, better, three-dimensionally like radii from the centre of a sphere.

(p. 135)

12.7   Relative Abundance: Dominance, Diversity and Evenness

So far, we have used species richness – the number of species in a quadrat or region – as a measure of diversity. This measure can be a misleading measure of diversity, since often a large proportion of the biomass is locked into the tissues of a few dominant species. The small fraction of biomass not locked up in the dominant species is then partitioned among a much larger number of relatively uncommon species. In a completely even community, each species would have an equal proportion of biomass (it would have an abundance of b/s where b is community biomass and s is the number of species). Such a situation maximizes a property of diversity called evenness (Peet 1974; Pielou 1975). The greater the departure from evenness, the greater the biomass concentration within a few species.

The number of species and the evenness of a community are best illustrated with a ranked abundance list – the abundance of each species (measured in biomass, number of individuals, cover or some similar measure) plotted against its order in a sequence from most abundant to least abundant.

Figure 12.24 shows an example from vascular plant communities in the Great Smoky Mountains and Figure 12.25 shows examples from wetlands. These kinds of ranked abundance lists are small-scale examples of a larger scale phenomenon, the canonical or log-normal distribution. The log-normal pattern is found in nearly all large samples: a few species are common (they dominate the sample) while most other species are uncommon (each is represented by only a few individuals). The analysis of the log-normal pattern is beyond the scope of this text, but you should know that there are many studies available to you in other sources such as Preston (1962a, b), Pielou (1975) and May (1981).

Sometimes one wants a single number to measure diversity. This means that one number is used to summarize all the information in a ranked abundance list, both the number of species overall and the relative abundance (or evenness) of the species. Two measures are commonly used: Simpson's index (Simpson 1949) and the Shannon–Weaver index (Shannon and Weaver 1949) (Box 12.1). More on the relative merits of these

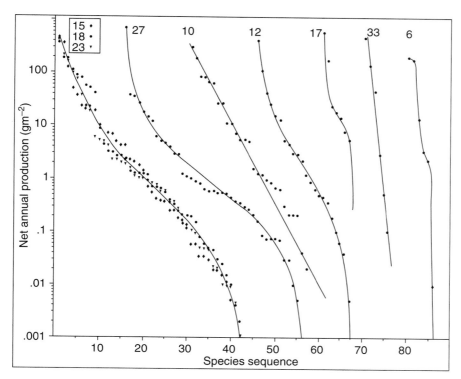

FIGURE 12.24
Dominance–diversity curves for vascular plant communities in the Great Smoky Mountains. For each curve, points represent species plotted by net annual above-ground production and order in the species sequence from most to least productive. For the sake of clarity, the curves have been arbitrarily spaced out, their origins being separated by 10 or 15 units along the sequence axis. The numbers across the top refer to sample sites and the first curve is based on data for three sites. (From Whittaker 1965)

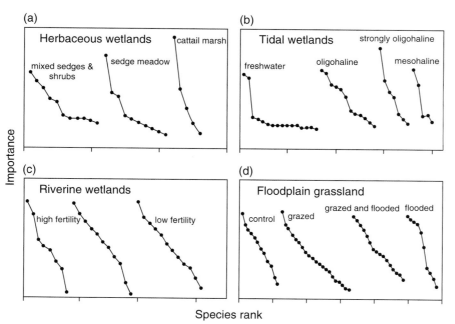

FIGURE 12.25 Ranked abundance lists (dominance–diversity curves) for four wetland types: (a) herbaceous (after Gosselink and Turner 1978); (b) tidal (after Latham et al. 1994); (c) riverine (after Weiher and Keddy unpublished); (d) floodplain grassland (after Chaneton and Facelli 1991). (From Keddy 2010 courtesy E. Weiher)

Box 12.1 Diversity Indices

The purpose of an index of diversity is to provide a single number that combines two different properties of a collection of species. The first property is simply the number of species (that is, species richness). The second is the relative abundance of the species. The diversity index will be highest for any number of species, s, when all the species occur in equal abundance, that is, when all elements p_i are identical.

The formula for Simpson's diversity index is:

$$C = \sum_{i=1}^{s} (p_i)^2$$

The formula for the Shannon–Weaver (or Shannon–Wiener) diversity index is:

$$H' = - \sum_{i=1}^{s} (p_i)(\ln p_i)$$

where C and H' are index numbers, s is the total number of species in the sample, and p_i is the proportion of all individuals in the sample that belong to species i.

(and other diversity measures) can be found in sources such as Peet (1974) and Pielou (1975).

Returning to ranked abundance lists, the traditional approach to exploring such patterns has been to develop statistical models that produce different relative abundance patterns (Whittaker 1965; May 1981), with the assumption that such models might suggest something about the processes generating these patterns in real data (Pielou 1975, 1977). Four models are commonly used:

1. Broken stick (MacArthur 1957). Assumes the simultaneous random division of resources among species, as if points were randomly assigned along a stick of wood that is then broken into pieces, the length of each piece representing the abundance of a species.

2. Geometric (Whittaker 1965, 1972). Assumes that each species takes a constant fraction of the available resources, availability being defined as resources not already allocated to the preceding set of dominant species. This assumes a strict hierarchy of competitive dominance.

3. Log-normal (Preston 1962a, b). The ranked abundance pattern that is associated with Preston's early descriptions of the canonical distribution of commonness and rarity.

4. Zipf–Mandelbrot (Frontier 1985). The most recent addition to the class of ranked abundance models, which assumes a successional process such that the entry of species into a community is dependent upon changes caused by those species already present.

Each of these models produces slightly different theoretical ranked abundance lists. Figure 12.26 shows an example of each fitted to real ecological data. The most notorious of such models is the broken-stick model that my doctoral supervisor, Chris Pielou, frequently criticized. In her opinion, it made the most unlikely assumptions of the entire set of such models, and she attributed its popularity and persistence to the fact that most ecologists could imagine breaking a stick, but few could do the calculations for, say, a geometric model. The relative merits of most of these models were extensively explored in the 1970s (e.g. Pielou 1977; May 1981), but it seems fair to conclude that all such modelling brought little understanding of the origin and persistence of differences in relative abundance. The tantalizing patterns remain.

Table 12.4 **The percentage of best-fits to four models of species' relative abundance in three grassland data sets.** (From Wilson et al. 1996)

Location	Broken stick	Geometric	General log–normal	Zipf–Mandelbrot
Monks Wood	1	29	43	27
Park Grass (1991)	2	28	46	24
Park Grass (1992)	2	28	22	48
Compton	0	58	42	0
Total	5	143	153	99

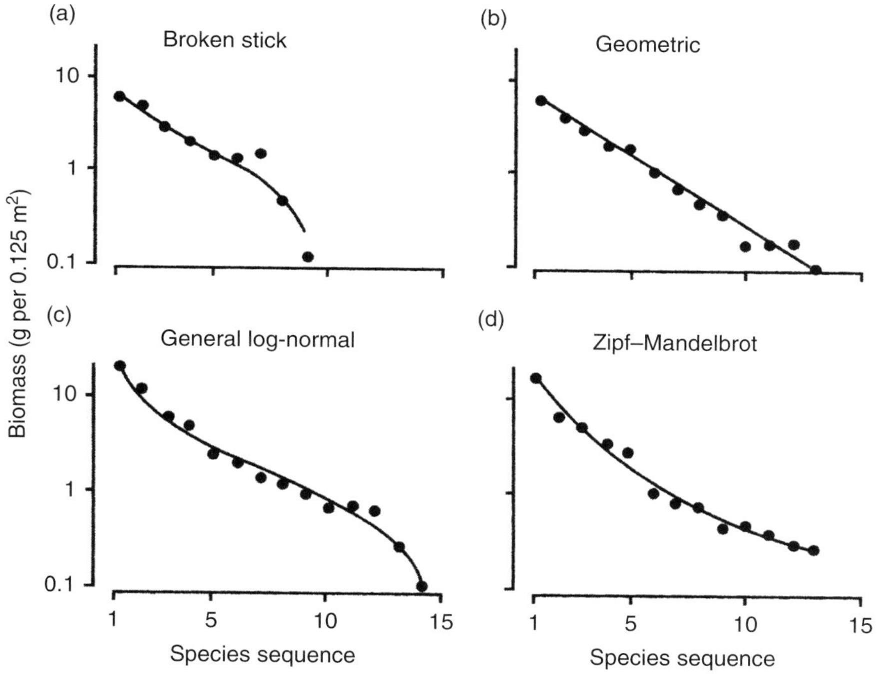

FIGURE 12.26 Four models of relative abundance fitted to data from British grasslands. (From Wilson et al. 1996)

The most comprehensive study of ranked abundance lists comes from three sets of British grasslands (Wilson et al. 1996). The Monks Wood experiment, established in 1978, allowed natural colonization and succession with annual hay mowing in mid August. The Park Grass experiment, established in 1856, includes different fertilization and mowing regimes (Box 12.2). The Compton experiment, initiated in 1987, explores factors thought to create grasslands with high species richness. Considering these grasslands collectively, the general log–normal model most frequently described the relative abundance of plant species (Table 12.4). A further analysis showed that it fit significantly better in plots fertilized with phosphorus. Changes in weather may also cause fluctuations in the patterns of ranked abundance. The data from the Park Grass experiment are presented for two separate sampling years. In 1991, the general log–normal model was most frequent ($n = 46$), whereas in 1992 the Zipf–Mandelbrot model was most frequent ($n = 48$).

Box 12.2   The Park Grass Experiment at Rothamsted

Most ecological experiments run for short periods of time, owing to practical constraints such as the length of a human career, the short nature of most grants and, in many cases, the duration of PhD research. As a consequence, long-term experiments have been rare in ecology, the Rothamsted experiments being a notable exception. Not only do they illustrate the value of long-term work, but they also provided the impetus for Sir R. A. Fisher's statistical studies that revolutionized thinking about variation in nature, and provided the quantitative arsenal that we now take for granted (Box 11.1). So let us examine these three monumental and intertwined stories, Rothamsted itself, the Park Grass plots and Ronald Fisher.

According to Johnston (1994), J. B. Lawes, later Sir John Lawes, inherited the Rothamsted estate, which included an old manor house and about 100 ha, in 1822, when he was only eight years old. His father had died, leaving family fortunes at a low ebb. After some time at Oxford, Lawes returned home and had a bedroom at the manor converted into a laboratory; here he worked on a variety of projects including medicinal plants. He finally took out a patent for the manufacture of superphosphate, and in 1843 starting commercial production at a factory in Deptford, London. In 1843, he appointed J. H. Gilbert (later Sir Henry Gilbert), a chemist by training, to assist him in field and laboratory experiments on nutrition of crops and animals. Today some of their experiments continue and are apparently the oldest continual agricultural experiments in the world. Their partnership continued for 57 years, and together they published some 150 scientific papers and some 300 popular articles for farmers.

Gilbert had earned a doctorate at Giessen in Germany as a student of Professor Liebig, a leading figure in the history of plant nutrition, still known for "Liebig's Law of the Minimum." The first edition of Professor Liebig's treatise on agriculture was published in 1840, and it rapidly went through several subsequent editions. Gilbert earned his doctorate in 12 months, returning to England to work at University College London and then in industry in Manchester before moving to Rothamsted. The principal field experiments he established there focused upon agricultural species: turnips (1843), winter wheat (1843), beans (1847), crops in rotation (1848), clover (1849), spring barley (1852), oats (1869) and, finally, of most interest to community ecologists, the only experiment involving plants in mixture, the Park Grass studies of permanent pasture (1856).

Here we shall largely pass over the details of the experiments themselves, particularly the ones on agriculture, except for Figure B12.1.1, which shows the dramatic improvement in yields of wheat through time as fertilizer, herbicides and fungicides were incorporated into agricultural practice.

Now to R. A. Fisher, who was hired at Rothamsted in 1919, nearly a century after the founding of the Statistical Society in 1834 (Barnett 1994). Over his period at Rothamsted, Fisher developed and introduced such essential concepts as factorial design, interaction of main effects, analysis of variance and blocking to account for the heterogeneity of fields.

The Park Grass plots themselves continue to be of interest to ecologists (Silvertown et al. 2016); one of the more interesting examples of this work (Silvertown 1980; Silvertown et al. 1994) shows that fertilization consistently led to steady declines in plant diversity (see ahead to Figure 12.29); in contrast, effects of rainfall were minor. Further, within any plot, there now appears to be some sort of botanical equilibrium at the coarse scale: the relative proportions of three groups (grasses, legumes and other species) show no trends over recent time. In contrast, however, individual species within each group have continued to fluctuate widely. There thus appeared to be a dynamic equilibrium at one scale, but not the other.

The Park Grass experiments, and others, continue at Rothamsted. Recognizing the value of this work, many other sites for long-term ecological research have been established over the last few decades. The valuable association of long-term experiments with large tracts of native vegetation also illustrates the close connection

Box 12.2 (cont.)

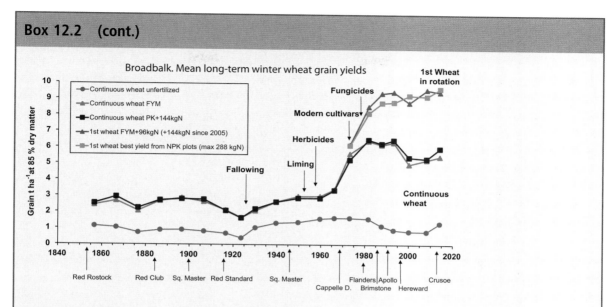

FIGURE B12.2.1 Yields of winter wheat grown at Broadbalk, Rothamsted from 1856 to 2014, with fertilizers and with farmyard manure, showing the effects of changing cultivars, the introduction of weed control and fungicides, and crop rotation to minimize effects of soil-borne pathogens. (© Rothamsted Research 2016)

possible between systems of protected areas and scientific research. There is a sort of symbiosis here, with the natural areas providing sites for research that otherwise would not be possible, while that research provides insight into management and protection of the reserve system.

Documenting the relative frequencies of these ranked abundance lists is an important contribution. The next step is to explore changes that occur with known factors, such as primary production, successional stage, soil fertility or rainfall. While mechanisms are still poorly understood, the ubiquity of competition and competitive hierarchies in vegetation (Chapter 4) suggests that competition plays a significant role in causing these dominance patterns. The sensitivity of the shape of the relationship to the addition of phosphorus also suggests a mechanism involving competition. However, the connection will be clearly established only when independent measures of relative competitive ability can be used to predict positions in ranked abundance lists or when dominant plants are removed to test for release of the subordinates.

12.8 Microcosm Experiments on Richness and Diversity

Small plant communities that are created artificially are called microcosms. Microcosm experiments are relatively artificial in the restricted number of species used, and in the few environmental factors that are

Table 12.5 The results of a five-year competition experiment using 20 species in 24 habitats (combinations of variables in column 1). F-values are shown for repeated-measures, three-way ANOVA of species richness, main effects and two-way interaction terms. (From Weiher and Keddy 1995a)

Environmental variable	Treatment	Fertility	Treatment × fertility	Year	Treatment × year	Fertility × year
Depth of water	195.60***	47.48***	9.67***	153.66***	46.11***	9.36***
Fluctuation of water level	73.75***	46.76***	5.28*	99.13***	48.44***	5.41**
Presence/ absence of litter	0.89	220.40***	8.65**	169.00***	4.98**	6.47**
Texture (sand, gravel, cobbles)	0.50	77.63***	0.35	296.58***	1.88	12.14***
Starting time	5.88**	56.96***	7.87**	268.52***	0.95	6.54**
Presence/absence of Typha	0.05	81.87***	2.55	84.58***	3.52*	4.06*

Notes:
* p <0.05; ** p <0.01; *** p <0.001.

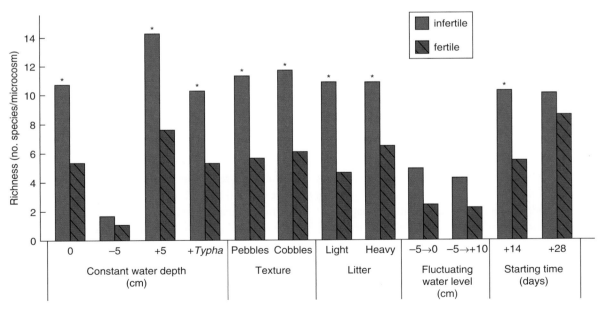

FIGURE 12.27 Species richness after two growing seasons for 12 artificially created wetland habitats at two levels of fertility. Depths are in centimetres. Significant differences in richness between fertility treatments at $p = 0.001$ are indicated by an asterisk. (From Moore and Keddy 1989b)

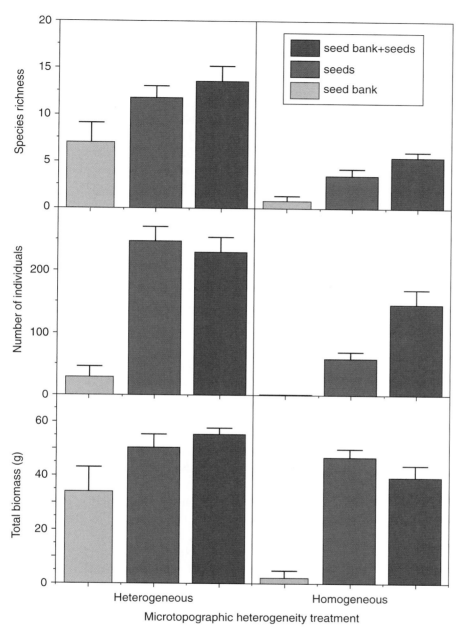

FIGURE 12.28 Three attributes of experimental plant communities (species richness, number of individuals and total biomass) shown in relation to substrate heterogeneity and seed source treatments (values shown are means +1 SE). (From Vivian-Smith 1997)

manipulated (Fraser and Keddy 1997). Unlike field experiments, microcosm experiments (which are also called laboratory experiments) sacrifice realism for precision (Diamond 1986). The advantage of microcosms for community-level research is the degree of

control they allow over the key environmental factors. We will look at three examples.

Grime et al. (1987) made artificial grassland communities to examine the effects of mycorrhizal infection, soil structure and grazing on plant species diversity. They found that mycorrhizae can increase diversity by increasing the biomass of the subordinate plant species relative to that of the dominant plant species. Perhaps the transfer of mineral nutrients and sugars through mycorrhizae reduces the intensity of competition and encourages species coexistence on fertile soils.

Weiher and Keddy (1995a) made artificial wetland communities to examine the effect of 24 different combinations of habitat factors on diversity. At the end of two growing seasons, species richness was significantly affected by fertility in eight different habitat types (Figure 12.27). After five years (Table 12.5), species richness was strongly affected by fertility level (column 3) and year (column 5) for all treatments. Beyond this, only two of the six

treatments (rows 1 and 2) strongly affected species richness – water depth and fluctuation of water level.

Vivian-Smith (1997) also made artificial wetland communities to study the effects of small-scale habitat variation on diversity. There is a good deal of evidence that heterogeneity can increase richness by allowing for differential establishment of seedlings (e.g. Harper 1965; Grubb 1977; Keddy and Constabel 1986). In this experiment, some containers had a flat peat–sand mixture, and some had five artificial peat hummocks raised 2.5 cm above the surrounding substrate. Three seed mixtures were then sown into these two treatments: an artificial seed mixture, a natural seed bank and a combination of the two. Figure 12.28 shows that heterogeneous treatments had significantly more species, regardless of the seed source used.

12.9   Field Experiments on Richness and Diversity

Experimental manipulation of independent variables in the field is more realistic, but often more difficult to interpret. Two of the most common field manipulations are based on the two fundamental forces that structure plant communities: fertility and disturbance (Chapters 3 and 5).

The longest running manipulative experiment is the Park Grass experiment at Rothamsted, England (Box 12.2). Begun in 1856, the treatments involve different fertilizer regimes on hay-meadow grassland. The original objective was to study the effects on hay yield, but ecologists have found it valuable for answering broader questions. Silvertown (1980) describes more of the history of the site and shows that over the years, species diversity has progressively declined in the nutrient-enriched treatments (with a corresponding increase in biomass). The pH of the soil was found to be another important factor controlling species richness, where soils with lower pH had lower species richness (Figure 12.29). The decline in species richness with nutrient enrichment has been found in many other studies (e.g. Grime 1973a, b, 1979; Huston 1979; Austin and Austin 1980; Tilman 1982),

including wetlands (e.g. Willis 1963; Moore et al. 1989; Verhoeven et al. 1996).

An experiment that combined both fertilization and the experimental removal of the dominant plant species (Gurevitch and Unnasch 1989) showed the close connections among fertilization, competitive dominance and reduced diversity. The sandy, nutrient-poor fields that they studied had some 15 herbaceous plant species, but this number declined to 6 with the application of NPK fertilizer. These effects could be entirely avoided by removing the one dominant species (Figure 12.30), in which case both fertilized and unfertilized plots had a mean of 18 species. This provides clear evidence that the decline in richness with fertilization is a result of competition from the dominant species. That is, it appears that increasing the resource supply increases monopolization. Presumably this is because the added nutrients allowed the dominant grass, *Dactylis glomerata*, to more effectively shade its neighbours (although Gurevitch was unable to detect a significant increase in *Dactylis* biomass after fertilization). Further, evidence for increased competition at higher resource levels comes from

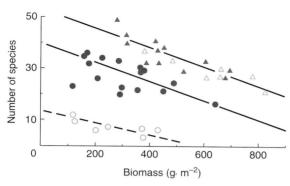

FIGURE 12.29 Changes in species richness with time in the Park Grass experiments at Rothamsted, UK. The plots in the top line represent the situation before fertilization; the bottom two lines illustrate the effects of fertilization (open symbols, acidified; closed symbols, not acidified when fertilized). The top line precedes the bottom lines by 80 years. In all three cases, across time and treatments, the slopes are nearly identical. (From Keddy 2001 after Silvertown 1980)

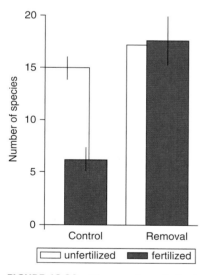

FIGURE 12.30 The number of plant species in an old field with sandy soil declined after addition of fertilizer (left bars), but this change did not occur if the dominant species, *Dactylis glomerata*, was also removed (right bars). Plots were 2 m × 0.5 m; responses were measured after two growing seasons; vertical lines show 1 SE for *n* = 5 replicates. (After Gurevitch and Unnasch 1989)

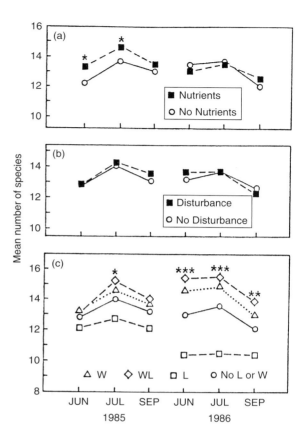

FIGURE 12.31 The effects of (a) increased nutrients, (b) disturbance and (c) increased water and light on species richness (mean number of species per 1.0 m × 0.5 m plot) in old fields. For (a) and (b) significant differences between treatments and within a date are indicated by asterisks. For (c), significant interactions for a given date are indicated by asterisks (* $p < 0.5$, ** $p < 0.01$, *** $p < 0.001$). (From Carson and Pickett 1990)

the greater difference between controls and treatments at high as opposed to low fertilities. This experiment, like that of Carson and Pickett (1990), tends to confirm the view that adding resources merely allows an already dominant species to further exclude the others by increasing its competitive effects upon neighbours.

In a revealing extension of the Gurevitch experiment, Carson and Pickett (1990) explored the effects of four different treatments on old fields in New Jersey: (1) added macronutrients; (2) added water; (3) added light (achieved by tying back tall plants to remove their shade from a plot; and (4) added disturbance (achieved by digging with a hand trowel early in the growing season). These fields are normally dominated by goldenrods and asters, with *Solidago canadensis* and *Aster pilosus* most common, and *S. graminifolia*, *S. juncea*, *S. rugosa* and *S. nemoralis* also found. The mean number of species was temporarily increased by the addition of nutrients (Figure 12.31a), but this effect disappeared in the second year. Disturbance had no effect upon the number of species in either year (Figure 12.31b). The principal factors increasing diversity were apparently the removal of the canopy and the addition of water (Figure 12.31c).

12.10   Implications for Conservation

Ecological management and conservation is a rapidly growing field. There are new journals dedicated to the topic, as well as clearly stated objectives for the maintenance of biodiversity (e.g. Noss and Cooperrider 1994; Table 12.6) and new global maps of where diversity is concentrated (Myers et al. 2000). There is a very important, and still frequently misunderstood, connection among local species diversity, the species pool and global biodiversity. Conserving "biodiversity" (e.g. Groombridge 1992; Reid et al. 1993; Noss 1995; Myers et al. 2000) requires one to think about the regional species pool for an area. It also requires setting clear priorities. Money, time and resources will always be limited, so it is imperative to concentrate upon those regions of the world that have the highest biological diversity.

While political boundaries are highly artificial, they do provide the administrative units within which conservation programs must be designed and implemented. Figure 12.32 shows that Brazil is therefore a global priority, being the only country with more than 50,000 plant species within its borders. Colombia, China and Mexico are next on the list, each having more than 25,000 plant species. At the global scale, the challenge is to maintain the species pool of the entire planet or at least of a region. Most managers, however, must focus on maintaining or increasing diversity at only one location. *It is entirely possible that attempts to increase local diversity can actually reduce diversity at a larger scale*, that is, reduce the regional species pool. We will explore two examples.

The New Jersey Pine Barrens support many rare plants, in part because the habitats are so infertile.

Table 12.6 Four fundamental land management objectives for maintaining native biodiversity of a region. (From Noss and Cooperrider 1994)

Represent, in a system of protected areas, all native ecosystem types and seral stages across their natural range of variation.

Maintain viable populations of all native species in natural patterns of abundance and distribution.

Maintain ecological and evolutionary processes, such as disturbance regimes, hydrological processes, nutrient cycles and biotic interactions.

Manage landscapes and communities to be responsive to short-term and long-term environmental change and to maintain the evolutionary potential of the biota.

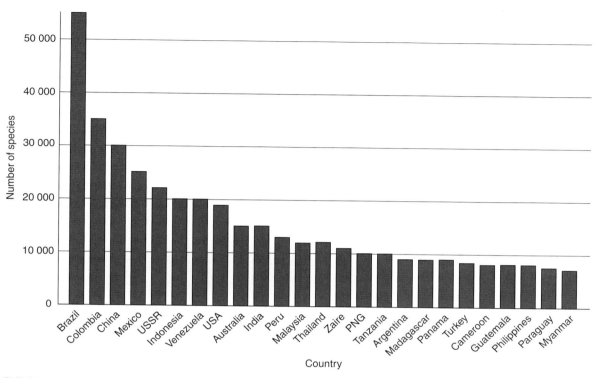

FIGURE 12.32   The 25 countries with the most species of plants. (From Groombridge 1992)

This infertility means that carnivorous plants are particularly well represented in the flora. Ehrenfeld (1983) found that experimentally nutrient-enriched sites supported nearly three times the number of species (73) as pristine sites (26). Enrichment increased local biodiversity by allowing the invasion of common, exotic species that were adapted to higher nutrient levels and already part of the regional species pool. It also resulted in the loss of native, uncommon carnivorous plants that occupy only infertile habitats, and thus reduced the regional species pool overall. The flora of infertile sites is 88 percent native species, 12 percent of which are carnivorous. The more development that occurs, the more fertility increases, and the more native species (particularly carnivorous plants) are replaced by invasive species.

Overgrazing of rangelands by domestic livestock and the subsequent loss of native plant diversity is a universal problem in semi-arid areas and has been a problem for several thousand years (Thirgood 1981; Hughes 1982). It has been a particularly contentious issue in southwestern North America, where ranchers pay a nominal fee to graze their cattle on publicly owned grasslands, and where overgrazing is implicated in conversion of grassland to shrubland, loss of the native biota, as well as increased erosion (e.g. Archer 1989; Fleischner 1994; Noss and Cooperrider 1994; McClaran 2003). The most obvious step seems to be to simply reduce grazing pressure – at least so long as critical thresholds have not been exceeded, a topic to which we shall return in the next chapter (Section 13.4.3). Other interventions may do more harm than good for biological diversity. Consider the Vail rangeland in eastern Oregon (Heady and Bartolome 1977; Heady 1988). In this case, efforts at "rehabilitation" primarily consisted of ploughing and seeding a monoculture of non-native grass

(*Agropyron pectiniforme*). Although marginal benefits to wildlife were reported (Heady 1988) little evidence exists to support such claims (Noss and Cooperrider 1994). For native plants, such programs can be quite destructive. For example, it has led to the invasion of non-native noxious weeds such as medusahead (*Taeniatherum asperum*) (Noss and Cooperrider 1994).

The general rule is that *if attempts at increasing local diversity are achieved by increasing populations of globally common species, local management is probably counterproductive.* Management for maintaining "bio-diversity" makes sense only if one takes a global perspective as the context for evaluating local actions.

CONCLUSION

There are, as you have seen, a large number of factors that control plant diversity, and Table 12.7 summarizes those we have explored as well as some for which there has been neither time nor space. One important topic we did not explore was the balance between rates of speciation and extinction, which you will have to read about elsewhere (Niklas

---

Table 12.7 A summary of the major factors that control diversity in plant communities at different temporal and spatial scales.

Large temporal and spatial scales (global species pool)

1. The isolation of continents allows increased plant diversity by permitting the evolution of new species and persistence of different floras on different continents

2. The angiosperm life cycle allows for greatly increased rates of speciation

3. The angiosperm method of pollination allows more species to coexist within a landscape

4. More than one major physical constraint on plant growth (e.g. salinity and flooding) reduces the species pool

Medium temporal and spatial scales (regional species pool)

1. Gradients allow for the coexistence of more species within a landscape, and the longer the gradients and more kinds of gradients, the more species occur

2. For any one gradient, the number of species is highest at intermediate levels of productivity

3. Within any segment of a gradient, disturbance and gap dynamics allow more species to coexist

4. There is little evidence that habitat specialization on narrower sections of gradients allows more species to coexist

Small temporal and spatial scales (local species pool)

1. A few species often dominate a site at the expense of most others

2. Increased amounts of productivity associated with factors such as higher soil fertility or longer growing season tend to increase the degree of dominance by these few species

3. Species richness reaches a maximum at intermediate levels of fertility and disturbance

4. Patches with extreme infertility or disturbance tend to support distinctive species, thereby increasing the size of the local species pool

et al. 1983, 1985). You see, however, that we have six useful models that address patterns in diversity: three that apply at the large scale and three that apply at more local scales.

Overall, we can conclude that global patterns in diversity are increasingly well documented (Gaston 2000; Myers et al. 2000; Willig et al. 2003; Mutke and Barthlott 2005). We have seen here that at the large scale, area, habitat variation and distance from the equator are important factors. At the smaller scale, factors such as competitive exclusion and intermediate disturbance become increasingly important. Overall, we are remarkably ignorant of factors controlling diversity. Willig et al. (2003) suggest that there are 30 hypotheses for possible causes of the latitudinal diversity gradient alone! We therefore need to approach the topic with a certain modesty.

In the short term, a top priority must be the establishment of nature reserve systems that efficiently represent the largest possible pool of native species (Scott et al. 1987; Myers 1988; Pressey et al. 1993; Noss and Cooperrider 1994). This will buy time while we refine management methods to ensure the persistence of biodiversity through time. Given that roughly one-third of the world's flora is at risk of extinction (Pitman and Jørgensen 2002) there remains a great deal of work for plant ecologists to accomplish.

Review Questions

1. What are the three main factors that control plant diversity at large scales?
2. What region of the world has the most species of gymnosperms? What regions of the world have the most species of angiosperms?
3. Explain how intermediate disturbance can maintain plant diversity. Go back to Chapter 5 and find some examples that illustrate this principle. Explain how fires and floods can both maintain plant diversity in wild habitats.
4. What is meant by centrifugal organization and what are peripheral habitats?
5. Collect data from a local habitat and construct a dominance–diversity curve. Compare it to Figures 12.24 and 12.25.
6. If nutrients limit plant growth, then adding nutrients might be expected to increase the diversity of plants in a habitat. What does Figure 12.27 show in this regard? Use the concept of competition to explain why fertilization usually decreases the diversity of plants.

Further Reading

Williams, C.B. 1964. *Patterns in the Balance of Nature.* London: Academic Press.

Whittaker, R.H. 1965. Dominance and diversity in land plant communities. *Science* 147: 250–260.

Pielou, E.C. 1975. *Ecological Diversity.* New York: John Wiley and Sons.

Huston, M.A. 1979. A general hypothesis of species diversity. *The American Naturalist* 113: 81–101.

Specht, A. and R. Specht. 1993. Species richness and canopy productivity of Australian plant communities. *Biodiversity and Conservation* 2: 152–167.

Rosenzweig, M.L. 1995. *Species Diversity in Space and Time*. Cambridge: Cambridge University Press.

Cowling, R.M. and M.J. Samways. 1995. Predicting global patterns of endemic plant species richness. *Biodiversity Letters* 2: 127–131.

Zobel, M. 1997. The relative role of species pools in determining plant species richness: an alternative explanation of species coexistence? *Trends in Ecology and Evolution* 12: 266–269.

Willig, M.R., D.M. Kaufman and R.D. Stevens. 2003. Latitudinal gradients of biodiversity: pattern, process, scale and synthesis. *Annual Review of Ecology, Evolution and Systematics* 34: 273–309.

Keddy, P.A. 2005. Putting the plants back into plant ecology: six pragmatic models for understanding, conserving and restoring plant diversity. *Annals of Botany* 96: 177–189.

13 Conservation and Management

Putting the pieces together. Five examples from history: ancient Assyria, Rome, Louisiana, Easter Island and the Galapagos. Principles for the design of reserve systems. Different levels of protection. Hotspots. Large forests. Large wetlands. Five advanced topics: services, buffers, thresholds, restoration, indicators. The thin red line. Where does one go from here?

FIGURE 13.1 A hillside with dipterocarp forest in the Ulu Gombak Forest Reserve (400 ha., granitic rock) in the Titiwangsa Range in Malaysia. Southeast Asian tropical forests are distinctive for their many hundreds of species of dipterocarps. Those on this hillside include *Shorea curtisii* (Seraya), with its distinctive silvery leaves. The understorey has an array of palms including *Eugeisonna tristis* (Bertam palm), *Oncosperma horridum* (Thorny palm), *Caryota maxima* (Giant mountain fishtail palm) and *Licuala* spp. (Fan palms). The tree near the bottom centre with the open crown is a species of *Ficus* (Strangler fig), that begins as an ephiphyte and then slowly replaces the host tree. Species composition changes with factors including altitude, slope and bedrock type. Tropical forests exemplify the problems facing plant conservation, as vast areas are annually lost to logging, fire, farmland and plantations. (Photo and description by Darrin Wu, www.rainforestjournal.com)

13.1   Introduction: It Is Time to Apply What We Know

It is now time to review the big picture and then bring the book to an end. We began the book by exploring how plants appeared and created the biosphere in which we now live. We have seen how plants and plant communities, such as the tropical forest in Figure 13.1, are a consequence of fundamental factors such as resources, competition, disturbance, herbivory and stress. When humans first migrated out of Africa some 100,000 years ago, the rest of the world was virgin territory covered in vast forests, wetlands and savannas. We have left behind us a trail of destruction. Many people cannot imagine the sheer scale of human alterations. In some parts of the world, more than 90 percent of the natural primary vegetation has already been cleared. The list includes: southern central China, Sri Lanka, Burma, Madagascar, West Africa and southeastern Brazil (the Atlantic Forest) (Myers et al. 2000). I am writing this book in Lanark County, which many people consider wild, yet there is not a single piece of unlogged forest left to illustrate what the first British settlers found only two centuries ago. Entire species and entire vegetation types are starting to disappear from the surface of the Earth. Not only are we losing unusual plants and beautiful plants, but we are losing the services they perform in the biosphere. My objective here is to review just how serious some of

these problems are, and then to offer you some positive and constructive responses.

Applied plant ecology and conservation biology require that we take the basic knowledge you have learned in the past 12 chapters, and apply it in many different circumstances. There are simple general principles that can be applied again and again. You can think of each chapter as illustrating one general principle. Now we will put them together. I will first share five stories of degradation of ecosystems. I have selected these for two reasons. First, these examples remind us that conservation problems are not something invented just a few decades ago: hundreds of years ago there were examples of abuse of species and ecosystems. (Of course, the Earth's population is vastly larger now, so the effects of humans are magnified. 12 billion people are bound to cause more harm than 100 million – about 120 times as much.) Second, these stories show how rather different cultures managed to create their own problems. Apparently the degradation of environments is a human problem, not just a problem for any one culture. We will then move from these examples to review the situation for plants and the Earth as a whole. And then you will learn some basic principles for managing plant communities wisely, for setting up reserve systems with protected landscapes and for restoring damaged ecosystems.

13.2   Some Historic Examples of Vegetation Degradation

13.2.1 Ancient Assyria

Let us begin with one of the earliest written human epics, Gilgamesh. This epic is also of interest because it pre-dates the Christian Bible, yet describes an ancient flood. Sandars (1972) recounts how in 1839 a young Englishman, Austen Henry Layard, spent some years excavating archaeological sites in Mesopotamia.

One of the most significant discoveries was thousands of broken tablets from the palace of Nineveh. Nineveh was an Assyrian city that fell in 612 BC to a combined army of Medes and Babylonians. The destruction was so complete that the city never rose again. Included in the ruins was the entire library of Assurbanipal "King of the World, King of Assyria." Over 25,000 broken tablets from this library were taken to the British

Museum, and eventual decipherment revealed an epic immeasurably older than Homer's *Iliad*, the fragments of which are now known as *The Epic of Gilgamesh*.

One section of this epic narrates how there was a flood: "The rider of the storm sent down the rain. . . a black cloud came from the horizon; it thundered within where Adad, lord of the storm, was riding. . . For six days and nights the winds blew, torrent and tempest and flood overwhelmed the world. . ." (Sandars 1972: pp. 110–111). This epic also has a boat full of survivors who come to rest on a mountain and who release a dove to search for land.

In an earlier part of the epic, Gilgamesh and his companion, Enkidu, travel to a mysterious cedar forest (which Sandars places in northern Syria or southwest Persia): "They gazed at the mountain of cedars, the dwelling-place of the gods. . . The hugeness of the cedar rose in front of the mountain, its shade was beautiful, full of comfort. . ." (p. 77). They encounter a monstrous guardian of the forest, Humbaba, whom they kill with their swords: "They attacked the cedars, the seven splendours of Humbaba were extinguished" (p. 83). *The Epic of Gilgamesh* therefore apparently records an early episode of deforestation followed by destructive flooding.

13.2.2 Ancient Rome

Moving into the realm of recorded history, the Roman civilization originated with the Etruscans, who "reclaimed Tuscany from forest and swamp" and built drainage tunnels to take the overflow from lakes (Durant 1944). The early history of Rome is little known, in part because the Gauls burned the city in 390 BC, presumably destroying most historical records. Although Rome was built on seven hills, it was not a healthy location: "rains, floods and springs fed malarial marshes in the surrounding plain and even in the lower levels of the city" (p. 12), but Etruscan engineers built walls and sewers for Rome, and "turned it from a swamp into a protected and civilized capital." One of the main sewers, the Cloaca Maxima, was large enough that wagons loaded with hay could pass beneath its arches; the city's refuse and

rain water passed through openings in the streets into these drains and then into the Tiber "whose pollution was a lasting problem in Roman life" (p. 81).

Meanwhile, deforestation accelerated to provide building materials and fuel. The famed Cedars of Lebanon, for example, now live only as literary references; the trees themselves vanished into the shipbuilding yards of the Egyptians, Phoenicians, Jews, Romans and Ottomans. Grazing by goats prevented regeneration of these forests and converted the Mediterranean countries into an increasingly barren landscape (Thirgood 1981; Hughes 1982). Twenty-five centuries ago, Plato was well aware of the consequences of these activities; he bemoans how "there has been a constant movement of soil away from the high ground and what remains is like the skeleton of a body emaciated by disease. All the rich soil has melted away, leaving a country of skin and bone" (Plato, *Critias* III, in Thirgood 1981, p. 36).

The effects of several thousand years of human settlement on Mediterranean forests are summarized in Figure 13.2. It illustrates how relatively small human populations can degrade entire landscapes. Over several thousand years, humans have changed the Mediterranean basin from a land of forests and streams to a land that is denuded and eroded. The great sea fleets of history – the sea-going ships of the Phoenician, Egyptian, Persian, Greek, Carthaginian, Roman, Muslim, Spanish and Napoleonic eras – were constructed from the trees that once covered the Mediterranean hillsides. Each ship took a toll.

The forests might have been able to withstand the pressures of cutting, except that grazing by livestock, particularly goats, killed any tree seedlings that regenerated. The present number of goats alone in this region likely exceeds 100 million. Until these great herds of grazing animals are removed, forests will be unable to re-establish (Thirgood 1981).

13.2.3 Louisiana Wetlands

One might try to explain away the degradation of the Middle East and Mediterranean forests by arguing that human beings a thousand years ago could not be

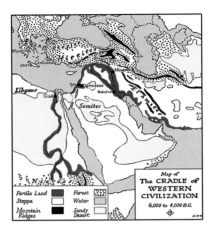

Phoenicia (ca. 3000 BC)

Some 5000 years ago, it is believed Semites from the east occupied the narrow coastal plain and established a series of settlements that developed into the Phoenician cities of Tyre, Sidon and Beirut. From the third millennium B.C., these were centers of trade and culture. According to early sources, Lebanon was "an oasis of green with running creeks" and "a vast forest whose branches hide the sky". The population grew and the Phoenicians were forced to migrate and colonize, to engage in manufacture and maritime trade; and to exploit the forest. These activities together formed the basis of their trading civilization. Phoenician achievements as cultivators in first breaking away from the hydraulic agriculture of the flatlands and establishing a permanent agriculture on slopes, are no less significant than their capacity as traders, while their maritime power and a good portion of their trade was founded on their exploitation of the forest. In classical times, Mount Lebanon appears an economic hinterland, its forests providing wood for export. Lebanese cedar became the first great commercial timber. The forests of the Lebanon range not only made Phoenicia a dominant seapower and trading nation, but, from an early date, supplied the civilizations of Egypt and Mesopotamia with choice timber for temples and palaces, and with essential oils and resins. Indeed, the fame and the rise and fall of the Phoenician civilization can be closely related to the availability of timber. As early as the Third Dynasty, 5000 years ago, timber is known to have been rafted from Lebanon to Egypt. (Thirgood 1981, pp. 95–96)

Greece (ca. 500 BC)

Contemporary Attica may accurately be described as a mere relic of the original country. There has been a constant movement of soil away from the high ground and what remains is like the skeleton of a body emasculated by disease. All the rich soil has melted away, leaving a country of skin and bone. Originally the mountains of Attica were heavily forested. Fine trees produced timber suitable for roofing the largest buildings; the roofs hewn from this timber are still in existence. The country produced boundless feed for cattle, there are some mountains which had trees not so very long ago, that now have nothing but bee pastures. The annual rainfall was not lost as it is now through being allowed to run over the denuded surface to the sea, it was absorbed by the ground and stored ... the drainage from the high ground was collected in this way and discharged into the hollows as springs and rivers with abundant flow and a wide territorial distribution. Shrines remain at the sources of dried up water sources as witness to this. (Plato, *Critias* III, in Thirgood 1981, p. 36)

Spain (ca. 1500)

The final destruction of the forests of Spain was initiated in medieval times. While the original extent of the forests of central Spain is uncertain ... the entire country was predominantly wooded. The Phoenicians, Greeks, Carthaginians and Romans all built ships on the Iberian coast. With the development of Catalanian and Castilian sea power, these encroachments on the forest cover of the coastal mountains became more extensive and by the fifteenth century, ship wood had to be imported through the Hanseatic League from as far away as Scandinavia. Around 1600, there was also considerable trade in barrel wood and boards between Hamburg and Spain. During the sixteenth century, naval dominance by Spanish shipbuilders attained its highest mark. At this time, Spain is said to have possessed almost 2000 seaworthy ships. Ships of more than 1000 tons were built and, with an average life of 15–20 years and 150 000 cubic feet of wood used in the building of a 1400 ton ship, the drain during the wooden ship era is evident. (Thirgood 1981, p. 49)

Lebanon (ca. 1960)

... In the Lebanon mountains ... the scene had to be witnessed to be believed for there one can see the most incredible scenes of wanton destruction of the last remnants of these beautiful trees. Not only are the last trees sought out and hacked down for timber and fuel, but one sees mature trees being lopped and actually felled in order to provide goat fodder. So heavy is goat grazing ... that the flocks have already consumed nearly all forms of vegetation within their reach. The shepherds, unperturbed, have therefore resorted to felling the last remnants of high forest in order to satisfy the empty bellies of their ravenous flocks. It is an astonishing sight to see a fine cedar or silver fir tree felled for this purpose and then to see hundreds of hungry goats literally pounce upon it the moment it falls to earth and devour every vestige of foliage from the branches. It does not take many minutes for such a flock to strip a tree of its foliage. The felled tree has then served the shepherd's purpose and is left to rot where it fell, he then turns his attention to the next tree and so on. (FAO 1961 in Thirgood 1981, p. 73)

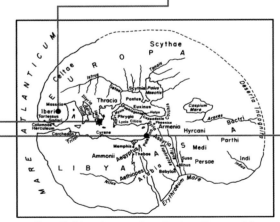

The world of the ancients according to Herodotus (fifth century BC)
(from Thirgood 1981)

FIGURE 13.2 Thousands of years of logging and grazing have removed forest and created the barren landscapes of the Middle East and the Mediterranean. (From Thirgood 1981)

expected to understand the consequences of their behaviour for the environment of their homelands. Let us therefore look at relatively recent human activity in the New World, using coastal Louisiana as the case study.

Louisiana is located on the Gulf of Mexico on the Gulf coastal plain, a region created by sedimentation in shallow coastal waters. The sediments that form contemporary Louisiana eroded from uplands far to the northwest (Rocky Mountains) and northeast (Appalachian Mountains), being carried here primarily by the Mississippi River (Figure 13.3). The central feature of southern Louisiana is the19,400 km² deltaic plain of the Mississippi River, an area that supports 41 percent of the coastal wetlands in the coterminous United States (Shaffer et al. 2005). Today the Mississippi River is considered scenic, but the region left a forbidding impression on some early visitors. In 1837, one traveller from Europe wrote:

> It is not like most rivers, beautiful to the sight. . . not one that the eye loves to dwell upon as it sweeps along. . . It is a furious, rapid, desolating torrent, loaded with alluvial soil. . . It sweeps down whole forests in its course, which disappear in tumultuous confusion, whirled away by the stream now loaded with the masses of soil which nourished their roots, often blocking up and changing the channel of the river, which, as if in anger at its being opposed, inundates and devastates the whole country round. . .
>
> *(Barry 1997: p. 96)*

The youngest parts of Louisiana have been newly built out into the Gulf of Mexico over just the past 5,000 years (Coleman et al. 1998). During flood periods, the river flows up and over older deposits, laying down new layers of sediment and forming natural levees. Thus the deltaic sediments build up above the level of the ocean, producing dry land (Figure 13.4a–c). As areas of the delta build up from accumulated sediment, water is naturally diverted to lower areas. The higher areas then begin to subside as the sediments settle, allowing the sea to creep inland (Figure 13.4d, e); eventually the delta may subside until only a small chain of islands or

even just a shoal remains (Boyd and Penland 1988). The natural vegetation of much of southern Louisiana would be fresh-or saltwater marshes, mixed with swamps of bald cypress and tupelo.

Long before European exploration, there were five dominant indigenous cultures (Caddo, Atakapa, Tunica, Natchez and Chitimacha). Typical villages had fewer than 500 people. The men hunted and the women farmed corn, beans, squash and sweet potatoes.

The history of European settlement begins with New Orleans, which was founded in 1717 when Louisiana was a French colony. New colonists were badly needed, but the city was so unpopular that when French convicts were offered release from prison if they would go to Louisiana, "Most preferred prison in France to freedom in America" (Taylor 1984: p. 9). A century and a half later, in the 1860s, New Orleans was considered to be the most unhealthy city in the nation. There were no sewers, except for ditches adjacent to streets. Frequent epidemics occurred (Taylor1984). The streets were lower than the Mississippi River, and could be submerged by spring floods, hurricanes or heavy rain. Hurricanes regularly caused flooding that inundated the lower areas of the city. The first recorded hurricane hit in 1719, two years after the city was founded.

Bald cypress (*Taxodium distichum*) is the most flood-tolerant tree on the delta and has been commercially exploited in Louisiana since around 1700 (Conner and Buford 1998). Because cypress swamps were flooded for much of the year, harvesting was at first a seasonal occupation. By 1725, loggers learned to girdle the trees during the late summer and winter to kill them, after which the wood dried enough to float during high water in the spring. Loggers would work from boats or rafts to fell the tree, trim the branches, cut the trunks to log lengths, and build them into rafts. These rafts could then be floated to mills for processing. Later, steam-powered pull boats appeared (Figure 13.5). Using cables and winches, they could pull in trees from as far away as 1.5 km. Canals could be dug at 3000-metre intervals, and entire forests systematically removed. In some cases, the mere dragging of large logs cut channels into the soft

(a)

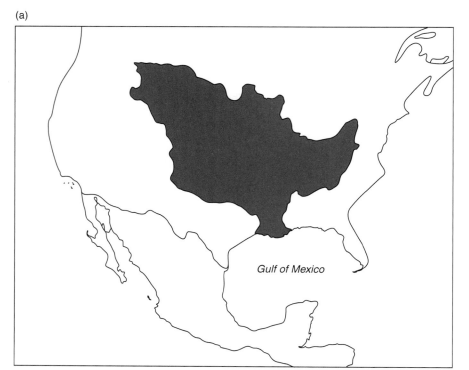

Gulf of Mexico

(b)

FIGURE 13.3 The Mississippi River drains all of central North America between the Rocky Mountains and the Appalachians. (a) Vast swamps of bald cypress and tupelo occupied the floodplain near the mouth of the Mississippi River. (b) This giant bald cypress was cut by the Lyon Lumber Company in the Maurepas Swamp around 1900. (Al Dranguet, Southeastern Louisiana University)

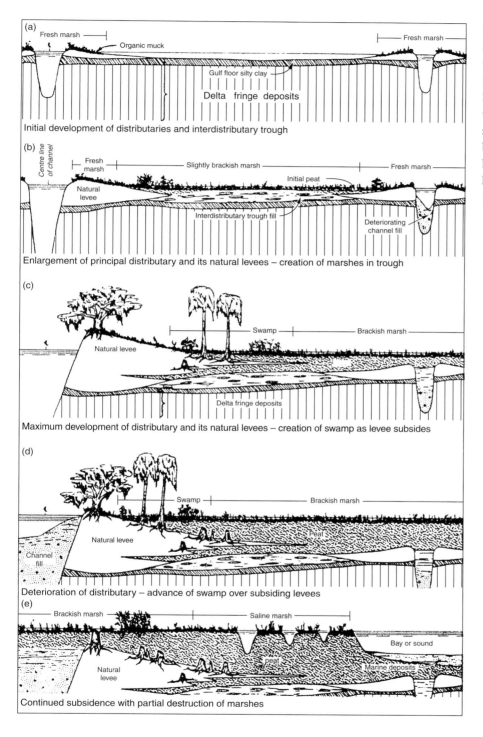

(a)
Fresh marsh
Organic muck
Fresh marsh
Gulf floor silty clay
Delta   fringe   deposits
Initial development of distributaries and interdistributary trough

(b)
Centre line of channel
Fresh marsh
Slightly brackish marsh
Initial peat
Fresh marsh
Natural levee
Interdistributary trough fill
Deteriorating channel fill
Enlargement of principal distributary and its natural levees – creation of marshes in trough

(c)
Swamp
Brackish marsh
Natural levee
Delta fringe deposits
Maximum development of distributary and its natural levees – creation of swamp as levee subsides

(d)
Swamp
Brackish marsh
Natural levee
Peat
Channel fill
Deterioration of distributary – advance of swamp over subsiding levees

(e)
Brackish marsh
Saline marsh
Bay or sound
Natural levee
peat
Marine deposits
Continued subsidence with partial destruction of marshes

FIGURE 13.4 The different kinds of wetland communities found in the Mississippi River delta depend upon the rate of sediment deposition (a, b, c) and the rate of sea level rise (d, e). (After Bernard and LeBlanc 1965)

FIGURE 13.5 Steam-powered pull boats allowed large-scale logging in cypress swamps. (From Williams 1989)

alluvial soil (Figure 13.6). You can still see these pull boat scars when flying in to New Orleans today. Early logging, then, not only removed the forest, but the spread of drainage ditches began to change water levels that were essential for the regrowth of cypress trees.

Logging reached a peak in the early 1900s, with a billion board feet being cut in 1913. One early logger said:

We just use the old method of going in and cutting down the swamp and tearing it up and bringing the cypress out. When a man's in here with all the heavy equipment, he might as well cut everything he can make a board foot out of; we're not ever coming back in here again.

(Conner and Buford 1998: p. 280)

It is doubtful that these forests will return. First, the drainage ditches have permanently altered the hydrology of the floodplain, and there are recurring surges of saline water. But more importantly, large dykes (levees) have been built along the length of the Mississippi River, so spring flooding no longer occurs in these floodplains, and new sediment is not deposited. The land is slowly sinking, no new soil is being left by floods and now the sea level is rising (Keddy et al. 2007). It is sobering to think that even if

bald cypress trees were replanted, it would be a thousand years until Louisiana recovered its forest.

Logging companies profited by cutting growth that began a thousand years ago, from trees that were already 500 years old when Columbus discovered the New World, trees that were already large before there were telescopes or printing presses, much less electricity and telephones. Ancient trees are a form of ecological capital that cannot easily be replaced. If we assume four human generations per century, it will be some 40 generations from now, around 3000 AD, before these marshlands might again support the ancient cypress trees that were common in Louisiana until the early part of the last century.

13.2.4 Easter Island

Easter Island, "the most remote habitable scrap of land in the world" (Diamond 1994: p. 365), is famous for its hundreds of giant stone statues of heads (Figure 13.7). When the Polynesians first arrived in the fifth century BC, the island was covered with palm forest. The human population density eventually grew to 60 people per square kilometre and the forests were cleared (Diamond 2004; Wright 2004). The large statues for which the island is known could no longer be transported and erected without logs to use as rollers and levers. For a generation or so after the last tree was felled, old lumber could still be used, but soon there was insufficient wood to build ocean-going canoes (cutting off the food supply) or to provide roof beams (cutting off the supply of housing). By the time Europeans arrived (the Dutch explorer, Jacob Roggeveen, arrived on Easter day in 1722), there had been destitution, multiple clan wars and many of the survivors lived in caves. The remaining sources of protein were humans and closely guarded chickens.

The decisive moment in this squalid story came when the last tree was cut. How is it that this island society was unable to appreciate the need for conservation? In the words of Wright (2004):

FIGURE 13.6 An aerial view of the wetlands south of Pass Manchac in Louisiana showing the scars left by pull-boat logging (Figure 13.5) that occurred about a hundred years ago. Each set of ditches radiates out from where a pull boat was anchored. These ditches have significantly altered the hydrology of these wetlands. (From Keddy et al. 2007)

FIGURE 13.7 Where have all the trees gone? These stone heads (moai) cannot speak, but appear puzzled. It is possible that the people who lived on Easter Island cleared the forests that provided their only source of house beams and shipbuilding materials, thereby leading to the collapse of their own civilization. (Photograph by Tomaz Kunst, Shutterstock)

We might think that in such a limited place, where from the height of Terevaka, islanders could survey their whole world at a glance, steps would have been taken to halt the cutting, to protect the saplings, to replant. . . The people who felled the last tree could see that it was the last, could know with complete certainty that there would never be another. And they felled it anyway.

(p. 60)

How indeed did they decide to cut the last trees on the island? That is a question to ask yourself as you lie awake at night. Of course, you should know that there are sceptics who question both the details of the story and the lesson (Young 2006; Hunt 2007; Hunt and Lipo 2010). Alternative explanations for the human decline on Easter Island might include European diseases, slave raiding or even introduced rats. Even the name, Easter Island, is questioned – the current residents call the island Rapa Nui. Future archaeological studies might help clarify details of the history and resolve some of this controversy – unless the evidence is eradicated by bulldozers. In the latest phase of Easter Island history, residents are planning a casino, golf course, hotels and resorts. Apparently the island's role as a modern morality tale continues.

13.2.5 The Galapagos: Pinta Island

Let us now consider an island group off the west coast of South America, one that the young Charles Darwin visited in 1835, and where he was stimulated to think about natural selection – the Galapagos Islands. The Galapagos Islands are of conservation interest because they already have a degree of protection in Galapagos National Park (created in 1968), and they have a rich endemic flora and fauna. The history of the Galapagos was surveyed by Rose (1924); Lack (1947) summarizes this history as being "mainly a tale of disaster, tempered by squalid crime."

Let us focus on one island, Pinta (Figure 13.8). Pinta makes up only 0.75 percent of the total land surface of the archipelago; it is somewhat isolated

even from the rest of the Galapagos, being one of the northern islands and one less often visited by tourists. This island is 59 km^2 and has an altitudinal range of 650 m. Because of the variation in elevation, the diversity of plants and animals is higher than that of the other Galapagos Islands (recall Chapter 12). No less than 180 taxa of higher plants are known, of which 59 are endemic to the Galapagos; two taxa *Scalesia baurii* ssp. *hopkinsii* (the Pinta sunflower-tree) and *Alternanthera filifolia* spp. *pintensis* are endemic to Pinta. There are no native land mammals and only three native reptiles – the giant tortoise, marine iguana and lava lizard (Hamann 1979, 1993; Whelan and Hamann 1989).

Many areas on the island are open forest known as dry season deciduous steppe forest in which *Bursera graveolens* and *Opuntia galapageia* form an open-canopied tree layer with a scattered shrub understorey. Grazing by tortoises can create a distinctive type of vegetation, in which forested areas are interspersed with open, meadow-like areas. The meadows contain herbaceous species that continually grow new tissue the tortoises consume; in the absence of grazing, tree-like forms predominate, often with foliage higher than tortoises can reach. On less fertile islands where droughts are particularly severe, tortoises have an elevated front on their carapace, which allows them to reach higher to consume the lower foliage of woody plants and cacti. These sorts of obvious differences in tortoises are one kind of evidence of the nature of natural selection (even if, as is so often the case, the birds get most of the attention).

As with other isolated islands, buccaneers and whalers took a heavy toll on the animal life. During the 1600s to 1800s, it is estimated that more than 100,000 tortoises were taken off the islands, often to be carried away alive and tied to the deck as a source of fresh meat on the long voyage across the Pacific. Two of the original 15 subspecies of tortoise were hunted to extinction. American whalers took 4,545 tortoises from Pinta during the period 1831 to 1868 alone; the last known female was taken in

FIGURE 13.8 Top: Pinta is a volcanic island on the northern edge of the Galapagos just above the equator. Middle: Grazing by tortoises once created open grassy meadows in this landscape. It is now increasingly covered by woody species such as cacti (*Opuntia galapageia*), small trees (*Bursera graveolens*) and shrubs (*Alternanthera filifolia*, *Castela galapageia*). Bottom: Lonesome George on the day scientists first found him in 1972. (Ole Hamann 1988, 2007, 1972)

1901, and the last three males were collected by the California Academy of Sciences in 1906! For some years, the Pinta Island tortoise existed as a single remaining individual, Lonesome George (Figure 13.8 inset), who lived at the Charles Darwin Research Station on Santa Cruz Island. Lonesome George was discovered in 1971 and caught in 1972. He died in 2012.

In the 1950s goats were introduced to Pinta, where they multiplied extremely rapidly and soon had a destructive impact on the vegetation. It is thought that a local fisherman introduced one male and two female goats, but by the early 1970s it was estimated that some 10,000 goats were present. Closed forest and scrub were opened, soil erosion started and natural regeneration of several plant species was prevented. The Galapagos National Park Service began a goat eradication campaign and the last goats were killed in 1990.

Scientists are now monitoring recovery of the vegetation. There is, however, a longer term issue: the destructive overgrazing by goats has ended, but now the island lacks a large herbivore, and the meadows once created by tortoises are gradually filling with woody plants (Figure 13.9). And what of those species that may require tortoises for processing and dispersing their seeds? To allow managers time to consider the options (discussed in Hansen et al. 2010), 30 sterilized hybrid tortoises have been stocked on the island to re-initiate natural disturbances and herbivory. Meanwhile goat eradication continues in the Galapagos. A much larger island, Santiago (>58,000 ha) was cleared of goats (>79,000 of them) between 2001 and 2006, and the vegetation is showing rapid recovery (Cruz et al. 2009). The island of Floreana is scheduled for eradication of rats, mice and cats (Nicholls 2013).

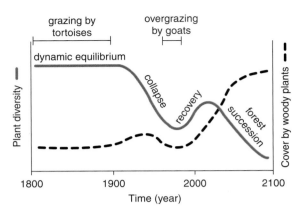

FIGURE 13.9   Historical and projected changes in the vegetation of Pinta Island. Projected forest succession in the absence of the native tortoise species remains a problem to be solved.

13.3   The World Needs Large Protected Areas

13.3.1 Designing a Protected Area System

As the twenty-first century begins, the first challenge for plant ecologists is to ensure that significant areas of landscape are protected from further degradation in composition, structure and function. Once these areas are protected, the next generation of managers will have to grapple with their wise management. The design of reserves, and of reserve systems, is a topic that deserves an entire book (e.g. Shrader-Frechette and McCoy 1993; Noss and Cooperrider 1994); a brief introduction has been provided by Noss (1995). One challenge in setting up networks of protected areas is to make future management as easy as possible. There are three basic steps to follow when designing a rep-resentative reserve system (Figure 13.10). These can be applied in any region of the world.

1. Classification of landscape based on physical fea-tures (topography, surficial geology, soils) and vegetation types (including seral stages) to enu-merate the habitat types that are present and that should be represented in the system of protected areas.

2. Identification of core protected areas that represent the habitats identified in step (1). Topics to be considered at this stage include area, number, redundancy and buffer zones.

3. Gap analysis, that is, re-inspection of step (2) to determine if any habitat types are missing.

The process begins with a classification that provides the list of habitats we are trying to include within the reserve system. You can now see why we spent time

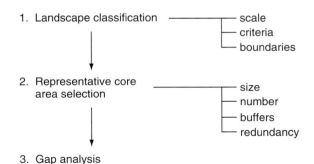

FIGURE 13.10   The first essential steps for designing a reserve system begin with a classification (step 1) and end with a gap analysis (step 3).

Table 13.1 **Some important factors for selecting and prioritizing natural areas for conservation. These are listed in approximate order of importance.**

Factor	Comments
Area	All important ecological values and functions increase with area
Naturalness	Minimal alteration to natural patterns and processes
Representation	Serves as an example of one or more important ecosystem types
Significance	Relative global importance: existing area of this habitat, rates of loss, percent of habitat type protected, better examples protected elsewhere?
Rare species	Globally and regionally significant species present
Diversity	Supports many native species
Productivity	Good production of commercial species (but high production may reduce rare species and diversity)
Hydrological functions	Flood reduction, groundwater recharge
Social functions	Education, tourism, recreation
Carrier functions	Contribution to global life support system: oxygen production, nitrogen fixation, carbon storage
Food functions	Harvesting of species for human consumption
Special functions	Spawning or nesting area, migratory stop over
Potential	Potential for restoration to recover lost values and functions
Prospects	Probability of long-term survival: future threats, buffer zones, possibilities for expansion, patrons, supporting organizations
Corridors	Existing connections to other protected areas or site itself is a corridor
Science function	Published work on site, existing use by scientists, existing research station, potential for future research

on classification systems (Section 11.4), and the further importance of maps of world floristic regions (Figure 2.12) and world ecoregions (Figure 2.14).

The size of each protected area of landscape should be large enough to retain the diversity of vegetation types and full array of species present. The well-documented relationship between species richness and area (Section 12.2) shows that, in general, the bigger the site the more species that are likely to be protected. Big areas have two particular merits. Big areas are important to maintain large mobile predators that need large territories (Weber and Rabinowitz 1996). As well, the bigger the site, the greater the possibility that natural processes can continue to generate or maintain habitat diversity. A coniferous forest reserve, for example, has to be large enough to allow wild fires to produce a mosaic of forest stands of different ages (Section 5.3.1). An alluvial wetland reserve ought to be large enough to allow flooding and bank erosion to continue unabated (Section 5.6.3). If these processes are missing, it may be impossible to retain the biological characteristics of the reserve, and it certainly will compound the difficulties and costs of management. Other factors can be used to select core protected areas including naturalness, significance, rare species, ecological functions and value for research (Table 13.1).

A protected area may represent a common vegetation type and serve as a protected example of a common landscape feature, or it may represent a rare vegetation type. Protection of both typical and rare features are complementary objectives for setting up reserve systems. When a reserve area is being considered one needs to ask careful questions. Are there similar examples already protected? Are there more important vegetation types that are not yet protected? This process is often termed "gap analysis" and algorithms now exist to evaluate reserve scenarios in order to maximize the ecological value of a reserve system (Scott et al. 1987; Pressey et al. 1993; Rodrigues et al. 2004). The objective is to define the smallest number of sites needed to achieve certain goals, such as providing one, two or three protected areas for each species or each community type.

13.3.2 There Are Different Levels of Protection

Most nations now have systems of protected areas. Some are quite strict and focused on protection of wild species and habitats, whereas others allow for a great deal of human activity. Some of the names you will likely encounter include greenspace, wildlife management areas, state parks, national forests, national parks, biosphere reserves and ecological reserves. The same name, such as national park, is used quite differently in different nations. The rules for national parks in England, for example, are very different from the stricter rules in Canada. But even in Canada you can find the occasional golf course or hydroelectric reservoir in a national park, along with fishing in most lakes and hunting by aboriginals.

In order to try to bring some semblance of order to the many categories, we need a set of names that are consistent around the world. To this end, the IUCN has recognized six categories that range from strictly protected areas (I) to sustainably used areas (VI) (Table 13.2). Hence when you read about the amount of land in a particular region that is protected, you need to immediately ask just what kind of protection they are talking about.

The topic of enforcement must also be considered. An area may be strictly protected on paper, but as the recent news headlines show that does not

Table 13.2 **How protected is "protected"? A classification for protected areas. (Note that the higher numbers allow increasing amounts of human activity and intervention.)**

Category I: Strict Nature Reserve/Wilderness Protection Area. An area possessing some outstanding or representative ecosystems, geological, or physiological features and/or species, which is protected and managed to preserve its natural condition.

Category II: National Park. An area designated to (a) protect the ecological integrity of one or more ecosystems for present and future generations, (b) exclude exploitation or occupation inimical to the purposes of designation of the area, and (c) provide a foundation for spiritual, scientific, educational, recreational and visitor opportunities, all of which must be environmentally and culturally compatible.

Category III: Natural Monument. An area containing specific natural or natural/cultural feature(s) of outstanding or unique value because of their inherent rarity, representativeness, or esthetic qualities or cultural significance.

Category IV: Habitat/Species Management Area. An area subject to active intervention for management purposes so as to ensure the maintenance of habitats to meet the requirements of specific species.

Category V: Protected Landscape/Seascape. An area where the interaction of people and nature over time has produced a landscape of distinct character with significant esthetic, ecological and/or cultural value.

Category VI: Managed Resource Protected Area. An area managed mainly for the sustainable use of natural resources – that is, an area containing predominantly unmodified natural systems, managed to ensure long-term protection and maintenance of biological diversity, while also providing a sustainable flow of natural products and services to meet community needs.

Source: Adapted from Anonymous. 1994. *Guidelines for Protected Area Management Categories*. Gland, Switzerland and Cambridge, UK: IUCN and the World Conservation Monitoring Centre.

stop AK-47 armed gangs from shooting elephants and rhinos in game reserves. Illegal logging is a significant threat to many forests, particularly in areas with unstable or corrupt governments. Significant numbers of humans may also live within some reserves, either legally or as squatters, and the resulting pressure from hunting bush meat (Box 13.1) may strip the forest of its wild creatures, some of which may be vital for pollinating trees or dispersing their seeds. Exotic species such as goats may be destroying habitat (Section 13.2.5; Box 13.2). In more developed areas, there are urban dwellers who think they are entitled to enter protected areas to claim their personal share of nature's wild harvest. Even in sedate downtown Ottawa, one can find bottles of wild leeks for sale in numbers that are probably unsustainable and likely originated in a nearby park.

A map of protected areas is just one step in a long-term process, and one must carefully consider both the

---

Box 13.1   The Plague of Bush Meat and the Road to Ruin

You know from the chapter on mutualism that small animals play a vital role in pollinating the flowers of tropical trees and dispersing the seeds of those trees. These small animals include a wide array of types of squirrels, bats and birds. Without these mutualistic animals, many tropical trees cannot successfully re-establish.

Overhunting is now putting these animals, and hence the trees, at risk. Although humans have hunted animals in tropical forests for millennia, the intensity and area of bush meat hunting is now unprecedented (Robinson and Bennett 2000; Harrison et al. 2013). It is a perfect storm: "Over the past 2–3 decades, the availability of firearms and affordable ammunition, on the one hand, and improved access to forests and markets, on the other, has resulted in widespread declines in wildlife populations throughout the tropics" (Harrison et al. 2013: p. 687). There is also the added factor of human population growth in tropical areas.

Consider just one example, involving just one road. Over 15 years a team tracked the performance of 470,000 trees for more than 1,100 species in a forest in Borneo. The onset of heavy hunting can be traced to a single access road that was upgraded in 1987. Within a decade (by 1998) hunters had exterminated the animals larger than 1 kg. These included important seed-dispersing species including hornbills, gibbons and flying foxes. Seed dispersal was seriously disrupted.

Bush meat hunting is an enormous threat to tropical forests, since overhunting affects far larger areas of forest than deforestation and logging combined. Every access road, then, can open a forest to defaunation. Roads are notorious for their negative impacts upon wild areas, and may be regarded as one of the most dangerous human activities in wild areas. Hence maintaining roadless areas is an important global conservation priority (Laurance et al. 2014). I have written myself about the negative effects of roads upon wetlands (Keddy 2010).

Shockingly, bush meat is not only being consumed locally. It is being smuggled into developed countries as a luxury, at high prices (Chaber et al. 2010). A single 17-day investigation at the Paris Roissy-Charles de Gaulle airport found 188 kg of bush meat being smuggled in personal baggage, which would mean 270 tonnes per year passing thorough just this one airport!

And quite apart from the destruction of forests, and the destruction of native animal populations, there are other direct negative consequences for humans. Bush meat is the source of the deadly Ebola virus, which, as I write in 2014, has killed thousands of people in West Africa this summer. One positive and simple step to help avoid future outbreaks of Ebola is to stop killing and eating wild bats.

Box 13.2   Goats and Desertification

Goats are one important single cause for pushing many species and landscapes across thresholds into decline. We saw in Chapter 6 how herbivores can shape the composition of vegetation and in Chapter 9 how herbivores reduce tree regeneration. One of the world's most dangerous herbivores, from the perspective of a plant, is the goat. There are many reasons why this is the case. First, goats have multiple impacts upon plants: they eat shoots, twigs and roots, and consume a wide array of species. Goats have sharp hooves, which damage the soil surface and increase soil erosion. Their numbers are growing exponentially, almost in lock step with the human population. Globally, there are now more than half a billion goats and about the same number of sheep (Aziz 2010). The biggest menace from goats is that in many cases they are released from the natural population limits that control most wild herbivores – by humans who tend them, and even cut trees and foliage during periods of drought that would otherwise limit the herds. As you saw in Figure 13.2, goat herds were probably the single largest contributor to the destruction of the Mediterranean forests of antiquity (along with boat building for warfare, true, but it was the goats that prevented new trees from regenerating). They are

FIGURE B13.2.1   What will happen to all the seedlings? A herd of goats in Kenya. Both human and goat populations in Africa are growing rapidly, putting significant areas of the continent at risk of deforestation and desertification. (Andrzei Kubik, Shutterstock)

Box 13.2 (cont.)

now doing the same in Africa, where human populations and livestock populations are climbing steadily (Brown 2012). As if this were not enough, many goat-dependent human societies occur in landscapes that are semi-arid, although it may be difficult to separate cause and effect here. Certainly, if you read more about landscapes at risk of desertification, picture a herd of goats (Figure B13.2.1).

In Lebanon, which was once famous for its forests, Thirgood (1981: p. 71) quotes a witness from 1961 describing how

> *the last trees are being sought out for destruction. . . The shepherds, unperturbed, have. . . taken to felling the last remnants of high forest in order to satisfy the empty bellies of their ravenous flocks. It is an astonishing sight to see a fine cedar or silver fir tree felled for this purpose, and then to see hundreds of hungry goats literally pounce upon it the moment it falls to earth and devour every vestige of foliage from the branches. It does not take many minutes for such a flock to strip a tree of its foliage. The felled tree has then served the shepherd's purpose, and is left to rot where it fell, he then turns his attention to the next tree and so on.*

Of course, one need not literally fell the last tree to feed a goat to eliminate a forest or a species of tree. If the trees cannot establish seedlings because of grazing, then it takes just a little longer to lose the forest, as each individual species of tree experiences reproductive failure (recall the dragon's blood tree in Section 9.4.3)

Even isolated islands are at risk from goats. Early global explorers had a bad habit of introducing goats to islands. The idea was to ensure fresh meat for the next time they visited. But many of these islands had plant species and entire vegetation types that had evolved without large herbivores and hence were poorly defended. Goat populations frequently grew rapidly, destroying native species and forests and landscapes (Campbell and Donlan 2005). There is now a concerted effort to remove goats from islands where species and ecosystems are at risk. Goats have now been entirely removed from at least 120 islands, totalling 132,867 ha. Examples include Flinders Island (Australia), Lana'i (Hawaii) and Pinta Island (Galapagos). "The successful eradication of introduced mammals from islands is no longer a rare event" (Campbell and Donlan 2005).

The bad news, from the point of view of plants, is that goat populations are still wreaking havoc in much of Africa. The good news is that they are now gone from some oceanic islands, and there are many more candidate islands where goat removal can restore forests and protect rare species from extinction.

level of protection and the amount of enforcement in evaluating the status of protected landscapes.

13.3.3 Biological Hotspots Are a Priority

Some areas of the Earth have inordinately large numbers of species and are called hotspots (Figure 13.11). The first publication on hotspots, prepared by Myers (1988), was based entirely upon plant species richness and identified 18 global hotspots. Later work expanded the list to 25. The criterion for hotspot designation is that the area must have at least 0.5 percent of all plant species worldwide as endemics. Table 13.3 supplements the map in Figure 13.11, listing the 25 hotspots, the number of plant species they contain in total and the percentage of endemics. Table 13.3 also shows the relative degree of risk, indicated by the percent of primary vegetation

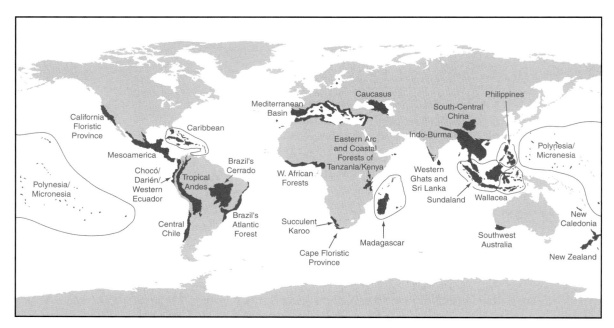

FIGURE 13.11   Twenty-five hotspots with high plant diversity. (From Myers et al. 2000) See Table 13.3.

that remains in each hotspot. Some of the areas most at risk according to this criterion are the Philippines (97 percent lost) and Indo-Burma (95 percent lost). The hotspots with the lowest risk seem to be central Chile, New Caledonia and the Succulent Karoo of South Africa – but even these have already lost more than 70 percent of their natural vegetation cover! Myers et al. (2000) estimate that the 25 hotspots in total contain more than 133,000 plant species and 9,645 vertebrates species – species that are confined to a total of only 2.1 million square kilometres or 1.4 percent of the Earth's land surface.

The richest hotspot for plants is the Tropical Andes of South America (Young et al. 2002; Conservation International 2006) – with approximately 45,000 plant species (Table 13.3). This region extends from elevations as low as 0.5 km to peaks exceeding 6 km, and spans a latitudinal range from Argentina to Venezuela. Recall that von Humboldt visited this area early in his career (Box 2.1) and set a world altitude record while climbing Mount Chimborazo (Figure B2.1.3). The hotspot includes tropical forests, montane cloud forests, thorn scrub, high elevation grasslands,

alpine communities (páramo) – and at the highest altitudes, Yareta (Figure 10.1). Overall, this hotspot contains roughly a sixth of all plant species in the world!

Such high plant diversity in the Andes is consistent with principles we covered in earlier chapters, the importance of gradients in general (Chapter 11), and altitude in particular, in generating plant diversity (Chapters 11 and 12). The final column in Table 13.3 shows the number of square kilometres of hotspot area that were protected when that table was compiled in 2000. The word "protected" must be used with caution since some protected areas may exist only on paper, and are still at risk from factors including deforestation, settlement and even changing climate. Even so, the good news is that an area slightly in excess of 800,000 km^2 has some form of protected status, and this makes up 38 percent of the land area in hotspots.

A gap analysis of the global protected areas system in 2004 found that many species remain unprotected, particularly those restricted to the mountains of South and Central America, and to islands in the Arabian

Table 13.3 **More information on the 25 hotspots of plant diversity mapped in Figure 13.11, ordered by the number of plant species found. (Myers et al. 2000)**

	Original extent of vegetation (km^2)	Plant species	Endemic plants (% of global plants, 300,000)	Percent of original extent remaining	Percent of area protected
New Zealand	270,500	2,300	0.60	22.0	87.7
New Caledonia	18,600	3,332	0.90	28.0	10.1
Eastern Arc and Coastal Forests of Tanzania/Kenya	30,000	4,000	0.50	6.7	100.0
Western Ghats/Sri Lanka	182,500	4,780	0.70	6.8	100.0
Succulent Karoo	112,000	4,849	0.60	26.8	7.8
SW Australia	309,850	5,469	1.40	10.8	100.0
Caucasus	500,000	6,300	0.50	10.0	28.1
Polynesia/Micronesia	46,000	6,557	1.10	21.8	49.0
Philippines	300,800	7,620	1.90	3.0	43.3
Cape Floristic Province	74,000	8,200	1.90	24.3	78.1
Western African Forests	1,265,000	9,000	0.80	10.0	16.1
Chocó/Darién/Western Ecuador	260,600	9,000	0.80	24.2	26.1
Brazil's Cerrado	1,783,200	10,000	1.50	20.0	6.2
Central Chile	300,000	10,000	0.50	30.0	10.2
California Floristic Province	324,000	10,000	0.70	24.7	39.3
Wallacea	347,000	10,000	0.50	15.0	39.2
South-Central China	800,000	12,000	1.20	8.0	25.9
Caribbean	263,500	12,000	2.30	11.3	100.0
Madagascar	594,150	12,000	3.20	9.9	19.6
Indo-Burma	2,060,000	13,500	2.30	4.9	100.0
Brazil's Atlantic Forest	1,227,600	20,000	2.70	7.5	35.9
Mesoamerica	1,155,000	24,000	1.70	20.0	59.9
Mediterranean Basin	2,362,000	25,000	4.30	4.7	38.3
Sundaland	1,600,000	25,000	5.00	7.8	72.0
Tropical Andes	1,258,000	45,000	6.70	25.0	25.3
Totals	17,444,300	N/A	44.0	12.2	37.7

Sea and South Pacific Ocean (Rodrigues et al. 2004). While 11.5 percent of the Earth's surface has been protected within reserves at the time of writing this book, the process of acquisition and protection will have to be accelerated and focused in the coming decades.

13.3.4 Large Forests Are a Priority

Tropical forests (Figure 13.1) provide important core areas for the protection of biodiversity. One global study concluded that the greatest threats to biodiversity occur in the losses of forests in Brazil and Colombia (Bryant et al. 1997). Brazil has by far the largest remaining area of natural forest, more than 2 million km^2, with some 36,000 plant species therein (Table 13.4). Note that while the area of forest at risk in Colombia is much smaller, it contains nearly as many plant species. The map in Figure 13.12 shows that vast

areas of this region are threatened. The need for protection of tropical forests is a problem that many people now understand and there are multiple programs for defining important areas of natural forests.

While tropical forests are priorities from the perspective of biodiversity, boreal coniferous forests stand out as another large area of comparatively natural forests (Bryant et al. 1997) now heavily exploited by humans. There is clear-cutting at massive scales, as is evident to anyone who has looked out of the aircraft windows while flying across northern areas of the world. These forests were once a fire-dominated ecosystem (Rowe and Scotter 1973; Suffling et al. 1988). Now, increasingly, areas of forest are being mechanically removed by equipment and transported south, mostly for pulp and paper (Wein et al. 1983). Here is where the issue of disturbance (Chapter 5) becomes important. Does logging merely replace fire, as some

Table 13.4 **Ten countries with the highest plant diversity in their frontier forests. (From Bryant et al. 1997)**

Global rank	Country	Frontier forest (1,000 km^2)	Estimated no. plant species within frontier forests (thousands)[a]	Approximate percentage of the country's plant species found within frontier forests
1	Brazil	2,284	36	65
2	Colombia	348	34	70
3	Indonesia	530	18	65
4	Venezuela	391	15	75
5	Peru	540	13	75
6	Ecuador	80	12	65
7	Bolivia	255	10	60
8	Mexico	87	9	35
9	Malaysia	47	8	50
10	Papua New Guinea	172	7	70

Note:
[a] Frontier forest plant species richness was estimated by multiplying the country's higher totals per unit area (standardized for size, using a species–area curve) by the country's total frontier forest area.

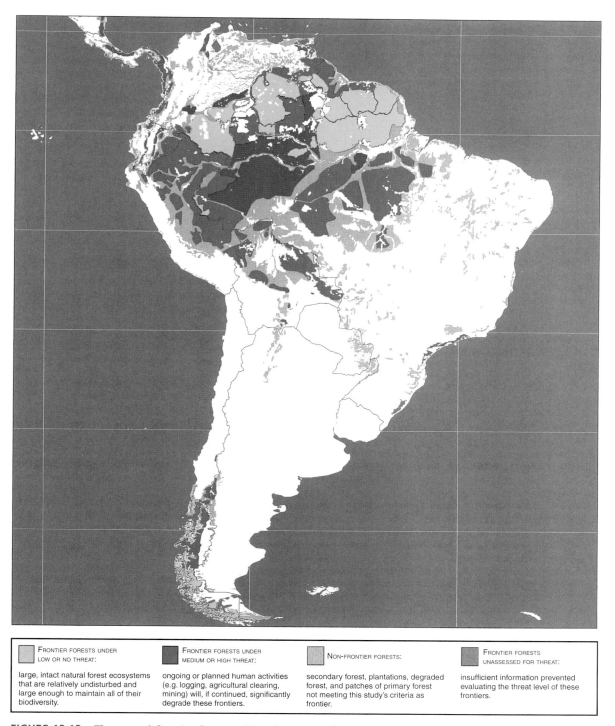

FRONTIER FORESTS UNDER LOW OR NO THREAT:	FRONTIER FORESTS UNDER MEDIUM OR HIGH THREAT:	NON-FRONTIER FORESTS:	FRONTIER FORESTS UNASSESSED FOR THREAT:
large, intact natural forest ecosystems that are relatively undisturbed and large enough to maintain all of their biodiversity.	ongoing or planned human activities (e.g. logging, agricultural clearing, mining) will, if continued, significantly degrade these frontiers.	secondary forest, plantations, degraded forest, and patches of primary forest not meeting this study's criteria as frontier.	insufficient information prevented evaluating the threat level of these frontiers.

FIGURE 13.12   Threatened frontier forests of South America (Bryant et al. 1997). For a global context, see Table 13.4.

claim, or does it lead to different forests? To study this issue Carleton and MacLellan (1994) collected data on plant composition and soil characteristics from 131 clear-cut stands and compared them with 250 stands that had regenerated naturally after fire. Multivariate analysis of these data showed that after accounting for site effects such as soil type and moisture availability, the kind of disturbance was very important in controlling the regenerating forest. They conclude "Comparison between postfire and postlogged woody vegetation data provides evidence of a massive conversion from needle-leaved, conifer-dominated ecosystems to broad-leaved, deciduous forest and shrub ecosystems as a result of logging activity."

13.3.5 Large Wetlands Are a Priority

Wetlands are another habitat with high biological diversity (Keddy 2010). In some areas, wetlands support extensive forest. Peatlands may support boreal forest, while tropical floodplains may support tropical forest. In such cases, large wetlands may, at times, belong in the preceding category of globally important forests. But other large areas of wetland do not support forest and instead have a wide array of herbaceous plants. These too may provide important core areas in a global natural areas network.

Large wetlands also provide a variety of ecological services including primary production, carbon sequestration, flood control, the removal of nutrients from water, food production and maintenance of biological diversity (de Groot 1992; Mitsch and Gosselink 2000; Keddy 2010). Often these services are directly related to the area of the wetland. For example, there is a direct relationship between the area of wetlands and fish production, in both Africa (Welcomme 1976, 1979) and the Gulf of Mexico (Turner 1977). Figure 13.13 and Table 13.5 show the world's largest remaining wetlands. The two largest are the West Siberian Lowland and the Amazon River basin. Much of the Siberian Lowland is a vast peatland, which is a globally important reservoir of stored carbon. It also provides an

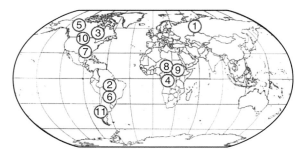

1 West Siberian Lowland	7 Mississippi River basin
2 Amazon River basin	8 Lake Chad basin
3 Hudson Bay Lowland	9 River Nile basin
4 Congo River basin	10 Prairie potholes
5 Mackenzie River basin	11 Magellanic moorland
6 Pantanal	

FIGURE 13.13 The world's largest wetlands. (From Keddy and Fraser 2005) See Table 13.5.

example of boreal forests and their biota (Solomeshch 2005). The Amazon has the largest flow of fresh water in the world, and supports some of the world's highest diversity of trees and fish (Junk and Piedade 2005).

13.3.6 A Global Assessment of Endemic Plant Conservation

Biodiversity hotspots are based on data including animals as well as plants; indeed, since birds and mammals are the best-known groups of animals, they may be weighted more heavily in the mapping. So what do we know about plants in particular? And what might this information tell us about the design of protected areas?

There are estimated to be 350,000 species of vascular plants, of which 96 percent are flowering plants, but there remains a large, but unknown, number of undescribed species (Joppa et al. 2011). Approximately 15 percent of the world's flowering plant species are thought to be undescribed, that is, with no formal name or description, and these are most likely to be species that are already rare and endemic. That is, the fact that they are rare is partly

Table 13.5 The world's largest wetlands, with areas rounded to the nearest 1,000 km². (From Fraser and Keddy 2005)

Rank	Continent	Wetland	Description	Area (km²)
1	Eurasia	West Siberian Lowland	Bogs, mires, fens	2,745,000
2	South America	Amazon River basin	Floodplain forest and savanna, marshes, mangrove	1,738,000
3	North America	Hudson Bay Lowland	Bogs, fens, swamps, marshes	374,000
4	Africa	Congo River basin	Swamps, riverine forest, wet prairie	189,000
5	North America	Mackenzie River basin	Bogs, fens, swamps, marshes	166,000
6	South America	Pantanal	Savannas, grasslands, riverine forest	138,000
7	North America	Mississippi River basin	Bottomland hardwood forest, swamps, marshes	108,000
8	Africa	Lake Chad basin	Grass and shrub savanna, marshes	106,000
9	Africa	River Nile basin	Swamps, marshes	92,000
10	North America	Prairie potholes	Marshes, meadows	63,000
11	South America	Magellanic moorland	Bogs	44,000

what has made them hard to find, and will make it difficult to find them in the future. Hence this undescribed portion may be at particular risk from habitat loss. What can be done? One approach is to use a simple model for the rate at which new species might be found to expand our existing knowledge, that is, to predict the number of unknown species from existing rates of discovery. What we want to know is the total number of species expected, S_T, in the following equation.

$$S_i = (a + b\ Y_i) * (T_i) * (S_T - \Sigma S_i)$$

where S_i is the number of species described per unit time, Y_i the time interval, T_i the number of taxonomists involved in the description, and ΣS_i the total number of species described to that time. The terms a, b, and S_T are constants to be estimated, the last one being the predicted total number of species.

Applying this equation to the world's botanic data bases, the two areas of the world with the most species yet to be discovered appear to be western South America (Ecuador to Peru, 29 percent) and southern Africa (Botswana and Namibia southward, 16 percent). In other words, if you are a young botanist eager to find and name new species, these are the two areas of the world in which you should work. The good news for conservation planning is that this analysis of unknown species is largely consistent with the map of biodiversity hotspots in Figure 13.11. The bad news is that a large number of the areas with unknown species are also areas in which habitat loss is extensive. This means that species may be going extinct before even being shown to exist.

Let us look at it another way. Suppose you were assigned the task of buying or otherwise protecting wild places to protect the most diversity with the least land. How might you go about doing so? Yes, you would start with biodiversity hotspots, but can we be more specific? Joppa et al. (2013) address this question by applying a hungry algorithm to existing plant data sets. This algorithm tries to protect as many endemic species as possible within a minimum land area. The first 43 regions to be selected were all islands, followed by Costa Rica. Costa Rica had

791 endemic plants. Panama entered next with 775 endemics. One can continue adding in areas indefinitely, keeping track of the cumulative additions of endemic species. One overall result is the generalization that "tropical and subtropical islands, moist tropical and subtropical forests (especially those in mountains), and Mediterranean ecosystems hold concentrations of plant endemics. The majority of as-yet undescribed plant species also live in these regions" (p. 1103). Using the arbitrary, but generally accepted, goal of protecting 17 percent of the Earth's land surface area (ca. 24.3 million km^2), it is possible to protect some two-thirds of the world's endemic plant species.

If we look at the actual protected areas in the world, how do they measure up? Many protected areas do not have high plant diversity. It is rather easier to set up large protected areas in the high arctic of Canada, for example, than on the smaller and more

densely populated island of Cuba. But Cuba has far more endemics. Hence if you compare the line of first choice made by the greedy algorithm with the cumulate diversity in areas that are already somewhat protected, as documented by the IUCN, the number of endemics actually being protected is rather lower. And if you use only those areas that are strictly protected, that is within IUCN classes 1 and 2, the percentage of endemic species protected falls further still. Nor do such calculations factor in the long-term viability of populations. On small islands (such as Socotra, Section 9.4.3, or Pinta, Section 13.2.5) the future of endemic species is precarious. The good news again, is that it is possible to removed exotic species such as goats from such islands, and thereby increase the likelihood of long-term survival of endemic plant species (Campbell and Donlan 2005). The simple lesson seems to be as follows. Protect land. On islands. In hot climates. And in the Andes Mountains.

13.4   Five Advanced Topics in Conservation Management

We have already covered a good deal of material in this chapter, from historical accounts of forest degradation, to contemporary efforts to design an effective system of reserves to protect the world's many kinds of plants. If these topics were mostly new to you, then it might be advisable to stop here, and move to the conclusion of this chapter, and this book. If, however, much of the preceding material was familiar, and you feel confident in your understanding, then let us turn to more advanced topics in the management of plant communities and nature reserve systems: (1) services, (2) buffers, (3) thresholds, (4) restoration and (5) indicators. In each case, we will look at a few principles that can take you into wider reading.

13.4.1 Communities and Ecosystems Provide Services

Plants and vegetation provide many **services** for humans (also sometimes called functions). Obvious

services include the production of oxygen, food and fresh water, but there are many more (Table 13.6). A future challenge for biologists is to document and better quantify these services.

These services are related, in a complex way, to the species that are present. If you lose enough species, you lose important functions. Extinction therefore has costs. Ehrlich and Ehrlich (1981) describe the loss of species from communities as analogous to the loss of rivets from the wings of an airplane. A certain number of rivets can be lost without the wings falling off because there is some redundancy but, eventually, if too many are lost, service is impaired and you crash! As a first approximation, one may assume that most services are carried out by more than one species; this is why species fall naturally into functional groups. If one species is lost, another may provide the same service, but if too many are lost, that service is lost. The degree of redundancy, and hence the safety margin, is still a huge unknown. The loss of animals that provide pollination and seed dispersal for tropical

Table 13.6 Services provided by ecosystems. (After de Groot 1992)

Regulation services
1. protection against harmful cosmic influences
2. regulation of the local and global energy balance
3. regulation of the chemical composition of the atmosphere
4. regulation of the chemical composition of the oceans
5. regulation of the local and global climate (including the hydrological cycle)
6. regulation of run-off and flood prevention (watershed protection)
7. water-catchment and groundwater recharge
8. prevention of soil erosion and sediment control
9. formation of topsoil and maintenance of soil fertility
10. fixation of solar energy and biomass production
11. storage and recycling of organic matter
12. storage and recycling of nutrients
13. storage and recycling of human waste
14. regulation of biological control mechanisms
15. maintenance of migration and nursery habitats
16. maintenance of biological (and genetic) diversity

Carrier services – providing space and suitable substrate for:
1. human habitation and (indigenous) settlements
2. cultivation (crop growing, animal husbandry, aquaculture)
3. energy conversion
4. recreation and tourism
5. nature protection

Production services
1. oxygen
2. water (for drinking, irrigation, industry, etc.)
3. genetic resources
4. medicinal resources
5. raw materials for clothing and household fabrics
6. raw materials for building, construction and industrial use
7. biochemicals (other than fuel and medicines)
8. fuel and energy
9. fodder and fertilizer
10. ornamental resources

Information services
1. aesthetic information
2. spiritual and religious information
3. historic information (heritage value)
4. cultural and artistic inspiration
5. scientific and educational information

forests is a vivid example of how quickly a service can be lost (Box 13.1).

Some writers use the word functions to describe services. Indeed, Table 13.6 comes from a book called *Functions of Nature*, and I have adjusted the table to use the word service. This dichotomy creates some problems if you are trying to read the work in this field. I prefer the word services for two reasons. First, it is a word that the general public appreciates, which makes it easier for us to communicate with others. Second, the concept of functional types is already well established in plant ecology, with the term meaning species that share similar types of life history traits. The Raunkiaer system is a classic example that we saw in Chapter 2. It is evident from work being written that even some scientists are confused about the distinction between functional types of plants (based on traits) and the services these plants provide. Keeping the words separate is an antidote to cloudy thinking.

Here I must provide a second caveat, again dealing with cloudy thinking. You should be warned that you can easily be confused and exhausted by the flood of papers written about function using artificial mixtures of small numbers of species (e.g. Tilman et al. 1996; Naeem 2002). In my opinion (Keddy 2005b) these papers illustrate work by a group of scientists knowing rather little about plants and caring rather little about conservation, pretending to have an interest in both (e.g. Tilman and Pacala 1993; Loreau 2000; Kinzig et al. 2002), while ignoring all the important issues of biodiversity conservation. Not only does this consume money needed for more important work, it deceives a new generation of students (Keddy 2004, 2005b). Many other plant ecologists have expressed serious concerns about both the experimental designs and the rather cavalier analysis of the data (e.g. Givnish 1994; Huston 1997; Wardle et al. 2000; Grime 2002), but there is no obvious end in sight.

For more on services, I suggest you start by reading Constanza et al. (1997) and then move on to de Groot (1992) and Pimm (2001). I too have also written a full chapter on the topic in *Wetland Ecology* (Keddy 2010).

13.4.2 A Full Protected Area System Has Buffers and Corridors

It is desirable to have a buffer zone surrounding a protected area (Figure 13.14 top). In this zone, land use practices such as logging and agriculture are regulated in certain ways to ensure the health of the protected core area. For example, since illegal hunting and logging closely follow roads, limitations on new road construction might be one general way of managing a buffer. Biosphere reserves are an example. A biosphere reserve (an international designation, recognized by UNESCO) typically surrounds a protected core area with a larger area of buffer landscape, a landscape that is still used by humans but also managed with consideration for the more strictly protected core area. The surrounding residents, then, become partners in protecting the core area. They, in turn, receive benefits. These might include income from ecotourism, or the increased yields from availability of wild pollinators for crops, or simply the pleasure of living near wild landscapes.

In some wild regions, we may wish to turn the buffer model inside out (Figure 13.14, bottom). That is, we may wish to treat our settlements as isolated units, and put a buffer around each settlement to ensure that the remaining landscape is protected. We could also look at this as a longer term goal for large-scale restoration, where cities, farms and plantations fit into a matrix, surrounded by wild places and supported by the services they provide.

Reserves can become increasingly smaller and more isolated from one another as intervening land is logged or developed for settlements and agriculture. The effects of such fragmentation are usually negative (Lovejoy et al. 1986; Pearce 1993a; Soulé and Terborgh 1999). Certainly, the proportion of edge habitats increases. Dispersal of seeds becomes increasingly constrained and species become increasingly broken into metapopulations with the dynamics typical of island species (MacArthur and Wilson 1967; Hanski and Gilpin 1991; Hanski 1994). Species with low dispersal abilities and long generation times are at

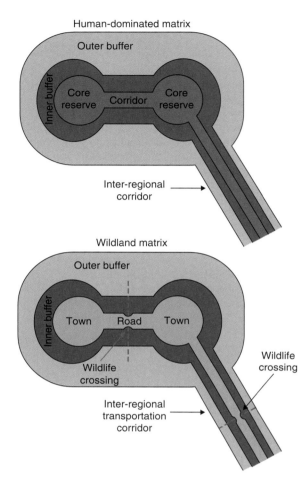

FIGURE 13.14 A protected areas system normally consists of a set of core areas surrounded by buffer zones and linked by corridors. In comparatively wild landscapes (bottom), the system can be reversed, with cities surrounded by buffer zones and linked by human corridors, all set within a wild landscape. (From Noss 1995)

particular risk. Think for example of ant-dispersed species (Section 7.5.3): ant dispersal works well in continuous areas of forest, but it is not going to move seeds across large agricultural fields or four-lane highways.

Landscape fragmentation is therefore a major problem for many protected areas. One solution is to ensure that corridors are protected to link the major core areas, that is, to maintain connectivity. Such corridors may follow land features, such as river valleys or escarpments, that have their own ecological values. In more extreme cases of heavily developed landscapes with a long history of human use, it may be necessary to reconstruct corridors.

13.4.3 There Are Thresholds in the Process of Degradation

We can define the process of degradation as the cumulative loss of services and species associated with human exploitation of landscapes. We have seen five examples of degradation in the first section of this chapter. It is important to understand that the process is rarely smooth and easily predicted. As with plant–herbivore models (Section 6.6.5), there may be periods of rapid and unexpected change. Let us look at grazing more closely, but with the aim of exploring critical points, or thresholds, in the degradation of ecosystems. Arid grasslands illustrate particularly well the sensitivity of stressed ecosystems and the concept of thresholds. First, some background. Overgrazing of grasslands may convert them to shrublands (Archer 1989; Sayre 2003). Overgrazing can threaten the survival of indigenous plants and animals (Noss and Cooperrider 1994; Milchunas et al. 1998). Overgrazing can threaten food production for indigenous human populations (Milton et al. 1994). In extreme cases of overgrazing, the vegetation cover is stripped from the land and erosion is greatly accelerated – recall Plato's complaint about the land "melting away" in Section 13.2.2.

The negative effects of overgrazing may not occur as smooth downward transitions, but as a series of steps. The edge of each step represents a threshold, a situation where a rather small amount of added grazing can make an enormous difference in species composition and ecological function. Figure 13.15 illustrates four stages in degradation. The stages are separated by two thresholds.

In the early stages of degradation, removing the grazing will allow recovery to occur. As the negative consequences of overgrazing accumulate, however, a transition threshold is crossed. At this point, cessation of overgrazing may be insufficient to allow recovery. The vegetation may remain in a degraded state unless expensive management is applied. This management might include reseeding, burning, herbicide treatment or selective-cutting. All of these may have other unwanted consequences. With further overgrazing, a second transition threshold is crossed. Even more costly intervention may be necessary. In extreme cases, it may be necessary to use heavy equipment to cut trenches to trap water and to artificially replant vegetation in such trenches. Needless to say, this kind of intensive management is a costly and undesirable substitute for ecologically sustainable grazing.

Although Figure 13.15 uses the example of rangeland, you should understand that all communities and ecosystems will degrade if humans interfere sufficiently with primary production and plant reproduction. In the extreme case, the plants disappear and even the soil is lost. But long before this stage typical changes include loss of larger species, replacement of slow-growing species by fast-growing ones, and lower rates of nutrient retention (Rapport et al.1985; Freedman 1995). You should think about how Figure 13.15 could be modified to represent the ecosystems around your own home. What are the pressures causing degradation and what are the thresholds?

Thresholds greatly complicate ecosystem management because it is often difficult to explain to land users how a small amount of added damage can lead to a great shift in ecological function – until that shift has actually occurred. At this point, it is not an easy matter to reverse the process. (Recall Easter Island's last tree.) Overall, the past centuries have seen a steady shift in the Earth's vegetation types from the left side to the right of Figure 13.15. Think, for example, of the Mediterranean cedar forests mentioned in Section 13.2.2. Who at the time would have believed that one could pass thresholds that would reduce these to semi-desert?

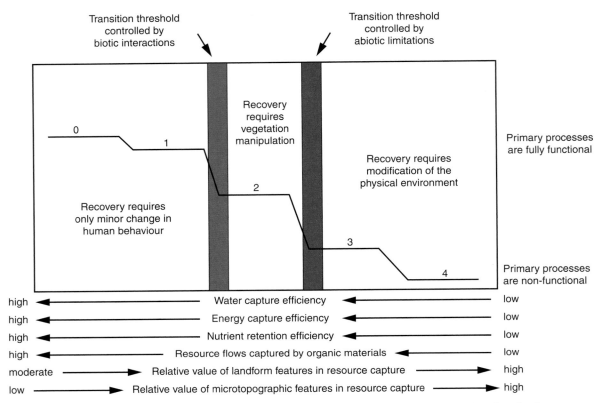

FIGURE 13.15 There are four stages in the degradation of arid land, with two thresholds. After the first threshold is passed, the vegetation must be manipulated to re-establish the original plant communities. After the second threshold is passed, the physical environment must also be modified for any recovery to occur. (After Whisenant 1999)

13.4.4 Restoration of Degraded Vegetation Types

Given the degradation described in the first part of this chapter, and the worldwide impacts of humans and their animals, it is increasingly unlikely that conservationists and managers will encounter a landscape with intact and pristine ecosystems. In many cases there will already have been considerable loss in the area of native vegetation, reductions in function and declines in biological diversity. Two of the principal challenges facing us are: (1) deciding to what degree it is possible to reverse these undesirable changes, and (2) implementing the programs to reverse them. The general name for this process is

restoration (Figure 13.16). More explicitly, restoration is the process of returning an area of landscape to a specified previously occurring ecological state. This is generally done by re-establishing the natural causal factors (often termed drivers) such as fire, flooding and grazing by native animals. There is a good deal of work needed to determine what the original system was, and which drivers will best provide for restoration (Cairns 1989; Noss 1995; Higgs 1997). Simultaneously, one can re-introduce species that have been lost during the process of degradation. As you now know from the chapter on herbivory, and the issue of top–down control, restoration ideally includes establishing natural populations of top carnivores.

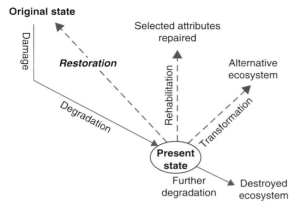

FIGURE 13.16 Restoration is one of several options for dealing with degraded ecosystems. (After Cairns 1989 and Magnuson et al. 1980)

Restoration requires explicit targets. Wherever possible, the target is the original composition that was present before the process of degradation began. Often a good deal of research is required to document the original composition. We should be very clear in using the term restoration that explicit and science-based targets are necessary. To plant non-native plantations of conifers to replace a clear-cut stand of tropical forest is **not** restoration. In cases where the landscape is so degraded that restoration is not possible, there are other alternatives such as rehabilitation or creating an alternative ecosystem (Figure 13.16); these admit that only selected species and functions can be re-established.

You will encounter other terms that can have similar objectives to restoration including "ecosystem management" (e.g. Christensen et al. 1996; Grumbine 1997), the "maintenance of ecological integrity" (e.g. Woodley et al. 1993; Noss 1995), "rehabilitation" (Cairns 1988), "reconstruction" (Allen 1988; Saunders et al. 1993) and "rewilding" (Soulé and Noss 1998; Foreman 2004). Whatever label one uses, the principles are similar: all involve the use of indigenous species and natural environmental factors to maintain and re-establish communities that will persist through time and provide services with minimal human input.

Restoration is often treated as a new discipline, but it has a long history. Clements (1935) discusses, in a paper that is still well worth reading, the degree to which research in ecology influenced landscape management. In this work, he draws in part upon his studies of carrying capacity in grazing lands in the American southwest (Sayre 2003). In these naturally arid grasslands, both shrubs and cacti occur naturally, but a combination of drought and over-grazing damaged the grasslands so severely that there was a die-off of livestock in the 1880s. As a consequence, livestock were excluded from much of a 21,000 ha experimental grassland, the Santa Rita Experimental Range, between 1903 and 1915 (McClaran 2003). Botanists then monitored the recovery of the vegetation and, after grazing was restored in 1915, carried out "hundreds of experiments and manipulations"(p. 17) to explore the interactions of livestock grazing, rodents and vegetation.

Further back still, Phipps (1883) wrote a treatise on forest restoration, and Larson (1996) describes a 107-year-old restoration experiment established by William Brown, an arboriculturalist from Scotland who was hired as a professor of agriculture at the University of Guelph in 1874. The site of Brown's restoration experiment, a former gravel pit, was planted with 14 species of trees, of which 10 are still present. Larson (1996) observes that this "represents an example. . . of an ecosystem that was planned and built based on the best information and principles available at the time."

While it is easy to be discouraged by the many ongoing cases of degradation, and the loss of natural habitats and plant species, restoration provides us with a positive and constructive response. It is possible to restore badly damaged habitats and rebuild successful systems of natural areas to protect the world's biological diversity and ecosystem services. There is now a scientific society devoted to restoration (the Society for Ecological Restoration, www.ser.org), which offers an online primer on the topic (Society for Ecological Restoration 2004).

13.4.5 Indicators Allow for Efficient Monitoring

In the section on restoration I used the term "target," emphasizing that these should be explicit and science-based. This brings us to the topic of indicators. Indicators are measurable attributes of communities and ecosystems that tell us about their state of health. The best analogy is gauges that occur in cars and airplanes: they monitor key aspects of vehicle performance. Engineers have already decided that certain indicators are essential: vehicle speed, fuel supply and engine temperature. We need similar indicators to monitor whether individual landscapes and entire protected area systems are working. We equally need such indicators to assess whether restoration is proceeding in the desired direction.

Indicators have another important role, that of simplifying the task of monitoring. It would be unduly expensive to measure the abundance of every species of plant (and animal) in each protected area. Hence the challenge is to pick a small number of measurements that contain a great deal of information. Returning to the gauges on a car, a great many engine malfunctions will cause an increase in engine temperature, so temperature is a valuable measurement. It also requires a simple tool: a thermometer.

We are slowly working toward lists of indicators for monitoring ecosystems. Consider the example of temperate deciduous forest. Table 13.7 shows some potential indicators and their critical values, extracted from published studies of old growth stands of deciduous forest in eastern North America. Simple

Table 13.7 A preliminary list of indicators for evaluating temperate deciduous forest and suggested critical values for them. (After Keddy and Drummond 1996)

Property	Measurement	Relative condition of forest		
		Good	Intermediate	Low
Stand indicators				
1. Tree size	Basal area (m^2/ha)	>29	20–29	<20
2. Canopy composition	Proportion of shade-tolerant tree species (American beech, sugar maple, basswood, hemlock)	>70%	30–70%	<30%
3. Coarse woody debris	Megagrams/ha, presence of large decaying logs (≥ 8 logs/ha)	>20, both firm and crumbling	10–20, either firm or crumbling	<10, no large logs present
4. Herbaceous layer	Number of ephemeral plant species	≥ 6	2–5	<2
5. Corticulous bryophytes	Number of bryophyte species	≥ 7	2–6	<2
6. Wildlife trees	Number of snags/10 ha (≥ 50.8 cm dbh)	≥ 4	1–3	<1
7. Fungi	No information available			
Landscape indicators				
8. Avian community	Number of forest interior species	>5	2–4	<2
9. Large carnivores	Number of species present	≥ 6	3–5	<3
10. Forest area	Hectares	>10^5	10^2–10^5	<10^2

factors such as tree size, amount of coarse woody debris and understorey plant species contain a great deal of information. Dead trees, called "snags," may be one of the most important predictors of habitat being provided by plants for forest animals (McComb and Muller 1983; Tubbs et al. 1987).

One might argue that there are overlooked elements – perhaps it is necessary to directly monitor other characteristics such as the activity of mycorrhizae that maintain tree growth, or the abundance of amphibians that form the base of many food webs. Another option is to monitor surrogates for these functions. For example, instead of monitoring mycorrhizal fungi or amphibians directly, one might measure the amount of dead wood on the forest floor (coarse woody debris, Table 13.7), since this provides a source of food for fungi and habitat for amphibians.

Sometimes individual species tell us a great deal about factors affecting entire vegetation types, and hence are termed indicator species. The presence of such species indicates that the rest of the ecosystem is functioning normally. Indicators might include selected plant species that are known to be sensitive to environmental change. For coastal plain wetlands *Dionaea muscipula* (Figure 12.13) would be an indicator of low fertility and low human perturbation. For deciduous forests, *Panax quinquefolius* (Figure 9.11) would serve as an indicator for deer grazing, logging and other forms of human interference. A few world floras now have lists of such sensitive, or conservative, species. In this approach, plants can be typically assigned to ten categories based upon expert opinion. Widespread and common species receive a score of 1, while species that depend upon small fragments of undisturbed habitat receive a score of 10. One can even calculate an index of conservatism, C, which is intended to measure how dependent the plant species

are upon natural vegetation types with minimal human alteration (Swink and Wilhelm 1979; Nichols 1999; Herman et al. 2001). The principal drawback of this system is that it requires a great deal of information on plants and their requirements. Such information is only available for limited areas of the world, which is why more general indicators may provide widespread immediate application.

In conclusion, indicators have two valuable uses:

1. Indicators allow us to monitor large pieces of vegetation, or entire reserve systems, to ensure that they continue to function in the desired manner. Since any project begins with less than complete information, Holling (1978a) introduced the concept of adaptive ecosystem management. You start with management decisions based upon the objectives and the best scientific information available. Over time, you monitor the results and compare them to the desired management outcome. If necessary, you adjust management practices accordingly, monitor the results of management and again compare them to the desired management outcome. And so on. This allows for management practices to evolve as human knowledge grows, enabling us to get closer and closer to the management goal as time passes. Along the way, we learn more about the details of the system, and about the underlying principles of science.

2. Indicators allow us to set specific measurable targets for restoration, and to determine how much recovery is actually occurring. This is no small matter. One review of 34 restoration projects found that only two had succeeded in creating the desired ecological community (Lockwood and Pimm 1997)! Hence the need for clearly stated ecological targets, carefully selected indicators and monitoring to ensure that targets are being met.

CONCLUSION

Congratulations! You now know something about the principles of plant ecology, and how they can be applied. You may already be preparing to put this book back on the shelf. Before you do so, we must ask one last question: what next? I hope that I have managed to convey to you the importance of plants in the biosphere, the remarkable diversity that exists within the plant kingdom and some of the excitement (and frustrations!) of research in plant ecology. I also hope that you feel like you have met some of the remarkable personalities that have shaped our view of plant ecology.

It would be entirely normal at this point to feel overwhelmed by the volume of data and the amount of work that has already been done. Sometimes this can be discouraging rather than inspiring! Keep in mind that it took me more than 30 years of working in plant ecology to prepare this book. My objective, overall, is to give you a bit of a shortcut. I have written for you the book that I wish I had owned when I was just starting my university studies. Now you have a head start! The first step is to learn the basic principles. Each chapter has introduced one basic principle and then explored its consequences. You can always come back to those basics: resource acquisition, competition, disturbance, herbivory and mutualism. The next step is to apply these principles wisely. That should keep you busy for an entire lifetime! The five examples of habitat degradation we saw early in this chapter illustrate a disturbing pattern that is being repeated in all landscapes around the globe. Vegetation is being degraded, forests are being cleared, species are vanishing and ecological services are at risk. Perhaps the world will end, as the poet T. S. Eliot put it, "not with a bang, but with a whimper."

Ecologists are like the legendary thin red line of British soldiers; we stand between our civilization and the ecosystems upon which it depends. These ecosystems and their many inhabitants are mute. We alone can provide them with voices and shape positive future action. Plant ecologists (Figure 13.17) have a particularly special role, since far more people are attracted to showy species like birds or whales or lions. Rather fewer students and professionals understand the importance of plants, or appreciate the services they provide.

If you look back over the many problems we have encountered, it would be easy to feel overwhelmed. No one can fix all these problems at once. But each of us is born with certain talents. If you learn the principles in this book, you can apply them wisely wherever you find yourself. You certainly won't have to look far. One way to orient yourself is to ask six simple questions.

1. In which world floristic region do you live?
2. In which world ecoregion do you live?
3. What is the appropriate flora that describes the species pool of this region?
4. What are the dominant species?
5. What are the most important factors that control plant abundance and distribution?
6. Which habitats and species are at risk?

Once you have answered these questions, it should help you decide what contributions you can make given your own circumstances. It may be to work on a local plant community, it may be to work abroad in one of the world's hotspots.

Too often writers will end with a plea for more research. Of course, more information is needed about certain species and certain processes. Information can also naturally accumulate as through adaptive management. But as you can see from the length of this book, we already know a great deal about plants. There is a tendency to forget the past,

FIGURE 13.17 Our challenge is to master what is already known about plants and then to apply this knowledge wisely. Here, a team of younger biologists, led by Daniel Laughlin (left), is introduced to a gymnosperm found only on the North Island of New Zealand: the giant kauri tree (*Agathis australis*).

or to ignore work already done, and then to spend time and money re-inventing the wheel – and then trying to take credit for it! I would like to encourage you to think differently. Start instead with what we already know and then try to apply it wisely. There are situations around the world desperately in need of input from plant ecologists, whether you are in the high Andes (Figure 10.1), on desert islands (Figure 9.6) or in tropical forest (Figure 13.1). The challenge is for you to master what is already known about plant communities, to discover new information where necessary, and then to apply this knowledge to ensure that wild species and wild places continue to thrive and provide the services we need for our own survival.

Review Questions

1. Write a short environmental history for your region of the world. You may use the section on Louisiana in Section 13.2.3 as a model.
2. What is a biological hotspot? Pick one that you would like to visit, explain its significance and list some species you would particularly like to see.
3. What are the three main steps in setting up a reserve system? Explain the terms core, buffer and corridor.
4. Obtain a list of "protected" areas in your region of the world. Where does each one fall in the IUCN classification in Table 13.2?

5. Read about current events in Madagascar, the island of Socotra, the Galapagos, Easter Island or the Congo River basin. Consider the likely consequences for the conservation of rare plants. Pick another area for consideration using Figures 12.10 and 12.11.

6. Explain how overgrazing can cause the degradation of landscapes. What are thresholds and why do they make management more difficult?

7. What is bush meat and what are the negative consequences for tropical forests?

8. You inherit 10 million dollars with the instructions that you must use the money to buy one of the most important vegetation types remaining in the world to donate to The Nature Conservancy. How might you go about making the decision rationally? What criteria would you consider? Do you remember the three guidelines from Section 13.3.6?

9. You are hired by a rich patron to restore a 10,000 acre property to its "natural state." Explain what is meant by restoration. Describe how you would decide what the desired natural state would be. Then list some of the techniques you could use to carry out that restoration.

10. This book mentions a number of historical figures in plant ecology. Who was overlooked and should be added to a future edition? Write a box for that person.

11. What topic do you wish had been discussed in this book that was not? Write a short piece that you think should have been included.

Further Reading

Ehrlich, A. and P. Ehrlich. 1981. *Extinction: The Causes and Consequences of the Disappearance of Species*. New York: Random House.

Cairns, J. 1989. Restoring damaged ecosystems: is predisturbance condition a viable option? *The Environmental Professional* 11: 152–159.

de Groot, R.S. 1992. *Functions of Nature*. The Netherlands: Wolters-Noordhoff.

Groombridge, B. 1992. *Global Biodiversity: Status of the Earth's Living Resources*. London: Chapman and Hall.

Pressey, R.L., C.J. Humphries, C.R. Margules, R.I. Vane-Wright and P.H. Williams. 1993. Beyond opportunism: key principles for systematic reserve selection. *Trends in Ecology and Evolution* 8: 124–128.

Noss, R.F. and A.Y. Cooperrider. 1994. *Saving Nature's Legacy*. Washington, DC: Island Press.

Bryant, D., D. Nielsen and L. Tangley. 1997. *The Last Frontier Forests: Ecosystems and Economies on the Edge*. Washington, DC: World Resources Institute.

Constanza, R., R. d'Arge, R. de Groot et al. 1997. The value of the world's ecosystem services and natural capital. *Nature* 387: 253–260.

Myers, N., R.A. Mittermeier, C.G. Mittermeier, G. A. B. da Fonseca and J. Kent. 2000. Biodiversity hotspots for conservation priorities. *Nature* 403: 853–858.

United Nations Environment Program. 2012. *Protected Planet Report 2012. Tracking Progress Towards Global Targets for Protected Areas*. Cambridge: United Nations Environment Programme World Conservation Monitoring Centre (UNEP-WCMC).

References

Aaviksoo, K., M. Ilomets and M. Zobel. 1993. Dynamics of mire communities: A Markovian approach (Estonia). pp. 23–43. In B.C. Patten (ed.) *Wetlands and Shallow Continental Water Bodies*. Vol. 2. The Hague: SPB Academic Publishing.

Abraham, K. F. and C. J. Keddy. 2005. The Hudson Bay lowland. pp. 118–148. In L. H. Fraser and P. A. Keddy (eds.) *The World's Largest Wetlands: Ecology and Conservation*. Cambridge: Cambridge University Press.

Abrahamson, W.G. and M. Gadgil. 1973. Growth form and reproductive effort in goldenrods (Solidago, Compositae). *The American Naturalist* 107: 651–661.

Adam, P. 1990. *Saltmarsh Ecology*. Cambridge: Cambridge University Press.

Adams J.M. 1997. *Global Land Environments Since the Last Interglacial*. Oak Ridge National Laboratory, TN, USA. www.esd.ornl.gov/ern/qen/nerc.html.

Adamus, P.R. 1992. Choices in monitoring wetlands. pp. 571–592. In D.H. McKenzie, D.E. Hyatt and V. J. McDonald (eds.) *Ecological Indicators*. London: Elsevier Applied Science.

Adl, S.M. 2003. *The Ecology of Soil Decomposition*. Wallingford: CABI Publishing.

Adolt, R. and J. Pavlis. 2004. Age structure and growth of *Dracaena cinnabari* populations on Socotra. *Trees* 18: 43–53.

Aerts, R. 1996. Nutrient resorption from senescing leaves of perennials: are there general patterns? *Journal of Ecology* 84:597–608.

Agnew. A.D.Q. 1961. The ecology of *Juncus effusus* L. in North Wales. *Journal of Ecology* 49: 83–102.

Alestalio, J. and J. Haikio. 1979. Forms created by the thermal movement of lake ice in Finland in winter 1972–73. *Fennia* 157: 51–92.

Allee, W.C. 1951. *Cooperation Among Animals with Human Implications*. Schuman, New York. (Revised edition of *Social Life of Animals*, Norton, New York, 1938)

Allee, W.C., A.E. Emerson, O. Park, T. Park and K.P. Schmidt. 1949. *Principles of Animal Ecology*. Philadelphia: Saunders.

Allen, E.B. 1988. *The Reconstruction of Disturbed Arid Ecosystems*. Boulder: Westview Press.

Allen, E.B. and M.F. Allen. 1990. The mediation of competition by mycorrhizae in successional and patchy environments. pp. 367–389. In J.B. Grace and D. Tilman (eds.) *Perspectives on Plant Competition*. San Diego: Academic Press.

Allison, S.K. 1995. Recovery from small-scale anthropogenic disturbances by northern California salt marsh plant assemblages. *Ecological Applications* 5: 693–702.

Al-Mufti, M.M., C.L. Sydes, S.B. Furness, J.P. Grime and S.R. Band. 1977. A quantitative analysis of shoot phenology and dominance in herbaceous vegetation. *Journal of Ecology* 65: 759–791.

Alvarez, W. 1998. *T. rex and the Crater of Doom*. New York: Vintage Books.

Alverson W.S., D.M. Waller and S.J. Solheim. 1988. Forests to deer: edge effects in northern Wisconsin. *Conservation Biology* 2: 348–358.

Amthor, J.S. and D.D. Baldocchi. 2001. Terrestrial higher plant respiration and net primary production, pp. 33–59. In J. Roy, B. Saugier and H. A. Mooney (eds.) *Terrestrial Global Productivity*. San Diego: Academic Press.

Anderson, R.C., J.S. Fralish and J.M. Baskin. 1999. *Savannas, Barrens, and Rock Outcrop Plant Communities of North America*. Cambridge: Cambridge University Press.

Angiosperm Phylogeny Group. 2003. An update of the Angiosperm Phylogeny Group classification for the orders and families of flowering plants: APG II. *Botanical Journal of the Linnean Society* 141: 399–443.

Angiosperm Phylogeny Group. 2016. An update of the Angiosperm Phylogeny Group classification for the orders and families of flowering plants: APG IV. *Botanical Journal of the Linnean Society* 181: 1–120.

Archer, S. 1989. Have southern Texas savannas been converted to woodlands in recent history? *The American Naturalist* 134: 545–561.

Archibold, O.W. 1995. *Ecology of World Vegetation.* London: Chapman and Hall.

Arimura, G., K. Matsui and J. Takabayashi. 2009. Chemical and molecular ecology of herbivore-induced plant volatiles: proximate factors and their ultimate functions. *Plant & Cell Physiology* 50: 911–923.

Arizaga, S. and E. Ezcurra. 1995. Insurance against reproductive failure in a semelparous plant: bulbil formation in *Agave macroacantha* flowering stalks. *Oecologia* 101: 329–334.

Arthur, W. 1982. The evolutionary consequences of interspecific competition. *Advances in Ecological Research* 12: 127–187.

1987. *The Niche in Competition and Evolution.* Chichester: Wiley.

Ashton, P.S. 1988. Dipterocarp biology as a window to the understanding of tropical forest structure. *Annual Review of Ecology and Systematics* 19: 347–370.

Ashton, P.S., T.J. Givnish and S. Appanah. 1988. Staggered flowering in the dipterocarpaceae: new insights into floral induction and the evolution of mast fruiting in the seasonal tropics. *The American Naturalist* 132: 44–66.

Atwood, E.L. 1950. Life history studies of the nutria, or coypu, in coastal Louisiana. *Journal of Wildlife Management* 14: 249–265.

Auclair, A.N.D., A. Bouchard and J. Pajaczkowski. 1976a. Plant standing crop and productivity relations in a *Scirpus–Equisetum* wetland. *Ecology* 57: 941–952.

1976b. Productivity relations in a *Carex*-dominated ecosystem. *Oecologia* 26: 9–31.

Augspurger, C.K. 1984. Seedling survival of tropical tree species: Interactions of dispersal distance, light-gaps, and pathogens. *Ecology* 65: 1705–1712.

Austin, M.P. 1968. An ordination study of a chalk grassland community. *Journal of Ecology* 56: 739–757.

1982. Use of a relative physiological performance value in the prediction of performance in multispecies mixtures from monoculture performance. *Journal of Ecology* 70: 559–570.

Austin, M.P. and B.O. Austin. 1980. Behaviour of experimental plant communities along a nutrient gradient. *Journal of Ecology* 68: 891–918.

Austin, M.P., J.G. Pausas and A.O. Nicholls. 1996. Patterns of tree species richness in relation to environment in southeastern New South Wales, Australia. *Australian Journal of Ecology* 21: 154–164.

Austin, M.P., J.G. Pausas and I.R. Noble. 1997. Modelling environmental and temporal niches of eucalypts. pp. 129–150. In J.E. Williams and J.C.Z. Woinarski (eds.) *Eucalypt Ecology: Individuals to Ecosystems.* Cambridge: Cambridge University Press.

Axelrod, D.I. 1970. Mesozoic paleogeography and early angiosperm history. *The Botanical Review* 36: 277–319.

Axelrod, D.I and P.H. Raven. 1972. Evolutionary biogeography viewed from plate tectonic theory. pp. 218–236. In J.A. Behnke (ed.) *Challenging Biological Problems: Directions Toward Their Solution.* Oxford: Oxford University Press.

Aziz, M.A. 2010. Present status of the world goat populations and their productivity. *Lohmann Information* 45(2): 42–52.

Bacon, P.R. 1978. *Flora and Fauna of the Caribbean.* Trinidad: Key Caribbean Publications.

Bailes, K.E. 1990. *Science and Russian Culture in an Age of Revolutions. V.I. Vernadsky and his Scientific School, 1863–1945.* Bloomington: Indiana University Press.

Bailey, R.G. 1998. *Ecoregions: The Ecosystem Geography of the Oceans and the Continents.* New York: Springer-Verlag.

Baker, H. 1937. Alluvial meadows: a comparative study of grazed and mown meadows. *Journal of Ecology* 25: 408–420.

Baker-Brosh, K. and R.K. Peet. 1997. The ecological significance of lobed and toothed leaves in temperate forest trees. *Ecology* 78: 1250–1255.

Bakker, R.T. 1978. Dinosaur feeding behaviour and the origin of flowering plants. *Nature* 274: 661–663.

Bakker, S.A., C. Jasperse and J.T.A. Verhoeven. 1997. Accumulation rates of organic matter associated with different successional stages from open water to carr forest in former turbaries. *Plant Ecology* 129: 113–120.

Baldwin, W.K.W. 1958. *Plants of the Clay Belt of Northern Ontario and Quebec.* National Museum of Canada, Bulletin No. 156.

Barbour, M.G. 2007. Closed cone pine and cypress forests. pp. 296–312. In M.G. Barbour, T. Keeler-Wolf and A.A. Schoenherr (eds.) *Terrestrial Vegetation of California.* 3rd edn. Berkeley: University of California Press.

Barnett, V. 1994. Statistics and the long-term experiments: past achievements and future challenges. pp. 165–183. In R.A. Leigh and A.E. Johnston (eds.) *Long-term Experiments in Agricultural and Ecological Sciences.* Proceedings of a conference to celebrate the 150th anniversary of Rothamsted Experimental Station, held at Rothamsted, 14–17 July, 1993. Wallingford: CAB International.

Barreda, V.D., N.R. Cúneo, P. Wilf, et al. 2012. Cretaceous/Paleogene floral turnover in Patagonia: drop in diversity, low extinction, and a *Classopollis* spike. *PLoS ONE* 7(12): e52455. doi:10.1371

Barrett, C.F., J.V. Freudenstein, D.L. Taylor and U. Kõljalg. 2010. Rangewide analysis of fungal associations in the fully mycoheterotrophic *Corallorhiza striata* complex (Orchidaceae) reveals extreme specificity on ectomycorrhizal *Tomentella* (Thelephoraceae) across North America. *American Journal of Botany* 97: 628–643.

Barron, G.L. 2003. Predatory fungi, wood decay, and the carbon cycle. *Biodiversity* 4: 3–9.

Barry, J.M. 1997. *Rising Tide: The Great Mississippi Flood of 1927 and How it Changed America.* New York: Simon and Schuster.

Barth, F.G. 1985. *Insects and Flowers: The Biology of a Partnership.* Princeton: Princeton University Press. Translated from 1982 German edition by M.A. Biederman-Thorson.

Barthlott, W., S. Porembski, E. Fischer and B. Gemmel. 1998. First protozoa trapping plant found. *Nature* 392: 447.

Bartram, W. 1791. *Travels Through North & South Carolina, Georgia, East & West Florida, the Cherokee Country, the Extensive Territories of the Muscogulges, or Creek Confederacy, and the Country of the Chactaws.* Philadelphia: Johnson. http://docsouth.unc.edu/nc/bartram.

Baskin, J. M. and C.C. Baskin. 1985. A floristic study of a cedar glade in Blue Licks Battlefield State Park, Kentucky. *Castanea* 50: 19–25.

Bauer, C.R., C.H. Kellogg, S.D. Bridgham and G.A. Lamberti. 2003. Mycorrhizal colonization across hydrological gradients in restored and reference freshwater wetlands. *Wetlands* 23: 961–968.

Baylis, G.T.S. 1980. Mycorrhizas and the spread of beech. *New Zealand Journal of Ecology* 3: 151–153.

Beard, J.S. 1944. Climax vegetation in tropical America. *Ecology* 25: 127–158.

1949. *The Natural Vegetation of the Windward and Leeward Islands.* Oxford: Clarendon Press.

1973. The physiognomic approach. pp. 355–386. In R.H. Whittaker (ed.) *Ordination and Classification of Communities. Handbook of Vegetation Science, Part V.* The Hague: W. Junk.

Beattie, A.J. and D.C. Culver. 1981. The guild of myrmecochores in the herbaceous flora of West Virginia forests. *Ecology* 62: 107–115.

Beeftink, W.G. 1977. The coastal salt marshes of western and northern Europe: an ecological and phytosociological approach. pp. 109–155. In V.J. Chapman (ed.) *Ecosystems of the World 1: Wet Coastal Ecosystems*. Amsterdam: Elsevier Scientific Publishing Company.

Bégin, Y., S. Arseneault and J. Lavoie. 1989. Dynamique d'une bordure forestière par suite de la hausse récente du niveau marin, rive sud-ouest du Golfe du Saint-Laurent, Nouveau-Brunswick. *Geographie Physique et Quaternaire* 43: 355–366.

Begon, M. and M. Mortimer. 1981. *Population Ecology: A Unified Study of Animals and Plants*. Oxford: Blackwell.

Beguin, P. 1990. Molecular-biology of cellulose degradation. *Annual Review of Microbiology* 44: 219–248.

Belcher, J., P.A. Keddy and P.M.C. Catling. 1992. Alvar vegetation in Canada: a multivariate description at two scales. *Canadian Journal of Botany* 70:1279–1291.

Belcher, J.W., P.A. Keddy and L. Twolan-Strutt. 1995. Root and shoot competition intensity along a soil depth gradient. *Journal of Ecology* 83: 673–682.

Bell, A.D. 1984. Dynamic morphology: a contribution to plant population ecology. pp. 48–65. In R. Dirzo and J. Sarukhán (eds.) *Perspectives on Plant Population Ecology*. Sunderland: Sinauer.

Bell, P.R. and A.R. Hemsley. 2000. *Green Plants: Their Origin and Diversity*. 2nd edn. Cambridge: Cambridge University Press.

Bell, R.A. 1993. Cryptoendolithic algae of hot semiarid lands and deserts. *Journal of Phycology* 29: 133–139.

Belsky, A.J. 1992. Effects of grazing, competition, disturbance and fire on species composition and diversity in grassland communities. *Journal of Vegetation Science* 3: 187–200.

Bender, E.A., T.J. Case and M.E. Gilpin. 1984. Perturbation experiments in community ecology: theory and practice. *Ecology* 65: 1–13.

Benecke, P. and R. Mayer. 1971. Aspects of soil water behavior as related to beech and spruce stands: some results of water balance investigations. pp. 153–168. In H. Ellenburg (ed.) *Integrated Experimental Ecology: Methods and Results of Ecosystem Research in the German Solling Project*, Vol. 2. Ecological Studies: Analysis and Synthesis. New York: Springer.

Benson, L. 1950. *The Cacti of Arizona*, 2nd edn. Tucson: University of Arizona Press.

1959. *Plant Classification*. Lexington: D.C. Heath and Company.

Bentz, B. and H. Nordhaus. 2009. *Bark Beetle Outbreaks in North America: Causes and Consequences*. Salt Lake City: University of Utah Press. ww.treesearch.fs.fed.us/pubs/43479.

Benzing, D.H. 1990. *Vascular Epiphytes: General Biology and Related Biota*. Cambridge: Cambridge University Press.

Berbee, M.L. and J.W. Taylor. 1993. Dating the evolutionary radiations of the true fungi. *Canadian Journal of Botany* 71: 1114–1127.

Berenbaum, M.R. 1991. Coumarins. pp. 221–249. In G.A. Rosenthal and M.R. Berenbaum (eds.) *Herbivores: Their Interactions with Secondary Plant Metabolites*. San Diego: Academic Press.

Berg, B. and C. McClaugherty. 2008. *Plant Litter: Decomposition, Humus Formation, Carbon Sequestration*. 2nd edn. Berlin: Springer-Verlag.

Berg, R.Y. 1975. Myrmecochorous plants in Australia and their dispersal by ants. *Australian Journal of Botany* 23: 475–508.

Bernard, H.A. and R.J. Leblanc. 1965. Résumé of the quaternary geology of the northwestern Gulf of Mexico province. pp. 137–185. In H.E. Wright and D.G. Frey (eds.) *The Quaternary of the United States*. Princeton: Princeton University Press.

Bernatowicz, S. and J. Zachwieja. 1966. Types of littoral found in the lakes of the Masurian and

Suwalki Lakelands. *Komitet Ekolgiezny-Polska Akademia Nauk* XIV: 519–545.

Bertness, M.D. and S.D. Hacker. 1994. Physical stress and positive associations among marsh plants. *The American Naturalist* 144: 363–372.

Bertness, M.D. and S.M. Yeh. 1994. Cooperative and competitive interactions in the recruitment of marsh elders. *Ecology* 75: 2416–2429.

Bidartondo, M.I. and T.D. Bruns. 2002. Fine-level mycorrhizal specificity in the Monotropoideae (Ericaceae): specificity for fungal species groups. *Molecular Ecology* 11: 557–569.

Billings, W.D. and H.A. Mooney. 1968. The ecology of arctic and alpine plants. *Biological Reviews* 43: 481–529.

Binford, M.W., M. Brenner, T.J. Whitmore, et al. 1987. Ecosystems, paleoecology and human disturbance in subtropical and tropical America. *Quaternary Science Reviews* 6: 115–128.

Björkman, E. 1960. *Monotropa hypopitys* L. an epiparasite on tree roots. *Physiologia Plantarum* 13: 308–327.

Black, D. (ed.) 1979. *Carl Linnaeus: Travels.* New York: Charles Scribner's Sons.

Bliss, L.C. and W.G. Gold. 1994. The patterning of plant communities and edaphic factors along a high arctic coastline: implications for succession. *Canadian Journal of Botany* 72: 1095–1107.

Blizard, D. 1993. *The Normandy Landings D-Day: The Invasion of Europe 6 June 1944.* London: Reed International Books.

Boesch, D.F., M.N. Josselyn, A.J. Mehta, et al. 1994. Scientific assessment of coastal wetland loss, restoration and management in Louisiana. *Journal of Coastal Research*, Special Issue No. 20.

Bohlen, P.J., S. Scheu, C.M. Hale, et al. 2004. Invasive earthworms as agents of change in north temperate forests. *Frontiers in Ecology and the Environment* 8: 427–435.

Bolan, N.S. 1991. A critical review on the role of mycorrhizal fungi in the uptake of phosphorus by plants. *Plant and Soil* 134: 189–207.

Bond, W.J. 1997. Functional types for predicting changes in biodiversity: a case study in Cape fynbos. pp. 174–194. In T.M. Smith, H.H. Shugart, and F.I. Woodward (eds.) *Plant Functional Types.* Cambridge: Cambridge University Press.

Bongers, T. and M. Bongers. 1998. Functional diversity of nematodes. *Applied Soil Ecology* 10: 239–251.

Boot, R.G.A. 1989. The significance of size and morphology of root systems for nutrient acquisition and competition. pp. 299–311. In H. Lambert et al. (eds.) *Causes and Consequences of Variation in Growth Rate and Productivity of Higher Plants.* The Hague: SPB Academic Publishing.

Booth, B. and D.W. Larson. 1999. Impact of history, language, and choice of system on the study of assembly rules. pp. 206–229. In E. Weiher and P. Keddy (eds.) *Ecological Assembly Rules: Perspectives, Advances, Retreats.* Cambridge: Cambridge University Press.

Borer, E.T., E.W. Seabloom, J.B. Shurin, et al. 2005. What determines the strength of a trophic cascade? *Ecology* 86: 528–537.

Borhidi, A. 1992. The serpentine flora and vegetation of Cuba. pp. 83–95. In A.J.M. Baker, J. Proctor, and R.D. Reeves (eds.) *The Vegetation of Ultramafic (Serpentine) Soils.* Andover: Intercept.

Bormann, B.T. and R.C. Sidle. 1990. Changes in productivity and distribution of nutrients in a chronosequence at Glacier Bay National Park, Alaska. *Journal of Ecology* 78: 561–578.

Bormann, F.H. and G.E. Likens. 1981. *Pattern and Process in a Forested Ecosystem.* Second corrected printing. New York: Springer-Verlag.

Boston, H.L. 1986. A discussion of the adaptation for carbon acquisition in relation to the growth strategy of aquatic isoetids. *Aquatic Botany* 26: 259–270.

Boston, H.L. and M.S. Adams. 1986. The contribution of crassulacean acid metabolism to the annual productivity of two aquatic vascular plants. *Oecologia* 68: 615–622.

Botkin, D.B. 1977. Life and death in a forest: the computer as an aid to understanding. pp. 213–233. In A.S. Hall and J.W. Day (eds.) *Ecosystem Modelling in Theory and Practice*. New York: John Wiley and Sons.

1990. *Discordant Harmonies: A New Ecology for the Twenty-first Century*. New York: Oxford University Press.

1993. *Forest Dynamics*. Oxford: Oxford University Press.

Boucher, D.H. (ed.) 1985a. *The Biology of Mutualism: Ecology and Evolution*. New York: Oxford University Press.

Boucher, D.H. 1985b. The idea of mutualism, past and future. pp. 1–28. In D.H. Boucher (ed.). *The Biology of Mutualism: Ecology and Evolution*. New York: Oxford University Press.

Boucher, D.H., S. James and K.H. Keeler. 1982. The ecology of mutualism. *Annual Review of Ecology and Systematics* 13: 315–347.

Bougoure, J.J., M.C. Brundrett and P.F. Grierson. 2010. Carbon and nitrogen supply to the underground orchid, *Rhizanthella gardneri*. *New Phytologist* 186: 947–956.

Boutin, C. and P.A. Keddy. 1993. A functional classification of wetland plants. *Journal of Vegetation Science* 4: 591–600.

Bowers, M.D. 1991. Iridoid glycosides. pp. 297–325. In G.A. Rosenthal and M.R. Berenbaum (eds.) *Herbivores: Their Interactions with Secondary Plant Metabolites*. San Diego: Academic Press.

Bowman, D.M.J.S. 2000. *Australian Rainforests: Islands of Green in a Land of Fire*. Cambridge: Cambridge University Press.

Bowman, V.C., J.E. Francis, R.A. Askin, J.B. Riding and G.T. Swindles. 2014. Latest Cretaceous-earliest Paleogene vegetation and climate change at the high southern latitudes: palynological evidence from Seymour Island, Antarctic Peninsula. *Palaeogeography, Palaeoclimatology, Palaeoecology* 408: 26–47.

Boyd, C.E. 1978. Chemical composition of wetland plants. pp. 155–168. In R.E. Good, D.F. Whigham and R.L. Simpson (eds.) *Freshwater Wetlands: Ecological Processes and Management Potential*. New York: Academic Press.

Boyd, R. and S. Penland. 1988. A geomorphologic model for Mississippi River delta evolution. *Transactions Gulf Coast Association of Geological Societies* 38: 443–452.

Bradford, J.B., W.K. Lauenroth and I.C. Burke. 2005. The impact of cropping on primary production in the U.S. Great Plains. *Ecology* 86: 1863–1872.

Braun, E.L. 1950. *Deciduous Forests of Eastern North America*. New York: Hafner.

Brewer, J.S. 1998. Effects of competition and litter on a carnivorous plant, *Drosera capillaris* (Droseraceae). *American Journal of Botany* 85: 1592–1596.

1999. Effects of fire, competition and soil disturbances on regeneration of a carnivorous plant (*Drosera capillaris*). *The American Midland Naturalist* 141: 28–42.

Bridges, E.M., N.H. Batjes and F.O. Nachtergaele (eds.) 1998. *World Reference Base for Soil Resources: Atlas*. Leuven: ACCO.

Brooks, R.R., R.D. Reeves and A.J.M. Baker. 1992. The serpentine vegetation of Goiás State, Brazil. pp. 67–81. In A.J.M. Baker, J. Proctor and R.D. Reeves (eds.) *The Vegetation of Ultramafic (Serpentine) Soils*. Andover: Intercept.

Brown, A., A. Batty, M. Brundrett and K. Dixon. 2003. *Underground Orchid* (Rhizanthella gardneri) *Interim Recovery Plan 2003–2008*. Australia: Department of Conservation and Land Management, Western Australian Threatened Species and Communities Unit.

Brown, J.F. 1997. Effects of experimental burial on survival, growth, and resource allocation of three species of dune plants. *Journal of Ecology* 85: 151–158.

Brown, J.H., D.W. Davidson, J.C. Munger and R.S. Inouye. 1986. Experimental community ecology: the desert granivore system. pp. 41–61. In J. Diamond and T.J. Case (eds.) *Community Ecology*. New York: Harper and Row.

Brown, J.K. and J.K. Smith. 2000. *Wildland Fire in Ecosystems: Effects of Fire on Flora*. USA:

Department of Agriculture, Forest Service, Rocky Mountain Research Station.

Brown, L.R. 2012. *Full Planet, Empty Plates: The New Geopolitics of Food Scarcity.* New York: W.W. Norton.

Browne, J. 1995. *Charles Darwin: Voyaging.* Princeton: Princeton University Press.

Brundrett, M.C. 2009. Mycorrhizal associations and other means of nutrition of vascular plants: understanding the global diversity of host plants by resolving conflicting information and developing reliable means of diagnosis. *Plant and Soil* **320**: 37–77.

Brussaard, L., V.M. Behan-Pelletier, D.E. Bignell et al. 1997. Biodiversity and ecosystem functioning in soil. *Ambio* **26**: 563–570.

Bryant, D., D. Nielsen, and L. Tangley. 1997. *The Last Frontier Forests: Ecosystems and Economies on the Edge.* Washington: World Resources Institute.

Buddenhagen, I.W. 2008. Bats and disappearing wild bananas. *Bats* **26**(4): 1–6.

Burch, W. Jr. 1999. *Daydreams and Nightmares: A Sociological Essay on the American Environment.* Madison: Social Ecology Press.

Burdon, J.J. 1982. The effect of fungal pathogens on plant communities. pp. 99–112. In E.I. Newman (ed.) *The Plant Community as a Working Mechanism.* Oxford: Blackwell.

Burger, J.C. and S.V. Louda. 1995. Interaction of diffuse competition and insect herbivory in limiting brittle prickly pear cactus, *Opuntia fragilis* (Cactaceae). *American Journal of Botany* **82**: 1558–1566.

Burgess, R.L. and D.M. Sharpe (eds.) 1981. *Forest Island Dynamics in Man-dominated Landscapes.* New York: Springer-Verlag.

Buss, L.W. 1988. *The Evolution of Individuality.* Princeton: Princeton University Press.

Bustard, L. 1990. The ugliest plant of the world: the story of *Welwitschia mirabilis. Kew Magazine* **7**: 85–90.

Cairns, J. (ed.) 1980. *The Recovery Process in Damaged Ecosystems.* Ann Arbor: Ann Arbor Science.

1988. *Rehabilitating Damaged Ecosystems.* Vol. 1 and 2. Boca Raton: CRC Press.

1989. Restoring damaged ecosystems: is predisturbance condition a viable option? *The Environmental Professional* **11**: 152–159.

Callaway, R.M. 2007. *Positive Interactions and Interdependence in Plant Communities.* Dordrecht: Springer.

Callaway, R.M. and L. King. 1996. Temperature-driven variation in substrate oxygenation and the balance of competition and facilitation. *Ecology* **77**: 1189–1195.

Campbell, B.D., J.P. Grime and J.M.L. Mackey. 1991. A trade-off between scale and precision in resource foraging. *Oecologia* **87**: 532–538.

1992. Shoot thrust and its role in plant competition. *Journal of Ecology* **80**: 633–641.

Campbell, K. and C.J. Donlan. 2005. Feral goat eradications on islands. *Conservation Biology* **19**: 1362–1374.

Canfield, R.H. 1948. Perennial grass composition as an indicator of condition of southwestern mixed grass ranges. *Ecology* **29**: 190–204.

Carleton, T.J. and P. MacLellan. 1994. Woody vegetation responses to fire versus clear-cutting logging: a comparative survey in the central Canadian boreal forest. *Ecoscience* **1**: 141–152.

Carpenter, S.R., S.W. Chisholm, C. J. Krebs, D.W. Schindler and R.F. Wright. 1995. Ecosystem experiments. *Science* **269**: 324–327.

Carroll, G. 1988. Fungal endophytes in stems and leaves: from latent pathogen to mutualistic symbiont. *Ecology* **69**: 2–9.

Carson, R. 1962. *Silent Spring.* Boston: Houghton Mifflin.

Carson, W.P. and S.T.A. Pickett. 1990. Role of resources and disturbance in the organization of an old-field plant community. *Ecology* **71**: 226–238.

Caruso, T., M.C. Rillig and D. Garlaschelli. 2012. On the application of network theory to arbuscular mycorrhizal fungi–plant interactions: the importance of basic assumptions. *New Phytologist* **194**: 891–894.

Caswell, H. 2001. *Matrix Population Models: Construction, Analysis, and Interpretation.* 2nd edn. Sunderland: Sinauer.

Catling, P.M. and V.R. Brownell. 1995. A review of the alvars of the Great Lakes region: distribution, floristic composition, biogeography and protection. *The Canadian Field-Naturalist* 109: 143–171.

1998. Importance of fire in alvar ecosystems: evidence from the Burnt Lands, Eastern Ontario. *The Canadian Field-Naturalist* 112: 661–667.

Catling, P.M., J.E. Cruise, K.L. McIntosh and S.M. McKay. 1975. Alvar vegetation in southern Ontario. *Ontario Field Biologist* 29: 1–25.

Cavers, P.B. 1983. Seed demography. *Canadian Journal of Botany* 61: 3578–3590.

Chaber A-L., S. Allebone-Webb, Y. Lignereux, A.A. Cunningham and J.M. Rowcliffe. 2010. The scale of illegal meat importation from Africa to Europe via Paris. *Conservation Letters* 3: 317–321.

Chaneton, E.J. and J.M. Facelli. 1991. Disturbance effects on plant community diversity: spatial scales and dominance hierarchies. *Vegetatio* 93: 143–156.

Chapin, III, F.S. 1980. The mineral nutrition of wild plants. *Annual Review of Ecology and Systematics* 11: 233–260.

Chapin, III, F.S., P.M. Vitousek and K. Van Cleve. 1986. The nature of nutrient limitation in plant communities. *The American Naturalist* 127: 48–58.

Chapman, C.A. and S.E. Russo. 2006. Primate seed dispersal: linking behavioral ecology with forest community structure. pp 510–525. In C.J. Campbell, A.F. Fuentes, K.C. MacKinnon, M. Panger and S. Bearder (eds.) *Primates in Perspective.* Oxford: Oxford University Press.

Charron, D. and D. Gagnon. 1991. The demography of northern populations of *Panax quinquefolium* (American ginseng). *Journal of Ecology* 79: 431–445.

Christensen, N.L., R.B. Burchell, A. Liggett and E.L. Simms. 1981. The structure and development of pocosin vegetation. pp. 43–61. In C.J. Richardson (ed.) *Pocosin Wetlands: An Integrated Analysis of Coastal Plain Freshwater Bogs in North Carolina.* Stroudsburg, Pennsylvania: Hutchinson Ross Publishing Company.

Christensen, N.L., A.M. Bartuska, J.H. Brown et al. 1996. The report of the Ecological Society of America Committee on the scientific basis for ecosystem management. *Ecological Applications* 6: 665–691.

Clarke, D. and N.J. Hannon. 1969. The mangrove swamp and salt marsh communities of the Sydney district. II. The holocoenotic complex with particular reference to physiography. *Journal of Ecology* 57: 213–234.

Clay, K. 1990. The impact of parasitic and mutualistic fungi on competitive interactions among plants. pp. 391–412. In J.B. Grace and D. Tilman (eds.) *Perspectives on Plant Competition.* San Diego: Academic Press.

Clements, F.E. 1916. *Plant Succession: An Analysis of the Development of Vegetation.* Pub. 242. Washington, DC: Carnegie Institute.

1933. Competition in plant societies. In *News Service Bulletin.* Washington: Carnegie Institution of Washington, 2 April 1933.

1935. Experimental ecology in the public service. *Ecology* 16: 324–363.

1936. Nature and structure of climax. *Journal of Ecology* 24: 254–282.

Clements, F.E., J.E. Weaver and H.C. Hanson. 1929. *Plant Competition.* Washington, D.C.: Carnegie Institution of Washington.

Cloud, P. 1976. Beginnings of biospheric evolution and their biogeochemical consequences. *Paleobiology* 2: 351–387.

Cloudsley-Thompson, J.L. 1996. *Biotic Interactions in Arid Lands.* Berlin: Springer-Verlag.

Clymo, R.S. and J.G. Duckett. 1986. Regeneration of *Sphagnum. New Phytologist* 102: 589–614.

Clymo, R.S. and P.M. Hayward. 1982. The ecology of *Sphagnum.* pp. 229–289. In A.J.E. Smith (ed.) *Bryophyte Ecology.* London: Chapman and Hall.

Cody, M.L. 1993. Do cholla cacti (*Opuntia* spp., subgenus *Cylindropuntia*) use or need nurse plants in the Mojave Desert? *Journal of Arid Environments* 24: 1–16.

Coe, M.J., D.L. Dilcher, J.O. Farlow, D.M. Jarzen and D.A. Russel. 1987. Dinosaurs and land plants. pp. 225–258. In E. Friis, W.G. Chaloner and P.R. Crane (eds.) *The Origins of Angiosperms and Their Biological Consequences.* Cambridge: Cambridge University Press.

Cole, L.C. 1949. The measurement of interspecific association. *Ecology* 30: 411–424.

Coleman, J.M., H.H. Roberts and G.W. Stone. 1998. Mississippi River delta: an overview. *Journal of Coastal Research* 14: 698–716.

Coleman, R.G. and C. Jove. 1992. Geological origin of serpentines. pp. 1–17. In A.J.M. Baker, J. Proctor and R.D. Reeves (eds.) *The Vegetation of Ultramafic (Serpentine) Soils.* Andover: Intercept.

Coley, P.D. 1983. Herbivory and defensive characteristics of tree species in a lowland tropical forest. *Ecological Monographs* 53: 209–233.

Colinvaux, P. 1978. *Why Big Fierce Animals are Rare: An Ecologist's Perspective.* Princeton: Princeton University Press.

1986. *Ecology.* Toronto: Wiley and Sons.

1993. *Ecology 2.* New York: Wiley and Sons.

Colinvaux, P.A., P.E. De Oliveira, J.E. Moreno, M.C. Miller and M.B. Bush. 1996. A long pollen record from lowland Amazonia: forest and cooling in glacial times. *Science* 274: 85–88.

Colinvaux, P.A., P.E. De Oliveira and M.B. Bush. 2000. Amazonian and neotropical plant communities on glacial time-scales: the failure of the aridity and refuge hypotheses. *Quaternary Science Reviews* 19: 141–169.

Colinvaux, P.A., G. Irion, M.E. Räsänen, M.B. Bush and J.A.S. Nunes de Mello. 2001. A paradigm to be discarded: geological and paleoecological data falsify the Haffer & Prance refuge hypothesis of Amazonian speciation. *Amazoniana* 16: 609–646.

Collinson, M.E. and J.J. Hooker. 1987. Vegetational and mammalian faunal changes in the Early Tertiary of southern England. pp. 259–304. In E. Friis, W.G. Chaloner and P. R. Crane (eds.) *The Origins of*

Angiosperms and Their Biological Consequences. Cambridge: Cambridge University Press.

Colwell, R.K. and E.R. Fuentes. 1975. Experimental studies of the niche. *Annual Review of Ecology and Systematics* 6: 281–309.

Connell, J.H. 1978. Diversity in tropical rain forests and coral reefs. *Science* 199: 1302–1310.

1980. Diversity and the coevolution of competitors, or the ghost of competition past. *Oikos* 35: 131–138.

1990. Apparent versus "real" competition in plants. pp. 9–26. In J.B. Grace and D. Tilman (eds.) *Perspectives on Plant Competition.* San Diego: Academic Press.

Connell, J.H. and R.O. Slatyer. 1977. Mechanisms of succession in natural communities and their role in community stability and organization. *The American Naturalist* 111: 1119–1144.

Conner, W.H. and M.A. Buford. 1998. Southern deepwater swamps. pp. 261–287. In M.G. Messina and W.H. Conner (eds.) *Southern Forested Wetlands. Ecology and Management.* Boca Raton: Lewis Publishers.

Connolly, J. 1986. On difficulties with replacement-series methodology in mixture experiments. *Journal of Applied Ecology* 23: 125–137.

Conservation International. 2006. Biodiversity Hotspots. Tropical Andes. (www.biodiversityhotspots.org/xp/Hotspots/andes) accessed 24 July 2006.

Constanza, R. (and 12 others). 1997. The value of the world's ecosystem services and natural capital. *Nature* 387: 253–260.

Corfield, T.F. 1973. Elephant mortality in Tsavo National Park, Kenya. *East African Wildlife Journal* 11: 339–368.

Cowling, R.M. 1990. Diversity components in a species-rich area of the Cape Floristic Region. *Journal of Vegetation Science* 1: 699–710.

Cowling, R.M. and M.J. Samways. 1995. Predicting global patterns of endemic plant species richness. *Biodiversity Letters* 2: 127–131.

Cowling, R.M., P.W. Rundel, B.B. Lamont, M.K. Arroyo and M. Arianoutsou. 1996. Plant diversity in Mediterranean-climate regions. *Trends in Ecology and Evolution* 11: 362–366.

Craighead, F.C. Sr. 1968. The role of the alligator in shaping plant communities and maintaining wildlife in the southern Everglades. *The Florida Naturalist* 41: 2–7, 69–74.

Crandell, D.R. and H.H. Waldron. 1969. Volcanic hazards in the Cascade Range. pp. 5–18. In R. Olson and M. Wallace (eds.) *Geologic Hazards and Public Problems. Conference Proceedings.* US Government Printing Office.

Crawley, M.J. and J.E. Harral. 2001. Scale dependence in plant biodiversity. *Science* 291: 864–868.

Crepet, W.L. and E.M. Friis. 1987. The evolution of insect pollination in angiosperms. pp. 181–201. In E. Friis, W.G. Chaloner and P.R. Crane (eds.) *The Origins of Angiosperms and Their Biological Consequences.* Cambridge: Cambridge University Press.

Crocker, R.L. and J. Major. 1955. Soil development in relation to vegetation and surface age at Glacier Bay, Alaska. *Journal of Ecology* 43: 427–448.

Cronquist, A. 1991. Asterales. pp. 721–726. In *Angiosperms: The Flowering Plants.* pp. 596–765, Vol. 13. *The New Encyclopaedia Britannica*, 15th edn. Chicago: Encyclopaedia Britannica Inc.

1993. A commentary on the general system of classification of flowering plants. pp. 272–293. In Flora of North America Editorial Committee. *Flora of North America*, Vol. 1. Introduction. New York: Oxford University Press.

Crowson, R.A. 1981. *The Biology of Coleoptera.* New York: Academic Press.

Cruz, F., V. Carrion, K.J. Campbell, C. Lavoie and C.J. Donlan. 2009. Bio-economics of large-scale eradication of feral goats from Santiago Island, Galápagos. *Journal of Wildlife Management* 73: 191–200.

Cyr, H. and M.L. Pace. 1993. Magnitude and patterns of herbivory in aquatic and terrestrial ecosystems. *Nature* 361: 148–150.

Dacey, J.W.H. 1980. Internal winds in water lilies: an adaptation for life in anaerobic sediments. *Science* 210: 1017–1019.

1981. Pressurized ventilation in the yellow water lily. *Ecology* 62: 1137–1147.

Dafni, A. 1992. *Pollination Ecology: A Practical Approach.* Oxford: Oxford University Press.

Dale, M. 1999. *Spatial Pattern Analysis in Plant Ecology.* Cambridge: Cambridge University Press.

Dansereau, P. 1959. Vascular aquatic plant communities of southern Quebec: a preliminary analysis. *Transactions of the Northeast Wildlife Conference* 10: 27–54.

Dansereau, P. and F. Segadas-Vianna. 1952. Ecological study of the peat bogs of eastern North America. *Canadian Journal of Botany* 30: 490–520.

Darwin, C. 1871. The descent of Man and selection in relation to sex. In M.J. Adler (ed.) 1990. *Great Books of the Western World.* 2nd ed., Vol. 49. Chicago: Encyclopaedia Britannica Inc.

1881. *The Formation of Vegetable Mould Through the Action of Worms, with Observations on Their Habits.* London: Murray.

1888. *Insectivorous Plants.* 2nd ed. London: John Murray. Revised by F. Darwin.

Darwin, F. (ed.) 1950. *Charles Darwin's Autobiography: With his Notes and Letters Depicting the Growth of the Origin of Species.* New York: Henry Schuman.

Daubenmire, R. 1978. *Plant Geography: With Special Reference to North America.* New York: Academic Press.

Davies, B.R. and K.F. Walker. 1986. *The Ecology of River Systems.* Dordrecht: W. Junk Publishers.

Davis, D.W. 2000. Historical perspective on crevasses, levees, and the Mississippi River. In C.E. Colten (ed.) *Transforming New Orleans and its Environs, Centuries of Change.* Pittsburgh: University of Pittsburgh Press.

Davis, M.B. 1976. Pleistocene biogeography of temperate deciduous forests. pp. 13–26. In R.C. West and W.G. Haag (eds.) *Geoscience and Man*, Vol. 13. *Ecology of the Pleistocene, a Symposium.* Baton Rouge: School of Geoscience, Louisiana State University.

Davis, S. and J. Ogden (eds.) 1994. *Everglades: The Ecosystem and its Restoration.* Del Ray Beach: St. Lucie Press.

Dawkins, R. 1976. *The Selfish Gene.* Oxford: Oxford University Press.

Day, R.T., P.A. Keddy, J. McNeill and T. Carleton. 1988. Fertility and disturbance gradients: a summary model for riverine marsh vegetation. *Ecology* **69**: 1044–1054.

Day, W. 1984. *Genesis on Planet Earth*, 2nd edn. New Haven: Yale University Press.

Dayton, P.K. 1979. Ecology: a science and a religion. pp. 3–18. In R.J. Livingston (ed.) *Ecological Processes in Coastal and Marine Systems.* New York: Plenum Press.

de Calesta, D.S. 1994. Effect of white-tailed deer on songbirds within managed forests in Pennsylvania. *Journal of Wildlife Management* **58**: 711–718.

Dearnaley, J. 2007. Further advances in orchid mycorrhizal research. *Mycorrhiza* **17**: 475–486.

Deckers, J.A., F.O. Nachtergaele and O.C. Spaargaren (eds.) 1998. *World Reference Base for Soil Resources: Introduction.* Leuven: ACCO.

de Duve, C. 1991. *Blueprint for a Cell: the Nature and Origin of Life.* Burlington: Neil Patterson.

de Groot, R.S. 1992. *Functions of Nature.* Groningen: Wolters-Noordhoff.

Delannoy, E., S. Fujii, C.C. des Francs-Small, M. Brundrett and I. Small. 2011. Rampant gene loss in the underground orchid *Rhizanthella gardneri* highlights evolutionary constraints on plastid genomes. *Molecular Biology and Evolution* **28**: 2077–2086.

Delcourt, H.R. and P.A. Delcourt. 1988. Quaternary landscape ecology: relevant scales in space and time. *Landscape Ecology* **2**: 23–44.

1991. *Quaternary Ecology: A Paleoecological Perspective.* London: Chapman and Hall.

del Moral, R. 1983. Competition as a control mechanism in subalpine meadows. *American Journal of Botany* **70**: 232–245.

del Moral, R. and L.C. Bliss. 1993. Mechanisms of primary succession: insights resulting from the eruption of Mount St. Helens. *Advances in Ecological Research* **24**: 1–66.

del Moral, R. and S.Y. Grishin. 1999. Volcanic disturbances and ecosystem recovery. pp. 137–160. In L.R. Walker (ed.) *Ecosystems of Disturbed Ground.* Ecosystems of the World Series. Amsterdam: Elsevier Science.

del Moral, R. and D.M. Wood. 1993. Early primary succession on the volcano Mount St. Helens. *Journal of Vegetation Science* **4**: 223–234.

del Moral, R., J.H. Titus and A.M. Cook. 1995. Early primary succession on Mount St. Helens, Washington, USA. *Journal of Vegetation Science* **6**: 107–120.

Denslow, J.L. 1987. Tropical rain forest gaps and tree species diversity. *Annual Review of Ecology and Systematics* **18**: 431–451.

Deshmukh, I. 1986. *Ecology and Tropical Biology.* Palo Alto: Blackwell Scientific.

Desmond, A. and J. Moore. 1991. *Darwin.* New York: Warner Books.

de Wit, C.T. 1960. On competition. *Verslagen van Landbouwkundige Onderzoekingen* **66**: 1–82.

Diamond, J.M. 1975. Assembly of species communities. pp. 342–444. In M.L. Cody and J.M. Diamond (eds.) *Ecology and Evolution of Communities.* Cambridge: Belknap Press of Harvard University Press.

1986. Overview: laboratory experiments, field experiments, and natural experiments. pp. 3–22. In J.M. Diamond and T.J. Case (eds.) *Community Ecology.* New York: Harper and Row.

Diamond, J. 1994. Ecological collapses of past civilisations. *Proceedings of the American Philosophical Society* **138**: 363–370.

2004. Twilight at Easter. *New York Review of Books* LI(5) (March 25). pp. 6–10.

Díaz, S., A. Acosta and M. Cabido. 1992. Morphological analysis of herbaceous communities under different grazing regimes. *Journal of Vegetation Science* **3**: 689–696.

Dickerson, R.E. 1969. *Molecular Thermodynamics*. New York: W. A Benjamin Inc.

Diestel, R. 2010. *Graph Theory*. 4th edn. New York: Springer.

Digby, P.G.N. and R.A. Kempton. 1987. *Multivariate Analysis of Ecological Communities*. London: Chapman and Hall.

Dilcher, D.L. and P.R. Crane. 1985. *Archaeanthus*: an early angiosperm from the Cenomanian of the western interior of North America. *Annals of the Missouri Botanical* 71: 351–383.

Dilcher, D.L. and W.L. Kovach. 1986. Early angiosperm reproduction: *Caloda delevoryana* gen. et sp. nov., a new fructification from the Dakota Formation (Cenomanian) of Kansas. *American Journal of Botany* 73: 1230–1237.

Dinerstein, E. 1991. Seed dispersal by greater one-horned rhinoceros (*Rhinoceros unicornis*) and the flora of *Rhinoceros* latrines. *Mammalia* 55: 355–362.

1992. Effects of *Rhinoceros unicornis* on forest structure in lowland Nepal. *Ecology* 73: 701–704.

Dirzo, R., C.C. Horvitz, H. Quevedo and M.A. Lopez. 1992. The effects of gap size and age on the understorey herb community of a tropical Mexican rain forest. *Journal of Ecology* 80: 809–822.

Dodson, C.H. 1991. Orchidales, pp. 738–746. In *Angiosperms: The Flowering Plants*. pp. 596–765, Vol. 13. *The New Encyclopaedia Britannica*, 15th edn. Chicago: Encyclopaedia Britannica Inc.

Douglas, R.J.W. 1972. *Geology and Economic Minerals of Canada*. Ottawa: Geological Survey of Canada.

Dowdeswell, J.A. 2006. The Greenland Ice Sheet and global sea-level rise. *Science* 311: 963–964.

Drew, M.C. 1975. Comparison of the effects of a localized supply of phosphate, nitrate, ammonium and potassium on the growth of the seminal root system, and the shoot, in barley. *New Phytologist* 75: 479–490.

Duchesne, L.C. and D.W. Larson. 1989. Cellulose and the evolution of plant life. *Bioscience* 39: 238–241.

Duddridge, J.A. and D.J. Read. 1982. An ultrastructural analysis of the development of

mycorrhizas in *Monotropa hypopithys* L. *New Phytologist* 92: 203–214.

Duffy, D.C. and A.J. Meier. 1992. Do Appalachian herbaceous understories ever recover from clearcutting? *Conservation Biology* 6: 196–201.

Dugan, P. (ed.) 1993. *Wetlands in Danger*. New York: Oxford University Press.

Durant, W. 1944. *Caesar and Christ*. New York: Simon and Schuster.

du Rietz, G.E. 1931. *Life-forms of Terrestrial Plants*. Acta Phytogeographica Suecia. III. Uppsala: Almqvist and Wiksells.

Dynesius, M. and C. Nilsson. 1994. Fragmentation and flow regulation systems in the northern third of the world. *Science* 266: 753–762.

Earth Impact Database. 2016. (http://www.passc.net/EarthImpactDatabase) accessed 27 October 2016.

Edmonds. J. (ed.) 1997. *Oxford Atlas of Exploration*. New York: Oxford University Press.

Egerton, F.N. 2007. A History of the Ecological Sciences, Part 25. American naturalists explore eastern North America: John and William Bartram. *Bulletin of the Ecological Society of America* 88: 253–268.

Ehrenfeld, J.G. 1983. The effects of changes in land-use on swamps of the New Jersey Pine Barrens. *Biological Conservation* 25: 353–375.

Ehrlich, A. and P. Ehrlich. 1981. *Extinction: The Causes and Consequences of the Disappearance of Species*. New York: Random House.

Ehrlich, P. and P.H. Raven 1964. Butterflies and plants: a study in coevolution. *Evolution* 18: 586–608.

Eissenstat, D.M. and E.I. Newman. 1990. Seedling establishment near large plants: effects of vesicular-arbuscular mycorrhizae on the intensity of plant competition. *Functional Ecology* 4: 95–99.

Ellenberg, H. 1985. Veränderungen der Flora Mitteleuropas unter dem Einfluss von Düngung und Immissionen. *Schweizerische Zeitschrift für Forstwesen* 136: 19–39.

1988a. Floristic changes due to nitrogen deposition in central Europe. In J. Nilsson and P. Grennfelt (eds.) *Critical Loads for Sulphur and Nitrogen.* Report from a workshop held at Skokloster, Sweden, 19–24 March 1988.

1988b. *Vegetation Ecology of Central Europe.* 4th edn. Cambridge University Press, Cambridge. Translated by G.K. Strutt.

Ellison, A.M. and E.J. Farnsworth. 1996. Spatial and temporal variability in growth of *Rhizophora mangle* saplings on coral cays: Links with variation in insolation, herbivory, and local sedimentation rate. *Journal of Ecology* 84: 717–731.

Elton, C. 1927. *Animal Ecology.* London: Sidgwick and Jackson Ltd.

Encyclopaedia Britannica. 1991a. Vol. 16. p. 500. Chicago: Encyclopaedia Britannica Inc.

1991b. Vol. 12. p. 41. Chicago: Encyclopaedia Britannica Inc.

1991c. Vol. 16. p. 481. Chicago: Encyclopaedia Britannica Inc.

Endress, P.K. 1996. *Diversity and Evolutionary Biology of Tropical Flowers.* Cambridge: Cambridge University Press.

Englert. S. 1970. *Islands at the Center of the World: New Light on Easter Island.* New York: Charles Scribner. Translated by W. Mulloy.

Eriksson, O. 1993. The species-pool hypothesis and plant community diversity. *Oikos* 68: 371–374.

Ernst, W. 1978. Discrepancy between ecological and physiological optima of plant species: a re-interpretation. *Oecologia Plantarum* 13: 175–188.

Estill, J.C. and MB. Cruzan. 2001. Phytogeography and rare plant species endemic to the southeastern United States. *Castanea* 66: 3–23.

Facelli, M., R., C. Leon and V.A. Deregibus. 1989. Community structure in grazed and ungrazed grassland sites in the flooding Pampa, Argentina. *The American Midland Naturalist* 121: 125–133.

Falls, J.B., E.A. Falls and J.M. Fryxell. 2007. Fluctuations of deer mice in Ontario in relation to seed crops. *Ecological Monographs* 77: 19–32.

Farjon, A. 1998. *World Checklist and Bibliography of Conifers.* Royal Botanical Gardens at Kew, Richmond, UK.

Farrow, E.P. 1917. On the ecology of the vegetation of Breckland. III. General effects of rabbits on the vegetation. *Journal of Ecology* 5: 1–18.

Faulkner, S.P. and C.J. Richardson. 1989. Physical and chemical characteristics of freshwater wetland soils. pp. 41–72. In D.A. Hammer (ed.) *Constructed Wetlands for Wastewater Treatment. Municipal, Industrial, and Agricultural.* Chelsea: Lewis Publishers.

Fedorov, A.V., P.S. Dekens, M. McCarthy et al. 2006. The Pliocene paradox (mechanisms for a permanent El Niño). *Science* 312: 1485–1489.

Feinsinger, P. 1976. Organisation of a tropical guild of nectivorous birds. *Ecological Monographs* 46: 257–291.

1993. Coevolution and pollination. pp. 282–310. In D. Futuyma and M. Slatkin (eds.) *Coevolution.* Sunderland: Sinauer.

Fernald, M.L. 1921. The Gray Herbarium expedition to Nova Scotia 1920. *Rhodora* 23: 89–111, 130–171, 184–195, 233–245, 257–278, 284–300.

1922. Notes on the flora of western Nova Scotia 1921. *Rhodora* 24: 157–164, 165–180, 201–208.

1935. Critical plants of the upper Great Lakes region of Ontario and Michigan. *Rhodora* 37: 197–222, 238–262, 272–301, 324–341.

Fernández-Armesto, F. 1989. *The Spanish Armada: The Experience of War in 1588.* Oxford: Oxford University Press.

Ferris, T. 1988. *Coming of Age in the Milky Way.* New York: Doubleday.

Fienberg, S.E. and DV. Hinkley (eds.). 1980. *R. A. Fisher: An Appreciation.* New York: Springer-Verlag.

Firbank, L.G. and A.R. Watkinson. 1985. On the analysis of competition within two-species mixtures of plants. *Journal of Applied Ecology* 22: 503–517.

Fischer, R., W. De Vries, W. Seidling, P. Kennedy and M. Lorenz. 2000. *Forest Condition in Europe.* 2000 Executive Report. United Nations Economic Commission for Europe/European Commission, Geneva and Brussels.

Fisher, R.A. 1925. *Statistical Methods for Research Workers*. London: Oliver and Boyd.

Fitter, A.H. and R.K.M. Hay. 1983. *Environmental Physiology of Plants*. London: Academic Press.

Fittkau, E.J. and H. Klinge. 1973. On biomass and trophic structure of the central Amazonian rain forest ecosystem. *Biotropica* 5: 2–14.

Flannery, T. 2001. *The Eternal Frontier: An Ecological History of North America and its Peoples*. Melbourne: Text Publishing.

2005. *The Weather Makers: How Man is Changing the Climate and What it Means for Life on Earth*. New York: Atlantic Monthly Press.

Fleischner, T.L. 1994. Ecological costs of livestock grazing in western North America. *Conservation Biology* 8: 629–644.

Flint, R.F. 1971. *Glacial and Quaternary Geology*. New York: John Wiley and Sons.

Fonteyn, P.J. and B.E. Mahall. 1978. Competition among desert perennials. *Nature* 275: 544–545.

1981. An experimental analysis of structure in a desert plant community. *Journal of Ecology* 69: 883–896.

Forde, B. and H. Zhang. 1998. Response: nitrate and root branching. *Trends in Plant Science* 3: 204–205.

Foreman, D. 2004. *Rewilding North America: A Vision for Conservation in the 21st Century*. Washington, D.C.: Island Press.

Forman, R.T.T. 1964. Growth under controlled conditions to explain the hierarchical distributions of a moss, *Tetraphis pellucida*. *Ecological Monographs* 34: 1–25.

Forman, R.T.T., D. Sperling, J. Bissonette et al. 2002. *Road Ecology: Science and Solutions*. Washington: Island Press.

Forsberg, R. 1982. A bilateral nuclear-weapon freeze. *Scientific American* 247(5): 52–61.

Forsyth, A. and K. Miyata. 1984. *Tropical Nature*. New York: Scribners.

Foster, A.S. and E.M. Gifford, Jr. 1974. *Comparative Morphology of Vascular Plants*. 2nd edn. San Francisco: W. H. Freeman and Company.

Foster, D.R. and P.H. Glaser. 1986. The raised bogs of south-eastern Labrador, Canada: classification, distribution, vegetation and recent dynamics. *Journal of Ecology* 74: 47–71.

Foster, D. R. and H.E. Wright, Jr. 1990. Role of ecosystem development and climate change in bog formation in central Sweden. *Ecology* 71: 450–463.

Fowler, N. 1981. Competition and coexistence in a North Carolina grassland. II. The effects of the experimental removal of species. *Journal of Ecology* 69: 843–845.

Fox, J.F. 1977. Alternation and coexistence of tree species. *The American Naturalist* 111: 69–89.

Fragoso, J.M.V. 1997. Tapir-generated seed shadows: scale-dependent patchiness in the Amazon rain forest. *Journal of Ecology* 85: 519–529.

Francis, R. and D.J. Read. 1984. Direct transfer of carbon between plants connected by vesicular-arbuscular mycorrhizal mycelium. *Nature* 307: 53–56.

Franco, A.C. and P.S. Nobel. 1989. Effect of nurse plants on the microhabit and growth of cacti. *Journal of Ecology* 77: 870–886.

Frank, B. 2005. On the nutritional dependence of certain trees on root symbiosis with belowground fungi (an English translation of A.B. Frank's classic paper of 1885). *Mycorrhiza* 15: 267–275.

Fraser, L.H. and P. Keddy 1997. The role of experimental microcosms in ecological research. *Trends in Ecology and Evolution* 12: 478–481.

Fraser, L.H. and P.A. Keddy (eds.). 2005. *The World's Largest Wetlands: Ecology and Conservation*. Cambridge: Cambridge University Press.

Freedman, B. 1995. *Environmental Ecology*. 2nd edn. San Diego: Academic Press.

Freedman, B., W. Zobens, T.C. Hutchinson and W.I. Gizyn. 1990. Intense, natural pollution affects arctic tundra vegetation at the Smoking Hills, Canada. *Ecology* 71: 492–503.

Freemark, K.E. and H.G. Merriam. 1986. The importance of area and habitat heterogeneity to bird assemblages in temperate forest fragments. *Biological Conservation* 36: 115–141.

French, B.M. 1998. *Traces of Catastrophe: A Handbook of Shock-Metamorphic Effects in Terrestrial Meteoric Impact Structures.* LPI Contribution No. 954. Houston: Lunar and Planetary Institute.

Fretwell, S.D. 1977. The regulation of plant communities by food chains exploiting them. *Perspectives in Biology and Medicine* 20: 169–185.

Frey, R.W. and P.B. Basan. 1978. Coastal salt marshes. pp. 101–169. In R.A. Davis (ed.) *Coastal Sedimentary Environments.* New York: Springer-Verlag.

Frey, T.E. 1973. The Finnish school and forest site types. pp. 403–433. In Whittaker, R.H. (ed.) *Ordination and Classification of Communities. Handbook of Vegetation Science, Part V.* The Hague: W. Junk.

Friedmann, E.I. 1982. Endolithic microorganisms in the Antarctic cold desert. *Science* 215: 1045–1053.

Friis, E.M., K.R. Pedersen and P.R. Crane. 2006. Cretaceous angiosperm flowers: innovation and evolution in plant reproduction. *Palaeogeography, Palaeoclimatology, Palaeoecology* 23: 251–293.

Frontier, S. 1985. Diversity and structure in aquatic ecosystems. *Oceanography and Marine Biology Annual Review* 23: 253–312.

Futuyma, D.J. and M. Slatkin. 1993. The study of coevolution. pp. 459–464. In D.J. Futuyma and M. Slatkin (eds.) *Coevolution.* Sunderland: Sinauer.

Galatowitsch, S.M. and A.G. van der Valk. 1994. *Restoring Prairie Wetlands: An Ecological Approach.* Ames: Iowa State University Press.

Gamble, T. 1921. Savannah as a naval stores port 1875–1920. pp. 59–62. In T. Gamble (ed.) *Naval Stores. History, Production, Distribution and Consumption.* Savannah, GA: Review Publishing and Printing Company.

Gardner, G. 1977. The reproductive capacity of *Fraxinus excelsior* on the Derbyshire limestone. *Journal of Ecology* 65: 107–118.

Gaston, K.J. 2000. Global patterns in biodiversity. *Nature* 405: 220–227.

Gaston, K.J., P.H. Williams, P. Eggleton and C.J. Humphries. 1995. Large scale patterns of biodiversity: spatial variation in family richness. *Proceedings of the Royal Society of London Series B-Biological Sciences* 260: 149–154.

Gauch, H.G. Jr. 1982. *Multivariate Analysis in Community Ecology. Cambridge Studies in Ecology.* Cambridge: Cambridge University Press.

Gauch, H.G. Jr. and T.R. Wentworth. 1976. Canonical correlation analysis as an ordination technique. *Vegetatio* 33: 17–22.

Gauch, H.G. Jr. and R.H. Whittaker. 1972. Comparison of ordination techniques. *Ecology* 53: 868–875.

Gauch, H.G. Jr., R.H. Whittaker and T.R. Wentworth. 1977. A comparative study of reciprocal averaging and other techniques. *Journal of Ecology* 65: 157–174.

Gaudet, C.L. 1993. *Competition in Shoreline Plant Communities: a Comparative Approach.* PhD thesis. Ottawa: University of Ottawa.

Gaudet, C.L. and P.A. Keddy. 1988. A comparative approach to predicting competitive ability from plant traits. *Nature* 334: 242–243.

Gause, G.F. and A.A. Witt. 1935. Behavior of mixed populations and the problem of natural selection. *The American Naturalist* 69: 596–609.

Geis, J.W. 1985. Environmental influences on the distribution and composition of wetlands in the Great Lakes basin. pp. 15–31. In H.H. Prince and F.M. D'Itri (eds.) *Coastal Wetlands.* Chelsea: Lewis Publishers.

Gentry, A.H. 1988. Changes in plant community diversity and floristic composition on environmental and geographical gradients. *Annals of the Missouri Botanical Garden* 75: 1–34.

Gibson, A.C. and P.S. Nobel. 1986. *The Cactus Primer.* Cambridge: Harvard University Press.

Gibson, C.W.D. and J. Hamilton. 1983. Feeding ecology and seasonal movements of giant tortoises on Aldabra atoll. *Oecologia* 56: 84–92.

Gibson, D.J. 2009. *Grasses and Grassland Ecology.* Oxford: Oxford Univerity Press.

Giosan, L., F. Filip and S. Constatinescu. 2009. Was the Black Sea catastrophically flooded in the early Holocene? *Quaternary Science Reviews* **28**: 1–6.

Given, D.R. and J. Soper. 1981. *The Arctic–Alpine Element of the Vascular Flora at Lake Superior.* Publications in Botany No. 10. Ottawa: National Museums of Canada.

Givnish, T.J. 1982. On the adaptive significance of leaf height in forest herbs. *The American Naturalist* **120**: 353–381.

　1984. Leaf and canopy adaptations in tropical forests. pp. 51–84. In E. Medina, H.A. Mooney and C. Vásquez-Yánes (eds.) *Physiological Ecology of Plants of the Wet Tropics.* The Hague: Dr. Junk.

　1987. Comparative studies of leaf form: assessing the relative roles of selective pressures and phylogenetic constraints. *New Phytologist* **106** (Suppl.): 131–160.

　1988. Ecology and evolution of carnivorous plants. pp. 243–290. In W.B. Abrahamson (ed.) *Plant–Animal Interactions.* New York: McGraw-Hill.

　1994. Does diversity beget stability? *Nature* **371**: 113–114.

Glaser, P.H. 1992. Raised bogs in eastern North America: regional controls for species richness and floristic assemblages. *Journal of Ecology* **80**: 535–554.

Glaser, P.H., J.A. Janssens and D.I. Siegel. 1990. The response of vegetation to chemical and hydrological gradients in the Lost River peatland, northern Minnesota. *Journal of Ecology* **78**: 1021–1048.

Gleason, H.A. 1917. The structure and development of the plant association. *Bulletin of the Torrey Botanical Club* **44**: 463–481.

　1926. The individualistic concept of the plant association. *Bulletin of the Torrey Botanical Club* **53**: 7–26.

Glenn-Lewin, D.C., R.K. Peet and T.T. Veblen (eds.) 1992. *Plant Succession: Theory and Prediction. Population and Community Biology.* No. 11. London: Chapman and Hall.

Glooschenko, W.A. 1980. Coastal salt marshes in Canada. pp. 39– 47. In C.D.A. Rubec and F.C.

Pollet (eds.) Proceedings of the Workshop on Canadian Wetlands. Saskatoon, Saskatchewan. Environment Canada, Lands Directorate, Ecological Land Class. Series No. 12.

Gnanadesikan, R. 1997. *Methods for Statistical Data Analysis of Multivariate Observations.* 2nd edn. New York: Wiley.

Goebel, K. 1905. Wilhelm Hofmeister. *The Plant World* **8**: 291–298.

Goldberg, D.E. 1982a. The distribution of evergreen and deciduous trees relative to soil type: an example from the Sierra Madre, Mexico, and a general model. *Ecology* **63**: 942–951.

　1982b. Comparison of factors determining growth rates of deciduous vs. broad-leaf evergreen trees. *The American Midland Naturalist* **108**: 133–143.

　1990. Components of resource competition in plant communities. pp. 27–49. In J.B. Grace and D. Tilman (eds.) *Perspectives on Plant Competition.* San Diego: Academic Press.

Goldberg, D.E. and K. Landa. 1991. Competitive effect and response: hierarchies and correlated traits in the early stages of competition. *Journal of Ecology* **79**: 1013–1030.

Goldberg, D.E. and P.A. Werner. 1983. Equivalence of competitors in plant communities: a null hypothesis and a field experimental approach. *American Journal of Botany* **70**: 1098–1104.

Goldsmith, F.B. 1973a. The vegetation of exposed sea cliffs at South Stack, Anglesey: I. The multivariate approach. *Journal of Ecology* **61**: 787–818.

　1973b. The vegetation of exposed sea cliffs at South Stack, Anglesey: II. Experimental studies. *Journal of Ecology* **61**: 819–829.

　1978. Interaction (competition) studies as a step towards the synthesis of sea-cliff vegetation. *Journal of Ecology* **66**: 921–931.

Goldsmith, F.B. and C.M. Harrison. 1976. Description and analysis of vegetation. pp. 85–155. In S.B. Chapman (ed.) *Methods in Plant Ecology.* Oxford: Blackwell Scientific.

Gopal, B. 1990. Nutrient dynamics of aquatic plant communities. pp. 177–197. In B. Gopal (ed.) *Ecology and Management of Aquatic Vegetation in the Indian Subcontinent*. Dordrecht: Kluwer Academic Publishers.

Gopal, B. and U. Goel. 1993. Competition allelopathy in aquatic plant communities. *The Botanical Review* 59: 155–210.

Gopal, B., J. Kvet, H. Loffler, V. Masing and B.C. Patten. 1990. Definition and classification. pp. 9–15. In B.C. Patten (ed.) *Wetlands and Shallow Continental Water Bodies*, Vol. 1. Natural and Human Relationships. The Hague: SPB Academic Publishing.

Gore, A.J.P. 1983. Introduction. In A.J.P. Gore (ed.) *Ecosystems of the World 4A. Mires: Swamp, Bog, Fen and Moor*. Amsterdam: Elsevier Scientific Publishing Company.

Gore, A. 2006. *An Inconvenient Truth: The Planetary Emergency of Global Warming and What We Can Do About It*. New York: Melcher Media/Rodale.

Goremykin, V.V., K. I. Hirsch-Ernst, S. Wölfl and F.H. Hellwig. 2003. Analysis of the *Amborella trichopoda* chloroplast genome sequence suggests that *Amborella* is not a basal angiosperm. *Molecular Biology and Evolution* 20: 1499–1505.

Gorham, E. 1953. Some early ideas concerning the nature, origin and development of peat lands. *Journal of Ecology* 41: 257–274.

1957. The development of peat lands. *The Quarterly Review of Biology* 32: 145–166.

1979. Shoot height, weight and standing crop in relation to density of nonspecific plant stands. *Nature* 279: 148–150.

1990. Biotic impoverishment in northern peatlands. pp. 65–98. In G.M. Woodwell (ed.) *The Earth in Transition*. Cambridge: Cambridge University Press.

1991. Northern peatlands role in the carbon cycle and probable responses to climatic warming. *Ecological Applications* 1: 182–195.

Gosselink, J.G. and R.E. Turner. 1978. The role of hydrology in freshwater wetland ecosystems. pp. 63–78. In R.E. Good, D.F. Whigham and R.L. Simpson (eds.) *Freshwater Wetlands:*

Ecological Processes and Management Potential. New York: Academic Press.

Gosselink, J.G., J.M. Coleman and R.E. Stewart, Jr. 1998. Coastal Louisiana. pp. 385–436. In M.J. Mac, P.A. Opler, C.E. Puckett Haecker and P.D. Doran (eds.) *1998. Status and Trends of the Nation's Biological Resources*, 2 Vols. Reston: US Department of the Interior, US Geological Survey.

Gotelli, N.J. and G.R. Graves. 1996. *Null Models in Ecology*. Washington, D.C.: Smithsonian Institution Press.

Gough, J. 1793. Reasons for supposing that lakes have been more numerous than they are at present; with an attempt to assign the causes whereby they have been defaced. *Memoirs of the Literary and Philosophical Society of Manchester* 4: 1–19. cited by D. Walker. 1970. Direction and rate in some British post-glacial hydoseres. pp. 117–139. In D. Walker, and R.G. West (eds.) *Studies in the Vegetational History of the British Isles*. Cambridge: Cambridge University Press.

Gough, L., J.B. Grace and K L. Taylor. 1994. The relationship between species richness and community biomass: the importance of environmental variables. *Oikos* 70: 271–279.

Gould, S.J. 1977. *Ever Since Darwin: Reflections in Natural History*. New York: W. W. Norton and Company.

Grace, J.B. 1993. The effects of habitat productivity on competition intensity. *Trends in Ecology and Evolution* 8: 229–230.

1999. The factors controlling species density in herbaceous plant communities: an assessment. *Perspectives in Plant Ecology, Evolution and Systematics* 2: 1–28.

2001. The roles of community biomass and species pools in the regulation of plant diversity. *Oikos* 92: 193–207.

2006. *Structural Equation Modeling and Natural Systems*. Cambridge: Cambridge University Press.

Grace, J.B. and B.H. Pugesek. 1997. A structural equation model of plant species richness and its application to a coastal wetland. *The American Naturalist* 149: 436–460.

Grace, J.B. and D. Tilman (eds.) 1990. *Perspectives on Plant Competition.* San Diego: Academic Press.

Grant, M.C. 1993. The trembling giant. *Discover* **4**(10): 82–89.

Green, P.T., D.J. O'Dowd and P.S. Lake. 1997. Control of seedling recruitment by land crabs in rain forest on a remote oceanic island. *Ecology* **78**: 2472–2486.

Greene, D.F. and E.A. Johnson. 1994. Estimating the mean annual seed production of trees. *Ecology* **75**: 642–647.

Greenslade, P.J.M. 1983. Adversity selection and the habitat templet. *Nature* **242**: 344–347.

Greig-Smith, P. 1952. Use of random and contiguous quadrats in the study of the structure of plant communities. *Annals of Botany* **16**: 293–316.

    1957. *Quantitative Plant Ecology.* London: Butterworths.

Grime, J.P. 1973a. Control of species density in herbaceous vegetation. *Journal of Environmental Management* **1**: 151–167.

    1973b. Competitive exclusion in herbaceous vegetation. *Nature* **242**: 344–347.

    1974. Vegetation classification by reference to strategies. *Nature* **250**: 26–31.

    1977. Evidence for the existence of three primary strategies in plants and its relevance to ecological and evolutionary theory. *The American Naturalist* **111**: 1169–1194.

    1979. *Plant Strategies and Vegetation Processes.* Chichester: John Wiley.

    1994. The role of plasticity in exploiting environmental heterogeneity. pp. 1–19. In M.M. Caldwell and R.W. Percy (eds.) *Exploitation of Environmental Heterogeneity by Plants: Ecophysiological Processes Above- and Belowground.* San Diego: Academic Press.

    1997. The humped-back model: a response to Oksanen. *Journal of Ecology* **85**: 97–98.

    2002. Declining plant diversity: empty niches or functional shifts? *Journal of Vegetation Science* **13**: 457–460.

Grime, J.P. and R. Hunt. 1975. Relative growth-rate: its range and adaptive significance in a local flora. *Journal of Ecology* **63**: 393–422.

Grime, J.P. and D.W. Jeffrey. 1965. Seedling establishment in vertical gradients of sunlight. *Journal of Ecology* **53**: 621–642.

Grime, J.P., G. Mason, A.V. Curtis et al. 1981. A comparative study of germination characteristics in a local flora. *Journal of Ecology* **69**: 1017–1059.

Grime, J.P., J.M.L. Mackey, S.H. Hillier and D.J. Read. 1987. Floristic diversity in a model system using experimental microcosms. *Nature* **328**: 420–422.

Grishin, S.Y., R. del Moral, P.V. Krestov and V.P. Verkholat. 1996. Succession following the catastrophic eruption of Ksudach volcano (Kamchatka, 1907). *Vegetatio* **127**: 129–153.

Groombridge, B. (ed.). 1992. *Global Biodiversity: Status of the Earth's Living Resources.* London: Chapman and Hall.

Grover, A.M. and G.A. Baldassarre. 1995. Bird species richness within beaver ponds in south-central New York. *Wetlands* **15**: 108–118.

Grubb, P.J. 1977. The maintenance of species-richness in plant communities: the importance of the regeneration niche. *Biological Reviews* **52**: 107–145.

    1987. Global trends in species-richness in terrestrial vegetation: a view from the Northern Hemisphere. pp. 99–118. In J.H.R. Gee and P.S. Giller (eds.) *Organization of Communities Past and Present.* Oxford: Blackwell Scientific Publications.

Grumbine, R.E. 1997. Reflections on "What is ecosystem management?" *Conservation Biology* **11**: 41–47.

Guariguata, M.R. 1990. Landslide disturbance and forest regeneration in the Upper Luquillo mountains of Puerto Rico. *Journal of Ecology* **78**: 814–832.

Gurevitch, J. and R.S. Unnasch. 1989. Experimental removal of a dominant species at two levels of soil fertility. *Canadian Journal of Botany* **67**: 3470–3477.

Haber, L.F. 1986. *The Poisonous Cloud. Chemical Warfare in the First World War.* Oxford: Clarendon Press.

Hairston, N.G., F.E. Smith and L.B. Slobodkin. 1960. Community structure, population control, and competition. *The American Naturalist* XCIV: 421–425.

Hamann, O. 1979. Regeneration of vegetation on Santa Fe and Pinta Islands, Galápagos, after the eradication of goats. *Biological Conservation* 15: 215–236.

1993. On vegetation recovery, goats and giant tortoises on Pinta Island, Galápagos, Ecuador. *Biodiversity and Conservation* 2: 138–151.

Hansen, D.M., C.J. Donlan, C.J. Griffiths and K.J. Campbell. 2010. Ecological history and latent conservation potential: large and giant tortoises as a model for taxon substitutions. *Ecography* 33: 272–284.

Hanski, I. 1994. Patch-occupancy dynamics in fragmented landscapes. *Trends in Ecology and Evolution* 9: 131–135.

Hanski, I. and M. Gilpin. 1991. Metapopulation dynamics: a brief history and conceptual domain. *Biological Journal of the Linnean Society* 42: 3–16.

Hare, J.D. 2011. Ecological role of volatiles produced by plants in response to damage by herbivorous insects. *Annual Review of Entomology* 56: 161–180.

Harper, J.L. 1965. The nature and consequence of interference amongst plants. *Genetics Today* 2: 465–482.

1967. A Darwinian approach to plant ecology. *Journal of Ecology* 55: 247–270.

1977. *Population Biology of Plants.* London: Academic Press.

1982. After description. pp. 11–25. In E. I. Newman (ed.) *The Plant Community as a Working Mechanism.* Oxford: Blackwell.

Harper, J.L. and J. Ogden. 1970. The reproductive strategy of higher plants. I. The concept of strategy with special reference to *Senecio vulgaris* L. *Journal of Ecology* 58: 681–698.

Harper, J.L. and J. White. 1974. The demography of plants. *Annual Review of Ecology and Systematics* 5: 419–463.

Harper, J L., B R. Rosen and J. White. 1986. *The Growth and Form of Modular Organisms.* London: The Royal Society.

Harris, L.D. 1984. *The Fragmented Forest: Island Biogeography Theory and the Preservation of Biotic Diversity.* Chicago: University of Chicago Press.

Harrison, R.D., S. Tan, J.B. Plotkin et al. 2013. Consequences of defaunation for a tropical tree community. *Ecology Letters* 16: 687–694.

Hartman, J.M. 1988. Recolonization of small disturbance patches in a New England salt marsh. *American Journal of Botany* 75: 1625–1631.

Harvey, P.H., R.K. Colwell, J.W. Silvertown and R.M. May. 1983. Null models in ecology. *Annual Review of Ecology and Systematics* 14: 189–211.

Hatch, A.B. 1937. The physical basis of mycotrophy in *Pinus. The Black Rock Forest Bulletin* 6:1–168.

Hawksworth, D.L. 1988. Coevolution of fungi with algae and cyanobacteria in lichen symbioses. pp. 125–148. In K.A. Pirozynski and D.L. Hawksworth (eds.) *Coevolution of Fungi with Plants and Animals.* London: Academic Press.

1990. The fungal dimension of biodiversity: magnitude, significance, and conservation. *Mycological Research* 95: 641–655.

Hawley, L.F. 1921. Lightwood, cut-over lands and the naval stores industry. pp. 237–239. In T. Gamble (ed.) *Naval Stores. History, Production, Distribution and Consumption.* Savannah, GA: Review Publishing and Printing Company.

Hayati, A.A. and M.C.F. Proctor. 1991. Limiting nutrients in acid-mire vegetation: peat and plant analyses and experiments on plant responses to added nutrients. *Journal of Ecology* 79: 75–95.

Heady, H.F. (ed.) 1988. *The Vale Rangeland Rehabilitation Program: An Evaluation.* USDA Forest Service, Resource Bulletin PNW-RB-157.

Heady, H.F. and J. Bartolome. 1977. *The Vale Rangeland Rehabilitation Program: The Desert Repaired in Southeastern Oregon.* USDA Forest Service, Resource Bulletin PHW-70.

Heckman, D.S., D.M. Geiser, B.R. Eidell, et al. 2001. Molecular evidence for the early colonization of land by fungi and plants. *Science* 293: 1129–1133.

Heinselman, M.L. 1973. Fire in the virgin forests of the Boundary Waters Canoe Area, Minnesota. *Quaternary Research* 3: 329–382.

1981. Fire and succession in the conifer forests of northern North America. pp. 374–405. In D.C. West, H.H. Shugart and D.B. Botkin (eds.) *Forest Succession: Concepts and Applications*. New York: Springer-Verlag.

Hemphill, N. and S.D. Cooper. 1983. The effect of physical disturbance on the relative abundances of two filter-feeding insects in a small stream. *Oecologia* 58: 378–382.

Herman, K.D., L.A. Masters, M.R. Penskar, et al. 2001. *Floristic Quality Assessment with Wetland Categories and Examples of Computer Applications for the State of Michigan. Revised*. 2nd edn. Lansing, MI: Natural Heritage Program, Michigan Department of Natural Resources.

Herre, E.A., L.C. Mejía, D.A. Kyllo et al. 2007. Ecological implications of anti-pathogen effects of tropical fungal endophytes and mycorrhizae. *Ecology* 88: 550–558.

Hett, J.M. 1971. A dynamic analysis of age in sugar maple seedlings. *Ecology* 52: 1071–1073.

Heywood, V.K., R.K. Brummitt, A. Culham and O. Seberg (eds.) 2007. *Flowering Plant Families of the World*. Richmond Hill, Ontario: Firefly Books.

Higgs, E.S. 1997. What is good ecological restoration? *Conservation Biology* 11: 338–348.

Hill, N.M. and P.A. Keddy. 1992. Predicting numbers of rarities from habitat variables: coastal plain plants of Nova Scotian lakeshores. *Ecology* 73: 1852–1859.

Hills, G.A. 1961. *The Ecological Basis for Land-Use Planning*. Report No. 46. Ontario: Ontario Department of Lands and Forests, Research Branch.

Hoagland, B.W. and S L. Collins. 1997. Gradient models, gradient analysis, and hierarchical structure in plant communities. *Oikos* 78: 23–30.

Hogenbirk, J.C. and R.W. Wein. 1991. Fire and drought experiments in northern wetlands: a climate change analogue. *Canadian Journal of Botany* 69: 1991–1997.

Hogg, E.H., V.J. Lieffers and R.W. Wein. 1992. Potential carbon losses from peat profiles: effects of temperature, drought cycles, and fire. *Ecological Applications* 2: 298–306.

Holechek, J.L., M. Vavra and R.D. Pieper. 1982. Botanical composition determination of herbivore diets: a review. *Journal of Range Management* 31: 309–315.

Holling, C.S. 1959. The components of predation as revealed by a study of small-mammal predation of the European pine sawfly. *Canadian Entomologist* 91: 293–320.

Holling, C.S. (ed.) 1978a. *Adaptive Environmental Assessment and Management*. New York: John Wiley and Sons.

Holling, C.S. 1978b. The spruce-budworm/forest-management problem. pp. 143–182. In C.S. Holling (ed.) *Adaptive Environmental Assessment and Management*. New York: John Wiley and Sons.

Holt, R.D. and J.H. Lawton. 1993. Apparent competition and enemy-free space in insect host-parasitoid communities. *The American Naturalist* 142: 623–645.

1994. The ecological consequences of shared natural enemies. *Annual Review of Ecology and Systematics* 25: 495–520.

Hook, D.D. 1984. Adaptations to flooding with fresh water. pp. 265–294. In T.T. Kozlowski (ed.) *Flooding and Plant Growth*. Orlando: Academic Press.

Hopkins, D.M. (ed.) 1967. *The Bering Land Bridge*. Stanford: Stanford University Press.

Horn, H.S. 1971. *The Adaptive Geometry of Trees*. Princeton: Princeton University Press.

1976. Succession. pp. 187–204. In R.M. May (ed.) *Theoretical Ecology: Principles and Applications*. Philadelphia: W. B. Saunders.

1981. Some causes of variety of patterns of secondary succession. pp. 24–35. In D.C. West, H.H. Shugart and D.B. Botkin (eds.) *Forest*

Succession: Concepts and Application. New York: Springer-Verlag.

Horn, H.S. and R.H. MacArthur. 1972. Competition among fugitive species in a harlequin environment. *Ecology* 53: 749–752.

Hsueh, Y.-P., P. Mahanti, F.C. Schroeder and P.W. Sternberg. 2013. Nematode-trapping fungi eavesdrop on nematode pheromones. *Current Biology* 23: 83–86.

Hubbell, S.P. and R.B. Foster. 1986. Biology, chance and history and the structure of the tropical rain forest tree communities. pp. 314–329. In J. Diamond and T.J. Case (eds.) *Community Ecology.* NewYork: Harper and Row.

Hubbell, S.P., F. He, R. Condit, et al. 2008. How many tree species are there in the Amazon and how many of them will go extinct? *Proceedings of the National Academy of Sciences* 105 (Supplement 1): 11,498–11,504.

Huber, H., S. Lukács and M.A. Watson. 1999. Spatial structure of stoloniferous herbs: an interplay between structural blue-print, ontogeny and phenotypic plasticity. *Plant Ecology* 141: 107–115.

Huempfner, R.A. and J.R. Tester. 1988. Winter arboreal feeding behavior of ruffed grouse in east-central Minnesota. pp. 122–157. In A.T. Bergerud and M.W. Gratson (eds.) *Adaptive Strategies and Population Ecology of Northern Grouse.* Minneapolis: University of Minnesota Press.

Hughes, F.M.R. (ed.) 2003. *The Flooded Forest: Guidance for Policy Makers and River Managers in Europe on the Restoration of Floodplain Forests.* FLOBAR2, Department of Geography, University of Cambridge, UK.

Hughes, J.D. 1982. Deforestation, erosion, and forest management in ancient Greece and Rome. *Journal of Forest History* 26: 60–75.

Humphries, C.J. 1981. Biogeographical methods and the southern beeches. pp. 283–297. In P.L. Forey (ed.) *The Evolving Biosphere.* British Museum (Natural History). Cambridge University Press.

Hungate, B.A., J.S. Dukes, M.R. Shaw, Y. Luo and C.B. Field. 2003. Nitrogen and climate change. *Science* 302:1512–1513.

Hunt, T.L. 2007. Rethinking Easter Island's ecological catastrophe. *Journal of Archaeological Science* 34: 485–502.

Hunt, T.L. and C.P. Lipo. 2010. Ecological catastrophe, collapse and the myth of "Ecocide" on Rapa Nui (Easter Island). pp. 21–44. In P.A. McAnany and N. Yoffee (eds.) *Questioning Collapse.* Cambridge: Cambridge University Press.

Hunt, R., D.W. Hand, M.A. Hannah and A.M. Neal. 1991. Response to CO_2 enrichment in 27 herbaceous species. *Functional Ecology* 5: 410–421.

Hunter, M.D. and P.W. Price. 1992. Playing chutes and ladders: heterogeneity and the relative roles of bottom–up and top–down forces in natural communities. *Ecology* 73: 724–732.

Huntley, B. 1990. European post-glacial forests: compositional changes in response to climatic change. *Journal of Vegetation Science* 1: 507–518.

Hurlbert, S.H. 1984. Pseudoreplication and the design of ecological field experiments. *Ecological Monographs* 54: 187–211.

1990. Spatial distribution of the montane unicorn. *Oikos* 58: 257–271.

Huston, M.A. 1979. A general hypothesis of species diversity. *The American Naturalist* 113: 81–101.

1994. *Biological Diversity. The Coexistence of Species on Changing Landscapes.* Cambridge: Cambridge University Press.

1997. Hidden treatments in ecological experiments: re-evaluating the ecosystem function of biodiversity. *Oecologia* 110: 449–460.

Hutchinson, G.E. 1959. Homage to Santa Rosalia; or, why are there so many kinds of animals? *The American Naturalist* 93: 145–159.

1961. The paradox of the plankton. *The American Naturalist* 95: 137–146.

1970. The biosphere. pp. 194–203. In E.O. Wilson (ed.) *1974. Ecology, Evolution, and Population Biology. Readings from Scientific American.* San Francisco: W.H. Freeman and Company.

1975. *A Treatise on Limnology*, Vol. 3. Limnological Botany. New York: John Wiley and Sons.

Hutchinson, H.N. 1892. *Extinct Monsters: A Popular Account of Some of the Larger Forms of Ancient Animal Life.* London: Chapman & Hall.

Huxley, C.R. 1980. Symbiosis between plants and epiphytes. *Biological Reviews* 55: 321–340.

Imbrie, J. (and 18 others). 1993. On the structure and origin of major glaciation cycles 2. The 100,000-year cycle. *Paleoceanography* 8: 699–736.

International Joint Commission. 1980. *Pollution in the Great Lakes Basin from Land Use Activities.* Washington, D.C.: International Joint Commission.

Irion, G.M., J. Müller, J.N. de Mello and W.J. Junk. 1995. Quaternary geology of the Amazon lowland. *Geo-Marine Letters* 15: 172–178.

IUCN. 2013. *Dracaena cinnabari* IUCN Red List of Threatened Species. Version 2013.1. (www.iucnredlist.org) accessed 6 August 2013.

Jackson, J.B.C. 1981. Interspecific competition and species distributions: the ghosts of theories and data past. *American Zoologist* 21: 889–901.

Jackson, J.B.C., L.W. Buss and R.E. Cook. 1985. *Population Biology and Evolution of Clonal Organisms.* New Haven: Yale University Press.

Jackson, M.B. and M.C. Drew. 1984. Effects of flooding on growth and metabolism of herbaceous plants. pp. 47–128. In T.T. Kozlowski (ed.) *Flooding and Plant Growth.* Orlando: Academic Press.

Jaksic, F.M. and E.R. Fuentes. 1980. Why are native herbs in the Chilean matorral more abundant beneath bushes: microclimate or grazing? *Journal of Ecology* 68: 665–669.

James, W. 1907. *Pragmatism.* Reprinted pp. xv–xvii, 1–89. In M. J. Adler (ed.) 1990. *Great Books of the Western World.* 2nd ed., Vol. 55. Chicago: Encyclopaedia Britannica Inc.

Janis, C. 1976. The evolutionary strategy of the Equidae and the origins of rumen and cecal digestion. *Evolution* 30: 757–774.

Janssens, F., A. Peeters, J.R.B. Tallowin, et al. 1998. Relationship between soil chemical factors and grassland diversity. *Plant and Soil* 202: 69–78.

Janzen, D.H. 1966. Coevolution of mutualism between ants and acacias in Central America. *Evolution* 20: 249–275.

1967. Interaction of the bull's-horn acacia (*Acacia cornigera* L.) with an ant inhabitant (*Pseudomyrmex ferruginea* F. Smith) in eastern Mexico. *The University of Kansas Science Bulletin* XLVII: 315–558.

1971. Seed predation by animals. *Annual Review of Ecology and Systematics* 2: 465–492.

1974. Epiphytic myrmecophytes in Sarawak: mutualism through the feeding of plants by ants. *Biotropica* 6: 237–259.

1976. Why bamboos wait so long to flower. *Annual Review of Ecology and Systematics* 7: 347–391.

1983. Dispersal of seeds by vertebrate guts. pp. 232–262. In D.J. Futuyma and M. Slatkin (eds.) *Coevolution.* Sunderland: Sinauer.

1985. The natural history of mutualisms. pp. 40–99. In D.H. Boucher (ed.) *The Biology of Mutualism: Ecology and Evolution.* New York: Oxford University Press.

Janzen, D.H. and P.S. Martin. 1982. Neotropical anachronisms: the fruits the gomphotheres ate. *Science* 215: 19–27.

Janzen, D.H., G.A. Miller, J. Hackforth-Jones, et al. 1976. Two Costa Rican bat-generated seed shadows of *Andira inermis* (Leguminosae). *Ecology* 57: 1068–1075.

Jarvis, P.G. 1964. Interference by *Deschampsia flexuosa* (L.) Trin. *Oikos* 15: 56–78.

Jefferies, R.L. 1977. The vegetation of salt marshes at some coastal sites in arctic North America. *Journal of Ecology* 65: 661–672.

Jenkins, D.L. (and 19 others). 2012. Clovis age western stemmed projectile points and human coprolites at the Paisley Caves. *Science* 337: 223–228.

Jenny, H. 1941. *Factors of Soil Formation.* New York: McGraw-Hill.

Jensen, T.S. 1985. Seed–seed predator interactions of European beech, *Fagus silvatica* and forest rodents, *Clethrionomys glareolus* and *Apodemus flavicollis*. *Oikos* 44: 149–156.

Jickells, T.D., S. Dorling, W.G. Deuser, et al. 1998. Air-borne dust fluxes to a deep water sediment trap in the Sargasso Sea. *Global Biogeochemical Cycles* 12: 311–320.

Johansson, M.E. and P.A. Keddy. 1991. Intensity and asymmetry of competition between plant pairs of different degrees of similarity: an experimental study on two guilds of wetland plants. *Oikos* 60: 27–34.

Johnston, A.E. 1994. The Rothamsted classical experiments. pp. 9–35. In R. A. Leigh and A. E. Johnston (eds.) *Long-term Experiments in Agricultural and Ecological Sciences.* Proceedings of a conference to celebrate the 150th anniversary of Rothamsted Experimental Station, held at Rothamsted, 14–17 July 1993. Wallingford: CAB International.

Johnson, L.A.S. and B.G. Briggs. 1975. On the Proteaceae: the evolution and classification of a southern family. *Botanical Journal of the Linnean Society* 20: 81–182.

Johnson, P.L. and W.D. Billings. 1962. The alpine vegetation of the Beartooth Plateau in relation to cryopedogenic processes and patterns. *Ecological Monographs* 32: 105–135.

Jones, C.G., J.H. Lawton and M. Shachak. 1994. Organisms as ecosystem engineers. *Oikos* 69: 373–386.

Jones, R.K., G. Pierpoint, G M. Wickware and J K. Jeglum. 1983a. A classification and ordination of forest ecosystems in the Great Claybelt of northeastern Ontario. pp. 83–96. In R.W. Wein, R.R. Riewe and I.R. Methven (eds.) *Resources and Dynamics of the Boreal Zone.* Proceedings of a Conference held at Thunder Bay, Ontario, August 1982. Ottawa: Association of Canadian Universities for Northern Studies.

Jones, R.K., G. Pierpoint, G.M. Wickware, et al. 1983b. *Field Guide to Forest Classification for the Clay Belt, Site Region 3E.* Toronto: Queen's Printer for Ontario.

Jones, W.G., K.D. Hill and J.M. Allen. 1995. *Wollemia nobilis,* a new living Australian genus and species in the Araucariaceae. *Telopea* 6: 173–176.

Joppa, L.N., D.L. Robert, N. Myers and S.L. Pimm. 2011. Biodiversity hotspots house most undiscovered plant species. *Proceedings of the National Academy of Sciences* 108: 13,171–13,176.

Joppa, L.N., P. Visconti, C.N. Jenkins and S.L. Pimm. 2013. Achieving the Convention on Biological Diversity's goals for plant conservation. *Science* 341: 1100–1103.

Jordan, C.F., F.B. Golley, J.D. Hall and J. Hall. 1980. Nutrient scavenging of rainfall by the canopy of an Amazonian rain forest. *Biotropica* 12: 61–66.

Jordan, W.R. III, M.E. Gilpin and J.D. Aber. 1987. *Restoration Ecology: A Synthetic Approach to Ecological Research.* Cambridge: Cambridge University Press.

Judd, W.S., C.S. Campbell, E. A. Kellogg, P.F. Stevens and M.J. Donoghue. 2002. *Plant Systematics: A Phylogenetic Approach.* 2nd edn. Sunderland: Sinauer.

Judson, S. 1968. Erosion of the land, or what's happening to our continents? *American Scientist* 56: 356–374.

Junk, W.J. 1983. Ecology of swamps on the middle Amazon. pp. 269–294. In A.J.P. Gore. (ed.) *Ecosystems of the World 4B: Mires: Swamp, Bog, Fen, and Moor.* Amsterdam: Elsevier Science.

Junk, W.J. and M.T.F. Piedade. 2005. The Amazon River basin. pp. 63–117. In L.H. Fraser and P.A. Keddy (eds.) *The World's Largest Wetlands: Ecology and Conservation.* Cambridge: Cambridge University Press.

Jutila, H.M. and J.B. Grace. 2002. Effects of disturbance on germination and seedling establishment in a coastal prairie grassland: a test of the competitive release hypothesis. *Journal of Ecology* 90: 291–302.

Kagan, D. 2003. *The Peloponnesian War.* New York: Viking.

Kalliola, R., J. Salo, M. Puhakka and M. Rajasilta. 1991. New site formation and colonizing vegetation in primary succession on the western Amazon floodplains. *Journal of Ecology* 79: 877–901.

Kaplan, D.R. and T.J. Cooke. 1996. The genius of Wilhelm Hofmeister: the origin of causal-analytical research in plant development. *American Journal of Botany* 83: 1647–1660.

Kastner, T.P. and M.A. Goñi. 2003. Constancy in the vegetation of the Amazon Basin during the late Pleistocene: evidence from the organic matter composition of Amazon deep sea fan sediments. *Geology* 31: 291–294.

Kay, S. 1993. Factors affecting severity of deer browsing damage within coppiced woodlands in the south of England. *Biological Conservation* 63: 524–532.

Kearney, T.H. and H.L. Shantz. 1912. The water economy of dry-land crops. pp. 351–362. *Yearbook of the United States Department of Agriculture-1911*. Washington: Department of Agriculture.

Keddy, P.A. 1980. Population ecology in an environmental mosaic: *Cakile edentula* on a gravel bar. *Canadian Journal of Botany* 58: 1095–1100.

1981a. Why gametophytes and sporophytes are different: form and function in a terrestrial environment. *The American Naturalist* 118: 452–454.

1981b. Vegetation with Atlantic coastal plain affinities in Axe Lake, near Georgian Bay, Ontario. *The Canadian Field-Naturalist* 95: 241–248.

1981c. Experimental demography of the sand dune annual, *Cakile edentula*, growing along an environmental gradient in Nova Scotia. *Journal of Ecology* 69: 615–630.

1982. Population ecology on an environmental gradient: *Cakile edentula* on a sand dune. *Oecologia* 52: 348–355.

1983. Shoreline vegetation in Axe Lake, Ontario: effects of exposure on zonation patterns. *Ecology* 64: 331–344.

1987. Beyond reductionism and scholasticism in plant community ecology. *Vegetatio* 69: 209–211.

1989. *Competition*. London: Chapman and Hall.

1990a. The use of functional as opposed to phylogenetic systematics: a first step in predictive community ecology. pp. 387–406. In S. Kawano

(ed.) *Biological Approaches and Evolutionary Trends in Plants*. London: Academic Press.

1990b. Competitive hierarchies and centrifugal organization in plant communities. pp. 265–289. In J.B. Grace and D. Tilman (eds.) *Perspectives on Plant Competition*. San Diego: Academic Press.

1991. Biological monitoring and ecological prediction: from nature reserve management to national state of the environment indicators. pp. 249–267. In F.B. Goldsmith (ed.) *Biological Monitoring for Conservation*. London: Chapman and Hall.

1992. Assembly and response rules: two goals for predictive community ecology. *Journal of Vegetation Science* 3: 157–164.

1994. Reflections on the 21st birthday of MacArthur's Geographical Ecology: applications of the Hertzsprung–Russel star diagram to ecology. *Trends in Ecology and Evolution* 9: 231–234.

1998. Review of *Null Models in Ecology* (N.J. Gotelli and G.R. Graves, 1996, Smithsonian Institution Press. Washington). *The Canadian Field Naturalist* 112: 752–754.

2001. *Competition*. 2nd edn. Dordrecht: Kluwer.

2004. Plants matter. Review of *The Ecology of Plants* (J. Gurevitch, S. Scheiner and G.A. Fox, 2002, Sinauer Associates, Sunderland, Massachusetts.) *The Quarterly Review of Biology* 79: 55–59.

2005a. Milestones in ecological thought: a canon for plant ecology. *Journal of Vegetation Science* 16: 145–150.

2005b. Putting the plants back into plant ecology: six pragmatic models for understanding, conserving and restoring plant diversity. *Annals of Botany* 96: 177–189.

2010. *Wetland Ecology: Principles and Conservation*. 2nd edn. Cambridge: Cambridge University Press.

Keddy. P.A. and P. Constabel. 1986. Germination of ten shoreline plants in relation to seed size, soil particle size and water level: an experimental study. *Journal of Ecology* 74: 122–141.

Keddy, P.A. and C.G. Drummond. 1996. Ecological properties for the evaluation, management. and restoration of temperate deciduous forest ecosystems. *Ecological Applications* 6: 748–762.

Keddy, P.A. and L.H. Fraser. 2005. Introduction: big is beautiful. pp. 1–10. In L.H. Fraser and P.A. Keddy (eds.) *The World's Largest Wetlands: Ecology and Conservation.* Cambridge: Cambridge University Press.

Keddy, P.A. and P. MacLellan 1990. Centrifugal organization in forests. *Oikos* 59: 75–84.

Keddy, P.A. and A.A. Reznicek. 1982. The role of seed banks in the persistence of Ontario's coastal plain flora. *American Journal of Botany* 69: 13–22.

1986. Great Lakes vegetation dynamics: the role of fluctuating water levels and buried seeds. *Journal of Great Lakes Research* 12: 25–36.

Keddy, P.A. and B. Shipley. 1989. Competitive hierarchies in plant communities. *Oikos* 49: 234–241.

Keddy, P.A. and I.C. Wisheu. 1989. Ecology, biogeography, and conservation of coastal plain plants: some general principles from the study of Nova Scotian wetlands. *Rhodora* 91: 72–94.

Keddy, P.A., L. Twolan-Strutt and I.C. Wisheu. 1994. Competitive effect and response rankings in 20 wetland plants: are they consistent across three environments? *Journal of Ecology* 82: 635–643.

Keddy, P.A., K. Nielsen, E. Weiher and L.R. Lawson. 2002. Relative competitive performance of 63 species of terrestrial herbaceous plants. *Journal of Vegetation Science* 13: 5–16.

Keddy, P.A., D. Campbell, T. McFalls et al. 2007. The wetlands of lakes Pontchartrain and Maurepas: past, present and future. *Environmental Reviews* 15: 1–35.

Keddy, P.A., L. Gough, J.A. Nyman, et al. 2009. Alligator hunters, pelt traders, and runaway consumption of Gulf coast marshes: A trophic cascade perspective on coastal wetland losses. pp. 115–133. In B.R. Silliman, E.D. Grosholz and M.D. Bertness (eds.) *Human Impacts on Salt Marshes. A Global Perspective.* Berkeley: University of California Press.

Keeler, K.H. 1985. Cost:benefit models of mutualism. pp. 100–127. In D.H. Boucher (ed.) *The Biology of Mutualism: Ecology and Evolution.* New York: Oxford University Press.

Keeley, J.E. 1998. CAM photosynthesis in submerged aquatic plants. *The Botanical Review* 64: 121–175.

2007. Chaparral and fire. *Freemontia* 35(4):16–21.

Keeley, J.E. and P.W. Rundel. 2003. Evolution of CAM and C_4 carbon concentrating mechanisms. *International Journal of Plant Science* 164 (Supplement): S55–S77.

Keeley, J.E., D.A. DeMason, R. Gonzalez and K.R. Markham. 1994. Sediment-based carbon nutrition in tropical alpine *Isoetes.* pp. 167–194. In P.W. Rundel, A.P. Smith and F.C. Meinzer (eds.) *Tropical Alpine Environments: Plant Form and Function.* Cambridge: Cambridge University Press.

Keller, G., T. Adatte, W. Stinnesbeck et al. 2004. Chicxulub impact predates the K–T boundary mass extinction. *Proceedings of the National Academy of Sciences* 101: 3753–3758.

Kellman, M. 1985. Forest seedling establishment in Neotropical savannas: transplant experiments with *Xylopia frutescens* and *Calophyllum brasiliense. Journal of Biogeography* 12: 373–379.

Kellman, M. and B. Delfosse. 1993. Effect of the red land crab (*Gecarcinus lateralis*) on leaf litter in a tropical dry forest in Vera Cruz, Mexico. *Journal of Tropical Ecology* 9: 55–65.

Kellman, M. and M. Kading. 1992. Facilitation of tree seedling establishment in a sand dune succession. *Journal of Vegetation Science* 3: 679–688.

Kellman, M. and N. Roulet. 1990. Nutrient flux and retention in a tropical sand-dune succession. *Journal of Ecology* 78: 664–676.

Kenrick, P. and P.R. Crane. 1997a. *The Origin and Early Diversification of Land Plants: A Cladistic Study.* Washington: Smithsonian Institution Press.

1997b. The origin and early evolution of plants on land. *Nature* **389**: 33–39.

Kershaw, K.A. 1962. Quantitative ecological studies from Landmannahellir, Iceland. *Journal of Ecology* **50**: 171–179.

1973. *Quantitative and Dynamic Plant Ecology.* 2nd edn. London: Edward Arnold.

Kershaw, K.A. and J.H.H. Looney. 1985. *Quantitative and Dynamic Plant Ecology.* 3rd edn. Victoria: Edward Arnold.

Kevan, P.G. 1975. Sun-tracking solar furnaces in high arctic flowers: significance for pollination and insects. *Science* **189**: 723–726.

Kidston, R. and W.H. Lang. 1921. On Old Red Sandstone plants showing structure, from the Rhynie Chert Bed, Aberdeenshire, Part 4: Restorations of the vascular cryptogams, and discussion of their bearing on the general morphology of the Pteridophyta and the origin of the organization of land-plants. *Transactions of the Royal Society of Edinburgh* **52**: 831–854.

Killingbeck, K.T. 1996. Nutrients in senesced leaves: keys to the search for potential resorption and resorption efficiency. *Ecology* **77**: 1716–1727.

King, J. 1997. *Reaching for the Sun: How Plants Work.* New York: Cambridge University Press.

Kingsland, S.E. 1995. *Modeling Nature: Episodes in the History of Population Ecology.* Chicago: University of Chicago Press.

Kinzig, A.P., S. Pacala and G.D. Tilman (eds.) 2002. *The Functional Consequences of Biodiversity: Empirical Progress and Theoretical Extensions.* Princeton: Princeton University Press.

Kleier, C. and P.W. Rundel. 2004. Microsite requirements, population structure and growth of the cushion plant *Azorella compacta* in the tropical Chilean Andes. *Austral Ecology* **29**: 461–470.

Knoll, A.H. 1992. The early evolution of eukaryotes: a geological perspective. *Science* **256**: 622–627.

Koch, P.L. and A.D. Barnosky. 2006. Late Quaternary extinctions: state of the debate. *Annual Review of Ecology, Evolution and Systematics* **37**: 215–250.

Koerselman, W. and A.F.M. Meulman. 1996. The vegetation N:P ratio: a new tool to detect the nature of nutrient limitation. *Journal of Applied Ecology* **33**: 1441–1450.

Koyama, H. and T. Kira. 1956. Intraspecific competition among higher plants. VIII. Frequency distributions of individual plant weight as affected by the interaction between plants. *Journal of the Institute of Polytechnics, Osaka City University Series D* **7**: 73–94.

Kozlowski, T.T. (ed.) 1984. *Flooding and Plant Growth.* Orlando: Academic Press.

Kozlowski, T.T. and S.G. Pallardy. 1984. Effect of flooding on water, carbohydrate, and mineral relations. pp. 165–193. In T.T. Kozlowski (ed.) *Flooding and Plant Growth.* Orlando: Academic Press.

Kramer, P.J. 1983. *Water Relations of Plants.* Orlando: Academic Press.

Krause, T and H. Zambonino. 2013. More than just trees: animal species diversity and participatory forest monitoring in the Ecuadorian Amazon. *International Journal of Biodiversity Science, Ecosystem Services & Management* **9**: 225–238

Krebs, C.J. 1978. *Ecology: The Experimental Analysis of Distribution and Abundance.* New York: Harper and Row.

1989. *Ecological Methodology.* New York: Harper and Row.

Kruckeberg, A.R. 1954. The ecology of serpentine soils. III. Plant species in relation to serpentine soils. *Ecology* **35**: 267–274.

Küchler, A.W. 1949. A physiognomic classification of vegetation. *Annals of the Association of American Geographers* **39**: 201–210.

1966. Analyzing the physiognomy and structure of vegetation. *Annals of the Association of American Geographers* **56**: 112–127.

Kuhry, P. 1994. The role of fire in the development of *Sphagnum*-dominated peatlands in western boreal Canada. *Journal of Ecology* **82**: 899–910.

Kuijt, J. 1969. *The Biology of Parasitic Flowering Plants*. Berkeley: University of California Press.

Kyte, F.T. 1998. A meteorite from the Cretaceous/Tertiary boundary. *Nature* 396: 237–239.

Lack, D. 1947. *Darwin's Finches: An Essay on the General Biological Theory of Evolution*. New York: Harper and Row.

Laing, H.E. 1940. Respiration of the rhizomes of *Nuphar advenum* and other water plants. *The American Journal of Botany* 27: 574–581.

1941. Effect of concentration of oxygen and pressure of water upon growth of rhizomes of semi-submerged water plants. *Botanical Gazette* 102: 712–724.

Lamont, B.B, D.C. Le Maitre, R.M. Cowling and N.J. Enright. 1991. Canopy seed storage in woody plants. *Botanical Review* 57: 277–317.

Langenheim, J.H. 2003. *Plant Resins: Chemistry, Evolution, Ecology, Ethnobotany*. Portland: Timber Press.

Langer, P. 1974. Stomach evolution in the *Artiodactyla*. *Mammalia* 38: 295–314.

Larcher, W. 1995. *Physiological Plant Ecology: Ecophysiology and Stress Physiology of Functional Groups*. 3rd edn. New York: Springer-Verlag.

2003. *Physiological Plant Ecology: Ecophysiology and Stress Physiology of Functional Groups*. 4th edn. Berlin: Springer-Verlag.

Larson, D.W. 1980. Patterns of species distribution in an *Umbilicaria*-dominated community. *Canadian Journal of Botany* 58: 1269–1279.

1982. Environmental stress and *Umbilicaria* lichens: the effect of subzero temperature pretreatments. *Oecologia* 55: 268–278.

1989. The impact of ten years at −20°C on gas exchange in five lichen species. *Oecologia* 78: 87–92.

1996. Brown's Woods: an early gravel pit forest restoration project, Ontario, Canada. *Society for Ecological Restoration* 4: 11–18.

2001. The paradox of great longevity in a short-lived tree species. *Experimental Gerontology* 36: 651–673.

Latham, P.J., L.G. Pearlstine and W.M. Kitchens. 1994. Species association changes across a gradient of freshwater, oligohaline, and mesohaline tidal marshes along the lower Savanna River. *Wetlands* 14: 174–183.

Latham, R.E. and R.E. Ricklefs. 1993a. Global patterns of tree species richness in moist forests: energy-diversity theory does not account for variation in species richness. *Oikos* 67: 325–333.

1993b. Continental comparisons of temperate zone tree species diversity. pp. 294–314. In R.E. Ricklefs and D. Schluter (eds.) *Species Diversity in Ecological Communities: Historical and Geographical Perspectives*. Chicago: The University of Chicago Press.

Latham, R.E., J. Beyea, M. Benner et al. 2005. *Managing White-tailed Deer in Forest Habitat from an Ecosystem Perspective: Pennsylvania Case Study*. Harrisburg: Audubon Pennsylvania and Pennsylvania Habitat Alliance.

Laughlin, D. and D. Laughlin. 2013. Advances in modelling trait-based plant community assembly. *Trends in Plant Science* 18: 584–593.

Laughlin, D.C., P.Z. Fulé, D.W. Huffman, J. Crouse and E. Laliberté. 2011. Climatic constraints on trait-based forest assembly. *Journal of Ecology* 99: 1489–1499.

Laurance, W.F. (and 11others). 2014. A global strategy for road building. *Nature* 513: 229–232.

Lavoisier, A.L. 1789. *Elements of Chemistry*. Translated by R. Kerr and reprinted in xi, xii and pp. 1–60. In M.J. Adler (ed.) 1990. *Great Books of the Western World*. 2nd edn., Vol. 42. Chicago: Encyclopaedia Britannica Inc.

Lawler, A. 2005. Reviving Iraq's wetlands. *Science* 307: 1186–1189.

Laws, B. 2010. *Fifty Plants that Changed the Course of History*. Richmond Hill, Ontario: Firefly.

Lechowicz, M.J. 1981. The effects of climatic pattern on lichen productivity: *Cetraria cucullata* (Bell.) Ach. in the arctic tundra of northern Alaska. *Oecologia* 50: 210–216.

Leck, M.A. and K.J. Graveline. 1979. The seed bank of a freshwater tidal marsh. *American Journal of Botany* 66: 1006–1015.

Leck, M.A., V.T. Parker and R.L. Simpson (eds.) 1989 *Ecology of Soil Seed Banks*. San Diego: Academic Press.

Lee, K.E. 1985. *Earthworms: Their Ecology and Relationships with Soils and Land Use*. Sydney: Academic Press.

Lefkovitch, L.P. 1965. The study of population growth in organisms grouped by stages. *Biometrics* 21: 1–18.

Legendre, L. and P. Legendre. 1983. *Numerical Ecology*. Amsterdam: Elsevier.

Leigh E.G. Jr. 1999. *Tropical Forest Ecology: A View From Barro Colorado Island*. Oxford: Oxford University Press.

Leopold, A. 1949. *A Sand County Almanac*. London: Oxford University Press.

Le Page, C. and P.A. Keddy. 1988. Reserves of buried seeds in beaver ponds. *Wetlands* 18: 242–248.

Leschine, S.B. 1995. Cellulose degradation in anaerobic environments. *Annual Review of Microbiology* 49: 399–426.

Leslie, P.H. 1945. The use of matrices in certain population mathematics. *Biometrika* 33: 183–212.

Levin, H.L. 1994. *The Earth Through Time*. 4th edn. Fort Worth: Saunders College Publishing; Harcourt Brace College Publishers.

Levins, R. 1968. *Evolution in Changing Environments*. Princeton: Princeton University Press.

Levitt, J. 1977. The nature of stress injury and resistance. pp. 11–21. In J. Levitt (ed.) *Responses of Plants to Environmental Stress*. New York: Academic Press.

1980. *Responses of Plants to Environmental Stresses*, Vols. I and II. 2nd edn. New York: Academic Press.

Lewis, III, R.R. (ed.) 1982. *Creation and Restoration of Coastal Plant Communities*. Boca Raton: CRC Press.

Lewis, D.H. 1987. Evolutionary aspects of mutualistic associations between fungi and photosynthetic organisms. pp. 161–178. In

A.D.M. Rayner, C.M. Brasier and D. Moore (eds.) *Evolutionary Biology of the Fungi*. Symposium of the British Mycological Society, held at the University of Bristol, April 1986. Cambridge: Cambridge University Press.

Leyser, O. and Fitter, A. 1998. Roots are branching out in patches. *Trends in Plant Science* 3: 203–204.

Li, X.-L., E. George and H. Marschner. 1991. Phosphorus depletion and pH decrease at the root–soil and hyphae–soil interfaces of VA mycorrhizal white clover fertilized with ammonium. *New Phytologist* 119: 397–404.

Lieth, H. 1975. Historical survey of primary productivity research. pp. 7–16. In H. Leith and R.H. Whittaker (eds.) *Primary Productivity of the Biosphere*. New York: Springer-Verlag.

Likens, G.E., F.H. Bormann, R.S. Pierce, J.S. Eaton and N.M. Johnson. 1977. *Biogeochemistry of a Forested Ecosystem*. New York: Springer-Verlag.

Lindroth, R.L. 1989. Chemical ecology of the luna moth: effects of host plant on detoxification enzyme activity. *Journal of Chemical Ecology* 15: 2019–2029.

Little, C.E. 1995. *The Dying of the Trees: The Pandemic in America's Forests*. New York: Penguin Books.

Liu, K. and M.L. Fearn. 2000. Holocene history of catastrophic hurricane landfalls along the Gulf of Mexico coast reconstructed from coastal lake and marsh sediments. pp. 38–47. In Z.H. Ning and K.K. Abdollhai (eds.) *Implications of Global Change for the Gulf Coast Region of the United States*. Baton Rouge, LA: Franklin Press, Gulf Coast Regional Climate Change Council.

Llewellyn, D.W., G.P. Shaffer, N.J. Craig et al. 1996. A decision-support system for prioritizing restoration sites on the Mississippi River alluvial plain. *Conservation Biology* 10: 1446–1455.

Lloyd, D.G. and S.C.H. Barrett (eds.) 1996. *Floral Biology: Studies on Floral Evolution in Animal-Pollinated Plants*. London: Chapman and Hall.

Lockwood, J.L. and S.L. Pimm. 1997. When does restoration succeed? pp. 363–392. In E. Weiher and P. Keddy (eds.) *Ecological Assembly Rules:*

Perspectives, Advances, Retreats. Cambridge, UK: Cambridge University Press.

Lodge, D.M. 1991. Herbivory on freshwater macrophytes. *Aquatic Botany* 41: 195–224.

Loehle, C. 1998a. Height growth rate tradeoffs determine northern and southern range limits for trees. *Journal of Biogeography* 25: 735–742.

1988b. Problems with the triangular model for representing plant strategies. *Ecology* 69: 284–286.

1995. Anomalous responses of plants to CO_2 enrichment. *Oikos* 73: 181–187.

Loreau, M. L. 2000. Biodiversity and ecosystem functioning: recent theoretical advances. *Oikos* 91: 3–17.

Louda, S.M. and S. Mole. 1991. Glucosinolates: chemistry and ecology. pp. 124–164. In G.A. Rosenthal and M.R. Berenbaum (eds.) *Herbivores: Their Interactions with Secondary Plant Metabolites*. San Diego: Academic Press.

Louda, S.M., K.H. Keller and R.D. Holt. 1990. Herbivore influence on plant performance and competitive interactions. pp. 413–444. In J.B. Grace and D. Tilman (eds.) *Perspectives on Plant Competiton*. San Diego: Academic Press.

Lovejoy, T.E., R.O. Bierregaard, Jr., A.B. Rylands et al. 1986. Edge and other effects of isolation on Amazon forest fragments. pp. 257–285. In M.E. Soulé (ed.) *Conservation Biology: the Science of Scarcity and Diversity*. Sunderland: Sinauer Associates.

Loveless, C.M. 1959. A study of the vegetation in the Florida Everglades. *Ecology* 40: 1–9.

Lovelock, J. 1979. *Gaia: A New Look at Life on Earth*. Oxford: Oxford University Press.

Lowman, M.D. 1992. Leaf growth dynamics and herbivory in five species of Australian rain-forest canopy trees. *Journal of Ecology* 80: 433–447.

Lowman, M.D. and H.B. Rinker (eds). 2004. *Forest Canopies*. 2nd edn. Burlington: Elsevier Academic Press.

Ludwig, D., D.D. Jones and C.S. Holling. 1978. Qualitative analysis of insect outbreak systems: the spruce budworm and forest. *Journal of Animal Ecology* 47: 315–332.

Lugo, A.E. and S.C. Snedaker. 1974. The ecology of mangroves. *Annual Review of Ecology and Systematics* 5: 39–64.

Lundberg, J. and K. Bremer. 2003. A phylogenetic study of the order Asterales using one morphological and three molecular data sets. *International Journal of Plant Sciences* 164: 553–578.

Lutman, J. 1978. The role of slugs in an *Agrostis-Festuca* grassland. pp. 332–347. In O.W. Heal and D.F. Perkins (eds.) *Production Ecology of British Moors and Montane Grasslands*. Ecological Studies, Vol. 27. Berlin: Springer-Verlag.

MacArthur, R.H. 1957. On the relative abundance of bird species. *Proceedings of the National Academy of Sciences* 43: 293–295.

1972. *Geographical Ecology*. New York: Harper and Row.

MacArthur, R.H. and E.O. Wilson. 1967. *The Theory of Island Biogeography*. Monographs in Population Biology, No. 1. Princeton: Princeton University Press.

MacDonald, P. (ed.) 1989. *The Solar System*. The World of Science, Vol. 7. Oxford: Equinox (Oxford) Ltd.

MacFarland, C.G., J. Villa and B. Toro. 1974. The Galápagos giant tortoises (*Geochelone elephantopus*). Part I: Status of the surviving populations. *Biological Conservation* 6: 118–133.

MacMahon, J.A. 1981. Successional processes: comparisons among biomes with special reference to probable roles of and influence on animals. pp. 277–305. In D.C. West, H.H. Shugart and D.B. Botkin (eds.) *Forest Succession: Concepts and Application*. New York: Springer-Verlag.

Magnuson, J.J., H.A. Regier, W.J. Christie and W.C. Sonzongi. 1980. To rehabilitate and restore Great Lakes ecosystems. pp. 95–122. In J. Cairns, Jr. (ed.) *The Recovery Process in Damaged Ecosystems*. Ann Arbour: Ann Arbour Science Publishers.

Magnusson, M. (ed.) 1990. *Chambers Biographical Dictionary*. 5th edn. Edinburgh: W & R Chambers.

Mains, G. 1972. *The Oxygen Revolution*. London: David and Charles.

Major, J. 1988. Endemism: a botanical perspective. pp. 117–146. In A.A. Myers and P.S. Giller (eds.) *Analytical Biogeography*. London: Chapman and Hall.

Margulis, L. 1993. *Symbiosis in Cell Evolution*. 2nd edn. New York: W. H. Freeman.

Margulis, L. and D. Sagan. 1986. *Microcosmos: Four Billion Years of Evolution from Our Microbial Ancestors*. Berkeley: University of California Press.

Marquis, R. 1991. Evolution of resistance in plants to herbivores. *Evolutionary Trends in Plants* 5: 23–29.

Marquis, R. and C. Whelan. 1994. Insectivorous birds increase growth of white oak through consumption of leaf-chewing insects. *Ecology* 75: 2007–2014.

Marschner, H. 1995. *Mineral Nutrition of Higher Plants*. 2nd edn. London: Academic Press.

Martin, D. and J. Bohlman. 2005. Molecular biochemistry and genomics of terpenoid defenses in conifers. pp. 29–56. In J.T. Romeo (ed.) *Chemical Ecology and Phytochemistry of Forest Ecosystems*. Oxford: Elsevier.

Martin, J.H. (and 43 others). 1994. Testing the iron hypothesis in ecosystems of the equatorial Pacific Ocean. *Nature* 371: 123–129.

Martin, P.S. and R.J. Klein. 1984. *Quaternary Extinctions: A Prehistoric Revolution*. Tucson: The University of Arizona Press.

Marx, K. 1867. In F. Engles (ed.) *Capital*. Translated from 3rd German edition by S. Moore and E. Aveling. Revised from 4th edition by M. Sachey and H. Lamm. pp. 1–411. In M.J. Adler (ed.) 1990. *Great Books of the Western World*. 2nd edn., Vol. 50. Chicago: Encyclopaedia Britannica Inc.

Maser, C., A.W. Claridge and J.M. Trappe. 2008. *Trees, Truffles, and Beasts: How Forests Function*. Piscataway: Rutgers University Press.

Matson, P.A., K. Lohse and S. Hall. 2002. The globalization of nitrogen deposition: consequences for terrestrial ecosystems. *Ambio* 31: 113–119.

Matthes-Sears, U., J. Gerrath and D. Larson. 1997. Abundance, biomass, and productivity of endolithic and epilithic lower plants on the temperate-zone cliffs of the Niagara Escarpment, Canada. *International Journal of Plant Science* 158: 451–460.

Maun, M.A. and J. Lapierre. 1986. Effects of burial by sand on seed germination and seedling emergence of four dune species. *American Journal of Botany* 73: 450–455.

May, E. 1982. *Budworm Battles: The Fight to Stop the Aerial Insecticide Spraying of the Forests of Eastern Canada*. Halifax: Four East Publications Ltd.

1988. How many species are there on Earth? *Science* 241: 1441–1449.

May, R.M. 1973. *Stability and Complexity in Model Ecosystems*. Princeton: Princeton University Press.

1977. Thresholds and breakpoints in ecosystems with a multiplicity of stable states. *Nature* 269: 471–477.

1981. Patterns in multi-species communities. pp. 197–227. In R.M. May (ed.) *Theoretical Ecology*. Oxford: Blackwell.

Mayr, E. 1982. *The Growth of Biological Thought: Diversity, Evolution, and Inheritance*. Cambridge: Belknap Press of Harvard University Press.

McCanny, S.J., P.A. Keddy, T.J. Arnason, et al. 1990. Fertility and the food quality of wetland plants: a test of the resource availability hypothesis. *Oikos* 59: 373–381.

McCarthy, K.A. 1987. *Spatial and Temporal Distributions of Species in Two Intermittent Ponds in Atlantic County, N.J.* MSc thesis. New Brunswick: Rutgers University.

McClaran, M.P. 2003. A century of vegetation change on the Santa Rita Experimental Range. pp. 16–33. In *USDA Forest Service Proceedings RMRS-P-30*. US Department of Agriculture.

McClure, J.W. 1970. Secondary constituents of aquatic angiosperms. pp. 233–265. In J.B. Harborne (ed.) *Phytochemical Phylogeny*. New York: Academic Press.

McComb, W.C. and R.N. Muller. 1983. Snag management in old growth and second-growth Appalachian forests. *Journal of Wildlife Management* 47: 376–382.

McFalls, T., P.A. Keddy, D. Campbell and G. Shaffer. 2010. Hurricanes, floods, levees, and nutria: vegetation responses to interacting disturbance and fertility regimes with implications for coastal wetland restoration. *Journal of Coastal Research* 26: 901–911.

McGraw, J.B. and M.A. Furedi. 2005. Deer browsing and population viability of a forest understory plant. *Science* 307: 920–922.

McIntosh, R.P. 1967. The continuum concept of vegetation. *The Botanical Review* 33: 130–187.

1981. Succession and ecological theory. pp. 10–23. In D.C. West, H.H. Shugart and D.B. Botkin (eds.) *Forest Succession: Concepts and Application.* New York: Springer-Verlag.

1985. *The Background of Ecology: Concept and Theory.* Cambridge: Cambridge University Press.

McKenzie, D.H., D.E. Hyatt and V.J. McDonald. 1992. *Ecological Indicators*, Vols. 1 and 2. London: Elsevier.

McNaughton, S.J. 1985. Ecology of a grazing ecosystem: the Serengeti. *Ecological Monographs* 55: 259–294.

McNaughton, S.J., R.W. Ruess and S.W. Seagle. 1988. Large mammal and process dynamics in African ecosystems. *Bioscience* 38: 794–800.

McNeill, J.R. and V. Winiwarter. 2004. Breaking the sod: humankind, history and soil. *Science* 304: 1627–1629.

McShea, W.J. and J.H. Rappole. 2000. Managing the abundance and diversity of breeding bird populations through manipulation of deer populations. *Conservation Biology* 14: 1161–1170.

McVaugh, R. 1943. The vegetation of the granitic flat-rocks of the southeastern United States. *Ecological Monographs* 13: 119–166.

Mead, R. 1988. *The Design of Experiments: Statistical Principles for Practical Application.* Cambridge: Cambridge University Press.

Meadows, D.H., D.L. Meadows, J. Randers and W.W. Behrens III. 1974. *The Limits to Growth: A Report for the Club of Rome's Project on the Predicament of Mankind.* 2nd edn. New York: The New American Library.

Meave, J. and M. Kellman. 1994. Maintenance of rain forest diversity in riparian forest of tropical savannas: implications for species conservation during Pleistocene drought. *Journal of Biogeography* 21: 121–135.

Merckx, V., M.I. Bidartondo and N.A. Hynson. 2009. Myco-heterotrophy: when fungi host plants. *Annals of Botany* 104: 1255–1261.

Meredith, E.T. 1920. The life of the naval stores industry as at present carried on in the south. Report of the Secretary of Agriculture Meredith to the United States Senate. pp. 89–90. In T. Gamble (ed.) *Naval Stores. History, Production, Distribution and Consumption.* Savannah, GA: Review Publishing and Printing Company.

Merrens, E.J. and D.R. Peart. 1992. Effects of hurricane damage on individual growth and stand structure in a hardwood forest in New Hampshire, USA. *Journal of Ecology* 80: 787–795.

Milchunas, D.G. and W.K. Lauenroth. 1993. Quantitative effects of grazing on vegetation and soils over a global range of environments. *Ecological Monographs* 63: 327–366.

Milchunas, D.G., W.K. Laurenroth and I.C. Burk. 1998. Livestock grazing: animal and plant biodiversity of shortgrass steppe and the relationship to ecosystem function. *Oikos* 83: 65–74.

Miller, G.R. and A. Watson. 1983. Heather moorland in northern Britain. pp. 101–117. In A. Warren and F.B. Goldsmith (eds.) *Conservation in Perspective.* Chichester: John Wiley and Sons Ltd.

Miller, K.G., M.A. Kominz, J.V. Browning et al. 2005. The Phanerozoic record of global sea-level change. *Science* 310: 1293–1298.

Miller, R.S. 1967. Pattern and process in competition. *Advances in Ecological Research* 4: 1–74.

Miller, S.L. 1953. A production of amino acids under possible primitive earth conditions. *Science* 117: 528–529.

Milliman, J.D. and R.H. Meade. 1983. World-wide delivery of river sediment to the oceans. *Journal of Geology* 91: 1–21.

Milton, S.J., W.R.J. Dean, M.A. du Plessis and W.R. Siegfried. 1994. A conceptual model of arid rangeland degradation. *Bioscience* **44**: 70–76.

Mitchley, J. 1988. Control of relative abundance of perennials in chalk grassland in southern England. II. Vertical canopy structure. *Journal of Ecology* **76**: 341–350.

Mitchley, J. and P.J. Grubb. 1986. Control of relative abundance of perennials in chalk grassland in southern England. I. Constancy of rank order and results of pot- and field-experiments on the role of interference. *Journal of Ecology* **74**: 1139–1166.

Mitsch, W.J. and J.G. Gosselink. 2000. *Wetlands*. 3rd edn. New York: John Wiley & Sons.

Mitsch, W.J. and S.E. Jørgensen. 1990. Modelling and management. pp. 727–744. In B.C. Patten (ed.) *Wetlands and Shallow Continental Water Bodies*, Vol. 1. The Hague: SPB Academic Publishing.

Moffett, M.W. 1994. *The High Frontier: Exploring the Tropical Rainforest Canopy*. Cambridge: Harvard University Press.

Moles, A.T., D.D. Ackerly, C.O. Webb, et al. 2005. A brief history of seed size. *Science* **307**: 576–580.

Molisch, H. 1937. *Der Einfluss einer Pflanze auf die andere. Allelopathie*. Jena: Gustav Fischer.

Moolman, H.J. and R.M. Cowling. 1994. The impact of elephant and goat grazing on the endemic flora of South African succulent thicket. *Biological Conservation* **68**: 53–61.

Mooney, H.A. and E.L. Dunn. 1970. Convergent evolution of Mediterranean climate evergreen sclerophyll shrubs. *Evolution* **24**: 292–303.

Mooney, H.A., B.G. Drake, R.J. Luxmoore, W.C. Oechel and L.F. Pitelka. 1991. Predicting ecosystem responses to elevated CO_2 concentrations. *Bioscience* **41**: 96–104.

Moore, B. and B. Bolin. 1987. The oceans, carbon dioxide and global climate change. *Oceanus* **29**: 9–15.

Moore, D.R.J. 1990. *Pattern and Process in Wetlands of Varying Standing Crop: the Importance of Scale*. PhD thesis. Ottawa: University of Ottawa.

1998. The ecological component of ecological risk assessment: lessons from a field experiment. *Human and Ecological Risk Assessment* **4**: 1103–1123.

Moore, D.R.J. and P.A. Keddy. 1989a. The relationship between species richness and standing crop in wetlands: the importance of scale. *Vegetatio* **79**: 99–106.

1989b. Infertile wetlands: conservation priorities and management. pp. 391–397. In M.J. Bardecki and N. Patterson (eds.) *Wetlands: Inertia or Momentum*. Proceedings of a Conference held in Toronto, Ontario, 21–22 October 1988. Federation of Ontario Naturalists.

Moore, D.R.J. and R.W. Wein. 1977. Viable seed populations by soil depth and potential site recolonization after disturbance. *Canadian Journal of Botany* **55**: 2408–2412.

Moore, D.R.J., P.A. Keddy, C.L. Gaudet and I.C. Wisheu. 1989. Conservation of wetlands: do infertile wetlands deserve a higher priority? *Biological Conservation* **47**: 203–217.

Moore, P.D., J.A. Webb and M.E. Collinson. 1991. *Pollen Analysis*. London: Blackwell Scientific.

Moreno, M.T., J.I. Cubero, D. Berner, D. Joel, L.J. Musselman and C. Parker (eds.) 1996. *Advances in Parasitic Plant Research*. Cordoba: Junta de Andalucia, Dirección General de Investigación Agraria.

Mori, S.A. 1992. The Brazil nut industry: past, present, and future. pp. 241–251. In M. Plotkin and L. Famolare (eds.) *Sustainable Harvest and Marketing of Rain Forest Products*. Washington: Island Press.

Morowitz, H.J. 1968. *Energy Flow in Biology: Biological Organization as a Problem in Thermal Physics*. New York: Academic Press.

Morris, E.C. and P.J. Myerscough. 1991. Self-thinning and competition intensity over a gradient of nutrient availability. *Journal of Ecology* **79**: 903–923.

Morris, W.F. and D.F Doak. 2002. *Quantitative Conservation Biology: Theory and Practice of Population Viability Analysis*. Sunderland: Sinauer.

Moss, B. 1983. The Norfolk Broadland: experiments in the restoration of a complex wetland. *Biological Reviews of the Cambridge Philosophical Society* **58**: 521–561.

    1984. Medieval man-made lakes: progeny and casualties of English social history, patients of twentieth century ecology. *Transactions of the Royal Society of South Africa* **45**: 115–128.

Mueller-Dombois, D. 1987. Natural dieback in forests. *Bioscience* **37**: 575–583.

Mueller-Dombois, D. and H. Ellenberg. 1974. *Aims and Methods of Vegetation Ecology*. New York: John Wiley and Sons.

Muller, C.H. 1966. The role of chemical inhibition (allelopathy) in vegetational composition. *Bulletin of the Torrey Botanical Club* **93**: 332–351.

    1969. Allelopathy as a factor in ecological process. *Vegetatio* **18**: 348–357.

Muller, R.A. and G.J. MacDonald. 1997. Glacial cycles and astronomical forcing. *Science* **277**: 215–218.

Müller, J., G. Irion, J.N. de Mello and W.J. Junk. 1995. Hydrological changes of the Amazon during the last glacial-interglacial cycle in Central Amazonia (Brazil). *Naturwissenschaften* **82**: 232–235.

Mutke, J. and W. Barthlott. 2005. Patterns of vascular plant diversity at continental to global scales. *Biologiske Skrifter* **55**: 521–531.

Myers, N. 1988. Threatened biotas: "hotspots" in tropical forests. *Environmentalist* **8**: 1–20.

Myers, N., R.A. Mittermeier, C.G. Mittermeier, G.A.B. da Fonseca and J. Kent. 2000. Biodiversity hotspots for conservation priorities. *Nature* **403**: 853–858.

Myers, R., J. O'Brien and S. Morrison. 2006. *Fire Management Overview of the Caribbean Pine* (Pinus caribaea) *Savannas of the Mosquitia, Honduras. GFI Technical Report 2006-1b.* Arlington: The Nature Conservancy.

Nabokov, P. 1993. Long threads. pp. 301–383. In B. Ballantine and I. Ballantine (eds.) *The Native Americans: An Illustrated History*. Atlanta: Turner Publishing Inc.

Naeem, S. 2002. Ecosystem consequences of biodiversity loss: the evolution of a paradigm. *Ecology* **83**: 1537–1552.

Naeem, S.L., J. Thompson, S.P. Lawler, J.H. Lawton and R.M. Woodfin. 1994. Declining biodiversity can alter the performance of ecosystems. *Nature* **368**: 734–737.

Naiman, R.J., C.A. Johnston and J.C. Kelley. 1988. Alteration of North American streams by beaver. *Bioscience* **38**: 753–762.

Nanson, G.C. and H.F. Beach. 1977. Forest succession and sedimentation on a meandering-river floodplain, northeast British Columbia, Canada. *Journal of Biogeography* **4**: 229–251.

Nantel, P. and P. Neuman. 1992. Ecology of ectomycorrhizal-basidiomycete communities on a local vegetation gradient. *Ecology* **73**: 99–117.

Nantel, P., D. Gagnon and A. Nault. 1996. Population viability analysis of American ginseng and wild leek harvested in stochastic environments. *Conservation Biology* **10**: 608–621.

Newman, E.I. 1993. *Applied Ecology*. Oxford: Blackwell Scientific Publications.

Newman, M.E.J. 2003. The structure and function of complex networks. *SIAM Review* **45**: 167–256.

Newman, S., J.B. Grace and J.W. Koebel. 1996. Effects of nutrients and hydroperiod on *Typha, Cladium* and *Eleocharis*: implications for Everglades restoration. *Ecological Applications* **6**: 774–783.

Nicholls, H. 2013. The 18-km^2 rat trap. *Nature* **497**: 306–308.

Nicholson, A. and P.A. Keddy. 1983. The depth profile of a shoreline seed bank in Matchedash Lake, Ontario. *Canadian Journal of Botany* **61**: 3293–3296.

Nichols, S.A. 1999. Floristic quality assessment of Wisconsin lake plant communities with example applications. *Journal of Lake and Reservoir Management* **15**: 133–141.

Nickrent, D.L. 2006. The parasitic plant connection. (www.parasiticplants.siu/index.htm) accessed 10 Nov. 2006.

Nicolson, M. and R.P. McIntosh. 2002. H. A. Gleason and the individualistic hypothesis revisited. *Bulletin of the Ecological Society of America* **83**: 133–142.

Nicotra, A.B., A. Leigh, C.K. Boyce et al. 2011. The evolution and functional significance of leaf shape in the angiosperms. *Functional Plant Biology* 38: 535–552.

Nielsen, K.B., R. Kjøller, P.A. Olsson, et al. 2004. Colonisation and molecular diversity of arbuscular mycorrhizal fungi in the aquatic plants *Littorella uniflora* and *Lobelia dortmanna* in southern Sweden. *Mycological Research* 108: 616–625.

Niering, W.A. and R.S. Warren. 1980. Vegetation patterns and processes in New England salt marshes. *Bioscience* 30: 301–307.

Niklas, K.J. 1994. Predicting the height of fossil plant remains: an allometric approach to an old problem. *American Journal of Botany* 81: 1235–1242.

Niklas, K.J., B.H. Tiffney and A.H. Knoll. 1983. Patterns in vascular land plant diversification. *Nature* 303: 614–616.

1985. Patterns in vascular plant diversification: an analysis at the species level. pp. 97–128. In J.W. Valentine (ed.) *Phanerozoic Diversity Pattern: Profiles in Macroevolution*. Princeton: Princeton University Press.

Nobel, P.S. 1976. Water relations and photosynthesis of a desert CAM plant, *Agave deserti*. *Plant Physiology* 58: 576–582.

1977. Water relations and photosynthesis of a barrel cactus, *Ferocactus acanthodes*, in the Colorado Desert. *Oecologia* 27: 117–133.

1985. Desert succulents. pp. 181–197. In B.F. Chabot and H.A. Mooney (eds.) *Physiological Ecology of North American Plant Communities*. London: Chapman and Hall.

Noble, I.R. and R.O. Slatyer. 1980. The use of vital attributes to predict successional changes in plant communities subject to recurrent disturbances. *Vegetatio* 43: 5–21.

Noss, R. 1995. *Maintaining Ecological Integrity in Representative Reserve Networks*. A World Wildlife Fund Canada/United States Discussion Paper, WWF.

Noss, R.F. and A.Y. Cooperrider. 1994. *Saving Nature's Legacy*. Washington, D.C.: Island Press.

Noy-Meir, I. 1973. Desert ecosystems: environment and producers. *Annual Review of Ecology and Systematics* 4: 25–51.

Noy-Meir, L. 1975. Stability of grazing systems: an application of predator–prey graphs. *Journal of Ecology* 63: 459–481.

Oakes, E.H. 2002. *Notable Scientists. A to Z of Chemists*. New York: Facts on File.

Ocampo, J.A. 1986. Vesicular-arbuscular mycorrhizal infection of "host" and "non-host" plants: effect on the growth responses of the plants and competition between them. *Soil Biology and Biochemistry* 18: 607–610.

Ogden, J.G. and M.J. Harvey. 1975. *Environmental Change in the Maritimes*. Halifax: Nova Scotia Institute of Science.

Okihana, H. and C. Ponnamperuma. 1982. A protective function of the coacervates against UV light on the primitive Earth. *Nature* 299: 347–349.

Oksanen, L. 1990. Predation, herbivory, and plant strategies along gradients of primary production. pp. 445–474. In J.B. Grace and D. Tilman (eds.) *Perspectives on Plant Competition*. San Diego: Academic Press.

Oksanen, L., S.D. Fretwell, J. Arruda and P. Niemelä. 1981. Exploitation ecosystems in gradients of primary productivity. *The American Naturalist* 118: 240–261.

Oksanen, L., M. Aunapuu, T. Oksanen et al. 1997. Outlines of food webs in a low arctic tundra landscape in relation to three theories on trophic dynamics. pp. 351–373. In A.C. Gange and V.K. Brown (eds.) *Multitrophic Interactions in Terrestrial Systems*. The 36th Symposium of The British Ecological Society. Oxford: Blackwell Science.

Olson, D.M., E. Dinerstein, E.D. Wikramanayake et al. 2001. Terrestrial ecoregions of the world: a new map of life on Earth. *Bioscience* 51: 933–938.

Ondok, J.P. 1990. Modelling ecological processes. pp. 659–89. In B.C. Patten (ed.) *Wetlands and Shallow Continental Water Bodies*, Vol. 1. The Hague: SPB Academic Publishing.

Oosting, H.J. 1956. *The Study of Plant Communities*, 2nd edn. San Francisco: W. H. Freeman.

Oparin, A.I. 1938. *The Origin of Life*. New York: The Macmillan Company. Translated by S. Morgulis.

Orloci, L. 1978. *Multivariate Analysis in Vegetation Research*. 2nd edn. The Hague: Junk.

Orson, R.A., R.L. Simpson and R.E. Good. 1990. Rates of sediment accumulation in a tidal freshwater marsh. *Journal of Sedimentary Petrology* 60: 859–869.

Ostrofsky, M.L. and E.R. Zettler. 1986. Chemical defenses in aquatic plants. *Journal of Ecology* 74: 279–287.

Oxford Atlas of the World. 1997. New York: Oxford University Press.

Paré, P.W. and J.H. Tumlinson. 1999. Plant volatiles as a defense against insect herbivores. *Plant Physiology* 121: 325–332.

Parker, V.T. 2007. Diversity and evolution of *Arctostaphylos* and *Ceanothus*. *Freemontia* 35 (4): 8–11.

2011. Diversity and management of rare *Arctostaphylos* and *Ceanothus* species in chaparral. pp. 233–238. In J.W. Willoughby, B.K. Orr, K.A. Schierenbeck and N.J. Jensen (eds.) *Proceedings of the California Native Plant Conservation Conference, 2009*. Sacramento: California Native Plant Society.

Parkinson, C.L., K.L. Adams and J.D. Palmer. 1999. Multigene analyses identify the three earliest lineages of extant flowering plants. *Current Biology* 9: 1485–1491.

Parks, J.C. and C.R. Werth. 1993. A study of spatial features of clones in a population of bracken fern, *Pteridium aquilinum* (Dennstaedtiaceae). *American Journal of Botany* 80: 537–544.

Pärtel, M., M. Zobel, K. Zobel and E. van der Maarel. 1996. The species pool and its relationship to species richness: evidence from Estonia plant communities. *Oikos* 75: 111–117.

Pearce, C.M. 1993a. Coping with forest fragmentation in southwestern Ontario. pp. 100–113. In S.F. Poser, W.J. Crins and T.J. Beechey (eds.) *Size and Integrity Standards for Natural Heritage Areas in Ontario*. Proceedings of a Seminar. Parks and Natural Heritage Policy Branch, Ontario Ministry of Natural Resources, Queen's Printer, Toronto.

Pearce, F. 1993b. Draining life from Iraq's marshes. *New Scientist* 1869: 11–12.

Pearsall, W.H. 1920. The aquatic vegetation of the English Lakes. *Journal of Ecology* 8: 163–201.

Peat, H.J. and A.H. Fitter. 1993. The distribution of arbuscular mycorrhizas in the British flora. *New Phytologist* 125: 845–854.

Peattie, D.C. 1922. The Atlantic coastal plain element in the flora of the Great Lakes. *Rhodora* 24: 50–70, 80–88.

Pedersen, O., K. Sand-Jensen and N.P. Revsbech. 1995. Diel pulses of O_2 and CO_2 in sandy lake sediments inhabited by *Lobelia dortmanna*. *Ecology* 76: 1536–1545.

Peel, M.C., B.L. Finlayson and T.A. McMahon. 2007. Updated world map of the Köppen–Geiger climate classification. *Hydrology and Earth System Sciences* 11:1633–1644.

Peet, R.K. 1974. The measurement of species diversity. *Annual Review of Ecology and Systematics* 5: 285–307.

1978. Forest vegetation of the Colorado Front Range: patterns of species diversity. *Vegetatio* 37: 65–78.

1992. Community structure and ecosystem function. pp. 103–151. In D.C. Glenn-Lewin, R.K. Peet and T.T. Veblen (eds.) *Plant Succession: Theory and Prediction*. London: Chapman and Hall.

Peet, R.K. and D.J Allard. 1993. Longleaf pine vegetation of the southern Atlantic and eastern Gulf coast regions: a preliminary classification. pp. 45–81. In S.M. Hermann (ed.) *Proceedings of the Tall Timbers Fire Ecology Conference. No. 18. The Longleaf Pine Ecosystem: Ecology, Restoration, and Management*. Florida: Tall Timbers Research Station.

Pennings, S.C. and R.M. Callaway. 1992. Salt marsh plant zonation: the relative importance of competition and physical factors. *Ecology* 73: 681–690.

Pennings, S.C., T.H. Carefoot, E.L. Siska, M.E. Chase and T.A. Page. 1998. Feeding preferences of a generalist salt-marsh crab: relative importance of multiple plant traits. *Ecology* 79: 1968–1979.

Percival, M.S. 1965. *Floral Diversity*. Oxford: Pergamon.

Pérez, J., J. Muñoz-Dorado, T. de la Rubia and J. Martínez. 2002. Biodegradation and biological treatments of cellulose, hemicellulose and lignin: an overview. *International Microbiology* 5: 53–63.

Peters, R.H. 1992. *A Critique for Ecology*. Cambridge: Cambridge University Press.

Petit, J.R. (and 19 others). 1999. Climate and atmospheric history of the past 420,000 years from the Vostok ice core, Antarctica. *Nature* 399: 429–436.

Petterson, B. 1965. Gotland and Öland: two limestone islands compared. *Acta Phytogeographic Suecica* 50: 131–140.

Phillips, D.L. and D.J. Shure. 1990. Patch-size effects on early succession in southern Appalachian forest. *Ecology* 71: 204–212.

Phipps, R.W. 1883. *On the Necessity of Preserving and Replanting Forests*. Toronto: Blackett and Robinson.

Pianka, E.R. 1981. Competition and niche theory. pp. 167–196. In R.M. May (ed.) *Theoretical Ecology*. Oxford: Blackwell.

1983. *Evolutionary Ecology*. 3rd edn. New York: Harper and Row.

Pickett, S.T.A. 1980. Non-equilibrium coexistence of plants. *Bulletin of the Torrey Botanical Club* 107: 238–248.

Pickett, S.T.A. and P.S. White. 1985. *The Ecology of Natural Disturbance and Patch Dynamics*. Orlando: Academic Press.

Pielou, E.C. 1975. *Ecological Diversity*. New York: John Wiley and Sons.

1977. *Mathematical Ecology*. New York: John Wiley and Sons.

1979. *Biogeography*. New York: John Wiley and Sons.

1984. *The Interpretation of Ecological Data: A Primer on Classification and Ordination*. New York: Wiley.

Pielou, E.C. and R.D. Routledge. 1976. Salt marsh vegetation: latitudinal gradients in the zonation patterns. *Oecologia* 24: 311–321.

Pietropaolo, J. and P. Pietropaolo. 1986. *Carnivorous Plants of the World*. Portland: Timber Press.

Pimm, S.L. 2001. *The World According to Pimm: A Scientist Audits the Earth*. New York: McGraw-Hill.

Pirozynski, D.W. and D.W. Malloch. 1988. Seeds, spores and stomachs: coevolution in seed dispersal mutualisms. pp. 228–244. In K.A. Pirozynski and D.L. Hawksworth (eds.) *Coevolution of Fungi with Plants and Animals*. London: Academic Press.

Pirozynski, K.A. and Y. Dalpé. 1989. Geological history of the Glomaceae with particular reference to mycorrhizal symbiosis. *Symbiosis* 7: 1–36.

Pitman, N.C.A. and P.M. Jørgensen. 2002. Estimating the size of the world's threatened flora. *Science* 298: 989.

Platt, W.J. 1999. Southeastern pine savannas. pp. 23–51. In R.C. Anderson, J.S. Fralish and J.M. Baskin (eds.) *Savannas, Barrens and Rock Outcrop Communities of North America*. Cambridge: Cambridge University Press.

Poelman E.H, M. Bruinsma, F. Zhu et al. 2012. Hyperparasitoids use herbivore-induced plant volatiles to locate their parasitoid host. *PLoS Biol* 10(11): e1001435.

Poljakoff-Mayber, A. and J. Gale. (eds.) 1975. *Plants in Saline Environments*. Berlin: Springer-Verlag.

Ponnamperuma, F.N. 1984. Effects of flooding on soils. pp. 9–45. In T.T. Kozlowski (ed.) *Flooding and Plant Growth*. Orlando: Academic Press.

Poole, R.W. and B.J. Rathcke. 1979. Regularity, randomness, and aggregation in flowering phenologies. *Science* 203: 470–471.

Porter, H. 1993. Interspecific variation in the growth response of plants to an elevated ambient CO_2 concentration. *Vegetatio* 104(105): 77–97.

Pound, R. 1893. Symbiosis and mutualism. *The American Naturalist* 27: 509–520.

Power, M.E. 1992. Top–down and bottom–up forces in food webs: do plants have primacy? *Ecology* 73: 733–746.

Power, M.J., C. Whitlock and P.J. Bartlein. 2011. Postglacial fire, vegetation, and climate history across an elevational gradient in the Northern Rocky Mountains, USA and Canada. *Quaternary Science Reviews* **30**: 2520–2533.

Prance, G.T. and D.M. Johnson. 1991. Plant collections from the plateau of Serra do Aracá (Amazonas, Brazil) and their phytogeographic affinities. *Kew Bulletin* **47**: 1–24.

Press, M.C. and J.D. Graves (eds.) 1995. *Parasitic Plants*. London: Chapman and Hall.

Pressey, R.L., C.J. Humphries, C.R. Margules, R.I. Vane-Wright and P.H. Williams. 1993. Beyond opportunism: key principles for systematic reserve selection. *Trends in Ecology and Evolution* **8**: 124–128.

Preston, F.W. 1962a. The canonical distribution of commonness and rarity: Part I. *Ecology* **43**: 185–215.

    1962b. The canonical distribution of commonness and rarity: Part II. *Ecology* **43**: 410–432.

Price, M.V. 1984. Alternative paradigms in community ecology. pp. 354–383. In P.W. Price, C.N. Slobodchikoff and W.S.A. Gaud (eds.) *A New Ecology: Novel Approaches to Interactive Systems*. New York: John Wiley and Sons.

Price, M.V. and O.J. Reichman. 1987. Distribution of seeds in Sonoran Desert soils: implications for heteromyid rodent foraging. *Ecology* **68**: 1797–1811.

Pringle, H. 1996. *In Search of Ancient North America*. New York: John Wiley and Sons.

Puerto, A., M. Rico, M.D. Matias and J.A. García. 1990. Variation in structure and diversity in Mediterranean grasslands related to trophic status and grazing intensity. *Journal of Vegetation Science* **1**: 445–452.

Putwain, P.D. and J.L. Harper. 1970. Studies in the dynamics of plant populations. III. The influence of associated species on populations of *Rumex acetosa* L. and *R. acetosella* L. in grassland. *Journal of Ecology* **58**: 251–264.

Putz, F.E. and C.D. Canham. 1992. Mechanisms of arrested succession in shrublands: root and shoot competition between shrubs and tree seedlings. *Forest Ecology and Management* **49**: 267–275.

Radford, A.E., H.E. Ahles and C.R. Bell. 1968. *Manual of the Vascular Flora of the Carolinas*. Chapel Hill: The University of North Carolina Press.

Raffa, K.F., B.H. Aukema, B.J. Bentz et al. 2008. Cross-scale drivers of natural disturbances prone to anthropogenic amplification: the dynamics of bark beetle eruptions. *Bioscience* **58**: 501–517.

Rapanarivo, S.H.J.V., J.J. Lavranos, A.J.M. Leeuwenberg and W. Röösli. 1999. Pachypodium *(Apocynaceae): Taxonomy, Habitats and Cultivation*. Rotterdam: A.A. Balkema.

Rapport, D.J. 1989. What constitutes ecosystem health? *Perspectives in Biology and Medicine* **33**: 120–132.

Rapport D.J., C. Thorpe and T.C. Hutchinson. 1985. Ecosystem behaviour under stress. *The American Naturalist* **125**: 617–640.

Raunkiaer, C. 1907. The life-forms of plants and their bearing on geography. Translated from Danish and republished in 1934. In *The Life Forms of Plants and Statistical Plant Geography*. pp. 2–104. Oxford: Clarendon Press.

    1908. The statistics of life-forms as a basis for biological plant geography. Translated from Danish and republished in 1934. In *The Life Forms of Plants and Statistical Plant Geography*. pp. 111–147. Oxford: Clarendon Press.

    1918. On the biological normal spectrum. Translated from German and republished in 1934. In *The Life Forms of Plants and Statistical Plant Geography*. pp. 425–434. Oxford: Clarendon Press.

    1934. *The Life Forms of Plants and Statistical Plant Geography: Being the Collected Papers of Raunkiaer*. Translated from the Danish, French and German. Preface by A.G. Tansley. Oxford: Clarendon Press.

Raven, P.H., R.F. Evert and S.E. Eichhorn. 2005. *Biology of Plants*. 7th edn. New York: W. H. Freeman.

Ravera, O. 1989. Lake ecosystem degradation and recovery studied by the enclosure method. pp. 217–243. In O. Ravera (ed.) *Ecological*

Assessment of Environmental Degradation, Pollution and Recovery. Amsterdam: Elsevier Science Publishers.

Ray, N. and J.M. Adams. 2001. A GIS-based vegetation map of the world at the last glacial maximum (25,000–15,000 BP). *Internet Archaeology* 11 (open access).

Raymond, A., P. Cutlip and M. Sweet. 2001. Rates and processes of terrestrial nutrient cycling in the Paleozoic: the world before beetles, termites, and flies. pp. 235–283. In W.D. Allmon and D.J. Bottjer (eds.) *Evolutionary Paleoecology. The Ecological Context of Macroevolutionary Change.* NY: Columbia University Press.

Read, D.J., H.K. Koucheki and J. Hodgson. 1976. Vesicular-arbuscular mycorrhizae in natural vegetation systems. 1. The occurrence of infection. *New Phytologist* **77**: 641–653.

Reader, R.J. and B.J. Best. 1989. Variation in competition along an environmental gradient: *Hieracium floribundum* in an abandoned pasture. *Journal of Ecology* **77**: 673–684.

Redford, K.H. 1992 .The empty forest. *Bioscience* **42**: 412–422.

Regal, P.J. 1977. Ecology and evolution of flowering plant dominance. *Science* **196**: 622–629.

Reid, D.M. and K.J. Bradford. 1984. Effect of flooding on hormone relations. pp. 195–219. In T.T. Kozlowski (ed.) *Flooding and Plant Growth.* Orlando: Academic Press.

Reid, W.V., J.A. McNeely, D.B. Tunstall, D.A. Bryant and M. Winograd. 1993. *Biodiversity Indicators for Policymakers.* Washington, D.C.: World Resources Institute.

Richards, P.W. 1996. *The Tropical Rain Forest. An Ecological Study.* 2nd edn. with contributions by R.P.D. Walsh, I.C. Baillie, and P. Greig-Smith. Cambridge: Cambridge University Press.

Richardson, C.J. (ed.) 1981. *Pocosin Wetlands: An Integrated Analysis of Coastal Plain Freshwater Bogs in North Carolina.* Stroudsburg: Hutchinson Ross Publishing Company.

Richardson, S.J., D.A. Peltzer, R.B. Allen and M.S. McGlone. 2005. Resorption proficiency along a chronosequence: responses among communities and within species. *Ecology* **80**: 20–25.

Rickerl, D.H., F.O. Sancho and S. Ananth. 1994. Vesicular-arbuscular endomycorrhizal colonization of wetland plants. *Journal of Environmental Quality* **23**: 913–916.

Ricklefs, R.E. 2000. *The Economy of Nature.* New York: W.H. Freeman and Company.

Rigler, F.H. 1982. Recognition of the possible: an advantage of empiricism in ecology. *Canadian Journal of Fisheries and Aquatic Sciences* **39**: 1323–1331.

Rigler, F.H. and R.H. Peters. 1995. *Science and Limnology.* Oldendorf/Lutie: Ecology Institute.

Ripple, W.J. and R.L. Beschta. 2012. Trophic cascades in Yellowstone: The first 15 years after wolf reintroduction. *Biological Conservation* **145**: 205–213.

Ritchie, J.C. 1987. *Postglacial Vegetation of Canada.* New York: Cambridge University Press.

Roberts, B.A. 1992. The serpentinized areas of Newfoundland, Canada: a brief review of their soils and vegetation. pp. 53–66. In A.J.M. Baker, J. Proctor and R.D. Reeves (eds.) *The Vegetation of Ultramafic (Serpentine) Soils.* Andover: Intercept.

Robinson, A.R. 1973. Sediment, our greatest pollutant? In R.W. Tank (ed.) *Focus on Environmental Geology.* London: University Press.

Robinson, D. 1996. Resource capture by localised root proliferation: why do plants bother? *Annals of Botany* **77**: 179–185.

Robinson, J.G. and E.L. Bennett (eds.). 2000. *Hunting for Sustainability in Tropical Forests.* New York: Columbia University Press.

Robinson, J.M. 1990. Lignin, land plants, and fungi: biological evolution affecting phanerozoic oxygen balance. *Geology* **18**: 607–610.

Rodman, J.E. 1974. Systematics and evolution of the genus *Cakile* (Cruciferae). *Contributions from the Gray Herbarium* **205**: 3–146.

Rodrigues, A.S.L., S.J. Andelman, M.I. Bakarr et al. 2004. Effectiveness of the global protected areas network in representing species diversity. *Nature* **428**: 640–643.

Roger, A.J. 1999. Reconstructing early events in eukaryotic evolution. *The American Naturalist* **154** (Suppl.): S146–S163.

Rohde, K. 1997. The larger area of tropics does not explain latitudinal gradients in species diversity. *Oikos* **79**: 169–172.

Rose, R. 1924. Man and the Galapagos Islands. pp. 332–417. In W. Beebe (ed.) *Galapagos: World's End.* New York: Putnam's Sons.

Rosenthal, G A. and M.R. Berenbaum. (eds.) 1991. *Herbivores: Their Interactions with Secondary Plant Metabolites.* San Diego: Academic Press.

Rosenzweig, M.L. 1995. *Species Diversity in Space and Time.* Cambridge: Cambridge University Press.

Rosgen, D.L. 1995. River restoration utilizing natural stability concepts. pp. 55–62. In J.A. Kusler, D.E. Willard and H.C. Hull, Jr. (eds.) *Wetlands and Watershed Management: Science Applications and Public Policy.* A collection of papers from a national symposium and several workshops at Tampa, Florida, April 23–26. New York: The Association of State Wetland Managers.

Rosman, K.J.R, W. Chisholm, S. Hong, J-P Candelone and C.F. Boutron. 1997. Lead from Carthaginian and Roman Spanish mines isotopically identified in Greenland ice dated from 600 B.C. to 300 A.D. *Environmental Science and Technology* **31**: 3413–3416.

Rowe, J.S. and G.W. Scotter. 1973. Fire in the boreal forest. *Quaternary Research* **3**: 444–464.

Rozan, T.F., K.S. Hunter and G. Benoit. 1994. Industrialization as recorded in floodplain deposits of the Quinnipiac River, Connecticut. *Marine Pollution Bulletin* **28**: 564–569.

Rubinstein, C.V., P. Gerrienne, G. S. de la Puente, R. A. Astini and P. Steemans. 2010. Early Middle Ordovician evidence for land plants in Argentina (eastern Gondwana). *New Phytologist* **188**: 365–369.

Rundel, P.W., R.M. Cowling, K.J. Esler, et al. 1995. Winter growth phenology and leaf orientation in *Pachypodium namaquanum* (Apocynaceae) in the succulent karoo of the Richtersveld, South Africa. *Oecologia* **101**: 472–477.

Russell, F.L., D.B. Zippin and N.L. Fowler. 2001. Effects of white-tailed deer (*Odocoileus virginianus*) on plants, plant populations and communities: a review. *The American Midland Naturalist* **146**: 1–26.

Russo, S.E., W.L. Cannon, C. Elowsky, S. Tan and S.J. Davies. 2010. Variation in leaf stomatal traits of 28 tree species in relation to gas exchange along an edaphic gradient in a Bornean rain forest. *American Journal of Botany* **97**: 1109–1120.

Ryan, W. and Pitman, W. 1999. *Noah's Flood: The New Scientific Discoveries About the Event That Changed History.* New York: Simon and Schuster.

Sala, O.E., L. Vivanco and P. Flombaum. 2013. Grassland ecosystems. pp. 1–7. In S. A. Levin (ed.). *Encyclopedia of Biodiversity.* 2nd edn. Vol. 4. Amsterdam: Elsevier.

Salisbury, E.J. 1942. *The Reproductive Capacity of Plants. Studies in Quantitative Biology.* London: G. Bell and Sons Ltd.

Salisbury, F.B. and C.W. Ross. 1988. *Plant Physiology.* 3rd edn. Belmont: Wadsworth Publishers.

Salisbury, S.E. 1970. The pioneer vegetation of exposed muds and its biological features. *Royal Society of London, Philosophical Transactions, Series B* **259**: 207–255.

Salo, J., R. Kalliola, I. Hakkinen, et al. 1986. River dynamics and the diversity of Amazon lowland forest. *Nature* **322**: 254–258.

Salzman, A.G. and M.A. Parker. 1985. Neighbours ameliorate local salinity stress for a rhizomatous plant in a heterogeneous environment. *Oecologia* **65**: 273–277.

Sandars, N.K. 1972. *The Epic of Gilgamesh.* An English version with an introduction by N. K. Sandars. Revised edition. London: Penguin Books.

Sargent, C.S. 1902. *The Silva of North America.* Vol. 14. Boston: Houghton Mifflin.

Sarthou, C. and J.-F. Villiers. 1998. Epilithic plant communities on inselbergs in French Guiana. *Journal of Vegetation Science* 9: 847–860.

Saugier, B.S., J. Roy and H.A. Mooney. 2001. Estimations of global terrestrial productivity: converging toward a single number? pp. 543–557. In J. Roy, B. Saugier and H. A. Mooney (eds.) *Terrestrial Global Productivity*. San Diego: Academic Press.

Saunders, D.A., R.J. Hobbs and P.R. Ehrlich (eds.) 1993. *Nature Conservation 3: Reconstruction of Fragmented Ecosystems – Global and Regional Perspectives*. Chipping Norton: Surrey Beatty and Sons Pty Limited.

Savile, D.B.O. 1956. Known dispersal rates and migratory potentials as clues to the origin of the North American biota. *The American Midland Naturalist* 56: 434–453.

1972. *Arctic Adaptations in Plants*. Monograph No. 6. Ottawa: Canada Department of Agriculture.

Sawyer, J., T. Keeler-Wolf and J. Evens. 2009. *A Manual of California Vegetation*. 2nd edn. Sacramento: California Native Plant Society.

Sayre, N.F. 2003. Recognizing history in range ecology: 100 years of science and management on the Santa Rita Experimental Range. pp. 1–15. In *USDA Forest Service Proceedings RMRS-P-30*. US Department of Agriculture.

Schatz, G.E. 2000. The rediscovery of a Malagasy endemic: *Takhtajania perrieri* (Winteraceae). *Annals of the Missouri Botanical Garden* 87: 297–302.

Schell, J. 1982. *The Fate of the Earth*. New York: Alfred A. Knopf.

Schnitzler, A. 1995. Successional status of trees in gallery forest along the river Rhine. *Journal of Vegetation Science* 6: 479–486.

Scholander, P.F., H.T. Hammel, B.D. Bradstreet, and E.A. Hemmingsen. 1965. Sap pressure in vascular plants. *Science* 148: 339–346.

Schopf, J. W. and E. S. Barghoorn. 1967. Alga-like fossils from the early Precambrian of South Africa. *Science* 156: 508–512.

Schwinning, S. and O.E. Sala. 2004. Hierarchy of responses to resource pulses in arid and semi-arid ecosystems. *Oecologia* 141: 211–220.

Scott, J.M., B. Csuti, J.D. Jacobi and J.E. Estes. 1987. Species richness: a geographic approach to protecting future biological diversity. *Bioscience* 37: 782–788.

Sculthorpe, C.D. 1967. *The Biology of Aquatic Vascular Plants*. Reprinted in 1985. London: Edward Arnold.

Seischab, F.K. and D. Orwig. 1991. Catastrophic disturbances in the presettlement forests of western New York. *Bulletin of the Torrey Botanical Club* 118: 117–122.

Selosse, M.-A., E. Baudoin and P. Vandenkoornhuyse. 2004. Symbiotic microorganisms, a key for ecological success and protection of plants. *Comptes Rendus Biologies* 327: 639–648.

Shaffer, G.P., J.G. Gosselink and S.S. Hoeppner. 2005. The Mississippi River alluvial plain. pp. 272–315. In L.H. Fraser and P.A. Keddy (eds.) *The World's Largest Wetlands: Ecology and Conservation*. Cambridge: Cambridge University Press.

Shannon, C.E. and W. Weaver. 1949. *The Mathematical Theory of Communication*. Urbana: University of Illinois Press.

Sheail, J. and T.C.E. Wells. 1983. The fenlands of Huntingdonshire, England: a case study in catastrophic change. pp. 375–393. In A.J.P. Gore (ed.) *Ecosystems of the World 4B. Mires: Swamp, Bog, Fen and Moor*. Amsterdam: Elsevier.

Shimwell, D.W. 1971. *The Description and Classification of Vegetation*. Seattle: University of Washington Press.

Shipley, B. 1993. A null model for competitive hierarchies in competition matrices. *Ecology* 74: 1693–1699.

2009. *From Plant Traits to Vegetation Structure: Chance and Selection in the Assembly of Ecological Communities*. Cambridge: Cambridge University Press.

Shipley, B. and J. Dion. 1992. The allometry of seed production in herbaceous angiosperms. *The American Naturalist* 139: 467–483.

Shipley, B. and P.A. Keddy. 1987. The individualistic and community-unit concepts as falsifiable hypotheses. *Vegetatio* 69: 47–55.

1994. Evaluating the evidence for competitive hierarchies in plant communities. *Oikos* 69: 340–345.

Shipley, B. and R.H. Peters. 1990. A test of the Tilman model of plant strategies: relative growth rate and biomass partioning. *The American Naturalist* 136: 139–153.

Shipley, B., P.A. Keddy and L.P. Lefkovitch. 1991. Mechanisms producing plant zonation along a water depth gradient: a comparison with the exposure gradient. *Canadian Journal of Botany* 69: 1420–1424.

Shrader-Frechette, K.S. and E.D. McCoy. 1993. *Method in Ecology: Strategies for Conservation.* Cambridge: Cambridge University Press.

Shugart, H.H., D.C. West and W.R. Emanuel. 1981. Patterns and dynamics of forests: an application of simulation models. pp. 74–106. In D.C. West, H.H. Shugart and D B. Botkin (eds.) *Forest Succession Concepts and Applications.* New York: Springer-Verlag.

Shure, D.J. 1999. Granite outcrops of the southeastern United States. pp. 99–118. In R.C. Anderson, J.S. Fralish and J.M. Baskin (eds.) *Savannas, Barrens, and Rock Outcrop Plant Communities of North America.* Cambridge: Cambridge University Press.

Silliman, B.R. and M.D. Bertness. 2002. A trophic cascade regulates salt marsh primary production. *Proceedings of the National Academy of Sciences* 99: 10,500–10,505.

Silvera, K., L.S. Santiago, J.C. Cushman and K. Winter. 2009. Crassulacean acid metabolism and epiphytism linked to adaptive radiations in the Orchidaceae. *Plant Physiology* 149: 1838–1847.

Silvertown, J. 1980. The dynamics of a grassland ecosystem: botanical equilibrium in the Park Grass experiment. *Journal of Applied Ecology* 17: 491–504.

1987. *Introduction to Plant Population Ecology.* 2nd edn. London: Longman.

Silvertown, J. and D. Charlesworth. 2001. *Introduction to Plant Population Biology.* 4th edn. Oxford: Blackwell Science.

Silvertown, J., M.E. Dodd, K. McConway, J. Potts and M. Crawley. 1994. Rainfall, biomass variation, and community composition in the Park Grass experiment. *Ecology* 75: 2430–2437.

Silvertown, J., P. Poulton, E. Johnston, et al. 2006. The Park Grass experiment 1856–2006: its contribution to ecology. *Journal of Ecology* 94: 801–814.

Simard, S.W., K.J. Beiler, M.A. Bingham, et al. 2012. Mycorrhizal networks: mechanisms, ecology and modeling. *Fungal Biology Reviews* 26: 39–60.

Simmons, I.G. 2003. *The Moorlands of England and Wales. An Environmental History 8000 BC–AD 2000.* Edinburgh: Edinburgh University Press.

Simon, L., J. Bousquet, R.C. Lévesque and M. Lalonde. 1993. Origin and diversification of endomycorrhizal fungi and coincidence with vascular land plants. *Nature* 363: 67–69.

Simpson, E.H. 1949. Measurement of diversity. *Nature* 163: 688.

Sinclair, A.R.E. 1983. The adaptations of African ungulates and their effects on community function. pp. 401–425. In F. Bouliere (ed.) *Tropical Savannas.* Amsterdam: Elsevier.

Sinclair, A.R.E. and J.M. Fryxell. 1985. The Sahel of Africa: ecology of a disaster. *Canadian Journal of Zoology* 63: 987–994.

Sinclair, A.R.E. and C.J. Krebs. 2001. Trophic interactions, community organization, and the Kluane ecosystem. pp. 25–48. In C.J. Krebs, S. Boutin and R. Boonstra (eds.) *Ecosystem Dynamics of the Boreal Forest: The Kluane Project.* New York: Oxford University Press.

Sinclair, A.R.E., D.S. Hik, O.J. Schmitz, et al. 1995. Biodiversity and the need for habitat renewal. *Ecological Applications* 5: 579–587.

Sinclair, A.R.E., C.J. Krebs, J.M. Fryxell, et al. 2000. Testing hypotheses of trophic level interactions: a boreal forest ecosystem. *Oikos* 89: 313–328.

Singh, D. and S. Chen. 2008. The white-rot fungus *Phanerochaete chrysosporium*: conditions for the production of lignin-degrading enzymes. *Applied Microbiology and Biotechnology* 81: 399–417.

Sioli, H. 1964. General features of the limnology of Amazonia. *Verhandlungen/ Internationale Vereinigung für Theoretische und Angewandte Limnologie* 15: 1053–1058.

Skellam, J.G. 1951. Random dispersal in theoretical populations. *Biometrika* 38: 196–218.

Sklar, F.H., R. Costanza and J.W. Day, Jr. 1990. Model conceptualization. pp. 625–658. In B.C. Patten (ed.) *Wetlands and Shallow Continental Water Bodies*, Vol. 1. The Hague: SPB Academic Publishing.

Sklar, F.H., M.J. Chimney, S. Newman et al. 2005. The ecological–societal underpinnings of Everglades restoration. *Frontiers in Ecology and the Environment* 3: 161–169.

Slade, A.J. and M.J. Hutchings. 1987. The effects of nutrient availability on foraging in the clonal herb *Glechoma hederacea*. *Journal of Ecology* 75: 95–112.

Sletvold, N. 2002. Effects of plant size on reproductive output and offspring performance in the facultative biennial *Digitalis purpurea*. *Journal of Ecology* 90: 958–996.

Smith, A. 1776. An enquiry into the nature and causes of the wealth of nations. In M.J. Adler (ed.) 1990. *Great Books of the Western World* 2nd edn. Vol. 36. Chicago: Encyclopaedia Britannica Inc.

Smith, C.C. 1970. The coevolution of pine squirrels (*Tamiasciurus*) and conifers. *Ecological Monographs* 40: 349–371.

Smith, C.C. and D. Follmer. 1972. Food preference of squirrels. *Ecology* 53: 82–91.

Smith, D.C. 1980. Mechanisms of nutrient movement between lichen symbionts. pp. 197–227. In C. B. Cook, P. W. Pappas, and E. D. Rudolph (eds.) *Cellular Interactions in Symbiosis and Parasitism*. Columbus: Ohio State University Press.

Smith, D.C. and A.E. Douglas. 1987. *The Biology of Symbiosis*. London: Edward Arnold.

Smith, L.M. and J.A. Kadlec. 1983. Seed banks and their role during the drawdown of a North American marsh. *Journal of Applied Ecology* 20: 673–684.

Smith, R.L. 1986. *Elements of Ecology*. New York: Harper and Row.

Smith, V.H. 1982. The nitrogen and phosphorus dependence of algal biomass in lakes: an empirical and theoretical analysis. *Limnology and Oceanography* 27: 1101–1112.

1983. Low nitrogen to phosphorus ratios favor dominance by blue-green algae in lake phytoplankton. *Science* 221: 669–671.

Smol, J.P. and B.F. Cumming. 2000. Tracking long-term changes in climate using algal indicators in lake sediments. *Journal of Phycology* 36: 986–1011.

Sneath, P.H.A. and R.R. Sokal. 1973. *Numerical Taxonomy*. San Francisco: W. H. Freeman.

Snell, T.W. and D.G. Burch. 1975. The effects of density on resource partitioning in *Chamaesyce hirta* (Euphorbiaceae). *Ecology* 56: 742–746.

Snow, A.A. and S.W. Vince. 1984. Plant zonation in an Alaskan salt marsh. II: An experimental study of the role of edaphic conditions. *Journal of Ecology* 72: 669–684.

Sobel, D. 1995. *Longitude: The True Story of a Lone Genius Who Solved the Greatest Scientific Problem of His Time*. New York: Penguin Books.

Society for Ecological Restoration. 2004. *SER International Primer on Ecological Restoration*. 2nd edn. Washington: SER Science & Policy Working Group.

Solomeshch, A.I. 2005. The West Siberian lowland. pp. 11–62. In L.H. Fraser and P.A. Keddy (eds.) *The World's Largest Wetlands: Ecology and Conservation*. Cambridge: Cambridge University Press.

Soper, J.H. and P.F. Maycock. 1963. A community of arctic–alpine plants on the east shore of Lake Superior. *Canadian Journal of Botany* 41: 183–198.

Sorrie, B.A. 1994. Coastal plain ponds in New England. *Biological Conservation* 68: 225–233.

Soulé, M. and R. Noss. 1998. Rewilding and biodiversity: complementary goals for continental conservation. *Wild Earth* 8(3): 18–28.

Soulé, M. E. and J. Terborgh (eds.) 1999. *Continental Conservation: Scientific Foundations of Regional Reserve Networks*. Washington, DC: Island Press.

Sousa, W.P. 1984. The role of disturbance in natural communities. *Annual Review of Ecology and Systematics* 15: 353–391.

Southwood, T.R.E. 1977. Habitat, the templet for ecological strategies? *Journal of Animal Ecology* 46: 337–365.

　1985. Interactions of plants and animals: patterns and processes. *Oikos* 44: 5–11.

　1988. Tactics, strategies, and templets. *Oikos* 52: 3–18.

Specht, A. and R. Specht. 1993. Species richness and canopy productivity of Australian plant communities. *Biodiversity and Conservation* 2: 152–167.

Specht, R.L. and A. Specht. 1989. Species richness of overstorey strata in Australian plant communities: the influence of overstorey growth rates. *Australian Journal of Botany* 37: 321–336.

Spence, D.H.N. 1982. The zonation of plants in freshwater lakes. *Advances in Ecological Research* 12: 37–125.

Sporne, K.R. 1956. The phylogenetic classification of the angiosperms. *Biological Reviews* 31: 1–29.

　1970. *The Morphology of Pteridophytes: The Structure of Ferns and Allied Plants*, 3rd edn. London: Hutchinson and Co.

Sprengel, C.K. 1793. Discovery of the secret of nature in the structure and fertilization of flowers. Vieweq, Berlin. Translation of title and first chapter by P. Haase. pp. 3–43. In D.G. Lloyd and S.C.H. Barrett (eds.) 1996. *Floral Biology. Studies on Floral Evolution in Animal-Pollinated Plants*. London: Chapman and Hall.

Starfield, A.M. and A.L. Bleloch. 1986. *Building Models for Conservation and Wildlife Management*. New York: Macmillan.

　1991. *Building Models for Conservation and Wildlife Management*. 2nd edn. Edina: MN Burgers International Group.

Stearn, W.T. 1979. Linnaean classification. pp. 96–101. In D. Black (ed.) *Carl Linnaeus: Travels*. New York: Scribner's Sons.

Steedman, R.J. 1988. Modification and assessment of an index of biotic integrity to quantify stream quality in southern Ontario. *Canadian Journal of Fisheries and Aquatic Sciences* 45: 492–501.

Steenbergh, W.F. and C.H. Lowe. 1969. Critical factors during the first years of life of the saguaro (*Cereus giganteus*) at Saguaro National Monument, Arizona. *Ecology* 50: 825–834.

Steila, D. 1993. Soils. pp. 47–54. In *Flora of North America*, Vol. 1. Introduction. New York: Oxford University Press.

Stellman, J.M., S.D. Stellman, R. Christian, T. Weber and C. Tomasallo. 2003. The extent and patterns of usage of Agent Orange and other herbicides in Vietnam. *Nature* 422: 681–687.

Steneck, R.S. and M.N. Dethier. 1994. A functional group approach to the structure of algal-dominated communities. *Oikos* 69: 476–498.

Stephenson, N.L. 1990. Climatic control of vegetation distribution: the role of the water balance. *The American Naturalist* 135: 649–680.

Stephenson, S.N. and P.S. Herendeen. 1986. Short-term drought effects on the alvar communities of Drummond Island, Michigan. *The Michigan Botanist* 25: 16–27.

Stevenson, J.C., L.G. Ward and M.S. Kearney. 1986. Vertical accretion in marshes with varying rates of sea level rise. pp. 241–259. In D.A. Wolfe (ed.) *Estuarine Variability*. San Diego: Academic Press.

Stewart, W.N. and G.W. Rothwell. 1993. *Paleobotany and the Evolution of Plants*. 2nd edn. Cambridge: Cambridge University Press.

Steyermark, J.A. 1982. Relationships of some Venezuelan forest refuges with lowland tropical floras. pp. 182–220. In G.T. Prance (ed.) *Biological Diversification in the Tropics*. New York: Columbia University Press.

Stokes, M.A. and T.L. Smiley. 1996. *An Introduction to Tree Ring Dating*. Tuscon: University of Arizona Press.

Stoltzenberg, D. 2004. *Fritz Haber: Chemist, Nobel Laureate, German, Jew.* Philadelphia: Chemical Heritage Press.

Strahler, A. 1971. *The Earth Sciences.* 2nd edn. New York: Harper and Row.

Street, F.A. and A.T. Grove. 1979. Global maps of lake-level fluctuations since 30,000 yr B.P. *Quaternary Research* 12: 83–118.

Strong, D.R. Jr., D. Simberloff, L.G. Abele and A.B. Thistle (eds). 1984. *Ecological Communities. Conceptual Issues and the Evidence.* Princeton: Princeton University Press.

Stupka, A. 1964. *Trees, Shrubs and Woody Vines of Great Smoky National Park.* Knoxville: The University of Tennessee Press.

Suffling, R., C. Lihou and Y. Morand. 1988. Control of landscape diversity by catastrophic disturbance: a theory and case study of fire in a Canadian boreal forest. *Environmental Management* 12: 73–78.

Sun, G., Q. Ji, D.L. Dilcher, et al. 2002. Archaefructaceae, a new basal angiosperm family. *Science* 296: 899–904.

Swink, F. and G. Wilhelm. 1979. *Plants of the Chicago Region: A Checklist of the Vascular Flora of the Chicago Region, With Keys, Notes on Local Distribution, Ecology, and Taxonomy, and a System for Evaluation of Plant Communities.* Lisle, IL: Morton Arboretum.

Syed, K. and J.S. Yadav. 2012. P450 monooxygenases (P450ome) of the model white rot fungus *Phanerochaete chrysosporium. Critical Reviews in Microbiology* 38: 339–363.

Szarek, S.R. and I.P. Ting. 1975. Photosynthetic efficiency of CAM plants in relation to C3 and C4 plants. pp. 289–297. In R. Marcelle (ed.) *Environmental and Biological Control of Photosynthesis.* The Hague: W. Junk.

Tabachnick, B.G. and L.S. Fidell. 2001. *Using Multivariate Statistics.* 4th edn. Boston: Allyn and Bacon.

Taiz, L. and E. Zeiger. 1991. *Plant Physiology.* San Francisco: Benjamin-Cummings.

Takhtajan, A. 1969. *Flowering Plants: Origin and Dispersal.* Edinburgh: Oliver and Boyd. Translated and revised from a Russian second edition published in Moscow in 1961.

1986. *Floristic Regions of the World.* Berkeley: University of California Press. Translated by T.J. Crovello.

Talbot, J.M., D.J. Yelle, J. Nowick and K.K. Treseder. 2012. Litter decay rates are determined by lignin chemistry. *Biogeochemistry* 108: 279–295.

Tansley, A.G. 1939. *The British Islands and their Vegetation.* Cambridge: Cambridge University Press.

Tansley, A. 1987. What is ecology? *Biological Journal of the Linnean Society* 32: 5–16.

Tansley, A.G. and R.S. Adamson. 1925. Studies of the vegetation of the English chalk. Part III. The chalk grasslands of the Hampshire–Sussex border. *Journal of Ecology* XIII: 177–223.

Tansley, A.G. and T.F. Chipp (eds.) 1926. *Aims and Methods in the Study of Vegetation.* London: The British Empire Vegetation Committee and Crown Agents for Colonies.

Taylor, D.R., L.W. Aarssen and C. Loehle. 1990. On the relationship between r/K selection and environmental carrying capacity: a new habitat template for life history strategies. *Oikos* 58: 239–250.

Taylor, J.G. 1984. *Louisiana: A History.* New York: W. W. Norton & Company.

Taylor, K.L. and J.B. Grace. 1995. The effects of vertebrate herbivory on plant community structure in the coastal marshes of the Pearl River, Louisiana, USA. *Wetlands* 15: 68–73.

Taylor, T.N. 1988. The origin of land plants: some answers, more questions. *Taxon* 37: 805–833.

1990. Fungal associations in the terrestrial paleoecosystem. *Trends in Ecology and Evolution* 5: 21–25.

Taylor, T.N., H. Kerp and H. Hass. 2005. Life history biology of early land plants: deciphering the gametophyte phase. *Proceedings of the National Academy of Sciences* 102: 5892–5897.

Teller, J.T. 2003. Controls, history, outbursts and impact of large late-Quaternary proglacial lakes in North America. pp. 45–61. In A. Gilespie, S. Porter, and B. Atwater (eds.) *The Quaternary Period in the United States.* Amsterdam: Elsevier.

Temple, S.A. 1977. Plant–animal mutualism: coevolution with dodo leads to near extinction of plant. *Science* 197: 886–887.

Terborgh, J. 1989. *Where Have All the Birds Gone?* Princeton: Princeton University Press.

Terborgh, J. (and 10 others). 2001. Ecological meltdown in predator-free forest fragments. *Science* 294: 1923–1926.

The Plant List. 2013. Version 1.1 Published on the Internet; www.theplantlist.org (accessed 26 October 2016).

Thirgood, J.V. 1981. *Man and the Mediterranean Forest: A History of Resource Depletion.* London: Academic Press.

Thompson, D.J. and J.M. Shay. 1988. First-year response of a *Phragmites* marsh community to seasonal burning. *Canadian Journal of Botany* 67: 1448–1455.

Thorn, R.G. and G.L. Barron. 1984. Carnivorous mushrooms. *Science* 224: 76–78.

Thorne, R.F. 1963. Some problems and guiding principles of angiosperm phylogeny. *The American Naturalist* 97: 287–305.

Tilghman, N.G. 1989. Impacts of white-tailed deer on forest regeneration in northwestern Pennsylvania. *Journal of Wildlife Management* 53: 524–532.

Tilman, D. 1982. *Resource Competition and Community Structure.* Princeton: Princeton University Press.

Tilman, D. and S. Pacala. 1993. The maintenance of species richness in plant communities. pp. 13–25. In R.E. Ricklefs and D. Schluter (eds.) *Species Diversity in Ecological Communities.* Chicago: University of Chicago Press.

Tilman, D., D. Wedin and J. Knops. 1996. Productivity and sustainability influenced by biodiversity in grassland ecosystems. *Nature* 379: 718–720.

Tinker, P.B., M.D. Jones and D.M. Durall. 1992. A functional comparison of ecto- and endomycorrhizas. pp. 303–310. In D.J. Read, D.H. Lewis, A.H. Fitter and I.J. Alexander (eds.) *Mycorrhizas in Ecosystems.* Wallingford: CAB International.

Tomlinson, P.B. 1986. *The Botany of Mangroves.* Cambridge: Cambridge University Press.

Trappe, J.M. 2005. A.B. Frank and mycorrhizae: the challenge to evolutionary and ecologic theory. *Mycorrhiza* 15: 277–281.

Tschudy, R.H., C.L. Pillmore, C.J. Orth, J.S. Gilmore and J.D. Knight. 1984. Disruption of the terrestrial plant ecosystem at the Cretaceous–Tertiary boundary, Western Interior. *Science* 225: 1030–1032.

Tubbs, C.H., R.M. DeGraff, M. Yamasaki and W.M. Healy. 1987. *Guide to Wildlife Tree Management in New England Northern Hardwoods.* United States Department of Agriculture and Forestry Service General Technical Report NE-118.

Turco, R.P., O.B Toon, T.P. Ackerman, J.B. Pollack and C. Sagan. 1983. Nuclear winter: global consequences of multiple nuclear explosions. *Science* 23: 1283–1292.

Turner, F. 1994. *Beyond Geography: The Western Spirit Against the Wilderness.* Fifth printing, first edition in 1983. New Brunswick: Rutgers University Press.

Turner, R.E. 1977. Intertidal vegetation and commercial yields of penaeid shrimp. *Transactions of the American Fisheries Society* 106: 411–416.

Turner, R.M. 1990. Long-term vegetation change at a fully protected Sonoran Desert site. *Ecology* 71: 464–477.

Turner, R.M., S.M. Alcorn, G. Olin and J.A. Booth. 1966. The influence of shade, soil, and water on saguaro seedling establishment. *Botanical Gazette* 127: 95–102.

Twolan-Strutt, L. and P. Keddy. 1996. Above- and belowground competition intensity in two contrasting wetland plant communities. *Ecology* 77: 259–270.

Uhl, C. and J.B. Kauffman. 1990. Deforestation, fire susceptibility and potential responses to fire in the eastern Amazon. *Ecology* 71: 437–449.

Underwood, A.J. 1978. The detection of non-random patterns of distribution of species along a gradient. *Oecologia* **36**: 317–326.

Urban, D.L. and H.H. Shugart. 1992. Individual based models of forest succession. pp. 249–292. In D.C. Glenn-Lewin, R.K. Peet and T.T. Veblen. (eds.) *Plant Succession.* London: Chapman and Hall.

Usher, M.B. 1992. Statistical models of succession. pp. 215–248. In D.C. Glenn-Lewin, R.K. Peet and T.T. Veblen. (eds.) *Plant Succession.* London: Chapman and Hall.

US Army Coastal Engineering Research Centre. 1977. *Shore Protection Manual,* Vol. 1, 3rd edn. Washington, D.C.: US Government Printing Office.

US Army Corps of Engineers. 2004. *The Mississippi River and Tributaries Project.* New Orleans District Office Website (www.mvn.usace.army.mil/pao/bro/misstrib.htm) accessed 26 Mar. 2006.

USDA. 1975. *Soil Taxonomy: A Basic System of Soil Classification for Making and Interpreting Soil Surveys.* Agricultural Handbook 436, Washington, D.C.: USDA.

2004. *Emerald Ash Borer; The Green Menace.* Animal and Plant Health Inspection Service. Program Aid No. 1769.

Vallentyne, J.R. 1974. *The Algal Bowl. Lakes and Man.* Miscellaneous Special Publication 22. Ottawa: Department of the Environment, Fisheries and Marine Service.

van Breemen, N. 1995. How *Sphagnum* bogs down other plants. *Trends in Ecology and Evolution* **10**: 270–275.

van der Heijden, M.G.A. and T.R. Horton. 2009. Socialism in soil? The importance of mycorrhizal fungal networks for facilitation in natural ecosystems. *Journal of Ecology* **97**: 1139–1150.

Vandermeer, J.B., B.A. Hazlett and B. Rathcke. 1985. Indirect facilitation and mutualism. pp. 326–343. In D.H. Boucher (ed.) *The Ecology of Mutualism.* New York: Oxford University Press.

van der Valk, A.G. 1981. Succession in wetlands: a Gleasonian approach. *Ecology* **62**: 688–696.

van der Valk, A.G. and C.B. Davis. 1976. The seed banks of prairie glacial marshes. *Canadian Journal of Botany* **54**: 1832–1838.

1978. The role of seed banks in the vegetation dynamics of prairie glacial marshes. *Ecology* **59**: 322–335.

van der Werf, A., A. Welschen, R. Welschen and H. Lambers. 1988. Respiration energy costs for the maintenance of biomass, for growth and for iron uptake in roots of *Carex diandra* and *Carex acutiformis. Physiologia Plantarum* **72**: 483–491.

Vasseur, D.A. and K.S. McCann. 2005. Resolution of Respect. Peter Yodzis. 1943–2005. *Bulletin of the Ecological Society of America* **85**: 203–205.

Venable, D.L. and C.E. Pake. 1999. Population ecology of Sonoran Desert annual plants. pp. 115–142. In R.H. Robichaux (ed.) *The Ecology of Sonoran Desert Plants and Plant Communities.* Tucson: University of Arizona Press.

Veneklaas, E.J., A. Fajardo, S. Obregon and J. Lozano. 2005. Gallery forest types and their environmental correlates in a Colombian savanna landscape. *Ecography* **28**: 236–252.

Vera, F.W.M. 2000. *Grazing Ecology and Forest History.* Wallingford: CABI Publishing.

Verhoeven, J.T.A. and W.M. Liefveld. 1997. The ecological significance of organochemical compounds in *Sphagnum. Acta Botanica Neerlandica* **46**: 117–130.

Verhoeven, J.T.A. and M.B. Schmitz. 1991. Control of plant growth by nitrogen and phosphorus in mesotrophic fens. *Biogeochemistry* **12**: 135–148.

Verhoeven, J.T.A., R.H. Kemmers and W. Koerselman. 1993. Nutrient enrichment of freshwater wetlands. pp. 33–59. In C.C. Vos and P. Opdam (eds.) *Landscape Ecology of a Stressed Environment.* London: Chapman and Hall.

Verhoeven, J.T.A., W. Koerselman and A.F.M. Meuleman. 1996. Nitrogen- or phosphorus-limited growth in herbaceous, wet vegetation: relations with atmospheric inputs and management regimes. *Trends in Ecology and Evolution* **11**: 494–497.

Vernadsky, V. 1929. *La Biosphère.* Paris: Felix Alcan.

Vernadsky, V.I. 1998. *The Biosphere.* New York: Copernicus, Springer-Verlag. Translated from the French and Russian, including a new foreword, introduction and appendices.

Vidal-Russell, R and D.L. Nickrent. 2008. Evolutionary relationships in the showy mistletoe family (Loranthaceae). *American Journal of Botany* 95: 1015–1029.

Vieira, S, S. Trumbore, P.B. Camargo et al. 2005. Slow growth rates of Amazonian trees: consequences for carbon cycling. *Proceedings of the National Academy of Sciences* 102: 18,502–18,507.

Vince, S.W. and A.A. Snow. 1984. Plant zonation in an Alaskan salt marsh I: Distribution, abundance, and environmental factors. *Journal of Ecology* 72: 651–667.

Vitousek, P.M. 1982. Nutrient cycling and nitrogen use efficiency. *The American Naturalist* 119: 553–572.

Vitousek, P., P.R. Ehrlich, A.H. Ehrlich and P. Matson. 1986. Human appropriation of the products of photosynthesis. *Bioscience* 36: 368–373.

(and 7 others). 1997. Human alteration of the global nitrogen cycle: causes and consequences. *Ecological Applications* 7: 737–750.

Vitt, D.H. and W. Chee. 1990. The relationship of vegetation to surface water chemistry and peat chemistry in fens of Alberta, Canada. *Vegetatio* 89: 87–106.

Vivian-Smith, G. 1997. Microtopographic heterogeneity and floristic diversity in experimental wetland communities. *Journal of Ecology* 85: 71–82.

Vogel, S. 1996. Christian Konrad Sprengel's theory of the flower: the cradle of floral ecology. pp. 44–62. In D.G. Lloyd and S.C.H. Barrett (eds.) *Floral Biology: Studies on Evolution in Animal-Pollinated Plants*. London: Chapman and Hall.

Vogl, R. 1969. One hundred and thirty years of plant succession in a southeastern Wisconsin lowland. *Ecology* 50: 248–255.

von Humboldt, A. 1845. *Cosmos: A Sketch of the Physical Description of the Universe, Vol. 1*. Translated by E.C. Otté. Foundations of Natural History. Baltimore: Johns Hopkins University

Press. 1997. (Originally produced in five volumes: 1845, 1847, 1850–51, 1858, and 1862.)

Waggoner, P.E. and G.R. Stephens. 1970. Transition probabilities for a forest. *Nature* 225: 1160–1161.

Walker, D. 1970. Direction and rate in some British post-glacial hydroseres. pp. 117–139. In D. Walker and R.G. West (eds.) *Studies in the Vegetational History of the British Isles*. Cambridge: Cambridge University Press.

Walker, J., C.H. Thompson, I.F. Fergus and B.R. Tunstall. 1981. Plant succession and soil development in coastal sand dunes of subtropical eastern Australia. pp. 107–131. In D.C. West, H.H. Shugart and D.B. Botkin (eds.) *Forest Succession. Concepts and Application*. New York: Springer-Verlag.

Wallin, I.E. 1927. *Symbioticism and the Origin of Species*. Baltimore: Williams and Wilkins.

Wang, J., H.W. Pfefferkorn, Y. Zhang and Z. Feng. 2012. Permian vegetational Pompeii from Inner Mongolia and its implications for landscape paleoecology and paleobiogeography of Cathaysia. *Proceedings of the National Academy of Sciences* 109: 4927–4932.

Wardle, D.A. 1995. Impacts of disturbance on detritus food-webs in agroecosystems of contrasting tillage and weed management practices. *Advances in Ecological Research* 26: 105–185.

2002. *Communities and Ecosystems: Linking the Aboveground and Belowground Components*. Princeton: Princeton University Press.

Wardle, D.A., M.A. Huston, J.P. Grime et al. 2000. Biodiversity and ecosystem function: an issue in ecology. *Bulletin of the Ecological Society of America* 81: 235–239.

Wardle, D.A., R.D. Bardgett, J.N. Klironomos, et al. 2004. Ecological linkages between aboveground and belowground biota. *Science* 304: 1629–1633.

Watkinson, A.R. 1985a. Plant responses to crowding. pp. 275–289. In J. White (ed.) *Studies in Plant Demography: A Festschrift for John L. Harper*. London: Academic Press.

1985b. On the abundance of plants along an environmental gradient. *Journal of Ecology* **73**: 569–578.

Watkinson, A.R. and R.P. Freckleton. 1997. Quantifying the impact of arbuscular mycorrhizae on plant competition. *Journal of Ecology* **85**: 541–545.

Watson, D.M. 2011. *Mistletoes of Southern Australia.* Collingwood: CSIRO.

Watt, A.S. 1919. On the causes of failure of natural regeneration in British oakwoods. *Journal of Ecology* **7**: 173–203.

1923. On the ecology of British beechwoods with special reference to their regeneration. Part I. The causes of failure of natural regeneration of the beech. *Journal of Ecology* **11**: 1–48.

Weaver, J.E. and F.E. Clements. 1929. *Plant Ecology.* New York: McGraw-Hill.

1938. *Plant Ecology.* 2nd edn. New York: McGraw-Hill.

Weber, W. and A. Rabinowitz. 1996. A global perspective on large carnivore conservation. *Conservation Biology* **10**: 1046–1054.

Weetman, G.F. 1983. Forestry practices and stress on Canadian forest land. pp. 260–301. In W. Simpson-Lewis, R. McKechnie and V. Neimanis (eds.) *Stress on Land in Canada.* Ottawa: Lands Directorate, Environment Canada.

Weiher, E. and P.A. Keddy. 1995a. The assembly of experimental wetland plant communities. *Oikos* **73**: 323–335.

1995b. Assembly rules, null models, and trait dispersion: new questions from old patterns. *Oikos* **74**: 159–165.

Weiher, E. and P.A. Keddy (eds.) 1999. *Ecological Assembly Rules: Perspectives, Advances, Retreats.* Cambridge: Cambridge University Press.

Weiher, E., A. van der Werf, K. Thompson, et al. 1999. Challenging Theophrastus: a common core list of plant traits for functional ecology. *Journal of Vegetation Science* **10**: 609–620.

Wein, R.W. 1983. Fire behaviour and ecological effects in organic terrain. pp. 81–95. In R.W. Wein and D.A. MacLean (eds.) *The Role of Fire in Northern Circumpolar Ecosystems.* New York: John Wiley and Sons Ltd.

Wein, R.W. and J.M. Moore. 1977. Fire history and rotations in the New Brunswick Acadian forest. *Canadian Journal of Forest Research* **7**: 285–294.

Wein, R.W., R.R. Riewe and I. R. Methven (eds.) 1983. *Resources and Dynamics of the Boreal Zone.* Proceedings of a Conference held at Thunder Bay, Ontario, August 1982. Ottawa: Association of Canadian Universities for Northern Studies.

Weinberg, G.M. 1975. *An Introduction to General Systems Thinking.* New York: John Wiley.

Weiner, J. 1985. Size hierarchies in experimental populations of annual plants. *Ecology* **66**: 743–752.

1986. How competition for light and nutrients affects size variablility in *Ipomea tricolor* populations. *Ecology* **67**: 1425–1427.

Weiner, J. and S.C. Thomas. 1986. Size variability and competition in plant monocultures. *Oikos* **47**: 221–222.

Weintraub, P. 2008. *Cure Unknown: Inside the Lyme Epidemic.* New York: St. Martin's Press.

Weisner, S.E.B. 1990. *Emergent Vegetation in Eutrophic Lakes: Distributional Patterns and Ecophysiological Constraints.* Lund, Sweden: Grahns Boktryckeri.

Welcomme, R.L. 1976. Some general and theoretical considerations on the fish yield of African rivers. *Journal of Fish Biology* **8**: 351–364.

1979. *Fisheries Ecology of Floodplain Rivers.* London: Longman.

Weldon, C.W. and W.L. Slauson. 1986. The intensity of competition versus its importance: an overlooked distinction and some implications. *The Quarterly Review of Biology* **61**: 23–44.

Weller, D.E. 1990. Will the real self-thinning rule please stand up? A reply to Osawa and Sugita. *Ecology* **71**: 1204–1207.

Weller, M.W. 1994. *Freshwater Marshes: Ecology and Wildlife Management.* 3rd edn. Minneapolis: University of Minnesota.

Werner, P.A. 1975. Predictions of fate from rosette size in teasel (*Dipsacus fullonum* L.). *Oecologica* 20: 197–201.

West, D.C., H.H. Shugart and D.B. Botkin (eds.) 1981. *Forest Succession: Concepts and Application.* New York: Springer-Verlag.

Westhoff, V. and E. van der Maarel. 1973. The Braun–Blanquet approach. pp. 617–707. In R.H. Whittaker (ed.) *Ordination and Classification of Communities. Handbook of Vegetation Science, Part V.* The Hague: Junk.

Westing, A.H. 1989. Herbicides in warfare: the case of Indochina. pp. 337–358. In P. Bourdeau, J.A. Haines, W. Klein and C.R. Krishna Murti (eds.) *Ecotoxicology and Climate.* Chichester: Wiley.

Westoby, M. 1984. The self-thinning rule. *Advances in Ecological Research* 14: 167–225.

1998. Leaf–height–seed (LHS) plant ecology strategy scheme. *Plant and Soil* 199: 213–227.

Westoby, M., M. Leishman and J. Lord. 1997. Comparative ecology of seed size and dispersal. pp. 143–162. In J. Silvertown, M. Franco and J.L. Harper (eds.) *Plant Life Histories: Ecology, Phylogeny and Evolution.* Cambridge: Cambridge University Press.

Wheeler, B. D. and K.E. Giller. 1982. Species richness of herbaceous fen vegetation in Broadland, Norfolk in relation to the quantity of aboveground plant material. *Journal of Ecology* 70: 179–200.

Whelan, P.M. and O. Hamann. 1989. Vegetation regrowth on Isla Pinta: a success story. *Noticias de Galápagos* 48: 11–13.

Whelan, R.J. 1995. *The Ecology of Fire.* Cambridge: Cambridge University Press.

Whisenant, S.G. 1999. *Repairing Damaged Wildlands.* Cambridge: Cambridge University Press.

White, I.D., D.N. Mottershead and S.J. Harrison. 1992. *Environmental Systems: An Introductory Text.* 2nd edn. London: Chapman and Hall.

White, P.S. 1979. Pattern, process and natural disturbance in vegetation. *The Botanical Review* 45: 229–299.

White, T.C.R. 1993. *The Inadequate Environment: Nitrogen and the Abundance of Animals.* Berlin: Springer-Verlag.

1994. Synthesis: vegetation pattern and process in the Everglades ecosystem. pp. 445–460. In S. Davis and J. Ogden (eds.) *Everglades: The Ecosystem and its Restoration.* Delray Beach: St. Lucie Press.

Whittaker, R.H. 1952. A study of summer foliage insect communities in the Great Smoky Mountains. *Ecological Monographs* 22: 1–44.

1954a. The ecology of serpentine soils. I. Introduction. *Ecology* 35: 258–259.

1954b. The ecology of serpentine soils. IV. The vegetational response to serpentine soils. *Ecology* 35: 275–288.

1956. Vegetation of the Great Smoky Mountains. *Ecological Monographs* 26: 1–79.

1960. Vegetation of the Siskiyou Mountains, Oregon and California. *Ecological Monographs* 30: 279–338.

1962. Classification of natural communities. *The Botanical Review* 28: 1–239.

1965. Dominance and diversity in land plant communities. *Science* 147: 250–260.

1967. Gradient analysis of vegetation. *Biological Reviews* 42: 207–264.

1972. Evolution and measurement of species diversity. *Taxon* 21: 213–251.

1973a. Direct gradient analysis: techniques. pp. 9–31. In R.H. Whittaker (ed.) *Ordination and Classification of Communities. Handbook of Vegetation Science, Part V.* The Hague: W. Junk.

Whittaker, R.H. (ed.) 1973b. *Ordination and Classification of Communities. Handbook of Vegetation Science, Part V.* The Hague: W. Junk.

1975. *Communities and Ecosystems.* 2nd edn. London: Macmillan.

Whittaker, R.H. and Likens, G.E. 1973. Carbon in the biota. pp. 281–302. In G.M. Woodwell and E.V. Pecan (eds.) *Carbon and the Biosphere.* Springfield, VA: National Technical Information Service.

Wiens, J.A. 1977. On competition and variable environments. *American Scientist* 65: 590–597.

Wilcove, D.S., C.H. McLellan and A.P. Dobson. 1986. Habitat fragmentation in the temperate zone. pp. 237–256. In M.E. Soulé (ed.) *Conservation Biology: The Science of Scarcity and Diversity.* Sunderland: Sinauer Associates.

Wilde, S.A. 1958. *Forest Soils: Their Properties and Relation to Silviculture.* New York: The Ronald Press Company.

Wilf, P., N.R. Cúneo, K.R. Johnson, et al. 2003. High plant diversity in Eocene South America: Evidence from Patagonia. *Science* 300: 122–125.

Williams, C.B. 1964. *Patterns in the Balance of Nature.* London: Academic Press.

Williams, E.J. 1962. The analysis of competition experiments. *Australian Journal of Biological Science* 15: 509–525.

Williams, G.C. 1975. *Sex and Evolution.* Princeton: Princeton University Press.

Williams, M. 1989. *Americans and Their Forests: A Historical Geography.* Cambridge: Cambridge University Press.

Williamson, G.B. 1990. Allelopathy, Koch's postulates and the neck riddle. pp. 143–162. In J.B. Grace and D. Tilman (eds.) *Perspectives on Plant Competition.* San Diego: Academic Press.

Willig, M.R., D.M. Kaufman and R.D. Stevens. 2003. Latitudinal gradients of biodiversity: pattern, process, scale and synthesis. *Annual Review of Ecology, Evolution and Systematics* 34: 273–309.

Willis, A.J. 1963. Braunton Burrows: the effects on the vegetation of the addition of mineral nutrients to the dune soils. *Journal of Ecology* 51: 353–374.

Wilson, J.B. 1988. Shoot competition and root competition. *Journal of Applied Ecology* 25: 279–296.

Wilson, J.B., T.C.E. Wells, I.C. Trueman et al. 1996. Are there assembly rules for plant species abundance? An investigation in relation to soil resources and successional trends. *Journal of Ecology* 84: 527–538.

Wilson, S.D. 1993. Competition and resource availability in heath and grassland in the Snowy Mountains of Australia. *Journal of Ecology* 81: 445–451.

1999. Plant interactions during secondary succession. pp. 629–650. In L.R. Walker (ed.) *Ecosystems of Disturbed Ground.* Amsterdam: Elsevier.

Wilson, S.D. and P.A. Keddy. 1986a. Measuring diffuse competition along an environmental gradient: results from a shoreline plant community. *The American Naturalist* 127: 862–869.

1986b. Species competitive ability and position along a natural stress/disturbance gradient. *Ecology* 67: 1236–1242.

Wing, S.L. 1997. Global warming and plant species richness: a case study of the Paleocene/Eocene boundary. pp. 163–185. In M.L. Reaka-Kudla, D.E. Wilson and E.O. Wilson (eds.) *Biodiversity II: Understanding and Protecting Our Biological Resources.* Washington, D.C.: Joseph Henry Press.

Wing, S.L. and B.H. Tiffney. 1987. Interactions of angiosperms and herbivorous tetrapods through time. pp. 203–224. In E. Friis, W.G. Chaloner and P.R. Crane (eds.) *The Origins of Angiosperms and Their Biological Consequences.* Cambridge: Cambridge University Press.

Wiser, S.K., R.K. Peet and P.S. White. 1996. High-elevation rock outcrop vegetation of the southern Appalachian Mountains. *Journal of Vegetation Science* 7: 703–722.

Wisheu, I.C. and P.A. Keddy. 1989a. Species richness – standing crop relationships along four lakeshore gradients: constraints on the general model. *Canadian Journal of Botany* 67: 1609–1617.

1989b. The conservation and management of a threatened coastal plain plant community in eastern North America (Nova Scotia, Canada). *Biological Conservation* 48: 229–238.

1991. Seed banks of a rare wetland plant community: distribution patterns and effects of human induced disturbance. *Journal of Vegetation Science* 2: 181–188.

1992. Competition and centrifugal organization of plant communities: theory and tests. *Journal of Vegetation Science* 3: 147–156.

Witmer, M.C. and A.S. Cheke. 1991. The dodo and the tambalacoque tree: an obligate mutualism reconsidered. *Oikos* 61: 133–137.

Wium-Anderson, S. 1971. Photosynthetic uptake of free CO_2 by the roots of *Lobelia dortmanna*. *Plantarum* 25: 245–248.

Wolbach, W.S., R.S. Lewis and E. Anders. 1985. Cretaceous extinctions: evidence for wildfires and search for meteoritic material. *Science* 230: 167–170.

Wolfe, J.A. 1991. Palaeobotanical evidence for a June "impact winter" at the Cretaceous/Tertiary boundary. *Nature* 352: 420–423.

Wolin, C.L. 1985. The population dynamics of mutualistic systems. pp. 248–269. In D.H. Boucher (ed.) *The Biology of Mutualism: Ecology and Evolution*. New York: Oxford University Press.

Wood, S.L. 1982. The bark and ambrosia beetles of North and Central America (Coleoptera: Scolytidae), a taxonomic monograph. *Great Basin Naturalist Memoirs* 6: 1–1356.

Woodbury, A.M. 1947. Distribution of pigmy conifers in Utah and northeastern Arizona. *Ecology* 28: 113–126.

Woodley, S., J. Kay and G. Francis. (eds.) 1993. *Ecological Integrity and the Management of Ecosystems*. Delray Beach: St. Lucie Press.

Woods, K.D. and R.H. Whittaker. 1981. Canopy-understory interaction and the internal dynamics of mature hardwood and hemlock-hardwood forests. pp. 305–323. In D.C. West, H.H. Shugart and D.B. Botkin (eds.) *Forest Succession: Concepts and Applications*. New York: Springer-Verlag.

Woodward, F.I. 1987. *Climate and Plant Distribution*. Cambridge: Cambridge University Press.

1992. Predicting plant responses to global environmental change. *New Phytologist* 122: 239–251.

Woodward, F.I. and C.K. Kelly. 1997. Plant functional types: towards a definition by environmental constraints. pp. 47–65. In T.M. Smith, H.H. Shugart and F.I. Woodward (eds.) *Plant Functional Types*. Cambridge: Cambridge University Press.

Woodwell, G.M. 1962. Effects of ionizing radiation on terrestrial ecosystems. *Science* 138: 572–577.

1963. The ecological effects of radiation. *Scientific American* 208: 42–47.

World Commission on Environment and Development. 1987. *Our Common Future*. Oxford: Oxford University Press.

Wright, D.H. and J.H. Reeves. 1992. On the meaning and measurement of nestedness of species assemblage. *Oecologia* 92: 416–428.

Wright, H.A. and A.W. Bailey. 1982. *Fire Ecology*. New York: Wiley.

Wright, I.J. (and 33 others). 2004. The worldwide leaf economics spectrum. *Nature* 428: 821–827.

Wright, J.P., C.G. Jones and A.S. Flecker. 2002. An ecosystem engineer, the beaver, increases species richness at the landscape scale. *Oecologia* 132: 96–101.

Wright, R. 2004. *A Short History of Progress*. Toronto: Anansi Press.

Yanko-Hombach, V., A.S. Gilbert and P. Dolukhanov. 2007. Controversy over the great flood hypotheses in the Black Sea in light of geological, paleontological and archaeological evidence. *Quaternary International* 167–168: 91–113.

Yoda, K., T. Kira, H. Ogawa and K. Hozumi. 1963. Self-thinning in overcrowded pure stands under cultivated and natural conditions. *Journal of Biology/Osaka City University* 14: 107–129.

Yodzis, P. 1986. Competition, mortality, and community structure. pp. 480–492. In J. Diamond and T.J. Case (eds.) *Community Ecology*. New York: Harper and Row.

1989. *Introduction to Theoretical Ecology*. New York: Harper and Row.

Young, E. 2006. Easter Island: a monumental collapse? *New Scientist* 2562: 30–34.

Young, K., C.U. Ulloa, J.L. Luteyn and S. Knapp. 2002. Plant evolution and endemism in Andean South America: an introduction. *The Botanical Review* 68: 4–21.

Young, T.P. 1990. Evolution of semelparity in Mount Kenya lobelias. *Evolutionary Ecology* 4: 157–191.

Young, T.P. and C.K. Augspruger. 1991. Ecology and evolution of long-lived semelparous plants. *Trends in Ecology and Evolution* 6: 285–289.

Zachos, J., M. Pagani, L. Sloan, E. Thomas and K. Billups. 2001. Trends, rhythms, and aberrations in global climate 65 Ma to present. *Science* 292: 686–693.

Zedler, J.B. and P.A. Beare. 1986. Temporal variability of salt marsh vegetation: the role of low-salinity gaps and environmental stress. pp. 295–306. In D.A. Wolfe (ed.) *Estuarine Variability*. San Diego: Academic Press.

Zedler, J.B. and C.P. Onuf. 1984. Biological and physical filtering in arid region estuaries: seasonality, extreme events, and effects of watershed modification. pp. 415–432. In V.S. Kennedy (ed.) *The Estuary as a Filter*. New York: Academic Press.

Zobel, M. 1997. The relative role of species pools in determining plant species richness: an alternative explanation of species coexistence? *Trends in Ecology and Evolution* 12: 266–269.

Zuidema, P.A. and R.G.A. Boot. 2002. Demography of the Brazil nut tree (*Bertholletia excelsa*) in the Bolivian Amazon: impact of seed extraction on recruitment and population dynamics. *Journal of Tropical Ecology* 18: 1–31.

Figure and Table Credits

A picture, it is said, is worth a thousand words. This book has over 300 such pictures. I have carefully selected them to be memorable, and illustrate important ideas or processes or consequences. This is not a coffee table book, but all the same, you should be able to flip through it and enjoy the images, and, more importantly, learn something from each one. Part of the history of ecology is the accumulation of such images. I have tried to give credit to the many scholars and artists and publishers who have contributed to the body of knowledge in plant ecology. This is no small task, involving probably several thousand pieces of correspondence. Each image is first given a source in the caption. For photographs, I have, where known, given the name of the photographer, and the location. In the case of historic plant illustrations, I have tried to give the name of the artist and the (approximate) year it was created, or more often, published. In the case of images from the literature, each is referenced in the caption to its original source in the literature cited, and then, in the list below, I provide the name of the rights-holding organization and the permission statement allowing reproduction of the image.

There are some gaps in figure numbers in this list. These include figures where the books are now out of print, and rights have reverted to the author. In most cases, I have received permission from the rights-holding author, while in other cases this was not possible, particularly when they were deceased. Images from original papers of my own, or created by me or by my spouse, Cathy Keddy, expressly for this book, are not included. Another body of images, owing to their age, or source, are in the public domain and not listed. Finally, images with rights-holders acknowledged in the captions are not below. Let me express again my gratitude to all the photographers, artists, and their publishers, who have provided the figures for this book. I also further thank Cathy Keddy for devoting more than a year to obtaining the permissions below.

Figure 1.6 with permission of the Botanical Society of America

Figure 1.8, Figure 4.9, Figure 5.11, Figure 6.5, Figure 6.15, Figure 6.16, Figure 10.29, Figure 13.11, Table 13.3 with permission of Macmillan Publishers Ltd.

Figure 1.10, Figure 1.19, Figure 3.14, Figure 7.15, Figure 9.7, Figure 9.12, Figure 10.20, Figure 10.22 with permission of AAAS

Figure 1.11 with permission of Yale University Press

Figure 1.15 with permission of Geological Survey of Denmark

Figure 1.16 with permission of Woods Hole Oceanographic Institution

Figure 1.18 with permission of The Geological Society of America

Figure 2.2, Figure 3.5, Figure 3.11, Figure 3.17, Figure 3.22, Figure 3.23, Figure 3.24, Figure 3.25, Figure 4.3, Figure 4.4, Figure 4.10, Figure 4.19, Figure 4.23, Figure 4.24, Figure 4.26, Figure 4.27, Figure 5.4, Figure 5.9, Figure 5.10, Figure 5.19, Figure 5.23, Figure 5.27, Figure 6.7, Figure 6.12, Figure 6.13, Figure 6.14, Figure 7.3, Figure 8.9, Figure 8.10, Figure 8.17, Figure 8.19, Figure 8.20, Figure 8.21, Figure 8.22, Figure 9.0, Figure 9.8, Figure 9.10, Figure 10.19, Figure 10.23, Figure 10.30, Figure 11.2, Figure 11.3, Figure 11.4, Figure 11.10, Figure 11.11b,c,d, Figure 11.25, Figure 12.9, Figure 12.15, Figure 12.17, Figure 12.26, Figure 12.28, Figure 12.31, Table 3.3, Table 6.1, Table 6.3, Table 6.4, Table 6.6, Table 6.7,

Figure 13.12, Table 13.4 with permission of World
    Resources Institute
Figure B1.1.1 with permission of Indiana
    University Press
Figure B2.2 with permission of journal *Takhtajania*
Table 1.5 courtesy Oxford University Press

Table 2.4 with permission of University of
    Washington Press
Table 5.3 with permission of Planetary and Space
    Science Centre, University of New Brunswick
Table 12.6 with permission of Island Press
Table 13.6 courtesy Wolters Kluwer

Glossary

Alvar. A kind of rock barren habitat with little or no soil over flat limestone (Figure 10.11, Section 10.2.4).

Angiosperms. Collective name for plants possessing flowers and fruits (Table 1.1, Section 1.1.3, Section 8.2).

ANPP. Annual net primary production. The rate at which carbon is stored in plant tissues per unit area of a vegetation type in a year. Normally expressed in grams of carbon fixed per square metre per year (g/m^2yr) (Section 3.5.1).

Apparent competition. The situation where a removal experiment yields a positive outcome that we interpret as competition, when the outcome is in fact caused by interactions with other species such as herbivores. More generally, apparent effects are those that arise out of indirect effects produced by species interactions in networks (Section 7.8).

Asymmetric competition. Competition that occurs when two plant species have different competitive abilities (Section 4.2.5). Often leads to competitive hierarchies in plant communities.

Block diagram. A drawing that shows the relationships between plants and environmental gradients using one or more cross-sections through plants and landforms (Section 11.2, Figure 11.1).

Bottom–up. A type of community in which the physical environment and plants control the abundance and composition of animals. Contrasted with top–down. (Section 6.6.1)

Bryophyte. Collective name for mosses and liverworts, that is, small terrestrial species in which the gametophyte phase is dominant over the sporophyte phase (Table 1.1).

CAM photosynthesis. See Crassulacean acid metabolism.

Carboniferous. The geological era dominated by seedless plants when coal was deposited, ca. 360 to 300 million years BP, between the Devonian and Permian (Section 1.2, Section 1.9, Figure 1.17).

Carnivorous. A term describing plants that capture animals in order to extract one or more nutrients, usually nitrogen, from the tissues of the animal (Section 3.7.3, Figure 3.29)

Carpel. The layer of tissue surrounding one or more seeds that forms an ovary in a flower and becomes a fruit. It is a defining feature of flowering plants. It likely originated as a leaf or bract that surrounded one or more ovules (Figure 8.4).

Cellulase. An enzyme that cleaves cellulose into smaller fragments (Box 3.3).

Cladogram. A branching diagram showing how species have diverged from common ancestors (Section 2.3.3, Figure 2.7).

Classification. A mathematical tool for constructing clusters of similar composition in data matrices comprised of species × species or species × environment observations (Section 11.4). Note that this automatically produces discontinuous variation, whether the underlying patterns are continuous or discontinuous. Contrast this with ordination.

Climax. The end point in a succession. Frequently, shade-tolerant species have become dominant (Section 8.4.1).

Commensalism. An interaction between two species where one benefits and the other has neither benefits nor costs (Section 7.1.1).

Competition. The negative effects that one organism has upon another, usually by consuming or controlling access to a resource that is limited in availability (Section 4.1).

Competitive dominance. An outcome of interactions where one species suppresses another through resource competition and/or interference competition (Section 4.2.5).

Competitive effect. The negative effect a plant has upon its neighbours.

Competitive response. The negative effect of neighbours upon a plant.

Crassulacean acid metabolism. A type of photosynthesis in which CO_2 is taken up at night and stored for use in photosynthesis during the day. Found in plant groups including epiphytes, orchids and cacti (Section 3.3.2).

CSR. A framework for plant evolution and physiology that divides plants into three main types: competitors, stress tolerators and ruderals (Section 10.8.6).

Decomposition. The chemical degradation of plant material by fungi and bacteria, releasing chemically bound nutrients (Section 3.5.3).

Degradation. A general term for the cumulative (and undesirable) changes in vegetation that result in the loss of species and a decline in **services** (Section 13.2, Section 13.4.3).

Detritivores. Organisms that mechanically break down plant material in litter and soil. These are mostly arthropods, particularly ants and termites (Section 3.5.3). The next step is **decomposition**, which leads to **remineralization**.

Dipterocarp. A family of trees with hundreds of species that is characteristic of tropical forests in south-east Asia (Figure 13.1).

Disjunct. A plant distribution with one or more significant gaps in the middle. Possible causes include long-distance dispersal or changing environments (Figures 8.8, 8.15).

Disturbance. A short-lived event that causes a measurable change in the properties of an ecological community. The most obvious change is a reduction in living biomass. Common examples include fallen trees, fires, animal burrows, sudden freezes, floods and landslides (Section 5.1). The four main properties of disturbance are described in Section 5.2.

Diversity. The number of species in a location weighted by their relative abundance (Section 12.7, Box 12.1). Sometimes used more generally to mean just how many species occur, but it is better to use the term **species richness** in such cases.

Ecoregion. A relatively large unit of land containing a distinct assemblage of natural communities and species, with boundaries that approximate the original extent prior to major human impacts (Figure 2.14).

Enclosure. An area of land fenced to manipulate herbivore density (Figure 6.12, Section 6.4). See **exclosure**. Recall the film *The Magnificent Seven* where the Mexican bandits are told, ominously "this fence was not built to keep you out – but to keep you *in*."

Exclosure. An area of land fenced specifically to prevent herbivores from feeding on the plants therein (Figure 6.12, Section 6.4). See **enclosure**.

Facilitation. A type of positive interaction in which one species of plant assists another in establishing. Often an early part of **succession** (Section 8.4). See also nurse plants (Section 7.2.1).

Flora. The list of all the plant species that can occur in a specified region. It is often part of a larger manual that includes information on identification and ecology, such as the *New Flora of the British Isles*. In a more technical sense, this list designates the species pool for the area (Box 10.2).

Floristic region. See **World Floristic Region**.

Gametophyte. The phase in plant life history that produces eggs and sperm (gametes). In some plants, such as Bryophytes, this is the dominant phase, but in most terrestrial plants this phase has been progressively reduced until it appears in the Angiosperms as a mere "embryo sac" (Box 8.1).

Gap dynamics. See **patch dynamics**.

Genet. An inclusive term for all the **ramets** that have arisen from one seedling. Large clonal plants such as cattails or poplars may consist of hundreds or thousands of **ramets**, often connected below ground, appearing to consist of many individuals but being in fact a single genet (Section 9.6).

Gondwana. The large continent that separated from Laurasia during the Jurassic era, producing the distinctive floras of the Southern Hemisphere (Section 8.2).

Gradient diagram. A graph showing one or more dependent variables, such as species composition or diversity, plotted along environmental gradients such as elevation or soil moisture (Section 11.2).

Grubbian. An adjective used to describe a quite clever kind of published study that criticizes another scholar's work without actually citing the work in question, so that only a few initiates know who is being simultaneously ignored and criticized. If the author also manages to take personal credit for ideas in that uncited work, it is known as a doubly Grubbian paper. As in, "he wrote a Grubbian (or doubly Grubbian) paper on plant competition".

Gymnosperm. The large group of terrestrial plants that have seeds, but neither flowers nor fruits. Instead most have cones. The largest sub-group is conifers, but lesser groups include gingkos and cycads (Table 1.1).

Haber process. An industrial process for removing nitrogen from the atmosphere and converting it to ammonia, named for Fritz Haber (Box 3.2).

Haustorium. An unusual kind of root that attaches to other plants, hence it is found in parasitic plants (Section 3.7.3, Parasites).

Hydrosere. A traditional term for a stage in wetland succession (Section 8.4.5). See also **sere**.

Inselberg. See *tepui*.

Interference competition. Competition between two individuals in which one damages the other. Probably much more common in animals than in plants. But below-ground chemical interactions and above-ground interference by shoots may occur (Section 4.1).

Iteroparous. An organism that reproduces many times after reaching sexual maturity. In contrast to **semelparous** (Section 9.9).

K–T boundary. The geological boundary that separates the Mesozoic era from the Cenozoic era (that is, the Cretaceous–Tertiary boundary). It is significant because mammals and flowering plants rapidly diversified after the event that created this boundary (Section 5.4.3, Figures 5.28 and 5.29). Note that some scholars are now using the newer term K–Pg boundary, for Cretaceous–Paleogene.

Same rock layer, same age, same event, but an update in nomenclature in the search for consistent nomenclature for geological history.

Lignin. A large molecule, common in plant tissues, particularly wood. It is a branched polymer synthesized from the amino acid phenylalanine. Lignin decays slowly and therefore is an important component of soil humus (Box 3.3).

Ligninase. The general term for an important but poorly understood group of molecules secreted by fungi that cut lignin into smaller fragments (Box 3.3).

LMA. Leaf mass area, the dry-mass investment per unit area of light interception, in units such as grams per cm^2. Higher values mean thicker and tougher leaves (Section 3.7.2). This is the inverse of SLA.

Loess. A deposit of wind-blown soil, often many metres thick (Section 8.3.2).

Macrophyte. A term used mostly by limnologists and wetland ecologists for vascular plants that grow in shallow water (e.g. water lilies). It may be further sub-divided into emergent, floating or submersed types (Section 6.5).

Mangal. A wooded wetland that occurs in coastal environments, often dominated by mangrove trees. Also known as mangrove swamp, which is an acceptable term, but for the large number of other tree species that can also occur in this habitat (Section 10.5.1).

Mast. An ancient and still popular name for the seeds and dry fruits produced by trees. Hence a year with high reproduction is still called a mast year (Section 9.4.2).

Meristem. The regions within a plant, usually at the tips of branches and roots, where cells are still undifferentiated and capable of further division. Meristems can produce new shoots, leaves, branches, roots or flowers depending upon their location. Since a majority of plants contain many meristems, plants are somewhat colonial, and it is often difficult if not impossible to determine what constitutes an individual (Sections 2.2.1, 3.3.5, 9.6, 10.2.2).

Mutualism. An interaction between two species in which both benefit, that is a +/+ interaction (Section 7.1.1).

Mycoheterotroph. A plant that obtains all of its carbohydrates (and perhaps other nutrients) from fungi (Section 3.7.3, Figure 3.32).

Mycorrhiza. The symbiotic relationship between plant roots and soil fungi (Figure 7.5).

Myrmecochory. The dispersal of seeds by ants (Section 7.5.3, Figure 7.14).

Ordination. A mathematical tool for finding simplified patterns of continuous variation in data matrices comprised of species × species or species × environment observations (Section 11.3).

Patch dynamics. The process that occurs when disturbances creates patches (or gaps) in vegetation, followed by the dispersal and establishment of new plants in those patches (Figure 4.26, Section 5.6).

Peat bog. A landscape with a high water table and deep accumulations of partially decayed plants, particularly (but not exclusively) species of *Sphagnum* moss. One of the four main types of wetlands (Section 8.4.2 Succession in Peat Bogs, Figures 8.23 and 8.24).

Peatland. A general term for a landscape that is largely covered in **peat bogs**, often mixed with coniferous forests rooted in peat.

Photosynthesis. The process by which plants use sunlight to build organic molecules from carbon dioxide and water (Section 3.3). If you are looking this term up, it is likely you would benefit from consulting a basic botany book before reading further in this book.

Phycobiont. The photosynthetic cells within a lichen, either cyanobacteria or algae (Figure 7.9).

Phylogeny. The evolutionary history of a set of taxonomic groups (Section 2.3.3, Figure 2.7). Hence a **phylogenetic** classification system for plants is one that is based upon a set of hypotheses about evolutionary relationships of plants.

Phytosociology. The detection and naming of plant communities (each called a phytocoenose) for purposes of classification (Section 11.4.1, Tables 11.1 and 11.2).

Pool. All the species that occur in a region and that therefore could potentially colonize a specified piece of habitat (Box 10.2).

Primary succession. See **succession**.

Princetonian. An erudite argument style based upon wishful thinking and cronyism rather that facts and experiments. As in, "He made a Princetonian model to show that angels carry atmospheric nitrogen into plants."

Profile diagram. A cross-sectional sketch of a piece of forest, often showing changes along a gradient (Section 11.2, Figure 11.2).

Propagule. A very general and inclusive term for a small (piece of) plant with a **meristem** that functions in population growth, or dispersal, or persistence in unfavourable periods. The term includes seeds, but also the many kinds of asexually produced structures such as turions, bulblets, gemmae or even **rhizome** fragments (Section 9.6).

Ramet. A ramet is the unit of clonal growth, the module that may often follow an independent existence if severed from the parent plant. Large clonal plants such as cattails or poplars may consist of many ramets connected below ground into a large **genet** (Section 9.6, Box 9.1). Each ramet has one or more **meristems.**

Remineralization. The process that deconstructs dead plant material and releases mineral nutrients such as nitrogen and phosphorus back into the soil for use by other plants. This has two stages, mechanical breakdown by detritivores and chemical breakdown by decomposers (Section 3.5.3).

Resorption. The removal of nutrients from dying tissues for use elsewhere in a plant. An important aspect of a plant's ability to conserve nutrients (Section 3.7.1).

Resource competition. **Competition** between two or more plants for the same pool of resources (Section 4.1). Also called exploitation competition in some books.

Restoration. A kind of management that aims to return an ecological community to a more natural composition and structure. In a sense, the reverse process of degradation (Section 13.4.4).

Rhizome. An underground organ of many plants, not a root but a buried horizontal stem with a leaf-producing meristem at the end. Found in some of the earliest land plants (Figure 1.5), often allowing regrowth after disturbance (Figure 5.17) and frequently important in growth of large clones (Section 9.6).

Richness. See **Species richness**.

Ruderal. A short-lived plant that typically colonizes disturbed areas. Often semelparous with a rapid growth rate (Section 9.9, Section 10.8.6).

Rumen. An enlarged fore-stomach found in certain groups of herbivores. Here, plant material is partially digested by microorganisms (Section 6.3.5).

Savanna. A kind of grassland containing scattered trees (Section 10.2.2, Figure 3.18b, Figure 6.1). More generally, grasslands that occur in the tropics, and are often maintained by grazing and fire (Figure 10.7).

Sclerenchyma. The supporting tissues in a plant, having thick cell walls comprised of cellulose and lignin, and therefore being resistant to decay (Section 1.2).

Sclerophyllous. A term describing shrubs with small, leathery, evergreen leaves that often closely overlap along the stem. Such shrubs are typical of Mediterranean climates (Section 10.2.3) and also occur in some peat bogs.

Secondary succession. See **succession**.

Semelparous. A type of life history in which an organism reproduces only once and then dies. In contrast to **iteroparous** (Section 9.9).

Sere. One recognizable stage of vegetation in a progressive sequence known as **succession** (Section 8.4). Also sometimes referred to as a seral stage.

Serotinous. A type of strobilus that is sealed shut by resin and opened by fire. They often remain attached to the parent tree for many years (Section 6.3.2) and provide long-term seed storage in the canopy.

Serpentine. The general term for rocks that are typically low in major nutrients, low in calcium and contain high levels of metals such as nickel or chromium. Distinctive plant communities often occur on such rocks, and they often have endemic species (Section 10.2.4).

Service. A benefit that humans obtain from an ecosystem. Such benefits may range from the production of food by selected tree species to the production of oxygen by entire forests (Section 13.4.1, Table 13.6).

SLA. Specific leaf area, the area of leaf constructed from a standard amount of dry mass, in units such as cm^2 per gram. Higher values mean thinner and more flimsy leaves. This is the inverse of LMA (Box 10.2).

Species richness. The number of species found in a designated area (Section 12.1).

Sphagnum. A genus of moss (**bryophyte**), significant because this genus covers vast areas of the Earth that are wet. It is one of the most common plants in **peat bogs** or **peatlands** (Figure 10.13).

Sporophyte. The diploid life-history stage in the alternation of generations. Now the dominant life-history stage in vascular plants (Box 8.1).

Stomata. The small openings in leaves and stems that allow a plant to take in carbon dioxide and release oxygen. Also an important source of water loss in plants (Section 1.2, Section 3.2).

Strain. The damage or constraints to individual physiology that arise from stress (Section 10.8.1).

Stress. One or more environmental factors that reduce the rate of production of biomass (Chapter 10).

Stress avoidance. The protection of the inner tissues of a plant from negative external factors. Examples include the waxy coatings on leaves of desert plants, or aerenchyma in aquatic plants (Section 10.8.1).

Stress tolerance. The modification of the physiology of a plant in response to external factors (Section 10.8.1).

Strobilus. The technically correct name for the cone found in plant groups including conifers and cycads. A strobilus consists of spirally arranged scales surrounding a central axis, the scales producing either pollen (microspores) or ovules (megaspores) or being sterile. The term is equally applied to the reproductive structures of seedless plants in the Lycopodiophyta and Pteridophyta (Figure 1.3, Section 6.3.2).

Succession. The sequence of changes in vegetation that occurs after a disturbance. Can be divided into primary succession (after a major disturbance) and secondary succession (after a minor disturbance) (Section 8.4).

Summary diagram. A visual interpretation of relationships among plants, vegetation types, environmental gradients and causal factors (Section 11.2, Figure 11.6).

Symbiosis. Two species occuring in close association, such as the ant plant in Figure 3.27 or the mycorrhizae in Figure B7.1.1. A very general term that implies that the nature of the exchanges between them is unclear. **Commensalism** and **mutualism** may both be considered symbiosis (Section 7.1.1).

Tepui. A high rock outcrop surrounded by forest. The top of the outcrop is somewhat like an island, with a distinctive flora and vegetation. Also known by the term **inselberg** (Section 10.2.4, Figure 10.12).

Terpene. A common molecule in plant resin, consisting of five carbon atoms attached to eight hydrogen atoms, often linked into multiple units.

Terpenes give many conifers, such as pines, their characteristic smell (Section 6.3.3).

Threshold. The stage at which a plant community or ecosystem is about to show an abrupt change (Section 13.4.3, Figure 13.15).

Tilmanesque. Grossly inflated and self-referential. As in, the perigynium of *Carex inflata* is tilmanesque.

Top–down. A type of community in which herbivores largely control the abundance and composition of plants, and in which the predators also control the number of herbivores. Contrasted with **bottom–up** (Section 6.6.1).

Trait. A measurable characteristic of a plant (individual or species) such as leaf shape, bark thickness, relative growth rate or photosynthetic type (Section 2.2).

Trophic cascade. A situation in which **top–down** effects extend through three or more trophic levels, such as alligators controlling nutria and thereby determining the abundance of marsh plants, or wolves controlling elk and thereby determining the abundance of trees (Section 6.6.1).

World Floristic Region. An area of the world having similar plants, based upon a phylogenetic (evolutionary) classification (Section 2.3.6, Figure 2.12).

Zonation. The patterns of plant distribution along natural gradients such as shorelines (Figure 8.31) or mountainsides (Figure B2.1.3). See also Section 11.5.3.

Index